The Physical Principles of Magnetism

WILEY SERIES ON THE SCIENCE AND TECHNOLOGY OF MATERIALS

Advisory Editors: J. H. Hollomon, J. E. Burke, B. Chalmers, R. L. Sproull, A. V. Tobolsky

---

THE PHYSICAL PRINCIPLES OF MAGNETISM
Allan H. Morrish

FRICTION AND WEAR OF MATERIALS
Ernest Rabinowicz

HANDBOOK OF ELECTRON BEAM WELDING
R. Bakish and S. S. White

PHYSICS OF MAGNETISM
Sōshin Chikazumi

PHYSICS OF III-V COMPOUNDS
Otfried Madelung (translation by D. Meyerhofer)

PRINCIPLES OF SOLIDIFICATION
Bruce Chalmers

APPLIED SUPERCONDUCTIVITY
Vernon L. Newhouse

THE MECHANICAL PROPERTIES OF MATTER
A. H. Cottrell

THE ART AND SCIENCE OF GROWING CRYSTALS
J. J. Gilman, editor

COLUMBIUM AND TANTALUM
Frank T. Sisco and Edward Epremian, editors

MECHANICAL PROPERTIES OF METALS
D. McLean

THE METALLURGY OF WELDING
D. Séférian (translation by E. Bishop)

THERMODYNAMICS OF SOLIDS
Richard A. Swalin

TRANSMISSION ELECTRON MICROSCOPY OF METALS
Gareth Thomas

PLASTICITY AND CREEP OF METALS
J. D. Lubahn and R. P. Felgar

INTRODUCTION TO CERAMICS
W. D. Kingery

PROPERTIES AND STRUCTURE OF POLYMERS
Arthur V. Tobolsky

MAGNESIUM AND ITS ALLOYS
C. Sheldon Roberts

PHYSICAL METALLURGY
Bruce Chalmers

FERRITES
J. Smit and H. P. J. Wijn

ZONE MELTING
William G. Pfann

THE METALLURGY OF VANADIUM
William Rostoker

INTRODUCTION TO SOLID STATE PHYSICS, SECOND EDITION
Charles Kittel

The Physical Principles of Magnetism

by Allan H. Morrish
University of Minnesota
Minneapolis, Minnesota

John Wiley & Sons, Inc. **New York · London · Sydney**

Library of Congress Catalog Card Number: 65-14248
Printed in the United States of America

To My Parents

Preface

Whatever else may be said concerning the two decades following World War II, they have been wonderful years for science. All of the scientific countries of the world have contributed to the great increase in knowledge. Although advances have been made in almost all areas, some have been favored over others. One of the subjects that has received the greatest attention is that of magnetism. Although the appearance of a vast amount of literature is in itself laudable, it has created educational problems. The scientist or engineer who desires or requires an integrated knowledge of magnetism has found it increasingly difficult to satisfy this need. This book was written with such readers primarily in mind.

The material between these covers has formed the basis for several solid state courses offered at the University of Minnesota since 1958. Some of the more elementary parts have been employed for the third quarter of a first course in solid state physics. However, the major use of this text has been as a three-quarter graduate course in magnetism. The students attending the lectures have come from chemistry, physics, electrical engineering, physical metallurgy, and geophysics departments. Since the backgrounds of the audience varied so much, the course was developed by starting as far as possible from first principles. However, it was necessary to begin somewhere, and consequently first courses in solid state physics and quantum mechanics were usually assumed as prerequisites.

Magnetic phenomena are discussed both from an experimental and theoretical point of view. The plan has been first to present the underlying physical principles and then to follow up with appropriate macroscopic or microscopic theories. Although quantum mechanical theories are given, a phenomenological approach is emphasized. More than half the book is devoted to a discussion of strongly coupled dipole systems, and in this area the molecular field theory is stressed. The principles and theories have been illustrated by selections from the experimental data, and various tables have been included in an attempt to give some idea of the scope of the literature. Very few references to 1964 papers are included because of publication deadlines.

The utter impossibility of completely surveying all aspects of the subject may be illustrated by considering the *1964 Magnetic Materials Digest*.[1] This publication, which attempts to summarize the 1963 literature pertaining to fundamental studies of ordered magnetic materials, contains more than 1700 references! The particular topics chosen to be discussed at greatest length in this book were determined, as it is usually put, by my own interests. This, of course, really means my own knowledge, experience, talents, and limitations as shaped by contact with colleagues, teaching assignments, and availability of research and travel grants. The coverage of some topics, for example alloys and superconductivity, inadequately reflects the number and extent of current investigations. Other topics, for example, magnetohydrodynamics, group theoretical analysis, and the magnetoelectric effect, have been completely omitted. Nevertheless, the book is rather long and indeed contains more material than can normally be covered in a one-year course.

The question of notation has been a vexing problem. In a book as comprehensive as this one, there are simply not enough symbols to go around. A partial solution has been sought in the use of different kinds of type for the same letter. For example, μ and μ represent the dipole magnetic moment and the permeability, H and $\mathcal{H}$ represent the magnetic field and Hamiltonian, and D, $\mathcal{D}$, and $\mathfrak{D}$ represent the demagnetizing factor, the crystal field spin parameter, and the Dzialoshinski vector, respectively. Even so, it was found necessary sometimes to employ the same symbol for more than one quantity; it is only hoped that the context will reduce, if not entirely remove, the danger of ambiguity. Although in general I have tried to adhere to the symbols commonly found in the literature, some deliberate departures have been made. In publications on nuclear magnetic resonance, and to some degree in electron paramagnetic resonance, T with some subscript denotes relaxation time. However, I prefer to reserve T for either temperature or period of time and have therefore taken the liberty of representing relaxation times by τ plus a subscript; this usage at least has the merit of being consistent with that in all other branches of science and engineering. Other, though less important examples, include the use of M_H for the magnetization

[1] A. H. Morrish, R. J. Prosen, and S. M. Rubens, editors, *1964 Magnetic Materials Digest*, M. W. Lads Publishing Co., Philadelphia, Pa. (1964).

along the direction of an applied field, H_1 for the amplitude of an alternating magnetic field, H_e for the effective magnetic field, H_m for the molecular field, J_e for an exchange interaction constant, T_{FN} for the Néel temperature of a ferrimagnet, and N with some subscript for the Weiss molecular field constants. For the Bohr magneton a choice between μ_B and β was possible; I have selected μ_B mainly because β also represents an angle or the relativistic quantity v/c. I have substituted the term "uncompensated poles" for "free poles," since the latter may be taken to imply the existence of monopoles, and they have never been observed in nature. The phrase "intensity of magnetization," common in the British literature, has been eliminated on the grounds that it is both inappropriate and inept. The symbol e for the elementary electric charge represents either type; that is, a negative number is to be substituted for an electron and a positive number for a proton or positron. Except as otherwise noted, the gaussian cgs system of units is employed.

There remains the pleasant task of thanking those who have played some role in the production of this book. I am very grateful to Professor A. J. Dekker, now at Groningen, The Netherlands, who first interested me in doing research in solid state physics and who encouraged me to write this book. I have been fortunate in having Professor W. F. Brown, Jr., as a colleague during part of my tenure at the University of Minnesota. The discussions we have had have left their imprint on a number of these pages. In addition, a critical reading of the first chapter by Dr. Brown has led to its improvement. The response and comments of my graduate students have greatly aided me in the removal of errors, inconsistencies, and obscurities. They have also demonstrated that almost all the problems are soluble. I am indebted to the many publishers and individuals who have given me permission to reproduce figures and tables; these sources are acknowledged at the pertinent places in the text. Finally, I wish to thank my wife for her help with the arduous tasks of proofreading and index preparation.

University of Manitoba
Winnipeg, Canada
April 1965

ALLAN H. MORRISH

Contents

I

The Magnetic Field

1. Historical

The writings of Thales, the Greek, establish that the power of loadstone, or magnetite, to attract iron was known at least as long ago as 600 B.C. It has been claimed that the Chinese used the compass sometime before 2500 B.C. That magnetite can induce iron to acquire attractive powers, or to become magnetic, was mentioned by Socrates. Thus permanent and induced magnetism represent two of man's earliest scientific discoveries. However, the only real interest in magnetism in antiquity appears to be concerned with its use in the construction of the compass. For example, it is illuminating that it was not until many centuries later that Gilbert (1540–1603) realized that the earth was a huge magnet, even though the operation of the compass depends on this very fact.

The discovery of two regions called magnetic poles, or sometimes just "poles," which attracted a piece of iron more strongly than the rest of the magnetite, was made by P. Peregrines about 1269 A.D. Coulomb (1736–1806), in accurate quantitative experiments with the torsion balance, investigated the forces between magnetic poles of long thin steel rods. His results form the starting point of this treatise on magnetism.

2. The Magnetic Field Vector H

Coulomb found that there were two types of poles, now called positive or north, and negative, or south. Like poles repel one another and unlike poles attract one another. This force of attraction or repulsion is proportional to the product of the strength of the poles and inversely proportional to the square of the distance between them. This is Coulomb's law, which can be stated mathematically as

$$\mathbf{F} = k\,\frac{m_1 m_2}{r^2}\,\mathbf{r}_0, \tag{1-2.1}$$

where $\mathbf{F}$ is the force,[1] m_1 and m_2 the pole strengths, r the distance between the poles, and $\mathbf{r}_0$ a unit vector directed along r. The constant of proportionality k that occurs permits a definition of pole strength. In the cgs system of units two like poles are of unit strength if they repel each other with a force of 1 dyne when they are 1 cm apart; that is, $k = 1$. Other systems of units and their relationship to the cgs system are discussed in Appendix I.

It is convenient to consider $\mathbf{F}$ as separated into two factors. One factor is just one of the poles, say m_2, usually called the test pole. The other factor depends on the other pole, called the source, and on the location with respect to it; it is called the magnetic field $\mathbf{H}$. This field is defined as the force the pole exerts on a unit positive pole, or

$$\mathbf{H} = \frac{m}{r^2}\,\mathbf{r}_0. \tag{1-2.2}$$

In addition to this use of $\mathbf{H}$ as the field at a point, we will employ the same symbol $\mathbf{H}$ as the set of values of the magnetic field at all points: no confusion should result, since the correct meaning will be clear from the context. The cgs unit of magnetic field is the oersted, although the term gauss is still frequently used. Should several poles be present, experiments show that the field is the vector sum of the forces on the test pole.

Instead of the vector field quantity $\mathbf{H}$, it is often convenient to use a scalar potential φ. The quantity φ is defined so that its negative gradient is the magnetic field

$$\mathbf{H} = -\nabla\varphi, \tag{1-2.3}$$

where the operator ∇ is

$$\nabla = \mathbf{i}\,\frac{\partial}{\partial x} + \mathbf{j}\,\frac{\partial}{\partial y} + \mathbf{k}\,\frac{\partial}{\partial z}.$$

Here $\mathbf{i}$, $\mathbf{j}$, $\mathbf{k}$, are the unit vectors of a Cartesian coordinate system, and (x, y, z) are the coordinates at the point where the field or potential is under consideration.

[1] Boldface type indicates a vector quantity.

It follows immediately that for an isolated pole

$$\varphi = \frac{m}{r}. \tag{1-2.4}$$

The work done in bringing a unit test pole from infinity to the point (x, y, z) a distance r from m is found by integrating (1-2.2) along the path.[2] This turns out to be equal to φ and provides a simple physical meaning of the scalar potential.

Consider the work done in taking a unit pole from point 1 to 2 in a field **H** given by the line integral

$$\int_2^1 \mathbf{H} \cdot d\mathbf{s} = \int_2^1 H_s \, ds. \tag{1-2.5}$$

Here ds is a line element of the path from point 1 to point 2. Now, since $H_s = -\partial\varphi/\partial s$, we get

$$\int_2^1 \mathbf{H} \cdot d\mathbf{s} = \varphi_2 - \varphi_1. \tag{1-2.6}$$

This integral has the same value for any path having the same first and final point; that is, the work done is independent of the path. Mathematically this is stated as

$$\oint \mathbf{H} \cdot d\mathbf{s} = 0. \tag{1-2.7}$$

3. The Magnetization Vector M

Isolated magnetic poles have never been observed in nature, but occur instead in pairs, one pole being positive, the other negative. Such a pair is called a dipole. The magnetic moment of a dipole is defined as

$$\boldsymbol{\mu} = m\mathbf{d}, \tag{1-3.1}$$

where **d** is a vector pointing from the negative to the positive pole and equal in magnitude to the distance between the poles assumed to be points. If **d** approaches zero and m increases so that $\boldsymbol{\mu} = m\mathbf{d}$ is a constant, then in the limit in which $\mathbf{d} = 0$ the dipole is said to be ideal.

Atomic theory has shown that the magnetic dipole moments observed in bulk matter arise from one or two origins: one is the motion of electrons about their atomic nucleus (orbital angular momentum) and the other is the rotation of the electron about its own axis (spin angular momentum).

[2] This is the work done *against* the magnetic force; to compute the work done *by* it, the limits of the integration are reversed.

The nucleus itself has a magnetic moment. Except in special types of experiments, this moment is so small that it can be neglected in the consideration of the usual macroscopic magnetic properties of bulk matter.

It turns out that a magnetic field **H** interacts with the electrons of an atom in such a way that a magnetic moment is induced. This phenomenon is called *diamagnetism*. Since all matter contains electrons moving in orbits, diamagnetism occurs in all substances.

Depending on the electronic structure of an atom, it may or may not have a *permanent* magnetic moment. All magnetic effects other than diamagnetism result because of permanent atomic magnetic moments. If the coupling between the moments of different atoms is small or zero, the phenomenon of *paramagnetism* results. In the absence of an applied field **H** such materials will exhibit no net magnetic moment. If the

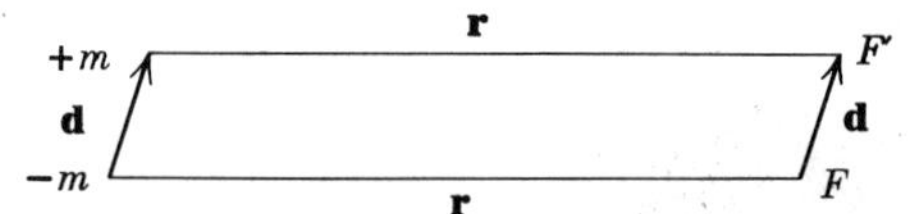

Fig. 1-3.1. F is the field point at which the potential produced by the dipole is calculated. F' is a point located a step **d** from F.

coupling between the atomic moments is very large, there are three important classifications. If the atomic moments are aligned parallel, the substance is said to be *ferromagnetic*. The magnetic moments may be aligned parallel within groups, usually two. If pairs of groups are aligned antiparallel and the atomic moments of the groups are equal, the substance is *antiferromagnetic*. However, the atomic moments of the groups may not be equal—for example, when two different elements are present; thus when they are aligned antiparallel there is a net moment. This phenomenon is called *ferrimagnetism*; some writers consider it to be just a special case of antiferromagnetism.

Because magnetic poles occur in pairs, it is of interest to calculate the magnetic field produced by such a combination. The dipole shown in Fig. 1-3.1 is considered to be almost but not quite ideal, so that $\mathbf{d} \ll \mathbf{r}$. The potential it produces at the field point F, φ_F, is due to both the positive and negative poles, that is,

$$\varphi_F = \varphi_F^{+} + \varphi_F^{-}.$$

Now let the point a step **d** from F be called F'. Except for sign, the potential $-m$ produces at F is the same $+m$ produces at F'. Hence

$$\begin{aligned}\varphi_F &= \varphi_F^{+} - \varphi_{F'}^{+}\\ &= -\mathbf{d}\cdot\nabla_F\varphi_F^{+}.\end{aligned}$$

Here ∇_F indicates differentiation with respect to the field coordinate (x, y, z) and not the source coordinate (x_i, y_i, z_i) since

$$\nabla_F \frac{1}{r} = -\nabla_s \frac{1}{r},$$

where

$$r = [(x - x_i)^2 + (y - y_i)^2 + (z - z_i)^2]^{1/2}.$$

Now, the point source $+m$ produces a potential at F given by

$$\varphi_F^{+} = \frac{m}{r}.$$

Therefore

$$\varphi_F = -m\mathbf{d} \cdot \nabla \frac{1}{r}.$$

Now suppose the dipole to be ideal, that is, $\mathbf{d} = 0$. In the limit we have

$$\varphi_F = -\boldsymbol{\mu} \cdot \nabla \frac{1}{r}.$$

Since

$$\nabla \frac{1}{r} = -\frac{1}{r^2}\left[\frac{(x - x_i)}{r}\mathbf{i} + \frac{(y - y_i)}{r}\mathbf{j} + \frac{(z - z_i)}{r}\mathbf{k}\right],$$

we get

$$\varphi_F = -\boldsymbol{\mu} \cdot \nabla \frac{1}{r} = \frac{1}{r^3}\boldsymbol{\mu} \cdot \mathbf{r} = \frac{1}{r^2}|\boldsymbol{\mu}| \cos \theta. \tag{1-3.2}$$

The magnetic field $\mathbf{H}_F$ is then

$$\mathbf{H}_F = -\nabla\left(-\boldsymbol{\mu} \cdot \nabla \frac{1}{r}\right)$$

$$= -\frac{\boldsymbol{\mu}}{r^3} + \frac{(3\boldsymbol{\mu} \cdot \mathbf{r})\mathbf{r}}{r^5}. \tag{1-3.3}$$

For substances that have a net magnetic moment it is usual to define a magnetization vector $\mathbf{M}$ as the ratio of the magnetic moment of a small volume at some point to that volume. The size of the volume chosen must be large enough so that a somewhat larger volume will still yield the same result for $\mathbf{M}$; in this way we ensure that atomic fluctuations are negligible. If $\mathbf{M}$ is constant for the specimen, the material is said to be uniformly magnetized. From the definition of $\mathbf{M}$ it is clear that it is also the pole strength for a unit area perpendicular to $\mathbf{M}$, that is

$$\sigma = \mathbf{M} \cdot \mathbf{n}, \tag{1-3.4}$$

where σ is the pole strength per unit area and $\mathbf{n}$ is a unit vector normal to the surface.

Introduction of the vector **M** permits generalization of equation 1-3.2 for bulk material. Summing over the dipoles gives the total potential at a point external to the specimen as

$$\varphi_F = -\sum \boldsymbol{\mu} \cdot \nabla \frac{1}{r}$$

$$= -\int \mathbf{M} \cdot \nabla_F \frac{1}{r}\, dv_s. \tag{1-3.5}$$

A special form of Green's theorem gives

$$\varphi_F = \int \frac{1}{r} \mathbf{M} \cdot \mathbf{n}\, dS - \int \frac{1}{r} \nabla_s \cdot \mathbf{M}\, dv \tag{1-3.6}$$

or

$$= \int \frac{\sigma\, dS}{r} - \int \frac{\rho}{r}\, dv,$$

where dS is an element of area. This result permits an interesting physical interpretation to be made. The magnetic potential can be considered to

(a) (b)

Fig. 1-3.2. In (*a*) each arrow represents a dipole, each with the same magnetic moment. The uncompensated charges for this dipole distribution are shown in (*b*). This illustrates the origin of volume charges for a simple case.

be due to two causes. One, the surface charge (or pole) density σ, and, two, a volume charge density ρ. The first of these can be easily pictured as arising from the uncompensated ends of the dipoles that end on the surface. The volume density may be pictured as the uncompensated poles that arise from an inhomogenity of the distribution of the moments, as illustrated in Fig. 1-3.2.

4. Magnetic Induction, the Vector B

The magnetic forces that must exist inside a ferro- or ferrimagnetic medium pose some special problems. Such forces have meaning only if it is possible to specify a method of measuring them. The approach adopted by Maxwell was to consider the medium as a continuum and to make a cavity around the point at which the force on the test pole was to be determined. However, the force per unit pole depends on the shape of the cavity, since this force depends partly on the pole distribution around the cavity, so that there exist an infinite number of ways that the field could be defined. In fact, two particular cavity shapes are chosen.

The field **H** is defined as the field vector in a needle-shaped cavity with an infinitesimally small diameter. The reason for this choice is that the field defined in this way satisfies equation 1-2.7. The field vector obtained when the cavity is a disk of infinitesimally small height is called the magnetic induction **B**; the reason for this choice is that Maxwell's equation $\nabla \cdot \mathbf{B} = 0$ is then satisfied. With the aid of Gauss's theorem, we can show that **B** and **H** are related by[3]

$$\mathbf{B} = \mathbf{H} + 4\pi\mathbf{M}; \tag{1-4.1}$$

B is said to be in the units of gauss in the cgs system.

The magnetic flux Φ is defined as the flux of the vector **B** through a surface of area A; that is,

$$\Phi = \int_A \mathbf{B} \cdot \mathbf{n}\, dS. \tag{1-4.2}$$

The unit of flux is called the maxwell. Thus the induction in gauss, at some field point, is equal to the flux density, the number of maxwells per square centimeter. The foregoing definition of flux is possible only because $\nabla \cdot \mathbf{B} = 0$, one of Maxwell's equations. Often a graphical meaning is given to the flux. It can be represented as lines or tubes whose density is equal to B and direction is along **B**.

When the vectors **B**, **H**, and **M** are parallel, it is useful to define the permeability μ by

$$B = \mu H \tag{1-4.3}$$

and the susceptibility χ by

$$M = \chi H. \tag{1-4.4}$$

The susceptibility per unit mass χ_ρ is defined as χ/ρ, where ρ is the density. The atomic or molar susceptibility χ_A or χ_m then is found by multiplying χ_ρ by the atomic or molecular weight. From (1-4.1) it follows immediately that

$$\mu = 1 + 4\pi\chi. \tag{1-4.5}$$

[3] A good treatment of this problem for the analogous electrical case may be found in C. J. F. Böttcher, *Theory of Electric Polarization*, Elsevier Publishing Co., New York (1952), Ch. II.

If the magnetic vectors are not parallel, the components of **B** and **H** relative to an arbitrarily chosen Cartesian coordinate system can be related by the set of equations:

$$B_x = \mu_{11}H_x + \mu_{12}H_y + \mu_{13}H_z,$$
$$B_y = \mu_{21}H_x + \mu_{22}H_y + \mu_{23}H_z,$$
$$B_z = \mu_{31}H_x + \mu_{32}H_y + \mu_{33}H_z.$$

The quantities μ_{ij} are components of the permeability tensor $\tilde{\boldsymbol{\mu}}$. Similarly, the relationship between M and H can be expressed with the aid of a susceptibility tensor $\boldsymbol{\chi}$.

It is an experimental result that χ is negative for diamagnetic materials and positive for the other types of magnetism, being very large for ferri- and ferromagnetic substances.

The magnetization of dia-, para-, and antiferromagnetic substances disappears if the applied field is removed. This is in contrast to the behavior of ferro- and ferrimagnetic materials, which usually retain at least part of their induced magnetic moment in the absence of an applied field. For these materials the susceptibility is a function of the applied field, the temperature, and the history of the samples. Discussion of the temperature and field dependence of the susceptibility is left until later.

5. The Demagnetization Factor D

The field H' inside a specimen is different from the applied field H because of the magnetization or equivalently, the poles. Consider a ferromagnet with ellipsoidal shape in a uniform external field **H**. As discussed later, the magnetization of the ellipsoidal specimen will also be uniform. The poles that appear on the surface, indicated in Fig. 1-5.1, produce a uniform internal field, $H' - H$, opposite in direction to **H**. For specimens with an ellipsoidal shape it is usual to write

$$\mathbf{H}' = \mathbf{H} - D\mathbf{M}, \tag{1-5.1}$$

where D is called the demagnetization factor. D depends on the geometry of the specimen. For diamagnets $H' > H$; for all other magnets $H' < H$. The difference in the field H' and H can usually be neglected for dia- and paramagnets, but it can be very large for ferro- and ferrimagnets. From the reasoning of Section 1-4, it can be seen that for a disk $D = 4\pi$ for the direction perpendicular to the plane of the disk. In general the demagnetizing factor is a tensor **D**.

In the easiest general case to calculate **M** is uniform. In equation 1-3.6,

$$\varphi = \int \frac{1}{r} \mathbf{M} \cdot \mathbf{n} \, dS - \int \frac{1}{r} \nabla \cdot \mathbf{M} \, dv, \tag{1-5.2}$$

the second term is zero, and the potential, and therefore the field, is due only to the surface pole distribution. Also, in expression 1-3.5 for φ, **M** can be taken outside the integral sign, and we have

$$\varphi = -\mathbf{M} \cdot \int \nabla \frac{1}{r} \, dv. \tag{1-5.3}$$

Now $-\int \nabla(1/r) \, dv$ is the gravitational force due to a volume of uniform unit mass density (the gravitational constant $G = 1$ here) or the electric force due to a volume of unit charge density. We therefore have the

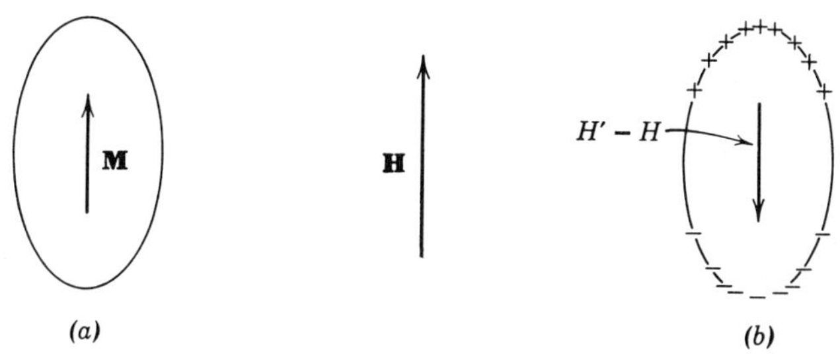

Fig. 1-5.1. (*a*) A uniformly magnetized ellipsoid and (*b*) the equivalent poles. The uncompensated surface poles produce a uniform internal field $H' - H$ and an external field that is identical to that of an equivalent dipole positioned at the ellipsoid's center. **H** is a uniform applied field.

important result that equations derived in potential theory may be used to find φ, next $H' - H$, and finally D.

It is instructive to consider the derivation of the foregoing result by a simple physical argument.[4] Let Ψ be the potential due to gravitation, or the electric charge of the body assumed of uniform density ρ. Now, if the body is moved a distance $-\delta x$ in the direction of x, the change of the potential at any point will be $-(d\Psi/dx)\,\delta x$. Instead, if we consider the body to be moved δx, and its original density ρ changed to $-\rho$, then $-(d\Psi/dx)\,\delta x$ is the resultant potential due to the two bodies (Fig. 1-5.2).

To any element of volume, mass, or charge $\rho \, dv$, there will correspond an element of the shifted body of $-\rho \, dv$ a distance $-\delta x$ away. Hence the dipole moment of these two elements is $\rho \, dv \, \delta x$, and the magnetization

[4] J. C. Maxwell, *Treatise on Electricity and Magnetism*, 3rd ed., Oxford University Press, Oxford (1891), vol. ii, p. 66. [Reprinted Dover Publications, New York (1954).]

is $\rho\,\delta x$. Therefore if $-(d\Psi/dx)\,\delta x$ is the magnetic potential of the body of magnetization $\rho\,\delta x$, then $-d\Psi/dx$ is the potential for a body of magnetization $\rho(= M)$.

In the volume common to the two bodies the density is effectively zero. A shell of positive charge or density resides on one side of the matter and one of negative on the other, each of density $\rho \cos \epsilon$, ϵ being the angle between the outward normal and the axis x. This then corresponds exactly to the first term of equation 1-5.2 and gives an immediate physical picture of the mathematics.

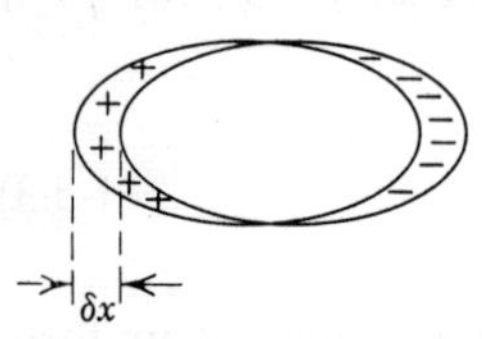

Fig. 1-5.2. Uncompensated charges or poles for two ellipsoids of opposite charge densities and a distance δx apart.

The x-component of $\mathbf{H}' - \mathbf{H}$ is $-(\partial\varphi/\partial x) = M_x(\partial^2\Psi/\partial x^2)$. Therefore, for $\mathbf{M}$ to be uniform, which implies that $\mathbf{H}'$ is uniform, Ψ must be a quadratic function of the coordinates. From potential theory[5] this occurs only when the body is bounded by a surface of second degree. The only physically possible body is then an ellipsoid.

The derivation of the expressions for Ψ is beyond the scope of this book and belongs to potential theory.[5] For completeness, and because of their practical usefulness, some of the important formulas for the demagnetization factor of ellipsoids of revolution are given.

We define a as the polar semiaxis and b as the equatorial semiaxis with $m = a/b$. Then for the prolate spheroid ($m > 1$)

$$D_a = \frac{4\pi}{(m^2 - 1)}\left\{\frac{m}{(m^2 - 1)^{1/2}} \ln\left[m + (m^2 - 1)^{1/2}\right] - 1\right\} \tag{1-5.4}$$

and

$$D_b = \tfrac{1}{2}(4\pi - D_a), \tag{1-5.5}$$

where D_a is the demagnetization factor for a and D_b along b.

In terms of the eccentricity $\epsilon^2 = 1 - (b/a)^2$

$$D_a = 4\pi\frac{(1 - \epsilon^2)}{\epsilon^2}\left(\frac{1}{2\epsilon}\ln\frac{1 + \epsilon}{1 - \epsilon}\right) - 1. \tag{1-5.6}$$

For the oblate spheroid ($m < 1$)

$$D_a = \frac{4\pi}{1 - m^2}\left[1 - \frac{m}{(1 - m^2)^{1/2}}\cos^{-1} m\right] \tag{1-5.7}$$

or

$$D_a = \frac{4\pi}{\epsilon^2}\left[1 - \frac{(1 - \epsilon^2)^{1/2}}{\epsilon}\sin^{-1}\epsilon\right]. \tag{1-5.8}$$

[5] W. Thomson and P. G. Tait, *Treatise on Natural Philosophy*, 2nd ed., Vol. i, Part ii, Cambridge University Press, Cambridge (1883).

For the sphere ϵ approaches zero and

$$D = \frac{4\pi}{3}. \tag{1-5.9}$$

Appendix II lists some values of D calculated by Stoner,[6] and Osborn[7] has calculated the demagnetization factor for the general ellipsoid.

Practical specimens are seldom ellipsoidal in shape but may be in the form of bars or rods, in which case the second term of (1-3.6) may be comparable with the first. The solution of the problem requires approximations and is very laborious. Bozorth and Chapin[8] have studied the demagnetization factors of rods.

6. Energy of Interaction

The potential energy of a dipole in a pre-existing magnetic field, sometimes called the mutual or interaction energy, is equal to the work done in bringing the dipole from infinity. To calculate this energy, consider the dipole as two poles, $-m$ and $+m$, distance d apart, brought to the position shown in Fig. 1-6.1. If the potential at the poles is φ_{-m} and φ_{+m} respectively, the work done is $-m\varphi_{-m}$ and $+m\varphi_{+m}$. The potential energy is then

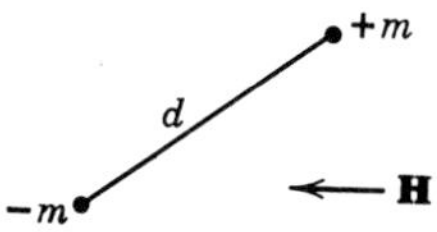

$$W = m(\varphi_{+m} - \varphi_{-m}) = md\,\frac{\partial \varphi}{\partial s},$$

Fig. 1-6.1. Dipole in an applied magnetic field **H**.

where s denotes the direction between $-m$ and $+m$. Then

$$W = \boldsymbol{\mu} \cdot \nabla\varphi = -\boldsymbol{\mu} \cdot \mathbf{H} \tag{1-6.1}$$

when **H** is the pre-existing magnetic field. The interaction energy of a permanent magnet is found by summing (1-6.1) over all the dipoles, giving

$$W = \int \mathbf{M} \cdot \nabla\varphi \, dv = -\int \mathbf{M} \cdot \mathbf{H} \, dv. \tag{1-6.2}$$

In addition, a permanent magnet will have potential energy because of its own field. To calculate this self-energy, consider the work done in bringing up a sequence of dipoles, $a, b, c \cdots n$, whose sum makes up the permanent magnet. No work is done in bringing a into position. Let us denote by $W_a(b)$, the work done in bringing dipole b into position in a field due to dipole a. Further, the work done in positioning dipole c due

[6] E. C. Stoner, *Phil. Mag.* **36,** 816 (1946).
[7] J. A. Osborn, *Phys. Rev.* **67,** 351 (1945).
[8] R. M. Bozorth and D. M. Chapin, *J. Appl. Phys.* **13,** 320 (1942).

to fields of a and b will be $W_a(c) + W_b(c)$. For the n dipoles the total work done W is

$$\begin{aligned} W = \; & W_a(b) \\ & + W_a(c) + W_b(c) \\ & + \cdots\cdots\cdots\cdots \\ & + W_a(n) + W_b(n) + W_c(n) + \cdots + W_{n-1}(n). \end{aligned}$$

On the other hand, if the dipoles had been brought up in the reverse order, we would get

$$\begin{aligned} W = \; & W_b(a) + W_c(a) + \cdots\cdots\cdots\cdots + W_n(a) \\ & \qquad + W_c(b) + \cdots\cdots\cdots\cdots + W_n(b) \\ & \qquad\qquad + \cdots\cdots \\ & \qquad\qquad\qquad + W_n(n-1) \end{aligned}$$

Adding, we get

$$\begin{aligned} 2W = \; & \qquad W_b(a) + W_c(a) + \cdots\cdots\cdots + W_n(a) \\ & + W_a(b) \qquad + W_c(b) + \cdots\cdots\cdots + W_n(b) \\ & + W_a(c) + W_b(c) \qquad + W_d(c) + \cdots\cdots + W_n(c) \\ & + \cdots\cdots\cdots\cdots \\ & + W_a(n) + W_b(n) + \cdots\cdots + W_{n-1}(n). \end{aligned}$$

Therefore

$$W = \tfrac{1}{2} \sum_{n=a}^{n} W(n), \tag{1-6.3}$$

where $W(a)$ is the energy of a dipole placed in a field due to the assembly of dipoles [$W_a(a) = 0$, etc.]. If the dipoles have a moment $\boldsymbol{\mu}$, then by (1-6.1)

$$W(n) = \boldsymbol{\mu} \cdot \nabla \varphi$$

and

$$\begin{aligned} W &= \tfrac{1}{2} \sum \boldsymbol{\mu} \cdot \nabla \varphi \\ &= -\tfrac{1}{2} \sum \boldsymbol{\mu} \cdot \mathbf{H}, \end{aligned} \tag{1-6.4}$$

where **H** is understood not to include the field of the dipole being summed.

We now wish to extend equation 1-6.4 for a permanent magnet. Direct passage from the sum to an integral over a volume element dv, in analogy to the step from equation 1-6.2 to equation 1-6.3, is not valid here.[9] The reason is that the field intensity which acts on a volume element dv, exclusive of its own field, depends on the shape of dv. One way out of this difficulty

[9] E. C. Stoner, *Phil. Mag.* **23,** 833 (1937); W. F. Brown, Jr., *Revs. Mod. Phys.* **25,** 131 (1953).

is to consider a cubic lattice of dipoles and to assume that the dipoles vary linearly (in direction) with position, to within negligible error, over distances containing many lattice spacings. It will also be assumed that there is no strain. The field acting on a dipole can then be evaluated by a method developed by Lorentz.

This method is as follows. A dipole on which the local field is to be calculated is considered as the center of a sphere of radius R that is large compared to the interdipole distance but small compared to macroscopic distances (Fig. 1-6.2). The difference between the field $\mathbf{H}_1$ of matter outside the sphere and the macroscopic field $\mathbf{H}$ is now computed. Both fields can be found from equation 1-3.6 by applying equation 1-2.3.

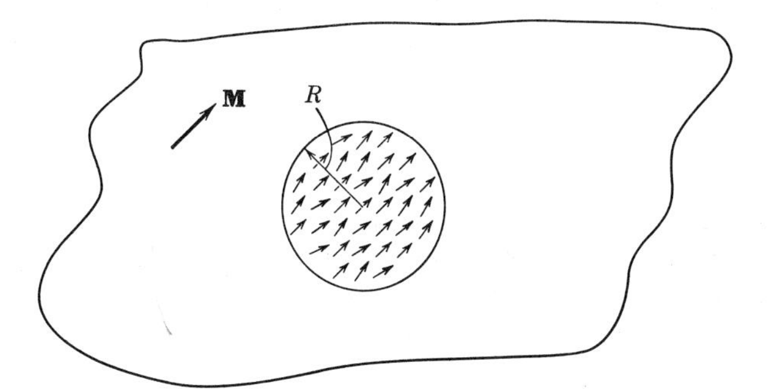

Fig. 1-6.2. The calculation of the Lorentz field.

However, $\mathbf{H}_1$ differs from $\mathbf{H}$ by omitting the volume integral over the interior of the sphere R and by including instead a surface integral over the surface of the sphere R. Now because of the assumption of the linear variation of μ, hence $\mathbf{M}$, with position $\nabla \cdot \mathbf{M} =$ constant, and the volume integral vanishes by symmetry. For the surface integral the expansion of $\mathbf{M}$ about the center contributes zero by symmetry, and $\mathbf{M}$ may be set equal to its value at the center of the sphere. Thus computation of the surface integral is elementary and gives for the difference $\mathbf{H}_1 - \mathbf{H}$ just $(4\pi/3)\mathbf{M}$. We must also compute the field $\mathbf{H}_2$ because of the dipoles inside the sphere. If the dipoles are parallel, it can be shown that $\mathbf{H}_2 = 0$; otherwise $\mathbf{H}_2 \neq 0$ in general. If $\mathbf{H}_2 = 0$, the total field $\mathbf{H}_1 + \mathbf{H}_2$ $(= \mathbf{H}_1)$ is known as the Lorentz field and is given by

$$\mathbf{H}_{\text{Lorentz}} = \mathbf{H} + \frac{4\pi}{3}\mathbf{M}. \tag{1-6.5}$$

If $\mathbf{H}_2 \neq 0$, the small energy term that results will not be considered as magnetic energy but will be lumped into other energy terms, such as anisotropy energy, to be discussed in Chapter 6.

In extending equation 1-6.4 to a permanent magnet, we make the following replacements:

$$\boldsymbol{\mu} \rightarrow \mathbf{M}\, dv \qquad \mathbf{H} \rightarrow \mathbf{H} + \frac{4\pi}{3}\mathbf{M} \qquad \Sigma \rightarrow \int$$

so that

$$W = -\tfrac{1}{2}\int \mathbf{M} \cdot \left(\mathbf{H} + \frac{4\pi}{3}\mathbf{M}\right) dv$$

$$= -\tfrac{1}{2}\int \mathbf{M} \cdot \mathbf{H}\, dv - \frac{2\pi}{3}\int M^2\, dv.$$

For a permanent magnet with ellipsoidal shape **M** is a constant, and the second term can be neglected by shifting the zero point of energy. Then we have

$$W = -\tfrac{1}{2}\int \mathbf{M} \cdot \mathbf{H}\, dv. \tag{1-6.6}$$

The result also holds for a system of permanent magnets, the integration then being over all the magnetic specimens. The $\frac{1}{2}$ factor of equation 1-6.6 is the usual one found in all self-energy expressions.

For a permanent magnet its self-field is given by (1-5.1) as

$$\mathbf{H} = -D\mathbf{M}.$$

Substituting in (1-6.5) gives

$$W = \tfrac{1}{2}DM^2 \tag{1-6.7}$$

per unit volume.

Expressions 1-6.5 and 1-6.6 indicate the energy to be localized at the magnetic particle. We now develop an equation which shows that the energy can be considered to be in the field, distributed throughout space. This result is in accordance with the ideas of Faraday and Maxwell.

First, we establish two relationships that will be useful in the derivation. By the divergence theorem

$$\int \nabla \cdot (\mathbf{H}\varphi)\, dv = \int \mathbf{n} \cdot \mathbf{H}\varphi\, dS.$$

Now consider the integration as being taken over the surface of an infinite sphere. Since $\varphi \propto 1/r^2$, $dS \propto r^2$, and $\mathbf{n} \cdot \mathbf{H} = 0$, we get

$$\int \nabla \cdot (\mathbf{H}\varphi)\, dv = 0. \tag{1-6.8}$$

By similar reasoning

$$\int \nabla \cdot (\mathbf{M}\varphi)\, dv = 0. \tag{1-6.9}$$

Now, returning to (1-6.5),

$$W = -\tfrac{1}{2}\int \mathbf{M}\cdot\mathbf{H}\,dv$$

$$= +\tfrac{1}{2}\int \mathbf{M}\cdot\nabla\varphi\,dv.$$

The integration can now be considered to be taken over all space, since **M** equals zero everywhere except in the permanent magnet. By virtue of equation 1-6.9

$$W = -\tfrac{1}{2}\int [\nabla\cdot(\mathbf{M}\varphi) - \mathbf{M}\cdot\nabla\varphi]\,dv$$

$$= -\tfrac{1}{2}\int (\nabla\cdot\mathbf{M})\varphi\,dv. \qquad (1\text{-}6.10)$$

Now, since

$$\mathbf{M} = \frac{1}{4\pi}(\mathbf{B} - \mathbf{H})$$

$$W = -\frac{1}{8\pi}\int [\nabla\cdot(\mathbf{B} - \mathbf{H})]\varphi\,dv.$$

It is one of Maxwell's fundamental relationships that $\nabla\cdot\mathbf{B} = 0$. By using equation 1-6.8,

$$W = -\frac{1}{8\pi}\int [\nabla\cdot(\mathbf{H}\varphi) - (\nabla\cdot\mathbf{H})\varphi]\,dv$$

$$= -\frac{1}{8\pi}\int \mathbf{H}\cdot\nabla\varphi\,dv$$

$$= \frac{1}{8\pi}\int H^2\,dv, \qquad (1\text{-}6.11)$$

and this is the equation we set out to develop.

7. Magnetic Effects of Currents. The Magnetic Shell. Faraday's Law

In 1820 Oersted discovered that an electric current produced a magnetic field. It was found by Ampère's experiments that the work per unit pole or the value of the line integral $\oint \mathbf{H}\cdot d\mathbf{s}$ (see equation 1-2.7) was no longer zero if the path taken enclosed the wire. Instead, it was directly proportional to the current i. Since the units of **H** have already been defined, this result permits a definition of the unit of current. For i to be in electromagnetic units, the constant of proportionality is taken as 4π so that

$$\oint \mathbf{H} \cdot d\mathbf{s} = 4\pi i. \tag{1-7.1}$$

The work done is no longer independent of the path. The use of the concept of the magnetostatic scalar potential to determine the magnetic field is thus limited to permanent magnet problems and situations in which the expression $\oint \mathbf{H} \cdot d\mathbf{s}$ is evaluated along paths that do not enclose current-carrying wires.

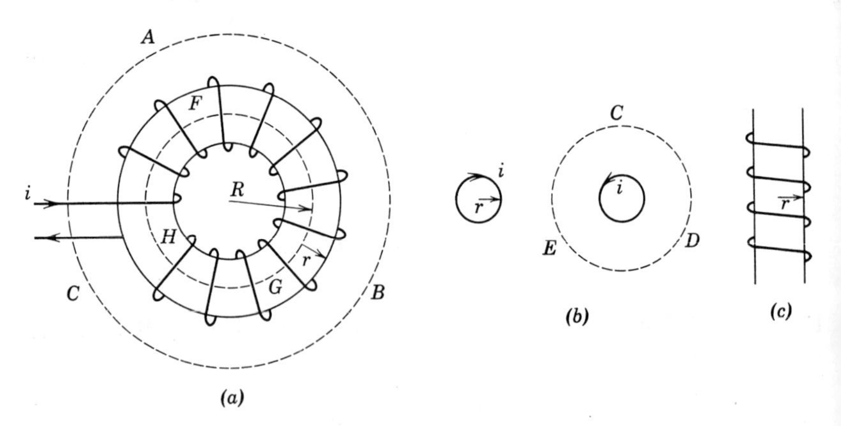

Fig. 1-7.1. The toroid is shown from above in (*a*) and in a vertical cross section in (*b*). The result when $R \to \infty$ is the solenoid shown in (*c*).

Application of (1-7.1) leads quickly to the calculation of **H** for a steady current if there is some symmetry to the problem. For example, if the path of integration for a straight wire of circular section is a concentric circle about the axis, and since by symmetry H is a constant, then $2\pi rH = 4\pi i$ and therefore $H = 2i/r$.

Consider now a toroid wound uniformly with n turns of wire per unit length. Figure 1-7.1*a* shows a view of the toroid from above a plane passing through its axis, whereas Fig. 1-7.1*b* shows a cross section for a plane perpendicular to the axis. Suppose the toroid is described by the radii R and r. We wish to determine the field when the toroid carries a current i. First we find the field outside the coil. On a circumferential path such as ABC of Fig. 1-7.1*a* no current is enclosed and $H = 0$. On a path such as CDE (Fig. 1-7.1*b*) one turn is crossed and $\oint \mathbf{H} \cdot d\mathbf{s} = 4\pi i$. Usually n is fairly large and i very small, so that $4\pi i$ can be neglected.

Hence the field outside the toroid is essentially zero. For a path inside the toroid and lying in a plane perpendicular to the axis no current is enclosed and therefore $H = 0$. For a path along the axis of the toroid (FGH) the field is

$$H = 4\pi ni. \tag{1-7.2}$$

Thus the only field is one inside the toroid and parallel to its axis. If now we let R approach infinity, in the limit the toroid becomes an infinitely long cylinder called a solenoid. The field inside the solenoid is parallel to the axis and uniform, equal to $4\pi ni$. If i is in amperes, the units are converted by dividing this result by 10; if it is in electrostatic units, the results are divided by $c = 3 \times 10^{10}$.

When the solenoid is of finite length and its length is appreciably larger than its radius, equation 1-7.2 is still accurate to a good degree at points not too close to the ends.

Often a solenoid is employed experimentally as a source of a uniform magnetic field. Practical difficulties limit the magnitude of the field that can be obtained when reasonable uniformity is required over a volume of several cubic centimeters. With only air cooling a rather husky coil is required to produce a continuous field in excess of 2000 oe. By passing cooling water through copper tubing, steady fields in the vicinity of 10,000 oe can be achieved.

Special winding arrangements, coupled with 2-Mw, high-current generators and large capacity cooling systems have permitted the attainment of static fields of over 100,000 oe[10]; the production of fields of 250,000 oe are planned for the future.[11] Besides water, organic fluids such as kerosene and orthodichlorobenzene, have been employed as coolants. Liquid N_2 and H_2 have also been used for cooling.[12] Low temperatures have the great advantage that the resistance of the electrical conductors is decreased by a factor of about 5. Hence the energy dissipated in the coil and the power source are also reduced by a factor of five.

[10] J. D. Cockroft, *Phil. Trans. Roy. Soc. (London)* **227,** 325 (1928); F. Bitter, *Rev. Sci. Instr.* **7,** 482 (1936), **10,** 373 (1939); J. M. Daniels, *Proc. Phys. Soc. (London)* **B-63,** 1028 (1950); F. Gaume, *J. Rech. Centre Natl. Rech. Sci.* **43,** 93 (1958); D. de Klerk, *Ned. Tijdschr. Natuurk.* **26,** 1 (1960), **26,** 345 (1960); S. Maeda, *High Magnetic Fields*, John Wiley and Sons, New York (1962), p. 406; R. S. Ingarden, *ibid.*, p. 427.

[11] B. Lax, *J. Appl. Phys.* **33,** 1025 (1962).

[12] T. W. Adair, C. F. Squire, and H. B. Utley, *Rev. Sci. Instr.* **31,** 416 (1960); C. E. Taylor and R. F. Post, *Advan. Cyrog. Eng.*, Plenum Press, New York (1960), Vol. 5; E. S. Borovik, F. I. Busel, and S. F. Grishin, *Zhur. Tekh. Fiz.* **31,** 459 (1961) [trans. *Sov. Phys.-Tech. Phys.* **6,** 331 (1961)]; J. R. Purcell, *High Magnetic Fields*, John Wiley and Sons, New York (1962) p. 166; H. L. Laquer, *ibid.*, p. 156.

Larger fields can be created by pulse techniques.[13] These methods consist of discharging a bank of condensers through a coil by means of some kind of switch. Fields of 150,000 oe are comparatively easy to make; with care, useful fields of close to 500,000 oe can be generated. Actually, fields in the neighborhood of 1 million oe have been produced, but the stresses on the coil are so great that it is usually destroyed. Pulsed fields have the disadvantage that special systems are required to detect the transient response.

Even larger fields have been produced by starting with a pulse field and then explosively reducing the volume of the flux container. The walls of the container act as perfect conductors; hence the total flux is conserved.[14] As a result the magnetic field is momentarily greatly increased. Fields in excess of 10 million oe lasting about 10^{-6} sec have been attained with this implosion technique.[15]

Solenoids constructed with hard superconductors are, at present, important sources of large continuous magnetic fields. They are likely, in the future, to be used even more widely. Hard superconductors are materials with zero electrical resistance that have the ability to remain superconducting in the presence of large magnetic fields. Hard superconductors are discussed at some length in Section 5-10. Finally, electromagnets, which are solenoids with soft iron cores, are also important sources of continuous magnetic fields; they will be considered in Section 1-9.

The magnetic shell. The magnetic effects of a current can be computed by using the concept of the equivalent magnetic shell. Such a shell is defined as a thin sheet of magnetic material, uniformly magnetized normal to its surface, the two surfaces having equal and opposite surface pole densities, $+\sigma$ and $-\sigma$. The boundary of the shell is taken to coincide with the current-carrying wire. If the surfaces of the shell are a small distance l apart, the moment or strength of the shell $\boldsymbol{\tau}$ is defined as

$$\boldsymbol{\tau} = \lim_{\substack{l \to 0 \\ \sigma \to \infty}} \sigma l \mathbf{n}$$

$$\sigma l = \text{constant},$$

[13] P. Kapitza, *Proc. Roy. Soc.* (*London*) **A-115,** 658 (1927); H. P. Furth and R. W. Waniek, *Rev. Sci. Instr.* **27,** 195 (1956); S. Foner and H. H. Kolm, *Rev. Sci. Instr.* **28,** 799 (1957); I. S. Jacobs and P. E. Lawrence, *Rev. Sci. Instr.* **29,** 713 (1958); D. H. Birdsall, *Rev. Sci. Instr.* **30,** 600 (1959); Y. B. Kim, *Rev. Sci. Instr.* **30,** 524 (1959); J. C. A. van der Sluijs, *High Magnetic Fields*, John Wiley and Sons, New York (1962), p. 290; L. W. Roeland and F. A. Muller, *ibid.*, p. 287; H. Zijlstra, *ibid.*, p. 281; M. A. Levine, *ibid.*, p. 277.

[14] Ia. P. Terletskii, *J. Exptl. Theoret. Phys.* (*USSR*) **32,** 387 (1957) [*trans. Soviet Phys.-JEPT*, 301 (1957)].

[15] C. M. Fowler, W. B. Garn, and R. S. Caird, *J. Appl. Phys.* **31,** 588 (1960).

where **n** is a unit normal vector drawn outward from the surface with positive pole density. Thus the moment or strength of a shell is the magnetic moment per unit area. A uniform shell is one for which τ is everywhere constant; hence a uniform shell need not be plane. Obviously the total dipole moment of the shell is given by the product of $\boldsymbol{\tau}$ and the area of the shell A.

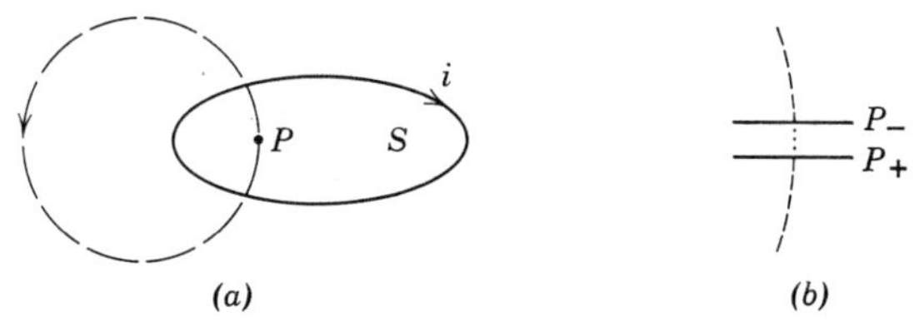

Fig. 1-7.2. S is the shell with wire-carrying current i as its boundary. For simplicity the shell is considered to be plane. The dotted curve is the path along which a unit pole is taken. (b) shows the points P_+, P_- at which the path crosses the shell of definite although infinitesimal thickness.

The potential at a point P due to a magnetic shell is easily calculated. By employing equation 1-3.5, we get

$$\varphi_P = \iint \boldsymbol{\tau}\, dS \cdot \boldsymbol{\nabla}_s\left(\frac{1}{r}\right),$$

where dS is an element of area of the shell so that its moment (strength) is $\tau\, dS$ and $\mathbf{r}$ is the vector from P to dS. Now, since $\boldsymbol{\nabla}_s(1/r) = -\mathbf{r}/r^3$

$$\varphi_P = \iint \tau \frac{\cos\theta}{r^2}\, dS,$$

where θ is the angle between $\boldsymbol{\tau}$ and $\mathbf{r}$. But $d\Omega = \pm(\cos\theta/r^2)\, dS$ is the solid angle of dS as seen at P. Therefore

$$\varphi_P = \iint \tau\, d\Omega.$$

Here $d\Omega$ is taken as positive if P faces the positive side of the shell. If τ is constant (uniform shell), then

$$\varphi_P = \tau\Omega. \tag{1-7.3}$$

Next we compute the work done on taking a unit pole around a path that crosses the magnetic shell once (Fig. 1-7.2).

By considering the potential, first of a plane surface distribution of poles, and then of a second plane with a surface distribution of poles of opposite sign, it may be shown[16] that the magnetic field between the layer

[16] M. Abraham and R. Becker, *Electricity and Magnetism*, Blackie and Son, London (1942), pp. 26–29.

is $4\pi\sigma$, which becomes infinite when $\sigma \to \infty$, plus a term that remains finite in the limit. The line integral of the field through the layer is $-4\pi\sigma l = -4\pi\tau$, a constant, plus a term that vanishes in the limit. The work done on the rest of the path is the difference in potential across the shell $\varphi_+ - \varphi_-$. Since the solid angle change is $\Omega = 4\pi$, we get $\varphi_+ - \varphi_- = 4\pi\tau$; hence the total work is zero. The field of a current differs from the field inside of the double layer by the singular term $4\pi\tau$; that is, the work done on traversing a circuit is $4\pi i$ (equation 1-7.1). Hence the field arising from a current-carrying circuit can be derived from a potential if complete traversals of the path are prevented by introducing a barrier surface. It then follows that a magnetic shell such as that defined above can be used to compute the field produced by a current if the magnetic moment τ is taken equal to the current i.

It is worth mentioning that other approaches are possible. A double layer can be defined as a surface distribution of dipoles, in which a dipole is defined as the limiting case of a pair of poles. Then, if a dipole is defined as the limiting case of a small circuit, we get a "double layer" with precisely the properties of a finite circuit. The little circuits side by side cancel each other everywhere except at the periphery.

Faraday's law. An effect, equally important as that investigated by Oersted and Ampère, was discovered by Faraday. He observed that a time-varying magnetic field caused a current to flow in a closed electrical circuit. Experimentally, it was found that

$$\oint_{\text{circuit}} \mathbf{E} \cdot d\mathbf{s} = -\frac{d}{dt}\int_A \mathbf{B} \cdot \mathbf{n}\, dS, \tag{1-7.4}$$

where A is a surface which has the circuit as its boundary. Here $\mathbf{E}$ is the electrostatic field, or force per unit charge, so that the left-hand side of equation 1-7.4 is just the work done on taking a unit electric charge about the path of the circuit (compare with equation 1-2.6). It is usual to refer to this term as the induced emf (electromotive force) $\mathcal{E}$. By employing equation 1-4.2 we obtain

$$\mathcal{E} = -\frac{d\Phi}{dt}. \tag{1-7.5}$$

The results of this section can be used to compute the energy required to change the magnetization of a material. Suppose that the material, of cross-sectional area A and length L, is placed in the solenoid, also of length L. If the solenoid is long and narrow, the magnetic field produced by a current i is given by equation 1-7.2,

$$H = 4\pi n i.$$

The induction in the solid is just (equation 1-4.1)

$$B = H + 4\pi M;$$

hence the total flux threading the solenoid is

$$\Phi = BnLA.$$

Suppose that this flux is changing with time; then an emf will be induced in the solenoid, given by equation 1-7.5 as

$$\mathcal{E} = -\frac{d\Phi}{dt}.$$

During this time work must be done to keep the current flowing; this energy is supplied by some power source such as a battery. If the amount of charge moved is dQ, the work that must be done is

$$\begin{aligned} dW &= \mathcal{E}\, dQ \\ &= -\frac{d\Phi}{dt}\, dQ \\ &= -i\, d\Phi \\ &= -\frac{H}{4\pi n}\, d[(H + 4\pi M)nLA] \\ &= -\frac{LA}{4\pi}(H\, dH + 4\pi H\, dM). \end{aligned}$$

Per unit volume of the material, the work done is

$$dW = -\frac{H\, dH}{4\pi} - H\, dM.$$

The first term is the work done to change the field H to $H + dH$; it is of no interest to us here. Let us take this term to the left-hand side of the equation and incorporate it into dW; for convenience this energy will still be called dW. This new dW is just the work done on increasing the magnetization of the material by an amount dM. Hence for a magnetic material we have[17]

$$dW = -H\, dM. \tag{1-7.6}$$

[17] This result can be obtained from a field, rather than a lumped parameter point of view. See, for example, E. A. Guggenheim, *Proc. Roy. Soc.* (*London*) **A-155,** 49, 70 (1936).

If the solenoid's current, hence its field, is increased or decreased slowly, the total amount of work done on the solid per unit volume is given by

$$W = -\int H\, dM, \tag{1-7.7}$$

where the limits of the integral are the initial and final values of M. If the material is isotropic so that equation 1-4.4 holds and the magnetic field is increased from 0 to H, the amount of work done per unit volume is

$$\begin{aligned} W &= -\int_0^H \chi H\, dH \\ &= -\tfrac{1}{2}\chi H^2. \end{aligned} \tag{1-7.8}$$

It can be shown that the form of equations 1-7.6 and 1-7.7 also holds if the body placed in the solenoid is ellipsoidal and so small that poles must be considered. By employing the equation for the self-field, $\mathbf{H} = -D\mathbf{M}$ (equation 1-5.1) we rederive equations 1-6.5 and 1-6.6.

8. Maxwell's and Lorentz's Equations

As a consequence of the experimental observations of Ampere and Faraday, Maxwell was led to the following equations:

$$\nabla \cdot \mathbf{D} = 4\pi\rho, \tag{1-8.1}$$

$$\nabla \cdot \mathbf{B} = 0, \tag{1-8.2}$$

$$\nabla \times \mathbf{E} = -\frac{1}{c}\frac{\partial \mathbf{B}}{\partial t}, \tag{1-8.3}$$

$$\nabla \times \mathbf{H} = \frac{1}{c}\left(4\pi\mathbf{j} + \frac{\partial \mathbf{D}}{\partial t}\right). \tag{1-8.4}$$

These equations are in gaussian units. Here $\mathbf{E}$ is the electric field, or force per unit charge, $\mathbf{D}$ is the displacement vector, $\mathbf{j}$ is the current density, and ρ is the electric charge volume density. $\mathbf{D}$ is related to $\mathbf{E}$ by the equation

$$\mathbf{D} = \mathbf{E} + 4\pi\mathbf{P}, \tag{1-8.5}$$

where $\mathbf{P}$ is the polarization of the electric dipole moment per unit volume. This relation is analogous to equation 1-4.1. $\mathbf{D}$ and $\mathbf{E}$ can also be related by

$$\mathbf{D} = \epsilon\mathbf{E}, \tag{1-8.6}$$

where ϵ is the dielectric constant. Often Maxwell's equations are considered as postulates instead of being derived from the foregoing experiments. The ability of the equations to predict vast numbers of experimental results correctly is then considered proof of their validity.

Because of equation 1-8.2 and since $\nabla \cdot \nabla \times \mathbf{A} \equiv 0$, a vector potential $\mathbf{A}$ can be defined as

$$\mathbf{B} = \nabla \times \mathbf{A}. \tag{1-8.7}$$

If the permeability μ is equal to 1, a dipole moment $\boldsymbol{\mu}$ will have a vector potential given by

$$\mathbf{A} = -\boldsymbol{\mu} \times \nabla\left(\frac{1}{r}\right) = \frac{1}{r^3}\boldsymbol{\mu} \times \mathbf{r}. \tag{1-8.8}$$

This follows since

$$\begin{aligned} \mathbf{H} = \mathbf{B} &= \nabla \times \mathbf{A} \\ &= \nabla \times \left[\boldsymbol{\mu} \times \nabla\left(\frac{1}{r}\right)\right] \\ &= (\boldsymbol{\mu} \cdot \nabla)\, \nabla\left(\frac{1}{r}\right) - \boldsymbol{\mu}\, \nabla^2\left(\frac{1}{r}\right). \end{aligned}$$

The term $\boldsymbol{\mu}\, \nabla^2(1/r) = 0$ except at $r = 0$. Therefore

$$\begin{aligned} \mathbf{H} &= \nabla\left(\boldsymbol{\mu} \cdot \nabla \frac{1}{r}\right) \\ &= -\nabla\varphi \end{aligned}$$

as in equation 1-2.3. The vector potential is used when currents that cannot be replaced by magnetic shells are present. It is also useful to introduce an electric scalar potential φ_E, defined by

$$\mathbf{E} = -\nabla\varphi_E - \frac{1}{c}\frac{\partial \mathbf{A}}{\partial t}. \tag{1-8.9}$$

Frequently the equation that expresses the force on a moving charge is required. Lorentz showed that this force is given by

$$\mathbf{F} = e\mathbf{E} + \frac{e}{c}\mathbf{v} \times \mathbf{H}, \tag{1-8.10}$$

where $\mathbf{v}$ is the velocity of the charge.

It is of interest to consider the transformation of Maxwell's equations to coordinate systems moving linearly with respect to the original frame of reference. According to the special theory of relativity, the length element

$$ds^2 = dx^2 + dy^2 + dz^2 - c^2\, dt^2$$

is an invariant quantity; that is $ds'^2 = dx'^2 + dy'^2 + dz'^2 - c^2\, dt'^2 = ds^2$. The linear transformations which keep ds^2 invariant are known as Lorentz transformations. If we let $x_1 = x$, $x_2 = y$, $x_3 = z$, and $x_4 = ict$, the

Lorentz transformation may be written

$$x_i' = \sum_{j=1}^{4} a_{ij} x_j, \qquad i = 1, 2, 3, 4. \tag{1-8.11}$$

For a frame of reference Σ' moving with a velocity $\mathbf{v}$ parallel to the y-axis of the Σ frame the transformation coefficients are given by

$$a_{ij} = \begin{pmatrix} 1 & 0 & 0 & 0 \\ 0 & \gamma_r & 0 & i\gamma_r\beta \\ 0 & 0 & 1 & 0 \\ 0 & -i\gamma_r\beta & 0 & \gamma_r \end{pmatrix}, \tag{1-8.12}$$

where $\beta = v/c$ and $\gamma_r = 1/\sqrt{1 - (v/c)^2}$.

It is desirable to rewrite Maxwell's equations so that their terms, although not invariant, transform with the same properties; when this is done, the equations are said to be covariant in form. Covariance may be achieved by writing the current $\mathbf{j}$ and charge density ρ as the four-vector

$$j_i = (\mathbf{j}, ic\rho), \tag{1-8.13}$$

the vector potential $\mathbf{A}$ and the electric scalar potential φ_E as the four-vector

$$A_i = (\mathbf{A}, i\varphi_E), \tag{1-8.14}$$

and the field quantities $\mathbf{E}$ and $\mathbf{H}$ by the antisymmetric tensor of rank two,

$$\mathscr{F}_{ij} = \begin{pmatrix} 0 & H_3 & -H_2 & -iE_1 \\ -H_3 & 0 & H_1 & -iE_2 \\ H_2 & -H_1 & 0 & -iE_3 \\ iE_1 & iE_2 & iE_3 & 0 \end{pmatrix}. \tag{1-8.15}$$

Equations 1-8.1 and 1-8.4 then take the form

$$\sum_{j=1}^{4} \frac{\partial \mathscr{F}_{ij}}{\partial x_j} = \frac{4\pi}{c} j_i, \qquad i = 1, 2, 3, 4. \tag{1-8.16}$$

To be covariant, this equation must take the form

$$\sum_{l=1}^{4} \frac{\partial \mathscr{F}_{kl}'}{\partial x_l'} = \frac{4\pi}{c} j_k' \tag{1-8.17}$$

in the Σ' reference frame. That this is indeed true is easily seen by applying the transformation of equation 1-8.11; equation 1-8.17 then becomes

$$\sum_{i=1}^{4} a_{ki} \left(\sum_{j=1}^{4} \frac{\partial \mathscr{F}_{ij}}{\partial x_j} - \frac{4\pi}{c} j_i \right) = 0.$$

Similarily, equations 1-8.2 and 1-8.3 reduce to

$$\frac{\partial \mathscr{F}_{ij}}{\partial x_k} + \frac{\partial \mathscr{F}_{ki}}{\partial x_j} + \frac{\partial \mathscr{F}_{jk}}{\partial x_i} = 0. \tag{1-8.18}$$

Finally, the Lorentz force (equation 1-8.10) has the covariant form

$$\mathbf{F}_i = \frac{1}{c} \mathscr{F}_{ij} j_j. \tag{1-8.19}$$

It is instructive to see how the field quantities themselves transform. For a coordinate system Σ' moving with a velocity $\mathbf{v}$ relative to the reference frame Σ, it follows from the general transformation relation $\mathscr{F}_{ij}' = a_{ik} a_{jl} \mathscr{F}_{kl}$ and equation 1-8.12 that

$$E_{\|}' = E_{\|}, \qquad H_{\|}' = H_{\|}$$

and

$$\mathbf{E}_{\perp}' = \gamma_r \left(\mathbf{E}_{\perp} + \frac{\mathbf{v}}{c} \times \mathbf{H} \right), \qquad \mathbf{H}_{\perp}' = \gamma_r \left(\mathbf{H}_{\perp} - \frac{\mathbf{v}}{c} \times \mathbf{E} \right), \tag{1-8.20}$$

where the subscripts $\|$ and $\perp$ refer to directions parallel and perpendicular, respectively, to $\mathbf{v}$. It is clear that $\mathbf{E}$ and $\mathbf{H}$ are not unique with respect to linear transformations. The physical origin of these results is closely related to the laws of Ampère and Faraday. Consider a charge or a permanent magnet at rest in the system Σ. In Σ' the charge or permanent magnet will be moving with respect to the axes. There will then be additional electric ($\mathbf{v}/c \times \mathbf{H}$) and magnetic ($\mathbf{v}/c \times \mathbf{E}$) fields present. Further details on the relativistic formulation of electrodynamics may be found in a number of standard textbooks.[18]

9. The Magnetic Circuit

The magnetic circuit, a concept analogous to the electric circuit, is of considerable use in solving many practical problems such as the design of electromagnets. We now develop this concept.

The work done on taking a unit pole around a closed path that goes through a solenoid of N turns is, by equation 1-7.1,

$$\oint \mathbf{H} \cdot d\mathbf{s} = \frac{4\pi N i}{10},$$

[18] P. G. Bergmann, *Introduction to the Theory of Relativity*, Prentice-Hall, Englewood Cliffs, N.J. (1942) Ch. 1–9; R. Becker, *Theorie der Elektrizität*, B. G. Teubner, Leipzig (1949), Sections 24–66; W. K. H. Panofsky and M. Phillips, *Classical Electricity and Magnetism*, Addison-Wesley, Cambridge, Mass. (1955), Ch. 14 and 15; J. D. Jackson, *Classical Electrodynamics*, John Wiley and Sons, New York (1962), Ch. 11.

where i is in amperes. The integral $\oint \mathbf{H} \cdot d\mathbf{s}$ is often called the magnetomotive force (mmf) since it is similar to the definition of the electromotive force (emf, equation 1-7.5). The mmf can be considered responsible for the production of the flux in a magnetic circuit, just as the emf causes current to flow in an electric circuit. In analogy to electrical resistance a magnetic reluctance is defined as

$$\mathcal{R} = \frac{\oint \mathbf{H} \cdot d\mathbf{s}}{\int \mathbf{B} \cdot \mathbf{n}\, dS} = \frac{\oint \mathbf{H} \cdot d\mathbf{s}}{\Phi}, \tag{1-9.1}$$

where Φ is the flux from equation 1-4.2. For the usual situation in which $\mathbf{B}$ is uniform,

$$\mathcal{R} = \frac{l}{\mu A}, \tag{1-9.2}$$

where l is the length of the circuit, A the cross-sectional area, and μ the permeability. When the magnetic circuit is made up of parts of length l_i, areas A_i, and permeabilities μ_i, equation 1-9.2 becomes

$$\mathcal{R} = \sum_i \frac{l_i}{\mu_i A_i}. \tag{1-9.3}$$

Usually a certain flux density in an air gap in the magnetic circuit is desired. Therefore for some particular system it is necessary to find the required value of the product Ni, or the ampere-turns. Since for air $\mu = 1$ and for a common ferromagnetic material such as iron $\mu \doteq 1000$, the reluctance in the short air gap is much greater than that in the iron part of the circuit. As stated in Section 1-4, μ is not a constant for iron but depends on the value of H or Ni. As a result, the solution of the magnetic circuit problem often involves a numerical successive approximation technique. This occurs most often in problems in which Ni is given and B is to be calculated.[19]

The magnetic flux in an air gap tends to spread out and have a larger value of cross-sectional area A_g than in the iron (A). A useful rule of thumb, based on experiment, is to increase each linear dimension of A by twice the length of the air gap in order to find A_g. If the air gap is very large, as, for example, in a "horseshoe" magnet, the reluctance can be

[19] R. A. Galbraith and D. W. Spence, *Fundamentals of Electrical Engineering*, Ronald Press, New York (1955), Ch. 9.

found by plotting the flux lines in the gap and drawing the corresponding magnetic equipotential lines.[20]

Examples of the design of some electromagnets are the following:
Uppsala University magnet. L. Dreyfus, *Arch. Electrotech.* **25**, 392 (1931); *ASEA J.* **12**, 8 (1935).
Leiden. G. Häden, *Siemens-Z.* **10**, 481 (1930).
Academie des Sciènces. S. Rosenblum, *Compt. rend.* (*Paris*) **188**, 1401 (1929).
Cambridge University. J. D. Cockcroft, *J. Sci. Instr.* **10**, 71 (1933).
Very large magnets for synchrocyclotrons, etc. For example, H. L. Anderson et al. *Rev. Sci. Instr.* **23**, 707 (1952).

10. Dipole in a Uniform Field

If a dipole of moment $\boldsymbol{\mu}$ is placed in a uniform field, it is subjected to a torque **L**, given by

$$L = -\frac{\partial W}{\partial \theta} = -\mu H \sin \theta, \qquad (1\text{-}10.1)$$

where θ is the angle between $\boldsymbol{\mu}$ and **H** and W is given by equation 1-6.1. In vectors, remembering that the cross product anticommutes, this can be rewritten as

$$\mathbf{L} = \boldsymbol{\mu} \times \mathbf{H}. \qquad (1\text{-}10.2)$$

The action of the uniform field **H** is thus to tend to rotate the dipole until it is parallel to **H**. This action, illustrated in Fig. 1-10.1, develops because we have assumed that the dipole acts like a conventional mechanical system, that is, that the dipole is a rigid body, capable only of rotation about an axis. In order to discuss the dynamical behavior of the dipole, we must consider how a dipole moment originates. For simplicity we

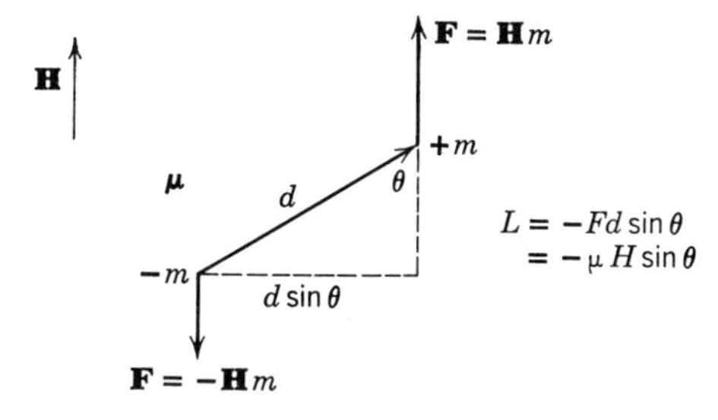

Fig. 1-10.1. Torque on a dipole produced by an external field **H**.

[20] For example, see W. B. Boast, *Principles of Electric and Magnetic Fields*, Harper and Bros, New York (1948), Ch. 14.

will assume that the dipole moment $\boldsymbol{\mu}$ arises from the motion of an electron in a circular orbit. Let r be the radius of the orbit, e its charge, and T the period of rotation. The moving electron can be considered essentially as a current flowing in a wire that coincides with the orbit. The magnetic effects can then be deduced by considering the equivalent magnetic shell. From Section 1-7 we have $\boldsymbol{\mu} = \boldsymbol{\tau} A$, in which $\boldsymbol{\mu}$ is the total dipole moment of the shell, $\boldsymbol{\tau}$ the strength of the shell, and A the shell area. Since $\tau = i$, and the current in this case is e/T, with e in gaussian units, we get

$$\boldsymbol{\mu} = \frac{eA}{\tau c}\mathbf{n} \quad (A = \pi r^2). \qquad (1\text{-}10.3)$$

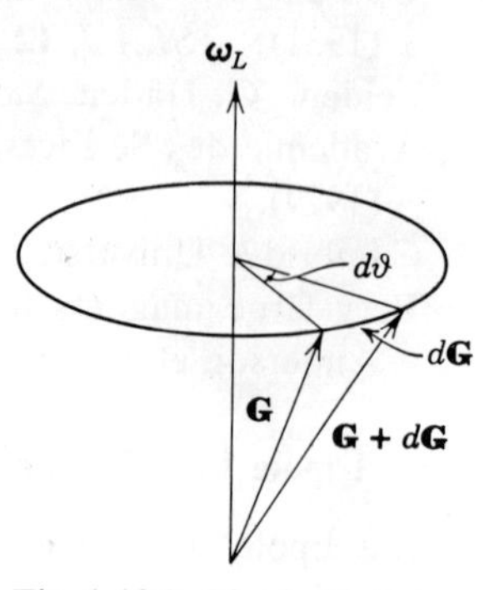

Fig. 1-10.2. Physical interpretation of equation 1-10.7.

Now, if m is the mass of the electron, the angular momentum is given by $\mathbf{G} = I\boldsymbol{\omega} = mr^2\boldsymbol{\omega} = mr^2(2\pi/T)\mathbf{n}$, in which I is the moment of inertia and ω $(= d\vartheta/dt)$ is the angular frequency. Substituting this into (1-10.3) gives

$$\boldsymbol{\mu} = \frac{e}{2mc}\mathbf{G}. \qquad (1\text{-}10.4)$$

Because the electron is negatively charged, the value to be inserted into this equation is $e = -4.80 \times 10^{-10}$; that is, the vectors $\boldsymbol{\mu}$ and $\mathbf{G}$ are antiparallel. Equation 1-10.4, together with Newton's law for angular momentum, $d\mathbf{G}/dt = \mathbf{L}$, and equation 1-10.2 give finally

$$\frac{d\mathbf{G}}{dt} = -\frac{e}{2mc}\mathbf{H} \times \mathbf{G}. \qquad (1\text{-}10.5)$$

Let us set

$$\boldsymbol{\omega}_L = \frac{-e}{2mc}\mathbf{H}; \qquad (1\text{-}10.6)$$

then (1-10.5) can be written as

$$d\mathbf{G} = \boldsymbol{\omega}_L \times \mathbf{G}\, dt. \qquad (1\text{-}10.7)$$

Consideration of Fig. 1-10.2 shows that this equation represents the equation of motion of a vector $\mathbf{G}$ precessing with an angular velocity $\omega_L = d\vartheta/dt$.

The action, therefore, of a uniform field $\mathbf{H}$ on such a dipole is to cause the moment $\boldsymbol{\mu}$ to precess about $\mathbf{H}$ with an angular frequency $eH/2mc$. This is in addition to the motion possessed by the electron before the application of the field. The precession is called the Larmor precession frequency, after the physicist who first derived it. These results also hold for noncircular orbits, provided only that the forces acting on the electron are central. In this derivation we have assumed that the angular momentum is left unchanged by the application of **H**. This is true to a first-order term in **H** (see problem 1-8). The precession phenomenon is the origin of diamagnetism and is discussed further in Chapter 2.

Problems

1-1. Show that for a uniformly magnetized sphere of radius a, the magnetic potential at a point distance $r > a$,

$$\varphi = \frac{4\pi}{3} Ma^3 \frac{\cos\theta,}{r^2}$$

where θ is the angle between **M** and **r**. Show that for $r < a$

$$\varphi = \frac{4\pi}{3} Mr\cos\theta.$$

1-2. From the result of problem 1-1, show that for a sphere $D = 4\pi/3$.

1-3. Show that the self-interaction energy of a permanent magnet is

$$\tfrac{1}{2}\int \rho\varphi \, \mathrm{dv} + \tfrac{1}{2}\int \sigma\varphi \, dS,$$

where the symbols have the meaning used in the text.

1-4. Use the result of problem 1-3 to calculate the energy of an ellipsoidal specimen, volume v_1, which contains an ellipsoidal cavity, volume v_2. [*Hint.* Assume first an ellipsoid, volume v_1, with uniform magnetization M, then another volume v_2, with uniform magnetization $-M$, and consider the interaction between these ellipsoids. See W. F. Brown Jr., and A. H. Morrish, *Phys. Rev.* **105**, 1198 (1957).]

1-5. Show that a permanent magnet, permeability μ, permanent magnetic moment M_0, has the field energy

$$W = \frac{1}{8\pi}\int \mu H^2 \, dV.$$

[*Hint.* $W = \frac{1}{2}DM_0^2$ and M_0 is defined by the equation
$M = M_0 + (\mu - 1)/4\pi H$.]

1-6. An electromagnet whose magnetization M is uniform has truncated pole pieces of base radius b and gap face radius a. The center of the air gap coincides with the cone from which the truncated pieces are fashioned. Show that the field at the center of the air gap is given by

$$4\pi M\left(1 - \cos\theta + \sin^2\theta\cos\theta\log_e\frac{b}{a}\right),$$

where θ is the angle of the cone. If the poles on the flat face of the pole pieces are neglected, show that the maximum field is achieved for $\theta = 54° 44'$.

1-7. Show that the energy in a magnetic field caused (equation 1-6.11) by circuits carrying currents is given by $\frac{1}{2}\Sigma i\Phi$ where Φ is the flux that threads the current i.

1-8. Consider an electron moving in a circular orbit about a nucleus of charge e. Use the Lorentz equation to set up a force equation when a magnetic field **H** is applied and show $\omega = \pm [(eH/2mc)^2 + (e^2/mr^3)]^{1/2} - (eH/2mc)$. Consider the magnitudes of the terms and make a suitable approximation to obtain the Larmor frequency.

Bibliography

Abraham, M., and R. Becker, *Classical Theory of Electricity and Magnetism*, Blackie and Sons, London (1937).

Brown, W. F., *Magnetostatic Principles in Ferromagnetism*, North-Holland Publishing Co., Amsterdam (1962).

Jackson, J. D., *Classical Electrodynamics*, John Wiley and Sons, New York (1962).

Jeans, J. H., *Electricity and Magnetism*, Cambridge University Press, Cambridge (1925).

Kolm, H., B. Lax, F. Bitter, and R. Mills, editors, *High Magnetic Fields*, M.I.T. Press and John Wiley and Sons, New York (1962).

Panofsky, W. K., and M. Philips, *Classical Electricity and Magnetism*, Addison-Wesley Publishing Co., Cambridge, Mass. (1955).

Stoner, E. C., *Magnetism and Matter*, Methuen and Company, London (1934).

Stratton, J. A., *Electromagnetic Theory*, McGraw-Hill Book Co., New York (1941).

2

Diamagnetic and Paramagnetic Susceptibilities

1. Introduction

Interest in the aspects of diamagnetism and paramagnetism described in this chapter dates back before 1900. In a sense, then, the material in this chapter can be considered as the classical aspects of the subject. This certainly does not imply that only nonquantum mechanical concepts are employed or that important experimental and theoretical investigations are not carried out today. Rather it separates experiments and theories concerning the bulk susceptibilities of matter from those that investigate the details of the level system and the relaxation processes. The first employ only a static magnetic field, whereas the second require the use of an alternating magnetic field, usually in addition to a static one. The last few years have seen a marked increase in the number of publications on static susceptibilities. On the experimental side this renewed activity has sprung from the easy availability of high purity chemicals, excellent quality single crystals, sometimes of unusual materials, and liquid helium. Theoretical work has been stimulated by the new experimental data, greater knowledge of wave functions, and improved computation techniques.

The chapter is begun with a review of results from quantum mechanics relating to atomic spectroscopy, together with a summary of the relationship between chemical binding and magnetic moments. The next section on diamagnetism presents a derivation of Langevin's equation and summarizes the pertinent experimental results. Paramagnetism is then treated, first in general, and second the various groups of paramagnets are considered in detail. The diamagnetism and paramagnetism of metals and alloys depend on the properties of the free electron gas and therefore form a somewhat special topic that is treated in Chapter 5.

2. Review of Quantum Mechanical and Other Results

It is assumed that the reader is acquainted with elementary quantum mechanics. Nevertheless, a brief summary of some pertinent results will be given in order to be available for ready reference.

The classical Hamiltonian $\mathcal{H}$ represents the energy of a system. For example, for a simple system such as a particle with linear momentum p, mass m, moving in a potential V, the Hamiltonian is

$$\begin{aligned} \mathcal{H} &= T + V \\ &= \frac{p^2}{2m} + V = E. \end{aligned} \tag{2-2.1}$$

Here E is the total energy, the sum of the kinetic energy T and the potential energy V. An operator for the Hamiltonian is obtained by replacing the components of momentum by $(\hbar/i)(\partial/\partial q)$ where $\hbar$ is Planck's constant h divided by 2π and q is the coordinate along which the component of p is determined (usually a Cartesian or some other orthogonal coordinate). By operating on a wave function ψ with this Hamiltonian, we obtain Schrödinger's equation,

$$\left[-\frac{\hbar^2}{2m}\left(\frac{\partial^2}{\partial x^2} + \frac{\partial^2}{\partial y^2} + \frac{\partial^2}{\partial z^2}\right) + V\right]\psi = E\psi. \tag{2-2.2}$$

Because the wave functions are required to satisfy certain conditions, such as being zero at infinity, solutions are obtained usually only for certain energies or eigenvalues of E. The value of the wave function for different values of E is said to represent different states of the system.

Application of Schrödinger's equation to the simplest atom, that of hydrogen, and the extension to other more complicated atoms, has led to information on the states that the electrons can occupy. These states are characterized by four quantum numbers. According to the Pauli principle, only one electron can occupy a state specified by some particular combination of quantum numbers. These numbers and their significance will now be itemized.

The Principal Quantum Number n. The value of this number primarily determines the energy of a particular shell or orbit. The number must be integral and nonzero. The shells are called K, L, M, N, O, P, and Q when n equals 1, 2, 3, 4, 5, 6, and 7, respectively.

The Orbital Angular Momentum Quantum Number l. For a particular value of l, the total angular momentum of an electron due to its orbital motion is given by

$$[l(l+1)]^{1/2}\,\hbar. \tag{2-2.3}$$

This quantum number is integral and is restricted to the values $l = 0, 1, 2, \ldots, (n-1)$. Electrons associated with such values of l are called s, p, d, f, and g electrons, respectively.

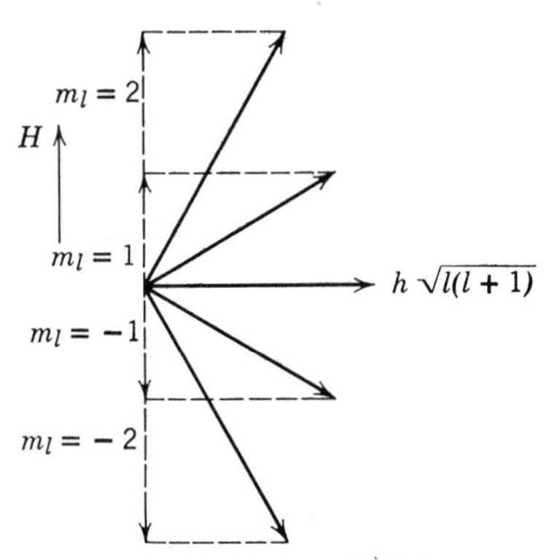

Fig. 2-2.1. Vector model of the atom.

The Magnetic Quantum Number m_l. This quantum number determines the component of the orbital angular momentum along a specified direction, usually that of an applied magnetic field. This number is also integral and for a given value of l may have $(2l + 1)$ possible values: $m_l = l, (l-1), \ldots, 0, \ldots, -(l-1), -l$. Thus a d electron has as permissible values of angular momentum along a field direction $2\hbar$, $\hbar$, 0, $-\hbar$, and $-2\hbar$. On the basis of the vector model of the atom, the plane of the electron's orbit can have only certain possible orientations; that is, the atom is spatially quantized (see Fig. 2-2.1).

The Spin Quantum Number m_s. An electron also rotates or spins about an internal axis and thus has an intrinsic angular momentum, called the spin. The quantum number m_s determines the component of the spin s along the direction of an applied field. The allowed values of m_s are $\pm\frac{1}{2}$, and thus the components of the spin angular momentum possible are $\pm\hbar/2$.

Magnetic moments. In accordance with equation 1-10.4, the existence of an electron with angular momentum leads to a magnetic moment. For the orbital angular momentum the magnetic moment is given by

$$|\mu| = \frac{|e|}{2mc}\,[l(l+1)]^{1/2}\hbar \tag{2-2.4}$$

and its projection along the direction of an applied field is

$$\mu_z = \mu_H = \frac{e\hbar}{2mc}\, m_l. \tag{2-2.5}$$

The quantity $e\hbar/2mc = -0.927 \times 10^{-20}$ erg/oe is known as the Bohr magneton μ_B.

The magnetic moment associated with the spin angular momentum is not given by equation 1-10.4. Instead, the right-hand side is multiplied by a factor g, called the spectroscopic splitting factor, the gyromagnetic

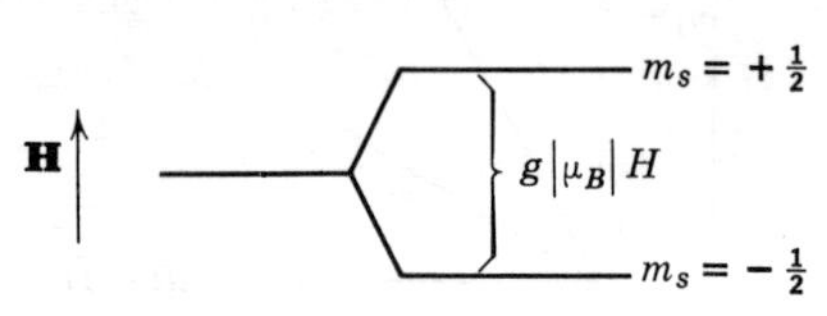

Fig. 2-2.2. Zeeman splitting of a twofold degenerate level by an applied field H.

factor, or simply the g-factor. The component of magnetic moment μ_{sz} along the field direction is thus given by

$$\mu_{sz} = g\left(\frac{e}{2mc}\right)\frac{\hbar}{2}, \tag{2-2.6}$$

where $g = 2.0023$ for a free electron.

When two or more electrons with different quantum numbers have the same energy, the state is said to be degenerate. For example, two electrons with $l = 0$ have spins of $+\frac{1}{2}$ and $-\frac{1}{2}$, respectively, and this energy level is twofold degenerate. Application of a magnetic field will split this level, for the electron whose spin is $+\frac{1}{2}$ has a magnetic moment antiparallel to the field, whereas that with $s = -\frac{1}{2}$ has μ_{sz} parallel to the field. Thus, by using the result of equation 1-6.1, the levels are split by

$$\begin{aligned} \Delta E &= 2\,|\mu_{sz}|\,H \\ &= g\left(\frac{|e|}{2mc}\right)\hbar H \\ &= g\,|\mu_B|\,H. \end{aligned} \tag{2-2.7}$$

This situation is illustrated in Fig. 2-2.2. Since the value of g determines the amount of the separation of the levels, it was natural that it be termed the splitting factor. The splitting was first observed in optical spectroscopy and is called the Zeeman effect.

In the vector model of the atom we introduce an additional quantum number j, which determines the total angular momentum due to vectoral combination of the orbital and spin angular momentum. Thus j is always half integral for a certain electron, being $l \pm \frac{1}{2}$.

Periodic table. The successive elements of the periodic table have an additional electron in the lowest possible energy state consistent with the Pauli principle. For a given value of n there are $2n^2$ possible values of l, m_l, and m_s. For a filled or closed shell the projection of the total angular momentum on a specified direction is zero. A table of the elements, together with the electron configuration of the lowest energy or ground state is given in Appendix III. The notation $1s^2$ means that there are two electrons (the superscript) in the $1s$ state ($n = 1, l = 0$) etc.

It is now necessary to consider how the orbital and spin momenta of the electrons of an atom are combined to form the atom's momentum. The method of combination that is important in magnetism is known as Russell-Saunders coupling. Here the l vectors of the various electrons are added (vectorially) to form a resultant $\mathbf{L}$, whereas the s vectors are summed to form a resultant $\mathbf{S}$. The resultants $\mathbf{L}$ and $\mathbf{S}$ are then combined to form the atomic angular momentum quantum number $\mathbf{J}$. J can assume the range of values

$$J = (L - S), \quad (L - S + 1), \ldots, (L + S - 1), \quad (L + S), \tag{2-2.8}$$

and such a group of levels is usually termed a *multiplet*. By definition, the multiplicity of the system is $2S + 1$. If $L \geqslant S$, there are indeed $2S + 1$ multiplets. However, if $L < S$, the full multiplicity is not developed, as there are then only $2L + 1$ multiplets. The multiplet spacing is determined by the spin-orbit coupling constant λ, defined so that the interaction energy is given by

$$\lambda \mathbf{L} \cdot \mathbf{S}. \tag{2-2.9}$$

Thus the Jth energy level is

$$\tfrac{1}{2}\lambda[J(J + 1) - L(L + 1) - S(S + 1)],$$

since $\mathbf{J}^2 = \mathbf{L}^2 + \mathbf{S}^2 + 2\mathbf{L} \cdot \mathbf{S}$ and $\mathbf{L}^2 = L(L + 1)$, $\mathbf{S}^2 = S(S + 1)$ and $\mathbf{J}^2 = J(J + 1)$.

The angular momentum is then given by

$$|\mathbf{G}| = [J(J + 1)]^{1/2}\hbar \tag{2-2.10}$$

and the splitting factor by the Landé formula

$$g = 1 + \frac{J(J + 1) + S(S + 1) - L(L + 1)}{2J(J + 1)}. \tag{2-2.11}$$

The magnetic moment is then $\mu = g(e/2mc)G$. The components of the magnetic moment parallel to the field may have $2J + 1$ values, namely $g\mu_B M_J$ where $M_J = J$, $(J - 1), \ldots, -(J - 1), -J$. Notice that in general μ is not codirectional with the angular momentum vector, since the value of g is different for an orbital and a spin magnetic moment. It is convenient to denote the components of L and S along the field direction by M_L and M_S.

The ground, or lowest energy, state of an atom is indicated by a capital letter, such as S[1], P, D, F, G, and represents the value of $L = 0$, 1, 2, 3, 4, . . . respectively. A prefix gives the value of $2S + 1$, the spin multiplicity of the energy levels. Prefixes of 1, 2, 3, . . . are read as singlets, doublets, triplets, etc. A suffix gives the J value. Examples are ${}^1S_0$ and ${}^5D_4$. In the first state $L = 0$, $S = 0$ and $J = 0$, and in the second $L = 2$, $S = 2$, and $J = 4$ (see Appendix III).

Hund's rules. From studies on spectra, Hund arrived at three rules that permit the prediction of the magnetic moment of free atoms, or ions, in their ground states. These rules follow:

1. $S = \Sigma m_{si}$ is the maximum allowed by the Pauli principle.
2. $L = \Sigma m_{li}$ is the maximum allowed by rule 1.
3. J for an incompletely filled shell is given by

$$J = L - S \text{ for a shell less than half occupied,}$$

$$J = L + S \text{ for a shell more than half occupied.}$$

These rules will be illustrated with a couple of examples. For Cr^{2+} the incompletely filled $3d$ shell has four electrons. Then $S = \frac{1}{2} + \frac{1}{2} + \frac{1}{2} + \frac{1}{2} = 2$, and $L = 2 + 1 + 0 - 1 = 2$ and $J = 2 - 2 = 0$, which gives a ${}^5D_0$ term. For Gd^{3+} there are seven $4f$ electrons. Here $S = 7 \times \frac{1}{2}$, $L = 3 + 2 + 1 + 0 - 1 - 2 - 3 = 0$, and $J = \frac{7}{2}$, for a ${}^8S_{7/2}$ ground term. For Er^{3+} there are eleven $4f$ electrons and $S = (7 \times \frac{1}{2}) - (4 \times \frac{1}{2}) = \frac{3}{2}$, $L = 0 + 3 + 2 + 1 + 0 = 6$, and $J = 6 + \frac{3}{2} = \frac{15}{2}$.

Nuclear angular momentum. The nucleus possesses a spin angular momentum and has associated a spin quantum number I such that the angular momentum is given by

$$G = [I(I + 1)]^{1/2}\, \hbar. \tag{2-2.12}$$

I can have half integral or integral values and, in projection along the direction of an applied magnetic field, can have the values $m_I = I$, $I - 1, \ldots, 0, \ldots, -(I - 1), -I$; that is, it is $(2I + 1)$-fold degenerate. This angular momentum gives rise to a nuclear magnetic moment given by

$$\mu_n = g_n \frac{e\hbar}{2M_p c}\, [I(I + 1)]^{1/2}, \tag{2-2.13}$$

[1] Not to be confused with the resultant spin quantum number S.

where M_p is the mass of the proton and g_n is a nuclear gyromagnetic ratio. The term $\mu_{Bn} = e\hbar/2M_pc = 5.05 \times 10^{-24}$ erg/oe is termed the nuclear Bohr magneton.

Weiss magneton. In the older literature a unit called the Weiss magneton μ_W is often met. It is related to the Bohr magneton by

$$\frac{\mu_B}{\mu_W} = 4.97. \tag{2-2.14}$$

Since it is purely an experimental unit, we shall not employ it in this book.

Energy units. The commonest energy unit is the electron volt. This is the energy an electron acquires when it goes through a potential difference of one volt. In terms of cgs units, 1 ev $= 1.60 \times 10^{-12}$ erg. In spectroscopy two other units are employed often to measure the energy between levels. These are the cm^{-1} and the degree, which are related to energy units through proportionality factors involving fundamental constants. They are derived from the relationships ev $= hc/\lambda = kT$. The conversion factors are given by 1 ev $= 8.065\ \text{cm}^{-1}$ and 1 $\text{cm}^{-1} = 1.436$°K.

Chemical binding and magnetic moments. Most free atoms and ions have an incompletely filled shell structure and thus have a magnetic moment. However, in practice, we are usually concerned with atoms in combination. In ionic compounds the positive ions lose as many electrons to the negative ions, so that each ion is left with a closed shell structure. Such *heteropolar binding* gives rise to molecules that have no magnetic dipole moment. In *homopolar* or *covalent binding*, for example H_2, N_2, electrons are usually shared so that no magnetic moment results. Exceptions are discussed under the section on paramagnetic gases. *Metallic binding* results from the interaction of free electrons and ion cores. It is these free electrons that give rise to a magnetic moment.

Nevertheless, a number of nonmetallic compounds have a permanent dipole moment. These compounds contain ions from certain parts of the periodic table. In some elements outer electrons may go into quantum states of higher principal quantum number n before all possible states of lower n are filled. For example, the $4s$ states are filled before the $3d$ and the $5s$, $5p$, and $6s$ states are filled before the $4f$. These outer electrons, and sometimes one or two electrons of the unfilled inner shell, are the valency electrons that partake in the bonding of compounds. The remaining inner electrons form an incomplete shell and give rise to a permanent magnetic moment. The *transition elements* that lead to magnetic ions are indicated in the periodic table in Appendix III.

DIAMAGNETISM

3. The Langevin Formula for Diamagnetic Susceptibility

The effect of the interaction of an external magnetic field on the electrons in orbitals in an atom can be understood qualitatively by considering Lenz's law of electricity. According to this law, when the magnetic flux enclosed by a current loop is changed, a current is induced in such a direction that its magnetic field opposes the original field. Thus the atom's electronic motion, considered as a current loop, will be changed in such a sense that a magnetic moment will be induced in a direction opposite to that of the applied field. Since the "resistance" in an electron orbit is zero, this diamagnetic moment will persist as long as the external field is applied.

A quantitative classical derivation of the diamagnetic susceptibility can be made by utilizing Larmor's precession theorem (see Chapter 1). From equation 1-10.4 the induced magnetic moment is given by

$$\begin{aligned}\boldsymbol{\mu} &= \left(\frac{e}{2mc}\right)\mathbf{G} \\ &= +\left(\frac{e}{2mc}\right)m\boldsymbol{\omega}_L\overline{\rho^2} \\ &= -\frac{e^2}{4mc^2}\mathbf{H}\overline{\rho^2},\end{aligned} \tag{2-3.1}$$

where $\boldsymbol{\omega}_L = -e\mathbf{H}/2mc$, the Larmor precession (equation 1-10.6), and $\overline{\rho^2} = \overline{x^2} + \overline{y^2}$ is the mean square radius of the electron's orbit when projected on a plane perpendicular to $\mathbf{H}$. The negative sign arises, of course, from the fact that the induced moment is opposite in direction to the applied field. If $\overline{r^2} = \overline{x^2} + \overline{y^2} + \overline{z^2}$ is the mean square distance from the nucleus, and on the average the distribution is spherically symmetrical so that $\overline{x^2} = \overline{y^2} = \overline{z^2}$, then $\overline{\rho^2} = \frac{2}{3}\overline{r^2}$. The atomic susceptibility is found by summing over the z electrons, and the susceptibility per cm^3 is obtained by multipling by N, the number of atoms per cm^3. Thus we get

$$\chi = -N\frac{e^2}{6mc^2}\sum_{i=1}^{z}\overline{r_i^2}. \tag{2-3.2}$$

This is Langevin's equation for the diamagnetic susceptibility, corrected for a numerical error by Pauli.[2]

[2] W. Pauli, *Z. Physik* **2**, 201 (1920).

Experimental values of χ can be used to give $\overline{r^2}$, or, conversely, values of $\overline{r^2}$ from other types of measurement or theory can be correlated with the observed χ. Taking $\overline{r_i^2} \simeq 10^{-16}$ cm^2, $N \simeq 5 \times 10^{22}$ cm^{-3}, and $Z \simeq 10$, we obtain the order of magnitude of χ as 10^{-6}.

To summarize, the following important conclusions may be drawn:

1. The diamagnetic susceptibility is always negative, since $\Sigma \overline{r_i^2} > 0$.
2. Diamagnetism is a property of all matter.
3. The temperature T does not enter explicitly into the susceptibility.
4. The amount of magnetization is proportional to $\Sigma \overline{r_i^2}$. Small variations of χ with temperature may be interpreted as a slight dependence of $\Sigma \overline{r_i^2}$ on T.

In the foregoing derivation it was assumed that the radius of a given orbit had only one particular size. J. H. van Leeuwen[3] pointed out that instead a continuous range of sizes was permitted for a given orbit. This classical picture led to the conclusion that the total susceptibility (diamagnetic plus paramagnetic) of matter was zero!

Application of even the old Bohr quantum theory, requiring, as it did, discrete quantum orbits instead of continuous ones, removed Miss van Leeuwen's objections and permitted a nonzero susceptibility. Quantum mechanics gives equation 2-3.2 exactly[4] for monatomic molecules, as shown in Section 2-8.

4. Susceptibility of Atoms and Ions

Satisfactory experimental evidence had been gathered for the existence of diamagnetism in some matter long before Langevin's work. In particular, Curie[5] made a comprehensive study of several substances over a wide temperature range and found that the susceptibility was independent of the applied magnetic field and almost independent of temperature. Since then, considerable further experimental work has been carried out. However, in the early literature, say prior to 1945, there is unfortunately considerable disagreement between the measurements. In the more recent investigations greatly increased precision is usually claimed.

Derivation of the Langevin equation implied that the electrons were under the influence of only one center of force. Strictly, then, equation 2-3.2 applies only to monatomic molecules. The rare gases are the ideal

[3] For a summary see J. H. Van Vleck, *Theory of Electric and Magnetic Susceptibilities*, Oxford University Press, Oxford (1932) Ch. IV.

[4] *Ibid*, Ch. VIII.

[5] P. Curie, *Ann. chim. et phys.* (7) **5**, 289 (1895).

examples of these systems, but polar salts such as Na^+Cl^- contain ions whose electrons are primarily controlled by one center of force. It is necessary to limit our discussion to the atoms and ions that have no orbital or spin moment ($^1S_0$ states), for otherwise a large paramagnetic contribution would exist.

On the theoretical side the problem of computing the diamagnetic susceptibility requires only the calculation of $\Sigma\,\overline{r^2}$. Although in principle quantum mechanics permits the calculation of the electronic charge distribution, in practice exact solutions have been obtained only for hydrogenlike atoms. For a hydrogenlike atom an electron under the influence of a nucleus of charge $+Ze$ has a mean square distribution given by

$$\overline{r^2} = a_0^2 \frac{n^2}{Z^2}\left[\frac{5}{2}n^2 - \frac{3l(l+1)-1}{2}\right], \tag{2-4.1}$$

where a_0 is the Bohr radius. This equation cannot be tested by experiment as hydrogenlike atoms have a nonzero spin and therefore a paramagnetic susceptibility. The equation can be extended to other electrons by considering the shielding of the nuclear charge by the inner electrons. This is achieved by the introduction of so-called screening constants σ, in which Z of equation 2-4.1 is replaced by $Z - \sigma$. Thus the contributions of the inner electrons may be approximately calculated and a value obtained for the susceptibility.

A more satisfying method of finding $\Sigma\,\overline{r^2}$ involves the use of approximate wave functions.[6] These are usually obtained by use of the Hartree[7] "self-consistent field" method. The years have seen the development of greatly improved Hartree-Fock wave functions, as they are called. The Thomas-Fermi statistical theory of the atom has also been employed[8] to calculate $\Sigma\,\overline{r^2}$. Theoretical values, obtained under various assumptions, are compared in Table 2-4.1 with the experimental ones.

Values of the ionic diamagnetic susceptibilities are obtained from the values for the salts, either in solution or in the solid form. According to the best available data, there are differences between the solution and the solid values. Some representative results are given in Table 2-4.2. The values quoted from Meyers are the means of some older experimental data.

It is usually assumed that the molecular susceptibility χ_M is equal to the sum of the kation and anion values, that is,

$$\chi_M = \chi_{\text{cation}} + \chi_{\text{anion}}. \tag{2-4.2}$$

[6] J. C. Slater, *Phys. Rev.* **36,** 57 (1930).
[7] D. R. Hartree, *Rept. Progr. Phys.* **11,** 113 (1946).
[8] T. Tietz, *Ann. Phys.* (*Leipzig*) **6,** 202 (1960).

Table 2-4.1. Diamagnetic Susceptibilities of Rare Gases

($\times 10^{-6}$ emu/mole)

Gas	Experiment	Theory
He	−2.02[a]	−1.86[c]
Ne	−6.96[a]	−5.8[d], −7.46[e], −11.4[f], −7.417[g], −7.48[h]
A	−19.23[b]	−18.8[f]

[a]C. Barter, R. G. Meisenheimer, and D. P. Stevenson, *J. Phys. Chem.* **64,**1312 (1960).
[b]G. G. Havens, *Phys. Rev.* **43,** 992 (1933).
[c]H. A. Bethe and E. E. Salpeter, *Quantum Mechanics of One- and Two-Electron Atoms*, Academic Press, New York (1962); R. Ya Damburg and E. M. Iolin, *J. Exptl. Theoret. Phys.* (*USSR*) **42,** 820 (1962) [trans. *Soviet Phys.-JEPT* **15,** 572 (1962)].
[d]K. E. Banyard, *J. Chem. Phys.* **34,** 338 (1961).
[e]B. H. Worsley, *Can. J. Phys.* **36,** 289 (1958).
[f]T. Tietz, *Ann. Phys.* (*Leipzig*) **6,** 202 (1960).
[g]L. M. Sachs, *Phys. Rev.* **124,** 1283 (1961).
[h]R. P. Hurst, M. Karplus, and T. P. Das, *J. Chem. Phys.* **36**, 2786 (1962).

This assumption is given some foundation when we notice that the susceptibility increments, as shown in Table 2-4.2, in going down the halide series are approximately constant. However, the results in Table 2-4.2 also show that the susceptibilities of the solids are generally 2 to 3%

Table 2-4.2. χ for Salts and Solutions

($\times 10^{-6}$ emu/mole)

	Solution				Solid		
Ion	H[a]	Li[a]	Na[a]	K[b]	Li[c]	Na[d]	K[b]
F	−9.0						
Cl	−21.8	−25.2	−30.3	−40.17	−24.3	−30.2	−38.71
Br	−32.8	−35.8	−42.0	−51.88	−34.7	−41.1	−49.60
I	−50.4	−53.7	−59.9	−68.23	−50	−57.0	−65.46

$MgCl_2$	−49.7
$CaCl_2$	−56

[a]W. R. Meyers, *Revs. Mod. Phys.* **24,** 15 (1952).
[b]V. C. G. Trew and S. F. A. Husain, *Trans. Faraday Soc.* **57,** 223 (1961).
[c]G. Foëx, *Diamagnetism and Paramagnetism*, Pergamon Press, London (1957).
[d]G. W. Brindley and F. E. Hoare, *Proc. Roy. Soc.* (*London*) **159,** 395 (1937).

less in absolute value then those of the solutions. From this we would suspect that the susceptibility of free ions would be larger than that indicated by equation 2-4.2. This is reasonable, since the nearest neighbors in a solid, and to a lesser extent in a liquid, may be expected to limit the electron cloud. The methods that have been used to deduce the ionic susceptibilities assume the correctness of equation 2-4.2 and provide values relative to some standard, for example, the H^+ ion. In addition,

Table 2-4.3. Ionic Susceptibilities

($\times 10^{-6}$ emu/mole)

	Weiss[a]	Joos[a]	Stoner[a]	Slater[a]	Pauling[a]	Courty[b]	Sachs[c]	Trew[d]	Tietz[e]
F^-		−12.1	−16.9	−8.1	−8.1		−12.665	−9.7	
Cl^-	−22.9	−18.3	−39.5	−25.2	−29			−24.3	
Br^-	−33.9	−30.0		−39.2	−54			−34.8	
I^-	−51.5	−47.7		−58.5	−80			−50.6	
Li^+	−3.0	−6.7	−0.7	−0.7	−0.6				
Na^+	−8.5	−12.5	−5.4	−4.1	−4.2	−4.522	−5.0816	−6.1	−8.3
K^+	−16.6	−21.1	−17.2	−14.1	−16.7	−13.687		−14.4	−12.1
Mg^{2+}	−7.3	−13.6	−4.3	−3.1	−3.2				
Ca^{2+}	−10.8	−18.5	−13.1	−11.1	−13.3	−10.553			

[a] W. R. Meyers, *Revs. Mod. Phys.* **24,** 15 (1952).
[b] C. Courty, *Compt. rend. (Paris)* **250,** 3293 (1960).
[c] L. M. Sachs, *Phys. Rev.* **124,** 1283 (1961).
[d] V. C. G. Trew and S. F. A. Husain, *Trans. Faraday Soc.* **57,** 223 (1961).
[e] T. Tietz, *Ann. Phys. (Leipzig)* **6,** 202 (1952).

studies of simple homopolar diatomic molecules can be used to indicate atomic susceptibilities. Values of ionic susceptibilities, as determined by various methods, experimental and theoretical, are given in Table 2-4.3.

5. Susceptibility of Molecules

Experiments show that many inorganic and almost all organic molecules are diamagnetic. The number of electrons in molecules is usually even, so that the spins balance in pairs. The orbital angular momentum either compensates in such a way to provide a zero value or else is otherwise rendered ineffective by a mechanism to be discussed later. Typical values for some homopolar molecules, as determined by Havens[9] and Honda and Owen,[10] are given in Table 2-5.1. Havens found the gas susceptibilities to be almost independent of pressure and temperature. Most of the Honda and Owen results are reported as atomic susceptibilities, because the molecular formula is not always certain.

[9] G. G. Havens, *Phys. Rev.* **43,** 992 (1933).
[10] K. Honda, *Ann. Phys. (Leipzig)* **32,** 1027 (1910); M. Owen, *Ann. Phys. (Leipzig)* **37,** 657 (1912).

Because there is no unique axis for the Larmor precession in a polyatomic molecule, consideration must be given to methods of calculating $\Sigma\ \overline{r^2}$ and thus the molecular susceptibility. If the nucleii are considered to be fixed, in simple cases the origin of axis may be taken at the center of gravity of the nucleii. This is reasonable, since the outer electrons make the largest contribution to the susceptibility. A better approach may be to take the precession axis at the nucleii for the inner electrons and on the line joining the nucleii of the molecule for the outer electrons. In general, the molecular susceptibility is expected to be less than the sum

Table 2-5.1. Atomic or Molecular Susceptibility

($\times 10^{-6}$ emu/mole)

	χ_M (Havens)		χ_A (Honda and Owen)		χ_A (Honda and Owen)
H_2	−4.00	B	−7.6	Se	−25.4
N_2	−12.0	C	−6.0	Te	−40.5
CO_2	−20.8			I	−45.7
	(Honda and Owen)	P	−22.9		
Cl_2	−40.2	S	−16.0		
Br_2	−64.0	As	−25.3		

of the atomic susceptibilities of the atoms making up the molecule, since the outer electrons are under the influence of a force center with a larger effective charge.

Further, even if a molecule has a resultant spin and orbital angular momentum of zero, there will be fluctuations of the angular momentum along the molecular axis that will contribute a temperature-independent term to the susceptibility. In the quantum mechanical treatment Van Vleck[11] has shown that the molecular susceptibility is

$$\begin{aligned}\chi_M &= -\frac{e^2N}{6mc^2}\sum\overline{r^2} + \tfrac{2}{3}N\left(\frac{e}{2mc}\right)^2 \sum_{n'} \frac{|\langle n'\ |G|\ n\rangle|^2}{\langle E_{n'} - E_n\rangle} \\ &= -\frac{e^2N}{6mc^2}\sum\overline{r^2} + N\,\alpha(n),\end{aligned} \tag{2-5.1}$$

where $\langle n'\ |G|\ n\rangle$ is the matrix element of the angular momentum

$$\left(\int \psi_{n'}{}^* G \psi_n \, d\tau\right)$$

[11] J. H. van Vleck, *Theory of Electric and Magnetic Susceptibilities*, Oxford University Press, Oxford (1932).

connecting the excited state n' with the ground state n, and $E_{n'} - E_n = h\nu_{n'-n}$ is the energy separation of the two states. The second term is, of course, the additional one, and it is to be noted that it is temperature-independent and positive. It is often called Van Vleck temperature-independent paramagnetism and is derived in section 2-8.

There must always be a contribution from the second term of equation 2-5.1 for a diatomic or polyatomic molecule, since the mean square of the angular momentum is not zero even though the mean angular momentum is. Even so, it would appear from the experimental data that this contribution to the molecular diamagnetic susceptibility is usually small, at least considerably smaller than the first term.

In most theoretical calculations the molecular wave functions are expressed as linear combinations of atomic orbitals (LCAO), and the Hartree-Fock atomic wave functions are used. The molecular susceptibility is then deduced by employing either perturbation or variational techniques. Most of the calculations have been for the hydrogen molecule,[12] for which the values of χ_M range from -3.93 to -4.21×10^{-6} emu/mole.[13] Other molecular susceptibilities that have been computed include (in units of $\times 10^{-6}$ emu/mole) $N_2(\chi_M = -13.539$,[14] $-24.6)$,[15] $F_2(\chi_M = -15.14)$,[16] $Cl_2(\chi_M = -40.67)$,[16] $Br_2(\chi_M = 58.84)$,[16] and $H_2O(\chi_M = -12.82)$.[16]

Extensive series of measurements, primarily on organic compounds in the liquid and solid state, have been carried out. Pascal and others[17] found that the data could be fitted with the additive formula

$$\chi_M = \sum_A n_A \chi_A + \lambda, \qquad (2\text{-}5.2)$$

where n_A is the number of atoms in the molecule with the atomic susceptibility χ_A, and λ is a constitutive constant that is determined by the

[12] J. H. Van Vleck and A. Frank, *Proc. Natl. Acad. Sci. U.S.* **15,** 539 (1929); G. C. Wick, *Phys. Rev.* **73,** 51 (1948); T. Tillieu and J. Guy, *Compt. rend. (Paris)* **240,** 1402 (1955); I. Espe, *Phys. Rev.* **103,** 1254 (1956); T. W. Marshall and J. A. Pople, *Molecular Phys. (G.B.)* **3,** 339 (1960); T. P. Das and R. Bersohn, *Phys. Rev.* **115,** 807 (1959); J. P. Auffray, *Phys. Rev.* **126,** 146 (1962).

[13] For a summary see H. F. Hameka, *Revs. Mod. Phys.* **34,** 87 (1962).

[14] H. F. Hameka, *J. Chem. Phys.* **34,** 366 (1961).

[15] J. V. Bonet and A. V. Bushkovitich, *J. Chem. Phys.* **21,** 2199 (1953).

[16] C. Courty, *Compt. rend. (Paris)* **249,** 2740 (1959).

[17] P. Pascal, *Ann. Chim. et Phys.* **19,** 5 (1910), **180,** 1596 (1925); A. Pacault, *Rev. Sci.* **86,** 38 (1948); P. Pascal, A. Pacault, and J. Hoaran, *Compt. rend. (Paris)* **233,** 1078 (1951); A. Pacault, N. Lumbroso, and J. Hoaran, *Cahiers Phys.* **43,** 54 (1953); C. Barter, R. G. Meisenheimer, and D. P. Stevenson, *J. Phys. Chem.* **64,** 1312 (1960); P. Pascal, F. Gallais, and J. F. Labane, *Compt. rend. (Paris)* **252,** 2644 (1961).

chemical bonding of the molecule's atoms. The atomic susceptibility is not, however, the value for the free ion, but it is a value characteristic of the atom when combined in a molecule. Pascal obtained his basic values by beginning with the halogen elements and then considering the change of the susceptibility with the addition of organic groups. Some of the values for χ_A and λ are given in Table 2-5.2. Use of these values gives the molecular susceptibility to about 1%. The effect of double or triple atomic bonds, as compared to single bonds, can be seen from the values of λ to decrease the molecular susceptibility. Physically, this probably

Table 2-5.2. Pascal's Values

	$\chi_A(\times 10^{-6})$			$\lambda(\times 10^{-6})$	
H	−2.95				
Li	−4.2	NO_3	−14.2	Benzene	−1.5
C	−6.0	SO_4	−33.6	Naphthalene	−6.1
N	−5.55	NH_3	−14.4	Ethylene	+5.5
O	−4.61	H_2O	−13.0	Acetylene	+0.8
F	−6.3				
Na	−9.2				
S	−15.0				
Cl	−20.0				
Br	−30.5				
K	−18.5				

arises from a greater concentration of charge in the double-bond region. In ring compounds, such as benzene, the effect is apparently to increase the electron spread. Recent attempts to derive Pascal's constants theoretically have met with some degree of success.[18]

From studies on single crystals of naphthalene, graphite, and other layerlike substances, it is found that the diamagnetic susceptibility is anisotropic, being abnormally large in the direction normal to the molecular plane. This must be due to the nonspherical symmetry of the electron cloud. Instead, the electronic distribution probably has a cylindrical symmetry about the bond directions. Such investigations obviously are important in chemistry, for they give information about the electron configuration of complex molecules.

[18] J. Baudet, *J. Chim. Phys.* **58,** 228 (1961); J. A. Pople, *J. Chem. Phys.* **37,** 53, 60 (1962).

PARAMAGNETISM

6. Curie's Law

In his 1895 paper Curie showed that a certain class of substance had a temperature-dependent and field-independent susceptibility given by

$$\chi = \frac{C}{T}, \tag{2-6.1}$$

where T is the absolute temperature and C is called Curie's constant. His conclusion was based on studies of oxygen gas (O_2), of solutions of some salts, and of some ferromagnetic metals at temperatures above a critical value, known as the Curie temperature. Substances that obey Curie's law (equation 2-6.1), at least to a first approximation, are usually called normal paramagnets. The type of susceptibility observed by Curie is now known to occur only for substances that contain permanent magnetic dipoles. Later experiments showed that not all paramagnetic materials obey Curie's law. Some materials have susceptibilities that can be fitted to the equation

$$\chi = \frac{C}{T - \theta}, \tag{2-6.2}$$

where θ is a constant. This equation is known as the Curie-Weiss law and is based on a classical theory developed by Weiss, which we shall consider in Chapter 6.

Atoms or ions of the transition group elements, or ions isoelectronic with these elements, form the largest class of substance to obey the Curie or Curie-Weiss law. It will be recalled that these elements have permanent moments because of uncompensated electrons in an inner shell. The rare earth or $4f$ transition ions and the iron or $3d$ group ions, when present in solution or in a solid salt, usually give rise to this type of paramagnetism. Sometimes complications, especially at very low temperatures, are related to such factors as the crystalline electric field of the diamagnetic neighbors and the details of the electronic level spacing. Ordering, either ferro- or antiferromagnetic, may occur for $T \leqslant 1°K$. The $4d$, $5d$, and $5f$ transition-group ions form compounds in which the magnetic behavior is complicated by the nature of the bonds.

In addition to the transition atoms and ions, some atoms and molecules have an odd number of electrons and thus a permanent moment. Free atoms of the alkali metals, for example, sodium vapor, are paramagnetic.

Examples of paramagnetic molecules include the gas, nitric oxide NO, and organic free radicals.

A few compounds have an even number of electrons, but their ground state happens to be such that there is a nonzero magnetic moment. The commonest example is the gas O_2, on which Curie made his original studies.

Certain metals and complex salts exhibit a form of paramagnetism that does not obey Curie's law even in an approximate way. Metals are discussed in Chapter 5.

Of course, in addition to a paramagnetic susceptibility, there must still be a diamagnetic one, so that the total susceptibility χ_T is the sum of the two; that is,

$$\chi_T = \chi_{\text{para}} + \chi_{\text{dia}}. \tag{2-6.3}$$

Experimentally, it turns out that the paramagnetic contribution is about 10^{-2} to 10^{-4} cgs units for a mole. This swamps the small diamagnetic contribution ($\sim 10^{-6}$). Nevertheless, in accurate experimental work a correction should be made in arriving at the paramagnetic susceptibility. Because of the size of χ_{dia}, the use of a theoretical value is normally sufficiently accurate. For solutions, or highly diluted salts, in which the number of diamagnetic ions is large compared to the number of paramagnetic ions, it is particularly important to apply a correction. In the literature on transition salts a correction is usually made for the non-paramagnetic ions, but not for the paramagnetic ion because the value of χ_{dia} for the transition ion is difficult to calculate with any degree of accuracy.

7. Theoretical Derivations of Curie's Law

Suppose we have a volume of gas containing N paramagnetic atoms per unit volume, each with a permament moment $\boldsymbol{\mu}$. The thermal agitation of the gas atoms will ensure that the moments are distributed at random over the possible orientations, so that there is no net magnetization of the gas. Now, let us apply a magnetic field. The dipoles will tend to be oriented in the direction of the field, whereas this effect will be opposed by the thermal agitation so that complete alignment will occur only at absolute zero. It will be assumed that thermal equilibrium is reached, probably by collision processes, but we shall not inquire how long it takes to achieve equilibrium. In addition, it will be assumed that no other interaction between the atomic dipoles exists, so that each dipole feels only the applied field **H**. The net magnetization per unit volume in the field direction M can then be calculated by the principles of elementary statistical mechanics.

Wide multiplets compared to kT. In the present derivation we assume that *all* atoms are in their ground state, characterized by the angular momentum quantum number J. This is equivalent to implying that the next level of the multiplet is widely spaced from the ground state, at least compared to the thermal energy kT; that is $E_{J'} - E_J \gg \mathrm{k}T$. According to Section 2-2 of this chapter, the permanent atomic magnetic moment can have the components $M_J g\mu_B$ in the field direction, where $M_J = J$, $(J-1), \ldots, -(J-1), -J$. From equation 1-6.1 the potential energy of a dipole is $-M_J g\mu_B H$. Then, according to statistical mechanics, the number of atoms with a particular orientation will be determined by the Boltzmann factor $e^{M_J g\mu_B H/\mathrm{k}T}$. We then get

$$M = N \frac{\sum_{-J}^{+J} M_J g\mu_B e^{M_J g\mu_B H/\mathrm{k}T}}{\sum_{-J}^{+J} e^{M_J g\mu_B H/\mathrm{k}T}}. \tag{2-7.1}$$

The quantity $M_J g\mu_B$ is likely to be the order of one Bohr magneton, that is, 10^{-20} erg/oe. At room temperature $\mathrm{k}T \approx 5 \times 10^{-14}$ erg, so that for any fields normally produced in the laboratory ($\approx 10^4$ oe) we have $M_J g\mu_B H/\mathrm{k}T \ll 1$. Under these circumstances we may write

$$M = Ng\mu_B \frac{\sum_{-J}^{+J} M_J(1 + M_J g\mu_B H/\mathrm{k}T)}{\sum_{-J}^{+J} (1 + M_J g\mu_B H/\mathrm{k}T)}.$$

Now terms of the type $\sum_{-J}^{+J} M_J$ equal zero, whereas

$$\sum_{-J}^{+J} 1 = 2J + 1 \quad \text{and} \quad \sum_{-J}^{+J} M_J^2 = \frac{J(J+1)(2J+1)}{3}.$$

We then get

$$M = \frac{Ng^2 J(J+1)\mu_B^2 H}{3\mathrm{k}T},$$

so that the susceptibility is

$$\chi = \frac{Ng^2 J(J+1)\mu_B^2}{3\mathrm{k}T} \tag{2-7.2}$$

$$= \frac{N\mu^2}{3\mathrm{k}T} \tag{2-7.3}$$

$$= \frac{N p_{\mathrm{eff}}^2 \mu_B^2}{3\mathrm{k}T}, \tag{2-7.4}$$

where the effective number of Bohr magnetons is given by

$$p_{\text{eff}} = g[J(J+1)]^{1/2} \tag{2-7.5}$$

$$= \frac{(3\chi kT)^{1/2}}{N\mu_B}. \tag{2-7.6}$$

We thus obtain Curie's law, with $C = N\mu^2/3k$. Equation 2-7.2 is often referred to as the Hund expression. Langevin (1905) first derived (2-7.3) from a classical theory in which it was assumed that all orientations of the dipoles were possible (see problem 2-3). Comparison with experiment is usually made by using equation 2-7.5. The order of magnitude of the molar susceptibility can be estimated from (2-7.3). For a moment of one Bohr magneton $\chi \approx 1/30T$. For $T = 300°K$, $\chi \approx 10^{-4}$, and for $T = 1°K$, $\chi \approx 10^{-2}$.

A rigorous quantum mechanical treatment shows that a temperature-independent term, similar to the second term of equation 2-5.1 must be added to (2-7.2). The complete expression for the susceptibility[19] is then

$$\chi = \frac{Ng^2J(J+1)\mu_B^2}{3kT} + \frac{N\mu_B^2}{6(2J+1)}\left[\frac{F(J+1)}{E_{J+1}-E_J} - \frac{F(J)}{E_J - E_{J-1}}\right],$$

$$= \sim +N\alpha(J) \tag{2-7.7}$$

where $F(J) = (1/J)[(S+L+1)^2 - J^2][J^2 - (S-L)^2]$. Either the first or second term is zero for the ground state. Generally this temperature-independent term is small and can be neglected. On the vector model (Fig. 2-7.1) the term $\alpha(J)$ arises from the perpendicular component bc, and is important if bc is larger than ac, that is, if the J value of the ground state is unusually small.

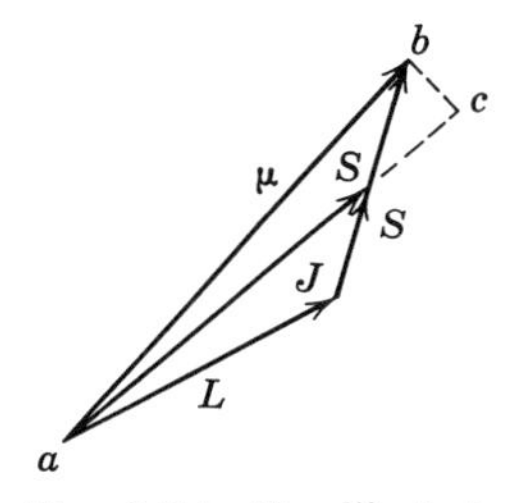

Fig. 2-7.1. To illustrate the origin of temperature independent paramagnetism.

Multiplet widths comparable to kT. When the multiplet widths are of the same order as kT, not all atoms will be in the ground state but will have values of J determined by equation 2-2.8. The atoms with a certain value of J will still contribute to the susceptibility according to equation 2-7.2. The total susceptibility will be obtained if we sum over all the atoms in different states. The number of atoms with a given J value will be proportional to the Boltzmann factor and the degree of degeneracy of

[19] J. H. Van Vleck, *Theory of Electric and Magnetic Susceptibilities*, Oxford University Press, Oxford (1932), p. 233.

the level, that is $(2J + 1)$. Then we get

$$\chi = N \frac{\sum_{J=|L-S|}^{L+S} \{[g_J^2 \mu_B^2 J(J+1)/3kT] + \alpha(J)\}(2J+1)e^{-E(J)/kT}}{\sum (2J+1)e^{-E(J)/kT}}. \quad (2\text{-}7.8)$$

Narrow multiplets with respect to kT**.** When the multiplets are very narrow, at least compared to kT, equation 2-7.7 can be simplified to give the susceptibility. Alternatively, on the basis of the vector model, the angular momenta represented by L and S can be considered decoupled and as contributing independently to the susceptibility. This situation requires the application of a strong magnetic field. However, the so-called principle of spectroscopic stability of quantum mechanics ensures that the result so obtained will hold for small fields. We then have

$$M = N \left[\frac{\sum_{-L}^{L} M_L g_L \mu_B e^{M_L g_L \mu_B H/kT}}{\sum e^{M_L g_L \mu_B H/kT}} + \frac{\sum_{-S}^{S} M_S g_S \mu_B e^{M_S g_S \mu_B H/kT}}{\sum e^{M_S g_S \mu_B H/kT}} \right].$$

Remembering that $g_L = 1$ and $g_S = 2$ and that M_L may take the values $L, (L-1), \ldots, -L$ and M_S the values $S, (S-1), \ldots, -S$, we easily obtain for the susceptibility

$$\chi = \frac{N\mu_B^2}{3kT}[L(L+1) + 4S(S+1)], \quad (2\text{-}7.9)$$

so that here

$$p_{\text{eff}} = [L(L+1) + 4S(S+1)]^{1/2}. \quad (2\text{-}7.10)$$

The term $\alpha(J)$ does not enter in the susceptibility here, since the contribution of bc of Fig. 2-7.1 is taken into account in equation 2-7.8.

The Curie-Weiss law. Weiss considered how the Curie law would be modified if there were interactions between the magnetic ions.[20] If the field due to the neighbors is equal to $N_W M$ where N_W is a constant, Curie's law can be written as

$$M = \frac{C(H + N_W M)}{T},$$

so that

$$M = \frac{CH}{T - CN_W} = \frac{CH}{T - \theta}$$

or

$$\chi = \frac{C}{T - \theta}. \quad (2\text{-}7.11)$$

[20] For a quantum mechanical derivation see L. Goldstein, *Ann. Phys.* (*USA*) **15,** 141 (1961).

Experimentally, the constants C and θ are determined by measurements of the susceptibility over a wide temperature range. Factors may exist, other than interactions, that modify Curie's law; θ then becomes just an additional parameter for data fitting purposes and lacks the above significance. The use of magnetically dilute compounds, in which most of the paramagnetic ions are replaced with similar diamagnetic ions so that interaction effects are greatly reduced, is a powerful experimental technique for exploring the origin of θ.

8. Quantum Mechanical Treatment

The derivations of the preceding section assumed that the presence of the magnetic field served only to orient the magnetic moments. However, the energy levels of the states of the atom are affected by this field. A complete solution to the problem is obtained by considering the additional terms introduced into the Hamiltonian as a perturbation. We suppose the Hamiltonian to be of the form

$$\mathcal{H} = \mathcal{H}_0 + H\mathcal{H}_1 + H^2\mathcal{H}_2 + \cdots, \tag{2-8.1}$$

where $\mathcal{H}_0$ is the unperturbed part and H is the magnetic field in which the perturbation part is developed as a series. We also develop the energy E for a particular state into a power series in H; that is

$$E = E_0 + HE_1 + H^2E_2 + \cdots, \tag{2-8.2}$$

where E_0 is the eigenvalue for the zero field.

It can be shown[21] from quantum mechanical perturbation theory that

$$E_1 = \int \psi_j{}^*\mathcal{H}_1\psi_j \, dv (= \langle j \,|\mathcal{H}_1|\, j\rangle)$$

and

$$E_2 = \sum_{j'}{}' \frac{|\langle j' \,|\mathcal{H}_1|\, j\rangle|^2}{E_{j'} - E_j} + \langle j \,|\mathcal{H}_2|\, j\rangle. \tag{2-8.3}$$

Here j' and j represent states of the unperturbed system, and the prime on the summation indicates that the term $j' = j$ is not included (it equals infinity). Our matrix notation is given in the first equation, in which the wave functions are for the unperturbed system (zero field) and the integration is over all configuration space.

[21] See, for example, J. H. Van Vleck, *Theory of Electric and Magnetic Susceptibilities*, Oxford University Press, Oxford (1932) Ch. VI, sec. 34, or H. Eyring, J. Walter, and G. E. Kimball, *Quantum Chemistry*, John Wiley and Sons, New York (1944).

Now the Hamiltonian of an atomic or molecular system can be shown to be[22]

$$\mathcal{H} = \sum_i \frac{1}{2m_i}\left(\mathbf{p}_i - \frac{e_i}{c}\mathbf{A}_i\right)^2 + V, \tag{2-8.4}$$

where the subscript i refers to a particular charge center, that is, an electron for example. **A** is the vector potential defined in equation 1-8.6. Let us assume the applied field H is along the z-direction, that is, $A_{x_i} = -\frac{1}{2}y_iH$, $A_{y_i} = \frac{1}{2}x_iH$, and $A_{z_i} = 0$. Then equation 2-8.4 becomes

$$\mathcal{H} = \sum_i \left[\frac{p_i^2}{2m_i} - H\frac{e_i}{2m_ic}(x_ip_{y_i} - y_ip_{x_i}) + H^2\left(\frac{e_i^2}{8m_ic^2}\right)(x_i^2 + y_i^2)\right] + V$$

or

$$= \sum_i \left[\frac{p_i^2}{2m_i} - H\mu_0 + H^2\frac{e_i^2}{8m_ic^2}(x_i^2 + y_i^2)\right] + V, \tag{2-8.5}$$

since the second term is $e/2mc$ times the angular momentum along the z-axis and is therefore equal to the projection of the magnetic moment in this direction in the absence of a magnetic field.[23] From equations 2-8.1 and 2-8.5 we have

$$\mathcal{H}_0 = \sum_i \frac{p_i^2}{2m_i} + V,$$

$$\mathcal{H}_1 = -\mu_0\left(= -\sum_i \mu_{0i}\right), \tag{2-8.6}$$

and

$$\mathcal{H}_2 = \sum_i \frac{e_i^2}{8m_ic^2}(x_i^2 + y_i^2).$$

Of course, the unperturbed Hamiltonian operator is obtained by making the usual substitution for p, as in equation 2-2.2.

Now, the average or expectation value of the magnetic moment in the field direction for a given state of energy E is deduced from equation 2-8.2 as[24]

$$\mu_H = -\frac{\partial E}{\partial H} = -E_1 - 2HE_2 \cdots. \tag{2-8.7}$$

[22] N. F. Mott and I. N. Sneddon, *Wave Mechanics and Its Applications*, Oxford University Press, Oxford (1948) p. 39.

[23] The subscript H which would normally indicate this direction is omitted here for convenience.

[24] Classically, this statement follows immediately from equation 1-6.1. For a quantum mechanical proof see Van Vleck, *op. cit.*, p. 145.

Substitution of equation 2-8.6 into (2-8.3) and then into (2-8.7) gives

$$\mu_H = \langle j\,|\mu_0|\,j\rangle + 2H\sum_i \frac{|\langle j'\,|\mu_0|\,j\rangle|^2}{E_{j'} - E_j} - H\left\langle j\,\middle|\,\sum_i \frac{e_i^2}{4m_ic^2}(x_i^2 + y_i^2)\,\middle|\,j\right\rangle. \tag{2-8.8}$$

The magnetic moment is then due to the permanent moment, plus two induced moments, induced since they vanish if the field is zero. The first induced moment is positive and therefore is a paramagnetic term, whereas the second is negative or a diamagnetic term. The last term in fact gives rise to the expression (2-3.2) for the diamagnetic susceptibility, since the matrix element gives the average value of $\Sigma\,(e^2/4mc^2)(x^2 + y^2)$ for the jth state, and when the average over all orientations is considered $(x^2 + y^2) = \frac{2}{3}r^2$. We will neglect this diamagnetic term in the remaining discussion.

The paramagnetic susceptibility will now be found. Since the atoms will be distributed between the states according to the Boltzmann factor, we have

$$\chi = \frac{N\bar{\mu}_H}{H} = \frac{N}{H}\frac{\sum \mu_H e^{-E/kT}}{\sum e^{-E/kT}}, \tag{2-8.9}$$

where the sum is over the states of the system.

Use of equation 2-8.2 and expansion of the exponential gives

$$e^{-E/kT} = e^{-E_0/kT}\left(1 - \frac{E_1H}{kT}\cdots\right). \tag{2-8.10}$$

Now we assume that the numerator of (2-8.9) has to be expanded only to the first power of H and that only the term independent of H in the denominator need be retained. Use of (2-8.7) gives

$$\chi = N\frac{\sum (E_1^2/kT - 2E_2)e^{-E_0/kT}}{\sum e^{-E_0/kT}} \tag{2-8.11}$$

since $\Sigma\, E_1e^{-E_0/kT} = \Sigma\,\langle j\,|\mu_0|\,j\rangle\, e^{-E_0/kT} = 0$; that is, in the absence of a field the net moment is zero. Substitution for E_1 and E_2 gives

$$\chi = \frac{N}{\sum e^{-E_0/kT}}\sum_j\left[\frac{|\langle j\,|\mu_0|\,j\rangle|^2}{kT} + \sum_{j'}{}' \frac{2\,|\langle j'\,|\mu_0|\,j\rangle|^2}{E_{j'} - E_j}\right]e^{-E_0/kT}. \tag{2-8.12}$$

Now, to derive the results of the preceding section, it is necessary only to specify the quantum numbers j for our atomic model. As an example, consider that the states of different n and J are spaced widely compared to kT, so that the contribution to the first term of equation 2-8.12 is negligible except for the lowest, or J, state. It is then only necessary to sum over the M_J states. The first term becomes

$$\frac{\overline{\sum |\langle M_J |\mu_0| M_J \rangle|^2}}{kT} = \frac{g^2\mu_B{}^2}{kT} \frac{\sum_{-J}^{+J} M_J{}^2}{2J+1}$$

$$= \frac{J(J+1)g^2\mu_B{}^2}{3kT} \qquad (2\text{-}8.13)$$

in agreement with equation 2-7.2. Here the bar means the average, which has been calculated in a fashion analogous to that used in deriving (2-7.2). The second term of (2-8.12) gives rise to the temperature-independent paramagnetic contribution quoted in equations 2-5.1 and 2-7.7. There may be two contributions, one due to the widely separated levels n and n' and another from the $J - J'$ separation. For the nn' levels we get

$$\alpha(n) = \sum_{n'}{}' \frac{2\,|\langle n' |\mu_0| n \rangle|^2}{E_{n'} - E_n},$$

as in (2-5.1), and for $\alpha(J)$ it can be shown[25] that the expression in (2-7.6) is obtained.

9. Susceptibility of Quasi-free Ions: the Rare Earths

The theory of the two preceding sections should be tested by comparison with experiments with monatomic paramagnetic gases. However, the only examples are the vapors of some elements, and experiments with vapors are difficult. For potassium vapor Lehrer[26] found Curie's law obeyed, with $p_{\text{eff}}^2 = 3.04$ to an accuracy of 10%. Because T was large in these experiments, we use equation 2-7.10. For the ground state of potassium $L = 0$ and $S = \frac{1}{2}$, and we get $p_{\text{eff}}^2 = 3.00$ theoretically in very satisfactory agreement.

Fortunately, the rare earth ions, when in the form of salts or solutions, are likely to behave as though they were almost free. This is so because the $4f$ electrons, responsible for the paramagnetism of the usually trivalent ions, are situated well to the interior of the ion, being surrounded by the $5s$ and $5p$ electrons. The insensitiveness of the $4f$ shell electrons to the effects of neighboring atoms is shown by the similarity of the chemical properties of the rare earths.

[25] J. H. Van Vleck, *Theory of Electric and Magnetic Susceptibilities*, Oxford University Press, Oxford (1932), Ch. IX, sec. 55.
[26] E. Lehrer, *Ann. Phys. (Leipzig)* **81**, 229 (1927).

The rare earths are indeed rare and difficult to purify. There is, nevertheless, agreement between different observers to within about 5%. These measurements were made on different salts, such as the hydrates, sulfates, and oxides, and on solutions, so that this good agreement is further evidence that the ion's $4f$ electrons were essentially free. Table 2-9.1 gives mean experimental[27] values of p_{eff} for the rare earth ions determined from equation 2-7.6 at room temperature. The table also shows the number of

Table 2-9.1. The Rare Earths

Ion	Number of $4f$ electrons	Ground State	Theory: $p_{\text{eff}} = g\sqrt{J(J+1)}$	Theory: Van Vleck and Frank	Experiment: $p_{\text{eff}} = \frac{\sqrt{3\chi kT}}{\mu_B}$
La^{3+}	0	$^1S_0$	0.00	0.00	dia.
$Ce^{3+}Pr^{4+}$	1	$^2F_{5/2}$	2.54	2.56	2.4
Pr^{3+}	2	$^3H_4$	3.58	3.62	3.6
Nd^{3+}	3	$^4I_{9/2}$	3.62	3.68	3.6
Pm^{3+}	4	$^5I_4$	2.68	2.83	—
Sm^{3+}	5	$^6H_{5/2}$	0.84	1.6	1.5
$Eu^{3+}Sm^{2+}$	6	$^7F_0$	0.00	3.5	3.6
$Gd^{3+}Eu^{2+}$	7	$^8S_{7/2}$	7.94	7.94	8.0
Tb^{3+}	8	$^7F_6$	9.72	9.7	9.6
Dy^{3+}	9	$^6H_{15/2}$	10.63	10.6	10.6
Ho^{3+}	10	$^5I_8$	10.60	10.6	10.4
Er^{3+}	11	$^4I_{15/2}$	9.59	9.6	9.4
Tu^{3+}	12	$^3H_6$	7.57	7.6	7.3
Yb^{3+}	13	$^2F_{7/2}$	4.54	4.5	4.5
$Lu^{3+}Yb^{2+}$	14	$^1S_0$	0.00	0.0	dia

$4f$ electrons for each ion, the spectroscopic ground state, and the theoretical value determined from equation 2-7.5. Except for Sm^{3+} and Eu^{3+}, the agreement between theory and experiment is satisfactory. Gd^{3+}, Tb^{3+}, Dy^{3+}, Ho^{3+} and Er^{3+} followed Curie's law.[28] For Ce^{3+}, Pr^{3+}, Nd^{3+}, Tm^{3+}, and Yb^{3+} there are deviations which can often be described by the Curie-Weiss law.[29]

[27] For an analysis of the experimental results see C. J. Gorter, *Arch. Mus. Tryler* **7,** 183 (1932); M. B. Cabrera, *Compt. rend. (Paris)* **205,** 400 (1937); W. Klemm and H. Bommer, *Z. anorg. u. allgem. Chem.* **231,** 138 (1937).

[28] Below 3°K dipole-dipole interactions are important in dysprosium ethyl sulfate, as shown by A. H. Cooke, D. T. Edmonds, F. K. McKim, and W. P. Wolf, *Proc. Roy. Soc. (London)* **A-252,** 246 (1959).

[29] R. P. Hudson and W. R. Hosler, *Phys. Rev.* **122,** 1417 (1961); K. H. Hellwege, W. Schembs, and B. Schneider, *Z. Physik* **167,** 477 (1962); K. H. Hellwege, S. H. Kusan, H. Lange, W. Rummel, W. Schembs, and B. Schneider, *Z. Physik* **167,** 487 (1962); J. Cohen and J. Ducloz, *J. Phys. radium* **20,** 402 (1959).

For Sm^{3+} and Eu^{3+} the deviations from Curie's law are large and the susceptibility is greater than that given by the Hund equation. Further investigation showed that the spacing of the multiplet levels was not large compared to kT, so that not all the atoms were in their ground state. Figure 2-9.1 shows the energy levels of Sm^{3+} and Eu^{3+} according to Judd,[30] together with the magnitude of 10 kT for room temperature. In order to calculate the susceptibility, it was necessary to use equation 2-7.8. Inclusion

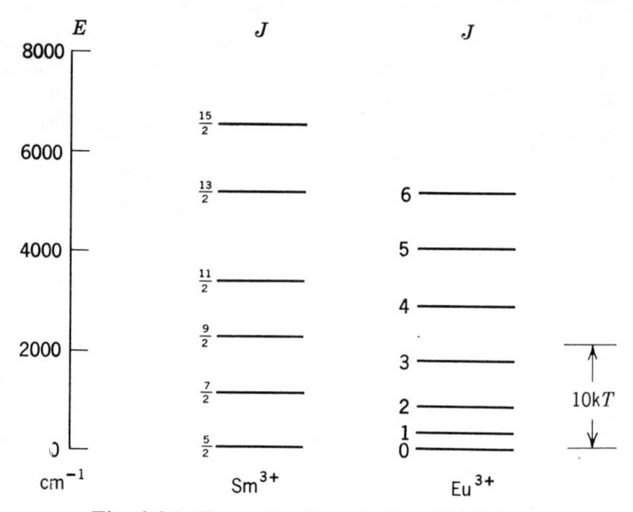

Fig. 2-9.1. Energy levels in Sm^{3+} and Eu^{3+} ions.

of the term $\alpha(J)$ is important, particularly for Eu^{3+}, since here J is small compared to S and L (see Fig. 2-7.1). The values for p_{eff} obtained by Frank,[31] who used intervals slightly too large for the multiplet structure, are shown in Table 2-9.1. The agreement is reasonable. Corrections for the other ions, due to multiplet structure, are small, as may be seen by comparing the Van Vleck and Frank results with the $g[J(J+1)]^{1/2}$ values in Table 2-9.1.

The temperature dependence of the susceptibility of Eu^{3+} and Sm^{3+} will be determined by equation 2-7.8 and will therefore be very different from Curie's law. Figure 2-9.2 shows Frank's calculated curves, together with some experimental results. Again the agreement is satisfactory.[32] Note the minimum in Sm^{3+} at about $T = 385°$K. The increase in χ at higher temperatures is, of course, due to the increase in the number of

[30] B. R. Judd, *Proc. Phys. Soc.* (*London*) **69**, 157 (1956).
[31] A. Frank, *Phys. Rev.* **39**, 119 (1932).
[32] See also N. Uryu, *J. Phys. Soc. Japan* **15**, 2041 (1960).

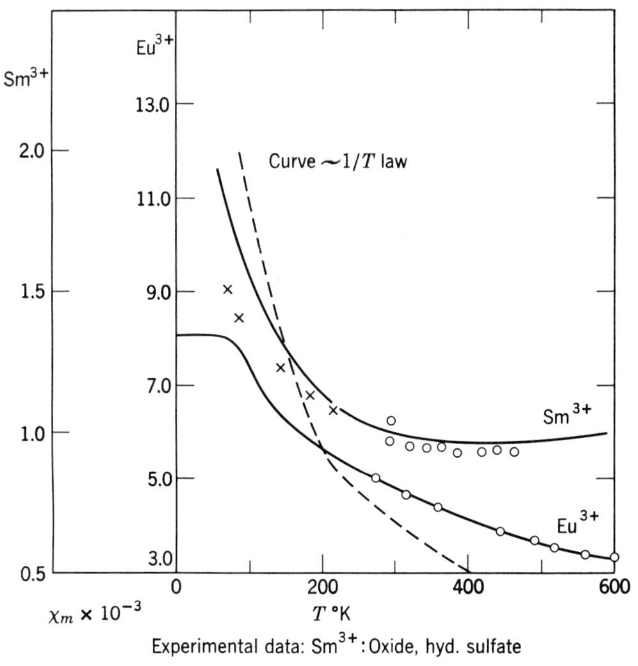

Fig. 2-9.2. $1/\chi$ versus T curves for Sm^{3+} and Eu^{3+} ions. From A. Frank, *Phys. Rev.* **39,** 119 (1932).

ions in states above the ground level, with the subsequent increase in the magnetic moment. The flat χ for Eu^{3+} at very low temperatures is due to the $\alpha(J)$ term, since $J = 0$ in the ground state. A Curie law curve is shown for the Eu^{3+} scale for comparison.

10. The Effect of the Crystalline Field

Up to this point, paramagnetic ions in solids and solutions have been considered as though they were free. They are, in fact, acted on by the inhomogeneous electric fields that are produced by their diamagnetic neighbors. We have already seen that the effect of these fields on the rare earth ions, at least at the more ordinary temperatures, is so small that the ions behave as though they were free. Departures from this behavior may be expected at very low temperatures and indeed do occur. In the iron-group ions the electrons responsible for the paramagnetism, the $3d$ shell electrons, will be affected by the crystalline fields to a much greater degree. This is partly due to the large radius of the $3d$ shell and partly to the lack of any outer electronic shells to screen the $3d$ shell.

The general effect of the crystalline field is to cause some splitting of the degenerate energy levels. The precise value of the electric field cannot be calculated by theory, since the charge distribution of the diamagnetic neighbors and the overlapping of the electron clouds are difficult to determine. However, the nearest neighbors of the magnetic ion usually have a high degree of symmetry which can be determined by X-ray analysis. For example, in the hydrated iron group salts the paramagnetic ion is usually at the center of a nearly regular octahedron, formed by the six negative oxygen ions of the H_2O dipoles. Bethe[33] considered how the

Table 2-10.1. Splitting of $(2N + 1)$-fold Degenerate Levels by a Crystalline Field

N	$(2N + 1)$	Cubic	Tetragonal	Trigonal	Rhombic
0	1	1	1	1	1
$\frac{1}{2}$	2	2	2	2	2
1	3	3	1,2	1,2	1,1,1
$\frac{3}{2}$	4	4	2,2	2,2	2,2
2	5	2,3	1,1,1,2	1,2,2	1,1,1,1,1
$\frac{5}{2}$	6	2,4	2,2,2	2,2,2	2,2,2
3	7	1,3,3	1,1,1,2,2	1,1,1,2,2	1,1,1,1,1,1,1
$\frac{7}{2}$	8	2,2,4	2,2,2,2	2,2,2,2	2,2,2,2

symmetry properties of the crystal field, when introduced into Schrödinger's equation as a perturbing potential, would affect the $(2N + 1)$ wave functions of the degenerate ground state. Here we let N be a general quantum number, either L, J, or S. By using group theoretical methods, Bethe showed how the degenerate levels would be split under the action of fields of various symmetries. However, the amount of the splitting cannot be determined by this method, although in some cases the relative distances between levels may be determinable. The general results for crystal fields of various symmetries are summarized in Table 2-10.1. Notice that in many cases a partial degeneracy remains. For example, for $N = 2$ in a trigonal field the fivefold level is split into a singlet and two doublet levels. The definitions of the field symmetries are given later.

In addition, two theorems of a general nature are of considerable value in determining the nature of the splittings. These theorems are due to Kramers[34] and Jahn and Teller.[35]

[33] H. A. Bethe, *Ann. Phys.* (*Leipzig*) **3,** 133 (1929) [trans. Consultants Bureau, Inc., New York].

[34] H. A. Kramers, *Proc. Acad. Sci. Amst.* **33,** 959 (1930).

[35] H. A. Jahn and E. Teller, *Proc. Roy. Soc.* (*London*) **A-161,** 220 (1937).

Kramers' theorem. If a system containing an *odd* number of electrons is placed in an external electric field, at least a twofold degeneracy in the levels will remain. The splittings for values of $N = \frac{1}{2}, \frac{3}{2}, \frac{5}{2}$, and $\frac{7}{2}$ of Table 2-10.1 are examples of Kramers' degeneracy.

Jahn-Teller theorem. Nonlinear complexes which have a degenerate ground state will spontaneously distort and the degeneracy will be removed. This occurs because the small displacement of the ions will reduce the energy. The theorem does not apply to Kramers' degeneracy, since here the minimum energy is already achieved.

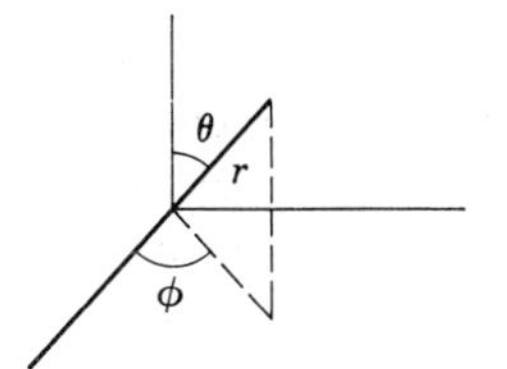

Fig. 2-10.1. Right-hand spherical coordinate system.

The crystal field potential V is assumed to satisfy Laplace's equation $\nabla^2 V = 0$. It can then be expanded as an infinite series of spherical harmonics, with terms of the type $A_n{}^m r^n P_n^{|m|}(\cos\theta)e^{im\phi}$. Here r, θ, and ϕ are the usual spherical coordinates (Fig. 2-10.1), $P_n^{|m|}(\cos\theta)$ are the associated Legendre polynomials, and $A_n{}^m$ are constants. Because of the orthogonality relationships for $3d$ electrons, matrix elements for the foregoing potential vanish for $n > 4$. Similarly, it can be shown that a polynomial ending at $P_6{}^6$ is sufficient for $4f$ electrons. Further reduction in the number of terms results when the crystal symmetry is considered.

For example,[36] for a regular octahedron in which the paramagnetic ion is surrounded by six nearest neighbors, with the Oz chosen as the polar axis, we find

$$V = A_4{}^0 r^4\{P_4{}^0(\cos\theta) - (\tfrac{5}{14})^{1/2}[P_4{}^4(\cos\theta)e^{i4\phi} + P_4{}^{-4}(\cos\theta)e^{-i4\phi}]\}. \tag{2-10.1}$$

In Cartesian coordinates this equation becomes

$$V = C(x^4 + y^4 + z^4 - \tfrac{3}{5}r^4), \tag{2-10.2}$$

where C is a numerical constant.

We are now in a position to define the field symmetry terms of Table 2-10.1. If $m = 0$, V does not depend on ϕ, and the symmetry is said to be axial. For $m = \pm 2$ the symmetry is rhombic, for $m = \pm 3$, trigonal, and for $m = \pm 4$, tetragonal. The symmetry as shown in equation 2-10.1 or 2-10.2 is termed cubic or octahedral.

Once the form of the potential V has been determined, on the basis of crystal structure, it is necessary to consider the effect on the magnetic

[36] B. Bleaney and K. W. H. Stevens, *Rept. Progr. Phys.* **16,** 108 (1953).

electrons, using a perturbation Hamiltonian given by

$$\mathcal{H} = \sum_i -eV(x_i, y_i, z_i).$$

The magnitude of this effect determines the point at which the perturbation calculation is made. Three different important cases arise.

1. The crystal field effect is relatively weak, being less than the spin-orbit interaction.
2. The field effects are larger than the spin-orbit interactions.
3. The effects are extremely large, in fact larger than the individual *l–l* and *s–s* Russell-Saunders coupling.

When this perturbation calculation is made, it is to be noted that the terms in V are unknown to the extent of the arbitrary A coefficients. These can be adjusted to fit the particular experimental results.

Case 1 refers to the rare earth salts and is discussed briefly in this section. Case 2 applies to the hydrated iron salts and is considered in Section 2-11. The third case occurs in covalent bonding and is considered briefly in Section 2-12.

The actual splitting of the ground state by the crystal field is found by computing the matrix elements of the secular equation; that is, integrals of the type $\int \psi_m^*(\sum_i -eV(x_i, y_i, z_i))\psi_m \, d\tau$, where the ψ's represent the appropriate angular momentum functions. This is extremely tedious, but Stevens[37] has found a way in which to shorten the labor by replacing the operators with the angular momenta, L_x, L_y, and L_z. The reader is referred to his original paper for details of this method. Once the splitting has been determined, the susceptibility is computed from equation 2-8.9.

Since the crystal field effect on the rare earth ion is considered after the effect of the larger spin-orbit interaction, it acts on the degenerate J eigenstates. It turns out that these levels are split by about 100 cm^{-1}. These splittings are so small that deviations from a Curie or Curie-Weiss law occur only at very low temperatures. Results for $Pr_2(SO_4)_3 \cdot 8H_2O$ and $Nd_2(SO_4)_3 \cdot 8H_2O$ are shown[38] in Fig. 2-10.2. Over the higher temperature range, the experimental curves are parallel to straight lines through the origin with the appropriate Hund values. For Pr^{3+}, $J = 4$ and there is an even number of magnetic electrons. The splitting due to the crystal field presumably places a singlet level lowest. At very low temperatures only this singlet is occupied and the only contribution to the susceptibility

[37] K. W. H. Stevens, *Proc. Phys. Soc. (London)* **A-65,** 209 (1952).

[38] C. J. Gorter and W. J. de Haas, *Comm. Kamerlingh Onnes Lab. Univ. Leiden* **2186** (1931).

arises from the temperature independent term $\alpha(J)$. For Nd^{3+}, $J = \frac{9}{2}$, and since there is an odd number of electrons the lowest level is at least doubly degenerate by Kramers' theorem. Penney and Schlapp[39] assumed a cubic potential $D\Sigma(x^4 + y^4 + z^4)$ and by adjusting D to produce a certain splitting obtained the curves in Fig. 2-10.2, in excellent agreement with the experimental results. However, Penney and Kynch[40] showed that these splittings were inconsistent with other experimental data from optical absorption and specific heat measurements. We may conclude that a

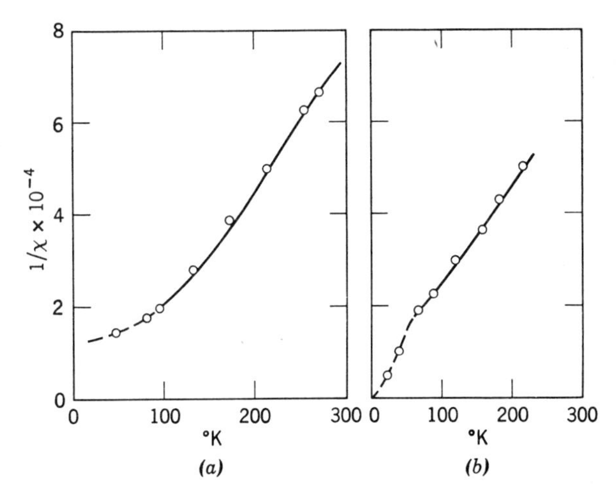

Fig. 2-10.2. Inverse susceptibility as a function of temperature for (*a*) $Pr_2(SO_4)_3{\cdot}8H_2O$ and (*b*) $Nd_2(SO_4)_3{\cdot}8H_2O$.

one-parameter theory, as used by Penney and Schlapp, is not sufficient. We will reconsider these questions in the discussion of paramagnetic resonance in a later chapter.

Single crystal anisotropy. Most of the old measurements of susceptibilities have been made on powders or polycrystalline solids. When a single crystal is employed, anisotropic behavior will occur if the crystal field lacks spherical symmetry. Then the electron orbits will be distorted along particular directions, and the magnetization will depend on the direction of an applied magnetic field. It is obvious that measurements on single crystals give valuable information on the symmetries of the crystal field.

In the general case the susceptibility can be described by a Cartesian

[39] W. G. Penney and R. Schlapp, *Phys. Rev.* **41,** 194 (1932), **42,** 666 (1932), **43,** 485 (1933).

[40] W. G. Penney and G. J. Kynch, *Proc. Roy. Soc.* (*London*) **A-170,** 112 (1939).

tensor of the type

$$\chi = \begin{pmatrix} \chi_{11} & \chi_{12} & \chi_{13} \\ \chi_{12} & \chi_{22} & \chi_{23} \\ \chi_{13} & \chi_{23} & \chi_{33} \end{pmatrix}. \tag{2-10.3}$$

The change in energy per unit volume is given by a generalized form of equation 1-7.8, namely,

$$E = -\tfrac{1}{2}\chi_{11}H_x^2 + \chi_{22}H_y^2 + \chi_{33}H_z^2 + 2\chi_{12}H_xH_y + 2\chi_{23}H_yH_z + 2\chi_{13}H_zH_x, \tag{2-10.4}$$

where H_x, H_y, and H_z are the components of the magnetic field along the axes x, y, and z. It is always possible to diagonalize[41] the matrix 2-10.3; that is, a set of orthogonal axes, called the *principal paramagnetic susceptibility axes*, exist such that $\chi_{12} = \chi_{13} = \chi_{23} = 0$. $\chi_{11} = \chi_1$, $\chi_{22} = \chi_2$, and $\chi_{33} = \chi_3$ are defined as the principal susceptibilities. For a field H with direction cosines (l, m, n) with respect to (x, y, z) the susceptibility for this direction χ, using equation 2-10.4, is

$$\chi_1 l^2 + \chi_2 m^2 + \chi_3 n^2 = \chi. \tag{2-10.5}$$

However, if the symmetry is isotropic, as, for example, for a cubic crystal field in which the diamagnetic ions form a regular octahedron, $\chi_1 = \chi_2 = \chi_3$, and the magnetic axes can be arbitrarily oriented. For trigonal or tetragonal symmetry $\chi_1 = \chi_2 = \chi_\perp$ and $\chi_3 = \chi_\parallel$, where $\chi_\parallel$ and $\chi_\perp$ are the susceptibilities parallel and perpendicular to the symmetry axis of the crystal, respectively.

As an example, we consider the measurements of Bogle et al.[42] on cerium ethylsulfate $[Ce(C_2H_5SO_4)_3 \cdot 9H_2O]$. When completely free, the ion Ce^{3+} has a $^2F_5$ ground state with $p_{\text{eff}} = 2.5$. The ethylsulfate crystals are hexagonal and the Ce ions have nine nearest neighbors, arranged so that the axis of symmetry is threefold, that is, trigonal symmetry. According to Elliott and Stevens,[43] the potential is given by

$$V = A_0 + A_2^0 r^2 P_2^0 + A_4^0 r^4 P_4^0 + A_6^0 r^6 P_6^0 + A_6^6 r^6 P_6^6 e^{i6\phi}$$

[41] See, for example, V. Rojanski, *Introduction to Quantum Mechanics*, Prentice-Hall, Englewood Cliffs, N.J. (1946), Ch. IX.

[42] G. S. Bogle, A. H. Cooke, and S. Whitley, *Proc. Phys. Soc.* (*London*) **A-64,** 931 (1951).

[43] R. J. Elliott and K. W. H. Stevens, *Proc. Phys. Soc.* (*London*) **A-65,** 205 (1951).

or in Cartesians by

$$V = A_0 + A_2^0(3z^2 - r^2) + A_4^0(35z^2 - 30r^2z^2 + 3r^4)$$
$$+ A_6^0(231z^6 - 315r^2z^4 + 105r^4z^2 - 5r^6)$$
$$+ A_6^6(x^6 - 15x^4y^2 + 15x^2y^4 - y^6), \qquad (2\text{-}10.6)$$

where P_6^6 is required because of the $4f$ wave functions of the rare earths. This field splits the ground state into three doublets, with $M_J = \pm\frac{1}{2}, \pm\frac{3}{2}$, and $\pm\frac{5}{2}$. The $\pm\frac{3}{2}$ state is approximately 125 cm^{-1} above the others and is populated only above 20°K. At low temperatures, following the reasoning of Section 2-7 of this chapter, the $\chi_\perp$ and $\chi_\parallel$ are expected to obey an equation of the type

$$\chi = \frac{N\mu_B^{\,2}}{4kT}\,\frac{g_{5/2}^2 + g_{1/2}^2\,e^{-\Delta/T}}{1 + e^{-\Delta/T}}\,. \qquad (2\text{-}10.7)$$

It was found that the $\pm\frac{5}{2}$ levels were below the $\pm\frac{1}{2}$ doublet by $\Delta = 6$ cm^{-1} (7.5°) and that a good fit with equation 2-10.7 resulted for $g_{5/2} = 3.8$ and $g_{1/2} = 1.0$ for H applied parallel to the hexagonal axis and $g_{5/2} = 0$ and $g_{1/2} = 2.25$ for H perpendicular to this axis.

11. The Iron Group Salts

In most of the salts for which the discussion of this section is pertinent a positive paramagnetic ion from the iron group is surrounded by six negatively charged diamagnetic ions or complexes arranged in a nearly regular octahedron. In the hydrated salts it is usually the negative oxygen ions of the water molecules that are the nearest neighbors of the magnetic cation. The iron group elements have degenerate $3d$ orbital wave functions which have appreciable values for much larger radii than the $4f$ wave functions of the rare earths. Moreover, since the iron group ion has lost its $4s$ electrons because of the ionic binding in these solids, there are no outer electrons to shield the $3d$ shell. Hence the splittings caused by the crystal field are about 10,000 cm^{-1} and thus are much larger than those produced by the spin-orbit coupling. The perturbation calculation then considers how the orbital levels are split by the crystal field. The usual Russell-Saunders coupling does not occur, and the states can no longer be specified by the quantum number J.

We will now give a simple physical explanation of the origin of the splitting of the orbital levels by the crystal field. For this purpose we consider an ion with only one $3d$ electron ($l = 2$) and form wave functions made up of certain linear combinations of the $3d$ hydrogenlike wave functions. If the hydrogen 3d wave functions are denoted by d_{+2}, d_{+1}, d_0, d_{-1}, and d_{-2}, where the numerical subscript to d represents the value of M_L, the linear combinations we choose are

$$\psi_1 = \frac{d_{+1} + d_{-1}}{\sqrt{2}} = xz\,f(r),$$

$$\psi_2 = -i\,\frac{d_{+1} - d_{-1}}{\sqrt{2}} = yz\,f(r),$$

$$\psi_3 = -i\,\frac{d_{+2} - d_{-2}}{\sqrt{2}} = xy\,f(r),$$

$$\psi_4 = \frac{d_{+2} + d_{-2}}{\sqrt{2}} = (x^2 - y^2)\,f(r),$$

$$\psi_5 = d_0 = (3z^2 - 1)\,f(r). \tag{2-11.1}$$

The electron charge cloud distributions of these orbitals are sketched in Fig. 2-11.1, together with circles to indicate the position of the diamagnetic neighbors. It is easier for a 3d electron cloud to avoid the charge cloud of the neighboring negative ions for the cases illustrated at (a) of Fig. 2-11.1. These orbitals (δ_ϵ) will then have a lower energy than for (b) or (c) (δ_γ orbitals), and splitting will have occurred. For a cubic field the splitting is shown in Fig. 2-11.1d.

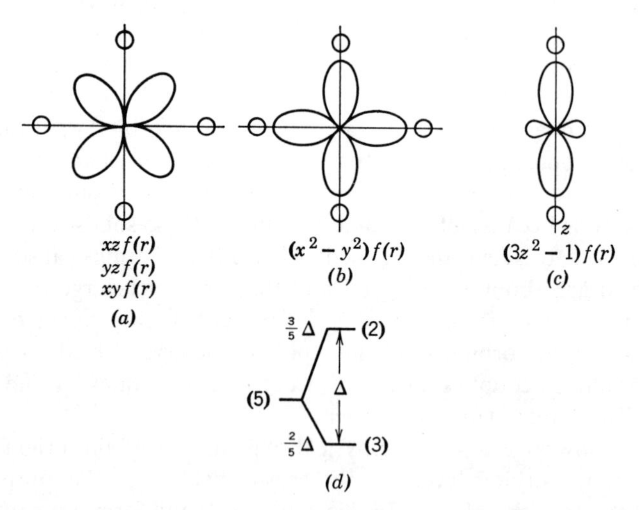

Fig. 2-11.1. Charge distributions for 3d orbitals (a, b, and c) and (d) the splitting of the energy level by an applied field.

Frequently the lowest orbital level, when split by a cubic field, is a singlet. Because of the large splitting, this is usually the only level populated. The orbital momentum is then said to be quenched, since it will make no contribution to the magnetic moment when a field is applied. When a doublet or triplet orbital level is lowest, there is further splitting by the spin-orbit coupling and by any crystal field of other than cubic symmetry that may be present. These other crystal fields may arise from a distortion in the position of the neighboring diamagnetic ions from an ideal octahedron or from the influence of diamagnetic ions farther away. These fields are usually of lower than cubic symmetry.

In quantum mechanical language, when a singlet level is lowest, the component of orbital momentum L_z is no longer a constant of the motion. Indeed, the average value of L_z will be zero, that is, $\int \psi^* L_z \psi \, d\tau = 0$, and this corresponds to quenching. We will now illustrate this behavior with an example. In order to simplify the mathematics, we consider one electron in a p-state ($L = 1$) subjected to a field of rhombic symmetry so that $V = Ax^2 + By^2 + Cz^2$, where (x, y, z) is the position of the electron. For a free ion with $L = 1$ the ground state is threefold degenerate. We denote the hydrogenlike wave functions by p_{+1}, p_0, and p_{-1} where again the subscripts indicate the value of M_L. We take as our threefold degenerate wave functions

$$\psi_1 = \frac{p_{+1} + p_{-1}}{\sqrt{2}} = x\,f(r),$$

$$\psi_2 = -i\,\frac{p_{+1} - p_{-1}}{\sqrt{2}} = y\,f(r), \tag{2-11.2}$$

$$\psi_3 = p_0 = z\,f(r).$$

In calculating[44] the effect of the crystal field as a perturbation, first we note that the nondiagonal matrix elements are zero, that is,

$$\int \psi_1^* V \psi_2 \, d\tau = \int \psi_1^* V \psi_3 \, d\tau = \int \psi_2^* V \psi_3 \, d\tau = 0.$$

For example, on considering

$$\int \psi_1^* V \psi_2 \, d\tau = \int_{-\infty}^{\infty} |f(r)|^2 \, xy(Ax^2 + By^2 + Cz^2) \, dx \, dy \, dz,$$

we see that the integrand is an odd power of x and thus the integral must be zero. The shift in the energy levels under the action of the crystal field is therefore given by the diagonal matrix elements of the type

[44] See for example H. Eyring, J. Walter, and J. H. Kimball, *Quantum Chemistry*, John Wiley and Sons, New York (1944).

$\int |\psi_1|^2 V\, d\tau$. Thus

$$\int \psi_1{}^2 V\, d\tau = \int |f(r)|^2 (Ax^4 + Bx^2y^2 + Cx^2z^2)\, dx\, dy\, dz,$$

$$\int \psi_2{}^2 V\, d\tau = \int |f(r)|^2 (Ax^2y^2 + By^4 + Cy^2z^2)\, dx\, dy\, dz,$$

and

$$\int \psi_3{}^2 V\, d\tau = \int |f(r)|^2 (Ax^2z^2 + By^2z^2 + Cz^4)\, dx\, dy\, dz.$$

It is not necessary to give $f(r)$ explicitly, since the nature of the splitting can be indicated by defining

$$S_1 = \int x^4\, |f(r)|^2\, d\tau = \int y^4\, |f(r)|^2\, d\tau = \int z^4\, |f(r)|^2\, d\tau$$

and

$$S_2 = \int x^2y^2\, |f(r)|^2\, d\tau = \int x^2z^2\, |f(r)|^2\, d\tau = \int y^2z^2\, |f(r)|^2\, d\tau.$$

The new energy levels are given by

$$\begin{aligned} E_1 &= AS_1 + (B + C)S_2, \\ E_2 &= BS_1 + (A + C)S_2, \\ E_3 &= CS_1 + (B + A)S_2. \end{aligned} \qquad (2\text{-}11.3)$$

This is an example of the splittings predicted by Table 2-10.1. Further, since the off-diagonal elements are zero, the wave functions of equation 2-11.2 are the wave functions of the perturbed state. Remembering that the angular momentum operator $L_z = (\hbar/i)[x(\partial/\partial y) - y(\partial/\partial x)]$, we determine that the average value of L_z for the three states is

$$\int \psi_1{}^* L_z \psi_1\, d\tau = \int \psi_2{}^* L_z \psi_2\, d\tau = \int \psi_3{}^* L_z \psi_3\, d\tau = 0; \qquad (2\text{-}11.4)$$

that is, the orbital angular momentum is quenched. Since, with a field applied along the z-direction, the magnetic moment is proportional to the average value of L_z, it also is zero. It is interesting to note that this result comes from the admixture of the two states with $M_L = +1$ and $M_L = -1$, which gives rise to new wave functions that are real.

Experimental measurements of the susceptibility can therefore not be expected to be in agreement with $p_{\text{eff}} = g[J(J + 1)]^{1/2}$. It is not zero, however, since the spin degeneracy of the levels still exists. As shown in Table 2-11.1 there is reasonably good agreement between the experiments and an effective magneton spin-only value of $p_{\text{eff}} = g[S(S + 1)]^{1/2}$ with $g = 2$. The experimental results are taken from a selection made from the literature by Gorter.[45] Some of the older measurements were made on solutions, but here also there may be large electrostatic forces, though not of a highly symmetric character.

The deviations from the spin-only value are usually due to incomplete quenching. This arises from the spin-orbit coupling, and the nature of the results are indicated by considering Hund's rules.[46] For elements

Table 2-11.1. The Iron Group Ions

Ion	No. of $3d$ Electrons	Ground State	Theory $p_{eff} = g[J(J+1)]^{1/2}$	Theory $p_{eff} = g[S(S+1)]^{1/2}$	Experiment p_{eff}
K^+, Ca^{3+}	0	$^1S_0$	0.00	0.00	Diamagnetic
Ti^{3+}, V^{4+}	1	$^2D_{3/2}$	1.55	1.73	1.7
V^{3+}	2	$^3F_2$	1.63	2.83	2.8
V^{2+}, Cr^{3+}, Mn^{4+}	3	$^4F_{3/2}$	0.77	3.87	3.8
Cr^{2+}, Mn^{3+}	4	$^5D_0$	0.0	4.90	4.9
Mn^{2+}, Fe^{3+}	5	$^6S_{5/2}$	5.92	5.92	5.9
Fe^{2+}	6	$^5D_4$	6.70	4.90	5.4
Co^{2+}	7	$^4F_{9/2}$	6.64	3.87	4.8
Ni^{2+}	8	$^3F_4$	5.59	2.83	3.2
Cu^{2+}	9	$^2D_{5/2}$	3.55	1.73	1.9
Cu^+, Zn^{2+}	10	$^2S_0$	0.00	0.00	Diamagnetic

with five or less $3d$ electrons the spin-orbit coupling favors antiparallel **L** and **S** momenta, and thus p_{eff} will tend to be less than that given by the spin-only value. For the other elements a parallel **L** and **S** is favored, leading to a larger p_{eff}. The spin-orbit coupling parameter is particularily large for the latter elements, and this is borne out in the larger disagreement with the spin-only value shown in Table 2-11.1.

In general, either the Curie or Curie-Weiss law is followed[47] for these salts, but low-temperature anomalies of the general type illustrated for the rare earths occur again, mainly for those ions with degenerate ground states in the presence of the crystalline field. For single crystals there is a large anisotropy for such elements. Thus Co^{2+} and Fe^{2+} may have an anisotropy of about 30%.[48] On the other hand, Ni^{2+}, which has a singlet

[45] C. J. Gorter, *Paramagnetic Relaxation*, Elsevier Publishing Co., New York (1947).

[46] J. H. Van Vleck, *Discussions Faraday Soc.* **26,** 96 (1958); A. S. Chakravarty and R. Chatterjie, *Indian J. Phys.* **33,** 531 (1959); A. S. Chakravarty, *Proc. Phys. Soc. (London)* **74,** 711 (1959); M. Kotani, *Prog. Theor. Phys. (Kyoto) Suppl.* **14,** 1 (1960).

[47] K. Kido and T. Watanabe, *J. Phys. Soc. Japan* **16,** 1217 (1959).

[48] A. Bose, *Indian J. Phys.* **22,** 74, 195, 276, 483 (1948); B. C. Guha, *Nature (London)* **184,** 50 (1959); L. C. Jackson, *Phil. Mag.* **4,** 269 (1959).

level lowest, has an anisotropy that is very small, sometimes less than 1%.[49]

12. Covalent Binding and the 3*d*, 4*d*, 5*d*, and 5*f*–6*d* Transition Groups

On the basis of the 3*d* wave functions illustrated in Fig. 2-11.1, covalent binding may be pictured as the sharing of electrons on the negative diamagnetic ions with the positive paramagnetic ions. The charge distribution is then smeared out so that it is more symmetric. The large overlapping of the orbitals can be considered as equivalent to the effect of a strong crystal field. Indeed, this effective field produces an interaction that is larger than the Russell-Saunders l–l or s–s coupling between the electrons and leads to a large splitting between the δ_ϵ triplet and the δ_γ

Table 2-12.1. Spin of the Iron Group Cyanides

Ion	Configuration	S
	$3d^1$	$\frac{1}{2}$
	$3d^2$	1
$Cr^{3+}CN_6$	$3d^3$	$\frac{3}{2}$
$Mn^{3+}CN_6$	$3d^4$	1
$Fe^{3+}CN_6$	$3d^5$	$\frac{1}{2}$
$Co^{3+}CN_6$, $Fe^{2+}CN_6$	$3d^6$	0

doublet. The result is that the susceptibilities are often quite different from those for ionic bound solids. The cyanides are the best-known covalently bound iron group substances. All octahedral complex compounds of the 4*d* and 5*d* transition groups exhibit covalent binding; other compounds are not understood at present. Covalent binding is also important in the oxides of the 5*f*–6*d* group. We shall discuss each of these systems briefly.

Because of the large orbital splitting, the electrons go into the δ_ϵ triplet in such a way that the maximum value of the total spin is achieved,[50] as illustrated in Table 2-12.1. Cyanides with more than six *d*-electrons are apparently not formed. Experiment indeed shows that $Co^{3+}CN_6$ and $Fe^{2+}CN_6$ are diamagnetic, whereas $Fe^{3+}CN_6$ has a susceptibility that can

[49] L. C. Jackson, *Proc. Roy. Soc.* (*London*) **A-140,** 695 (1933); T. Haseda, H. Kobayashi, and M. Date, *J. Phys. Soc. Japan* **14,** 1724 (1959); M. A. Lasheen, J. van den Broek, and C. J. Gorter, *Physica* **24,** 1061 (1958); A. Bose, S. C. Mitra, and S. K. Datta, *Proc. Roy. Soc.* (*London*) **A-248,** 153 (1958).

[50] J. H. Van Vleck, *J. Chem. Phys.* **3,** 807 (1935); J. B. Howard, *J. Chem. Phys.* **3,** 813 (1935).

be explained by $S = \frac{1}{2}$ together with a small contribution from the presumably incompletely quenched orbital momentum. There is fair agreement with the susceptibility measurements on other cyanides.[51]

Data on the palladium and platinum group compounds are scarce and often inconsistent. Some experimental results, mainly from Cabrera and Duperier and Bose,[52] are tabulated in Table 2-12.2. The spin-only values of p_{eff}, calculated for values of S from Hund's rules and from Table 2-12.1, are also given. The latter values show the better agreement, but it is still not good.

Table 2-12.2. Palladium and Platinum Group Ions

	Ion	Configuration	$p_{eff} = 2[S(S+1)]^{1/2}$ (Hund)	$p_{eff} = 2[S(S+1)]^{1/2}$ (Table 2-12.1)	Experimental p_{eff}
Palladium	Mo^{4+}	$4d^2$	2.83	2.83	0.2
	Mo^{3+}	$4d^3$	3.87	3.87	3.6
	Ru^{4+}	$4d^4$	4.90	2.83	0.6
	Ru^{3+}	$4d^5$	5.92	1.73	2.1
	Rh^{3+}	$4d^6$	4.90	0.00	0.06
	Pd^{2+}	$4d^8$	2.83	—	0.1
Platinum	Ir^{4+}	$5d^5$	5.92	1.73	1.9
	Os^{2+}	$5d^6$	4.90	0.00	0.3
	Ir^{3+}				0.04
	Pt^{2+}	$5d^8$	2.83	—	diamagnetic

It became clear, late in the 1940's, when sufficient quantities of the transuranic elements had been produced, that a second "rare earth" series existed. It appears that the series begins with actinium, the analog of lanthanum. However, the precise electronic configuration of some of the ions is uncertain. Thus in Table 2-12.3 only the number of electrons with unpaired spins is given for the various compounds whose susceptibilities have been reliably determined.[53]

The experimental values for four or more magnetic electrons are close to that expected for rare earth ions, whereas for one and two magnetic electrons the spin-only values give the best fit. This led Dawson to suggest that the magnetic electrons were in $6d$ states in ions of the latter type and

[51] M. Kotani, *J. Phys. Soc. Japan* **4,** 293 (1949); B. C. Guha, *Proc. Roy. Soc.* (*London*) **A-206,** 353 (1951).

[52] B. Cabrera and A. Duperier, *Proc. Phys. Soc.* **51,** 845 (1939); D. M. Bose, *Z. Physik* **48,** 716 (1928); B. N. Figgis, J. Lewis, R. S. Nyholm, and R. D. Peacock, *Discussions Faraday Soc.* **26,** 103 (1958); H. Selig, F. A. Cafasso, D. M. Gruen, and J. G. Malm, *J. Chem. Phys.* **36,** 3440 (1962).

[53] J. K. Dawson, *Nucleonics* **10,** 39 (1952); D. M. Gruen and C. Hutchinson, *J. Chem. Phys.* **22,** 386 (1954).

in $5f$ states for the former. Pryce,[54] however, has developed a theory which shows that the spin-only values obtained are fortuitous and that in fact the neptunyl and plutonyl ions' magnetic electrons are in the $5f$ shell.

Pryce's theory depends on the observation that the actinides, in contrast to the lanthanides, form stable complexes such as $(UO_2)^{2+}$, $(NpO_2)^{2+}$, and $(PuO_2)^{2+}$. The U^{2+} ion may be considered as having an electron configuration $5f6d7s^2$ (neutral U is $5f^36d7s^2$). The orbitals of these four

Table 2-12.3. Transuranic Ions

Compound	No. of Magnetic Electrons	Theory: $p_{eff} = g[J(J+1)]^{1/2}$	Theory: $p_{eff} = 2[S(S+1)]^{1/2}$ d-electrons	Theory: $p_{eff} = 2[S(S+1)]^{1/2}$ f-electrons	Experimental p_{eff}
$Na(NpO_2)(C_2H_3O_2)_3$	1	2.54	1.73	1.73	1.84
$Na(PuO_2)(C_2H_3O_2)_3$	2	3.58	2.83	2.83	2.83
U^{3+} (soln.)	3	3.62	3.87	3.87	3.14
PuF_4	4	2.68	4.90	4.90	2.98
PuF_3 (conc.)	5	0.84	5.92	5.92	1.2
$PuCl_3$					1.0
AmF_3	6	0.00	4.90	6.92	1.3
CmF_3	7	7.94	3.83	7.92	7.9

electrons can be combined to form highly directional hybrid bonds with the orbitals of the two oxygens. This produces a linear $(UO_2)^{2+}$ ion with strong covalent binding. No electrons remain to produce a paramagnetic effect, and indeed $(UO_2)^{2+}$ is diamagnetic; $(NpO_2)^{2+}$ has one more electron than the uranyl complex. Under the assumption that the magnetic electron is in the $5f$ shell, it turns out that because of the electronic field of the bonding electrons and the spin-orbit coupling the ground state would be a doublet. Only this doublet is occupied at room temperature, and, with an effective angular momentum represented by $J = S = \frac{1}{2}$, it gives rise to the observed paramagnetism. The theory for $(PuO_2)^{2+}$ is somewhat more complicated, but gives $J = S = 1$. Paramagnetic resonance[55] experiments have confirmed Pryce's theory.

It is obvious that our understanding of the transition-group compounds, described in this section, is far from satisfactory. A clearer picture must await the collection of more data and further theoretical studies.

13. Saturation in Paramagnetic Substances

In the derivation of Curie's law in Section 2-7 it was assumed that $M_J g \mu_B H/kT \ll 1$. At low temperature, and in high fields, this inequality

[54] J. C. Eisenstein and M. H. L. Pryce, *Proc. Roy. Soc.* (*London*) **A-229**, 20 (1955).
[55] B. Bleaney, *Discussions Faraday Soc.* **19**, 112 (1955).

no longer holds. We start with equation 2-7.1,

$$M = N \frac{\sum_{-J}^{+J} M_J g \mu_B e^{M_J g \mu_B H/kT}}{\sum_{-J}^{+J} e^{M_J g \mu_B H/kT}}$$

and substitute $x = g\mu_B H/kT$. The result may be written

$$M = Ng\mu_B \frac{d}{dx} \ln \sum_{-J}^{+J} e^{M_J x}.$$

By using the formula for the sum of the geometric progression

$$e^{-Jx}(1 + e^x + e^{2x} + \cdots + e^{2Jx}),$$

this becomes

$$M = Ng\mu_B \frac{d}{dx}\left(\ln e^{-Jx} \frac{1 - e^{(2J+1)x}}{1 - e^x}\right).$$

Remembering $\sinh x = (e^x - e^{-x})/2$, we obtain

$$M = Ng\mu_B \frac{d}{dx} \ln \frac{\sinh [(2J + 1)/2]x}{\sinh (x/2)}$$

so that finally

$$\begin{aligned} M &= Ng\mu_B J\left[\frac{2J + 1}{2J} \coth \left(\frac{2J + 1}{2J}\right)y - \frac{1}{2J} \coth \frac{y}{2J}\right] \qquad (2\text{-}13.1) \\ &= Ng\mu_B J B_J(y), \end{aligned}$$

where

$$y = \frac{Jg\mu_B H}{kT}.$$

The expression $B(y)$ is usually called the Brillouin function.[56] Equation 2-13.1 was derived for wide multiplets. For very narrow multiplets it can be shown in a similar fashion that $M = 2NS\mu_B B_S(2S\mu_B H/kT) + NL\mu_B B_L(L\mu_B H/kT)$. This result is not of much use, since at low temperatures the multiplet levels are not small with respect to kT.

When $J \to \infty$, that is, when all orientations of the moment become possible when a field is applied, we get

$$\underset{J\to\infty}{B_J(y)} = \coth y - \frac{1}{y} = L(y), \qquad (2\text{-}13.2)$$

where $L(y)$ is the classical Langevin function, derived in 1905 by Langevin. This is an example of the Bohr correspondence principle, which relates quantum mechanical results to the corresponding classical case. Since

[56] L. Brillouin, *J. Phys. radium.* **8**, 74 (1927).

$B(y)$ and $L(y)$ approach unity as $y \to \infty$, the dipoles will, in sufficiently large fields, be directed so that the magnetization along H is the maximum possible; that is, there will be saturation. In the classical case Langevin considered that the dipole moments would align themselves along H,

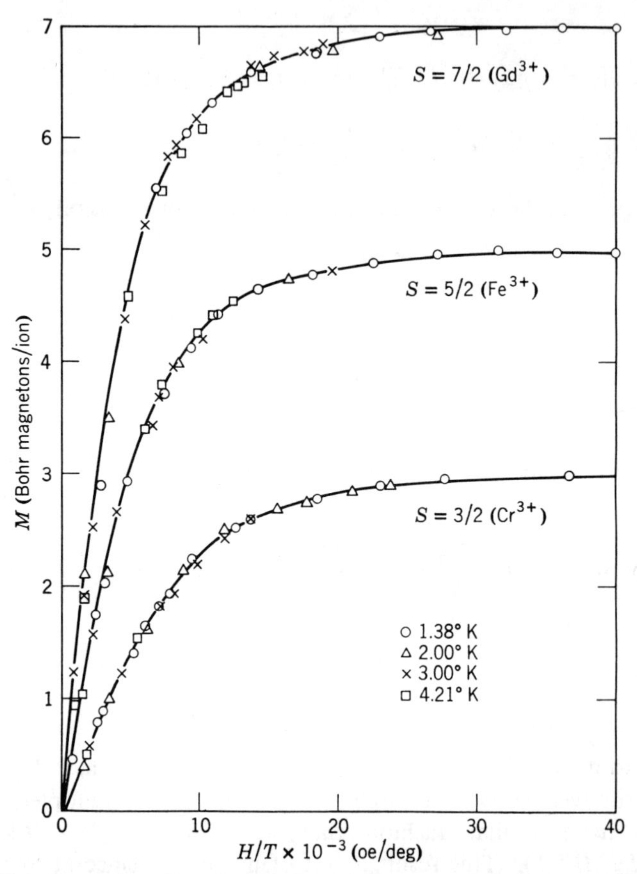

Fig. 2-13.1. Saturation effects in high fields and at low temperatures for various paramagnetic ions. (After W. E. Henry.)

whereas in the quantum mechanical case the maximum component of the moment $\mu_B g[J(J+1)]^{1/2}$ along H is $Jg\mu_B$ (see problem 2-6). Thus, classically, $M/N\mu \to 1$ whereas quantum mechanically $M/N\mu \to \sqrt{J/(J+1)}$ at saturation.

Experimental saturation was first observed[57] for hydrated gadolinium sulfate, $Gd_2(SO_4)_3{\cdot}8H_2O$. This is a particularly favorable case, since the

[57] H. R. Woltjer and H. Kamerlingh Onnes, *Commun. Kamerlingh Onnes Lab. Univ. Leiden*, **167c** (1923).

Gd^{3+} ion has $L = 0$, and therefore no complications caused by the crystalline field. It is also well diluted magnetically because of the eight water molecules. Since $J = S = \frac{7}{2}$ for this ion, the difference between the Brillouin and the Langevin function is not large; nevertheless, the Brillouin function was favored by experiment. Further work at higher values of field has been carried out by Henry[58] on potassium chromium alum [$KCr(SO_4)_2 \cdot 12H_2O$] and ferric ammonium alum [$FeNH_4(SO_4)_2 \cdot 12H_2O$], as well as hydrated gadolinium sulfate. Since L is quenched for iron group ions, the value of S should be substituted for J in the Brillouin function. The Brillouin functions for $J = \frac{7}{2}(Gd^{3+})$, $J = \frac{5}{2}(Fe^{3+})$, and $J = \frac{3}{2}(Cr^{3+})$ together with Henry's results are shown in Fig. 2-13.1. The agreement is satisfactory. For the special case of two levels, that is $J = \frac{1}{2}$, equation 2-13.1 reduces to

$$M = \frac{Ng\mu_B}{2} \tanh \frac{g\mu_B H}{2kT}. \qquad (2\text{-}13.3)$$

A curve with this shape has been obtained for dysprosium ethylsulfate, using a magnetorotation technique.[59] The free Dy^{3+} ion has $J = \frac{15}{2}$. The crystalline field splits the levels, so that at very low temperatures only the lowest state, a doublet, is occupied.

14. Paramagnetic Molecules

Some molecules have an odd number of electrons and will therefore exhibit a paramagnetism. The orbital angular momentum is usually completely quenched, so that the paramagnetism is due to the spin (i.e., $p_{eff} = 2[S(S + 1)]^{1/2}$). Organic free radicals are common examples; in fact, the paramagnetism of α-naphthyl diphenylmethyl was discovered in 1925. Other examples investigated are the ketyls, diphenyl-trinitro phenyl hydrazyl, and triphenylmethyls. Compounds containing paramagnetic inorganic free radicals are rare; an example is $(KSO_3)_2NO$. Most molecules, of course, have an even number of electrons and are diamagnetic. There are a few, however, such as organic biradicals, that are normal paramagnets.

Some complex molecules exhibit a feeble paramagnetism that is usually independent of temperature. It is believed that the resultant spin of the complex is zero and that Van Vleck temperature-independent paramagnetism is being observed[60] (see equation 2-5.1). Possible examples are $KMnO_4$,[61] K_2CrO_4, $K_2Cr_2O_7$, MoO_3, WO_3, and TiO_2.

[58] W. E. Henry, *Phys. Rev.* **88,** 559 (1952).
[59] J. Becquerel, W. J. de Haas, and J. van den Handel, *Physica* **3,** 1133 (1936).
[60] For a critical review see J. Baudet, *J. Chim. Phys.* **58,** 845 (1961).
[61] O. P. Singhal, *Proc. Phys. Soc.* (*London*) **79,** 389 (1962).

Very few molecules in the gaseous state at ordinary temperatures are paramagnetic. Those that are include the diatomic gases oxygen and nitric oxide and the polyatomic gases NO_2 and ClO_2.

It is appropriate to make a few remarks concerning the spectroscopic notation used to describe diatomic molecules. The molecule has symmetry about the axis formed by the two nuclei, and as a result the total orbital momentum **L** is not a constant of the motion. Instead, the axial component $|M_L| = \Lambda$ is a good quantum number; Λ takes the values 0, 1, 2, . . . , L, and such states are denoted by Σ, Π, Δ, . . . , respectively. The different values of the spin components,[62] $M_S = \Sigma$, for a certain Λ, give rise to a multiplet whose multiplicity is indicated by a superscript on the symbol for the state, such as $^2\Sigma$ and $^3\Pi$. The sum $\Omega = \Lambda + \Sigma$ is also a constant of the motion and is given as a subscript. When the rotational angular momentum of the molecular axis is large, the electronic rotational and electronic orbital momentum are added to give a resultant **K**, which is then added to the spin to form the total momentum represented by **J**, that is, $\mathbf{S} + \mathbf{K} = \mathbf{J}$.

When the multiplets are widely spaced compared to kT, only the lowest state is occupied. The component of the magnetic moment along the nuclear axis is given by $\mu_{\parallel} = \mu_B(\Lambda + 2\Sigma)$, so that the susceptibility will be

$$\chi = \frac{\mu_B^{\,2}}{3kT}(\Lambda + 2\Sigma)^2. \tag{2-14.1}$$

For narrow multiplets the set of states with $\Sigma = S, S - 1, \ldots, -S$ must be considered. The mean value of the effective magneton number over these states is given by

$$\begin{aligned}\overline{p_{\text{eff}}^2} &= \overline{(\Lambda + 2\Sigma)^2}\\ &= \Lambda^2 + 4\overline{\Sigma}\Lambda + 4\overline{\Sigma^2}\\ &= \Lambda^2 + 4S(S + 1),\end{aligned}$$

since $\overline{\Sigma} = 0$. This gives a susceptibility of

$$\chi = \frac{\mu_B^{\,2}}{3kT}[\Lambda^2 + 4S(S + 1)]. \tag{2-14.2}$$

Although oxygen has an even number of electrons, its ground state is $^3\Sigma$ ($\Lambda = 0$, $S = 1$). Since its multiplet structure is very small, of the order of 4 cm^{-1}, equation 2-14.2 applies for temperatures above the boiling point (90°K). Calculation gives $p_{\text{eff}} = 2.83$ and is in good agreement with the experimental result[63] $p_{\text{eff}} = 2.85$. By trapping individual

[62] Not to be confused with the state symbol Σ.

[63] E. Bauer and A. Piccard, *J. Phys. radium* **1**, 97 (1920).

oxygen molecules in the large and well-separated holes of the clathrate compound β-quinol, it has been possible to show that Curie's law is valid down to 10°K.[64] Below this temperature the measurements can be explained if it is assumed that the oxygen molecules are partly frozen in by an axially symmetric crystal field due to the neighboring atoms of the clathrate compound.[65] Susceptibility measurements of sulfur vapor have established that diatomic sulfur, S_2, also has a $^3\Sigma$ ground state.[66]

The ground level of the NO molecule is $^2\Pi_{1/2}$, and it is separated by 120 cm^{-1} from the next $^2\Pi_{3/2}$ level. For the $^2\Pi_{1/2}$ level equation 2-14.1 shows that $p_{\text{eff}} = 0$, so that for very low temperatures the gas is diamagnetic. At very high temperatures equation 2-14.2 should be valid, and $p_{\text{eff}} = 2$. For ordinary temperatures an intermediate equation must be used. Good agreement is then obtained with the experimental results.[67] For the polyatomic gases the orbital moment is quenched, and the susceptibility is due only to the spin and a small temperature independent term.

15. Paramagnetic Susceptibility of the Nucleus

The nucleus has a spin **I** and therefore a magnetic moment, as given by equation 2-2.13. The vector **I** combines with **J** to form a resultant **F**, and this produces a hyperfine structure. The spacing of these hyperfine levels is small compared with kT, except at very low temperatures. Further, the coupling of **I** and **J** is small, so that application of even a small magnetic field decouples the vectors, hence they contribute to the susceptibility independently. This is analogous to the situation discussed in Section 2-7. The susceptibility of the nucleus is then given by

$$\chi_{\text{nuc}} = \frac{N g_n^2 I(I+1)}{3kT} \mu_{Bn}^2, \qquad (2\text{-}15.1)$$

where g_n is the nuclear gyromagnetic ratio and μ_{Bn} is the nuclear Bohr magnetron. Since the difference between a nuclear and an electronic magnetic moment is of the order of 10^{-3}, χ_{nuc} is about 10^{-9} emu/mole at room temperature. Even diamagnetic susceptibilities are at least 10^3 times

[64] A. H. Cooke, H. Meyer, W. P. Wolf, D. F. Evans, and R. E. Richards, *Proc. Roy. Soc.* (*London*) **A-225,** 112 (1954).

[65] H. Meyer, M. C. M. O'Brien, and J. H. Van Vleck, *Proc. Roy. Soc.* (*London*) **A-243,** 414 (1957).

[66] L. Néel, *Compt. rend.* (*Paris*) **194,** 2035 (1932); A. B. Scott, *J. Am. Chem. Soc.* **71,** 3145 (1949); J. A. Poulis, C. H. Massen, and P. van der Leeden, *Trans. Faraday Soc.* **58,** 52 (1962).

[67] J. H. Van Vleck, *Theory of Electric and Magnetic Susceptibilities*, Oxford University Press, Oxford (1932), p. 270.

larger than this, and it is clear that the nuclear susceptibility can usually be neglected.

At very low temperatures, however, χ_{nuc} may be comparable with χ_{dia}. It has been found possible to measure the nuclear susceptibility of solid hydrogen, which is diamagnetic in the electron system, at a temperature close to absolute zero.[68] The results were consistent with the value of the proton's magnetic moment as determined by other methods. The contribution of the protons to the susceptibility of water has also been measured.[69] Evans[70] has succeeded in detecting the static nuclear susceptibility of hexamethylbenzene ($C_6(CH_3)_6$) by using a torque method and utilizing the different time effects of the paramagnetism and the larger diamagnetism.

Problems

2-1. From first principles, calculate the value of $\overline{r^2}$ for atomic hydrogen in its ground state [$\psi = (\pi a_0^3)^{-1/2} e^{-r/a_0}$]. Check the result, using equation 2-4.1. Compare it with the square of the mean value of the electronic radius. Next, compute the diamagnetic susceptibility of the hydrogen atoms. (*Ans.* $\overline{r^2} = 3a_0^2$, $\bar{r}^2 = a_0^2$, where $a_0 = 0.529$ Å, the Bohr radius.)

2-2. Using the data of Table 2-5.2, compute the susceptibility of $NH_3C_6H_5$, C_2H_4, C_2H_2, C_6H_5Cl. Compare the values for LiCl, LiBr, KCl, and KBr with those given in Table 2-4.2 and the ionic values of Table 2-4.3.

2-3. Consider a monatomic paramagnetic gas whose molecules each have a permanent magnetic moment $\boldsymbol{\mu}$. Further, when a magnetic field H is applied, assume that all orientations of the moment with respect to the field are possible. Show that the magnetization is given by replacing the summation of equation 2-7.1 by integration so that

$$M = \frac{\int_0^{\pi} \mu \cos\theta e^{\boldsymbol{\mu}\cdot\mathbf{H}/kT}(2\pi \sin\theta \, d\theta)}{\int_0^{\pi} e^{\boldsymbol{\mu}\cdot\mathbf{H}/kT}(2\pi \sin\theta \, d\theta)},$$

where $2\pi \sin\theta \, d\theta = d\Omega$, the solid angle, and θ is the angle between $\boldsymbol{\mu}$ and $\mathbf{H}$. By assuming that $\mu H/kT \ll 1$, expand the exponentials to the usual approximation and derive the classical analogue of equation 2-7.3.

2-4. Show that for two doublet levels separated by an energy Δ, in a zero field, the susceptibility is given by equation 2-10.7. Using the values of the

[68] B. Lasarew and L. Schubnikow, *Phys. Z. Sowjet* **11**, 445 (1937).

[69] J. A. Poulis, C. H. Massen, and P. van der Leeden, *Magnetic and Electric Resonance and Relaxation*, J. Smidt, editor, North-Holland Publishing Co., Amsterdam (1963), p. 743.

[70] D. F. Evans, *Phil. Mag.* **1**, 370 (1956).

parameters found by Bogle et al., plot χ as a function of temperature up to $T = 20°K$.

2-5. Find the new energy levels of a $3d$ electron when under the action of a rhombic field, using the wave functions of equation 2-11.1. Check the result with the predictions of Table 2-10.1. Show that the angular momentum is quenched.

2-6. When the approximation used in problem 2-3 is not valid, show that $M = N\mu L(\mu H/kT)$, where $L(\mu H/kT) = \coth(\mu H/kT) - (kT/H)$. *Hint.* Change the integration parameter by substituting $x = \cos\theta$. For $\mu H/kT \ll 1$ show that $L(y) = y/3 - y^3/45 + 2y^5/945 \cdots$ and thus rederive the result of problem 2–3.

2-7. The principle susceptibilities of a crystal are χ_1, χ_2, and χ_3. Show that a powder of randomly orientated crystals would give a susceptibility of $\chi = (\chi_1 + \chi_2 + \chi_3)/3$.

2-8. For a paramagnetic gas of N atoms/cm³ and with $L = 0$, $S = \frac{1}{2}$, calculate the number of atoms in the two levels at the temperature T and in the field H. Also calculate the resultant magnetization and thus rederive equation 2-13.3. Calculate values for the populations for $N = 10^{22}$, $H = 25{,}000$ oersteds, and $T = 300$ and $4°K$.

2-9. For the gas of problem 2-8 calculate the free energy $F = E - TS$, where S is the entropy, in terms of the excess number of atoms in the lowest energy state. Using the condition that F is a minimum for thermodynamic equilibrium, find the susceptibility.

2-10. Consider matrix elements of the crystal field with $3d$ wave functions of the type $\int \psi_2{}^* eV\psi_1\, d\tau$ and let V be expanded in spherical harmonics of the type $A_n{}^m r^n P_n^{|m|}(\cos\theta)e^{im\phi}$. First, expand $\psi_2{}^*\psi_1$ in spherical harmonics and show that there are no terms for $n > 4$. Thus show that the matrix elements for $n > 4$ vanish. Also show that for n odd the matrix elements are zero by considering the change of sign of $\psi_2{}^*\psi_1$ and V on substituting $-x$, $-y$, and $-z$ for x, y, z.

Bibliography

Bates, L. F., *Modern Magnetism*, 4th ed. Cambridge University Press, Cambridge (1962).

Foëx, G., *Diamagnetism and Paramagnetism*, Pergamon Press, London (1957).

Griffith, J. S., *Theory of Transition Metal Ions*, Cambridge University Press, Cambridge (1961).

Meyers, W. R., "Diamagnetism of Ions," *Revs. Mod. Phys.* **24,** 15 (1952).

Selwood, P. W., *Magnetochemistry*, 2nd ed. Interscience Publishers, New York (1956).

Spedding, F. H. and A. H. Duane, editors, *The Rare Earths*, John Wiley and Sons, New York (1961).

Stoner, E. C., *Magnetism and Matter*, Methuen and Co., London (1934).

van den Handel, J., "Paramagnetism," *Advan. Electron. and Electron Phys.* **6,** 463, New York (1954).

Van Vleck, J. H., *Theory of Electric and Magnetic Susceptibilities*, Oxford University Press, Oxford (1932).

3

Thermal, Relaxation, and Resonance Phenomena in Paramagnetic Materials

1. Introduction

It is clear, from the preceding chapter, that the population of energy levels of paramagnetic ions depends on the temperature. Calorimetric studies, such as specific heat measurements, can be expected to yield information concerning these levels; conversely, paramagnetic salts are likely to have applications in low-temperature work. The first part of this chapter, which includes a section in which pertinent thermodynamic equations are reviewed, is concerned with thermal effects.

Next, the time required to reach an equilibrium state, overlooked in Chapter 2, is considered. The use of alternating magnetic fields provides a method of investigating the various mechanisms in relaxation effects. These effects are analyzed primarily from the thermodynamic point of view and provide, in addition to information on time effects, data on the specific heats of the magnetic system.

The lowest energy levels of paramagnetic ions, either free or in a crystal, are studied by the method of paramagnetic resonance. The splitting of levels is measured by finding the frequency of an alternating field that induces transitions between these levels. In addition, the width of these absorption lines or resonances is related to the relaxation processes. Electron paramagnetic resonance phenomena are treated in some detail in this chapter, and nuclear paramagnetic resonance is treated in Chapter 4.

Other techniques for the study of paramagnetic materials, such as optical spectroscopy, are not discussed in this book.

THERMAL PHENOMENA

2. Summary of Thermodynamic Relationships

The equilibrium states of a system can usually be described by three variables, of which only two are independent. Two examples of groups of such variables are PVT (pressure, volume, temperature) and HMT (magnetic field, magnetization, temperature). If xyz represents any three thermodynamic variables, with y the dependent variable so that $y = y(x, z)$, then

$$dy = \left(\frac{\partial y}{\partial x}\right)_z dx + \left(\frac{\partial y}{\partial z}\right)_x dz,$$

where the subscript to the partial derivatives emphasizes the variable that is kept constant. For the case of y being constant $dy/dx = 0$, and it follows immediately that

$$\left(\frac{\partial x}{\partial y}\right)_z \left(\frac{\partial y}{\partial z}\right)_x \left(\frac{\partial z}{\partial x}\right)_y = -1. \tag{3-2.1}$$

Also, if $P = (\partial y/\partial x)_z$ and $Q = (\partial y/\partial z)_x$, then

$$\left(\frac{\partial P}{\partial z}\right)_x = \left(\frac{\partial Q}{\partial x}\right)_z, \tag{3-2.2}$$

which is the condition for an exact differential. Equations 3-2.1 and 3-2.2 are useful in establishing some thermodynamic relationships.

The first law of thermodynamics states that the heat dQ added to a system is equal to the increase in the internal energy dU, plus the work dW done by the system. When the system is a solid whose magnetization is changed, work must be done on the system, so that $dW = -H\,dM$, according to equation 1-7.7. We then have

$$dU = dQ + H\,dM. \tag{3-2.3}$$

Now considering U as a function of T and M, so that

$$dU = \left(\frac{\partial U}{\partial T}\right)_M dT + \left(\frac{\partial U}{\partial M}\right)_T dM$$

(3-2.3) becomes

$$dQ = \left(\frac{\partial U}{\partial T}\right)_M dT + \left[\left(\frac{\partial U}{\partial M}\right)_T - H\right] dM. \tag{3-2.4}$$

The heat capacity C of a system is defined in general as dQ/dT and when measured per unit mass or per mole is termed the specific heat. The specific heat depends on the conditions under which the temperature rises when heat is added to the system. For magnetic systems C_H and C_M, the specific heats at constant field and constant magnetization, respectively, are of interest. From (3-2.4) we have

$$C_M = \left(\frac{dQ}{dT}\right)_M = \left(\frac{\partial U}{\partial T}\right)_M = T\left(\frac{\partial S}{\partial T}\right)_M, \tag{3-2.5}$$

remembering that the entropy S is related to Q by $T\,dS = dQ$.

The enthalpy E is defined as

$$E = U - HM, \tag{3-2.6}$$

so that

$$dE = dQ - M\,dH. \tag{3-2.7}$$

It is clear that the enthalpy is an internal energy, of greatest utility in processes where H is held constant. Some writers prefer to define their thermodynamic systems so that the internal energy does not include the interaction energy $\mathbf{H} \cdot \mathbf{M}$, in which case the roles of internal energy and enthalpy are interchanged. The choice made is unimportant as long as one is consistent. Expressions for the specific heat at constant H are then

$$C_H = \left(\frac{dQ}{dT}\right)_H = \left(\frac{\partial E}{\partial T}\right)_H = T\left(\frac{\partial S}{\partial T}\right)_H. \tag{3-2.8}$$

Now (3-2.3) can be written as

$$dU = T\,dS + H\,dM, \tag{3-2.9}$$

so that by using (3-2.2) we have

$$\left(\frac{\partial T}{\partial M}\right)_S = \left(\frac{\partial H}{\partial S}\right)_M. \tag{3-2.10}$$

Further, from (3-2.7) we have

$$dE = T\,dS - M\,dH, \tag{3-2.11}$$

so that

$$\left(\frac{\partial T}{\partial H}\right)_S = -\left(\frac{\partial M}{\partial S}\right)_H. \tag{3-2.12}$$

The free energy or Helmholtz function F is defined as

$$F = U - TS, \tag{3-2.13}$$

so that

$$dF = H\,dM - S\,dT \tag{3-2.14}$$

and

$$\left(\frac{\partial H}{\partial T}\right)_M = -\left(\frac{\partial S}{\partial M}\right)_T. \tag{3-2.15}$$

The Gibbs free energy function or thermodynamic potential G is defined as

$$G = E - TS; \tag{3-2.16}$$

hence

$$dG = -M\,dH - S\,dT \tag{3-2.17}$$

and

$$\left(\frac{\partial M}{\partial T}\right)_H = \left(\frac{\partial S}{\partial H}\right)_T. \tag{3-2.18}$$

Equations 3-2.10, 3-2.12, 3-2.15, and 3-2.18 are known as Maxwell's relations.

It is useful to derive two relationships for $T\,dS$. Consider S to be a function of T and M; then

$$T\,dS = T\left(\frac{\partial S}{\partial T}\right)_M dT + T\left(\frac{\partial S}{\partial M}\right)_T dM.$$

Using (3-2.5) and (3-2.15), we get the first relation

$$T\,dS = C_M\,dT - T\left(\frac{\partial H}{\partial T}\right)_M dM. \tag{3-2.19}$$

Now consider S as a function of T and H; then

$$T\,dS = T\left(\frac{\partial S}{\partial T}\right)_H dT + T\left(\frac{\partial S}{\partial H}\right)_T dH.$$

From (3-2.8) and (3-2.18), we have

$$T\,dS = C_H\,dT + T\left(\frac{\partial M}{\partial T}\right)_H dH, \tag{3-2.20}$$

the second relationship.

Substitution of (3-2.19) into (3-2.9) gives

$$dU = C_M\,dT - \left[T\left(\frac{\partial H}{\partial T}\right)_M - H\right] dM.$$

But, since

$$dU = \left(\frac{\partial U}{\partial T}\right)_M dT + \left(\frac{\partial U}{\partial M}\right)_T dM,$$

it follows that

$$\left(\frac{\partial U}{\partial M}\right)_T = H - T\left(\frac{\partial H}{\partial T}\right)_M. \tag{3-2.21}$$

Similarly, combining (3-2.20) and (3-2.11) leads to

$$\left(\frac{\partial E}{\partial H}\right)_T = T\left(\frac{\partial M}{\partial T}\right)_H - M. \tag{3-2.22}$$

Finally, it is of interest to have expressions for the difference and ratio of the two specific heats. Equating (3-2.19) to (3-2.20) and solving for dT gives

$$dT = \frac{-T(\partial H/\partial T)_M \, dM}{C_H - C_M} - \frac{T(\partial M/\partial T)_M \, dM}{C_H - C_M}.$$

Since we may write

$$dT = \left(\frac{\partial T}{\partial M}\right)_H dM + \left(\frac{\partial T}{\partial H}\right)_M dH,$$

then

$$\left(\frac{\partial T}{\partial M}\right)_H = \frac{-T(\partial H/\partial T)_M}{C_H - C_M}$$

or

$$C_H - C_M = -T\left(\frac{\partial H}{\partial T}\right)_M \left(\frac{\partial M}{\partial T}\right)_H. \tag{3-2.23}$$

Further, since

$$\left(\frac{\partial H}{\partial T}\right)_M = -\left(\frac{\partial M}{\partial T}\right)_H \left(\frac{\partial H}{\partial M}\right)_T,$$

from (3-2.1), we get

$$C_H - C_M = T\left(\frac{\partial M}{\partial T}\right)_H^2 \left(\frac{\partial H}{\partial M}\right)_T. \tag{3-2.24}$$

To find an expression for the ratio, we consider first (3-2.19) and solve for dT, that is,

$$dT = \frac{T\,dS}{C_M} + \frac{T}{C_M}\left(\frac{\partial H}{\partial T}\right)_M dM.$$

Since we may write

$$dT = \left(\frac{\partial T}{\partial S}\right)_M dS + \left(\frac{\partial T}{\partial M}\right)_S dM,$$

it follows on equating the coefficients of dM that

$$C_M = T\left(\frac{\partial H}{\partial T}\right)_M \left(\frac{\partial M}{\partial T}\right)_S.$$

Similarly, from (3-2.20) we obtain

$$C_H = -T\left(\frac{\partial H}{\partial T}\right)_S \left(\frac{\partial M}{\partial T}\right)_H,$$

so that

$$\frac{C_H}{C_M} = -\frac{(\partial H/\partial T)_S(\partial M/\partial T)_H}{(\partial H/\partial T)_M(\partial M/\partial T)_S}$$

$$= -\frac{(\partial H/\partial T)_S(\partial T/\partial M)_S}{(\partial H/\partial T)_M(\partial T/\partial M)_H}.$$

On using (3-2.1) for the denominator, we have finally

$$\frac{C_H}{C_M} = \frac{(\partial H/\partial M)_S}{(\partial H/\partial M)_T}. \tag{3-2.25}$$

3. The Magnetocaloric Effect: The Production and Measurement of Low Temperatures

For an adiabatic process, that is, when there is no change in entropy S, equation 3-2.20 gives

$$C_H\, dT = -T\left(\frac{\partial M}{\partial T}\right)_H dH.$$

A finite adiabatic change in field thus produces a temperature change given by

$$\Delta T = -\frac{T}{C_H}\left(\frac{\partial M}{\partial T}\right)_H \Delta H. \tag{3-3.1}$$

Since for a normal paramagnetic salt $(\partial M/\partial T)_H$ is negative, an increase in field produces a heating and conversely decreasing H gives rise to a temperature drop. This is often called the *magnetocaloric effect.*

The important application of this effect is in the production of temperatures below 1°K.[1] This magnetic cooling is accomplished in the following way. By pumping on liquid helium, in order to lower its vapor pressure, a temperature in the neighborhood of 1°K is achieved. A paramagnetic salt is situated in an evacuated container that is immersed in the helium bath. Since the salt is thermally insulated from the walls of the container, thermal contact with the helium bath is made by introducing He gas (exchange gas) at a reduced pressure into the container. A field is then applied, and the salt is magnetized. This magnetization is essentially accomplished isothermally, since the heat developed is transferred to the liquid helium by the exchange gas. After thermal equilibrium is established the exchange gas is pumped out so that the salt is thermally isolated. The magnet is now switched off, and the temperature falls in accordance with equation 3-3.1. In this way temperatures in the region of 10 millidegrees Kelvin are obtained.

The decrease in entropy in the isothermal magnetization may be calculated by using equation 3-2.20 with $dT = 0$ or (3-2.18), so that

$$S_H - S_{H=0} = \int \left(\frac{\partial M}{\partial T}\right)_H dH. \tag{3-3.2}$$

[1] First suggested by P. Debye, *Ann. Phys.* (*Leipzig*) **81,** 1154 (1926) and independently by W. F. Giauque, *J. Am. Chem. Soc.* **49,** 1864 (1927).

If the salt obeys Curie's law $M = CH/T$, we have

$$S_H = S_{H=0} - \frac{C}{2}\left(\frac{H}{T}\right)^2. \tag{3-3.3}$$

For a normal paramagnet with excellent thermal insulation this is also the entropy of the salt after the adiabatic demagnetization.

The entropy is a measure of the order of a system; the greater the ordering, the smaller the entropy. It is of value to consider the entropy changes of the salt when treated as an assembly of noninteracting dipoles. In the absence of a field the dipoles are randomly orientated. On application of a field the moments are partly aligned, the amount depending also on the temperature T. Indeed, the lower the temperature at which the process is carried out, the greater the degree of dipole alignment, or ordering, hence the lower the entropy (see Section 2-7). At temperatures below 1°K the entropy of the system of lattice vibrations is so small that it can be neglected, so that the entropy is due wholly to the ordering of the magnetic spin system. Hence on the removal of the magnetic field adiabatically, so that the entropy of the magnetic system remains unchanged, the ordering will appear to be that of the system at a much lower temperature.

However, Curie's law is not valid at the lowest temperatures, since the lowest energy levels of the ions are split by the crystalline field and by other interaction effects even in a zero applied field. The lowest temperatures attainable are limited by this effect and may be estimated[2] by a simple method. Let $\delta = \mathrm{k}\theta_0$ be the difference in energy between levels due to the natural splitting. At an initial temperature T_1 high enough so that $T_1 \gg \theta_0$ the $2J + 1$ ground-state levels of an ion in the presence of a field will be separated by $g\mu_B H$, provided $g\mu_B H \gg \mathrm{k}\theta_0$. The distribution of the dipoles, hence the ordering and entropy, will be determined by the factor $g\mu_B H/\mathrm{k}T_1$. On removal of the field, so that the temperature falls to T_2, the occupation of the levels, and therefore the entropy, will be a function of $\mathrm{k}\theta_0/\mathrm{k}T_2$. Since the entropy depends only on the occupation of the levels, it has the same functional relationship for the two states. Therefore, since this entropy is constant, we may equate these ratios so that

$$\frac{T_2}{T_1} = \frac{\mathrm{k}\theta_0}{g\mu_B H}. \tag{3-3.4}$$

Since the final temperature is directly proportional to θ_0, only salts with $\theta_0 < 1$°K are of value in the production of very low temperatures.

The conventional methods for the measurement of temperature fail below 1°K. It is usual to use a magnetic temperature T^*, defined by

[2] N. Kurti and F. E. Simon, *Proc. Roy. Soc. (London)* **A-149,** 152 (1935).

assuming Curie's law is valid and determined by measuring the susceptibility of the paramagnetic salt, since $T^* = c/\chi$, where c is Curie's constant per cubic centimeter. Because of demagnetizing effects, the magnetic temperature depends on the shape of the specimen. The magnetic temperature for a spheroidal specimen $T^⊛$ is often employed. The magnetic temperature of ellipsoidal specimens is related to that for spheres by $T^⊛ = T^* + (D_s - D)c$, where D and D_s are the demagnetization factors of the ellipsoid and sphere, respectively.

The magnetic temperature will be equal to the absolute temperature in regions where Curie's law is still valid. Zero field splitting of the lowest levels ultimately produces deviations from this law and at temperatures of the order of θ_0 and below T^* will not equal T. One salt that is of great value as a thermometer is cerium magnesium nitrate

$$[2Ce(NO_3)_3 \cdot 3Mg(NO_3)_2 \cdot 24H_2O].$$

Its zero-field-level splitting is so small that $T^⊛ = T$ down to 0.006°K.[3] Other salts that have been studied in adiabatic demagnetization experiments are discussed by Ambler and Hudson.[4]

The absolute temperature can be determined by a calorimetric method. Since $T = dQ/dS$, then for $H = 0$

$$T = \frac{(dQ/dT^*)_{H=0}}{(dS/dT^*)_{H=0}} = \frac{C^*_{H=0}}{(dS/dT^*)_{H=0}}. \tag{3-3.5}$$

The relationship between S and T^* is found by measuring T^* after a demagnetization and calculating S by using equation 3-3.3. The specific heat can be determined if the change of T^* is measured when a known quantity of heat dQ is added. The difficulty lies in heating the sample uniformly and allowing for stray heating. One ingenious method heats the specimen by irradiating with gamma rays[5]; another uses an alternating magnetic field.[6] Neither method is entirely satisfactory.[7]

Because of zero field splitting, there is an anomaly in the specific heat C_H. At high temperatures ($T \gg \theta_0$) the levels are essentially equally populated, whereas when $T \ll \theta_0$ only the lowest level is occupied. As the temperature is increased toward θ_0, energy is required to produce a

[3] J. M. Daniels and F. N. H. Robinson, *Phil. Mag.* **44,** 630 (1953).
[4] E. Ambler and R. P. Hudson, *Rept. Progr. Phys.* **18,** 251 (1955).
[5] F. E. Simon, *Nature (London)* **135,** 763 (1935).
[6] W. F. Giauque and D. P. MacDougall, *Phys. Rev.* **43,** 768 (1935).
[7] M. J. Steenland, D. de Klerk, and C. J. Gorter, *Physica* **17,** 149 (1949).

redistribution of the populations, and as a result the specific heat has a maximum at $T = \theta_0$. To calculate $C_{H=0}$, consider a system of N ions (dipoles) which has a zero level spacing δ and take the energy of the lowest level zero. Then, according to the Boltzmann distribution law, the ratio of the number of ions n_1 in the upper level to the total number of ions $n_1 + n_2 = N$ is given by $e^{-\delta/kT}/(1 + e^{-\delta/kT})$. It follows that the average total energy of the system is

$$\bar{E} = \frac{N\,\delta e^{-\delta/kT}}{1 + e^{-\delta/kT}}.$$

By applying equation 3-2.8, we get for the specific heat C, after simplifying,

$$C = \left(\frac{\partial E}{\partial T}\right)$$

$$= \frac{\delta^2 N}{kT^2}\frac{e^{-\delta/kT}}{(1 + e^{-\delta/kT})^2}. \tag{3-3.6}$$

At high temperatures this tails off to

$$\frac{C}{R} = \frac{\delta^2}{4k^2T^2}. \tag{3-3.7}$$

Determination of the slope of the plot of C/R versus $1/T^2$ leads to a value of δ or θ_0. Measurements of the specific heat which show this anomaly have been made[8] for temperatures close to θ_0. Figure 3-3.1 shows the

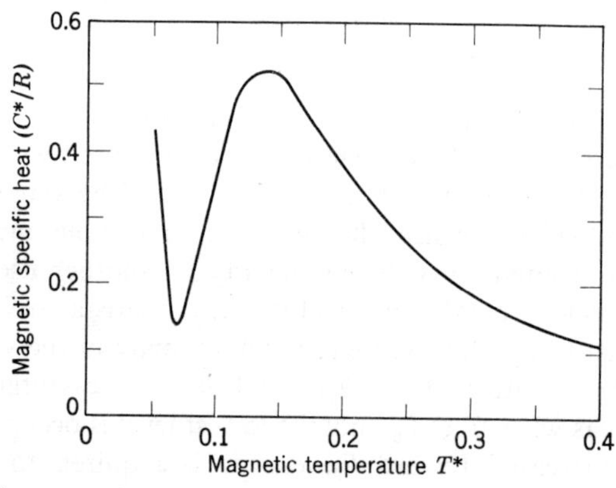

Fig. 3-3.1. The magnetic specific heat of chromium methylamine alum. (After W. E. Gardner and N. Kurti.)

[8] For example, W. E. Gardner and N. Kurti, *Proc. Roy. Soc.* (*London*) **A-223,** 542 (1954).

experimental results for chromium methylamine alum.[8] At higher temperatures the specific heat of the magnetic system is too small compared with that of the lattice to be measured directly. Fortunately, the relaxation measurements, to be described in the next sections, permit C_H to be determined at higher temperatures.

Before 1945 adiabatic demagnetization experiments gave the best information concerning zero-field-level splitting. Since then the paramagnetic resonance technique has provided a much better method for the determination of these splittings.

Finally, it should be mentioned that cooling also can be accomplished by demagnetizing polarized nucleii. A two-stage process, in which an electronic demagnetization is followed by a nuclear demagnetization, has produced a temperature of about 10 microdegrees Kelvin.[9]

PARAMAGNETIC RELAXATION

4. The Susceptibility in an Alternating Magnetic Field

We have already treated the behavior of paramagnetic substances when placed in a static magnetic field. Only equilibrium states were considered; the time required to establish equilibrium when the magnetic field was changed was neglected. In fact, it turns out that these times are very small. However, when an alternating magnetic field of sufficiently high frequency is applied, the magnetization of the material is unable to follow the changes of the field, but instead lags in phase.

It is convenient to describe this phenomena macroscopically in a manner analogous to that used for the static case. If the applied field is of the form

$$H(t) = H + H_1 \cos \omega t, \tag{3-4.1}$$

the magnetization may be represented by

$$\begin{aligned} M(t) &= M_0 + M_1 \cos (\omega t - \phi) \\ &= M_0 + M_1 \cos \omega t \cos \phi + M_1 \sin \omega t \sin \phi, \end{aligned} \tag{3-4.2}$$

where H is a static field, M_0 is the equilibrium value of the magnetization in this static field, and ϕ is the phase angle by which the magnetization lags the field. The equations have been written in scalar form, since the constant field H is usually, although not always, parallel to the oscillating one.

[9] N. Kurti, F. N. H. Robinson, F. E. Simon, and D. A. Spohr, *Nature* (*London*) **178,** 450 (1956).

Then, in analogy to equation 1-4.4, $\chi_0 = M_0/H$, we define

$$\chi' = \frac{M_1 \cos \phi}{H_1} \quad \text{and} \quad \chi'' = \frac{M_1 \sin \phi}{H_1}, \tag{3-4.3}$$

so that $\chi''/\chi' = \tan \phi$. Equation 3-4.2 then becomes

$$M(t) = \chi_0 H + \chi' H_1 \cos \omega t + \chi'' H_1 \sin \omega t. \tag{3-4.4}$$

It is convenient to use complex notation, so that

$$H(t) = H + H_1 e^{i\omega t}$$

and

$$M(t) = M_0 + \chi H_1 e^{i\omega t},$$

where

$$\begin{aligned} \chi &= \chi' - i\chi'' \\ &= |\chi| \, e^{-i\phi}. \end{aligned} \tag{3-4.5}$$

With the aid of Euler's theorem ($e^{i\omega t} = \cos \omega t + i \sin \omega t$), it is easy to show that the real parts of these equations are identical with the preceding ones.

Both χ' and χ'' depend on the frequency ω as well as the magnitude of the static field H: χ' is called the *high frequency susceptibility*, and the variation of χ' with ω is called the *paramagnetic dispersion*; χ'' is proportional to the energy absorbed by the substance from the high frequency field. This is shown by the following considerations. According to equation 1-7.7, the work done on the specimen is given by $dW = -H\, dM$. Then the energy A that is absorbed per second per unit volume is given by

$$\begin{aligned} A &= \frac{\omega}{2\pi} \int_{\text{cycle}} H \, dM \\ &= \frac{\omega}{2} \chi'' H_1^2. \end{aligned} \tag{3-4.6}$$

At low frequencies $\chi' = \chi_0$, the static susceptibility, and $\chi'' = 0$, that is, M and H are in phase.

Kramers and Kronig[10] have shown that χ' and χ'' are not independent of

[10] H. A. Kramers, *Atti. congr. fis. (Como)* 545 (1927), and R. Kronig, *J. Opt. Soc. Am.* **12,** 547 (1926). Simple proof may be found in H. Fröhlich, *Theory of Dielectrics,* Oxford University Press, Oxford (1949), and C. Kittel, *Elementary Statistical Physics,* John Wiley and Sons, New York (1958). More elaborate and rigorous proof is given in A. Abragam, *The Principles of Nuclear Resonance,* Oxford University Press, Oxford (1961).

each other but are related by the equations

$$\chi'(\omega_0) = \frac{2}{\pi}\int_0^\infty \frac{\omega\chi''(\omega)\, d\omega}{\omega^2 - \omega_0^2} + \chi'(\infty),$$

$$\chi''(\omega_0) = -\frac{2}{\pi}\int_0^\infty \frac{\omega_0\chi'(\omega)\, d\omega}{\omega^2 - \omega_0^2}. \tag{3-4.7}$$

We shall not give the proof of these relations. Usually $\chi'(\infty) = 0$.

Most experiments have been made on paramagnetic ions of the $3d$ group or on Gd of the $4f$ group. For these ions the dipole moments are a result of the electron spin only. Because of this, it has become common to refer to their magnetic system as the *spin system*; we shall follow this nomenclature here, with the understanding that the term does not preclude the possibility that the electronic orbital momentum contributes to the magnetic moment. Waller[11] was the first to point out that two different time effects, or relaxations, were involved. One relaxation refers to the spin system only and depends on the interaction between the spins; the other involves the interaction between the spin system and the system of lattice vibrations of the crystal.

In order to gain some conception of the two relaxation effects, consider a system of dipoles in an applied static magnetic field H. As discussed in Section 1-10, these dipoles will precess about the field direction. Let us suppose that the system of dipoles has reached an equilibrium state so that their energy levels are populated according to the Boltzmann distribution. Now, if H is increased in magnitude, but not direction, to $H + \Delta H$, the Boltzmann distribution predicts that the population of the levels will shift so that the net moment parallel to the field is increased. This requires the release of energy. If the spin system were thermally isolated, no energy would be released, the level populations would remain constant, and the temperature of the spin system would rise from T to T_s so that $H/T = (H + \Delta H)/T_s$ (compare Section 3-3). However, interaction of the spin system with the lattice provides a mechanism by which this energy may be released. Since it takes time for the magnetization M_z to reach its equilibrium value M_0, the process can be described phenomenologically by the equation

$$\frac{\partial \mathbf{M}_z}{\partial t} = \frac{\mathbf{M}_0 - \mathbf{M}_z}{\tau_1}. \tag{3-4.8}$$

Here the subscript z is used to indicate that H is applied along the z direction and τ_1 is called the spin-lattice relaxation time. Experimentally,

[11] I. Waller, *Z. Physik* **79,** 370 (1932).

it is found that τ_1 varies inversely with temperature. It is often about 10^{-6} sec at liquid nitrogen temperatures, increasing to about 10^{-2} sec in the liquid helium region.

The spin-spin relaxation time τ_2 arises because of interaction between the dipoles of the system and is a measure of the time taken for the spin system to come to equilibrium. It is much shorter than τ_1, being about 10^{-10} sec, and is temperature-independent. Because of this, the two relaxation processes can usually be studied independently, since for frequencies less than 10^7 cps the spin-spin relaxation can be neglected.

The relaxations also depend on whether the applied field H is larger or smaller than the fluctuating internal fields H_i produced by the dipoles. If $H \ll H_i$, the main effect of the field is to change the direction but not the magnitude of the field that each dipole sees. As a result, the dipoles undergo a slight reorientation, since they will now precess about a new field direction. This polarization of the spin system will require no interaction with the lattice but instead occurs by a small change in the energy levels of the dipoles. However, if $H \gg H_i$, a small increase of H changes the magnitude but not the direction of the field that the dipole sees. The change of magnetization then occurs by level population changes and requires interaction with the lattice.

As a consequence of the preceding considerations, it follows that spin-lattice relaxation can be studied by using a relatively large static field applied parallel to an alternating field whose frequency is so low that spin-spin relaxation is unimportant. On the other hand, spin-spin effects are studied at a higher frequency. If an applied static field has a component parallel to the oscillating field, making this component small will ensure that the spin-lattice relaxation will not affect the spin-spin results.

5. Spin-Lattice Relaxation

The redistribution of the populations of energy levels of a paramagnetic substance which occurs through the agency of the spin-lattice relaxation will now be treated microscopically. For simplicity, consider a system of N spins with $S = \frac{1}{2}$ which will have only two energy levels. In a magnetic field H the levels will be separated by an energy $g\mu_B H$ (Fig. 3-5.1). Let N_1 and N_2 be the equilibrium populations of the upper and lower energy levels, respectively ($N = N_1 + N_2$).

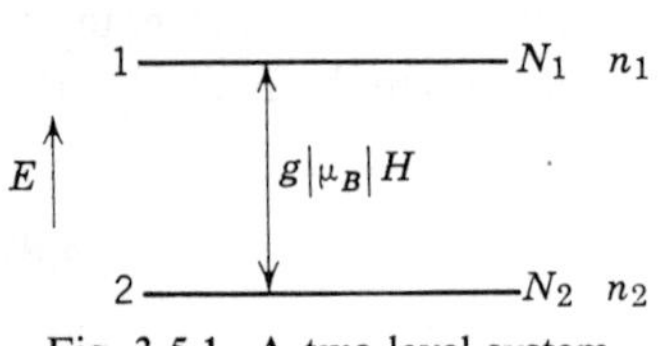

Fig. 3-5.1. A two-level system.

Even at equilibrium there is a certain probability that transitions will occur between the states. In order to preserve equilibrium, a detailed balancing of the transitions must occur, so that

$$N_1 P_{12} = N_2 P_{21}, \tag{3-5.1}$$

where P_{12} is the probability for a single transition per second from level 1 to level 2. Further, if we define the time for a transition as $\tau_{12} = 1/P_{12}$ and $\tau_{21} = 1/P_{21}$ and employ the Boltzmann formula, we get

$$\begin{aligned}\frac{N_2}{N_1} &= \frac{\tau_{21}}{\tau_{12}} \\ &= e^{+g\mu_B H/kT} \\ &= 1 + g\mu_B H/kT,\end{aligned} \tag{3-5.2}$$

where the last equation holds if $g\mu_B H \ll kT$.

Now consider the same system before equilibrium and let n_1 and n_2 represent the instantaneous populations of the upper and lower energy levels, respectively. In order to approach and reach the equilibrium distributions, transitions from the upper to lower energy levels must occur more often than transitions in the opposite direction. Assuming that the probability for a transition is independent of the level populations, we may write

$$\begin{aligned}+\frac{dn_2}{dt} &= -\frac{dn_1}{dt} \\ &= \frac{n_1}{\tau_{12}} - \frac{n_2}{\tau_{21}}.\end{aligned} \tag{3-5.3}$$

It is easy to show that this expression can be written as

$$\begin{aligned}\frac{dn_2}{dt} &= \left(\frac{n_1 + n_2}{2}\right)\left(\frac{1}{\tau_{12}} + \frac{1}{\tau_{21}}\right)\left(\frac{\tau_{21} - \tau_{12}}{\tau_{12} + \tau_{21}} - \frac{n_2 - n_1}{n_1 + n_2}\right) \\ &= \frac{1}{2}\left(\frac{1}{\tau_{12}} + \frac{1}{\tau_{21}}\right)[(N_2 - N_1) - (n_2 - n_1)],\end{aligned} \tag{3-5.4}$$

since $N = n_1 + n_2 = N_1 + N_2$ and $(N_2 - N_1)/(N_1 + N_2) = (\tau_{21} - \tau_{12})/(\tau_{12} + \tau_{21})$. Now the net moment of the system is just the difference between the projections of the moments of the two energy states, so that

$$M_0 = \tfrac{1}{2} g\mu_B (N_2 - N_1) \quad \text{and} \quad M_z = \tfrac{1}{2} g\mu_B (n_2 - n_1). \tag{3-5.5}$$

By differentiating and using (3-5.3) we obtain

$$\begin{aligned}\frac{dM_z}{dt} &= \tfrac{1}{2} g\mu_B \left(\frac{dn_2}{dt} - \frac{dn_1}{dt}\right) \\ &= g\mu_B \frac{dn_2}{dt}.\end{aligned}$$

Substitution of equations 3-5.4 and 3-5.5 leads to

$$\frac{dM_z}{dt} = \left(\frac{1}{\tau_{12}} + \frac{1}{\tau_{21}}\right)(M_0 - M_z).$$

If the relaxation time τ_1 is defined by $1/\tau_1 = 1/\tau_{12} + 1/\tau_{21}$, we get

$$\frac{dM_z}{dt} = \frac{M_0 - M_z}{\tau_1},$$

which is identical with equation 3-4.8. Thus, by considering a microscopic picture, we are able to gain some insight into the origin of the phenomenological equation used to describe spin-lattice relaxation.

Some indication of the reason for the difference between the probabilities $1/\tau_{12}$ and $1/\tau_{21}$ can be understood on the basis of the following simple model. Suppose the lattice vibrations can be represented by a set of harmonic oscillators whose equally spaced energy levels are separated by an amount $g\mu_B H$, the energy difference between the levels of the spin system. An upward transition from the energy levels 2 to 1 will then occur by absorption of the energy $g\mu_B H$ from a certain oscillator. This may take place only if the oscillator is in some state other than the ground state. Since the probability that the oscillator is in the ground state is $-g\mu_B H/kT$, the probability that it is in some other state is $1 + g\mu_B H/kT$. A downward transition from the levels 1 to 2 of the spin system will occur when the oscillator receives the energy $g\mu_B H$; here the oscillator may be in any energy level. For this model the ratio of the upward $1/\tau_{21}$ to the downward $1/\tau_{12}$ probability is just the ratio of the probability that the oscillator will give a quantum of energy $g\mu_B H$ to the probability that it will receive a quantum of the same energy. This ratio is simply $1 + g\mu_B H/kT$, in agreement with equation 3-5.2.

It is now necessary to obtain a relationship between the susceptibility at high frequencies and the spin-lattice relaxation time; thermodynamic considerations may be used. Since the spin-spin relaxation time τ_2 is short compared to the spin-lattice relaxation time τ_1, the spin system may be assumed to be in thermodynamic equilibrium at all times for the experimental conditions considered. The spin system therefore may be treated as a separate system with its own specific heat and a certain spin temperature T_s. As a consequence, when an oscillating magnetic field, parallel to a static field, is applied to a paramagnetic salt, the spin temperature will also oscillate and will, in general, be different from the constant temperature of the lattice T_l. The difference between the spin and lattice temperature will be reduced because of the thermal contact between the spin and lattice systems, that is, by the spin-lattice interaction. The larger the interaction, of course, the smaller this temperature difference will be. The period $\tau(=2\pi/\omega)$ of the oscillatory field will also influence this temperature difference. At low frequencies ($\tau \gg \tau_1$) the spin temperature will always be essentially equal to the lattice temperature; the susceptibility of the salt will have its static or *isothermal* value $\chi_0 = \chi_T$. On the other hand, if $\tau \ll \tau_1$, there is essentially no heat exchange between the spin and lattice systems and the susceptibility is said to have its *adiabatic* value, χ_S. There will, of course, be a gradual change from χ_T to χ_S as the frequency of the oscillations is increased. Application of the thermodynamical equation 3-2.25 gives a relationship between these two susceptibilities that is independent of τ_1, namely

$$\frac{C_H}{C_M} = \frac{\chi_T}{\chi_S}, \tag{3-5.6}$$

since by definition

$$\chi_T = \left(\frac{\partial M}{\partial H}\right)_T \quad \text{and} \quad \chi_S = \left(\frac{\partial M}{\partial H}\right)_S. \tag{3-5.7}$$

In order to proceed, some relationship that describes the effect of the spin-lattice interaction is now required. Casimir and du Pré[12] made the assumption that the rate of exchange of heat dQ/dt from the lattice system to the spin system was proportional to the difference in temperature between the systems, that is

$$\begin{aligned} \frac{dQ}{dt} &= -\alpha(T_s - T_l) \\ &= -\alpha\theta, \end{aligned} \tag{3-5.8}$$

where α is a constant that describes the degree of thermal contact between the spin system and the lattice. Use of equation 3-2.19 gives

$$-\alpha\theta\, dt = C_M\, dT - T\left(\frac{\partial H}{\partial T}\right)_M dM,$$

since $dQ = T\,dS$. Now, since the applied field and magnetization, given by equations 3-4.5, oscillate with an angular frequency ω, the temperature difference θ will oscillate with the same frequency, and we may write $\theta = \theta_1 e^{i\omega t}$. By substitution into the above equation, we have

$$-\alpha\theta_1 = i\omega\left[C_M\theta_1 - T\left(\frac{\partial H}{\partial T}\right)_M \chi H_1\right].$$

Also, if M is considered a function of H and T,

$$dM = \left(\frac{\partial M}{\partial H}\right)_T dH + \left(\frac{\partial M}{\partial T}\right)_H dT,$$

[12] H. B. G. Casimir and F. K. du Pré, *Physica* **5**, 507 (1938).

so that

$$\chi H_1 = \left(\frac{\partial M}{\partial H}\right)_T H_1 + \left(\frac{\partial M}{\partial T}\right)_H \theta_1.$$

Elimination of θ_1 from these two equations leads to

$$\chi = \left(\frac{\partial M}{\partial H}\right)_T \frac{C_M + \alpha/i\omega}{C_M - T(\partial H/\partial T)_M(\partial M/\partial T)_H + \alpha/i\omega}$$

$$= T\frac{C_M + \alpha/i\omega}{C_H + \alpha/i\omega}, \tag{3-5.9}$$

where use has been made of equation 3-2.23. Equating the real and

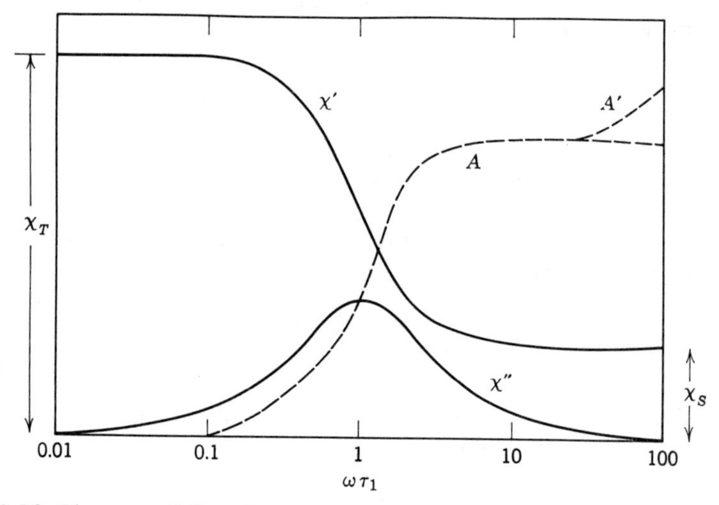

Fig. 3-5.2. The susceptibility of a paramagnetic salt in the frequency region in which spin-lattice relaxation is important. The absorption A is also shown in arbitrary units. A' shows the additional absorption at high frequencies caused by the spin-spin relaxation.

imaginary parts gives

$$\chi' = \chi_S + \frac{\chi_T - \chi_S}{1 + \omega^2\tau_1^2}$$

and

$$\chi'' = \frac{(\chi_T - \chi_S)\omega\tau_1}{1 + \omega^2\tau_1^2}, \tag{3-5.10}$$

provided the spin-lattice relaxation time is taken as $\tau_1 = C_H/\alpha$. For low frequencies $\chi' = \chi_T$, $\chi'' = 0$, whereas for high frequencies $\chi' = \chi_S$, $\chi'' = 0$; χ'' has its maximum value at $\omega = 1/\tau_1$. Figure 3-5.2 shows the variation of χ' and χ'' with frequency for $\chi_S = 0.25\chi_T$ as well as the absorption A (equation 3-4.6).

The spin-lattice interaction has its origin in the thermal vibrations that the atoms and ions of the lattice undergo. These lattice vibrations are often called Debye waves or phonons; the number of phonons increases with the frequency of vibration to some maximum frequency, as indicated in Fig. 3-5.3. There are three basic types of spin-lattice relaxation mechanisms, known as the direct, Raman, and Orbach processes. In the direct process the reversal of an electronic spin is accomplished by the absorption or emission of a single phonon with the correct frequency. In the Raman process a phonon of one energy is absorbed, followed by the emission of a phonon of another energy; the difference in energy of the two phonons equals the energy of the spin transition. The Orbach process[13] also involves two phonons. However, the absorbed phonon has precisely the energy necessary to lift the spin to a higher electronic energy level. The emitted phonon then corresponds in energy to the transition of the spin to the ground state. Hence the Orbach and direct processes are similar in the sense that they are both resonance processes, that is, they are associated with transitions between the spin energy levels.

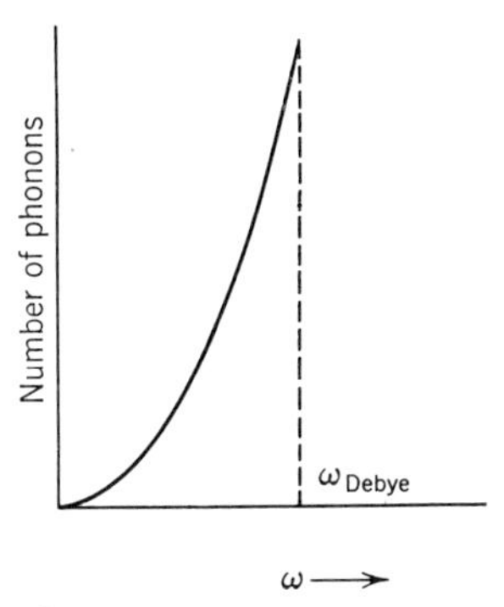

Fig. 3-5.3. Frequency distribution of phonons (lattice vibrations); ω_{Debye} is the upper frequency limit for phonons.

Waller considered the effect of the lattice vibrations on varying the distance between the magnetic ions of the salt. As a result, the magnetic field produced by a dipole at a neighboring dipole varies approximately as $(\mu/r^3)/(dr/r)\cos\omega t$ (see equation 1-3.2), where r is the nearest neighbor distance between dipoles and dr is the amplitude of the variation of this distance. This mechanism is insufficient to explain the relaxation times observed except perhaps when the ions have large magnetic moments and are present in large concentrations[14] or when the ions are in S states and have only relatively small splitting of their energy levels by the crystalline field.[15]

[13] R. Orbach, *Proc. Phys. Soc. (London)* **77,** 821 (1961), *Proc. Roy. Soc. (London)* **A-264,** 458 (1961).

[14] S. A. Al'tshuler, *Izvest. Akad. Nauk. SSSR, Ser. Fiz.* **20,** 1207 (1956) [trans. Columbia Technical Translations, White Plains, N.Y.].

[15] Sh. Sh. Bashkirov, *Fiz. Metall. i Metalloved.* **6,** 577 (1958); A. M. Leushin, *Fiz. Tverd. Tela* **4,** 1564 (1962).

Kronig[16] and Van Vleck[17] have considered the effect of variations in the crystalline electric field that arise from the lattice vibrations. These variations will modulate the electronic orbits of the magnetic ions, and the orbital changes will interact with the spins via the spin-orbit coupling. This mechanism corresponds to a field 10^3 to 10^4 greater than that provided by Waller's theory. To find the spin-lattice relaxation time for the one phonon direct process, the crystalline field potential is expanded in powers of the strain ϵ. The first-order term $\epsilon V_{cf}'$ is responsible for the spin-lattice relaxation, since the zeroth order term generates only the static

Table 3-5.1. Dependence of $(1/\tau_1)$ on T and H
$(T \ll \theta_D)$

Spin	Direct	Raman	Orbach ($T \ll \Delta/\mathrm{k} \ll \theta_D$)
$S = \frac{1}{2}$	H^4T	T^9 and H^2T^7	$e^{-\Delta/\mathrm{k}T}$
$S > \frac{1}{2}$	H^2T	T^7	$e^{-\Delta/\mathrm{k}T}$

wave functions. The probability for transitions between the upper (1) and lower (2) energy levels according to a well-known result of time-dependent perturbation theory is

$$P_{21} = \frac{2\pi}{\hbar} |\langle 1 |\epsilon V_{cf}'| 2\rangle|^2 \rho(E), \tag{3-5.11}$$

where $\rho(E)$ is the energy density of the final states. Then P_{21} may be evaluated by summing over all phonon modes considered as a continuum and assuming $T \ll \theta_D$, the Debye temperature, so that $\omega_q = vq$, where ω_q is the phonon frequency with wave number q and v is the velocity. The relaxation time τ_1 is then obtained by relating it to the transition probability through equations 3-5.1 *et seq.* For the two-phonon process a similar calculation is made, except that now second-order time-dependent perturbation theory is employed. The functional dependence of the reciprocal of the spin-lattice relaxation time $(1/\tau_1)$ on the temperature T and the static field H is given[18] in Table 3-5.1. Here Δ in the Orbach process is the energy separation between the low-lying levels (1 and 2) and the next higher level. In the limit $\mathrm{k}T \gg \Delta$ the Orbach process goes over to the direct one. When appreciable interactions occur between the dipoles, the dependence of $1/\tau_1$ on the static magnetic field is more complicated.[19]

[16] R. de L. Kronig, *Physica* **6,** 33 (1939).

[17] J. H. Van Vleck, *Phys. Rev.* **57,** 426 (1940), **59,** 724 (1941).

[18] R. Orbach, "Spin-Lattice Relaxation in Solids" in *Fluctuations, Relaxation and Resonance in Magnetic Systems*, Oliver and Boyd, Edinburgh (1962).

[19] J. H. Van Vleck, "The Theory of Paramagnetic Relaxation" in *Magnetic and Electric Resonance and Relaxation*, J. Smidt, Editor, North-Holland Publishing Co., Amsterdam (1963).

The direct process, since it is of first order, has the highest probability for occurrence. However, the phonons must have the right frequency, whereas in the Raman process almost all the phonons can take part. Hence, at low temperatures at which there are only a limited number of phonons present, the direct or Orbach process will be most important. At higher temperatures, say in the vicinity of $T \approx 20°K$, there will be many more high-frequency phonons present and the Raman process will predominate. These physical arguments are clearly in accord with the calculated temperature dependences given in Table 3-5.1. It is therefore expected that τ_1 will vary as T^{-1} when $T \approx 4°K$ and then as T^{-7} or T^{-9} as the temperature is raised. For $T \gg \theta_D$ it can be shown that $\tau_1 \propto T^{-2}$.

At still lower temperatures, in the neighborhood of $T \approx 1°K$, there is a further complication. The phonons in direct contact with the spin system have a relatively small conductivity to the helium bath. As a result these phonons are unable to transfer the energy received from the spins fast enough to maintain the lattice temperature at that of the bath. Van Vleck has called this the bottleneck problem;[20] it is also known in modern parlance as the hot phonon problem.[21] Theory predicts that τ_1 will then vary as T^{-2}. Experiment shows the effect to be less important than expected from the theory, perhaps because cracks or defects either decrease the path length to the bath or scatter energy into other parts of the phonon spectrum.

The extensive investigations in recent years of materials useful for masers (Section 3-12) have shown that the foregoing relaxation theories are inadequate to describe the data. All individual processes in the spin-lattice relaxation theories involve a transition in energy of only one spin. Now, however, let us assume that a material may possess two separate spin systems, each characterized by its own spin temperature and different values of τ_1. An obvious example would be a paramagnet with two types of magnetic ions, each with a different g-value and moment. Another would be a compound with magnetic ions all of the same kind but located in two different crystallographic sites. A less obvious example would be a single spin system with more than two available energy levels. The transfer of energy between these two spin systems has become known as *cross relaxation*;[22] the corresponding relaxation time is denoted by τ_{12}.

Probably the simplest example of a cross-relaxation process is the following: consider two dipoles, A and B, with almost the same transition

[20] J. H. Van Vleck, *Phys. Rev.* **59,** 724 (1941).

[21] J. A. Giordmaine, L. E. Alsop, F. R. Nash, and C. H. Townes, *Phys. Rev.* **109,** 302 (1958); P. W. Anderson, *Phys. Rev.* **114,** 1002 (1959).

[22] N. Bloembergen, S. Shapiro, P. S. Persham, and J. O. Artman, *Phys. Rev.* **114,** 445 (1959); A. Kiel, *Phys. Rev.* **120,** 137 (1960); D. N. Klyshko, V. S. Tumanov, and L. A. Ushakova, *Ah. Eksperim. Tech. Fiz.* **43,** 25 (1962).

frequency; that is, each dipole has the energy levels separated by the same energy. Suppose dipole A is in its ground state and dipole B is in its excited state. The dipoles will then be antiparallel to each other and will precess with nearly the same frequency. As a result, each dipole will produce at the other dipole a rotating field whose frequency is almost equal to the precession frequency. It is then possible that both dipoles will flip over, with no change in the total energy of the system. This process, by which a rotating field of the right frequency produces a transition between energy levels, is basic to paramagnetic resonance; it is discussed in detail in Section 3-8. Even though the transition frequency for the A and B dipoles differ somewhat, a simultaneous double spin flip is still possible either because of the finite width of the energy levels or because additional energy is given to or received from the dipole or other interaction energy. Cross relaxation that involves more than two dipole transitions may also occur. For example, in a three spin-flip process two dipoles may undergo an upward transition which is just balanced by the downward transition of the third dipole. If the two upward transitions are equal in energy, the process is called harmonic cross relaxation. Cross relaxation may also take place in nuclear systems.[23]

The question of principal concern here is how the cross-relaxation process influences the spin-lattice relaxation time. Suppose spin system A has a long spin-lattice relaxation time τ_{1A} and that system B has a relatively short one τ_{1B}. This could occur, for example, if the ground state of an A ion were nondegenerate, whereas that of a B ion was degenerate. Further, the B ions may be able to transfer energy by the Orbach process. Also, assume $\tau_{1B} < \tau_{1A}$. As a consequence, the A system, instead of relaxing directly to the lattice, will transfer energy to the B system, which in turn relaxes to the lattice. It is reasonable to expect that the spin-lattice relaxation time will be directly proportional to the concentration, and indeed this is predicted by a quantitative theory.[24] The temperature variation will depend on the details of the systems. As an example, suppose the B ions have a large number of energy levels. At very low temperatures most of the population is in the lowest level. The chance of cross relaxation is then reduced, since the population of those levels with the proper transition frequency is low. Also τ_{1B} is smaller because there are fewer high-frequency phonons available for the Orbach process. At higher temperatures the higher levels of the B system become populated quickly and the cross relaxation and τ_{1B} increase rapidly. Cross relaxation also may permit an escape from the phonon bottleneck.

[23] A. Abragam and W. G. Proctor, *Phys. Rev.* **109,** 1441 (1958).

[24] U. Kh. Kopvillem, *Fiz. Tverd. Tela.* **2,** 1829 (1960) [trans. *Soviet Phys.-Solid State* **2,** 1653 (1961)]; S. Shapiro and N. Bloembergen, *Phys. Rev.* **116,** 1453 (1959).

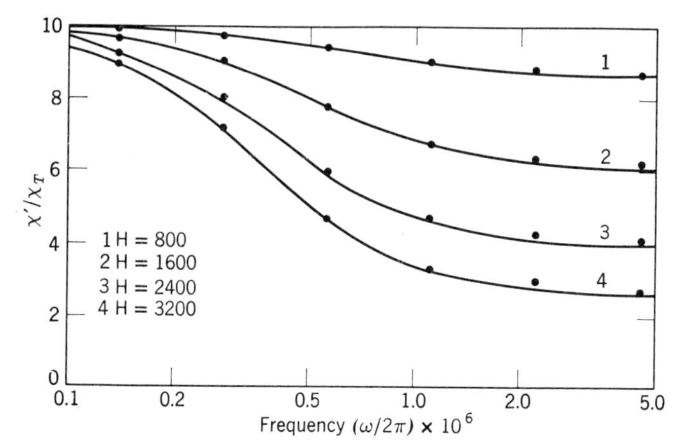

Fig. 3-5.4. Paramagnetic dispersion of $Gd_2(SO_4)_3{\cdot}8H_2O$ at 77°K. **H** in oersteds. (After L. J. F. Broer and C. J. Gorter.)

We now turn to a consideration of the experimental results. The dispersion[25] of the salt $Gd_2(SO_4)_3{\cdot}8H_2O$ at 77°K as a function of the magnitude of the applied static field H is shown in Fig. 3-5.4. The results have been normalized to the isothermal susceptibility. The experimental points are shown and the curves are calculated from equation 3-5.10. The agreement is clearly satisfactory. Further, the influence of H on the results is as predicted. (To determine how H enters in equation 3-5.10, see problem 3-5.) The results for the absorption[26] in the same salt are shown in Fig. 3-5.5. Again agreement with the theoretical equations 3-5.10 and

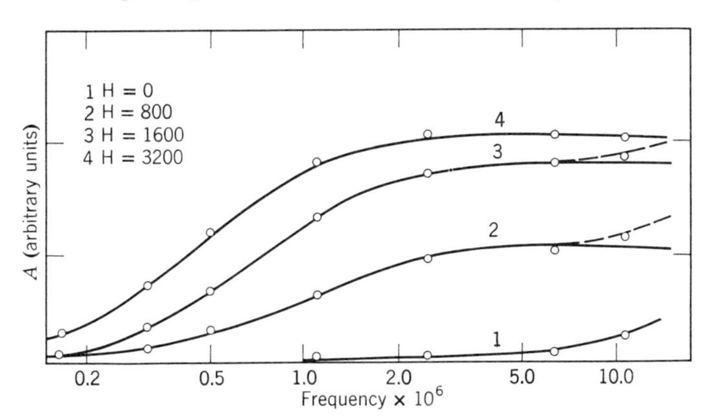

Fig. 3-5.5 Paramagnetic absorption in $Gd_2(SO_4)_3{\cdot}8H_2O$ at 77°K for various values of the static field H. (After F. W. DeVrijer, J. Volger, and C. J. Gorter.)

[25] L. J. F. Broer and C. J. Gorter, *Physica* **10,** 621 (1943).
[26] F. W. De Vrijer, J. Volger, and C. J. Gorter, *Physica* **11,** 412 (1946).

3-4.6 is good. The increase at high frequencies is interpreted as the effect of the spin-spin relaxation.

For both the iron and rare earth groups $1/\tau_1$ varies as T^x where $5 < x < 9$ in the region in which the Raman process is expected to be important. Any disagreement with the T^7 or T^9 terms predicted by theory may be caused by violation of the condition $T \ll \theta$, deviations from the Debye phonon spectrum, and saturation effects.[27] At low temperatures relaxation via the Orbach process has been observed in cerium magnesium nitrate[28] and dysprosium ethyl sulfate.[29] There is evidence that the phonon bottleneck is sometimes important.[30] The variation of $1/\tau_1$ with H has the expected behavior in dilute samples. However, in nondilute samples there are frequently deviations from the theoretical predictions.[31] Information obtained from paramagnetic resonance measurements concerning the dependence of the relaxation time with temperature is discussed in Section 3-8.

The magnetic specific heats may be determined by measurements of spin-lattice relaxation. From equation 3-2.24 we have

$$C_H - C_M = T\left(\frac{\partial M}{\partial T}\right)_H^2 \left(\frac{1}{\chi_T}\right),$$

so that together with equation 3-5.6 we get

$$C_M = T\left(\frac{\partial M}{\partial T}\right)_H^2 \frac{\chi_S}{\chi_T(\chi_T - \chi_S)}. \tag{3-5.12}$$

If Curie's law is obeyed, $M = CH/T$, and this leads to

$$\frac{b}{C} = \frac{H^2\chi_S}{\chi_T - \chi_S}, \tag{3-5.13}$$

where $b = C_M T^2$. Since all the quantities on the right-hand side of this equation can be determined experimentally, values of b or C_M can be calculated. In this way magnetic specific heats can be determined without

[27] C. J. Gorter, in *Fluctuations, Relaxation and Resonance in Magnetic Systems*, D. ter Haar, Editor, Oliver and Boyd, Edinburgh (1961), p. 87. A summary of early results may be found in C. J. Gorter, *Paramagnetic Relaxation*, Elsevier Publishing Co. Amsterdam (1947).

[28] C. B. P. Finn, R. Orbach, and W. P. Wolf, *Proc. Phys. Soc.* (*London*) **77,** 261 (1961).

[29] A. H. Cooke, C. B. P. Finn, K. W. Magum, and K. L. Orbach, *J. Phys. Soc. Japan Suppl. B*-1 **17,** 462 (1962).

[30] C. J. Gorter, L. C. van der Marcel, and B. Bölger, *Physica* **21,** 103 (1955).

[31] L. C. van der Marcel, J. van den Broek, and C. J. Gorter, *Physica* **23,** 361 (1957), **24,** 101 (1958), **27,** 661 (1961); J. van den Broek, L. C. van der Marcel, and C. J. Gorter, *Physica* **25,** 371 (1959); K. P. Sitnikov, *J. Exptl. Theoret. Phys.* (*USSR*) **34,** 1090 (1958) [trans. *Soviet Phys.-JEPT* **34,** 755 (1958)]; P. H. Fang, *Physica* **27,** 681 (1961); P. H. Fang in *Magnetic and Electric Resonance and Relaxation*, J. Smidt, Editor, North-Holland Publishing Co., Amsterdam (1963), p. 123.

recourse to low-temperature calometric methods. The quantity b/C is independent of temperature at temperatures well above that for the specific heat maximum (see equation 3-3.7 and Section 3-3). Numerous publications from the Leiden group report on relaxation measurements of the specific heats. In addition, Benzie and Cooke[32] have studied salts containing various concentrations of the magnetic ions. In this way the specific heat could be analyzed into a part due to dipole interaction and a part due to the crystalline field and nuclear interaction, since the former depends on the concentration, whereas the latter does not.

Finally, it is important to mention that the spin-lattice interaction has recently been investigated by a reverse procedure. In this method a piezoelectric transducer is used to generate acoustic vibrations, or, in other words, phonons, with microwave frequencies in the paramagnetic crystal.[33] When the phonons have the same frequency as that required for a spin transition, they will tend to be absorbed. Both the decrease in the transmitted phonons[34] and the increase in the population of the upper spin energy levels[35] then expected have been observed. Spin transitions produced by the combined absorption of a phonon plus a photon have also been detected.[36] The experiments are in general agreement with the theory of the direct process of the spin-lattice relaxation.[37]

6. Spin-Spin Relaxation

Experimentally it is found that as the frequency of the oscillatory magnetic field is increased an absorption additional to the spin-lattice absorption occurs. This absorption, attributed to spin-spin interaction effects, is indicated by A' in Fig. 3-5.2 and by the dashed curves in Fig. 3-5.5. An expression for the spin-spin absorption may be obtained by substituting the contribution to χ'' by the spin system into equation 3-4.6. The value of χ''_{spin} in analogy to equation 3-5.10, is

$$\chi''_{\text{spin}} = \frac{\chi_s \omega \tau_2}{1 + \omega^2 \tau_2^{\,2}},$$

[32] R. J. Benzie and A. H. Cooke, *Proc. Phys. Soc. (London)* **A-63,** 213 (1950).

[33] H. E. Bömmel and K. Dransfeld, *Phys. Rev.* **117,** 1245 (1960); E. H. Jacobsen, *Phys. Rev. Letters* **2,** 249 (1959).

[34] E. B. Tucker, *Phys. Rev. Letters* **6,** 183 (1961).

[35] N. S. Shiren and E. B. Tucker, *Phys. Rev. Letters* **2,** 206 (1959); E. H. Jacobsen, N. S. Shiren, and E. B. Tucker, *Phys. Rev. Letters* **3,** 81 (1959); R. D. Mettuck and M. W. P. Strandberg, *Phys. Rev.* **119,** 1204 (1960); E. B. Tucker, *Phys. Rev. Letters* **6,** 547 (1961).

[36] N. S. Shiren, *Phys. Rev. Letters* **6,** 168 (1961).

[37] N. S. Shiren in *Magnetic and Electric Resonance and Relaxation*, J. Smidt, Editor, North-Holland Publishing Co., Amsterdam (1963), p. 114.

where τ_1, the spin-lattice relaxation time, has been replaced by τ_2, the spin-spin relaxation time, and $(\chi_T - \chi_S)$, the additional susceptibility associated with the spin-lattice interaction, has been replaced by χ_s, the susceptibility appropriate to the spin system alone. At the time paramagnetic relaxation studies were made it was not easy to produce extremely high frequencies, and as a result only the initial tail of the spin-spin absorption was studied. For this region it is legitimate to neglect $\omega^2\tau_2^2$ compared to 1, so that $\omega\tau_2/(1 + \omega^2\tau_2^2)$ may be replaced by $\omega\tau_2$ in the expression for χ''_{spin}. Then for the spin-spin absorption we get

$$A_{\text{spin}} = \tfrac{1}{2}\omega^2\tau_2\chi_s H_1^2. \tag{3-6.1}$$

This equation is of historical importance, since the spin-spin relaxation time was first determined by colorimetric measurements of the absolute absorption. This method is commonly restricted to the frequency range 10 to 10^2 Mc. The spin relaxation time can also be determined by dispersion measurements or by measuring the absorption as a function of frequency. These methods have the advantage of requiring only relative measurements; they have been employed for frequencies of 10^2 to 10^3 Mc.

The spin-spin relaxation time is a measure of the time required to establish thermal equilibrium in the spin system. The origin of τ_2 on a microscopic basis must be sought by considering the different mechanisms by which the spins may interact. First, consider the local or internal magnetic field at a dipole produced by neighboring dipoles. These neighboring dipoles will have their dipole moments arranged over the possible spatial orientations, and, moreover, as transitions between the different energy levels occur, the arrangement will change. As a result, the internal magnetic field will vary from dipole to dipole and also with time at a given dipole. Since the average value of the internal field is zero, we calculate a root mean square value H_i by employing equation 1-3.2. The result is[38]

$$H_i = \left(2\mu^2 \sum_j \frac{1}{r_{ij}^{\,6}}\right)^{1/2}. \tag{3-6.2}$$

For a dipole moment of one Bohr magneton this is of the order of 1000 oe. Should a static field H be applied to the dipoles, a given dipole would see an effective field that is the resultant of the internal and static fields. However, whether or not a static field is applied, the dipoles precess with a range of frequencies. This spread of frequencies is given approximately by (see equation 1-10.6)

$$\begin{aligned}\Delta\omega &\approx \frac{ge}{2mc} H_i \\ &\approx 10^{10}\ \text{sec}^{-1}.\end{aligned} \tag{3-6.3}$$

[38] J. H. Van Vleck, *J. Chem. Phys.* **5**, 320 (1937).

Consequently, the relaxation time due to this mechanism is approximately $\tau_2 \approx 1/\Delta\omega = 10^{-10}$ sec, since we expect it to be the time that a dipole will require to reorient itself about a different precession axis.

For some paramagnetic salts a quantum mechanical interaction between the dipoles, called the exchange interaction, is important.[39] The origin and form of the exchange interaction are discussed in Section 6-3. Since this interaction favors either parallel or antiparallel arrangements of two adjacent spins, the neighboring spins are likely to become reoriented. Further, it tends to propagate through the crystal and indeed it may be said that a spin wave is created (see Section 6-6). If this motion in the spin system is rapid enough, the fields produced by the dipoles are averaged out. A similar effect occurs for the molecular motion in a liquid, discussed in Section 4-4. As a result $\Delta\omega$ predicted by equation 3-6.2 is reduced or equivalently τ_2 is increased. The exchange interaction can also play a role in the cross-relaxation process.

In the vicinity of imperfections the crystalline field will be somewhat different from other regions of the crystal. Both the zero-energy-level splitting and the g-value of a paramagnetic ion located at or near the imperfection will also be changed. As a result, there will be a random variation in the frequencies required for resonance and the lines will be broadened.

According to the Heisenberg uncertainty principle, $\Delta E\, \Delta t \sim \hbar$, where ΔE is the spread in the energy levels and Δt is the mean lifetime of a state. Therefore the dispersion in precession frequencies will give rise to a broadening of the system's energy levels of

$$\Delta E \sim \hbar\, \Delta\omega. \tag{3-6.4}$$

On this basis, the spin system may be pictured as attaining thermodynamic equilibrium by small energy changes over the energy range ΔE.

We may also consider these mechanisms as phase-destruction processes, for if two dipoles were precessing in phase at a given instant we would expect them to lose this phase relationship in about a time τ_2 later. If the dipoles were precessing about a field in the z-direction, the phase destruction would occur for the x and y components of the magnetization M_x and M_y. To describe this phase destruction, Bloch[40] introduced the phenomological equations

$$\frac{dM_x}{dt} = -\frac{M_x}{\tau_2} \quad \text{and} \quad \frac{dM_y}{dt} = -\frac{M_y}{\tau_2}. \tag{3-6.5}$$

[39] J. H. Van Vleck, *Phys. Rev.* **74,** 1168 (1948); A. Wright, *Phys. Rev.* **76,** 1826 (1949); P. W. Anderson and P. R. Weiss, *Revs. Mod. Phys.* **25,** 269 (1953).

[40] F. Bloch, *Phys. Rev.* **70,** 460 (1946).

Although these equations defining the spin-spin relaxation time have not been employed for the interpretation of paramagnetic relaxation experiments, they have been widely used in the theory of paramagnetic resonance, both for electron and nuclear dipoles.

By extending Waller's treatment of the effect of magnetic dipole interaction, Broer[41] has calculated the spin-spin relaxation time from the absorption from theoretical considerations. This absorption of energy is proportional to the square of the nondiagonal matrix elements of the magnetic moment for states with a difference in energy given by $\hbar\omega$, ω being the angular frequency of the alternating field.[42] It is also proportional to the square of the alternating field and to the population difference of the energy levels, namely $\hbar\omega/2kT$, provided $\hbar\omega \ll kT$. It can then be shown that the spin absorption is given by

$$A_{\text{spin}} = \frac{\omega^2 H_1{}^2}{8kT} \sum |\langle i\,|M|\,j\rangle|^2,$$

where the summation extends over the range of energy levels given by equation 3-6.4. By equating this value for A_s with that of equation 3-6.1, we obtain an expression for τ_2,

$$\tau_2 = \frac{\Sigma\,|\langle i\,|M|\,j\rangle|^2}{4\chi_s kT}.$$

Broer then applies certain conditions and assumptions to the calculation of $\Sigma\,|\langle i\,|M|\,j\rangle|^2$. As a result, he finds for the spin-spin relaxation time

$$\tau_2 = \frac{1}{8\pi}\,\frac{h}{g\mu_B H_i}. \tag{3-6.6}$$

It is interesting to note that except for a small numerical factor this is the same result that we get from the order of magnitude equation (3-6.3), namely

$$\tau_2 \approx \frac{1}{\Delta\omega}$$
$$\approx \frac{h}{2\pi g\mu_B H_i}.$$

Wright has calculated the effect of the exchange interaction; the result can be expressed by the addition of a correction term to equation 3-6.6.

Let us now turn to a discussion of paramagnetic spin-spin experiments.

[41] L. J. F. Broer, *Physica* **10,** 801 (1943), **13,** 353 (1947).

[42] The reader familiar with time-dependent perturbation theory will recognize this statement as reasonable.

Table 3-6.1 lists the internal field H_i for a few paramagnetic salts, calculated by using equation 3-6.2. Then the values of τ_2 obtained from equation 3-6.6 are given. The agreement with the experimental values determined by Volger et al.[43] is generally satisfactory for most of the salts. The large discrepancy observed for the free radical, diphenylpicryl hydrazyl[44] $[(C_6H_5)_2N—NC_6H_2(NO_2)_2]$ undoubtedly derives from the presence of a strong exchange interaction. The smaller differences found

Table 3-6.1. Spin-Spin Relaxation Data

Salt	H_i (oersteds)	τ_2 Theory ($\times 10^{10}$ sec)	τ_2 Experiment ($\times 10^{10}$ sec)	H_{perp} (oersteds)
$Gd_2(SO_4)_3 \cdot 8H_2O$	1380	0.47	0.57	1600
$CrK(SO_4)_2 \cdot 12H_2O$	310	2.55	2.39	870
$FeNH_4(SO_4)_2 \cdot 12H_2O$	450	1.6	1.1	870
$MnSO_4 \cdot 4H_2O$	1200	0.64	0.64	1200
$CuSO_4 \cdot 5H_2O$	200	2.4	6.7	50
$CuCl_2 \cdot 2H_2O$	590	1.4	9.6	72

for some of the other materials probably have the same origin. Appreciable exchange effects also occur in $Cu(NH_4)_2(SO_4)_2 \cdot 6H_2O$.[45]

Experimentally, it is found that τ_2 is independent of temperature, as expected on the basis of theory. When a static field is applied parallel to the oscillatory field, one theory predicts that τ_2 will shift to higher values,[46] whereas another theory predicts that τ_2 will shift downward.[47] The latter theory also predicts absorption in three different frequency regions. Experiments show that τ_2 for one absorption band does increase in a number of salts. On the other hand, two additional absorptions were also observed at higher frequencies.[48] A theory by Caspers[49] has clarified these inconsistencies.

[43] J. Volger, F. W. De Vrijer, and C. J. Gorter, *Physica* **13,** 621 (1947).

[44] J. C. Verstelle, G. W. J. Drewes, and C. J. Gorter, *Physica* **26,** 520 (1960).

[45] H. Hadders, P. R. Locher, and C. J. Gorter, *Physica* **24,** 839 (1958).

[46] R. Kronig and C. J. Bouwkamp, *Physica* **5,** 521 (1938).

[47] L. J. F. Broer, *Physica* **10,** 801 (1943), **17,** 531 (1951).

[48] P. R. Locher and J. C. Verstelle, *Proc. VII Conf. on Low-Temperature Phys.*, University of Toronto Press, Toronto, **56,** (1960); P. R. Locher and C. J. Gorter, *Physica* **27,** 997 (1961); A. I. Kurushin, *J. Exptl. Theoret Phys. (USSR)* **32,** 938 (1957) [trans. *Soviet Phys.-JEPT* **5,** 766 (1957)]; V. A. Kutuzov, *J. Exptl. Theoret. Phys. (USSR)* **35,** 1304 (1958) [trans. *Soviet Phys.-JEPT* **8,** 910 (1959)]; N. S. Garif'yanov, *J. Exptl. Theoret. Phys. (USSR)* **37,** 1551 (1959) [trans. *Soviet Phys.-JEPT* **10,** 110 (1960)].

[49] W. J. Caspers, *Physica* **25,** 43 (1959), **26,** 778, 798 (1960), *Theory of Spin Relaxation*, Interscience Publishers, New York (1964).

If a static field is applied perpendicular to the alternating one, it is observed that the spin absorption, and therefore τ_2, is decreased. This occurs because the static field increases the precession frequency of the dipoles. As a result, the frequency at which maximum absorption takes place is shifted upward. Consequently, there is less absorption at the frequency of the applied alternating field. According to a theory by Broer, the absorption (and τ_2) will decrease to half its value when the perpendicular field is the same order of magnitude as H_i. Values of the perpendicular field required to halve the absorption, H_{perp} are given in Table 3-6.1. Broer's theory is generally borne out by experiment except for salts in which the exchange interaction is important.

A relaxation effect, often called third relaxation, has been found for salts with $J > \frac{1}{2}$ and crystal field splittings of the order of the dipole-dipole interaction.[50] This new relaxation time is of the order of 10^{-7} sec and is temperature-independent. It has been observed in potassium chrome alum [$CrK(SO_4)_2 \cdot 12H_2O$] and other chrome alums. The third relaxation has been attributed to simultaneous double spin-flip transitions in two neighboring Cr^{3+} ions, that is, to one of the cross-relaxation processes (see Section 3-5).

To summarize, the description of relaxation phenomena is usually carried out phenomologically. Although much progress has been made toward an understanding of the mechanisms operating, many details remain to be clarified before the theory of paramagnetic relaxation can be considered to be satisfactory.

PARAMAGNETIC RESONANCE

7. Conditions for Paramagnetic Resonance

Consider an ion with a permanent magnetic moment. If the ion is free and has a resultant angular momentum quantum number J, the application of a static field H will produce Zeeman splitting of the $2J + 1$ states into levels with energies of $M_J g \mu_B H$; M_J is the magnetic quantum number and g is the Landé splitting factor given by equation 2-2.11. If the ion forms part of a crystalline solid and is not free, its angular momentum may be described by a spin quantum number S. In a field H the energies of the split $2S + 1$ levels are given again by the equation $M_S g \mu_B H$, except that here g is a splitting factor that makes the energy difference between levels,

[50] F. W. De Vrijer and C. J. Gorter, *Physica* **14,** 617 (1948), **18,** 549 (1952); L. J. Smits, H. E. Derksen, J. C. Verstelle, and C. J. Gorter, *Physica* **22,** 773 (1956); J. C. Verstelle, G. W. J. Drewes, and C. J. Gorter, *Physica* **24,** 632 (1958).

$g\mu_B H$, come out correctly. In general, g is not equal to the Landé value. According to quantum mechanics, a selection rule operates so that for magnetic dipole radiation only the transitions between adjacent levels for which $\Delta M_S = \pm 1$ are possible. Transitions between levels can be induced by the application of an oscillating field whose frequency is

$$\hbar\omega = -g\mu_B H$$

or

$$\omega = -\frac{ge}{2mc} H. \tag{3-7.1}$$

This is just the Larmor precession frequency of equation 1-10.7 modified to include the effect of the spin by the inclusion of the factor g. For a system of ions in thermal equilibrium the lowest energy levels will have the greatest populations. Consequently, even though individual transitions for emission or absorption have equal probability, there is a net absorption of energy from the radiation field that can be observed experimentally. Since the absorption occurs only for frequencies at or near the Larmor frequency, the phenomenon is called *paramagnetic resonance* or sometimes *electron spin resonance.*

The principle of paramagnetic resonance may also be illustrated by considering the equations of motion. If the dipole moment of an ion $\boldsymbol{\mu}$ is related to the angular momentum $\mathbf{G}$ by the equation $\boldsymbol{\mu} = (ge/2mc)\mathbf{G}$, where g is the generalized splitting factor just introduced, the classical equation of motion (1-10.5) may be written

$$\frac{d\mathbf{G}}{dt} = \frac{ge}{2mc}\mathbf{G} \times \mathbf{H}$$

or, in terms of the dipole moment,

$$\frac{d\boldsymbol{\mu}}{dt} = \frac{ge}{2mc}\boldsymbol{\mu} \times \mathbf{H}. \tag{3-7.2}$$

If a static field $\mathbf{H}$ is applied along the z-axis, then by writing equation 3-7.2 in component form it is easily seen by substitution that a solution for $\boldsymbol{\mu}$ is

$$\mu_x = \mu \sin\theta \cos\omega_L t,$$
$$\mu_y = \mu \sin\theta \sin\omega_L t,$$
$$\mu_z = \mu \cos\theta = \text{constant},$$

if

$$\boldsymbol{\omega}_L = -\gamma\mathbf{H}$$
$$= -\frac{ge}{2mc}\mathbf{H}.$$

That is, $\boldsymbol{\mu}$ (and also **G**) precesses about the applied field at the Larmor frequency, a result already derived in Chapter 1 (equation 1-10.6). The dipole makes an angle θ with the applied field, so that μ_z is a constant, by quantum mechanics, equal to $M_S\hbar$, where M_S is the magnetic quantum number.

Now, let an alternating field $2H_1 \cos \omega t$ be applied in a direction perpendicular to H; for example, along the x-axis. This alternating field may be

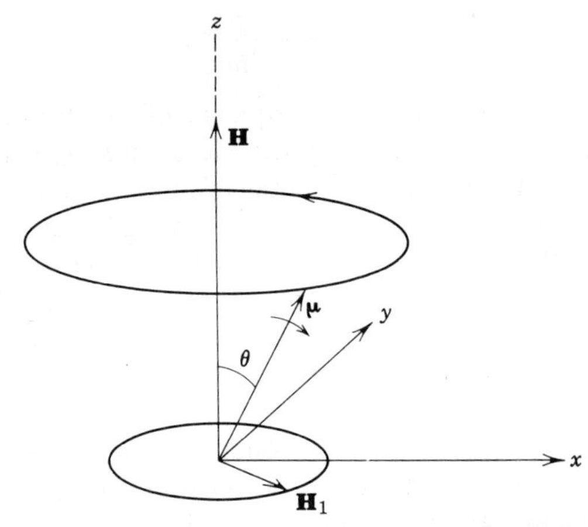

Fig. 3-7.1. Torque on dipole precessing about the field H.

considered as made up of two circularly polarized fields, one rotating clockwise, the other counterclockwise, namely,

clockwise	$H_x = H_1 \cos \omega t$	$H_y = H_1 \sin \omega t$	$H_z = 0,$
counterclockwise	$H_x = H_1 \cos \omega t$	$H_y = -H_1 \sin \omega t$	$H_z = 0.$

Since the counterclockwise field rotates in the same direction as the precessing electronic dipole moment, a torque $\boldsymbol{\mu} \times \mathbf{H}_1$ will tend to tip the dipole moment away from the direction of **H**, in the situation indicated in Fig. 3-7.1. Now, if ω is considerably different from ω_L, the phase between $\boldsymbol{\mu}$ and $\mathbf{H}_1$ will be continuously changing and the direction of the torque $\boldsymbol{\mu} \times \mathbf{H}_1$ will change, so that on the average the torque is zero. For example, when $\mathbf{H}_1$ is 180° to the direction shown in Fig. 3-7.1, for the same location of $\boldsymbol{\mu}$, the torque will tend to turn $\boldsymbol{\mu}$ toward the direction of **H**.

However, if $\omega = \omega_L$, $\boldsymbol{\mu}$ and $\mathbf{H}_1$ will rotate in synchronism, so that the torque will always be constant. Therefore in time $\boldsymbol{\mu}$ will turn either toward **H**, in which case energy will be emitted, or away from **H**, in which case it will be absorbed; again there will be a net absorption because of the difference in level population. The clockwise rotating polarized field will act as a frequency far from the Larmor frequency and its effect can be neglected. Hence the foregoing considerations again lead to equation 3-7.1 as the condition for paramagnetic resonance and, in addition, show why it is necessary to have the alternating field perpendicular to the static one.

It is usual, in treating this principle quantitatively, to sum the dipole moments vectorially, so that equation 3-7.2 becomes

$$\frac{d\mathbf{M}}{dt} = \gamma \mathbf{M} \times \mathbf{H}, \tag{3-7.3}$$

where $\gamma = ge/2mc$ and $\mathbf{M}(= \Sigma\, \boldsymbol{\mu})$ is the magnetization, or the dipole moment per unit volume. Writing equation 3-7.3 in component form and neglecting the clockwise rotating part of the alternating field, we get

$$\frac{dM_x}{dt} = \gamma(M_y H - M_z H_1 \sin \omega t),$$

$$\frac{dM_y}{dt} = \gamma(M_z H_1 \cos \omega t - M_x H),$$

and

$$\frac{dM_z}{dt} = \gamma(M_x H_1 \sin \omega t - M_y H_1 \cos \omega t).$$

It is convenient to use the complex magnetizations, defined as

$$M_+ = M_x + iM_y$$

and

$$M_- = M_x - iM_y. \tag{3-7.4}$$

It follows immediately that these quantities satisfy the differential equations

$$\frac{dM_+}{dt} = -i\gamma M_+ H + i\gamma M_z H_1 e^{i\omega t},$$

$$\frac{dM_-}{dt} = +i\gamma M_- H - i\gamma M_z H_1 e^{-i\omega t}, \tag{3-7.5}$$

$$\frac{dM_z}{dt} = \gamma(M_+ e^{-i\omega t} - M_- e^{i\omega t})H_1.$$

If $M_\pm$ have a time dependence of $e^{\pm i\omega t}$, substitution into equation 3-7.5 gives the solution

$$M_\pm = \frac{\gamma H_1 M_z}{\gamma H + \omega} e^{\pm i\omega t}$$

and

$$\frac{dM_z}{dt} = 0, \quad \text{that is, } M_z = \text{constant}, \tag{3-7.6}$$

and further

$$\left|\frac{M_\pm}{M_z}\right| = \tan\theta$$

$$= \frac{H_1}{\gamma H + \omega}$$

$$= \frac{H_1}{H + \omega/\gamma}, \tag{3-7.7}$$

where θ is the angle M makes with the z-axis. Since $\omega_L = -\gamma H$, when $\omega = \omega_L$, we find $\theta = \pi/2$. The susceptibility, defined as the ratio of $M_\pm$ to the alternating field, will tend to infinity at this frequency. Of course, damping factors, to be discussed in the next section, modify this result. It is also to be noted that the magnitude of $M_\pm$ depends on the amplitude H_1, that is, on the power of the alternating radiation.

This derivation has been made, assuming the classical equation of motion (3-7.2) holds in atomic systems, wherein $\mathbf{G}$ (or $\boldsymbol{\mu}$) is replaced by its expectation value. That this is so may be seen from the following considerations. From a postulate of quantum mechanics the commutation of the angular momentum with the Hamiltonian $[\mathcal{H}, \mathbf{G}]$ is given by

$$\mathcal{H}\mathbf{G} - \mathbf{G}\mathcal{H} = -i\hbar\frac{d\mathbf{G}}{dt},$$

where $\mathcal{H} = -\boldsymbol{\mu}\cdot\mathbf{H}$
$= -\gamma\mathbf{G}\cdot\mathbf{H}.$

This gives

$$(\mathbf{G}\cdot\mathbf{H})\mathbf{G} - \mathbf{G}(\mathbf{G}\cdot\mathbf{H}) = \frac{i\hbar}{\gamma}\frac{d\mathbf{G}}{dt}.$$

Considering only the x-component, the left-hand side gives

$$[(G_zH_z)G_x - G_x(G_zH_z)] + [(G_yH_y)G_x - G_x(G_yH_y)]$$
$$+ [(G_xH_x)G_x - G_x(G_xH_x)] = i\hbar(G_yH_z - G_zH_y),$$

since the commutation relationships for G_x, G_y, and G_z are

$$[G_xG_x] = 0, \qquad [G_zG_x] = i\hbar G_y, \qquad \text{and } [G_xG_y] = i\hbar G_z.$$

This gives then

$$\gamma(G_xH_z - G_zH_y) = \frac{dG_x}{dt}.$$

The other components give analogous results, so that we get

$$\frac{d\mathbf{G}}{dt} = \gamma \mathbf{G} \times \mathbf{H}$$

in agreement with the classical result. Now, when the expectation value of the operator **G**, or the operator $\boldsymbol{\mu}$, is taken, using the appropriate electronic wave functions, the same form of the equation results. Since this expectation value corresponds to the observable quantity, we conclude that equation 3-7.2 holds for atomic systems.

Paramagnetic resonance was first observed by Zavoiskey and Cummerow and Halliday.[51] When numerical values are substituted into equation 3-7.1, we obtain for the resonance condition

$$\nu = 1.401gH, \tag{3-7.8}$$

where ν is in megacycles and H in oersteds.[52] For a g value of about 2 and a field of several kilo-oersteds the resonance frequency lies in the microwave region. Since the radiation sources in this region are operated only over a limited frequency range, it is customary to carry out paramagnetic resonance experiments at a fixed frequency and by varying the magnetic field H. However, some experiments have been made at radio frequencies.[53] The major difficulty is that the width of the absorption line may be much larger than the field required for resonance. For example, at 20 Mc/sec the value of H at resonance is only about 7 oe. A further advantage in working at microwave frequencies occurs because the difference in the populations of levels is greater when larger magnetic fields are employed and therefore the intensity of power absorbed is greater. The main disadvantage is the high cost and complexity of the apparatus.

8. Line Widths: the Effect of Damping

Under the heading of paramagnetic relaxation (see p. 87), spin-lattice and spin-spin interactions and the determination of τ_1 and τ_2 by non-resonance dispersion and absorption measurements were considered. These interactions must also influence paramagnetic resonance. Further, the interactions must originate, at least for solids, from the same basic processes already discussed. However, theoretical treatments of the effect

[51] E. Zavoiskey, *J. Phys.* (*USSR*) **9,** 211 (1945); R. L. Cummerow and D. Halliday, *Phys. Rev.* **70,** 433 (1946).

[52] In the literature it is common to call the units of H, gauss, perhaps because the field is produced between the pole faces of an electromagnet.

[53] T. R. Caver and C. P. Slichter, *Phys. Rev.* **92,** 212 (1953), for example.

of the interactions on resonance commonly utilize a phenomenological approach. Although this leads to a satisfactory understanding of experiments, the detailed similarities and differences between the role played by the interactions in nonresonance and resonance phenomena is obscured.[54]

The internal fields, then, are taken into account in the equation of motion (3-7.3) by adding, as damping terms, equations 3-4.8 and 3-6.5 to the right-hand side. For a static field H applied along the z-axis, and considering only the counterclockwise rotating part of the alternating field, we get the Bloch equations.[55]

$$\begin{aligned}\frac{dM_x}{dt} &= \gamma(M_yH - M_zH_1\sin\omega t) - \frac{M_x}{\tau_2},\\ \frac{dM_y}{dt} &= \gamma(M_zH_1\cos\omega t - M_xH) - \frac{M_y}{\tau_2},\\ \frac{dM_z}{dt} &= \gamma(M_xH_1\sin\omega t - M_yH_1\cos\omega t) + \frac{M_0 - M_z}{\tau_1}.\end{aligned} \tag{3-8.1}$$

In terms of $M_{\pm} = M_x \pm iM_y$, these equations become

$$\begin{aligned}\frac{dM_+}{dt} + i\gamma HM_+ + \frac{M_+}{\tau_2} &= i\gamma M_zH_1e^{i\omega t},\\ \frac{dM_-}{dt} - i\gamma HM_- + \frac{M_-}{\tau_2} &= -i\gamma M_zH_1e^{-i\omega t},\\ \frac{dM_z}{dt} + \frac{M_z}{\tau_1} &= \frac{i\gamma}{2}(M_+e^{-i\omega t} - M_-e^{i\omega t})H_1 + \frac{M_0}{\tau_1}.\end{aligned} \tag{3-8.2}$$

Again, assuming that $M_{\pm}$ has a time variation $e^{\pm i\omega t}$, respectively, we get, after substitution into the above equations,

$$M_{\pm} = \frac{\gamma M_zH_1e^{\pm i\omega t}}{\omega + \gamma H \mp i/\tau_2}. \tag{3-8.3}$$

In evaluating M_z, let us assume $dM_z/dt = 0$. Physically, this means it is assumed that equilibrium exists in the spin system; that is, the spin temperature has reached an equilibrium value determined by the power absorption, by the relaxations, and by the lattice temperature. Thus, although the spin temperature may be equal to the lattice temperature,

[54] A discussion of the relationship between the relaxation times in nonresonance and resonance experiments may be found, for example, in G. E. Pake, *Paramagnetic Resonance*, W. A. Benjamin, New York (1962), pp. 112–120.

[55] F. Bloch, *Phys. Rev.* **70,** 460 (1946).

it may also be at some higher temperature. On substitution for M_+ and M_- into the last equation of (3-8.2), we obtain the result

$$\frac{M_z}{M_0} = \frac{1 + (\omega + \gamma H)^2\tau_2^2}{1 + (\omega + \gamma H)^2\tau_2^2 + \gamma^2 H_1^2\tau_1\tau_2}. \tag{3-8.4}$$

Further, on substitution of this value of M_z into equation 3-8.3, we have

$$\frac{M_\pm}{M_0} = \frac{\gamma H_1 e^{\pm i\omega t}\,\tau_2[(\omega + \gamma H)\tau_2 \pm i]}{1 + (\omega + \gamma H)^2\tau_2^2 + \gamma^2 H_1^2\tau_1\tau_2}. \tag{3-8.5}$$

In analogy with equation 3-4.5, we may define as the complex susceptibility in the x, y plane $\chi' - i\chi'' = M_\pm/H_1 e^{\pm i\omega t}$ remembering that there is no component of M_0 in this plane. From equation 3-8.5 we obtain, on equating reals and imaginaries,

$$\frac{\chi'}{\chi_0} = \frac{\gamma H\tau_2^2(\omega + \gamma H)}{1 + (\omega + \gamma H)^2\tau_2^2 + \gamma^2 H_1^2\tau_1\tau_2}$$

and (3-8.6)

$$\frac{\chi''}{\chi_0} = \frac{-\gamma H\tau_2}{1 + (\omega + \gamma H)^2\tau_2^2 + \gamma^2 H_1^2\tau_1\tau_2},$$

where $\chi_0 = M_0/H$. Notice that, since $\gamma H = -\omega_L$, χ'' is positive.

The results are simplified if the condition

$$\gamma^2 H_1^2\tau_1\tau_2 \ll 1 \tag{3-8.7}$$

is satisfied, so that this term may be neglected in the foregoing equations. This is a condition that essentially places a limit on the amplitude H_1 of the alternating field for many salts. Since $\gamma^2 \approx 10^{14}$, $\tau_2 \approx 10^{-10}$, and above liquid nitrogen temperatures $\tau_1 \approx 10^{-6}$ or smaller for transition ion salts, it is difficult *not* to satisfy this condition in the paramagnetic resonance of these materials with the radiation sources available at microwave frequencies. When $\gamma^2 H_1^2\tau_1\tau_2$ is small, equation 3-8.4 shows that M_z approaches M_0; that is, in spite of the absorption of the microwave energy, which tends to increase the population of the upper levels at the expense of the lower levels, the level populations tend toward their equilibrium distributions. This is achieved because the spin-lattice relaxation time τ_1 is short enough so that the energy absorbed by the spin system is conducted quickly to the lattice, and consequently the spin temperature is always essentially equal to the lattice temperature. The angle of precession is given by

$$\tan\theta = \frac{|M_{\pm}|}{M_z}$$

$$= \frac{\gamma H_1\tau_2}{[1 + (\omega + \gamma H)^2\tau_2^{\,2}]^{1/2}}$$

$$= \frac{H_1}{[1/(\gamma\tau_2)^2 + (H + \omega/\gamma)^2]^{1/2}}. \tag{3-8.8}$$

This should be compared to equation 3-7.7. Equation 3-8.6 reduces to

$$\frac{\chi'}{\chi_0} = \frac{\gamma H\tau_2^{\,2}(\omega + \gamma H)}{1 + (\omega + \gamma H)^2\tau_2^{\,2}}$$

and (3-8.9)

$$\frac{\chi''}{\chi_0} = \frac{-\gamma H\tau_2}{1 + (\omega + \gamma H)^2\tau_2^{\,2}}.$$

These equations are plotted in Fig. 3-8.1 as functions of $\tau_2(\omega + \gamma H)$.

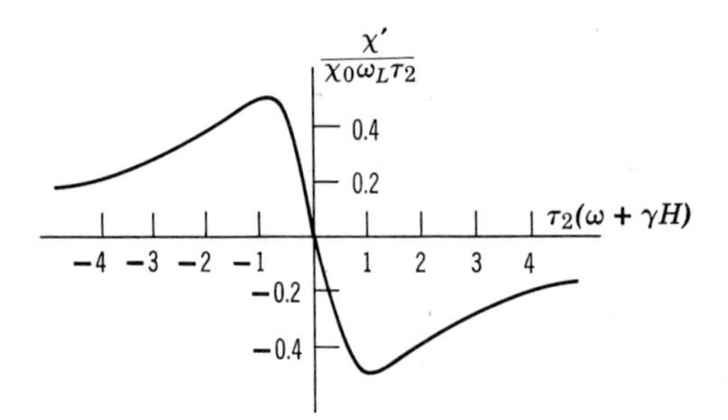

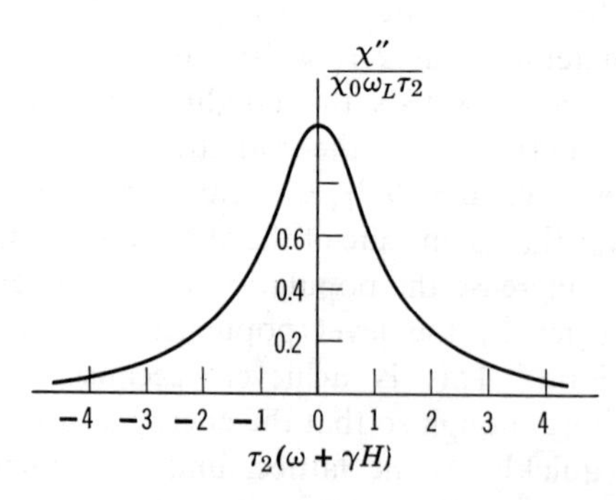

Fig. 3-8.1. Paramagnetic susceptibilities for $\gamma^2H_1^{\,2}\tau_1\tau_2 \ll 1$.

At resonance, we have simply

$$\chi''/\chi_0 = \omega_L \tau_2. \qquad (3\text{-}8.10)$$

The maximum absorption per cycle occurs at the Larmor frequency. The half-width at half maximum of χ'' occurs at

$$\tau_2 \, |\omega + \gamma H| = 1. \qquad (3\text{-}8.11)$$

In terms of the angular frequency $\Delta\omega = \omega - \omega_L$, a convenient experimental quantity, the interesting result is that the width of the absorption line depends only on the spin-spin relaxation time, that is, $\Delta\omega = 1/\tau_2$.

In certain situations $\gamma^2 H_1^{\,2} \tau_1 \tau_2$ may not be negligible compared to unity. This may occur, for example, for paramagnetic salts, at liquid helium temperatures at which τ_2 may be about 10^{-2} sec, for the paramagnetic free radicals, or whenever an extremely high powered radiation generator is employed. Then $M_z < M_0$, in fact, at resonance $\omega = \omega_L$, we have from equation 3-8.4

$$\frac{M_z}{M_0} = \frac{1}{1 + \gamma^2 H_1^{\,2} \tau_1 \tau_2}.$$

The system is then said to be saturated, that is, the spin-lattice relaxation is unable to maintain an equilibrium distribution. Both χ' and χ'' are reduced in magnitude, and in addition the absorption lines become broader. At resonance we have from equation 3-8.6 for the ratio of the magnetization associated with χ'' to the value M_0

$$\frac{\chi'' H_1}{M_0} = \frac{\gamma H_1 \tau_2}{1 + \gamma^2 H_1^{\,2} \tau_1 \tau_2}.$$

It is easily shown that this has a maximum for varying H_1 at

$$H_1 = \frac{1}{|\gamma|} (\tau_1 \tau_2)^{-1/2}. \qquad (3\text{-}8.12)$$

When H_1 is very small, the absorption of radiation energy is small and difficult to detect, whereas when H_1 is very large saturation occurs which reduces and broadens the line. Consequently, equation 3-8.12 is usually considered as expressing the optimum value for the magnitude of the alternating field.

In order to avoid power saturation of the line by meeting the condition of equation 3-8.7, it helps if the spin-lattice relaxation time τ_1 is less than 10^{-5} sec. On the other hand if τ_1 approaches τ_2, that is, for $\tau_1 < 10^{-7}$ sec, the spin-lattice interaction acts to broaden the line more than the spin-spin interaction. Phenomenologically, τ_1 replaces τ_2, since it is more important in destroying the phase relationships of the magnetization in the plane perpendicular to H. The broadening produced is about $\Delta\nu \approx 1/\tau_1$. It frequently happens that at room temperature the paramagnetic absorption lines are so broad they cannot be observed. However, since τ_1 increases rapidly with decrease of temperature (Section 3-5), use of liquid nitrogen or, in extreme cases, liquid helium temperatures usually makes the lines narrow enough to be observed. It may be shown that the spin-lattice relaxation gives a line shape similar to collision broadening.[56] Also it can be shown that τ_1 varies as δ, the separation of the lowest orbitals. Experiments[57] confirm this, since it is found that ions with small δ-values have broad absorption lines.

Many investigations by resonance methods of relaxation in paramagnetic materials have been made in recent years. To a large extent, these studies have been stimulated by an interest in masers (to be discussed in Section 3-11). Two methods for the determination of the spin-lattice relaxation time are in common use. In one the microwave power is held continuously at a high enough level to cause saturation. In the other a pulse of microwave power large enough to produce saturation is applied to the specimen, and the subsequent recovery of the resonance line is followed on an oscilloscope screen.

The experimental variation of τ_1 with temperature for both iron and rare earth group ions is in remarkably good accord with theory. The operation of the direct process at low temperatures and the Raman process at higher temperatures has been established for Cu and Fe ions in salts and in various host crystals.[58] The direct Orbach and Raman processes have been observed for several rare earth ions in ethylsulfates and double nitrate crystals.[59] For the Raman process the T^7 dependence was found for ions with a Kramers doublet (Sm, Nd, Dy) and the T^9 dependence when $S \neq \frac{1}{2}$(Pr).[59] Evidence has been obtained for the phonon bottleneck,[60] and, in particular, the predicted variation of $1/\tau_1$ versus T^2 has been observed.[59]

Cross relaxation has been demonstrated to be important in ruby and

[56] J. H. Van Vleck and V. Weisskopf, *Revs. Mod. Phys.* **17,** 227 (1945).

[57] D. M. S. Bagguley and J. H. E. Griffiths, *Proc. Phys. Soc.* (*London*) **A-65,** 594 (1952).

[58] D. H. Paxman, *Proc. Phys. Soc.* (*London*) **78,** 180 (1961); J. G. Castle, P. F. Chester, and P. E. Wagner, *Phys. Rev.* **119,** 953 (1960); J. G. Castle and P. W. Feldman, *Phys. Rev.* **121,** 1349 (1961); B. Bolger, *Physica* **26,** 761 (1960); B. Bolger, J. M. Noothoven van Goor, R. J. van Damme, and C. J. Gorter, *Physica* **27,** 18, 277 (1961); J. M. Daniels and K. E. Rieckhoff, *Can. J. Phys.* **38,** 604 (1960).

[59] P. L. Scott and C. D. Jeffries, *Phys. Rev.* **127,** 32 (1962).

[60] R. H. Ruby, H. Benoit, and C. D. Jeffries, *Phys. Rev.* **127,** 51 (1962); G. V. Marr and P. Swarup, *Can. J. Phys.* **38,** 495 (1960); F. R. Nash, *Phys. Rev. Letters* **7,** 59 (1961).

certain other materials. Ruby, which has been much used as a maser material, is just an α-Al_2O_3 crystal doped with Cr^{3+} ions. Even though the concentrations c of Cr^{3+} ions is relatively small, an appreciable percentage of Cr^{3+} ion pairs are nearest neighbors (actually zc^2, in which z is the number of nearest-neighbor sites). There is appreciable exchange interaction between the ions of nearest neighbor pairs, and as a consequence their energy levels differ considerably from those of the single ions.[61] The Cr^{3+} pairs, strongly coupled to the lattice because of the Orbach process,[62] play the role of the B spin system (p. 98). On the other hand, the single Cr^{3+} ions are coupled more weakly to the lattice and constitute the A spin system. Since some energy-level separations of the A and B system ions are comparable, the coupling between these systems is appreciable. As a result, the single ions transfer energy to the lattice via the exchange-coupled pairs.[63] Both harmonic and nonharmonic cross relaxation involving three or more transitions have also been observed in ruby[64] and other paramagnetic materials.[65] When $\tau_{12} < \tau_{1B}$, the use of pulse saturation methods has shown the existence of the two relaxation times, τ_{12} and τ_1.

The mechanisms that produce line broadening[66] and therefore determine τ_2 have already been discussed in Section 3-6. For the purpose of spectroscopy, in which the main interest is the accurate determination of the energy differences between levels, it is an advantage to have the narrowest lines possible. Consequently, a great deal of effort in the early 1950's was made to reduce line widths.

Under the usual experimental operating conditions, the line breadth is determined by the dipole-dipole interaction. The broadening caused by this interaction can be reduced only by increasing the distance of separation between the ions and is achieved by diluting the salt by replacing some

[61] A. L. Schawlow, D. L. Wood, and A. M. Clogston, *Phys. Rev. Letters* **3,** 271 (1959); L. Rimai, H. Statz, M. J. Weber, G. A. de Mars, and G. F. Koster, *Phys. Rev. Letters* **4,** 125 (1960); J. Owen, *J. Appl. Phys.* **32,** 213 (1961).

[62] J. C. Gill, *Nature (London)* **190,** 619 (1961).

[63] G. S. Bogle and F. F. Gardner, *Australian J. Phys.* **14,** 381 (1961).

[64] W. B. Mimms and J. D. McGee, *Phys. Rev.* **119,** 1233 (1960); R. W. Roberts, J. H. Burgess, and H. D. Tenney, *Phys. Rev.* **121,** 997 (1961); J. E. Geusic, *Phys. Rev.* **118,** 129 (1960); R. A. Armstrong and A. Szabo, *Can. J. Phys.* **38,** 1304 (1960); M. Soutif, *J. Phys. radium* **22,** 841 (1961); W. S. C. Chang, *Quantum Electronics*, C. H. Townes, Editor, Columbia University Press, New York (1960), p. 346.

[65] J. A. Giordmaine, L. E. Alsop, F. R. Nash, and C. H. Townes, *Phys. Rev.* **109,** 302 (1958); P. Sorokin, G. J. Lasher, and I. L. Gilles, *Phys. Rev.* **118,** 939 (1960); F. R. Nash and E. Rosenvasser, *Quantum Electronics*, C. H. Townes, Editor, Columbia University Press, New York (1960), p. 302.

[66] See, for example, C. MacLean and G. J. W. Kor, *Appl. Sci. Research* **B4,** 425 (1956).

paramagnetic ions with isomorphous diamagnetic ones.[67] A limit to the line width on further dilution is set by the interaction of the paramagnetic ion with the dipole moments of the nuclei of the diamagnetic neighbors. For hydrated salts this width can be reduced by substituting heavy water, since the deuteron's dipole moment is less than the proton's moment.

In certain paramagnetic salts, mainly for those that are undiluted, the line width is affected by exchange interaction. As discussed in Section 3-6, the dipole fields tend to be averaged out and hence the line widths are narrowed.[68] At very low temperatures the condition $h\nu < kT$ is no longer valid, and as a result τ_2 (and ΔH) are no longer temperature-independent. This follows because the populations of the pertinent energy levels are then changing rapidly with temperature.[69] It should also be mentioned that the line shapes of polycrystalline samples[70] and of liquids[71] have also been studied.

Although χ'' determines the absorption per second A, the intensity of the line depends on the transition probabilities and the difference of level populations as well as on the power. By using considerations similar to those that led to equation 3-6.6 it can be shown that for transitions from M to $M - 1$, when the levels are equally spaced,

$$\chi'' = \pi\chi_0\nu^2 f(\nu),$$

where $f(\nu)$ is the frequency distribution determined by the matrix elements of the magnetic moment and is given by the shape of the absorption line.[72] Considering the noise in the usual experimental conditions, it turns out that the minimum quantity necessary to observe resonance corresponds

[67] J. H. Van Vleck, *Phys. Rev.* **74,** 1168 (1948); M. H. L. Pryce and K. W. H. Stevens, *Proc. Phys. Soc. (London)* **A-63,** 36 (1950); P. W. Anderson and P. R. Weiss, *Revs. Mod. Phys.* **25,** 269 (1953).

[68] C. Kittel and E. Abrahams, *Phys. Rev.* **90,** 238 (1953).

[69] M. McMillan and W. Opechowski, *Can. J. Phys.* **38,** 1168 (1960), **39,** 1369 (1961); G. Seidel and I. Svare, *Paramagnetic Resonance*, Vol. II, W. Low, Editor, Academic Press, New York (1963), p. 468.

[70] J. P. Lloyd and G. E. Pake, *Phys. Rev.* **94,** 579 (1954); D. Kivelson, *J. Chem. Phys.* **33,** 1094 (1961); K. Hausser, *Z. Elektrochem.* **65,** 636 (1961); R. N. Rogers and G. E. Pake, *J. Chem. Phys.* **33,** 1107 (1960); J. G. Powles and M. H. Mosley, *Proc. Phys. Soc. (London)* **75,** 729 (1960); B. M. Kozyrev, *Discussions Faraday Soc.* **19,** 135 (1955).

[71] R. H. Sands, *Phys. Rev.* **99,** 1222 (1955); R. Neiman and D. Kivelson, *J. Chem. Phys.* **35,** 156 (1961); J. W. Searl, R. C. Smith, and S. J. Wyard, *Proc. Phys. Soc. (London)* **78,** 1174 (1961); F. K. Kneubühl and B. Natterer, *Helv. Phys. Acta* **34,** 710 (1961); T. Vänngård and R. Aasa, *Paramagnetic Resonance*, Vol. II, W. Low, Editor, Academic Press, New York (1963), p. 509.

[72] See B. Bleaney and K. W. H. Stevens, *Rept. Progr. Phys.* **16,** 120 (1953); B. Bleaney, *Phil. Mag.* **42,** 441 (1951).

to about 10^{-11} gm of paramagnetic material. Thus paramagnetic resonance can be used to study smaller numbers of paramagnetic ions than any other physical or chemical technique.

9. Fine and Hyperfine Structure: the Spin-Hamiltonian

In Sections 2-10 and 2-11 the effect of the crystalline field on the magnetic ions in paramagnetic salts was discussed. The present discussion will be limited to $3d$ transition ions, with the crystalline field effects that are larger than the spin-orbit interaction. As a result of the crystalline field, the orbital levels are split so that the orbital magnetic moment may be partly and perhaps completely quenched. Further, with partial quenching the orbital moment will have an angular variation determined by the symmetry of the crystal field. Consequently, the level splitting will depend on the direction the applied field makes with the crystal axes. This means the splitting factor g will be anisotropic, the amount depending on the degree of quenching and of asymmetry of the crystal field.

Now, consider an ion whose orbital levels are split by the crystal field in such a way that the lowest level is a singlet. The spin-degeneracy of this level is often reduced in zero applied magnetic fields, provided $S \geqslant 1$. As an example, the sevenfold orbital degeneracy of a Cr^{3+} ion ($3d^3$) splits into $1 + 3 + 3$-fold degenerate levels in a cubic field (Table 2-10.1), with the singlet level lowest. In an axially symmetric crystalline field the fourfold lowest level ($S = \frac{3}{2}$) splits into two doublets, with $M_S = \pm\frac{1}{2}$ and $M_S = \pm\frac{3}{2}$, and an energy difference of $2D$ say. Application of a static magnetic field H parallel to the axis of the crystal field will cause the energy levels to diverge linearly, as indicated in Fig. 3-9.1*a*. An alternating magnetic field of constant frequency applied at right angles to H induces transitions for $\Delta M_S = \pm 1$. Three transitions are possible, occurring at different values of H as indicated by the arrows. Hence in the paramagnetic resonance absorption of Cr^{3+} a triplet is observed (Fig. 3-9.1*b*). This is called the *fine structure* of the spectrum. Note that the lines are of different intensity, the central line being the strongest. This comes about because the absorption is greatest for states which have the largest component of magnetic moment in the plane perpendicular to H, since this condition gives the maximum value of $M_{\pm}$, hence χ''. Actually, in quantum mechanical terms, the absorption intensity is proportional to the square of the matrix element of the perpendicular magnetic moment. It should also be noted that when the magnetic field is applied at some angle to the crystal axis the separation of the levels with field is more complicated. There is competition between the crystal and magnetic field directions, so that the axis of precession of the magnetic moment in low fields changes from the crystal axis to the direction of the magnetic field as H is increased. In low fields the splitting is nonlinear, but it becomes linear again at high fields.

These problems will now be treated quantitatively for an iron group transition ion whose lowest orbital state is a singlet as a result of the crystalline field interaction. The effect of the spin-orbit coupling and of the applied field H will be considered by using as the perturbation Hamiltonian

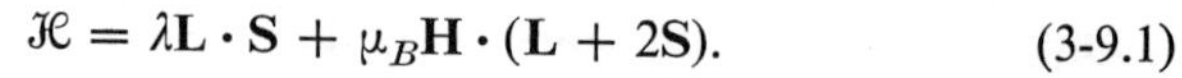

$$\mathcal{H} = \lambda \mathbf{L} \cdot \mathbf{S} + \mu_B \mathbf{H} \cdot (\mathbf{L} + 2\mathbf{S}). \tag{3-9.1}$$

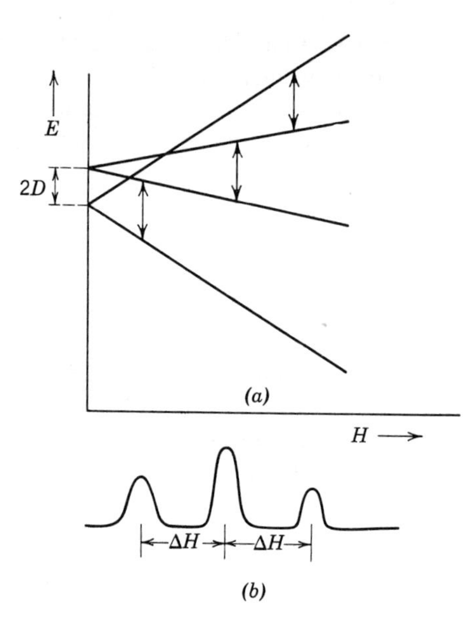

Fig. 3-9.1. Fine structure in paramagnetic resonance.

The perturbation calculation will be carried out by a method developed by Pryce.[73] In this method the usual operator representation is employed for **L**, whereas no representation is chosen for **S**. Then, instead of getting the change of energy of the levels produced by the perturbation as usual, an expression called the *spin-Hamiltonian*,[74] which is an explicit function of **S**, is obtained.

The matrix notation in Chapter 2 is employed in which $|0\rangle$ and $|n\rangle$ represent the wave functions for the ground and excited orbital states,

[73] M. H. L. Pryce, *Proc. Phys. Soc.* (*London*) **A-63,** 25 (1950).

[74] A. Abragram and M. H. L. Pryce, *Proc. Roy. Soc.* (*London*) **A-205,** 135 (1951). For another formulation see G. F. Koster and H. Statz, *Phys. Rev.* **113,** 445 (1959), **115,** 1568 (1959).

respectively. Since the ground state is nondegenerate, first-order perturbation theory gives $\langle 0 |\mathcal{H}| 0\rangle = \langle 0 |\lambda \mathbf{L} \cdot \mathbf{S} + \mu_B \mathbf{H} \cdot (\mathbf{L} + 2\mathbf{S})| 0\rangle$. For the term $\langle 0 |\mu_B \mathbf{H} \cdot 2\mathbf{S}| 0\rangle$, since no operation is made on the wave functions and the functions are normalized, ($\langle 0 | 0\rangle = 1$), we get just $2\mu_B \mathbf{H} \cdot \mathbf{S}$. The term $\langle 0 |\mu_B \mathbf{H} \cdot \mathbf{L} + \lambda \mathbf{L} \cdot \mathbf{S}| 0\rangle$ on expansion gives

$$(\mu_B H_z + \lambda S_z)\langle 0 |L_z| 0\rangle + (\mu_B H_x + \lambda S_x)\langle 0 |L_x| 0\rangle + (\mu_B H_y + \lambda S_y)\langle 0 |L_y| 0\rangle = 0,$$

since the matrix elements such as $\langle 0 |L_z| 0\rangle$ vanish because a singlet orbital ground state has zero orbital moment. Proceeding to the second order in the perturbation requires the computation of

$$-\sum{}' \frac{\langle 0 |\lambda \mathbf{L} \cdot \mathbf{S} + \mu_B \mathbf{H} \cdot (\mathbf{L} + 2\mathbf{S})| n\rangle^2}{E_n - E_0}.$$

The term $\langle 0 |\mu_B \mathbf{H} \cdot 2\mathbf{S}| n\rangle = 0$, since $\langle 0 | n\rangle = 0$ for nonzero values of n. The matrix elements $\langle 0 |L_x| n\rangle$, $\langle 0 |L_y| n\rangle$, and $\langle 0 |L_z| n\rangle$ are nonzero, however, and have values that depend on the particular level configuration in any given case. It is necessary, then, to compute the values of the tensor Λ_{ij}, defined as

$$\Lambda_{ij} = \sum_n{}' \frac{\langle 0 |L_i| n\rangle \langle n |L_j| 0\rangle}{E_n - E_0},$$

where $i, j = x, y, z$. The perturbation calculation then leads to the spin-Hamiltonian operator,

$$\mathcal{H}_s = 2\mu_B(\delta_{ij} - \lambda\Lambda_{ij})S_i H_j - \lambda^2\Lambda_{ij}S_i S_j - \mu_B{}^2\Lambda_{ij}H_i H_j, \quad (3\text{-}9.2)$$

where δ_{ij} is the Kronecker delta, equal to 1 for $i = j$ and equal to zero for $i \neq j$. Note that the spin-Hamiltonian contains terms for each value of the subscript i and j, that is, repeated indices indicate summation.

The significance of the various terms of equation 3-9.2 will now be considered. The magnetic moment is given by

$$\mu_i = -\frac{\partial \mathcal{H}_s}{\partial H_i} = -2\mu_B S_i + 2\mu_B \lambda \Lambda_{ij} S_j + 2\mu_B^2 \Lambda_{ij} H_j.$$

The first term is the permanent moment originating from the spin, whereas the second is the permanent moment originating from the orbit via the spin-orbit coupling. The last term is the field-induced paramagnetic moment. It is the temperature-independent term that occurs in the magnetic susceptibility (Chapter 2). The corresponding term in the spin-Hamiltonian is frequently written $-\mu_B{}^2\mathbf{H} \cdot \boldsymbol{\Lambda} \cdot \mathbf{H}$.

For $H = 0$ we have just the permanent moment

$$\mu_i = -2\mu_B(\delta_{ij} - \lambda\Lambda_{ij})S_j$$
$$= -\mu_B g_{ij} S_j,$$

where

$$g_{ij} = 2(\delta_{ij} - \lambda\Lambda_{ij}) \tag{3-9.3}$$

is the definition of the g-factor tensor. The first term of the spin-Hamiltonian (3-9.2) can be written in general as $\mu_B \mathbf{H} \cdot \mathbf{g} \cdot \mathbf{S}$, and it describes the level splitting caused by the application of the field $\mathbf{H}$. In general, the elements g_{ij} will be nonzero, but if the coordinate axes are chosen to be the principal axes of the tensor Λ_{ij}, then g_{ij} is a diagonal tensor. It has only the principal values $g_{xx} = g_x$, $g_{yy} = g_y$, and $g_{zz} = g_z$, the other elements being zero. When the symmetry is axial, $g_z = g_{\|}$, and $g_x = g_y = g_{\perp}$, whereas if it is cubic $g_x = g_y = g_z$. For a field $\mathbf{H}$ applied with direction cosines (l, m, n) with respect to the principal axes (x, y, z), the g-value for this direction is given by

$$g = (l^2 g_x^2 + m^2 g_y^2 + n^2 g_z^2)^{1/2}. \tag{3-9.4}$$

Note that for the ions with $3d$-shells less than half filled $g < 2$ since $\lambda > 0$, whereas if the $3d$ shell is more than half filled $g > 2$ since $\lambda < 0$.

Finally, consider the term $-\lambda^2\Lambda_{ij}S_iS_j$ of equation 3-9.2. This is the only term remaining in the spin-Hamiltonian for $H = 0$ and is the term that causes the zero-field splitting of the spin multiplet. It is usually written as $\mathbf{S} \cdot \mathfrak{D} \cdot \mathbf{S}$, where $\mathfrak{D}_{ij} = -\lambda^2\Lambda_{ij}$. For axially symmetric crystal fields, with choice of the principal axes of Λ_{ij}, we can have $\mathfrak{D}_{zz} = \mathfrak{D}_{\|}$ and $\mathfrak{D}_{xx} = \mathfrak{D}_{yy} = \mathfrak{D}_{\perp}$; the other components of the tensor will be zero. The term then gives

$$\mathfrak{D}_{\|}S_z^2 + \mathfrak{D}_{\perp}(S_x^2 + S_y^2),$$

which, using the identity $S_x^2 + S_y^2 = S(S+1) - S_z^2$, simplifies to

$$(\mathfrak{D}_{\perp} - \mathfrak{D}_{\|})S_z^2 - \mathfrak{D}_{\perp}S(S+1).$$

It is usual practice to rewrite this expression as

$$\mathfrak{D}[S_z^2 - \tfrac{1}{3}S(S+1)] + \tfrac{1}{3}S(S+1)[2\mathfrak{D}_{\perp} + \mathfrak{D}_{\|}], \tag{3-9.4}$$

where

$$\mathfrak{D} = (\mathfrak{D}_{\|} - \mathfrak{D}_{\perp}).$$

The first term in (3-9.4) gives the zero field splitting of the lowest spin levels, since the last term represents just a constant shift of all these levels. The splitting described by equation 3-9.4 is caused primarily by the crystal field entering via the spin-orbit coupling. There may also be a contribution to this term because of the spin-spin interaction between the ion's electrons.

In general, the spin-Hamiltonian (3-9.2) can be written

$$\mathcal{H}_s = \mu_B \mathbf{H} \cdot \mathbf{g} \cdot \mathbf{S} + \mathbf{S} \cdot \mathfrak{D} \cdot \mathbf{S} - \mu_B{}^2 \mathbf{H} \cdot \Lambda \cdot \mathbf{H}, \qquad (3\text{-}9.5)$$

where $\mathbf{g}$, $\mathfrak{D}$, and Λ are tensors. It is to be recalled that this equation was derived under the assumption that the lowest orbital level was a singlet. We now consider how this restriction can be removed. By Kramers' and Jahn-Teller's theorem (Section 2-10), we are lead to the conclusion that the orbital-plus-spin ground state of a system of electrons will be a singlet for an even number of electrons and a doublet for an odd number of electrons. When the lowest orbital level is a singlet, the splitting of the spin levels is generally small enough so that equation 3-9.3 will hold with a value of S equal to that for the free atom. However, when the lowest orbital level is degenerate, the splittings, because of the spin-orbit coupling and distortions of the crystal field, are larger. As a result, the lowest level is either a doublet or a singlet, depending on whether the number of electrons is odd or even. For the latter situation, no resonance will be observed and need not be treated further. For the former, the splittings are of the order of 100 cm^{-1}, so that only the doublet ground state need be considered. The magnetic system can then be described, provided that it is treated as having an *effective* spin, equal to $\frac{1}{2}$.

Therefore the foregoing conditions can be covered by a generalization of the spin-Hamiltonian of equation 3-9.5, provided that the quantum number S is considered as the effective spin, determined by equating $2S + 1$ to the number of electronic levels in the lowest group. Experimentally, it turns out that this spin-Hamiltonian can be used to describe the fine structure of the $3d$-transition ions in an axially symmetric crystalline field with one exception, namely the ions in S states. This case is discussed in the next section.

The actual energies of the electronic levels are the eigenvalues E of the Schrödinger equation $\mathcal{H}_s\psi = E\psi$, in which $\mathcal{H}_s$ is the spin-Hamiltonian (3-9.5) and the wave functions ψ are for the states with effective spin S. It is customary to employ the spin-Hamiltonian to describe the paramagnetic resonance spectrum of a transition ion situated in some particular crystal. The main advantage of this method is the small number of constants, such as S, $\mathfrak{D}$, $g_{\parallel}$, $g_{\perp}$, that are required to describe the often complicated behavior of the lowest energy levels of a salt. Indeed, in the early days of the subject, various involved formulas, together with energy-level diagrams, shown as a function of H, were given. Although the spin-Hamiltonian is a compact way of describing the resonance behavior, it is at the same time often a difficult task to determine the corresponding eigenvalues. The principle of the method is illustrated with a simple example; more complicated examples may be found in the literature.[75]

First, by assuming axial symmetry for the crystal field and omitting most of the terms that merely shift all the levels equally, we get from (3-9.5)

$$\mathcal{H}_s = \mu_B g_{\|} H_z S_z + \mu_B g_{\perp}(H_x S_x + H_y S_y) + \mathfrak{D}[S_z^2 - \tfrac{1}{3}S(S+1)]. \tag{3-9.6}$$

For a field H applied parallel to the z-direction this reduces to

$$\mathcal{H}_s = \mu_B g_{\|} H_z S_z + \mathfrak{D}[S_z^2 - \tfrac{1}{3}S(S+1)] \tag{3-9.7}$$

Since $\mathcal{H}_s$ is only a function of S_z, the eigenvalues are M_S, where M_S has the values $S_z, S_z - 1 \cdots -S_z$. Consider again Cr^{3+}, in which $S_z = \frac{3}{2}$. There we have (for strong fields)

$$\begin{aligned}
\mathcal{H}_s\,|\tfrac{3}{2}\rangle &= \tfrac{3}{2}g_{\|}\mu_B H + \mathfrak{D}|\,\tfrac{3}{2}\rangle,\\
\mathcal{H}_s\,|\tfrac{1}{2}\rangle &= \tfrac{1}{2}g_{\|}\mu_B H - \mathfrak{D}|\,\tfrac{1}{2}\rangle,\\
\mathcal{H}_s\,|-\tfrac{1}{2}\rangle &= -\tfrac{1}{2}g_{\|}\mu_B H - \mathfrak{D}|\,-\tfrac{1}{2}\rangle,\\
\mathcal{H}_s\,|-\tfrac{3}{2}\rangle &= -\tfrac{3}{2}g_{\|}\mu_B H + \mathfrak{D}|\,-\tfrac{3}{2}\rangle.
\end{aligned}$$

The energy levels are found by solving the secular determinant formed from the energy matrix for E, that is,

$$\begin{vmatrix}
(\tfrac{3}{2}g_{\|}\mu_B H + \mathfrak{D}) - E & 0 & 0 & 0\\
0 & (\tfrac{1}{2}g_{\|}\mu_B H - \mathfrak{D}) - E & 0 & 0\\
0 & 0 & (-\tfrac{1}{2}g_{\|}\mu_B H - \mathfrak{D}) - E & 0\\
0 & 0 & 0 & (-\tfrac{3}{2}g_{\|}\mu_B H + \mathfrak{D}) - E
\end{vmatrix} = 0.$$

Thus there are four levels, with energies $E_{3/2,-3/2} = \pm\frac{3}{2}g_{\|}\mu_B H + \mathfrak{D}$ and $E_{1/2,-1/2} = \pm\frac{1}{2}g_{\|}\mu_B H - \mathfrak{D}$. In zero applied field there are two doublets, with $M_S = \pm\frac{3}{2}$ and $M_S = \pm\frac{1}{2}$, separated by the energy $2\mathfrak{D}$. Remembering the selection rule $\Delta M_S = \pm 1$, we determine immediately that the separation of the lines of the fine structure is given by $2\mathfrak{D}/g_{\|}\mu_B$. The value of $g_{\|}$ can be determined from the value of H for the central line, since here the level energy separation is $\Delta E = h\nu$, so that we have $H_{\text{res}} = h\nu/g_{\|}\mu_B$ (see Fig. 3-9.1). From the foregoing it is obvious that zero-field splitting occurs only when $S \geqslant 1$; for $S = \frac{1}{2}$ there is only one line, and $\mathfrak{D} = 0$.

It should be remarked that in the literature it is common to describe the zero field splitting with the notation

$$\mathcal{H}_s = \mu_B \mathbf{H}\cdot\mathbf{g}\cdot\mathbf{S} + \mathfrak{D}[S_z^2 - \tfrac{1}{3}S(S+1)] + E_c(S_x^2 - S_y^2) \tag{3-9.8}$$

[75] For example, K. D. Bowers and J. Owen, *Rept. Progr. Phys.* **18**, 304 (1955).

where $E_c(S_x^2 - S_y^2)$ represents the zero field splitting caused by fields of lower than axial symmetry.

The preceding discussion has assumed that the ion had no nuclear moment, that is, $I = 0$. For a nucleus with spin I there will be $2I + 1$ values for the projection of the nuclear moment when a field H is applied. Each of these values is proportional to the quantum number m_I, where

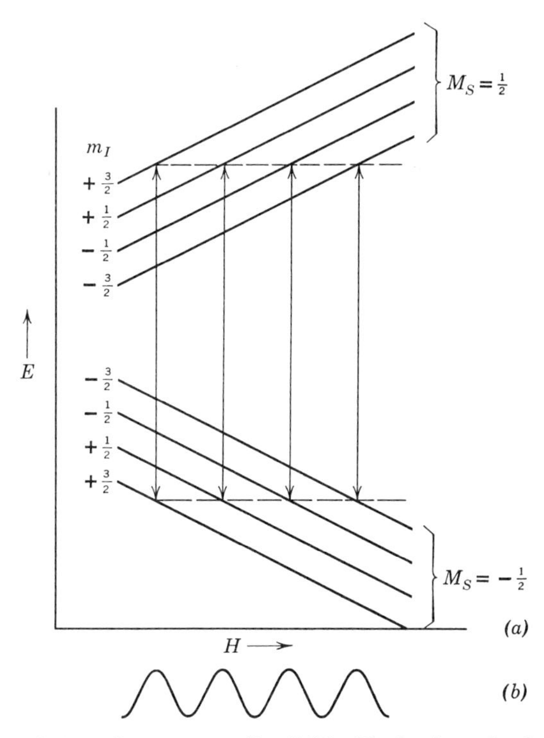

Fig. 3-9.2. Isotropic hyperfine structure (for Cu^{2+}). The level spacing is not linear for small magnetic fields. Hyperfine splittings vary from a few oersteds to a few hundred oersteds. [From B. Bleaney, *Physica* **17**, 175 (1951).]

$m_I = I, I - 1 \cdots -I$ (see Section 2-2). As a result, the electronic moment will see a net field, which has $2I + 1$ equally spaced values (in a strong applied field). Therefore electronic transitions will occur for a fixed frequency when the net field satisfies the resonance condition. On an energy level picture the nuclear moment produces an additional $(2I + 1)$-fold degeneracy to each electron energy level; this degeneracy is removed by application of a magnetic field. Figure 3-9.2*a* shows the level splitting for a Cu^{2+} ion, in which $S = \frac{1}{2}$ and $I = \frac{3}{2}$. The allowed transitions obey the selection rules $\Delta M_S = \pm 1$ and $\Delta m_I = 0$. For a fixed frequency, four equally spaced absorption lines called the hyperfine structure (hfs) of the spectrum are observed as a function of H. They were first observed by Penrose.[76] At the usual temperatures employed the nuclear levels have equal populations, and as a result the hyperfine lines have equal intensities. This forms an important criterion in distinguishing the hyperfine from the fine structure. The $2I + 1$ hyperfine lines can be used to determine nuclear spins. Indeed, the nuclear spins of a number of transition elements, including Cr^{53}, V^{50}, Ru^{99}, Ru^{101}, Nd^{143}, Nd^{145}, Sm^{147}, Sm^{149}, Er^{167}, and Pu^{241}, have been determined by this method.

The interaction energy between the electronic and nuclear moment can be represented often by $\mathcal{A}\mathbf{I}\cdot\mathbf{S}$, where $\mathcal{A}$ depends on the orbital radius, the nuclear and electronic g and magneton values, and the wave functions for the ground state.

In order to derive the operator representing the hyperfine interaction in the spin-Hamiltonian, we proceed as before and treat $\mathcal{A}\mathbf{I}\cdot\mathbf{S}$ as a perturbation. The first-order perturbation simply gives $\langle 0\,|\mathcal{A}\mathbf{I}\cdot\mathbf{S}|\,0\rangle = \mathcal{A}\mathbf{I}\cdot\mathbf{S}\langle 0\,|\,0\rangle = \mathcal{A}\mathbf{I}\cdot\mathbf{S}$ and the second order vanishes, since $\langle 0\,|\mathcal{A}\mathbf{I}\cdot\mathbf{S}|\,n\rangle = 0$. For the applied magnetic field parallel to the symmetry axis z the hyperfine spacing is given by $\Delta H = \mathcal{A}/g_{\|}\mu_B$. In the general case the hyperfine splitting depends on the direction of H, and $\mathcal{A}$ is therefore a tensor. For the principal axes this term usually is written

$$\mathcal{A}I_zS_z + \mathcal{B}(I_xS_x + I_yS_y). \tag{3-9.9}$$

in the spin-Hamiltonian.

If the nucleus has an electric quadrupole moment, it will interact with an electric field gradient at the nucleus to produce a shifting of the hyperfine energy levels. To first order the levels are shifted equally, but in second order there are small differences in the separation of the lines. Also, forbidden transitions are now allowed so that additional weak lines may be observed for $\Delta m_I = \pm 1$ and ± 2. The quadrupole interaction can be described by a term $\mathfrak{I}\cdot\mathcal{P}\cdot\mathfrak{I}$ in the spin-Hamiltonian. Referred to the principal axes, this term is usually written

$$\mathcal{P}[I_z^2 - \tfrac{1}{3}I(I+1)] + \mathcal{P}'(I_x^2 - I_y^2). \tag{3-9.10}$$

The quadrupole interaction was discovered experimentally by Ingram.[77] It has been observed in only a few salts; for example, in Cu^{2+} $6H_2O$ complexes, erbium ethylsulfate, and a neptunyl complex.

The interaction of H with the nuclear moment introduces an additional term $-g_N\mu_{B_n}\mathbf{H}\cdot\mathbf{I}$ in the spin-Hamiltonian. For $\Delta m = 0$ transitions

[76] R. P. Penrose, *Nature* (*London*) **163,** 992 (1949). For a review see B. Bleaney, *Physica* **17,** 175 (1951).

[77] D. J. E. Ingram, *Proc. Phys. Soc.* (*London*) **A-62,** 664 (1949).

there will be a small constant shift of the spectrum lines; usually this shift is small, if not negligible.

In summary, the paramagnetic resonance spectrum is usually described by the spin-Hamiltonian

$$\mathcal{H}_s = \mathbf{S} \cdot \mathbf{\mathcal{D}} \cdot \mathbf{S} + \mu_B \mathbf{H} \cdot \mathbf{g} \cdot \mathbf{S} + \mathbf{S} \cdot \mathbf{\mathcal{A}} \cdot \mathbf{I} + \mathbf{I} \cdot \mathbf{\mathcal{T}} \cdot \mathbf{I} - g_N \mu_B \, \mathbf{H} \cdot \mathbf{I} - \mu_B{}^2 \mathbf{H} \cdot \Lambda \cdot \mathbf{H}. \quad (3\text{-}9.11)$$

For a crystalline field with axial symmetry this becomes

$$\mathcal{H}_s = \mathcal{D}[S_z{}^2 - \tfrac{1}{3}S(S+1)] + \mu_B[g_{\parallel} H_z S_z + g_{\perp}(H_x S_x + H_y S_y)] + \mathcal{A} I_z S_z + \mathcal{B}(I_x S_x + I_y S_y) + \mathcal{T}[I_z{}^2 - \tfrac{1}{3}I(I+1)] + \mathcal{T}'(I_x{}^2 + I_y{}^2) - g_{\parallel}\mu_{B_n} \mathbf{H} \cdot \mathbf{I}. \quad (3\text{-}9.12)$$

For strong fields **H**, Bleaney[78] has calculated a general formula for the energy levels as a function of angle. The values of $g_{\parallel}$, $g_{\perp}$, $\mathcal{D}$, $\mathcal{A}$, $\mathcal{B}$, and $\mathcal{T}$ of the spin-Hamiltonian can be obtained immediately from the observed spectra once the principal axes have been determined.[79] For some spectra a more complicated spin-Hamiltonian is required to fit the results. Whether these terms represent real crystal field distortions or give a better fit with the data because of the additional parameters involved may be open to question.

10. The Spectra of the Transition Group Ions

Paramagnetic resonance studies of the transition group ions in various crystals are extensive and are tabulated in various reviews.[80] At various places in Chapters 2 and 3 some paramagnetic salts have been discussed. For convenience, the name, chemical formula, and crystal structure of the salts which have been investigated the most thoroughly are listed in Table 3-10.1. The list is not exhaustive, and the resonance spectra of some other crystal types have been studied. In addition, paramagnetic resonance of some of the transition ions present as impurities in several host crystals has been observed. A general outline of the information obtained first from studies of paramagnetic salts and then from doped crystals is presented in that order in this section. The paramagnetic resonance of other systems than the transition ions in crystals, such as paramagnetic molecular gases and free radicals, are discussed in the next section.

[78] B. Bleaney, *Phil. Mag.* **42,** 441 (1951).

[79] B. Bleaney and D. J. E. Ingram, *Proc. Roy. Soc.* (*London*) **A-205,** 336 (1951), show how to determine the relative signs of $\mathcal{A}$, $\mathcal{B}$, and $\mathcal{D}$.

[80] K. D. Bowers and J. Owen, *Rept. Progr. Phys.* **18,** 304 (1955); J. W. Orton, *ibid.* **22,** 204 (1959).

Table 3-10.1. Paramagnetic Crystal Types

Name	Chemical Formula	Crystal Structure
Tutton salts	$M^{+}_{2}M^{2+}(SO_4)_2 \cdot 6H_2O$ or $M^{+}_{2}M^{2+}(SeO_4)_2 \cdot 6H_2O$, $M^+ = K^+, Rb^+, NH_4^+$, etc.; M^{2+} = divalent ion from 3*d* group	Monoclinic; M^{2+} ion in distorted octahedron of six molecules of H_2O
Alums	$M^+M^{3+}(SO_4)_2 \cdot 12H_2O$, $M^+ = K^+, NH_4^+$, etc.; M^{3+} = trivalent ion from 3*d* group	M^{3+} at center of almost regular octahedron of six H_2O molecules
Flurosilicates	$M^{2+}SiF_6 \cdot 6H_2O$; M^{2+} = divalent 3*d* ion	Orthorhombic; M^{2+} surrounded by almost regular octahedron of six H_2O molecules
Bromates	$M^{2+}(BrO_3)_2 \cdot 6H_2O$; M^{2+} = divalent 3*d* ion	Cubic; M^{2+} surrounded by almost regular octahedron of six H_2O molecules
Sulfates	$M^{2+}SO_4 \cdot 7H_2O$; $M^{2+} = Ni^{2+}$, etc.	Orthorhombic; M^{2+} surrounded by distorted octahedron of six H_2O molecules
Cyanides	$K_3M^{3+}(CN)_6$, $M^{3+} = Cr^{3+}, Mn^{3+}, Fe^{3+}, Co^{3+}$, or $K_4M^{2+}(CN)_6 \cdot 3H_2O$; $M^{2+} = V^{2+}, Mn^{2+}, Fe^{2+}$	Orthorhombic / Tetragonal; M^{2+} or M^{3+} in octahedron formed by six CN groups
Halides	$M^{+}_{2}M^{4+}X_6$, X = Cl, Br.; $M^+ = K^+, NH_4^+$, etc., or $Na_2M^{4+}X_6 \cdot 6H_2O$; $M^{4+} = Pt^{4+}, Ir^{4+}$	Cubic / Triclinic; M^{4+} surrounded by octahedron of six halogens
Double Nitrates	$M^{2+}_{3}M^{3+}_{2}(NO_3)_{12} \cdot 24H_2O$; $M^{2+} = Mg^{2+}$ or divalent 3*d* ion; $M^{3+} = Bi^{3+}$ or trivalent 4*f* ion	Trigonal; M^{2+} in almost regular octahedron of six H_2O molecules; M^{3+} surrounded by NO_3 groups
Ethylsulfates	$M^{3+}(C_2H_5SO_4) \cdot 9H_2O$; M^{3+} = trivalent ion from 4*f* group	Hexagonal; M^{3+} ion at center of triangle of three water molecules; in parallel planes above and below are two more triangles of water molecules, rotated 60° with respect to the center one (trigonal symmetry)

The 3*d* group ions in salts. From Table 3-10.1 it is clear that the paramagnetic ion of the 3*d* group is located at the center of a more or less regular octahedron of negatively charged diamagnetic ions. The splitting of the orbital levels by an octahedral or cubic crystalline field for an ion with an electron configuration $3d^1$ has already been considered in Section 2-11. For $3d^6$ the five electrons with parallel spins have a spherical charge distribution and zero orbital angular momentum. Hence the electron charge distribution is similar for $3d^1$ and $3d^6$ ions, and their fivefold degenerate orbitals split into two- and threefold degenerate levels as in Fig. 3-10.1*a* (also Fig. 2-11.1*d*). The ions $3d^4$ and $3d^9$ are also in *D*-states and are complementary to the $3d^1$ and $3d^6$ ions. They can be considered as having a positive hole in the spherically symmetric distribution $3d^5$ and $3d^{10}$. As a result, the splitting is similar, but inverted, as in Fig. 3-10.1*b*. Distortions from octahedral symmetry and spin-orbit coupling split the levels further. This splitting turns out to be smaller for $3d^1$ and $3d^6$ than

for $3d^4$ and $3d^9$ ions. Therefore for the former ions there is an important contribution from the orbital momentum. The g-value and zero field splitting of these ions tend to vary considerably from salt to salt.

For ions in F-states there is sevenfold orbital degeneracy. In a cubic crystalline field the $3d^3$ and $3d^8$ configurations have an orbital singlet lowest in energy (Fig. 3-10.1c), whereas $3d^2$ and $3d^7$ have a triplet lowest.

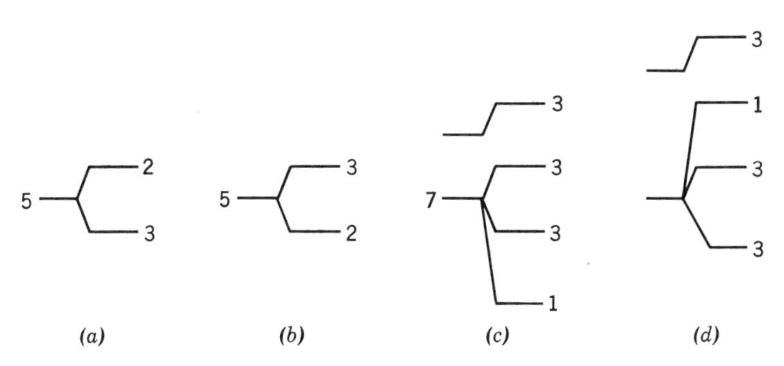

Fig. 3-10.1. The splitting in a cubic field of ions with the electron configurations (a) $3d^1$ and $3d^6$, (b) $3d^4$ and $3d^9$, (c) $3d^3$ and $3d^8$, and (d) $3d^2$ and $3d^7$. The degeneracy of the levels is indicated by the numbers.

In the former case only simple spin-degenerate levels result. In the latter, even though there is further splitting, the orbital momentum is far from being quenched; this leads to g-values that vary from salt to salt and are often large. Table 3-10.2 lists the $3d$ group ions that have been studied by paramagnetic resonance and includes the effective spin and the range of g-values observed for different salts. It may be seen that the g-values

Table 3-10.2. Data on 3d group ions in salts, obtained by paramagnetic resonance

Ion	Electron Configuration	Free Ion Spin	Effective Spin	Range of g-Values (experimental)
Ti^{3+}	$3d^1$	$\frac{1}{2}$	$\frac{1}{2}$	1.1–1.3
not detected	$3d^2$		not detected	
V^{2+}, Cr^{3+}	$3d^3$	$\frac{3}{2}$	$\frac{3}{2}$	2.0
Cr^{2+}	$3d^4$	2	2	1.9–2.0
Mn^{2+}, Fe^{3+}	$3d^5$	$\frac{5}{2}$	$\frac{5}{2}$	2.0
Fe^{2+}	$3d^6$	2	$\frac{1}{2}$	0–9
Co^{2+}	$3d^7$	$\frac{3}{2}$	$\frac{1}{2}$	1.4–7
Ni^{2+}	$3d^8$	1	1	2.2–2.3
Cu^{2+}	$3d^9$	$\frac{1}{2}$	$\frac{1}{2}$	2.0–2.5

are in line with those expected from the foregoing discussion of the splitting of the orbital levels.

The ions with $3d^5$ form a special case. Since $L = 0$, neither the spin-orbit coupling nor the crystalline field will produce splitting, at least to the first order. Experimentally, it is found that there is splitting of the sixfold degenerate spin levels so that the spin-Hamiltonian can be written

$$\mathcal{H} = g\mu_B \mathbf{H} \cdot \mathbf{S} + \frac{a}{6}\left(S_x^{\,4} + S_y^{\,4} + S_z^{\,4} - \frac{707}{16}\right) + \mathcal{D}\left(S_z^{\,2} - \frac{35}{12}\right). \quad (3\text{-}10.1)$$

The term in S^4 enters in a fifth-order perturbation calculation,[81] depending on the crystalline field to first order and the spin-orbit interaction to fourth order. The last term develops from the presence of an axially symmetric crystalline field. It is apparently caused by a small distortion from a spherical charge distribution brought about by a small admixture of $3d^44s$ to the $3d^5$ configuration.[82]

We refer the reader to the literature for details on the hyperfine structure.

Covalent binding and the 3*d*, 4*d*, 5*d*, and 5*f*–6*d* groups. The orbital level splitting into δ_ϵ and δ_γ levels for the iron group cyanides and for salts of the 4*d* and 5*d* ions have been discussed in Section 2-12. In principle, the large separation of the δ_ϵ and δ_γ levels for these compounds could arise from either a large crystalline field or from covalent bonding. It has been possible to show that the latter factor is important. Because of the sharing of the *d*-electrons, two important effects result. The first is a small reduction of the orbital magnetic moment. The second is the appearance of an additional hyperfine structure because of the interaction of the electron with the magnetic moment of the nucleus which shares this electron. Both effects have been observed.[83] For example, for ammonium chloroiridate $[(NH_4)_2(IrCl)_6]$, the hyperfine structure is produced both by the magnetic moment of the iridium nucleus and by the neighboring chlorine nuclei.

Measurements on some of the 3*d* cyanides and on salts of the 4*d* and 5*d* groups are summarized in Table 3-10.3. The results may be correlated with the data of Table 2-12.1 and 2-12.2.

For the 5*f*-6*d* group elements resonance has been observed in neptunyl and plutonyl ions. The results are summarized in Table 3-10.4. The results are in accord with the theory of Pryce described in Section 2-12.

4*f* rare earth ions in salts. Because the 4*f* electrons are well shielded by outer electrons and have, in addition, a small average radius, the perturbation caused by the crystalline field is considered after the spin-orbit

[81] J. H. Van Vleck and W. G. Penney, *Phil. Mag.* **17,** 961 (1934).

[82] A. Abragam and M. H. L. Pryce, *Proc. Roy. Soc.* (*London*) **A-205,** 135 (1951).

[83] J. Owen, *Proc. Roy. Soc.* (*London*) **A-227,** 183 (1955).

coupling has been accounted for. Hence J, representing the total angular momentum, is a good quantum number, and the perturbation of the field is taken on its states. The crystalline field for the ethylsulfates, the rare earth salt that has been the most extensively investigated, is given by equation 2-10.6. These salts have trigonal symmetry, with nine water molecules surrounding the rare earth ion. The H_2O molecules are usually

Table 3-10.3. Results for Some Covalently Bound Ions of the 3*d*, 4*d*, and 5*d* Groups

Ion	Configuration	Free Ion Spin	Effective Spin	Range of g-Values
Cr^{3+}, V^{2+}	$3d^3$	$\frac{3}{2}$	$\frac{3}{2}$	1.992
Mn^{3+}	$3d^4$	Resonance not yet observed		
Fe^{3+}	$3d^5$	$\frac{5}{2}$	$\frac{1}{2}$	0.7–2.6
Mo^{5+}	$4d^1$	$\frac{1}{2}$	$\frac{1}{2}$	1.95
Mo^{3+}	$4d^3$	$\frac{3}{2}$	$\frac{3}{2}$	1.90–2.0
Ru^{3+}	$4d^5$	$\frac{5}{2}$	$\frac{1}{2}$	1.0–3.24
Ag^{2+}	$4d^9$	$\frac{1}{2}$	$\frac{1}{2}$	2.04–2.18
Re^{4+}	$5d^3$	$\frac{3}{2}$	$\frac{3}{2}$	1.8
Ir^{4+}	$5d^5$	$\frac{5}{2}$	$\frac{1}{2}$	0.75–2.20

arranged in three equilateral triangles, stacked in parallel planes, with the triangle in the center plane twisted 60° with respect to the others. It turns out that the electronic levels are doublets when the number of electrons is odd and either singlets or doublets when the number is even. A further splitting, because of the Jahn-Teller effect, occurs in the latter doublets, but this splitting is very small. The g-values may vary considerably from

Table 3-10.4. Results for the 5*f*-6*d* group ions

Ion	Configuration	Effective Spin	Range of g-Values
$(NpO_2)^{2+}$	$5f^1$	$\frac{1}{2}$	0.20–3.40
$(PuO_2)^{2+}$	$5f^2$	$\frac{1}{2}$	0–5.9

$g = 2$ and frequently are very anisotropic. The $4f^7$ ions in S-states have properties similar to the $3d^5$ ions; that is, the splittings are small and the g-values are almost isotropic. The experiments are conducted at low temperatures because of the line broadening arising from the rapid spin-lattice relaxation time at higher temperatures. Table 3-10.5 summarizes some of the experimental observations. Another salt of the rare earths that has been well studied is the double nitrate.[84]

[84] B. R. Judd, *Proc. Roy. Soc. (London)* **A-232,** 458 (1955).

Transition ions in various host lattices. Transition ions may be present in various host materials, either as an accidental impurity or as the result of deliberate doping. In the term host materials we arbitrarily exclude the very dilute paramagnetic salts. Most of the crystals that serve as hosts have high melting points and are very hard. They include refractory materials such as MgO and gems such as sapphire and emerald. Frequently the transition ion enters the lattice substitutionally, although sometimes it occupies an interstitial site. Paramagnetic resonance studies provide

Table 3-10.5. Data on the Rare Earth Ions

Ion	Configuration	Free Ion Spin	Effective Spin	Range of g-Values
Ce^{3+}	$4f^1$	$\frac{1}{2}$	$\frac{1}{2}$	0.95–2.18[a]
Pr^{3+}	$4f^2$	1	$\frac{1}{2}$	0.3–1.7
Nd^{3+}	$4f^3$	$\frac{3}{2}$	$\frac{1}{2}$	2.0–3.6
Sm^{3+}	$4f^5$	$\frac{5}{2}$	$\frac{1}{2}$	0.6
Eu^{2+}, Gd^{3+}	$4f^7$	$\frac{7}{2}$	$\frac{7}{2}$	1.99
Tb^{3+}	$4f^8$	3	$\frac{1}{2}$	0.3–17.8
Dy^{3+}	$4f^9$	Not observed		
Er^{3+}	$4f^{11}$	$\frac{3}{2}$	$\frac{1}{2}$	1.47–8.9
Yb^{3+}	$4f^{13}$	Not observed		

[a] This g-value is for the lowest doublet in the diluted crystal only. See Section 2-10, equation 2-10.7.

important information concerning the strength and symmetry of the crystalline fields in the hosts. Often it is possible to change the valency state of the transition ion by irradiation or other treatment. For example, the spectra of monovalent, divalent, and trivalent iron has been observed in MgO. Further, because of the requirement of charge balance, additional vacancies may be incorporated into the crystal. Again by irradiation, color centers of different types (Section 3-11) may be produced and their paramagnetic resonance spectra studied. Besides being of fundamental interest, doped crystals are technically important because of their widespread use as maser materials (Section 3-12).

A list of the transition ions whose spectra has been observed in various host lattices is given in Table 3-10.6. This list is far from exhaustive. Isotropic g-values are also listed. The crystal field and hyperfine interaction parameters of the spin-Hamiltonian may be found in the literature.[85]

[85] An excellent source of references is *Paramagnetic Resonance*, Vol. I, W. Low, Editor, Academic Press, New York (1963).

Table 3-10.6. Transition Ions in Various Host Materials

Host	Transition Ions and Their *g*-Value (in brackets)
α-Al_2O_3 (sapphire or corundum)	Cr^{3+}[1.985](ruby), Fe^{3+}[2.00], Gd^{3+}, Mn^{4+}[1.9937], Cu^{3+}[2.078], Ti^{3+}[1.067]
TiO_2 (rutile)	V^{4+}[1.914], Cr^{3+}[1.97], Mn^{4+}[1.99], Mn^{3+}[2.00], Fe^{3+}[2.00], Ni^{3+}, Co^{2+}, Ni^{2+}[2.1], Cu^{2+}, Nb^{4+}, Mo^{5+}, Ta^{4+}, Ce^{3+}, Er^{3+}[<0.1]
$Al_2O_3 \cdot 6SiO_2 \cdot 3BeO$ (beryl or emerald)	Cr^{3+}[1.97], Fe^{3+}
MgO	Cr^{3+}[1.980], Mn^{2+}[2.0014], Fe^{3+}[2.004], Co^{2+}[4.278], Co^{+}[2.173], Ni^{2+}[2.214], Ni^{+}[2.169], Fe^{2+}, Fe^{+}[4.15], V^{2+}[1.980], Cu^{2+}[2.190], Ru^{3+}[2.170], Rh^{3+}[2.171], Pd^{2+}[2.170], Gd^{3+}
CaO	Cr^{3+}[1.9732], Co^{2+}[4.372], Mn^{2+}[2.001], V^{2+}[1.9683], Fe^{3+}[2.006], Ni^{2+}[2.327], Eu^{2+}[1.994], Gd^{3+}[1.992]
Ga_2O_3	Cr^{3+}[1.96]
$CaWO_4$	Gd^{3+}
$SrTiO_3$	Cr^{3+}[1.978], Mn^{4+}[1.994], Yb^{3+}, Nd^{3+}[2.61], Eu^{2+}[1.990], Gd^{3+}[1.992], Ni^{+}, Ni^{2+}[2.21], Ni^{3+} [2.180], Ce^{3+}
ThO_2	Gd^{3+}
CaF_2	Mn^{2+}[1.998], Eu^{2+}[1.989], Gd^{3+}[1.991], Tm^{3+}[3.453], Ho^{2+}[5.975], Yb^{3+}[1.328], Tb^{3+}, Tb^{4+}, U^{3+}, U^{4+}, H
ZnF_2	Cr^{3+}[1.97]

11. The Spectra of Paramagnetic Molecules and Other Systems

Any system that has one or more unpaired electrons may be expected to exhibit paramagnetism. Besides the transition elements, systems that have been studied by paramagnetic resonance techniques include (*a*) certain gases, notably O_2, NO, and NO_2, as well as some composed of free atoms, (*b*) free radicals, (*c*) donors and acceptors in semiconductors, (*d*) *F*-centers and other types of trap, and (*e*) defects produced by radiation damage. These classifications are not mutually exclusive; for example, irradiation can produce both *F*-centers and free radicals. We shall discuss each of these systems briefly.

Paramagnetic gases. The paramagnetic susceptibility of the gases O_2 and NO was treated in Section 2-14. In addition, a brief outline of the notation used to describe the quantum states of diatomic molecules was given. The oxygen molecule has zero axial orbital momentum ($\Lambda = 0$)

and two unpaired electron spins ($S = 1$). There is coupling between the rotation of the molecule and the spin to give levels with $J = K$ and $J = K \pm 1$. The latter levels are almost degenerate and separated by about 2 cm^{-1} from $J = K$. The selection rule is $\Delta J = \pm 1$; these transitions have been observed with an oscillatory magnetic field and zero steady field. Application of a small steady magnetic field splits the J levels linearly, but as the applied field is increased, the coupling between K and S begins to break down. As a result, a large number of resonance absorption lines are observed.[86] A number of these transitions have been accounted for theoretically.[87] For the O^{16} nucleus, $I = 0$, and there is no hyperfine structure.

The paramagnetic doublet level $^2\Pi_{3/2}$ of the NO molecule is split into components because of the strong coupling with the molecular rotation. A fine structure of three lines has been observed from the four $J = \frac{3}{2}$ levels[88] (the selection rule is $\Delta M_J = \pm 1$). There is also a hyperfine structure because of the interaction with the nuclear moment of nitrogen ($I = 1$).

The spectra of the polyatomic molecule NO_2 can be interpreted if complete decoupling of the electronic spin moment ($\Sigma = \frac{1}{2}$), the nuclear moment ($I = 1$), and the moment arising from the molecular rotation is assumed.[89] The spectra of free atoms of hydrogen, oxygen, nitrogen, and phosphorus, produced by a gaseous discharge method, have also been observed.[90] By using photodissociation, atomic iodine (I^{127}) and an isotope of oxygen (O^{17}) have been investigated.[91] The dependence of the free atom concentration on various conditions has been studied also for the latter atoms.

Free radicals. Susceptibility measurements on some organic and inorganic free radicals were mentioned in Section 2-14. The paramagnetic resonance of free radicals was observed first in the late 1940's.[92] Because of the covalent binding, the orbital momentum is strongly quenched, and as a result the g-values observed are close to the free spin value, $g \approx 2$, although usually slightly larger. The g-values are very nearly isotropic,

[86] R. Beringer and J. G. Castle, *Phys. Rev.* **81,** 82 (1951).

[87] A. F. Henry, *Phys. Rev.* **80,** 396 (1950); M. Tinkham and M. W. P. Strandberg, *Phys. Rev.* **99,** 537 (1955).

[88] R. Beringer, E. B. Rawson, and A. F. Henry, *Phys. Rev.* **94,** 343 (1954).

[89] J. G. Castle and R. Beringer, *Phys. Rev.* **80,** 114 (1950).

[90] R. Beringer and M. A. Heald, *Phys. Rev.* **95,** 1474 (1954); H. G. Dehmelt, *Phys. Rev.* **99,** 527 (1955).

[91] K. D. Bowers, R. A. Kamper, and C. D. Lustig, *Proc. Phys. Soc. (London)* **B-70,** 1176 (1957).

[92] B. M. Kosirev and S. G. Salekov, *Dokl. Akad. Nauk SSSR* **58,** 1023 (1947); A. N. Holden, C. Kittel, F. R. Merritt, and W. A. Yager, *Phys. Rev.* **75,** 614 (1949).

but in some materials a small anisotropy has been observed.[93] For most free radicals there is only one unpaired electron present, and consequently there is no fine structure.

Because the spin-orbit coupling is small, the spin-lattice interaction is weak and this results in relaxation times that are considerably longer than those of the transition ions; indeed τ_1 may be of the order of seconds. For some free radicals the spin-lattice interaction probably arises primarily in an exchange interaction in a solid or in motion of the molecules in a liquid rather than by the spin-orbit coupling mechanism. For any of these cases it appears that the spin-lattice interaction is too weak to produce appreciable line broadening by shortening the lifetime of the states.

On the other hand, the long relaxation times mean that saturation may occur, since then the spins in the upper energy level may not be able to return to the lower level quickly enough to maintain the Boltzmann population distribution. Saturation was discussed in Section 3-8; the condition for no saturation was given as $\gamma^2 H_1^2 \tau_1 \tau_2 \ll 1$ (equation 3-8.7). The effect of saturation is to reduce the power absorption. Further, this reduction is greater for the center than for the leading and trailing edge of the line, since it is proportional to the amount of power absorbed. As a result the line shape is flattened, and hence the apparent line width is increased. This is termed homogeneous broadening to contrast it from inhomogeneous broadening caused either by overlapping hyperfine structure or by inhomogeneities in the applied magnetic field. In inhomogeneous broadening the line shape is the envelope of the unresolved lines. Saturation will reduce each individual line because of homogeneous broadening. As a result the peak of each line will be reduced by the same factor. Therefore, although the absorption is less, the envelope retains the same shape, so that the line width remains the same. This difference in behavior with power saturation thus permits differentiation between homogeneous and inhomogeneous broadening.[94]

Exchange interaction is often important for free radicals in the solid state; the effect is to cause considerable narrowing of the line width. For example, for diphenylpicryl hydrazyl[$(C_6H_5)_2N—NC_6H_2(NO_2)_3$] the line width expected from the dipole-dipole interaction is about 30 oersteds, whereas because of the exchange interaction the actual width observed is only 2.7 oe.[95] This material has $g = 2.0038$, extremely close to the free

[93] For example, in diphenylpicryl hydrazyl, by L. S. Singer and C. Kikuchi, *J. Chem. Phys.* **23,** 1738 (1955).

[94] A. M. Portis, *Phys. Rev.* **91,** 1071 (1953).

[95] J. P. Goldsborough, M. Mandel, and G. E. Pake, *Phys. Rev. Letters* **4,** 13 (1960); J. R. Singer, *Paramagnetic Resonance*, W. Low, Editor, Academic Press, New York (1963), p. 577.

spin-value and also gives an intense line owing to the high magnetic concentration. Because of these properties hydrazyl is frequently employed as a g-marker standard. The g-value of other salts can then be determined accurately, provided they are close to 2.

On dissolving a free radical compound in some solvent, such as benzene, the line is broadened because of the reduction of the exchange interaction. On further dilution the dipole-dipole interaction is also reduced (for the reason see Section 4-4), and, provided the unpaired electron spends some time in the vicinity of one or more nuclei possessing magnetic moments, a resolved hyperfine structure is frequently obtained.[96] The magnitude of the splitting of the hyperfine structure depends on the amount of coupling between the unpaired electron and a nucleus. Also there will be hyperfine structure resulting from the interaction with each magnetic nucleus embraced by the electron's orbit. Studies of the hyperfine splitting are thus fruitful in providing information on the molecular orbital of the odd electron.

When more than one magnetic nucleus is present, two cases occur most frequently. In the one case the interaction with one nucleus with a spin I_1 is much stronger than with a second nucleus with a spin I_2. Here the stronger interaction splits the electronic line into $(2I_1 + 1)$ lines, and then each of these is split into $(2I_2 + 1)$ lines by the weaker interaction. In the other case the couplings with two or more identical nuclei of spin I are all equal. Then the structure consists of $(2nI + 1)$ lines, where n is the number of identical nuclei. Further, the intensity is a maximum for the central line or lines and is symmetrical for the off-center lines. For example, consider an electron coupling equally with two protons ($m_{I_1} = m_{I_2} = \pm\frac{1}{2}$). The first proton will split the two electronic levels into two, and the second proton will split each of these into two more. Two of these sublevels will coincide in energy, as indicated in Fig. 3-11.1, so that each electronic level is split into three equally spaced components. As a result, there will be three allowed transitions (the selection rule is $\Delta m_{I_1} = \Delta m_{I_2} = 0$) giving rise to three equally spaced lines, with relative intensities 1:2:1.

The spectrum can be represented by the spin-Hamiltonian

$$\mathcal{H} = g\mu_B \mathbf{S} \cdot \mathbf{H} + \mathcal{A}_1 \mathbf{I}_1 \cdot \mathbf{S} + \mathcal{A}_2 \mathbf{I}_2 \cdot \mathbf{S}, \ldots, \tag{3-11.1}$$

where $\mathcal{A}_1$ and $\mathcal{A}_2$ are a measure of the interaction with the two nuclei.

It is now pertinent to consider the interaction energy between the

[96] For example, see G. E. Pake, J. Townsend, and S. I. Weissman, *Phys. Rev.* **85**, 682 (1952); K. H. Hanser, *Z. Naturforsch.* **14a**, 425 (1959).

magnetic moment of the nucleus and the field at the nucleus set up by the electron. It is given by

$$gg_n\mu_B\mu_{B_n}\left\{\left[\frac{(\mathbf{l}-\mathbf{s})\cdot\mathbf{I}}{r^3}+\frac{3(\mathbf{r}\cdot\mathbf{s})(\mathbf{r}\cdot\mathbf{I})}{r^5}\right]+\frac{8\pi}{3}\,\delta(r)\mathbf{s}\cdot\mathbf{I}\right\} \quad (3\text{-}11.2)$$

where r is the distance between the nucleus and the electron and $\delta(r)$ is the Dirac δ-function, equal to zero everywhere except at the nucleus.

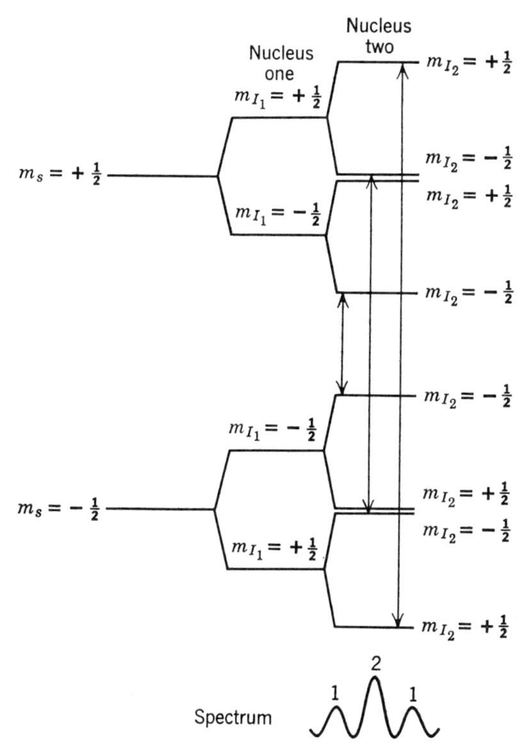

Fig. 3-11.1. Electronic energy levels when two magnetic nuclei are present.

The equation applies both for free radicals and transition ions; for the radicals $l = 0$, for the ions the expression should be summed over all the magnetic electrons. The first two terms arise because of the dipole-dipole interaction and can be derived from the classical equation 1-3.3. These terms give rise to an anisotropic hyperfine splitting. For free radicals in the liquid state it may be shown that these terms average out to zero (see Section 4-4). The last (contact) term is obtained from the Dirac theory of the electron[97]; it gives rise to an isotropic splitting. This term is zero unless the electron wave function is nonzero at the nucleus. For atoms this nonzero condition is fulfilled only for *s*-electrons; indeed, this is the reason for the configurational mixing required to explain the hyperfine splitting of the spectra of $3d^5$ and $4f^7$ transition ions, discussed in Section 3-10. For free radicals it is necessary to consider molecular orbitals. Only σ molecular orbitals (corresponding to *s*-atomic orbitals) have a finite electron density at the nucleus. On the other hand, delocalization of the unpaired electron in a planar free radical requires that it be in a π molecular orbital (corresponding to a *p*-atomic orbital). The fact that hyperfine structure is observed is explained by some configurational mixing of the π-orbital with the σ-orbital of an excited state.

Table 3-11.1. Data on some Free Radicals

Radical	*g*-Value	Line Width in Solid	Hyperfine Structure in Solution
Diphenylpicryl hydrazyl	2.0036	1.9 to 2.7	5 lines
Picryl aminocarbazyl	2.0036	0.5	7 lines
Wurster's blue perchlorate	2.00	2.7	39 lines
Napthalene	2.00	5	19 lines
p-Benzosemiquinone			5 lines
Pi-*p*-anisyl nitric oxide	2.0063	31.4	3 lines

The stable free radicals have permitted detailed studies. These results have been employed in the identification and analysis of the kinetics of short-lived active free radicals, which may be produced by irradiation, during polymerization, and in the charring of organic materials. Free radicals may occur in natural organic matter, so that the use of paramagnetic resonance techniques has become important in biology and medicine.[98] Biradicals, or free radicals with two unpaired electrons, also exist.

Table 3-11.1 summarizes some of the paramagnetic resonance data for a few representative free radicals.

Donors and acceptors in semiconductors. The common elemental semiconductors are silicon and germanium; they are in the fourth column of the periodic table and have four valence electrons. In the solid they are covalently bound. An element from the fifth column such as P, As,

[97] E. Fermi, *Z. Physik* **60**, 320 (1930); F. J. Milford, *Am. J. Phys.* **28**, 521 (1960).

[98] D. J. E. Ingram, *Paramagnetic Resonance*, Vol. II, W. Low, Editor, Academic Press, New York (1963), p. 809.

and Sb, has five valence electrons. When these elements are present as substitutional impurities in Ge or Si, four of the valence electrons take part in the covalent bonding, whereas the additional electron is bound very weakly. It moves in a hydrogenlike $1s$ orbit of large radius. This radius is large for two reasons. First, the electron behaves as though it had an effective mass m^* of the order of $0.3m$. This arises because the electron moves in the periodic potential of the lattice as well as the Coulomb potential of the impurity nucleus. Second, because the orbit encompasses many host atoms, the electron may be considered to be moving in matter with a dielectric constant ϵ of the host material. On a simple Bohr model this orbital radius is $R = \epsilon(m/m^*)a_0$, where a_0 is the first Bohr radius for hydrogen. The unpaired electron has a small ionization energy (about 0.02 ev) and may be donated to the conduction band to produce negative or n-type material; hence the term donor atom. Similarly, atoms from the third column, such as B or Al, with a deficiency of one electron for the covalent bonding, act as acceptors for electrons from the top of the filled valence band and thus produce holes or p-type material.

Resonance has been observed for donors in both silicon[99] and germanium[100] at low temperatures; it has also been observed for acceptors in silicon,[101] but not in germanium. For silicon the g-value is close to the free spin value, although slightly less. Below 2°K the spin system relaxes to the lattice via the direct process[102] with $\tau_1 \approx 5 \times 10^{-4}$ sec. Above 2°K, Raman processes become important and $1/\tau_1 \propto T^7$. This follows because of the small spin-orbit coupling. Hyperfine structure resulting because of the magnetic moment of the impurity nucleus has been observed for Li, P, As, and Sb. The hyperfine line separation confirms the large orbit radius. At higher donor concentrations the electrons are no longer localized, the local hyperfine fields are averaged out, and the lines become narrower. For intermediate concentrations some electrons interact in pairs and act as a magnetic unit with $S = 1$ to give half the hyperfine spacing. For donors in germanium local strains greatly broaden the line unless the static field is applied along the [100]-direction.

Iron group and other transition ions have also been incorporated into silicon and germanium. However, the unpaired electrons are now tightly bound, and the donor states are called *deep* ones. Resonance of these

[99] R. C. Fletcher, W. A. Yager, G. L. Pearson, and F. R. Merritt, *Phys. Rev.* **95,** 844 (1954); G. Feher, *Phys. Rev.* **103,** 834 (1956).

[100] G. Feher, D. K. Wilson, and E. A. Gere, *Phys. Rev. Letters* **3,** 25 (1959).

[101] G. Feher, J. C. Hensel, and E. A. Gere, *Phys. Rev. Letters* **5,** 309 (1960).

[102] A. Honig and E. Stupp, *Phys. Rev. Letters* **1,** 275 (1958); G. Feher and E. A. Gere, *Phys. Rev.* **114,** 1245 (1959); L. Roth, *Phys. Rev.* **118,** 1534 (1960); H. Hasegawa, *Phys. Rev.* **118,** 1523 (1960); D. K. Wilson and G. Feher, *Phys. Rev.* **124,** 1068 (1961).

deep impurities has also been observed.[103] The reader is referred to the literature for further details concerning resonance in semiconductors.[104]

Traps, *F*-centers, etc. We shall restrict the discussion to traps in alkali halide crystals. These traps are indicated schematically in Fig. 3-11.2. An *F*- or color center (from *Fabre*, the German for color) is an unpaired electron trapped at a negative ion vacancy. A V_1-center is a hole trapped at a positive ion vacancy; it is the result of an electron being missing from

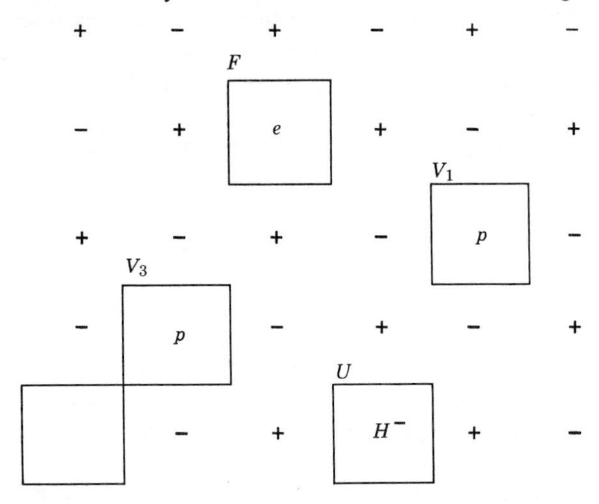

Fig. 3-11.2. Illustrating various types of traps, as labeled: + indicates alkali ion, — indicates halide ion, *e* is a trapped electron, *p* is a trapped hole, and H^- is a negative hydrogen ion.

one of the neighboring ions. A V_2-center consists of two adjacent V_1-centers; it is nonmagnetic. A V_3-center is formed by two adjacent positive ion vacancies with one trapped hole, and a *U*-center is a negative ion vacancy that contains a negative hydrogen ion. These centers are commonly produced by irradiation with X- or γ-rays or neutrons. Resonance has been observed for *F*-, V_3-, and *U*-centers.

The *g*-values of *F*-centers are slightly less than for free spins,[105] ranging

[103] H. H. Woodbury and G. W. Ludwig, *Phys. Rev.* **117,** 102 (1960), **126,** 466 (1960); G. W. Ludwig and H. H. Woodbury, *Phys. Rev. Letters* **5,** 98 (1960).

[104] For a survey see G. W. Ludwig and H. H. Woodbury, *Solid State Physics*, F. Seitz and D. Turnbull, Editors, Academic Press, New York, **13,** 223 (1962); and G. Feher, *Paramagnetic Resonance*, Vol. II, W. Low, Editor, Academic Press (1963), p. 715.

[105] C. A. Hutchinson, *Phys. Rev.* **75,** 1769 (1949); E. E. Schneider and T. S. England, *Physica* **17,** 221 (1951); M. Tinkham and A. F. Kip, *Phys. Rev.* **83,** 657 (1951); C. P. Slichter, *Phys. Rev. Letters* **5,** 197 (1960).

from about 1.98 to 2.00. Above 90°K the spin-lattice interaction contributes to the line width. For high enough concentrations ($\sim 10^{17}/cm^3$) there is appreciable line broadening from spin-spin interaction. For low concentrations and at low temperatures, the line is broadened because of unresolved hyperfine structure.[106] There is good agreement with the theory of the molecular orbital of the *F*-center.[107]

V_3-centers observed by Kanzig,[108] were identified from V_1-centers by their lack of cubic symmetry. The *g*-values were anisotropic, ranging from 2.0023 to 2.024. The hyperfine structure in KCl resulted from the interaction with Cl^{35} nuclei ($I = \frac{3}{2}$). *U*-centers were observed in mixed crystals of KCl—KH and KCl—KD irradiated with ultraviolet light.[109]

Defects from radiation damage. Irradiation of a material may produce a defect with an associated unpaired electron. The production of color centers as the result of trapping in a crystal vacancy in an alkali halide crystal has just been discussed. Impurities may act also as a trapping center; for example,[110] Al impurities in quartz (SiO_2). In some organic materials the covalent carbon-to-carbon bonds are believed to be broken by irradiation. Normally the two electrons of the bond have antiparallel spins. Breaking the bond by irradiation may occur by removal of one electron to some point far from the bond, by one electron being excited so that the electrons will have parallel spins, or by displacement of the bond's atoms so that the bond is no longer formed. These processes are believed to occur in plastics such as Teflon.[111] Irradiation by high-energy particles may cause lattice damage by knocking lattice atoms into interstitials and producing vacancies. This apparently occurs in quartz and diamond irradiated by neutrons.[112] When lanthanum magnesium nitrate,[113] potassium nitrate,[114] and potassium azite with nitrate impurities[115] are irradiated with γ-rays, the NO_3 group has an oxygen atom removed to leave the paramagnetic NO_2 molecule. Irradiation of LiH and AgCl

[106] A. M. Portis, *Phys. Rev.* **91,** 1071 (1953); G. Feher, *Phys. Rev.* **105,** 1122 (1957).

[107] A. F. Kip, C. Kittel, R. A. Levy, and A. M. Portis, *Phys. Rev.* **91,** 1066 (1953).

[108] W. Kanzig, *Phys. Rev.* **99,** 1890 (1955).

[109] C. J. Delbecq, B. Smaller, and P. H. Yuster, *Phys. Rev.* **104,** 599 (1956).

[110] J. H. E. Griffiths, J. Owen, and I. M. Ward, *Nature (London)* **173,** 439 (1954).

[111] E. E. Schneider, *Discussions Faraday Soc.* **19,** 158 (1955); H. N. Rexroad and W. Gordy, *J. Chem. Phys.* **30,** 399 (1959).

[112] J. H. E. Griffiths, J. Owen, and I. M. Ward, *Defects in Crystalline Solids*, Physical Society (London) (1955), p. 81.

[113] B. Bleaney, W. Hayes, and P. M. Llewellyn, *Nature (London)* **179,** 149 (1957).

[114] H. Zeldes, *Paramagnetic Resonance*, Vol. II, W. Low, Editor, Academic Press, New York (1963), p. 764.

[115] D. Mergerian and S. A. Marshall, *Phys. Rev.* **127,** 2015 (1962).

Table 3-11.2. Data for Some Materials with Defects Produced by Radiation Damage

Substance	Irradiation	Defect Type	g-Value	Hyperfine Structure
SiO_2 with Al impurity	X at 290°K	0^-	2.00–2.06	From Al^{27} ($I = \frac{5}{2}$)
$LaMg_3(NO_3)\cdot 24H_2O$	α, β, or γ at 290°K	NO_2 molecule	1.993–2.005	From N^{14} ($I = 1$)
KNO_3	γ-ray	NO_2 and NO_3^{2-}	1.993–2.007	From N^{14}
KN_3 plus nitrate impurity	γ-ray	NO and NO_2	2.00	From K^{19}
H_2SO_4 (solid)[a]	γ at 77°K	Atomic H	2.0024	From H^1 ($I = \frac{1}{2}$)
H_2O (solid)[b]	γ at 77°K	H and OH	2.00	From H^1
Teflon	X at 290°K	Broken C—C bond	2.02	From F^{19} ($I = \frac{1}{2}$)
CF_3CONH_2[c]	γ at 290°K	Free radical	2.0025–2.0045	From F^{19} ($I = \frac{1}{2}$)
$CF_3CF_2CONH_2$[c]	γ at 290°K	Free radical	2.00	From F^{19}
Diamond	Neutron at 290°K	C vacancies and interstitials Perhaps C_2 molecules	2.0028	None
LiH	Ultraviolet	Colloidal Li	2.0	None
AgCl	Ultraviolet and X	Colloidal Ag	2.0	None

[a] R. Livingston, H. Zeldes, and E. H. Taylor, *Discussions Faraday Soc.* **19,** 166 (1955).
[b] B. Smaller, M. S. Matheson, and E. L. Yasaitis, *Phys. Rev.* **94,** 202 (1954).
[c] R. Lontz and W. Gordy, *Paramagnetic Resonance*, Vol. II, W. Low, Editor, Academic Press, New York (1963), p. 795.

leaves colloidal lithium[116] and silver[117] metal, respectively.[118] Table 3-11.2 lists some of the resonance data on irradiated materials.

12. The Three-Level Maser and Laser

A maser is a device that acts as an amplifier or oscillator for electromagnetic radiation in the microwave region by employing in some manner the energy levels of an atomic or molecular system. The word is derived[119] from the first letter of the key words in the phrase "*m*icrowave *a*mplification by the *s*timulated *e*mission of *r*adiation." A laser is the optical analog of the maser (*l*ight). A maser that employs three electronic levels of a paramagnetic salt has been proposed by Bloembergen.[120] This scheme is an interesting application of principles developed earlier in this chapter and is now described.

Suppose that for a certain strength of an applied steady magnetic field

[116] W. T. Doyle, D. J. E. Ingram, and M. J. A. Smith, *Proc. Phys. Soc.* (*London*) **74,** 540 (1959).
[117] M. J. A. Smith and D. J. E. Ingram, *Proc. Phys. Soc.* (*London*) **80,** 310 (1962).
[118] Resonance of conduction electrons is discussed in Section 5-7.
[119] J. P. Gordon, H. J. Zeiger, and C. H. Townes, *Phys. Rev.* **99,** 1264 (1955).
[120] N. Bloembergen, *Phys. Rev.* **104,** 324 (1956).

the three levels of the salt have energies $E_3 > E_2 > E_1$ (see Fig. 3-12.1). Let the populations of these levels be denoted by N_3, N_2, and N_1, respectively, if the system is in thermal equilibrium, and by n_3, n_2, and n_1 otherwise. Now transitions between the levels can be produced by alternating fields of angular frequencies given by

$$\hbar\omega_{13} = E_3 - E_1, \qquad \hbar\omega_{32} = E_3 - E_2, \qquad \hbar\omega_{21} = E_2 - E_1. \quad (3\text{-}12.1)$$

Suppose that the salt is initially in thermal equilibrium and then consider the application of an alternating field, with frequency ω_{13} and of sufficiently large amplitude that saturation is approached. As a result, the equilibrium distribution population will be destroyed and n_1 will approach

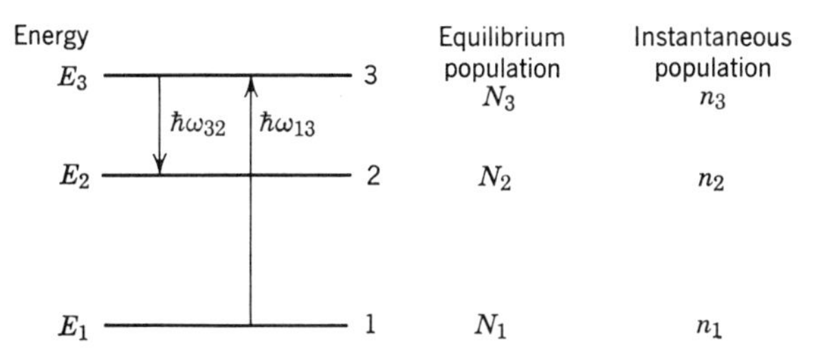

Fig. 3-12.1. The three-level maser principle.

and, in the limit of complete saturation, equal n_3. It follows immediately that either the condition $n_3 > n_2$ or $n_2 > n_1$ will hold [since in the limit $n_1 = n_3 = (N_1 + N_3)/2$]. If $n_3 > n_2$, then, in the presence of radiation of frequency ω_{32}, stimulated emission instead of absorption of energy will occur, hence amplification will result. It is usual to call ω_{13} the pump frequency and ω_{32} the signal frequency. These masers are operated at liquid helium temperatures in order to reduce the spin-lattice relaxation and permit saturation. As a result, they have very low noise, mainly caused by cavity losses and spontaneous transitions. The first three-level maser to operate[121] employed the salt $Gd_{.005}La_{.995}(C_2H_5SO_4)_3 \cdot 9H_2O$. However, it is an advantage mechanically if the crystal is hard rather than soft and water soluble. Ruby, which is α-Al_2O_3 doped with Cr^{3+} ions, is well suited as a maser and laser material and is in widespread use. Gases are often employed for continuously operating lasers because of the reduced power requirements.[122]

[121] H. E. D. Scovil, G. Feher, and H. Seidel, *Phys. Rev.* **105,** 762 (1957).

[122] Fuller details may be found in the books on masers and quantum electronics listed in the bibliography.

Problems

3-1. Show for a paramagnetic material that obeys Curie's law that the difference in the specific heats is given by

$$C_H - C_M = \frac{CH^2}{T^2},$$

and therefore that the specific heat of the spin system C_H is given by an expression

$$C_H = \frac{b + CH^2}{T^2},$$

where C is Curie's constant and $b = C_M T^2$.

3-2. The zero field splitting into two doublets of the ground state of chromic potassium alum ($Cr_2(SO_4)_3K_2SO_4 \cdot 24H_2O$) is $\theta_0 = 0.25°K$. Plot the magnetic specific heat as a function of temperature below 10°K. Compare the specific heat of the salt with that of some metal in this temperature range. If the initial temperature of the salt is 1.5°K in a field of 15,000 oe, calculate the final temperature on removal of the field.

3-3. Show that equations 3-5.10 can be written in the form

$$\left(\frac{\chi'}{\chi_T} - \frac{\chi_S}{\chi_T} - \frac{1}{2}\right)^2 + \left(\frac{\chi''}{\chi_T}\right)^2 = \left(\frac{1}{2} - \frac{\chi_S}{2\chi_T}\right)^2$$

or

$$\left[\frac{\chi'}{\chi_T - \chi_S} - \frac{\chi_S + \chi_T}{2(\chi_T - \chi_S)}\right]^2 + \left(\frac{\chi''}{\chi_T - \chi_S}\right)^2 = \frac{1}{4}.$$

Plot χ'/χ_T as a function of χ''/χ_T. Show that a chord with its origin at the intersection of the circle with the χ'/χ_T axis has the equation $\tan \phi = \omega\tau_1$ where ϕ is the acute angle the chord makes with the χ'/χ_T axis.

3-4. Show that equations 3-5.10 satisfy the Kramers-Kronig relationships (equations 3-4.7).

3-5. Generalize the results of Section 3-3, in which $1/\tau_1 = 1/\tau_{12} + 1/\tau_{21}$, to show that when many energy levels are involved the relaxation time is given by

$$\frac{1}{\tau_1} = \frac{n\sum P_{ij}E_{ij}^2}{\sum E_{ij}^2},$$

where the summation is for all values of i and j, P_{ij} being the transition probability, E_{ij} the energy difference between levels i and j, and n is the number of levels populated.

3-7. Consider a system of N noninteracting spins with $S = \frac{1}{2}$ and equilibrium populations N_1 and N_2 in the upper and lower energy levels in a static applied field H. Now, when a field $H_{(t)} = H + H_1 e^{i\omega t}$ is applied, assume (contrary to that for equation 3-5.3) that the transition probability becomes $P_{12} = (P_{12})_{dc} + (\partial P_{12}/\partial H)_T H_1 e^{i\omega t}$. Now find dn_1/dt and dn_2/dt and then the magnetization arising from the ac field. Thus derive the Debye relaxation equations,

$$\chi' = \frac{\chi_T}{1 + \omega^2\tau_1^2}$$

and

$$\chi'' = \frac{\chi_T\omega\tau_1}{1 + \omega^2\tau_1^2}.$$

[*Hint*. Note that the energy levels vary with a frequency ω. See C. J. Gorter and R. de L. Kronig, *Physica* **3,** 1009 (1936).]

3-7. For $S = 1$, calculate the level energies and the zero field splitting for the spin-Hamiltonian of equation 3-9.7.

3-8. By using the results $S_\pm = S_x \pm iS_y$, so that

$$S_+ \mid M> = [S(S+1) - M(M+1)]^{1/2} \mid M+1>$$

and

$$S_- \mid M> = [S(S+1) - M(M-1)]^{1/2} \mid M-1>,$$

show that the energy levels for the spin-Hamiltonian of equation 3-9.8 for $H \parallel z$ and $S = 1$ are

$$E_{1,2} = \tfrac{1}{3}\mathfrak{D} \pm (g_z^2\mu_B^2H^2 + E_c^2)^{1/2}$$

and

$$E_3 = -\tfrac{2}{3}\mathfrak{D}.$$

3-9. The Cu^{2+} ion situated in an octahedron formed by six molecules of H_2O has its ground state orbitals split by the crystal field as shown in the figure. The corresponding angular part of the wave function for each state is also listed. By using equation 3-9.3 show that

$$g_\parallel = 2\left(1 - \frac{4\lambda}{E_3 - E_1}\right)$$

and

$$g_\perp = 2\left(1 - \frac{\lambda}{E_{4,5} - E_1}\right).$$

Ref. D. Polder, *Physica* **9,** 709 (1942).

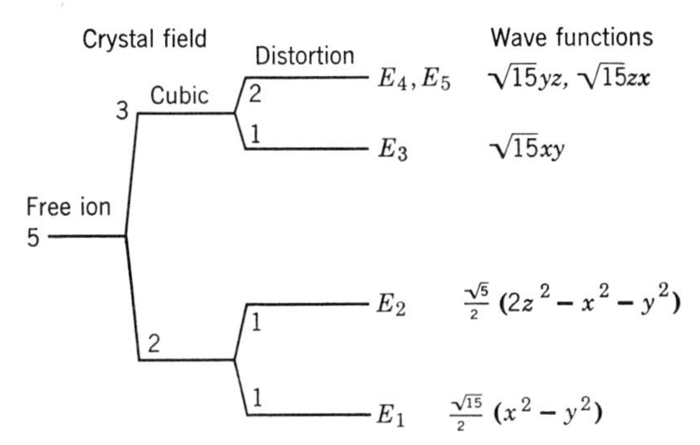

Fig. P3-9. The numbers indicate the degree of degeneracy of each orbital.

3-10. The spin-Hamiltonian for a paramagnetic salt is assumed to be

$$\mathcal{H}_s = g_{\|}\mu_B H_z S_z + g_{\perp}\mu_B(H_x S_x + H_y S_y) + \mathcal{A} S_z I_z + \mathcal{B}(S_x I_x + S_y I_y),$$

since both electron spin and nuclear spin operators are present. Suppose H is applied along the z-direction (the axis of quantization). For $S = 1$ and $I = 1$ find the secular determinant and evaluate the energy levels. [*Hint.* The operations with the nuclear spin operator are analogous to those for the electron spin operator; the operations may be performed separately.]

3-11. Show that in a paramagnetic resonance experiment the hyperfine splitting produced when an unpaired electron interacts with a nucleus of spin $I = \frac{1}{2}$ is equal to $2\Delta H$, in which ΔH is the field produced at the electron by the nuclear moment. Show that the anisotropic part of equation 3-11.2 can be written in scalar form as

$$g g_n \mu_B \mu_{B_n} m_s m_I \left(\frac{3\cos^2\theta - 1}{r^3}\right)$$

if $l = 0$ and θ is the angle between the direction of the nuclear dipole axis (assumed the z-direction, the same as the direction of an applied steady field) and the line joining the nucleus and the unpaired electron. Show that this corresponds to $\Delta H = g_n \mu_{B_n} m_I[(3\cos^2\theta - 1)/r^3]$. Discuss the anisotropy expected in the hyperfine splitting.

3-12. For a two-electron-level system the change in the level populations, $n = n_2 - n_1$, is the sum of the change because of (a) absorption of microwave radiation and (b) the spin-lattice interaction; that is,

$$\frac{dn}{dt} = \left(\frac{dn}{dt}\right)_{\text{rf}} + \left(\frac{dn}{dt}\right)_{\text{sl}}.$$

Radiation theory shows that $(dn/dt)_{\text{rf}} = -\frac{1}{4}\gamma^2 H_1^2 \tau_2 n$.
Using the Bloch-like equation

$$\left(\frac{dn}{dt}\right)_{\text{sl}} = \frac{N_0 - n}{\tau_1},$$

where $N_0 = N_2 - N_1$, show that at equilibrium

$$n = N_0(1 + \tfrac{1}{4}\gamma^2 H_1^2 \tau_1 \tau_2)^{-1}.$$

Since the rate of absorption of energy from the microwave field is given by $A = -\hbar\omega(dn/dt)_{\text{rf}}$,
show that

$$\chi'' = \frac{\chi_0 \omega_0 \tau_2}{1 + \frac{1}{4}\gamma^2 H_1^2 \tau_1 \tau_2},$$

where $\chi_0 = \gamma^2 \hbar N_0 / 2\omega_0$ and ω_0 is the resonant frequency. [*Hint.* Use equation 3-4.6.]

Bibliography

Al'tshuler, S. A., and B. M. Kozyrev, *Electron Paramagnetic Resonance* (translated from the Russian), Academic Press, New York (1964).

Ambler, E., and R. P. Hudson, "Magnetic Cooling," *Rept. Progr. Phys.* **18,** 251 (1955).

Bagguley, D. M. S., and J. Owen, "Microwave Properties of Solids," *Rept. Progr. Phys.* **20,** 304 (1957).

Benedek, G., *Magnetic Resonance at High Pressure*, Interscience Publishers, New York (1963).

Bleaney, B., and K. W. H. Stevens, "Paramagnetic Resonance," *Rept. Progr. Phys.* **16,** 304 (1955).

Bowers, K. D., and J. Owen, "Paramagnetic Resonance II," *Rept. Progr. Phys.* **18,** 304 (1955).

Casimir, H. B. G., *Magnetism and Very Low Temperatures*, Cambridge University Press, Cambridge (1940).

Caspers, W. J., *Theory of Spin Relaxation*, Interscience Publishers, New York (1964).

Cooke, A. H., "Paramagnetic Relaxation Effects," *Rept. Progr. Phys.* **13,** 276 (1953).

Garrett, C. G. B., *Magnetic Cooling*, Harvard University Press, Cambridge (1954).

Gorter, C. J., *Paramagnetic Relaxation*, Elsevier Publishing Co. Amsterdam (1947).

Gorter, C. J., Editor, *Progress in Low Temperature Physics*, Interscience, New York (1955).

Gorter, C. J., "Paramagnetic Relaxation," *Nuovo cimento Suppl.* **6,** 887 (1957).

Heine, V., *Group Theory in Quantum Mechanics*, Pergamon Press, New York (1960).

Hellwege, K. H., and A. M. Hellwege, *Magnetic Properties I*, Springer-Verlag, Berlin (1962).

Herzfield, C. M., and P. H. E. Meijer, "Group Theory and Crystal Field Theory," *Solid State Phys.*, **12,** 1 (1961).

Hutchinson, C. A., Jr., "Magnetic Resonance," *Ann. Rev. Phys. Chem.* **7,** 359 (1956); H. M. McConnell, *ibid* **8,** 105 (1957); J. E. Wertz, *ibid.* **9,** 93 (1958).

Ingram, D. J. E., *Spectroscopy at Radio and Microwave Frequencies*, Butterworth Publications, London (1955).

Ingram, D. J. E., *Free Radicals*, Butterworth Publications, London (1958).

Low, W., Editor, *Paramagnetic Resonance*, Vols. 1 and 2, Academic Press, New York (1963).

Low, W., *Paramagnetic Resonance in Solids*, Academic Press, New York (1960).

Kurti, N., "Magnetism at Very Low Temperatures and Nuclear Orientation," *Nuovo Cimento, Suppl.* **6,** 1101 (1957).

Orton, J. W., "Paramagnetic Resonance Data," *Rept. Progr. Phys.* **22,** 204 (1959).

Pake, G. E., *Paramagnetic Resonance*, W. A. Benjamin, New York (1962).

Pryce, M. H. L., "Paramagnetism in Crystals," *Nuovo Cimento, Suppl.* **6,** 817 (1957).

Siegman, A. E., *Microwave and Optical Masers*, McGraw-Hill Book Co., New York (1964).

Singer, J. R., *Masers*, John Wiley and Sons, New York (1959).

Singer, J. R., Editor, *Advances in Quantum Electronics*, Columbia University Press, New York (1961).

Smidt, J., Editor, *Magnetic and Electric Resonance and Relaxation*, North-Holland Publishing Co., Amsterdam (1963).

ter Haar, D., Editor, *Fluctuations, Relaxation and Resonance in Magnetic Systems* Oliver and Boyd, Edinburgh (1963).

van den Handel, J., "Paramagnetism," *Advan. Electron. Electron Phys.* **6**, 463 (1954).
Vuylsteke, A. A., *Elements of Maser Theory*, D. Van Nostrand, Princeton (1960).
Wilson, A. H., *Thermodynamics and Statistical Mechanics*, Cambridge University Press, Cambridge (1957).
Zemansky, M. W., *Heat and Thermodynamics*, 4th ed., McGraw-Hill Book Co. New York (1955).

4

Nuclear Magnetic Resonance

1. Introduction

Nuclear paramagnetic resonance is basically analogous to electron paramagnetic resonance. As a result of the spatial quantization of the nuclear magnetic moments when a static magnetic field H is applied, the $2I + 1$ states are split into levels with energies $m_I g_n \mu_{B_n} H$. Then transitions between the levels can be induced by the application of an alternating field with a frequency given by

$$\hbar\omega = -g_n \mu_{Bn} H$$

or

$$\omega = \frac{-g_n e}{2M_p c} H. \tag{4-1.1}$$

Equation 4-1.1 is similar to equation 3-7.1 for the electron, except that the electronic moment, g-value, and mass are replaced by the corresponding nuclear quantities. Because the protonic mass is about 1830 times greater than the electronic mass, the nuclear resonance frequency occurs at a much lower frequency than that of the electronic resonance. Numerically the resonance condition is

$$\nu = 2.13 g_n H, \tag{4-1.2}$$

where ν is in kilocycles and H in oersteds. In other words, for fields in the kilo-oersted region, resonance absorption occurs at radio frequencies (rf). As a result, the nuclear resonance apparatus is much simpler, since lumped parameter circuit methods may be employed.

Nuclear resonance was observed first in 1945 by Bloch et al.[1] and by Purcell et al.,[2] although Gorter had attempted earlier experiments unsuccessfully. The method of detecting the resonance was different for these two experiments. A schematic diagram of Bloch's apparatus is shown in Fig. 4-1.1. The alternating magnetic field $2H_1 \cos \omega t$ is applied to the specimen at right angles to the steady field (the z-direction) H via coil 1.

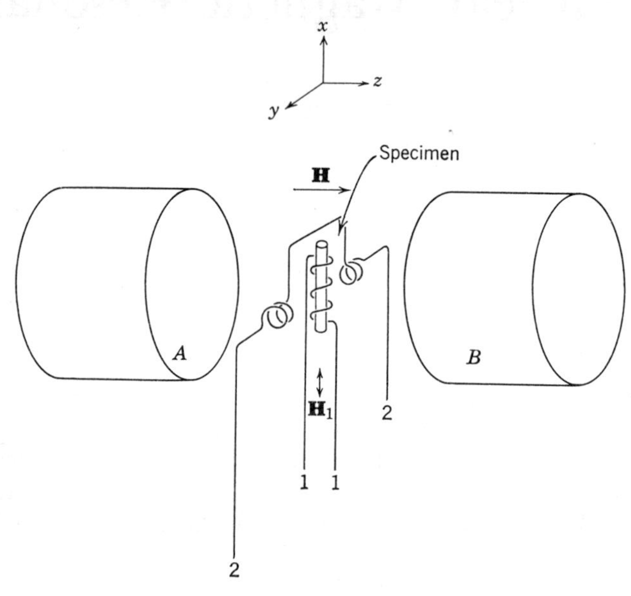

Fig. 4-1.1. Illustrating nuclear induction resonance apparatus. A–B are the pole pieces of the electromagnet. The rf is applied via coil 1 (the transmitter), and the induced signal is picked up in coil 2 (the receiver). For so-called nuclear absorption coil 2 is missing and the signal is detected in coil 1.

The solutions of the equation of motion (3-7.3, etc.) show that there is a component of the magnetization rotating in the x, y plane, given by equation 3-7.6

$$M_{\pm} = \frac{\gamma H_1 M_z}{\gamma H + \omega} e^{\pm i\omega t}$$

or by equation 3-8.3 if damping is included. As a result, a voltage will be induced in coil 2, which has its axis along the y-direction, given by

$$\mathcal{E} = -\frac{4\pi A}{c} \dot{M}_y,$$

where A is the effective cross-sectional area of the coil, and $\dot{M}_y$ is the rate of

[1] F. Bloch, W. W. Hansen, and M. E. Packard, *Phys. Rev.* **69**, 127 (1946).
[2] E. M. Purcell, H. C. Torrey, and R. V. Pound, *Phys. Rev.* **69**, 37 (1946).

change of the y-component of the magnetization. At resonance $M_{\pm}$, M_y, hence $\mathcal{E}$, have their maximum value. Placing the axes of coils 1 and 2 at right angles has the advantage that no voltage will be directly induced in coil 2 because of the alternating current in coil 1. This method of detection is known as *nuclear induction.*

It is not necessary that the two coils be perpendicular; a signal will be induced with the axis of coil 2 at any angle to coil 1. Indeed the axes of the two coils could lie along the same direction; in this case they could be the same coil. This is the arrangement Purcell used; the term nuclear magnetic resonance usually implies use of this experimental setup. The nuclear resonance may be studied by employing bridge circuits,[3] marginal oscillators,[4] super regeneration,[5] and other methods.

The solution of the equation of motion for the nuclear dipoles (equation 3-7.3)

$$\frac{d\mathbf{M}}{dt} = \gamma \mathbf{M} \times \mathbf{H}$$

is identical with that for the electron, given in Section 3-7, except that nuclear quantities are used for $\gamma(=g_n e/2M_p c)$. The same remarks hold for the Bloch equations (3-8.1), in which the damping effects are taken into account by adding terms involving the nuclear spin-lattice and spin-spin relaxation times τ_1 and τ_2. Indeed, these equations and solutions were employed originally by Bloch for the nuclear resonance and not for the electron resonance. The line shapes of the Bloch susceptibilities (equation 3-8.10), usually called Lorentzian, are observed to fit the results for liquids better than for solids. The reason for this behavior lies in the details of the mechanisms involved in the nuclear spin-lattice and spin-spin interactions. These mechanisms are somewhat different from the electron case and are discussed in later sections.

There is little point in repeating in detail the solutions of Sections 3-7 and 3-8. Instead we shall treat the phenomenon of resonance from a different point of view by using the method of rotating coordinates.[6] Besides giving some additional insight into the earlier equations of magnetic resonance, the method is useful for the discussion of transient effects, to be treated later in this chapter.

In a fixed coordinate system the equation of motion of a nuclear moment $\boldsymbol{\mu}$ with angular momentum $\mathbf{G}$ in a field $\mathbf{H}$ is given by

$$\frac{d\mathbf{G}}{dt} = \boldsymbol{\mu} \times \mathbf{H}. \qquad (4\text{-}1.3)$$

[3] N. Bloembergen, E. M. Purcell, and R. V. Pound, *Phys. Rev.* **73,** 679 (1948).
[4] R. V. Pound, *Prog. Nuc. Phys.* **2,** 21 (1952).
[5] A. Roberts, *Rev. Sci. Instr.* **18,** 845 (1947).
[6] I. I. Rabi, N. F. Ramsey, and J. Schwinger, *Revs. Mod. Phys.* **26,** 167 (1954).

Consider now a second coordinate system, rotating about the z-axis with an angular velocity $\boldsymbol{\omega}$. Then the time rate of change of $\mathbf{G}$ in the rotating system, denoted by $(d\mathbf{G}/dt)^*$, is related to $d\mathbf{G}/dt$ by[7]

$$\frac{d\mathbf{G}}{dt} = \left(\frac{d\mathbf{G}}{dt}\right)^* + \boldsymbol{\omega} \times \mathbf{G}. \tag{4-1.4}$$

These two results give

$$\left(\frac{d\mathbf{G}}{dt}\right)^* = \boldsymbol{\mu} \times \mathbf{H} - \boldsymbol{\omega} \times \mathbf{G}.$$

Using $\boldsymbol{\mu} = \gamma \mathbf{G}$, we obtain

$$\begin{aligned}\left(\frac{d\mathbf{G}}{dt}\right)^* &= \boldsymbol{\mu} \times \left(\mathbf{H} + \frac{\boldsymbol{\omega}}{\gamma}\right) \qquad (4\text{-}1.5)\\ &= \boldsymbol{\mu} \times \mathbf{H}_e,\end{aligned}$$

where

$$\mathbf{H}_e = \mathbf{H} + \frac{\boldsymbol{\omega}}{\gamma}. \tag{4-1.6}$$

Hence the rotation of G in the rotating system can be considered as equivalent to the rotation in a stationary system with an effective field given by equation 4-1.6. When the only applied field is a steady one $\mathbf{H}$ along the z-direction and, in addition, when $\boldsymbol{\omega} = -\gamma\mathbf{H}$, the effective field $\mathbf{H}_e$ is zero. Then, since there is no torque on $\boldsymbol{\mu}$ in the rotating frame, the magnetic moment will remain fixed with time in this system.

Next consider the application of a field of magnitude H_1, lying along the x^*-axis and rotating about the z- (or z^*-) axis with frequency ω, in addition to the steady field $\mathbf{H}$. The new equation of motion is

$$\begin{aligned}\left(\frac{d\mathbf{G}}{dt}\right)^* &= \gamma\mathbf{G} \times (\mathbf{H}_{ez} + \mathbf{H}_1),\\ &= \gamma\mathbf{G} \times \mathbf{H}_e\end{aligned}$$

where the effective field has the components

$$\begin{aligned} H_{ez^*} &= H + \frac{\omega}{\gamma}\\ H_{ex^*} &= H_1, \qquad (4\text{-}1.7)\\ \text{and}\quad H_{ey^*} &= 0,\end{aligned}$$

[7] Consider any vector $\mathbf{A}$ that is stationary in the rotating coordinate system. The change of $\mathbf{A}$, $\delta\mathbf{A}$, with respect to the fixed coordinate system, in a rotation through an angle $\delta\mathbf{n} = \delta t\boldsymbol{\omega}$ is $\delta\mathbf{A} = \delta t\boldsymbol{\omega} \times \mathbf{A}$. It follows immediately that $\delta\mathbf{A}/\delta t = \boldsymbol{\omega} \times \mathbf{A}$. See, for example, J. L. Synge and B. A. Griffith, *Principles of Mechanics*, 2nd ed., McGraw-Hill Book Co., New York (1945), p. 347.

and is indicated in Fig. 4-1.2. The effective field is constant, with the magnitude

$$H_e = \left[\left(H + \frac{\omega}{\gamma}\right)^2 + H_1^{\,2}\right]^{1/2} \tag{4-1.8}$$

and makes an angle θ with the z-direction given by

$$\tan\theta = \frac{H_1}{H + \omega/\gamma}. \tag{4-1.9}$$

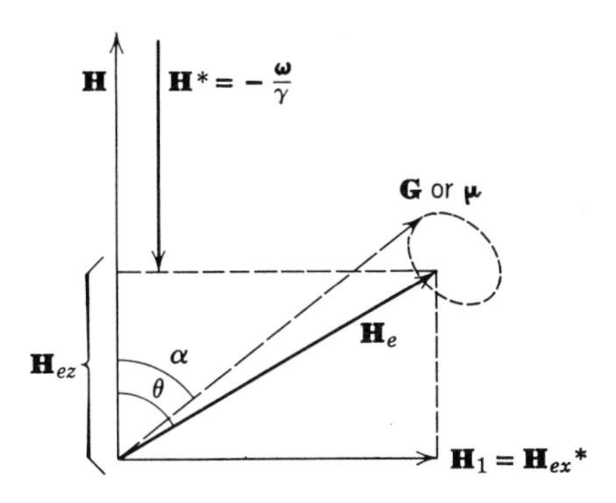

Fig. 4-1.2. Precession of the magnetic moment in the rotating coordinate system.

If the angular momentum vector **G** or the dipole moment $\boldsymbol{\mu} = \gamma\mathbf{G}$ makes some angle, say α, with the z-direction, it will precess about the effective field with an angular frequency of $\Omega = -\gamma H_e$ in the rotating system. This motion may be described with the aid of a third frame of reference (double-starred system) rotating about the axis of H_e in the single-starred frame with a frequency Ω so that $\mathbf{H}_e + \boldsymbol{\Omega}/\gamma = 0$. There will be zero torque on $\boldsymbol{\mu}$ in this third frame, and consequently the moment will retain its initial orientation in the double-starred frame.

If the single-starred system is rotating at the Larmor frequency $\boldsymbol{\omega} = -\gamma\mathbf{H}$, the effective field $\mathbf{H}_e$ equals $\mathbf{H}_1$. Suppose a dipole $\boldsymbol{\mu}$ is inserted into the system along the z-direction ($\alpha = 0$). The moment will oscillate between $+z$ and $-z$, in precessing about the H_1- axis, at a rate $\Omega_r = -\gamma H_1$; this is the phenomena of magnetic resonance.

It is not, however, the usual experimental situation in magnetic resonance. Normally we start with a steady field H somewhat below (or above) the resonance value and then change the field gradually through the resonance. The solutions of the equation of motion given in Chapter 3 (with or without damping) are for steady-state conditions. These solutions will describe the situation only if the steady field is changed under the

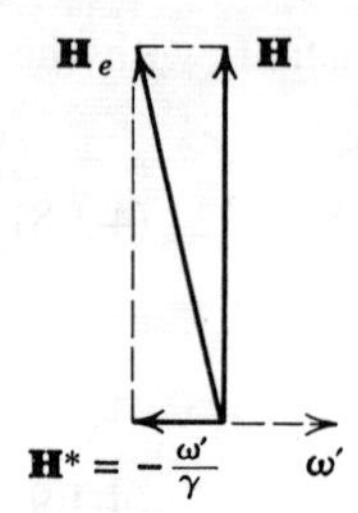

Fig. 4-1.3. The effective field $\mathbf{H}_e$ when $\mathbf{H}$ is rotated with the angular frequency $\boldsymbol{\omega}'$.

condition necessary for the maintenance of the steady state. Maintenance of the steady state is achieved if the angle the magnetic dipole moment makes with the effective field is preserved as the steady field is changed. Stated alternatively in quantum mechanical terms, the magnetic quantum number is conserved; this is known as the *adiabatic* condition, a term first introduced by Ehrenfest in another connection.

Use of rotating coordinates permits a simple derivation of the adiabatic criterion. Let only a steady field **H** be applied to a dipole of moment **M**. At equilibrium **M** will lie along **H**. Now let **H** be varied by causing it to rotate about a perpendicular axis with an angular velocity $\boldsymbol{\omega}'$. In the rotating system the effective field will be

$$\mathbf{H}_e = +\frac{\boldsymbol{\omega}'}{\gamma} + \mathbf{H}$$

as indicated in Fig. 4-1.3. The dipole moments will preserve their orientation with respect to **H**, provided that $\mathbf{H}_e$ remains approximately equal to **H**, that is if

$$H^* \ll H$$

or

$$|\omega'| \ll \gamma H. \tag{4-1.10}$$

Further, since

$$\omega' = \frac{d\theta}{dt} = \frac{1}{H}\frac{dH}{dt},$$

we obtain

$$\frac{dH}{dt} \ll \gamma H^2 \tag{4-1.11}$$

as the criterion for the rate at which H can be changed, for adiabatic conditions. If, in addition, the change of H with time is slow compared to the relaxation times τ_1 and τ_2, the condition is known as *adiabatic slow passage*.

As an example, let us consider the solution to the equation of motion given by equation 3-3.7,

$$\left|\frac{M_\pm}{M_z}\right| = \tan\theta = \frac{H_1}{H + \omega/\gamma}$$

on the rotating coordinate basis. The angle θ is also the angle the effective field H_e makes with the z-axis. Hence $\mathbf{M}$ is constant in the rotating system and parallel to the effective field $\mathbf{H}_e$. Suppose that initially $H \gg H^*$ $(= -\omega/\gamma)$, then $\mathbf{M}$ would be almost parallel to $\mathbf{H}$. Now if H is reduced through resonance slowly enough that the adiabatic condition is fulfilled, $\mathbf{M}$ and also $\mathbf{H}_e$ will turn from parallel to antiparallel to $\mathbf{H}$. This particular problem neglects the relaxation times, which would be the case if both τ_1 and τ_2 were very large, so that $1/\tau_1 \to 0$ and $1/\tau_2 \to 0$. Although this condition is probably never satisfied in nature in general, it may be satisfied temporarily if the change of H with time is rapid compared to τ_1 and τ_2. If, in addition, the passage is slow enough to be adiabatic, the relationship between $\mathbf{M}$ and $\mathbf{H}_e$ of this problem will be achieved temporarily. This sweep condition is called *adiabatic rapid passage*; it is important for the creation of inverted populations of levels, that is, negative temperatures, a process discussed in Section 4-7.

Let us consider equation 3-8.8, obtained when damping is included, also on a rotating coordinate basis. Here we have

$$\left|\frac{M_\pm}{M_z}\right| = \tan\theta$$

$$= \frac{H_1}{[1/(\gamma\tau_2)^2 + (H + \omega/\gamma)^2]^{1/2}}\,.$$

It is easily seen that $\mathbf{M}$ no longer lies in the $H - H_1$ plane, but lags by an angle ϵ given by

$$\tan\epsilon = \frac{-1/\gamma\tau_2}{H + \omega/\gamma}\,. \qquad (4\text{-}1.12)$$

For adiabatic slow passage conditions, M always has a component along the direction of H, since $M_\pm/M_z > 0$ always, that is, the level populations are never inverted. The situation in which slow passage conditions are not met is discussed later in this chapter (Section 4-6).

One of the most important applications of nuclear resonance has been the accurate determination of the magnetic moments of nuclei. From equation 4-1.1

$$\hbar\omega = -g_n \mu_{B_n} H$$

it is obvious that location of the resonance frequency in a certain field H yields the g_n-value for the moment. We then obtain the maximum value of the component of the magnetic moment from the relation $\mu_z = g_n \mu_{B_n} I$, provided, of course, that the spin quantum number I is known. The magnetic field H, which is very difficult to measure to better than one part in 10^3, sets the limit to the accuracy with which the magnetic moment can be determined absolutely. The frequency, however, can be determined to better than one part in 10^6. Consequently, most measurements determine the ratio of two moments in the same field from the relation

$$\frac{(\omega_0)_a}{(\omega_0)_b} = \frac{(\mu_z/I)_a}{(\mu_z/I)_b}. \tag{4-1.13}$$

The proton is frequently used as the reference nucleus, partly because the proton's magnetic moment has been determined accurately in several precision experiments. In one of these experiments the magnetic field is measured by accurately determining the force on current carrying conductors.[8] In two others the moment is obtained by comparing the magnetic resonance frequency with the cyclotron resonance of either the proton[9] or an electron.[10]

Even though the applied field is the same, the actual field at the two nuclei in question may be different; if so, certain corrections must be made to the frequencies before insertion into equation 4-1.13. These corrections are as follows.

1. If two specimens are employed, one for each type of nucleus, the diamagnetic susceptibility will be different for each. The reduction in the field at the molecule or ion may be calculated, provided χ is known. If one specimen containing a mixture of the two nuclei is employed, the reduction in the field is the same and the frequency ratio is not affected.

2. If paramagnetic ions are present, a similar correction should be made. Experimentally, it turns out that a mixed specimen does not eliminate this correction[11]; it is best to avoid specimens that contain paramagnetic ions.

3. The orbital electrons of an ion or molecule produce a diamagnetic screening effect on the field at the nucleus (see problem 4–1). An approximate calculation can be made for monatomic atoms or ions by using the Thomas-Fermi statistical model of the atom.[12] For complex ions or molecules a second-order paramagnetic term enters as well as the diamagnetic.[13] This paramagnetic term has been calculated only for molecular hydrogen. The term plays an important part in the so-called chemical shift of the nuclear resonance frequency, as discussed later.

[8] H. A. Thomas, R. L. Driscoll, and J. A. Hipple, *Phys. Rev.* **78,** 787 (1950).

[9] H. Sommer, H. A. Thomas, and J. A. Hipple, *Phys. Rev.* **82,** 697 (1951); C. D. Jeffries, *Phys. Rev.* **81,** 1040 (1951).

[10] J. H. Gardner, *Phys. Rev.* **83,** 996 (1951).

[11] W. C. Dickinson, *Phys. Rev.* **81,** 717 (1951).

[12] W. E. Lamb, *Phys. Rev.* **60,** 817 (1941); W. G. Proctor and F. C. Yu, *Phys. Rev.* **81,** 20 (1951).

[13] N. F. Ramsey, *Phys. Rev.* **77,** 567 (1950), **86,** 243 (1952), *Physica* **17,** 303 (1951).

The nuclear moment may be positive or negative, depending on the sign of g_n. The sign determines the sense of precession of the moment. It may be determined by using the nuclear induction technique by comparing the phase of the induced signal to the applied radio frequency, which also induces a signal in the receiving coil because of flux leakage.

Table 4-1.1. Nuclear Magnetic Moments, Spins, and Quadrupole Moments[a]

Isotope	Magnetic Moment (in units of μ_{B_n})	Spin Quantum Number I	Electric Quadrupole Moment (if any) ($\times 10^{-24}$ cm^2)
n^1	-1.9130	$\frac{1}{2}$	
H^1	2.79270	$\frac{1}{2}$	
H^2	0.85738	1	2.77×10^{-3}
He^3	-2.1274	$\frac{1}{2}$	
Li^7	3.2560	$\frac{3}{2}$	-4.2×10^{-2}
Be^9	-1.1774	$\frac{3}{2}$	2×10^{-2}
B^{10}	1.8006	3	0.111
B^{11}	2.6880	$\frac{3}{2}$	3.55×10^{-2}
C^{13}	0.70216	$\frac{1}{2}$	
N^{14}	0.40357	1	2×10^{-2}
F^{19}	2.6273	$\frac{1}{2}$	
Na^{23}	2.2161	$\frac{3}{2}$	0.1
Cl^{35}	0.81089	$\frac{3}{2}$	-7.97×10^{-2}
Cu^{63}	2.2206	$\frac{3}{2}$	-0.15
Ba^{137}	0.936	$\frac{3}{2}$	

[a] Tabulated by Varian Associates, Palo Alto, Calif.

When the spin is unknown, only the g-value can be determined. There are ways in which nuclear resonance can be used to determine the spin. For example, if the nucleus possesses a quadrupole moment and interacts with a static electric field gradient, the nuclear levels are displaced unequally. Since the energy level intervals are now unequal, $2I + 1$ levels will produce $2I$ individual absorption lines; hence I is obtained by simply counting the number of lines. The spin can also be determined by methods which (a) compare the maximum amplitudes or the areas under the curves of the resonance lines of two nuclei,[14] (b) consider the resonance line width,[15] or (c) employ the fine structure.[16]

Table 4-1.1 lists the magnetic moments and spins of a few representative nuclei. Mention should be made of the application of nuclear resonance

[14] G. D. Watkins and R. V. Pound, *Phys. Rev.* **89**, 658 (1953).
[15] H. S. Gutowsky, R. E. McClure, and C. J. Hoffman, *Phys. Rev.* **81**, 635 (1951).
[16] R. A. Ogg and J. D. Ray, *J. Chem. Phys.* **22**, 147 (1954).

techniques to the measurement of steady magnetic fields. Protons are used as a rule, since their magnetic moments are known with great accuracy. The resonance frequency is measured and the field H is determined with the aid of equation 4-1.1. In addition, the field of electromagnets may be stabilized by utilizing the phenomena of magnetic resonance.

2. Line Shapes and Widths

Examination of the shape and width of magnetic resonance lines yields much information concerning certain magnetic effects; for example, relaxation phenomena. In this section the main models and methods that are employed for the analysis of line shapes are outlined. This section is included in Chapter 4 rather than Chapter 3 because to date there have been more investigations of line shapes for nuclear magnetic systems. However, important work has been carried out for electron magnetic systems, and the subject is likely to receive more attention in the future. In any case, the contents of this section refer to electron as well as nuclear systems.

The shape of the magnetic absorption line, as a function of frequency ω in a fixed steady field H, is denoted by the normalized function $g(\omega)$, so that

$$\int_0^\infty g(\omega)\, d\omega = 1. \tag{4-2.1}$$

It is now necessary to discuss the expressions that have been employed for $g(\omega)$.

Lorentz considered an ensemble of harmonic oscillators interacting by means of collisions of very short duration. At each impact he assumed there was a sudden change of phase but that immediately after a collision one phase was as likely as another. The collision-broadened line shape given by this model is called Lorentzian[17]; it has the same mathematical form as that given by the Bloch equations with no saturation (equation 3-8.10). In order to satisfy the normalization condition, the line-shape function for the Bloch susceptibilities is defined as

$$g(\omega) = \frac{\tau_2/\pi}{1 + (\omega - \omega_L)^2\tau_2{}^2}. \tag{4-2.2}$$

At the maximum of the curve we have

$$g(\omega)_{\max} = \frac{\tau_2}{\pi}, \tag{4-2.3}$$

[17] H. E. White, *Introduction to Atomic Spectra*, McGraw-Hill Book Co., New York (1934), Ch. 21; also, F. G. Breene, Jr., *Revs. Mod. Phys.* **29**, 94 (1957).

whereas the half-width is

$$\Delta\omega = \frac{1}{\tau_2} \tag{4-2.4}$$

and the width at the point of maximum slope is $1/\sqrt{3}\tau_2$. It is not so strange that a line shape determined by collisions does not fit the results for solids too well, since here the atoms are essentially fixed in space. The model may be expected to be a better fit for gases and liquids.

For solids, with either nuclear or electronic moments, a given atom experiences a local magnetic field produced by the moments of many surrounding atoms. This field is quasi-static, in contrast to the sudden impact type of interaction in the collision theory. The field is the origin of the spin-spin relaxation for electron systems, as discussed in Section 3-6. The line shape arises because the different atoms experience a different field. Because the total field at an atom is the sum of a large number of small fields, it is reasonable to assume that the distribution of resultant fields will be approximately Gaussian. Then the normalized line shape can be written as

$$g(\omega) = \frac{\tau_2}{\pi} \exp\left[-(\omega - \omega_L)^2 \frac{\tau_2{}^2}{\pi}\right] \tag{4-2.5}$$

so that, as for Lorentzian lines,

$$g(\omega)_{\max} = \frac{\tau_2}{\pi} \tag{4-2.6}$$

and the half-width is

$$\Delta\omega = \frac{1}{\tau_2}(\pi \log 2)^{1/2}. \tag{4-2.7}$$

Clearly, for absorption lines of similar shape the value of $g(\omega)_{\max}$ is larger for narrow lines and small for broad lines. Hence $1/g(\omega)_{\max}$ is a measure of the line width as well as of τ_2, as shown by equations 4-2.3 and 4-2.6.

The calculation of the line shape for systems in which many nuclear moments significantly contribute to the local field is difficult and often not possible. It turns out, however, that the moments of the spectrum can be calculated fairly easily. The nth moment is defined as

$$\overline{(\omega - \omega_L)^n} = \overline{(\Delta\omega)^n} = \int_{-\infty}^{\infty} (\omega - \omega_L)^2\, g(\omega)\, d\omega. \tag{4-2.8}$$

Since $g(\omega)$ is an even function for line broadening caused by magnetic dipoles, the odd moments are zero. Usually interest is centered on the second moment. The quantum mechanical formula for the second moment is

$$\overline{(\Delta\omega)^2} = \frac{\sum_{nn'} \omega_{nn'}^2 \, |\langle n \, |I| \, n' \rangle|^2}{\sum_{nn'} |\langle n \, |I| \, n' \rangle|^2} - \omega_L^2.$$

The result of the calculation[18] for a single crystal, expressed in the corresponding magnetic field units, is

$$\overline{(\Delta H)^2} = \frac{\frac{3}{2} I(I+1) g_n^2 \mu_{B_n}^2}{N} \sum_{i>k} \left(\frac{3 \cos^2 \theta_{ik} - 1}{r_{ik}^3} \right)^2, \tag{4-2.9}$$

where N is the total number of nuclei that make important contributions to the local field at a nucleus, r_{ik} is the length of the vector from nucleus i to nucleus k, and θ_{ik} is the angle between r_{ik} and the applied static magnetic field H. The factor in the sum is just the angular and radial dependence in a dipole-dipole interaction (see problem 3-11). Hence, if the location of the dipoles in the lattice is known, $\overline{(\Delta H)^2}$ can be calculated; conversely if the second moment is determined from the spectrum, some information on the location of the nuclei can be obtained. For example, the bond length for N—H in the NH_4^+ ion has been measured with this method.[19] Equation 4-2.9 has been verified experimentally[20] with a single crystal of CaF_2.

If a polycrystalline sample is employed and the grains can be assumed to be oriented isotropically, the average of equation 4-2.9 gives

$$\overline{(\Delta H)^2} = \tfrac{6}{5} I(I+1) g_n^2 \mu_{B_n}^2 \sum_{i>k} \left(\frac{1}{r_{ik}^6} \right).$$

If there is more than one type of nuclear moment present in the crystal, an additional term

$$\frac{\mu_{B_n}^2}{3N} \sum_{jf} I_f(I_f+1) g_{nf}^2 \frac{(3 \cos \theta_{jf} - 1)^2}{r_{jf}^6} \tag{4-2.10}$$

must be added, where I_f and g_{nf} describe the other nuclear species. The numerical coefficient in (4-2.9) is $\frac{9}{2}$ times greater than in equation 4-2.10. A trivial factor of 2 occurs because in equation 4-2.9 each pair of nuclei is counted only once. The real difference is a factor $\frac{9}{4} = (\frac{3}{2})^2$. This factor arises because of the spin exchange process, discussed in Section 3-6, which can occur only for identical nuclei.

[18] J. H. Van Vleck, *Phys. Rev.* **74,** 1168 (1948).

[19] H. S. Gutowsky and G. E. Pake, *J. Chem. Phys.* **16,** 1164 (1948).

[20] G. E. Pake and E. M. Purcell, *Phys. Rev.* **74,** 1184 (1948).

Finally, it is interesting to note that the second moment for the Lorentzian line shape (equation 4-2.2) is infinite. This result occurs because of the very slow convergence of the edges of the curve. The second moments for experimentally observed spectra are always finite.

3. Resonance in Nonmetallic Solids

Only nonmetallic solids whose atoms are rigidly fixed in the lattice, except, of course, for thermal lattice vibrations, are considered in this section. Solids within which the atoms (or molecules) have considerable freedom either to rotate or to diffuse within the crystal are considered in the next section, together with liquids. For the solids we are concerned with here the spin-lattice relaxation time is generally several orders of magnitude larger than the spin-spin relaxation time.

The field at a nuclear moment differs from the applied steady field because of the field produced by a neighboring dipole by the amount (equation 1-3.2)

$$-\frac{\boldsymbol{\mu}}{r^3} + \frac{3(\boldsymbol{\mu} \cdot \mathbf{r})\mathbf{r}}{r^5}.$$

The component of this local field parallel to H is

$$\pm \frac{\mu_z(3\cos^2\theta - 1)}{r^3},$$

where μ_z is the component of the moment of the dipole in the field direction, r is the separation of the dipoles, and θ is the angle between **H** and **r**. The occurrence of the plus or minus sign depends on whether the dipole in question is parallel or antiparallel to **H**, respectively. Now consider a solid in which the nuclear dipoles occur in pairs; that is, two dipoles are much closer than the next nearest dipoles. The field at a given dipole, then, is essentially just

$$H_{\text{eff}} = H \pm \frac{\mu_z(3\cos^2\theta - 1)}{r^3} \qquad (4\text{-}3.1)$$

and in a nuclear resonance experiment we would expect to find two resonance lines, symmetrically spaced. In gypsum, $CaSO_4 2H_2O$, for example, this situation exists for pairs of protons, with the additional complication that there are two types of proton pairs, with different values of θ. In addition, these lines are broadened because of the smaller local fields produced by the neighbors farther away. Finally, there is another complication because of the process of spin exchange. As a result it can be shown[21] that the local field is greater by a factor of $\frac{3}{2}$, so that equation 4-3.1 becomes

$$H_{\text{eff}} = H \pm \tfrac{3}{2}\,\frac{\mu_z(3\cos^2\theta - 1)}{r^3}. \qquad (4\text{-}3.2)$$

The actual spectrum observed for a single crystal depends on the orientation of the crystal with the field H. The data obtained gives information on the angular disposition of the proton pairs, which may be compared with X-ray data.

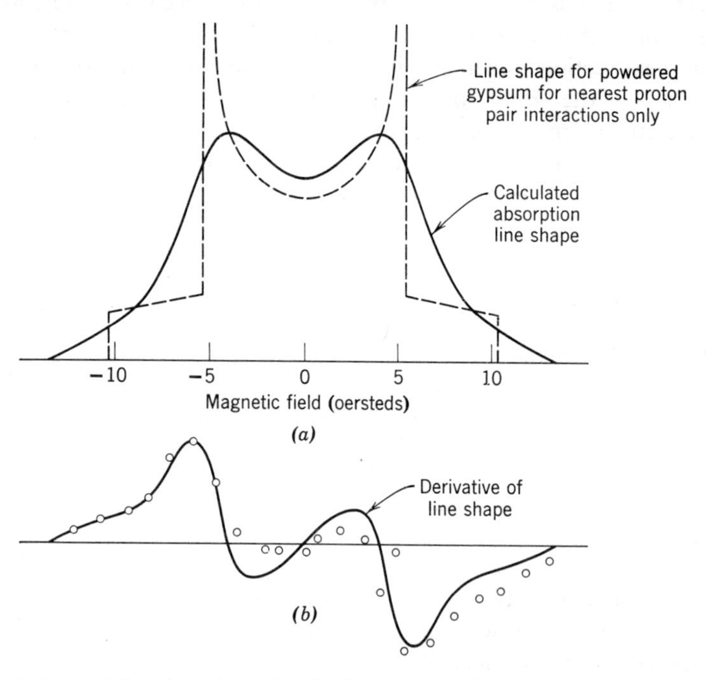

Fig. 4-3.1. Calculated line shape for a powder in which the dipoles are located in close pairs. In (*a*) the calculated line shape is shown and in (*b*) the derivative of this line shape, together with experimental points, is given. (After G. E. Pake.)

For a powdered specimen, if the crystal grains are distributed at random, we must average over the spectra for different orientations. The result for the proton pairs of gypsum is indicated by the broken line of Fig. 4-3.1*a*. When the line broadening caused by all the other neighbors is considered, the full curve of this figure is obtained. The derivative of this line shape, together with some experimental points is shown in Fig. 4-3.1*b*. Theory and experiment are in good agreement.[21]

[21] G. E. Pake, *J. Chem. Phys.* **16,** 327 (1948).

Several other hydrated salts have been studied.[22] Nuclear moments grouped in clusters of three located at the vertices of an equilateral triangle have also been studied theoretically and experimentally.[23] Groups of four identical nuclei located at the vertices of a tetrahedron have been treated theoretically,[24] but, in general, as more nuclei are involved the calculation becomes very difficult. For these cases it is usual to employ the second and fourth moment method, defined and discussed in the preceding section.[25]

We will now consider the mechanism of the spin-lattice relaxation in these solids. Because of the lattice vibrations, the local magnetic fields at a nucleus will vary simply because the internuclear spacing is changing. This is just the mechanism Waller considered for electron spins (see Section 3-5). The interaction may occur by the same two processes here, namely either by direct interaction with a phonon with the nuclear magnetic resonance frequency or via a Raman-like process. This mechanism gives a value of $\tau_1 \approx 10^4$ sec at room temperature; experimentally τ_1 is at most 10^3 sec, and usually it is much shorter. At lower temperatures the difference between the predicted value of τ_1 and the observed value is even greater, being several orders of magnitude different. It is believed that the discrepancy arises because of the presence of minute amounts of paramagnetic impurities (ions or *F*-centers) in any actual solid.[26] The excess energy in the spin system is considered to diffuse toward the paramagnetic magnetic moment by the process of spin-exchange and then to be transferred to the lattice via the fluctuating field of the electronic moment. The energy exchange with the lattice then depends on the paramagnetic spin-lattice relaxation time, which is very short. For concentrations of paramagnetic ions greater than 1% no diffusion of the spin energy by spin exchange is necessary. The electronic energy levels may then be broadened sufficiently so that transfer of the quanta from the nuclear dipoles requires only a reorientation of the electronic moments. Then τ_1 is expected to be independent of temperature, a result that has been confirmed experimentally by Bloembergen.[26]

[22] H. S. Gutowsky, G. B. Kistiakowsky, G. E. Pake, and E. M. Purcell, *J. Chem. Phys.* **17,** 972 (1949); J. Itoh, R. Kusaka, Y. Yamagata, R. Kiriyama, and H. Ibamota, *J. Phys. Soc. Japan* **8,** 293 (1953).

[23] E. R. Andrew and R. Bersohn, *J. Chem. Phys.* **18,** 159 (1953); R. E. Richards and J. A. S. Smith, *Trans. Faraday Soc.* **47,** 1261 (1951).

[24] K. Tomita, *Phys. Rev.* **89,** 429 (1953); R. Bersohn and H. S. Gutowsky, *J. Chem. Phys.* **22,** 651 (1954).

[25] See, for example, J. S. Waugh, F. B. Humphrey, and D. M. Yost, *J. Phys. Chem.* **57,** 486 (1953); E. R. Andrew and D. Hyndman, *Proc. Phys. Soc.* (*London*) **A-66,** 1187 (1953); E. R. Andrew and R. G. Eades, *Proc. Roy Soc.* (*London*) **A-216,** 398 (1953).

[26] B. V. Rollin and J. Hatton, *Phys. Rev.* **74,** 346 (1948); N. Bloembergen, *Physica* **15,** 386 (1949).

4. The Influence of Nuclear Motion on Line Widths and Relaxations

It is found experimentally that the resonance line widths for liquids and gases are usually much narrower than for solids. It turns out that this narrowing is the result of the rotational and translational motion of the molecules of the liquid and gas. The effect is observed for paramagnetic as well as nuclear systems, as, for example, for solutions of free radicals discussed in Section 3-11.

The explanation of *motional narrowing* may be found by considering the angular and radial dependence of the field produced at a dipole i by another dipole j, given by expressions of the type

$$\left(\frac{3\cos^2\theta_{ij} - 1}{r_{ij}^{\ 3}}\right). \tag{4-4.1}$$

For a rigid crystal lattice this expression, which occurs in equations 4-2.9, 4-2.10, and 4-3.2, for example, is just a constant depending on geometry. However, when there is motion of the dipoles, θ_{ij} and r_{ij} are random functions of time. If θ_{ij} and r_{ij} are changing sufficiently rapidly with time, the average of expression 4-4.1 will tend to zero; this is essentially the origin of the narrowing observed.

It is necessary to consider how fast these fluctuations must occur for line narrowing, and for this purpose a correlation time τ_c is introduced.[27] This is approximately the time a molecule takes to rotate through a radian or to move a distance of the order of its dimensions. For times much longer than τ_c, the value of the local field is considered to be completely uncorrelated. It is usual in statistical theory to introduce the correlation time by defining a correlation function $g(\tau)$ by

$$g(\tau) = \overline{H_1(t)\, H_1(t - \tau)},$$

where $H_1(t)$ is the local field at the time (t). Since $g(\tau) \ll 1$ for $t \gg \tau_c$, a suitable simple function to assume for $g(\tau)$ would be $g(\tau) = g(0)e^{-\tau/\tau_c}$. This equation defines τ_c.

Consider a rigid lattice of dipoles and let the line width be $\Delta\omega = 1/\tau_2^{\ 0}$, where $\tau_2^{\ 0}$, the spin-spin relaxation time, is a measure of the time the precessing dipoles take to lose their initial phase relationship. Next, suppose that motion of the dipoles throughout the lattice occurs. A

[27] N. Bloembergen, E. M. Purcell, and R. V. Pound, *Phys. Rev.* **73,** 679 (1948); R. Kubo and K. Tomita, *J. Phys. Soc. Japan* **9,** 888 (1954); R. K. Wangness and F. Bloch, *Phys. Rev.* **89,** 728 (1953); F. Bloch, *Phys. Rev.* **102,** 104 (1956); A. G. Redfield, *IBM J. Res. Dev.* **1,** 19 (1957).

particular dipole will see a number of different lattice sites and therefore will sometimes precess more rapidly and at other times more slowly. If the motion is rapid enough so that a dipole sees a number of sites in the time τ_2^0, its phase difference with the other nuclei will be reduced and narrowing will occur. Hence narrowing occurs when $\tau_c \approx \tau_2^0$ and will increase as τ_c approaches zero.

The dependence of τ_2 on τ_c for $\tau_c \ll \tau_2$ may be obtained from the following simple argument. Suppose that the frequency changes from the mean value by $\pm\Delta\omega$ for a time τ_c. During this time the phase is shifted by an angle $\pm\tau_c\,\Delta\omega$, and for n such time intervals the mean square phase displacement will be $\overline{(\Delta\phi)^2} = n(\tau_c\,\Delta\omega)^2$. The initial phase relationship may be considered lost when $\overline{(\Delta\phi)^2} \approx 1$, and this is the time $\tau_2 = n\tau_c$. Then $1 \approx (\tau_2/\tau_c)(\tau_c\,\Delta\omega)^2$ or

$$\frac{1}{\tau_2} = \tau_c(\Delta\omega)^2, \tag{4-4.2}$$

where this equation is valid only in the region $\tau_c(\Delta\omega) < 1$, that is, $\tau_c < \tau_2^0$, in which τ_2^0 is the spin-spin relaxation time for the rigid lattice.

The problem may be treated quantitatively by considering the time average of equation 4-4.1 expanded as a Fourier integral, namely

$$\overline{\sum_j \left| \frac{3\cos^2\theta_{ij}(t) - 1}{r_{ij}(t)^3} \right|^2} = \int_{-\infty}^{+\infty} J_0(\nu)\,d\nu. \tag{4-4.3}$$

However, only frequencies of the spectrum close to zero will contribute to the line width, so that the infinite limits of the integral may be replaced by a frequency of the order of $\Delta\nu$, the line width itself. It may be shown that

$$J_0(\nu) = \overline{\sum_j \left| \frac{3\cos^2\theta_{ij} - 1}{r_{ij}^{\,3}} \right|^2} \; \frac{2\tau_c}{1 + 4\pi^2\nu^2\tau_c^{\,2}} \tag{4-4.4}$$

if it is assumed that the correlation time of each dipole is the same. Equation 4-4.4 may then be substituted into (4-4.3) and the integration performed between the limits $\pm\Delta\nu$. The result gives the dependence of the line width on τ_c. For $\tau_2\,\Delta\omega < 1$ we obtain the same dependence between τ_2 and τ_c given in the simplified equation (4-4.2). There is another source of line broadening in the region $\tau_2\,\Delta\omega < 1$, which is usually of the same order of magnitude. This broadening arises from the limitation of the lifetime of the nucleus in certain states because of spin-lattice relaxation. The dependence of τ_2 on τ_c is illustrated in Fig. 4-4.1, in which both contributions to τ_2 have been included.

With narrow lines, there is an experimental source of line broadening that is often important. If the steady magnetic field is somewhat inhomogeneous, different parts of the specimen will have slightly different resonant frequencies, and as a result the resonance line will be broadened. Nonuniformity of the magnetic field frequently sets the limit to the line width in liquid samples.

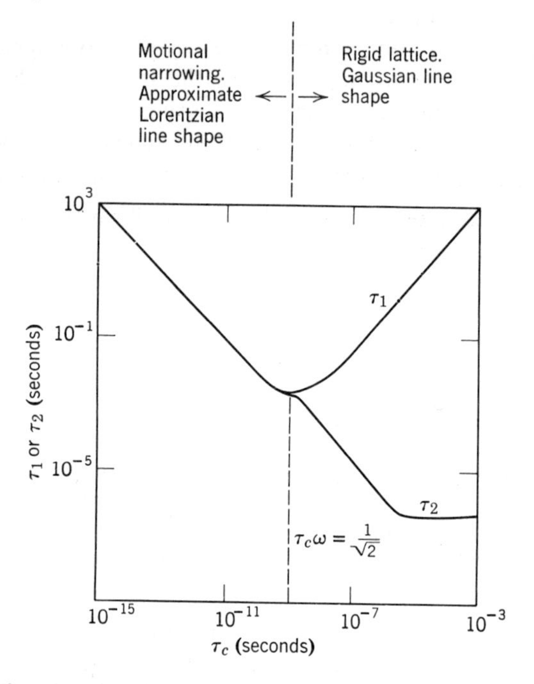

Fig. 4-4.1. The relaxation times τ_1 and τ_2 as a function of the correlation time τ_c for a given frequency. Included in τ_2 is a contribution that arises from the limit on the lifetime of a state because of τ_1. [From N. Bloembergen, E. M. Purcell, and R. V. Pound, *Phys. Rev.* **73,** 679 (1948).]

The mechanisms that give rise to the spin-lattice interaction for rigid lattices, discussed in the preceding section, are of comparatively little importance for liquids or gases and need not be considered. Instead, the spin-lattice time τ_1 is primarily the result of the molecular motion; the mechanism arises in the following way. Because of the molecular motion, the local field fluctuates and has a frequency spectrum whose intensity is denoted by $J(\nu)$ (see equation 4-4.2). This spectrum has an appreciable intensity for frequencies at or near the resonance frequency ν_L. This component of the local field induces transitions between the energy levels in the same way that an applied alternating field does, hence tends to bring the spin system into equilibrium with the lattice. In addition,

Bloembergen et al.[28] have shown that frequencies of $2\nu_L$ can also produce changes of the level populations.

It follows, then, that the transition probability will be proportional to $J(\nu_L)$ and $J(2\nu_L)$. In Section 3-5 $1/\tau_1$ has been defined as the sum of the transition probabilities P_{12} and P_{21}; hence we have

$$\frac{1}{\tau_1} \propto [J(\nu_L) + J(2\nu_L)].$$

The next step is to express the Fourier component in terms of the correlation time τ_c. When this is done, we obtain[28]

$$\frac{1}{\tau_1} = C\left(\frac{\tau_c}{1 + 4\pi^2\nu^2\tau_c^{\ 2}} + \frac{2\tau_c}{1 + 16\pi^2\nu^2\tau_c^{\ 2}}\right), \tag{4-4.5}$$

where C is independent of temperature and frequency. For $\omega\tau_c \ll 1$, equation 4-4.5 reduces to $1/\tau_1 \approx 3C\tau_c$, whereas, for $\omega\tau_c \gg 1$, $1/\tau_1 \approx 3C/2\omega^2\tau_c$. There is therefore a minimum in τ_1; this minimum occurs at $\omega\tau_c = 1/\sqrt{2}$. The variation of τ_1 with τ_c is shown in Fig. 4-4.1 also.

It is important to note that for $\omega\tau_c \ll 1$ the two relaxation times τ_1 and τ_2 are approximately equal. Both τ_1 and τ_2 originate in the same fluctuating local field, but τ_1 depends on frequencies close to ν_L and $2\nu_L$, whereas τ_2 depends on frequencies close to zero. The equality of τ_1 and τ_2 results because the intensity of the frequency spectrum is almost the same at these frequencies for short correlation times. For a hypothetical liquid we take

$$\tau_1 = \tau_2 = \tau \tag{4-4.6}$$

and this leads to a simplified form of the Bloch equations of motion. The situation is more complicated when either $I > \frac{1}{2}$ or the molecule consists of more than two spins. If, however, the molecular motion is sufficiently rapid, equation 4-4.6 will still apply.[29]

Bloembergen et al.[30] performed experiments on several liquids in an attempt to confirm the foregoing theory; as an example, we shall consider their experiments on glycerine. The correlation time depends on the viscosity of the liquid which in turn depends on the temperature. By varying the temperature they were able to investigate a large range of values of τ_c, extending from well below to well above the minimum in τ_1. The fit between theory and experiment was very good, as indeed it was also for the other liquids investigated.

The effect of dissolving in the sample a salt containing paramagnetic

[28] N. Bloembergen, E. M. Purcell, and R. V. Pound, *Phys. Rev.* **73,** 679 (1948).

[29] P. S. Hubbard, *Phys. Rev.* **128,** 650 (1962); P. M. Richards, *Phys. Rev.* **132,** 27 (1963).

[30] N. Bloembergen, E. M. Purcell, and R. V. Pound, *loc. cit.*

ions as an impurity is to produce a larger fluctuating local field because of the larger electron moment. As a result, the relaxation time τ_1 is shortened, the amount depending on the paramagnetic ion concentration and the size of the dipole moment.[31] Use of a paramagnetic impurity is a useful way to obtain a desired value of spin-lattice relaxation time.

Narrow resonance lines have been observed for hydrogen and other gases. It is common to employ high pressures in order to increase the gas density and thus increase the signal strength. Since τ_c will be of the order of the mean time between collisions, the correlation time is expected to be a function of pressure. Therefore we would expect, as for liquids, that $(1/\tau_1)\alpha\tau_c$ and $\tau_1 \approx \tau_2$; this has been confirmed by various experimenters.[32] Measurements of τ_1 as a function of pressure have provided information concerning the interaction between pairs of H_2 molecules.[33]

For a solid we expect to observe a broad resonance line, consistent with the rigid lattice theory of the preceding section. However, for many solids a narrowing of the absorption line is observed as the temperature is increased. This is interpreted as being caused by molecular motion within the crystal. Certain atomic or molecular groups have the property that they undergo rotation about one or more axes with a frequency that increases with temperature. This hindered rotation produces a situation in between the rigid lattice and the liquid cases, for although there are now fluctuations in the local field they are not so rapid as they are for liquids. An example is the variation of the line width with temperature for 1,2 dichloroethane,[34] shown in Fig. 4-4.2. Below 160°K the molecules are presumably at rest, and the line is very broad. The melting point is 240°K; hence we observe the narrow width characteristic of liquids in this region. Between these two temperatures the line narrowing is believed caused by the rotation of the molecule about the *H-H* axis. Above 175°K the reorientation of the molecules occurs so quickly that the maximum narrowing possible occurs. Other solids in which rotational narrowing has been observed include benzene, methane, cyclohexane, ammonium inorganic salts, and long-chain paraffins and polymers.[35] Cyclohexane brings in another

[31] I. Solomon and N. Bloembergen, *J. Chem. Phys.* **25,** 261 (1956); N. Bloembergen and L. O. Morgan, *J. Chem. Phys.* **34,** 842 (1961); J. S. Dohnanyi, *Phys. Rev.* **127,** 1980 (1962).

[32] E. M. Purcell, R. V. Pound, and N. Bloembergen, *Phys. Rev.* **70,** 988 (1946); M. E. Packard and H. E. Weaver, *Phys. Rev.* **88,** 163 (1952); M. Bloom, *Physica* **23,** 237 (1957); M. Lipsicas and M. Bloom, *Can. J. Phys.* **39,** 881 (1961).

[33] I. Oppenheim and M. Bloom, *Can. J. Phys.* **39,** 845 (1961); M. Lipsicas and A. Hartland, *Phys. Rev.* **131,** 1187 (1963).

[34] H. S. Gutowsky and G. E. Pake, *J. Chem. Phys.* **18,** 162 (1950).

[35] For references see E. R. Andrew, *Nuclear Magnetic Resonance,* Cambridge University Press, Cambridge (1956), pp. 165–175.

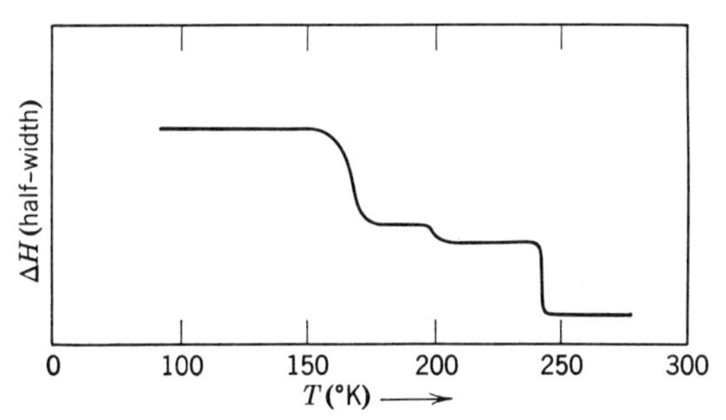

Fig. 4-4.2. Variation of the half-width with temperature for 1,2-dichloroethane. (After H. S. Gutowsky and G. E. Pake.)

mechanism for line narrowing in solids. The variation of the second moment of cyclohexane with temperature is illustrated in Fig. 4-4.3. Below 150°K the crystal is essentially rigid. The line narrowing between 150 and 180°K is probably the result of reorientation of the molecules about their triad axis. At 186°K there is a phase change, and in the expanded crystal the molecular reorientation occurs about another axis. A new feature occurs between 220 and 240°K, where the line narrows to a value characteristic of a liquid. To account for this narrowing, it is necessary to

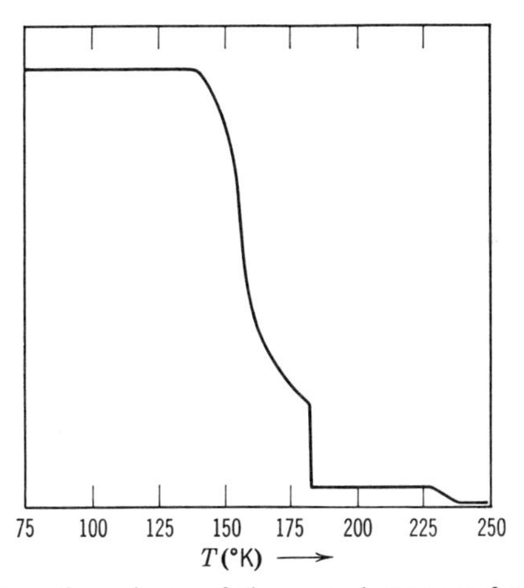

Fig. 4-4.3. Temperature dependence of the second moment for cyclohexane, C_6H_{12}. [From E. R. Andrew and R. G. Eades, *Proc. Roy. Soc.* (*London*) **A-216,** 398 (1953).]

assume that the molecules can now *diffuse* through the lattice. Self-diffusion of atoms or molecules in crystals has been observed, for example, in solid hydrogen, methane, methyl alcohol, and neopentane.[36]

[36] For references see E. R. Andrew, *Nuclear Magnetic Resonance*, Cambridge University Press, Cambridge (1956), pp. 165–175.

5. The Chemical Shift: Fine Structure

Experimentally, it has been observed[37] that a nucleus may have a different resonant frequency for a given applied field in different chemical compounds; this difference in resonant frequencies is called the *chemical shift*. The shift is believed to arise because of the differences in shielding of the nuclear moment when the chemical environment is changed, since the electronic configuration of a molecule depends on the chemical binding. Because the shift is observed only for molecular compounds and not for ions or free atoms, it appears that it is the second-order paramagnetic term,

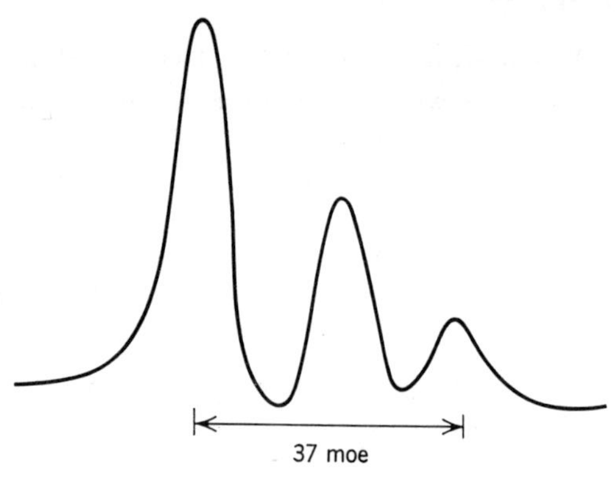

Fig. 4-5.1. Fine structure in methyl alcohol arising from different chemical shifts in the proton resonance. (After J. T. Arnold, S. S. Dharmetti, and M. Packard.)

referred to in Section 4-1, that gives rise to the effect. In general, it is not possible to calculate a theoretical value for this shift, but significant progress has been made in correlating the shift with the electronic structure of molecules.[38]

When the same nuclear species occupies sites with different electronic environments in a certain material, there will be different chemical shifts, hence different resonant frequencies for each type of site. As an example, the fine structure observed in the spectrum of ethyl alcohol is shown[39] in Fig. 4-5.1. The three lines correspond to protons in the CH_3, CH_2, and OH groups. Such fine structure has an important application in chemical analysis.

With the improvement of the homogeneity of the applied magnetic

[37] W. D. Knight, *Phys. Rev.* **76,** 1259 (1949); W. C. Dickinson, *Phys. Rev.* **77,** 736 (1950); W. G. Proctor and F. C. Yu, *Phys. Rev.* **77,** 717 (1950), etc.

[38] H. S. Gutowsky, D. W. McCall, B. R. McGarvey, and L. H. Meyer, *J. Am. Chem. Soc.* **74,** 4809 (1952); N. F. Ramsey, *Phys. Rev.* **86,** 243 (1952); J. A. Pople, *Proc. Roy. Soc.* (*London*) **A-239,** 541 (1957); H. M. McConnell, *J. Chem. Phys.* **27,** 226 (1957).

[39] J. T. Arnold, S. S. Dharmetti, and M. Packard, *J. Chem. Phys.* **19,** 507 (1951).

field, additional fine structure was resolved; for example, 28 lines have been observed with ethyl alcohol. This fine structure is believed to be the result of dipole-dipole interaction between the resonating nucleus and the moments of its nearest neighbors. Since, in a liquid, this interaction averages to zero, it is postulated[40] that the nuclear moments interact via the spin moments of the molecule's bonding electrons.

Fine structure may also result from the quadrupole interaction for nuclei with $I > \frac{1}{2}$ (see Section 4-8).

6. Transient Effects: the Spin-Echo Method

When the conditions of adiabatic slow passage are not met, or when saturation conditions are not avoided, a number of transient effects may occur. Transient conditions depend on the relationship between the various times involved in magnetic resonance, such as the spin-lattice relaxation time τ_1, the spin-spin relaxation time τ_2, the time spent on passing through the resonance line, the period of the alternating field $1/\nu$, and the time $(1/\gamma H_1)$, which depends on the amplitude of the alternating field. The solutions for transient effects have never been worked out in general. In this section only a few of the interesting transient effects will be considered. Since τ_1 and τ_2 are involved in transient phenomena, it is perhaps not surprising that studies of such phenomena provide the most convenient and important methods for the determination of relaxation times.

For a system of nuclear dipoles in the liquid state the spin-lattice relaxation time is sufficiently long that it is not too difficult to produce radio frequency power of a high enough level to produce saturation. As a result, the amplitude of the resonance line will be decreased, the amount depending on the degree of saturation. If, now, the radio frequency power is suddenly reduced, the level populations will gradually return to the Boltzmann equilibrium values. Therefore the resonance line will gradually increase in amplitude at a rate depending on the value of τ_1; this rate is in fact given by $(1 - e^{-t/\tau_1})$. Hence measurement of the rate of growth of the line permits a very direct determination of τ_1. For values of $\tau_1 < 10$ sec, it is usual to employ a cine camera to take pictures of the recovery of the resonance line as displayed on an oscilloscope.

If τ_1 is very short, it may be determined by a method known as progressive saturation, which depends on the fact that τ_1 is a function of the saturation; the reader is referred to the literature for details of the method.[41]

[40] N. F. Ramsey and E. M. Purcell, *Phys. Rev.* **85,** 143 (1952).

[41] N. Bloembergen, *Nuclear Magnetic Relaxation,* W. A. Benjamin, New York (1961); G. Suryan, *Proc. Indian Acad. Sci.* **A-30,** 107 (1951); C. Sherman, *Phys. Rev.* **93,** 1429 (1954).

We now consider a situation for which the conditions for adiabatic slow passage (Section 4-1) are not fulfilled. Suppose τ_1 and τ_2 are long compared to $1/\gamma H_1$ and that passage through the resonance is short compared to τ_1 and τ_2 but long compared to $1/\gamma H_1$. The solution to the Bloch equations of motion still shows that the magnetic moment precesses, although the magnitude and orientation of the moment may vary with time. It may be shown from problem 4-3 that the dispersion component of the susceptibility has a shape proportional to

$$\frac{M(t)}{(1 + \delta^2)^{1/2}},$$

where

$$\delta = \frac{(H + \omega/\gamma)}{H_1}. \tag{4-6.1}$$

In other words, the shape of the dispersion curve for fast passage conditions is more like the absorption than the dispersion shapes for slow passage (see Fig. 3-8.1); this shape has been observed experimentally. Further, it may be shown that $M(t)$ depends primarily on the value it attains before the passage through resonance, since it remains almost constant during the passage. This value depends on δ; it is positive if $H > \omega/\gamma$ and negative for $H < \omega/\gamma$ initially. Hence, when the direction of passage is changed, the resonance line observed will be inverted. A value of τ_1 can be deduced by studying the line inversions and by varying the speed of passage.[42]

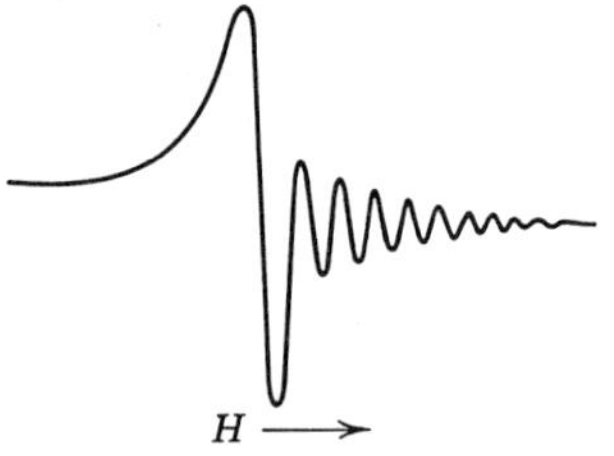

Fig. 4-6.1. Phenomena of the "wiggles." The derivative of the absorption line is shown.

An interesting transient effect results when the resonance is swept through in a time short compared to τ_1, τ_2, and $1/\gamma\delta H$, where δH is the inhomogeneity of the steady field over the sample. We then often observe wiggles that occur after the resonance line, as illustrated in Fig. 4-6.1. Note that the times τ_1, τ_2, and $1/\gamma\delta H$ determine the line width. The wiggles can be explained physically by considering a nuclear-induction type of experiment. When the resonant field is reached, the nuclei are brought into phase by the alternating field of frequency ω and a signal is induced in the receiving coil. When the field is quickly moved above the resonant value, the moment continues to precess, but at the frequency γH which is no longer equal to ω. There are then two signals induced in the receiving coil at frequencies of γH and ω; these frequencies

[42] F. Bloch, W. W. Hansen, and M. E. Packard, *Phys. Rev.* **70,** 474 (1946); L. E. Drain, *Proc. Phys. Soc. (London)* **A-62,** 301 (1949); S. D. Gvozdover and A. A. Magazanik, *J. Exptl. Theoret. Phys. (USSR)* **20,** 705 (1950).

beat together and produce the alternate maxima and minima observed. The beats will continue as long as the nuclear moments remain essentially in phase, which will occur for a time of the order of τ_2. Further, the amplitude of the beats will decay exponentially,[43] as the field is swept, with the characteristic time τ_2. Hence τ_2 may be determined from the wiggles. Other transient effects that involve nonpulse techniques are not discussed here; instead the reader is referred to the literature.[44]

When pulses of radio frequency are employed, a number of interesting transient effects occur. A physical explanation of these effects can be given most simply by utilizing the rotating coordinate analysis outlined in Section 4-1.

Let us discuss first the behavior of a nuclear spin system during the time the pulse is applied.[45] Just before the pulse is applied, the moment **M** will be precessing about the steady field **H** with a frequency $\boldsymbol{\omega}_L = -\gamma\mathbf{H}$. Now consider the application of a pulse of radio frequency of frequency ω and amplitude H_1 and neglect relaxation effects at present. The result, in the rotating frame of reference, is that **M** will precess about the effective field $\mathbf{H}_e$ (equation 4-1.8) with a frequency

$$\Omega = -\gamma\left[\left(H + \frac{\omega}{\gamma}\right)^2 + H_1^{\,2}\right]^{1/2}.$$

Near resonance $\Omega \ll \omega$, so that in the laboratory frame of reference the motion of **M** consists of a nutation of frequency Ω superimposed on the precessional frequency ω_L. The effect of the relaxations is to damp the nutation. Since the nutational motion will modulate the amplitude of the radio frequency pulse, an oscillation with an exponential decay is observed during the period in which the pulse is applied. It may be shown[45] that this decay time is given by

$$2\left(\frac{1}{\tau_1} + \frac{1}{\tau_2}\right)^{-1}. \qquad (4\text{-}6.2)$$

In these experiments pulse amplitude is large enough so that some saturation sets in by the end of the pulse. By varying the time between pulses, it is possible to observe whether the system has recovered from the saturation produced by the preceding pulse. From this information

[43] B. A. Jacobsohn and R. K. Wangness, *Phys. Rev.* **73,** 942 (1948).

[44] For example, R. Gabillard, *Compt. rend.* (*Paris*), **232,** 324, 1477, 1551 (1951), **233,** 39, 307 (1951), *Phys. Rev.* **85,** 694 (1952); C. Manus, R. Mercier, P. Denis, G. Béné, and R. Extermann, *Compt. rend.* (*Paris*) **238,** 1315 (1954).

[45] H. C. Torrey, *Phys. Rev.* **76,** 1059 (1949).

the spin-lattice relaxation time may be found; equation 4-6.2 can then be employed to determine τ_2. If $\tau_1 \gg \tau_2$, it is not necessary to know τ_1 in order to determine τ_2 from this equation.

Now let us consider the behavior of the nuclear spin system after the application of the radio frequency pulse. It will be assumed that the nuclear induction method is used. Suppose two pulses are applied to the specimen, one at $t = 0$ and the other at the time $t = t_0$. Then, in addition to the two signals induced in the receiving coil at $t = 0$ and $t = t_0$, a third signal is observed at the time $t = 2t_0$, as indicated in Fig. 4-6.2. Since this third pulse occurs in the absence of applied radio frequency radiation, it is appropriately called a *spin echo.*[46]

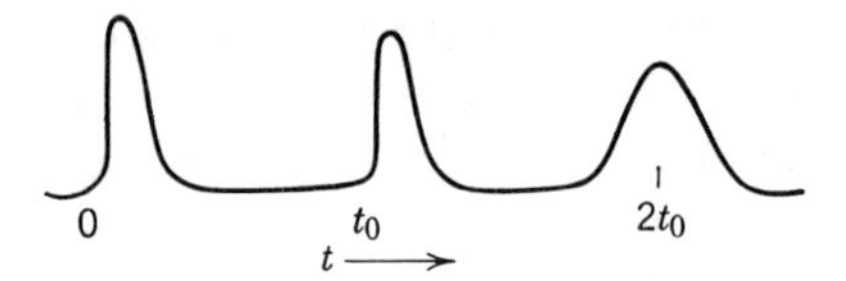

Fig. 4-6.2. The third pulse is the spin echo.

A physical explanation of the spin-echo phenomenon will now be given for one special situation; a rotating coordinate system (x^*, y^*, z^*) will be employed. Suppose that the magnetic moment **M** has reached an equilibrium value $\mathbf{M}_0$ along the direction of the steady applied field **H** (Fig. 4-6.3*a*). Next suppose a pulse of amplitude $\mathbf{H}_1$ and a frequency equal to the resonant frequency $-\gamma H$ is applied at right angles to the steady field **H**, say along the x^*-axis. In the rotating coordinate system the effective field will then be just $\mathbf{H}_e = \mathbf{H}_1$, and **M** will precess about H_1 at the frequency $-\gamma H_1$ (Fig. 4-6.3*b*). Suppose that the pulse lasts for a time t_p so that

$$\gamma H_1 t_p = \frac{\pi}{2}; \tag{4-6.3}$$

such a pulse is usually called a 90° pulse. Then, at the instant of the removal of the pulse, the moment **M** will have been rotated so that it lies in the x^*, y^* plane (Fig. 4-6.3*c*). A signal will be induced in the receiving coil because of the rotation of **M** in the laboratory system. Now, because of inhomogeneities in the applied field **H** and the distribution in local fields throughout the sample, certain regions of the sample will have slightly higher and other regions slightly lower precession frequencies. For the purposes of illustration, suppose that the sample is composed of

[46] E. L. Hahn, *Phys. Rev.* **80**, 580 (1950).

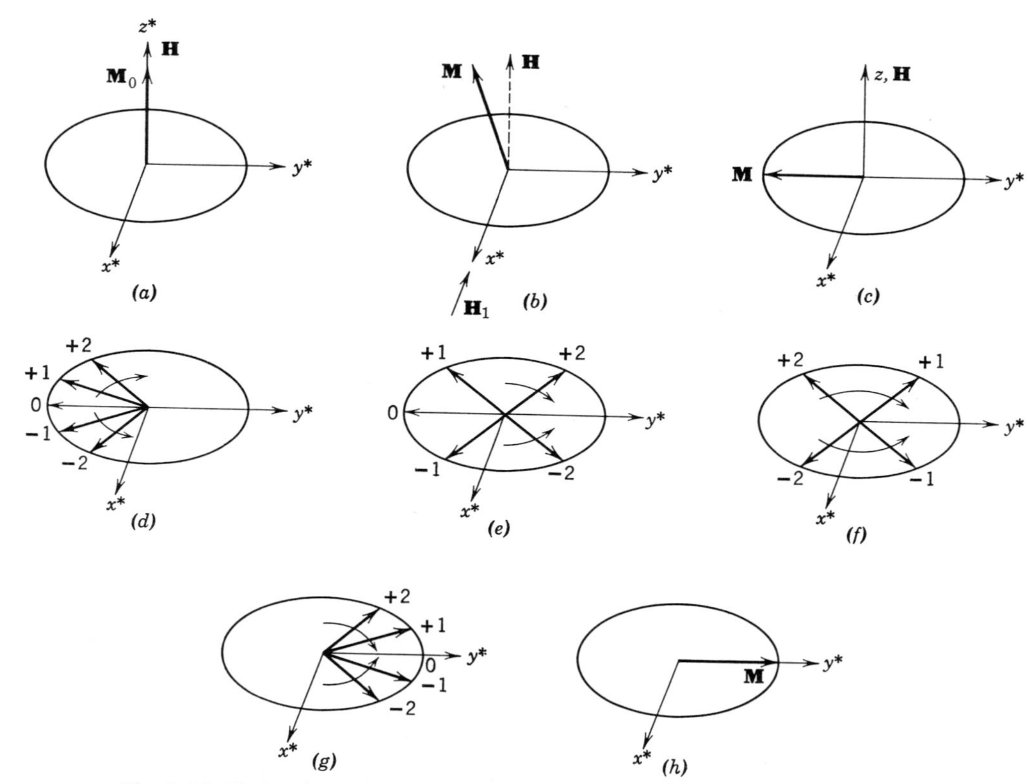

Fig. 4-6.3. Illustrations of the spin-echo phenomena (see text). (After E. L. Hahn.)

five regions, each called a spin isochromat.[47] Suppose one isochromat (labeled 0) precesses with the frequency $\omega(= -\gamma H)$, whereas two (labeled +1 and +2) precess faster and the other two slower (labeled −1 and −2) than ω. In time the isochromats will fan out, as indicated first in Fig. 4-6.3*d*. Note that there will still be a net moment in the $-y^*$-direction; however it is smaller than before, hence the induced signal will have begun to decay. At a later time, t_0, the spin isochromats will be fanned out further (Fig. 4-6.3*e*); there is no net moment in the x, y plane and the induced signal will have decayed to zero.

Consider that a 180° pulse (with $\gamma H_1 t_p = \pi$) is applied to the system along the x^*-axis when the system is as described in Fig. 4-6.3*e*. As a result, each isochromat will be rotated through 180° about the direction of H_1, as indicated in Fig. 4-6.3*f*. An induced signal will appear in the receiver at the time of this rotation. The isochromats will still be processing in the same sense, hence as time goes on the isochromat will tend to become "unfanned" (Fig. 4-6.3*g*). A resultant moment will be produced, and a signal will begin to be induced in the receiver. At $t = 2t_0$ the isochromats will be in phase again and the echo signal will reach its maximum value. Then, as time goes on, the spins will again fan out and the phase coherence will gradually be lost.

The use of a 90–180° pulse sequence is not essential; induced spin echoes are observed for any pair of pulses. When more than two pulses are applied, a series of echoes is observed. From an analysis of the Bloch equations[48] it can be shown that studies of the amplitude of the echo pulse as well as the decay of the pulses provide a direct method for the measurement of relaxation times and of the spin diffusion constant.

The result of rotating the sample itself in a spin-echo experiment has been studied.[49] It should also be mentioned that the spin-echo technique has been used in electron resonance investigations.[50] Indeed, it has been used extensively for the determination of these quantities. Particularly noteworthy examples are experiments on both gaseous and solid He^3, materials with interesting quantum mechanical properties. For the gas the self-diffusion constant was found to be inversely proportional to the density, and the spin-lattice relaxation time was found to increase with the

[47] E. L. Hahn, *Phys. Rev.* **80,** 580 (1950).

[48] *Ibid.* E. L. Hahn and D. E. Maxwell, *Phys. Rev.* **88,** 1070 (1952); T. P. Das and D. K. Roy. *Phys. Rev.* **98,** 525 (1955); E. T. Jaynes, *Phys. Rev.* **98,** 1099 (1955); A. L. Bloom, *Phys. Rev.* **98,** 1105 (1955); I. J. Lowe and R. E. Norberg, *Phys. Rev.* **107,** 46 (1957).

[49] I. J. Lowe, *Phys. Rev. Letters* **2,** 522 (1959).

[50] R. J. Blume, *Phys. Rev.* **109,** 1867 (1958); L. K. Wanlass and J. Wakabayashi, *Phys. Rev. Letters* **6,** 271 (1961); D. Cutler and J. G. Powles, *Proc. Phys. Soc.* (*London*) **80,** 130 (1962).

density for moderate gas densities.[51] For the solid it is concluded[52] that two mechanisms, diffusion and coupling between the Zeeman and exchange energy, make the predominant contributions to τ_1. The exchange term is appreciable because of the large overlap of the atomic orbital functions.

7. Negative Temperatures

A scheme to provide a net emission of radiation by redistribution of the populations of a three-level system was discussed in Section 3-12. Systems that have only the two lowest levels appreciably populated will be discussed now. Let N_1 and N_2 be the populations of the ground and excited levels, respectively, and let $\Delta E = \hbar\omega_{21}$ be the energy difference between the levels. Then, as usual, the level populations at thermal equilibrium are related by the Boltzmann factor

$$\frac{N_2}{N_1} = e^{-\Delta E/kT}. \tag{4-7.1}$$

For positive temperatures N_2 is always less than N_1 (except for $T = +\infty$, when $N_2 = N_1$). Under this condition we will always get a net absorption of energy when the system is stimulated with radiation. In order to obtain a net emission of energy, it is necessary to have $N_2 > N_1$; if this situation is achieved, it will be necessary to ascribe a negative temperature to the spin system. Although the idea of a negative absolute temperature may at first seem strange, it is clear that the physical meaning is simply that the total energy of the system is higher rather than lower than that for the limiting case in which $N_1 = N_2 (T = \pm\infty)$. A system with $T < 0$ is very hot, rather than cold, for as it cools its temperature will fall first to $-\infty$, then switch to $+\infty$, and finally drop to the positive temperature that corresponds to thermal equilibrium with the lattice.

Three techniques for the inversion of populations of two-level systems will be outlined next. First,[53] consider the application of a steady magnetic field **H** to a system and suppose that equilibrium has been reached, so that a magnetic moment **M** lies along the field direction. Now suppose that the field **H** is so quickly reversed in direction that the level populations do not have time to readjust. Then, for a short time at least, depending on the relaxation time, **M** will be antiparallel to **H** (see Fig. 4-7.1*a*). Since the

[51] K. Luszczynski, R. E. Norberg, and J. E. Opfer, *Phys. Rev.* **128,** 186 (1962).

[52] J. M. Goodkind and W. M. Fairbank, *Phys. Rev. Letters* **4,** 458 (1960); R. L. Garwin and H. A. Reich, *Phys. Rev.* **115,** 1478 (1959); H. A. Reich, *Phys. Rev.* **129,** 630 (1963); S. R. Hartmann, *Phys. Rev.* **133,** A17 (1964).

[53] E. M. Purcell and R. V. Pound, *Phys. Rev.* **81,** 279 (1951).

angle **M** makes with **H** is not conserved in the reversal, this method is called nonadiabatic reversal.

In a second method[54] radio frequency radiation of frequency ω and amplitude $\mathbf{H}_1$ is applied at right angles to the steady field **H**. It is convenient to utilize the rotating coordinate method of analysis. Far off resonance the effective field will be almost parallel to **H**. It will be assumed that equilibrium has been reached so that either the condition of equation 3-7.7 or equation 3-8.8 will apply; that is, **M** lies either along the effective field $\mathbf{H}_e$, or else merely lags $\mathbf{H}_e$ by a phase angle ϵ (equation 4-1.12). Now

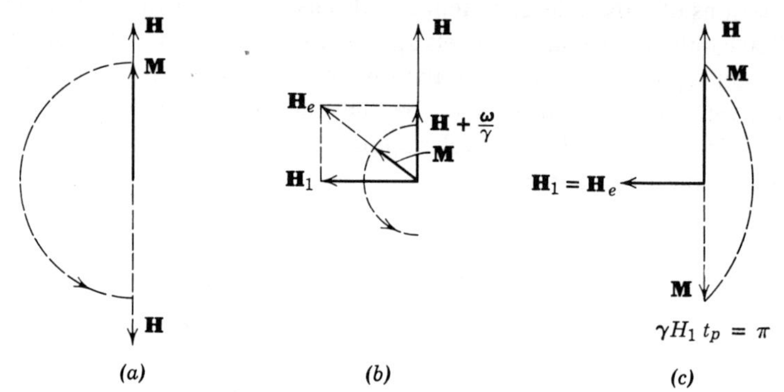

Fig. 4-7.1. Negative temperatures achieved by (*a*) nonadiabatic fast reversal, (*b*) adiabatic fast passage, and (*c*) 180° pulse.

suppose that **H** is reduced under the conditions of adiabatic rapid passage. These conditions are that the rate of change of H is slow enough that **M** will retain its orientation with respect to $\mathbf{H}_e$ but fast enough that the level populations will remain unchanged. On sweeping H through resonance, the behavior of the system is described by equation 3-7.7, as discussed in Section 4-1. Hence $\mathbf{H}_e$ and also **M** turn from parallel to antiparallel to **H**, as indicated in Fig. 4-7.1*b*. In the laboratory system **M** is now reversed with respect to **H**.

In the third method[55] **M** is assumed to have reached its equilibrium value, so that it lies along the direction of the steady field **H**. A pulse with the resonant frequency $-\gamma H$, applied at right angles to **H**, will cause **M** to rotate about the effective field (equal to $\mathbf{H}_1$). If $\gamma H_1 t_p = \pi$, that is, for a 180° pulse, **M** will be rotated until it is antiparallel to **H**, as indicated in Fig. 4-7.1*c*. Hence the level populations will be inverted.

As an example, we shall describe an interesting experiment by Purcell and Pound[56] in which the method of nonadiabatic fast reversal was

[54] J. Combrisson, A. Honig, and C. H. Townes, *Compt rend.* (*Paris*) **241,** 245 (1956).
[55] M. W. P. Strandberg, *Phys. Rev.* **106,** 617 (1957).
[56] E. M. Purcell and R. V. Pound, *Phys. Rev.* **81,** 279 (1951).

employed. The experiment was carried out on the nuclear spin system of a very pure single crystal of LiF. The system was studied by observing the nuclear spin resonance of the lithium (Li^7) nucleus. Pound had observed earlier that the spin-lattice relaxation time for the Li and F nuclear spins was very large—about 15 sec at 1°K and 300 sec at room temperature. This fact permitted a crystal to be moved about in a time short compared to that required for the spin and lattice system to come to equilibrium. The spin-spin relaxation time of the system was found to be[57] about 10^{-5} sec, even in zero field. Hence, in order to invert the populations, the magnetic field had to be reversed in a time less than 10^{-6} sec.

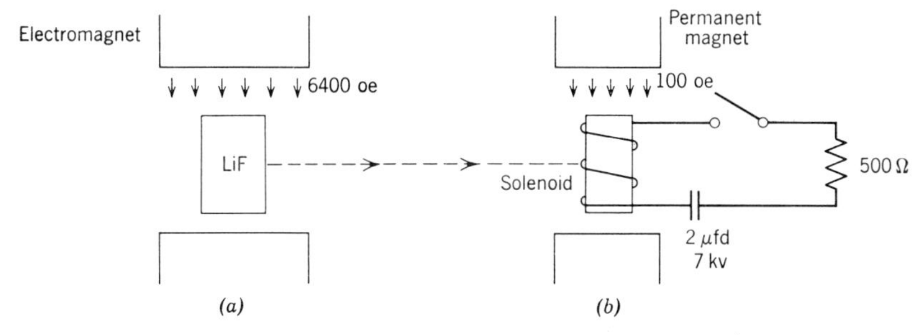

Fig. 4-7.2. The Pound-Purcell negative temperature experiment.

The experiment consisted of the following sequence of operations. First the LiF crystal was placed at room temperature into a large field of 6400 oe produced by an electromagnet (Fig. 4-7.2*a*). The specimen was left in the field for some time in order to establish thermal equilibrium. Then the sample was removed quickly and placed in a small solenoid whose axis was parallel to a field of 100 oe produced by a permanent magnet (Fig. 4-7.2*b*). This transplanting took a time short compared to τ_1 but long compared to τ_2. The effect was essentially an adiabatic demagnetization from 6400 to 100 oe, with the spin system temperature dropping from 293°K to about 293 (100/6400) $\approx$ 5°K. A 2-μfd condensor, initially charged to 7 kv, was then discharged through the solenoid (Fig. 4-7.2*b*), so that the solenoid produced a field of -200 oe in the very short time of about 10^{-7} sec. Thus the specimen was exposed to a net field of -100 oe instead of $+100$ oe in a time short compared to the spin-spin relaxation time; as a result the level populations were reversed. The field produced by the solenoid then decayed in a time of about 10^{-3} sec, and the specimen again was in a field of about $+100$ oe.

[57] N. F. Ramsey and R. V. Pound, *Phys. Rev.* **81,** 278 (1951); N. F. Ramsey, *Phys. Rev.* **103,** 20 (1956).

In order to show that a negative temperature had been achieved, the specimen was returned to the original field of 6400 oe and the Li^7 resonance was inspected. The lines observed as a function of time are illustrated in Fig. 4-7.3. The peak at the extreme left is an absorption curve taken before the sequence of operations was begun. The next peak was observed when the sample had been returned to the 6400-oe field; it is negative, indicating an emission of energy. The emission lines decrease in height and finally become zero corresponding to a spin temperature of $-\infty$, when the level populations are equal. Thereafter an absorption occurs. The positive peaks increase, hence show the cooling of the sample from $+\infty$ to room temperature. The decay of the signal occurs in the characteristic

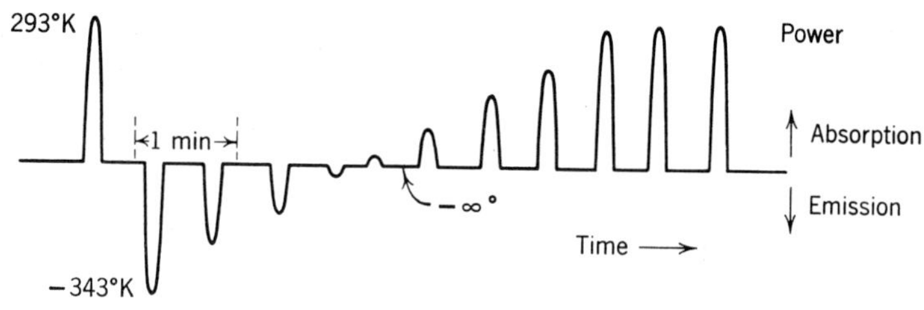

Fig. 4-7.3. Emission and absorption lines in the Pound-Purcell experiment. [From E. M. Purcell and R. V. Pound, *Phys. Rev.* **81,** 279 (1951).]

spin-lattice time τ_1. The first negative peak is not so large as the initial positive one because of some irreversible heat flow in the spin system.

It is of interest to consider the changes that occur in the spin system when the solenoid is energized.[58] The Hamiltonian for the spin system consists of two terms, one for the dipole-dipole or spin-spin coupling, given by the quantum mechanical analog of equation 1-3.3

$$\mathcal{H}_{ss} = \mu_{B_n}{}^2 \sum_{j>i} \frac{g_{ni}g_{nj}}{r_{ij}{}^3}\left[\mathbf{I}_i \cdot \mathbf{I}_j - \frac{3(\mathbf{r}_{ij} \cdot \mathbf{I}_i)(\mathbf{r}_{ij} \cdot \mathbf{I}_j)}{r_{ij}{}^2}\right] \qquad \text{(4-7.2)}$$

and the other for the Zeeman energy

$$\mathcal{H}_z = -\mu_{B_n} H \sum_i g_{ni} m_{Ii}. \qquad \text{(4-7.3)}$$

In the expression or the Boltzmann factor two temperatures can be formally introduced to describe the spin system, a dipole or spin-spin temperature T_{ss} and a Zeeman temperature T_z that is,

$$\exp\left(-\frac{\mathcal{H}_{ss}}{kT_{ss}} - \frac{\mathcal{H}_z}{kT_z}\right).$$

[58] A. Abragam and W. G. Proctor, *Phys. Rev.* **109,** 1441 (1958); J. H. Van Vleck, *Z. Phys. Chem.* **16,** 363 (1958).

The spin-spin temperature is positive if there are more parallel pairs than antiparallel pairs of dipoles, whereas the Zeeman temperature is positive if there is a net number of spins aligned parallel to the field H. For an initial spin temperature of $+5°$K the sudden reversal of the field from $+100$ to -100 oersteds will produce a Zeeman temperature of $-5°$K, whereas the spin-spin temperature will remain at $+5°$K, since the spin-spin Hamiltonian is independent of the field. As the solenoid's field decays, these two temperatures gradually come to an equilibrium value (see problem 4–6). The success of the Purcell-Pound experiment depends on the difference in behavior of these two energies. It should be noted that the rapid field reversal must occur in a time not only rapid compared to τ_2 but also compared to the period of the Larmor precession in the constant magnetic field. The concept of two temperatures for the spin system has also found use in the analysis of spin-spin relaxation and in other problems.[59]

8. Quadrupole Effects and Resonance

Only nuclei with a spin quantum number I greater than $\frac{1}{2}$ can have an electric quadrupole moment.[60] Experiment shows that most nuclei with $I > \frac{1}{2}$ do indeed possess such a moment. The quadrupole moment arises when the charge distribution over the nucleus is not spherically symmetric. The interaction of the quadrupole moment with the electronic energy levels has already been discussed briefly in Section 3-9. The quadrupole moment Q is defined by

$$eQ = \int \rho(3z^2 - r^2)\, dv, \tag{4-8.1}$$

where ρ is the electric charge density and dv is a volume element in the nucleus whose location is given by r, the distance from the center of the nucleus, and z, the coordinate along the symmetry axis of the nucleus. Values of Q of a few nuclei have been listed in Table 4-1.1.

The nuclear quadrupole moment interacts with the gradient of the crystalline electrostatic field of a crystal, provided that this field is of less than cubic symmetry. The energy of interaction depends on the orientation of the nucleus in such a way that the energy levels of the nucleus are shifted by an amount proportional to m_I^2.[61] Hence the $2I$ intervals between the energy levels are unequal, so that the resonance spectrum has a fine

[59] A. G. Anderson and S. R. Hartmann, *Magnetic and Electric Resonance and Relaxation*, J. Smidt, Editor, North-Holland Publishing Co., Amsterdam (1963), p. 157; B. N. Provotoroff, *J. Exptl. Theoret. Phys.* **42,** 882 (1962) [trans. *Soviet Phys.-JEPT*, **15,** 611 (1962)]; J. Jeener, H. Eisendarth, and R. Van Steenwinkel, *Phys. Rev.* **133,** A-478 (1964); J. Philippott, *Phys. Rev.* **133,** A-471 (1964).

[60] G. E. Pake, "Nuclear Magnetic Resonance," *Solid State Phys.* **2,** 50–54 (1956).

[61] *Ibid.*

structure of $2I$ lines. For many solids the quadrupole interaction is small compared to the interaction with the applied magnetic field and can be treated with a perturbation method. There is good agreement between theory and experiment.[62]

The interaction between the quadrupole moment and the electric field gradient also depends on the nuclear charge distribution. The nuclear radius differs for different isotopes (same A, different mass) and isomers (same Z and mass but different nuclear state). The shift of the energy levels from these effects has also been observed in nuclear resonance experiments.[63]

The origin of the spin-lattice relaxation in solids for nuclei without a quadrupole moment was discussed in Section 4-3. It was concluded that the mechanism considered by Waller, wherein the lattice vibrations cause the local magnetic field to vary, could not account for the spin-lattice relaxation times observed. When the nuclei have a quadrupole moment, we must consider an additional mechanism for the exchange of energy between the lattice and spin systems. Lattice vibrations also cause the electric field gradient to vary. The electric field gradient components of frequency ν_L and $2\nu_L$ interact with the quadrupole moment and induce transitions between the energy levels. This quadrupole interaction is usually considerably greater than the dipole interaction and indeed does predict values of τ_1 comparable to those observed.[64] Experiments by Pound[64] show that for some crystals at least the quadrupole interaction is the predominant source of the spin-lattice relaxation.

In liquids or gases the local magnetic field fluctuates rapidly because of the molecular motion. The interaction of the nuclear magnetic dipoles and this field is the origin of the spin-lattice relaxation, discussed in Section 4-4. The local electric field gradient also fluctuates rapidly, and when an electric quadrupole moment exists the interaction provides an additional relaxation mechanism. A formula similar to equation 4-4.5 applies[65] for τ_1. Experiments on deuterons and protons have shown the importance of this mechanism.[66]

[62] R. V. Pound, *Phys. Rev.* **79,** 1410 (1950); G. M. Volkoff, H. E. Petch, and D. W. L. Smellie, *Can. J. Phys.* **30,** 270 (1952); G. Burns, *Phys. Rev.* **127,** 1193 (1962).

[63] L. R. Walker, G. K. Wertheim, and V. Jaccarino, *Phys. Rev. Letters* **6,** 98 (1961).

[64] R. V. Pound, *loc. sit.*; J. H. Van Kranendonk, *Physica* **20,** 781 (1954); J. Kondo and J. Yamashita, *J. Phys. Chem. Solids* **10,** 245 (1959); E. G. Wikner, W. E. Blumberg, and E. L. Hahn, *Phys. Rev.* **118,** 631 (1960); R. L. Michen, *Phys. Rev.* **125,** 1537 (1962).

[65] N. Bloembergen, *Nuclear Magnetic Relaxation*, W. A. Benjamin, New York (1961).

[66] N. Bloembergen, E. M. Purcell, and R. V. Pound, *Phys. Rev.* **73,** 679 (1948). See also P. Diehl, *Helv. Phys. Acta* **29,** 219 (1956); H. A. Christ, *Helv. Phys. Acta* **33,** 572 (1960); H. A. Christ, P. Diehl, H. R. Schneider, and H. Dahn, *Helv. Chim. Acta* **44,** 865 (1961).

In some solids the splitting of the energy levels by the quadrupole interaction is larger than that produced by the magnetic fields usually employed in the laboratory. This situation develops because of the presence of large electric field gradients and is most common for molecular solids. Because of the large splitting, resonance can be observed[67] when a radio frequency field is applied without the application of a steady magnetic field. This phenomena is called *nuclear quadrupole resonance* or, more usually, *pure quadrupole resonance*.[68] Since the frequency of resonance is fixed by the internal electric field gradient, it is necessary to be able to vary the radio frequency over a wide range of frequencies. Further, because the separations between the successive levels are unequal more than one resonance line is observed. In addition to the spin-spin interaction and to small values of τ_1, line broadening occurs because of variations in the electric field gradient throughout the crystals. These variations may be caused by crystal defects, strains, or torsional vibrations of the molecules.

9. Nuclear Orientation

Magnetic cooling produced by the adiabatic demagnetization of polarized nuclei was mentioned briefly at the end of Section 3-3. It is now appropriate to outline techniques for the orientation of nuclei. Four static methods have been proposed and all have been realized experimentally. A necessary condition for polarization by these methods is that the nuclear energy levels must be split by an amount greater than kT. Then most of the nuclei will occupy the lowest level, hence there will be a preferred spin orientation. In order to satisfy this condition the temperature must be 1°K or lower.

The first method,[69] often called the "brute-force" method, involves simply the application of a very large magnetic field. The interaction of this field with the nuclear dipole moments will split the energy levels. Because of the size of $\boldsymbol{\mu}$ the fields required are of the order of 50 koe for temperatures about 0.01°K. The experimental difficulties are considerable; the first successful experiment was achieved in 1955.[70] This method is the only one that can give nuclear magnetic cooling, since it is the only one in which the disorder of the spin system can be altered by varying an external parameter, namely the applied magnetic field.

[67] The first successful experiment is described by H. G. Dehmelt and H. Krüger, *Naturwiss.* **37,** 111 (1950).

[68] For a review see H. G. Dehmelt, *Am. J. Phys.* **22,** 110 (1954); T. P. Das and E. L. Hahn, *Nuclear Quadrupole Resonance Spectroscopy*, Academic Press, New York (1958).

[69] C. J. Gorter, *Phys. Z.* **35,** 923 (1934); N. Kurti and F. E. Simon, *Proc. Roy. Soc. (London)* **A-149,** 152 (1935).

[70] J. W. T. Dabbs, L. D. Roberts, and S. Bernstein, *Phys. Rev.* **98,** 1512 (1955).

The other three methods make use of internal interactions present in the solid state. In the second method[71] the magnetic fields produced by the electronic dipole moments are employed. The magnetic field at the nucleus of a paramagnetic ion can be of the order of 100 koe or higher. A modest external magnetic field is applied to polarize the electronic dipoles and nuclear polarization follows. The first experiments with this method were carried out by Ambler et al.[72] Nuclear orientation has also been achieved in magnetically ordered materials, for example ferro- and antiferromagnets.[73]

In the first two methods the axis of quantization was provided by an applied magnetic field. In the next two methods this axis is determined by the crystalline electric field. In the third method[74] the splitting of the levels occurs because of the interaction between the nuclear quadrupole moment and the electric field gradient. Dabbs et al.[75] have produced nuclear orientation with this method. In the fourth method[76] the electric field interacts with the nuclear moment via the electronic magnetic hyperfine interaction. This method is the one that has been used most widely, and it is also the method whereby the first nuclear orientation was achieved.[77]

In methods three and four a doublet series of levels degenerate in $\pm m_I$ are produced; that is, in the ground state there are equal numbers of nuclei with spins pointing in opposite directions. This type of orientation is called *alignment*. In methods one and two all the spins in the ground state point the same way and the term *polarization* is employed to describe the nuclear orientation. In many nuclear physics experiments alignment is sufficient to produce the desired effect. These experiments include studies of (a) γ-ray anisotropy, (b) polarization of γ-rays, (*c*) absorption of polarized neutrons, and (d) nonconservation of parity.[78]

Nuclear orientation can also be produced by dynamical methods that do not require the use of very low temperatures.[79] One utilizes the Overhauser

[71] C. J. Gorter, *Physica* **14,** 504 (1948); M. E. Rose, *Phys. Rev.* **75,** 213 (1949).

[72] E. Ambler, M. A. Grace, H. Halban, N. Kurti, H. Durand, C. E. Johnson, and H. R. Lemmer, *Phil. Mag.* **44,** 216 (1953).

[73] J. M. Daniels, J. C. Giles, and M. A. R. Le blanc, *Can. J. Phys.* **39,** 53 (1962); J. M. Daniels and J. M. Cotignola, *Can. J. Phys.* **40,** 1342 (1962).

[74] R. V. Pound, *Phys. Rev.* **76,** 1410 (1949).

[75] J. W. T. Dabbs, L. D. Roberts, and G. W. Parker, *Bull. Am. Phys. Soc.* **1,** 207 (1956).

[76] B. Bleaney, *Proc. Phys. Soc.* (*London*) **A-64,** 315 (1951).

[77] J. M. Daniels, M. A. Grace, and F. N. H. Robinson, *Nature* (*London*) **168,** 780 (1951).

[78] C. S. Wu, E. Ambler, R. W. Hayward, D. D. Hoppes, and R. P. Hudson, *Phys. Rev.* **105,** 1413 (1957).

[79] C. D. Jeffries, "Dynamic Nuclear Polarization," *Progress in Cryogenics*, Heywood and Co., London (1961).

effect,[80] another depends on the hyperfine electronic line structure.[81] The Overhauser effect is discussed in Chapter 5.

10. Double Resonance

A double resonance technique in which both electron and nuclear resonance are involved has been developed by Feher and others.[82] The nuclear resonance is observed via the electron resonance line. The method can be understood easily by the following example.

For a system with an electron spin **S** and a nuclear spin **I** in a magnetic field **H**, the spin-Hamiltonian of equation 3-9.11 becomes

$$\mathcal{H} = g\mu_B \mathbf{H} \cdot \mathbf{S} + \mathcal{A}\mathbf{S} \cdot \mathbf{I} - g_n \mu_{B_n} \mathbf{H} \cdot \mathbf{I}, \tag{4-10.1}$$

provided that there is no crystalline field, no quadrupole nuclear moment, and that g, g_n and $\mathcal{A}$ are constants. The terms in equation 4-10.1 arise from the direct interaction between the applied magnetic field and the electron's magnetic moment, the hyperfine interaction between the electron and nuclear spins, and the direct interaction between the applied field and the nuclear moment, respectively. These energies are in order of decreasing size. It will be assumed that $S = \frac{1}{2}$ and $I = \frac{1}{2}$. The energy levels are then given, to a good approximation, by

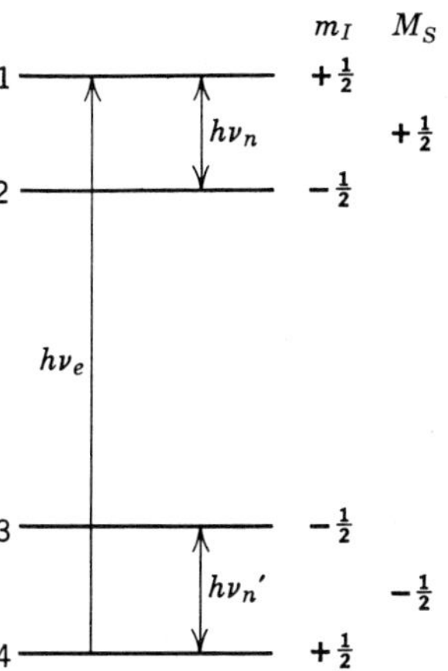

Fig. 4-10.1. The energy levels of a system with $S = \frac{1}{2}$, $I = \frac{1}{2}$.

$$\begin{aligned}
&(1) && \tfrac{1}{2}g\mu_B H + \tfrac{1}{4}\mathcal{A} - \tfrac{1}{2}g_n\mu_{B_n}H,\\
&(2) && \tfrac{1}{2}g\mu_B H - \tfrac{1}{4}\mathcal{A} + \tfrac{1}{2}g_n\mu_{B_n}H,\\
&(3) && -\tfrac{1}{2}g\mu_B H + \tfrac{1}{4}\mathcal{A} + \tfrac{1}{2}g_n\mu_{B_n}H,\\
&(4) && -\tfrac{1}{2}g\mu_B H - \tfrac{1}{4}\mathcal{A} - \tfrac{1}{2}g_n\mu_{B_n}H,
\end{aligned} \tag{4-10.2}$$

and are illustrated in Fig. 4-10.1. Note that the energy difference between levels 1 and 2 is slightly different from that between 3 and 4. The electron spin transition will be either between levels 4 and 1 or 3 and 2, since the

[80] A. W. Overhauser, *Phys. Rev.* **92,** 411 (1953).
[81] A. Honig, *Phys. Rev.* **96,** 234 (1954).
[82] G. Feher, *Phys. Rev.* **103,** 834 (1956); J. Lambe, N. Laurance, E. C. McIrvine, and R. W. Terhune, *Phys. Rev.* **122,** 1161 (1961).

condition $\Delta m_I = 0$ must be fulfilled. Let us assume that the transition is between levels 4 and 1; the resonance frequency is then given by

$$h\nu_e = g\mu_B H + \tfrac{1}{2}\mathcal{A} \quad . \tag{4-10.3}$$

Let us suppose now that this transition is saturated, so that the populations of these levels are made equal and the absorption line disappears. A slowly varying radio frequency ν_n is applied next. When this frequency satisfies the condition

$$h\nu_n = \tfrac{1}{2}\mathcal{A} - g_n\mu_{B_n}H \tag{4-10.4}$$

stimulated transitions between levels 1 and 2 will occur and level 2 will be populated from level 1. Hence the saturation of the electron transition is removed and the electron resonance line is restored. Removal of the saturation also occurs for a radio frequency given by

$$h\nu_{n'} = \tfrac{1}{2}\mathcal{A} + g_n\mu_{B_n}H \tag{4-10.5}$$

for transitions will then be stimulated between levels 3 and 4.

The technique is important for two reasons. One, the observation of the frequencies ν_n and $\nu_{n'}$ permit an accurate determination of g_n and $\mathcal{A}$. Two, hyperfine structure which is obscured by the width of the electron line can be resolved. An example of the first application[83] is the determination of A and g_n for the donor electrons of P in Si. Examples of the other application are the resolution of the hyperfine structure of the electron resonance line (a) of F-centers[84] in KCl and (b) of donors in semiconductors.[85]

A double nuclear resonance technique has also been developed. An ensemble of nuclear spins, which by themselves produce a weak signal, are detected by observing their effect on another abundant ensemble of nuclear spins.[86]

11. Beam Methods

No account of electronic and nuclear magnetic moments would be complete without at least a brief summary of beam methods. These methods employ beams either of neutral atoms or of neutral molecules; beams of electrically charged particles properly belong to another area of study. The source of the beam is usually a heated oven which produces

[83] G. Feher, *Phys. Rev.* **103,** 834 (1956).

[84] G. Feher, *Phys. Rev.* **105,** 1122 (1957); W. T. Doyle, *Phys. Rev.* **126,** 1421 (1962).

[85] G. Feher, *Phys. Rev.* **114,** 1219 (1959).

[86] D. F. Holcomb, B. Pedersen, and T. R. Sliker, *Phys. Rev.* **123,** 1951 (1961); B. N. Provotorov, *Phys. Rev.* **128,** 75 (1962); S. R. Hartmann and E. L. Hahn, *Phys. Rev.* **128,** 2042 (1962); F. M. Lurie and C. P. Slichter, *Phys. Rev.* **133,** A1108 (1964).

the vapor of the material to be studied. The molecules emerge from a small aperture in the oven, hence form a collimated beam. The beam traverses a path in an apparatus that is so highly evacuated that few of the molecules suffer a collision with another molecule. The beam of particles is studied by applying certain types of magnetic fields. The force on an atom or molecule that possesses a magnetic moment $\boldsymbol{\mu}$ is found by differentiating the quantum mechanical analog of equation 1-6.1, that is

$$F_r = -\frac{dW}{dr}$$

$$= gM_J\mu_B \frac{dH}{dr}, \qquad (4\text{-}11.1)$$

where $gM_J\mu_B$ is the component of $\boldsymbol{\mu}$ along the direction of H and F_r is the force parallel to the r-direction. Hence, in order to deflect the beam, the magnetic field must be inhomogeneous.

Since 1938 almost all beam experiments have employed the radio-frequency resonance method introduced by Rabi and associates.[87] Nevertheless, it is of interest to outline briefly some of the earlier beam experiments. A molecular beam was produced as early as 1911.[88] However, the first important experiment was the one carried out by Stern and Gerlach on beams of silver atoms.[89] For such atoms $J = S = \frac{1}{2}$, and therefore only two orientations of the magnetic moment are possible, namely, a component of μ_B either parallel or antiparallel to the applied field **H**. When these atoms were sent through an inhomogeneous magnetic field, the deflecting force was either $+\mu_B(dH/dr)$ or $-\mu_B(dH/dr)$, so that the beam was split into two components. The observation of this effect gave convincing evidence for the existence of spatial quantization. Experiments have also been performed on other metal atoms, on hydrogen atoms, and on the paramagnetic diatomic gases O_2 and NO. The effect of an inhomogeneous field on nuclear moments was studied by employing molecules that had no electronic moment.[90] The experimental difficulties are greater, since much longer paths through the field as well as more sensitive detection methods are necessary. For hydrogen molecules the results are complicated further because the two protons may have their spins either antiparallel (parahydrogen) or parallel (orthohydrogen). With this method, reasonably accurate values of the moment of the proton and deutron were obtained. Nuclear moments can also be measured with

[87] I. I. Rabi, J. R. Zacharias, S. Millman, and P. Kusch, *Phys. Rev.* **53,** 318 (1938).
[88] L. Dunoyer, *Compt. rend.* (*Paris*) **152,** 594 (1911).
[89] W. Gerlach and O. Stern, *Ann. Phys.* (*Leipzig*) **74,** 673 (1924).
[90] R. Frisch and O. Stern, *Z. Physik* **85,** 4 (1933).

beams of atoms that possess an electronic moment because of the hyperfine interaction.[91] The *zero moment* method[92] is one way in which such a measurement can be made. In this method the beam passes first through a strong and then through a weak inhomogeneous field onto a receiver placed to collect the maximum number of atoms in zero field. The weak field is varied continuously, starting from zero, and the number of atoms collected is observed to fall and then to rise; that is, a series of maxima is observed for which the effective value of the magnetic moment is zero.

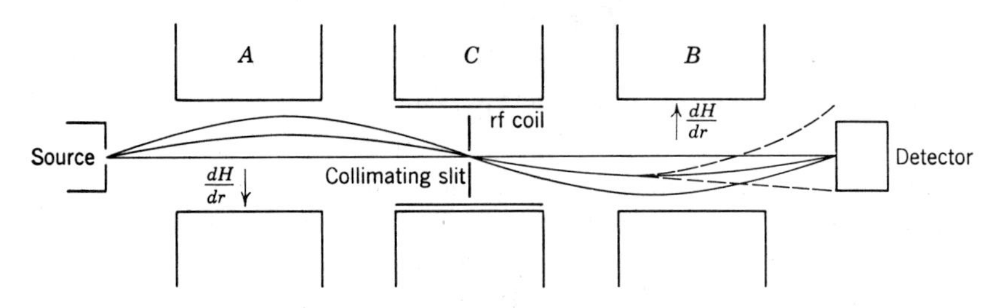

Fig. 4-11.1. Beam resonance apparatus.

Analysis of the number of maxima and their spacing yields values of the nuclear spin and moment. For atoms with $I = \frac{1}{2}$ another zero moment method was used.[93]

It is now appropriate to describe the beam resonance method. A schematic sketch of a typical apparatus is shown in Fig. 4-11.1. Magnets A and B produce inhomogeneous fields with gradients dH/dr in opposite directions. The molecules, which in the absence of any field would have gone in a straight-line path to the detector, are deflected by magnet A. These deflected molecules would then miss the detector if it were not for the presence of magnet B, which refocuses the beam on the detector. The deflection and refocusing of the beam is independent of the velocity of the molecules, since a fast molecule is deflected less than a slow one (see the diagram). The refocusing action requires that the magnitude and orientation of the nuclear magnetic moments be the same in both fields A and B. If in the region between A and B the nuclear magnetic moment changes, the beam collected will be reduced. A magnet C which produces a uniform field H is located between the A and B magnets. A coil is placed between the poles of magnet C in order to produce an rf field of frequency ν at right angles to the steady field. When the resonance

[91] I. I. Rabi, J. M. B. Kellog, and J. R. Zacharias, *Phys. Rev.* **46,** 157 and 163 (1934).

[92] V. W. Cohen, *Phys. Rev.* **46,** 713 (1934); M. Fox and I. I. Rabi, *Phys. Rev.* **48,** 746 (1935).

[93] J. M. B. Kellog, I. I. Rabi, and J. R. Zacharias, *Phys. Rev.* **50,** 472 (1936).

condition of equation 4-1.2 is fulfilled, transitions will occur and there will be a reorientation of nuclear spins. As a result, the paths in the region of magnet B will change as indicated, for example, by the dotted lines in Fig. 4-11.1. For molecular beams the frequency ν is usually held constant and the field H is varied; when the resonance condition is achieved, the beam collected by the detector will be reduced to a minimum. Clearly, a beam resonance experiment is simply another way of observing the phenomena of nuclear magnetic resonance; historically, this method preceded the absorption or induction methods described earlier in this chapter.

The magnetic moments of the neutron and the μ-meson have been determined by variations of the foregoing beam resonance method. Even though neutrons have no charge, they have a magnetic moment. Beams of neutrons can be obtained either from a reaction in a cyclotron or from nuclear reactors. On passing through some soft iron, the neutron beam becomes polarized as a result of the interaction with the electronic moment of the iron atoms (see Section 8-2). A plate of iron at the exit of the apparatus serves as an analyzer. When the static homogeneous field for a given frequency ν satisfies the resonance condition, the neutron beam is partly depolarized. This is detected by a decrease in the intensity of the beam emerging from the analyzer.[94] The value of the static field H required for resonance of the neutrons has been determined accurately by calibration with a proton resonance spectrometer of the type discussed at the beginning of the chapter[95,96] (Section 4-1). The magnetic moment of the neutron was found to be[97] $-1.91307 \pm 0.0006\ \mu_{Bn}$.

A beam of π-mesons can be produced in a high-energy accelerator. The π-meson decays spontaneously to a μ-meson and a neutrino. Because parity is not conserved,[97] there is a correlation between the spin angular momentum of the μ-meson and the linear momentum of the π-meson, and hence the μ-meson beam is polarized. A μ-meson in turn decays with a half-life of 2.2×10^{-6} sec to an electron and two neutrinos. Again, because of parity nonconservation, the decay electrons are emitted anisotropically with respect to the spin angular momentum of the μ-meson. Thus the asymmetry in the angular distribution of the electron is a measure of the polarization of the μ-meson. For a given counter position the counting rate will then be a function of the strength of an applied static field. This experiment was first performed to verify the nonconservation

[94] L. W. Alvarez and F. Bloch, *Phys. Rev.* **57,** 111 (1940).

[95] W. R. Arnold and A. Roberts, *Phys. Rev.* **71,** 878 (1947).

[96] F. Bloch, D. Nicodemus, and H. H. Staub, *Phys. Rev.* **74,** 1025 (1948); H. H. Staub and E. H. Rogers, *Helv. Phys. Acta* **23,** 63 (1950).

[97] T. D. Lee and C. N. Yang, *Phys. Rev.* **104,** 254 (1956).

of parity, but it also provides a measurement of the μ-meson's magnetic moment.[98] More accurate experiments have employed radiofrequency as well as a static field.[99] However, the greatest accuracy has been achieved by measuring the difference between the orbital cyclotron frequency (see Section 5-8) and the spin precession frequency.[100] The g-value for the positive μ-meson was determined to be[100] $g = 2.002324$. Muonium, or an atom consisting of an electron and a positive μ-meson, has also been studied with beam resonance techniques.[101]

Nuclear moments with beams of atoms possessing an electronic magnetic moment are also studied by observing transitions between the hyperfine levels. Because the hyperfine spacing depends on whether the applied field H is weak or strong, it is convenient to be able to vary the frequency ν in these experiments. Besides yielding accurate values of g_n, analysis of the data gives the spin I, hence the nuclear moment.[102]

Additional important information obtained from beam experiments include the following.

1. If a nuclear quadrupole moment exists, the interaction with an electric field gradient produced by the electronic configuration will affect the splitting of the hyperfine levels. From a study of the weak and zero field transitions the values of the quadrupole moment may be determined.[103]
2. Isotopes of atoms have been observed to have a hyperfine structure that is different from that expected on the basis of a different nuclear mass. The effect is explained by considering the structure of the nucleus.[104] When an electron is close to the nucleus, the interaction depends on the spatial distribution of the magnetic moment of the nucleons that comprise the nucleus.
3. Precise measurements[105] of the hyperfine splitting for atoms of hydrogen, Ga, In, and Na can be explained only if the g-value of the

[98] R. L. Garwin, L. M. Lederman, and M. Weinrich, *Phys. Rev.* **105,** 1415 (1957).

[99] T. Coffin, R. L. Garwin, L. M. Lederman, S. Penman, and A. M. Sachs, *Phys. Rev.* **106,** 1108 (1957); R. A. Lundy, J. C. Sens, R. A. Swanson, V. L. Telegdi, and D. D. Yovanovitch, *Phys. Rev. Letters* **1,** 38 (1958); D. P. Hutchinson, J. Menes, G. Shapiro, A. M. Patlach, and S. Penman, *Phys. Rev. Letters* **7,** 129 (1961).

[100] G. Charpak, F. J. M. Farley, R. L. Garwin, T. Muller, J. C. Sens, V. L. Telegdi, and A. Zichichi, *Phys. Rev. Letters* **6,** 128 (1961).

[101] V. W. Hughes, D. W. McColm, K. Ziock, and R. Prepost, *Phys. Rev. Letters* **5,** 63 (1960).

[102] N. F. Ramsey, *Molecular Beams*, Oxford University Press, Oxford (1956), Ch. IX.

[103] For example, see A. K. Mann and P. Kusch, *Phys. Rev.* **77,** 427 (1950).

[104] A. Bohr, *Phys. Rev.* **73,** 1109 (1948); A. Bohr and V. F. Weisskopf, *Phys. Rev.* **77,** 94 (1950).

[105] J. E. Nafe and E. B. Nelson, *Phys. Rev.* **73,** 718 (1948); P. Kusch and H. M. Foley, *Phys. Rev.* **74,** 250 (1948); S. H. Koenig, A. G. Prodell, and P. Kusch, *Phys. Rev.* **88,** 191 (1952).

electron spin is not exactly 2.0 but slightly larger. With the assumption that $g_l = 1$, the best value for g for an electron spin after a certain relativistic correction is

$$g_s = 2.002292 \pm 0.000025. \tag{4-11.2}$$

The magnetic moment of the electron is said to be anomalous, since it is slightly larger than one Bohr magneton.

4. An experiment on a beam of excited atoms of hydrogen[106] has shown that the $2\,^2S_{1/2}$- and $2\,^2P_{1/2}$-states are not degenerate; the displacement of the energy levels is known as the Lamb shift. This result contradicts a prediction of the Dirac theory of the electron. The information obtained from experiments (3) and (4) has led to a new formulation of quantum electrodynamics.[107] According to this theory, the electron interacts with the electromagnetic field and with electrons in the negative energy states of the Dirac theory. The theory predicts values of the anomalous magnetic moment of the electron and the Lamb shift in excellent agreement with experiment.

It is outside the scope of this book to give a more detailed discussion of beam methods. The fact that the Nobel prizes awarded in physics for discoveries made with beam methods are second in number only to those made with the Wilson cloud chamber is mute evidence of the great importance of beam techniques.

Problems

4-1. Suppose an atom with a spherically symmetric charge distribution is placed in an external field H. Show that the induced diamagnetic current produces a field at the nucleus given by

$$\Delta H = -\left(\frac{eH}{3mc^2}\right)\varphi_E(0),$$

where $\varphi_E(0)$ is the electrostatic potential at the nucleus.

4-2. From the solutions to the Bloch equations (equations 3-8.4 and 3-8.5) deduce the value of M_x and M_y. Recall that the alternating field $2\mathbf{H}_1 \cos \omega t$ is applied along the x-direction and compute the susceptibility for this direction by using equation 3-4.3. Note that the results are one half the values obtained for equation 3-8.6 or 3-8.10. These are usually the susceptibilities of interest experimentally.

[106] W. E. Lamb and R. C. Retherford, *Phys. Rev.* **79,** 549 (1950), **81,** 222 (1951); E. S. Dayhoff, S. Triebwasser, and W. E. Lamb, *Phys. Rev.* **89,** 106 (1954).

[107] J. Schwinger, *Phys. Rev.* **73,** 416 (1948), **76,** 790 (1949); R. Karplus and N. Kroll, *Phys. Rev.* **77,** 536 (1950).

4-3. Suppose that the magnetic field is swept through a resonance in a time short compared to τ_1 and τ_2 and long compared to $1/\gamma H_1$; this is the condition of adiabatic rapid passage. Assume solutions of the Bloch equations

$$M_x = M_1(t) \cos \omega t \qquad M_y = M_1(t) \sin \omega t,$$

so that the resultant moment $M(t)$ is given by

$$M^2 = M_x^2 + M_y^2 + M_z^2 = M_1^2 + M_z^2.$$

Show that the conditions by which these equations are satisfied are

$$\dot{M}_1 + \frac{M_1}{\tau_2} = 0,$$

$$M_1\left(H + \frac{\omega}{\gamma}\right) = M_z H_1,$$

$$\dot{M}_z + \frac{M_z}{\tau_1} = \frac{M_0}{\tau_1},$$

by substitution into the Bloch equations of motion (3-8.1). Define $\delta = M_z/M_1$ so that

$$\delta = \frac{H + \omega/\gamma}{H_1}$$

$$= \frac{\omega - \omega_L}{\gamma H_1}.$$

Hence show $M_x = M/(1 + \delta^2)^{1/2} \cos \omega t$, $M_y = M/(1 + \delta^2)^{1/2} \sin \omega t$, $M_z = M\delta/(1 + \delta^2)^{1/2}$. Further show that M is determined by the solution of the differential equation

$$\dot{M} + \frac{M}{\tau_1}\frac{(\delta^2 + \tau_1/\tau_2)}{(1 + \delta^2)} = \frac{M_0 \delta}{\tau_1(1 + \delta^2)^{1/2}}.$$

This solution is

$$M(t) = \int_{-\infty}^{t} dt' \frac{M_0(t')\,\delta(t')}{\tau_1[1 + \delta^2(t')]^{1/2}} \exp\left[\int_t^{t'} \frac{\delta^2(t'') + \tau_1/\tau_2}{\tau_1\{1 + \delta^2(t'')\}} dt''\right],$$

where $M_0(t') = \chi_0 H(t')$.

4-4. In considering the effects of an alternating magnetic field $2H_1 \cos \omega t$ in a resonance experiment, only the counterclockwise rotating part was retained. If H_1 is comparable to the steady field H, the other rotating component cannot be ignored. It has been shown by F. Bloch and A. Siegert, *Phys. Rev.* **57**, 522 (1940), that the resonance condition $\omega = -\gamma H$ then becomes to first order

$$\omega = -\gamma H\left(1 + \frac{1}{4}\frac{H_1^2}{H^2}\right).$$

Plot the resonant frequencies given by these equations as a function of H and also as a function of H_1/H for constant H.

4-5. The interaction of a quadrupole moment with an electric field gradient $\partial^2\varphi_E/\partial z^2$ shifts the energy levels by an amount

$$E = \frac{eQ}{4I(2I-1)}[3m^2 - I(I+1)]\frac{\partial^2\varphi_E}{\partial z^2},$$

as calculated by R. V. Pound, *Phys. Rev.* **79,** 685 (1950), with first-order perturbation theory; φ_E is the electrostatic potential. Plot the energy levels (in the absence of an applied field) for both integer and half-integer I. Note that Q cannot be determined unless $\partial^2\varphi_E/\partial z^2$ is known, which is seldom the case.

4-6. Assume that the spin-spin interaction in a LiF crystal corresponds to an effective field of 30 oe. As in the Purcell-Pound experiment, assume that in the nonadiabatic reversal of a field from +100 to −100 oe the spin-spin temperature is +5°K and the Zeeman temperature is −5°K. Assume that the field of −100 oe is maintained and then show that the final equilibrium spin temperature is −6°K. Neglect the spin-lattice interaction. If the field is then removed, show that the final spin temperature in the field-free state is −1.7°K. Finally, what is the temperature when the field returns to +100 oe?

Bibliography

Abragam, A., *The Principles of Nuclear Resonance*, Oxford University Press, Oxford (1961).

Andrew, E. R., *Nuclear Magnetic Resonance*, Cambridge University Press, Cambridge (1956).

Cohen, M. H., and F. Reif, "Quadrupole Effects in Nuclear Magnetic Resonance Studies of Solids," *Solid State Phys.* **5,** 321 (1957).

Das, T. P., and E. L. Hahn, *Nuclear Quadrupole Resonance Spectroscopy*, Academic Press, New York (1958).

Pake, G. E., "Nuclear Magnetic Resonance," *Solid State Phys.* **2,** 1 (1956).

Pople, J. A., W. G. Schneider, and H. J. Bernstein, *High-Resolution Nuclear Magnetic Resonance*, McGraw-Hill Book Co., New York (1959).

Powles, J. G., "Nuclear Magnetism in Pure Liquids," *Rept. Progr. Phys.* **22,** 433 (1959).

Purcell, E. M., "Nuclear Magnetism and Nuclear Relaxation," *Nuovo cimento, Suppl.* **3,** 961 (1956).

Ramsey, N. F., *Molecular Beams*, Oxford University Press, Oxford (1956).

Roberts, J. D., *Nuclear Magnetic Resonance*, McGraw-Hill Book Co., New York (1959).

Saha, A. K., and T. P. Das, "Theory and Applications of Nuclear Induction," *Saha Inst. Nuc. Phys. (Calcutta)* (1957).

Slichter, C. P., *Principles of Magnetic Resonance*, Harper and Row, New York (1963).

Smidt, J., Editor, *Magnetic and Electric Resonance and Relaxation*, North-Holland Publishing Co., Amsterdam (1963).

Van Vleck, J. H., "Line Breadths and Theory of Magnetism," *Nuovo cimento, Suppl.* **3,** 993 (1956).

5

The Magnetic Properties of an Electron Gas

1. Statistical and Thermodynamic Functions for an Electron Gas

Metals, except for those few that are ferromagnetic, exhibit either diamagnetism or a feeble paramagnetism. The magnetic properties of metals are quite different from those of the systems discussed in Chapters 2 and 3. Many of the properties of a metal can be understood on the basis of the following model. It is assumed that the valence electrons of the atoms that constitute the metal are not localized at the atoms but instead are able to wander throughout the volume of the metal. In a first approximation the interaction of these electrons with each other and with the ion cores is neglected, and the electrons are said to form a *free-electron gas.* This approximation works best for metals composed of monovalent atoms. Consideration of the interaction with the ion cores leads to the band picture; the mobile electrons are said to occupy the conduction band and are called conduction electrons.

For a monovalent metal the number of conduction electrons per unit volume is equal to the number of atoms per unit volume. Because of this high density, together with the small mass of the electron, an electron gas at ordinary temperatures obeys Fermi-Dirac statistics rather than the usual Boltzmann statistics. In this section we give an outline of the statistical and thermodynamic relationships required for a discussion of the magnetic properties of the electron gas.

In order to discuss the statistics of the electron gas, it is necessary to know the quantum states and energies that are possible. The one-dimensional Schrödinger equation for an electron is

$$\left[\left(-\frac{\hbar^2}{2m}\frac{\partial^2}{\partial x^2} + V\right)\right]\psi = E\psi.$$

First the interaction with the ion cores will be neglected; that is, the electron will be considered to be free and then $V = 0$. The general solution is given by

$$\psi(x) = Ae^{ikx} + Be^{-ikx}$$

where

$$k^2 = \frac{2m}{\hbar^2}E.$$

The quantity k is called the wave vector, and for this case $|\mathbf{k}| = 2\pi/\lambda$, where λ is the de Broglie wavelength. The electron is confined to the interior of the metal; that is $0 < x < L$, where L is the x-dimension of the metal. As a result, the boundary conditions on the wave function are $\psi = 0$ for $x = 0$ and for $x = L$. From the first condition $A = -B$ and $\psi = C \sin kx$. From the second condition

$$\sin kL = 0$$

that is, $k_n = n\pi/L$, where $n = 1, 2, 3, \ldots$. Hence the wave functions are standing waves, given by $\psi_n = C \sin n\pi x/L$. The corresponding energies are

$$E_n = \frac{\hbar^2 k_n^{\,2}}{2m} = \frac{\hbar^2 \pi^2 n^2}{2mL^2}.$$

Analogously, for an electron in a three-dimensional metal, with edges of length L, the solutions are the standing waves

$$\psi(x, y, z) = C \sin\frac{n_x \pi x}{L}\sin\frac{n_y \pi y}{L}\sin\frac{n_z \pi z}{L}, \tag{5-1.1}$$

where n_x, n_y, and n_z are positive nonzero integers. The energy levels are given by

$$E_n = \frac{\hbar^2 k_n^{\,2}}{2m} = \frac{\hbar^2 \pi^2}{2mL^2}(n_x^{\,2} + n_y^{\,2} + n_z^{\,2}). \tag{5-1.2}$$

The allowed values of the wave vector **k** are given by

$$\mathbf{k} = \frac{\pi}{L}\mathbf{n} \tag{5-1.3}$$

where $\mathbf{n} = n_x\mathbf{i} + n_y\mathbf{j} + n_z\mathbf{k}$. In general, the states are degenerate, since different combinations of n_x, n_y, and n_z may have the same energy. The expression $\hbar^2\pi^2/2mL^2$ is a measure of the difference between successive energy levels. For a macroscopic sized piece of metal, with say $L = 1$ cm, we have $\hbar^2\pi^2/2mL^2 \approx 10^{-16}$ ev. Hence for this case the levels are so closely spaced they may be considered to be quasi-continuous.

It is useful to have an expression for the number of possible states, or wave functions, that correspond to a momentum between p and $p + dp$. Equation 5-1.2 can be rewritten as

$$n_x^2 + n_y^2 + n_z^2 = R^2,$$

where $R^2 = p^2L^2/\hbar^2\pi^2$ since $E = p^2/2m$. Now, since each set of positive integers n_x, n_y, n_z represents one state, the number of states in a shell between R and $R + dR$ is just the volume of this shell. This volume is given by

$$\tfrac{1}{8}4\pi R^2\, dR = \frac{4\pi p^2\, dpV}{h^3},$$

where $V = L^3$, and this is the required expression.

Electrons obey the Pauli exclusion principle, so that each possible state can be occupied at most by one electron. However, the foregoing discussion neglected the spin of the electron. Since the spin can be either up or down, the number of possible states in the momentum range dp at p is given by

$$C(p)\, dp = \frac{8\pi p^2\, dpV}{h^3}. \tag{5-1.4}$$

The number of states between the energies E and $E + dE$ is

$$C(E)\, dE = KE^{1/2}\, dE,$$

where

$$K = \frac{4\pi V(2m)^{3/2}}{h^3}, \tag{5-1.5}$$

which is easily shown from equation 5-1.4 with the aid of $E = p^2/2m$.

The interaction of the free electron with the positive ion cores is treated by considering Schrödinger's equation when the potential V is periodic; that is,

$$V(\mathbf{r}) = V(\mathbf{r} + \mathbf{d}),$$

where **d** is a vector that describes the periodicity of the lattice. It can then be shown that the solutions of the Schrödinger equation are of the form

$$\psi_k = u_k(\mathbf{r})e^{i\mathbf{k}\cdot\mathbf{r}} \tag{5-1.6}$$

where

$$u_k(\mathbf{r}) = u_k(\mathbf{r} + \mathbf{d}).$$

Hence the solutions, known as Bloch functions, are just plane waves modulated by a function that has the same periodicity as the lattice. Only certain ranges of energy, known as bands, are now possible. The highest energy band containing electrons is known as the conduction band because the electrons do not fill all its states, and hence are mobile. The next lowest band in energy is known as the valence band. When each state of this band is occupied by electrons, an external field cannot change the state of an electron. Hence this band then will not contribute to the conductivity, susceptibility, and other properties of a metal. The energy of the electrons can, as a rule, still be represented by a quadratic function of k, as for the free electron, but it is necessary to replace the mass m of the electron by the effective mass m^*. For example, it may be possible to write the energy as

$$E = \frac{\hbar^2 k^2}{2m^*}. \tag{5-1.7}$$

The concept of the effective mass perhaps may be best understood by considering the motion of a conduction electron under the action of an applied field, either magnetic or electric. It is customary to express the motion explicitly in terms of this external field only. Then, in order to take account of the interaction between the electron and the ion cores, the effective mass must be used instead of the mass. From these considerations it can be shown that the effective mass is given by

$$m^* = \frac{\hbar^2}{d^2E/dk^2} \tag{5-1.8}$$

and this is consistent with the special case of (5-1.7). When the energy surface is anisotropic, the effective mass may be represented by a tensor with components given by

$$\left(\frac{1}{m^*}\right)_{ij} = \frac{1}{\hbar^2}\frac{d^2E}{dk_i\,dk_j}, \tag{5-1.9}$$

where $i, j = x, y, z$. The ratio m^*/m may be greater or smaller than unity for different metals. For a detailed discussion of band theory the reader is referred to a standard textbook.[1] For our purposes we are usually content to extend the free electron treatment to include the ion core interactions by replacing m by m^*.

Let us now consider the statistics of an electron gas. Because of the Pauli exclusion principle, each possible state available to the electron gas is either unoccupied or occupied by at most one electron. Let C_i denote the number of states (energy levels) with energies between E_i and $E_i + dE_i$; this range will be assumed small enough so that these levels may be treated as though each had the energy E_i. If $\mathcal{N}_i$ represents the number of particles in the C_i levels, then $C_i - \mathcal{N}_i$ is the number of unoccupied levels. Since the electrons are indistinguishable, the number of possible arrangements W_i is given by

$$W_i = \frac{C_i!}{N_i!(C_i - \mathcal{N}_i)!}.$$

If altogether there are $\mathcal{N}$ electrons, the total number of arrangements in which the $\mathcal{N}$ electrons can be distributed over the different energy levels E_i for a given set of values of N_i is equal to

$$W = \prod_i \frac{C_i!}{\mathcal{N}_i!(C_i - \mathcal{N}_i)!}.$$

According to the principles of statistical mechanics, the average properties of the gas when in equilibrium will be described by the most probable distribution. Thus W must be a maximum for this case, subject to the conditions

$$\sum \mathcal{N}_i = \mathcal{N}$$

and

$$\sum \mathcal{N}_i E_i = E,$$

where E is the total energy of the electron gas. It is convenient to employ $\log W$ rather than W, so that

$$\begin{aligned}\log W &= \sum_i \log W_i \\ &= \sum_i [C_i \log C_i - \mathcal{N}_i \log \mathcal{N}_i - (C_i - \mathcal{N}_i) \log (C_i - \mathcal{N}_i)],\end{aligned}$$

where Stirling's formula $\log \mathcal{N}! \approx \mathcal{N} \log \mathcal{N}$ has been used. For W to be a maximum for small variation $\delta\mathcal{N}_i$ in $\mathcal{N}_i$, we must have $\delta \log W = 0$ together with $\delta\mathcal{N} = 0$ and $\delta E = 0$. When the method of undetermined

[1] For example, A. J. Dekker, *Solid State Physics*, Prentice-Hall, Englewood Cliffs, New Jersey (1957), Ch. 10; C. Kittel, *Solid State Physics*, John Wiley and Sons, New York (1956), Ch. 11; F. Seitz, *Modern Theory of Solids*, McGraw-Hill Book Co., New York (1940), Ch. 8.

multipliers of Lagrange is applied, we obtain

$$\delta \log W - \alpha \sum_i \delta \mathcal{N}_i - \beta \sum_i E_i \, \delta \mathcal{N}_i = 0,$$

where α and β are undetermined constants. Since this equation must be satisfied for all values of i, we have

$$\log \left(\frac{C_i - \mathcal{N}_i}{\mathcal{N}_i} \right) - \alpha - \beta E_i = 0;$$

hence

$$\mathcal{N}_i = \frac{C_i}{e^{\alpha + \beta E_i} + 1}$$

$$= C_i \, F(E_i). \tag{5-1.10}$$

The function $F(E_i)$ is known as the Fermi-Dirac distribution function, or the Fermi function, for short.

When the density of electrons $\mathcal{N}_i/C_i$ is small, a situation that would be expected to occur at high temperatures, then $e^{\alpha + \beta E_i}$ must be large compared to 1 and the unity in the denominator may be neglected. Hence we have

$$\mathcal{N}_i = C_i e^{-\alpha} e^{-\beta E_i}.$$

By comparison with the Boltzmann distribution, we see that

$$\beta = \frac{1}{kT}.$$

For this case, in which $e^{\alpha} \gg 1$, the electron gas is said to be nondegenerate. When $0 < e^{\alpha} < 1$, that is, for $\alpha < 0$, the gas is degenerate and the unity term in the denominator of equation 5-1.10 must be retained. The latter situation applies for the electron gas of a metal.

The parameter α is determined by the condition $\Sigma \, \mathcal{N}_i = \mathcal{N}$. It is convenient to define the energy E_F, called the *Fermi energy*, by

$$\alpha = - \frac{E_F}{kT}.$$

Equation 5-1.10 may then be rewritten as

$$\mathcal{N}(E) \, dE = C(E) \, F(E) \, dE$$

with

$$F(E) = \frac{1}{e^{(E - E_F)/kT} + 1}, \tag{5-1.11}$$

where $\mathcal{N}(E) \, dE$ is the number of electrons occupying $C(E) \, dE$ states with energies between E and $E + dE$ and $F(E)$ is the Fermi function (see also equation 5-1.10).

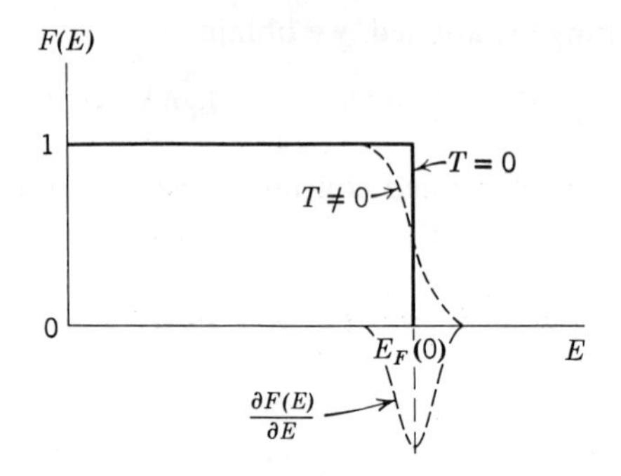

Fig. 5-1.1. The Fermi function $F(E)$ at $T = 0$ and at $T \neq 0$.

At $T = 0°\text{K}$, absolute zero, the Fermi function has the property that $F(E) = 1$ for $E < E_F(0)$ and $F(E) = 0$ for $E > E_F(0)$. Thus all states with energies less than $E_F(0)$ are occupied and all those above $E_F(0)$ are empty, as indicated in Fig. 5-1.1. As the temperature is increased, states within about $\mathrm{k}T$ below the Fermi level begin to depopulate and states within about $\mathrm{k}T$ above the Fermi level begin to be populated. For energies below $E_F(T)$ by more than a few $\mathrm{k}T$ the value of $F(E)$ remains almost equal to unity. This behavior is illustrated also in Fig. 5-1.1. The position of the Fermi energy level is determined by the value of E for which $F(E) = \frac{1}{2}$, as follows from equation 5-1.11.

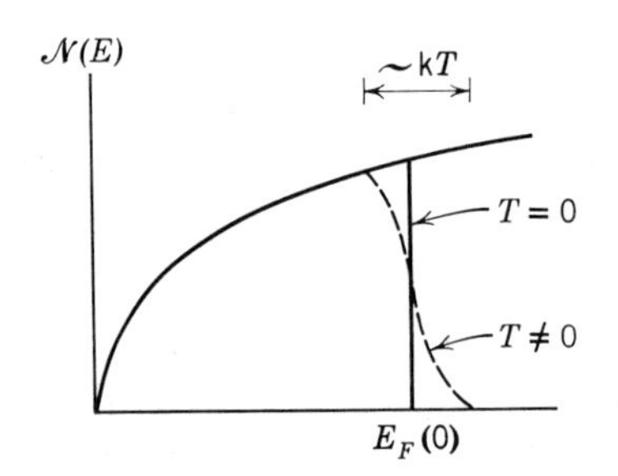

Fig. 5-1.2. The density of states $N(E)$ for $T = 0$ and for $T \neq 0$. For $T = 0$ the states are filled up to the Fermi energy level $E_F(0)$.

For a free electron gas the density of states $C(E)\,dE$ is given by equation 5-1.5. This function is shown in Fig. 5-1.2 with the energy distribution $\mathcal{N}(E)$, found by multiplying $C(E)$ by $F(E)$, both for $T = 0°\text{K}$ and $T \neq 0$. The Fermi energy is easily calculated for this case from the condition

$$\int_0^\infty \mathcal{N}(E)\,dE = \int_0^\infty C(E)\,F(E)\,dE$$

$$= \mathcal{N}.$$

At absolute zero this gives

$$\mathcal{N} = K \int_0^{E_F(0)} E^{1/2}\,dE$$

or

$$E_F(0) = \frac{h^2}{2m}\left(\frac{3N}{8\pi}\right)^{2/3}, \tag{5-1.12}$$

where $N = \mathcal{N}/V$, the number of electrons per unit volume. $E_F(0)$ is of the order of a few electron volts. Hence below the melting point of metals $kT \ll E_F(T)$, and $E_F(T)$ is close to its value at 0°K. It is also of interest to calculate the average energy of the electrons at absolute zero. Here

$$\overline{E(0)} = \frac{1}{\mathcal{N}}\int_0^{E_F(0)} E\, C(E)\, dE,$$

which gives

$$\overline{E(0)} = \frac{3}{5} E_F(0). \tag{5-1.13}$$

For statistical calculations it is often convenient to employ the partition function $\mathfrak{Z}$, defined as

$$\mathfrak{Z} = \sum_i e^{-Ei/kT} \tag{5-1.14}$$

where the summation is over *all* states of the system. For the quantum theory E_i is the eigenvalue of the Hamiltonian $\mathcal{H}$, whereas for classical theory E is equal to the Hamiltonian $\mathcal{H}$. For a quasi-continuous distribution of states, as for the free electron gas, the number of states between x and $x + dx$, y and $y + dy$, z and $z + dz$, p_x and $p_x + dp_x$, p_y and $p_y + dp_y$, and p_z and $p_z + dp_z$ is given by equation 5-1.4 as

$$\frac{2}{h^3}\, dx\, dy\, dz\, dp_x\, dp_y\, dp_z.$$

Hence the partition function can be approximated by the integral

$$\mathfrak{Z} = \frac{2}{h^3}\iint \cdots \int \exp\left[-\frac{E(p,q)}{kT}\right] dx\, dy\, dz\, dp_x\, dp_y\, dp_z, \tag{5-1.15}$$

where $E(p, q)$ expresses the relationship of the energy to the momenta and coordinates. The magnetization M is given in terms of the partition function by

$$M = NkT\frac{\partial \log \mathfrak{Z}}{\partial H}, \tag{5-1.16}$$

since by differentiating we get

$$M = \frac{N\sum(\partial E_i/\partial H)e^{-E_i/kT}}{\sum e^{-E_i/kT}}$$

$$= \frac{N\sum \mu_i e^{-E_i/kT}}{\sum e^{-E_i/kT}}, \tag{5-1.17}$$

where equation 2-8.7 has been used. This result is just the magnetic moment when the moments are distributed over the states according to the Boltzmann factor (compare with equation 2-8.9). An expression for the magnetization may also be given in terms of the Gibbs free-energy function, since by equation 3-2.17

$$dG = -M\,dH - S\,dT$$

and then, for constant temperature,

$$M = -\left(\frac{\partial G}{\partial H}\right)_T. \tag{5-1.18}$$

Consequently G must be related to the partition function by

$$G = -N\mathrm{k}T \log \mathfrak{Z}. \tag{5-1.19}$$

Many writers prefer to employ the free energy F in place of the Gibbs function G in the foregoing equations, in which case the thermodynamic system must be defined so that the internal energy U does not include the interaction energy HM. Our definition requires the use of enthalpy E where these other writers employ the internal energy U.

The partition function is of the greatest utility when Boltzmann statistics apply. When it is necessary to consider cases in which the Fermi-Dirac statistics apply, it is extremely convenient to allow the number of particles of the system (actually the subsystem) to vary. It is then more suitable to employ the *grand* canonical ensemble[2] in the treatment of the electron gas rather than the method employed earlier in this section. This treatment leads to the definition of a *grand* partition function. We shall not go into these matters here. However, it is appropriate to consider some of the thermodynamic equations that result when $\mathcal{N}$ is allowed to vary. For a one-component system (all particles the same) we have

$$\begin{aligned} dG &= -M\,dH - S\,dT + \frac{\partial G}{\partial \mathcal{N}}\,d\mathcal{N} \\ &= -M\,dH - S\,dT + \mu\,d\mathcal{N}, \end{aligned} \tag{5-1.20}$$

where the chemical potential μ is defined as

$$\mu = \left(\frac{\partial G}{\partial \mathcal{N}}\right)_{H,T}. \tag{5-1.21}$$

[2] R. C. Tolman, *Principles of Statistical Mechanics*, Oxford University Press, Oxford (1938); C. Kittel, *Elementary Statistical Physics*, John Wiley and Sons, New York (1958).

Clearly equation 5-1.18 still applies, provided $\mathcal{N}$ is a constant, a condition already implied for equation 3-2.17. Since $G = E - TS$, we have

$$dE = T\,dS - M\,dH + \mu\,d\mathcal{N}, \tag{5-1.22}$$

which is simply a statement of the first law of thermodynamics. It is left to the reader to show that μ is equal to the Fermi energy E_F (problem 5-2) and that the Fermi levels of two systems in thermal equilibrium are equal (problem 5-3).

The use of the grand partition function leads directly to an expression for the Gibbs function G in terms of the Fermi-Dirac distribution function. This expression is

$$G = \mathcal{N}E_F - \mathrm{k}\sum_i \log\left(1 + e^{(E_F - E_i)/\mathrm{k}T}\right). \tag{5-1.23}$$

Since $G = E - TS$, it follows immediately that the entropy is

$$S = \frac{E}{T} + \mathrm{k}\sum_i \log\left(1 + e^{(E_F - E_i)/\mathrm{k}T}\right) - \frac{\mathcal{N}E_F}{T}. \tag{5-1.24}$$

The summation in these two equations is over all the individual states. To make this clear, consider equation 5-1.10 rewritten as

$$\begin{aligned} n_i &= \frac{\mathcal{N}_i}{C_i} \\ &= \frac{1}{e^{(E_i - E_F)/\mathrm{k}T} + 1}, \end{aligned}$$

where n_i is the average number of electrons in each state. Because of the Pauli principle, n_i is some number lying between 0 and 1. Then, on summing over all the individual states, we get

$$\sum_i n_i = \mathcal{N} = \sum_i \frac{1}{e^{(E_i - E_F)/\mathrm{k}T} + 1} \tag{5-1.25}$$

and

$$\sum_i n_i E_i = E = \sum_i \frac{E_i}{e^{(E_i - E_F)/\mathrm{k}T} + 1}. \tag{5-1.26}$$

We shall be content here to prove equation 5-1.24, hence (5-1.23), by showing that the differential dS leads to equation 5-1.22. We obtain

$$\begin{aligned} dS = d\left(\frac{E}{T}\right) &- \mathrm{k}\sum_i \frac{1}{e^{(E_i - E_F)/\mathrm{k}T} + 1} \\ &\times \left[E_i d\left(\frac{1}{\mathrm{k}T}\right) + \frac{1}{\mathrm{k}T}dE_i - d\left(\frac{E_F}{\mathrm{k}T}\right)\right] - d\left(\frac{\mathcal{N}E_F}{T}\right). \end{aligned} \tag{5-1.27}$$

Now the second term on the right-hand side is just equal to $-Ed(1/T)$ because of equation 5-1.26. This result combines with the first term $d(E/T)$ to give dE/T. The third term may be written

$$-\sum_i \frac{n_i}{T} dE_i = -\sum_i \frac{n_i}{T} \frac{\partial E_i}{\partial H} dH,$$

for which equation 5-1.25 has been used. But $\partial E_i/\partial H = -\mu_i$, the magnetic moment for the state with energy E_i (equation 2-8.7). Hence for this term we get

$$+\frac{M\, dH}{T}.$$

The fourth term on the right-hand side of (5-1.27) gives

$$\sum_i n_i\, d\left(\frac{E_F}{T}\right) = \mathcal{N}\, d\left(\frac{E_F}{T}\right).$$

When combined with the last term $-d(\mathcal{N}E_F/T)$, the result is $-(E_F/T)\, d\mathcal{N}$ or $-(\mu/T)\, d\mathcal{N}$. Therefore we have finally

$$dS = \frac{dE}{T} + \frac{M\, dH}{T} - \frac{\mu\, d\mathcal{N}}{T}$$

as required by the first law of thermodynamics.

2. The Spin Paramagnetism of the Electron Gas

The susceptibility of an electron gas is composed of a paramagnetic contribution from the electron spin and a diamagnetic contribution from the electron's translational or orbital motion. The spin part is discussed in this section; it is usually the larger portion and, moreover, is simpler to treat. The separation of the spin and orbital parts is permissible because there is no spin-orbit coupling to a first order of approximation.

If the electron gas obeyed Boltzmann statistics, the susceptibility would be given by Hund's formula (equation 2-7.2) with $J = S$; that is,

$$\chi = \frac{Ng^2\mu_B{}^2 S(S+1)}{3kT} \tag{5-2.1}$$

where N is the number of conduction electrons per unit volume and $S = \frac{1}{2}$. The result gives $\chi \approx 10^{-4}$ per cm^3, whereas for metals we find by experiment that $\chi \approx 10^{-6}$ per cm^3. Also, the observed susceptibility depends only slightly on temperature, instead of being inversely proportional to it. This discrepancy between theory and experiment is removed by the application of Fermi-Dirac statistics, as first shown by Pauli.[3]

Suppose that the electron gas is at temperature $T = 0°K$. It is convenient to consider the electron gas as two subsystems, one with electrons of a given spin direction and the other with electrons of the opposite spin direction. In the absence of an applied magnetic field the electrons of each subsystem will fill all the possible energy states below $E_F(0)$, whereas

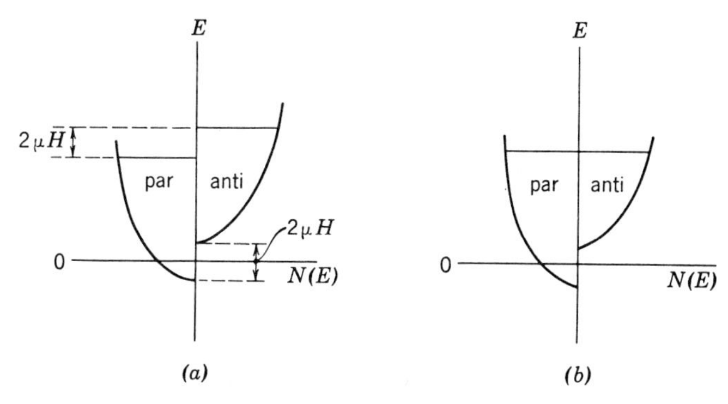

Fig. 5-2.1. The distribution of states for a free electron gas when a magnetic field H is applied. In (a) the situation is unstable; in (b) equilibrium between the parallel and antiparallel subsystems has been achieved.

those above $E_F(0)$ will be empty. The net magnetic moment will then be zero, since there will be equal numbers of electrons in each subsystem. If a field **H** is applied, the energy of an electron in the subsystem with a magnetic moment component parallel to **H** becomes[4] $E - \mu_B H$, whereas that for an electron with an antiparallel moment component in the other subsystem becomes $E + \mu_B H$, where E is the energy when $H = 0$. If the number of electrons in each subsystem remained the same, the situation would be as illustrated in Fig. 5-2.1a. This is an unstable situation, since energy is gained when electrons with antiparallel spins enter the parallel spin subsystem. Equilibrium is reached when the two subsystems are filled to the same energy level, that is, when the Fermi levels of the two systems are equal, as illustrated in Fig. 5-2.1b. This is a special case of the general result of problem 5-3. Since $\mu_B H \ll E_F$, only a few electrons near the Fermi level change their spin direction; this is the reason for the small magnetic moment observed for metals compared to that predicted by Boltzmann statistics.

[3] W. Pauli, *Z. Physik* **41**, 81 (1927).

[4] It is assumed that $g = 2$, $S = \frac{1}{2}$; otherwise $\boldsymbol{\mu} \cdot \mathbf{H}$ should be written $g\mu_B m_s H$.

The number of electrons per unit volume, ΔN, that change their spin direction is equal to the number of electrons in the energy range $\mu_B H$ at the top of the distribution. To a good approximation this is

$$\Delta N = \mu_B H \frac{C(E_F)}{2V},$$

where $C(E_F)/2$ is the density of states for one of the subsystems at the Fermi level. If the electrons are completely free, $C(E_F)$ is given by equation 5-1.5, and we have

$$\Delta N = \frac{2\pi(2m)^{3/2}}{h^3} \mu_B H [E_F(0)]^{1/2}.$$

The magnetic moment is then given by

$$M = \mu_B(2\Delta N)$$
$$= \frac{4\pi}{h^3}(2m)^{3/2}[E_F(0)]^{1/2}\mu_B{}^2 H,$$

since $2\Delta N$ is the difference in the number of electrons in the two subsystems. By substituting for $E_F(0)$ from equation 5-1.12 we get

$$M = \frac{4\pi m}{h^2}\left(\frac{3N}{\pi}\right)^{1/3} \mu_B{}^2 H.$$

The susceptibility is given by

$$\chi = \frac{4\pi m}{h^2}\left(\frac{3N}{\pi}\right)^{1/3} \mu_B{}^2$$
$$= \frac{3N\mu_B{}^2}{2kT_F}$$

where

$$kT_F = E_F. \tag{5-2.2}$$

The latter relation is obtained by multiplying top and bottom by $E_F(0)$. For $N \approx 10^{22}/\text{cm}^3$ we get $\chi \approx 10^{-6}/\text{cm}^3$, in agreement with experiment. Further, χ is almost temperature-independent, for the influence of temperature on the Fermi distribution and on N is small. The interaction of the electrons with the ion cores has been neglected. It may be taken into account by replacing m in equation 5-2.2 by the effective mass m^*.

In order to treat the general case, we start with the equation for the magnetization

$$M = \mu_B \int [F(E - \mu_B H) - F(E + \mu_B H)] \frac{C\ E)}{2V}\, dE.$$

Now, a Taylor's series expansion gives for the Fermi function

$$F(E - \mu_B H) = F(E) - \mu_B H \frac{\partial F(E)}{\partial E} + \frac{1}{2}(\mu_B H)^2 \frac{\partial^2 F(E)}{\partial E^2} - \cdots.$$

and a similar result for $F(E + \mu_B H)$. Hence we have

$$M = -\frac{\mu_B{}^2 H}{V} \int \frac{\partial F(E)}{\partial E} C(E)\, dE.$$

Now let us make a Taylor expansion of $C(E)$ about the Fermi energy E_F:

$$C(E) = C(E_F) + (E - E_F)C'(E_F) + \tfrac{1}{2}(E - E_F)^2 C''(E_F) + \cdots,$$

where

$$C'(E_F) = \left.\frac{\partial C(E)}{\partial E}\right|_{E=E_F} \quad \text{and} \quad C''(E_F) = \left.\frac{\partial^2 C(E)}{\partial E^2}\right|_{E=E_F}.$$

Then by substituting in the equation for M we have

$$M = -\frac{\mu_B{}^2 H}{V}\left[C(E_F)\int_0^\infty \frac{\partial F(E)}{\partial E}\, dE + C'(E_F)\int_0^\infty (E - E_F)\frac{\partial F(E)}{\partial E}\, dE + \tfrac{1}{2}C''(E_F)\int_0^\infty (E - E_F)^2 \frac{\partial F(E)}{\partial E}\, dE + \cdots\right].$$

Now the function $\partial F(E)/\partial E$ is symmetrically shaped and has a maximum at the Fermi energy E_F. Further, since $F(E)$ is influenced by temperature only in a narrow range of the order of kT around E_F, $\partial F(E)/\partial E$ vanishes, except close to the Fermi energy, as illustrated in Fig. 5-1.1. Hence the lower limit of zero on the integrals may be replaced by $-\infty$. Then for the first integral we get

$$\int_{-\infty}^\infty \frac{\partial F(E)}{\partial E}\, dE = F(E)\big|_{E=\infty} - F(E)\big|_{E=-\infty} = -1.$$

Next, since $\partial F(E)/\partial E$ is an even power of $(E - E_F)$, the second integral is just zero. Finally, by writing $x = (E - E_F)/kT$ we find for the third integral

$$-\tfrac{1}{2}(kT)^2 \int_{-\infty}^\infty \frac{x^2 e^x}{(1 + e^x)^2}\, dx = -\frac{\pi^2}{6}(kT)^2,$$

a result that may be seen by expanding the integrand in powers of e^{-x}. Therefore we have

$$M = \frac{\mu_B{}^2 H}{V}\left[C(E_F) + \frac{\pi^2}{6}(kT)^2 C''(E_F) + \cdots\right]. \qquad (5\text{-}2.3)$$

Since this series converges rapidly, it is usually not necessary to include higher order terms.

It is left to the reader (problem 5-4) to show that the Fermi energy at $T \neq 0$ is

$$E_F(T) = E_F(0)\left\{1 - \frac{\pi^2}{12}\left[\frac{kT}{E_F(0)}\right]^2\right\}. \qquad (5\text{-}2.4)$$

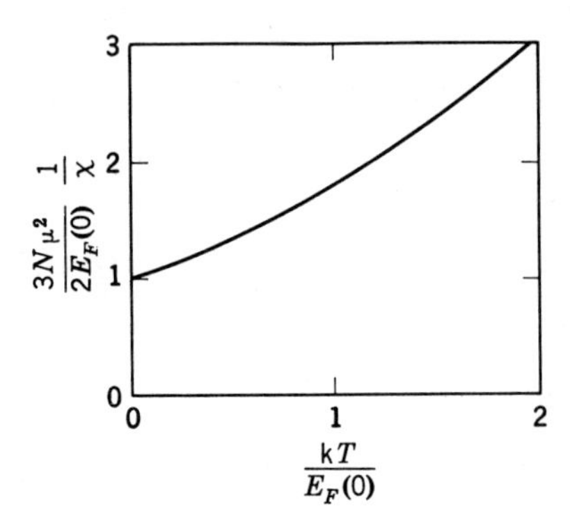

Fig. 5-2.2. The inverse of the susceptibility as a function of temperature for a free electron gas. Exchange and correlation effects are neglected.

When this result is substituted into equation 5-2.3, we find that the variation of the susceptibility with temperature is

$$\chi = \frac{4\pi m}{h^2}\left(\frac{3N}{\pi}\right)^{1/3}\mu_B^{\,2}\left\{1 - \frac{\pi^2}{12}\left[\frac{kT}{E_F(0)}\right]^2\right\}$$

$$= \frac{3N\mu_B^{\,2}}{2E_F(0)}\left\{1 - \frac{\pi^2}{12}\left[\frac{kT}{E_F(0)}\right]^2\right\}, \qquad (5\text{-}2.5)$$

provided $kT \ll E_F(0)$. A plot of this equation is shown in Fig. 5-2.2. For $T = 0$ the result is identical to equation 5-2.2. Again, the interaction with the ion cores may be taken into account by replacing m [which appears implicitly in $E_F(0)$] by m^*.

There is another important interaction to consider, namely that between the conduction electrons themselves. This interaction is not the result of the classical Coulomb forces, since these forces are independent of the spin direction, and hence do not influence the susceptibility. Instead, the interaction is the result of the *exchange* and *correlation* forces which do depend on the number of electrons that have parallel and antiparallel spin directions. These forces have a quantum mechanical origin. They arise because the wave function of the electron gas must have the antisymmetric properties required by the Pauli principle.

The exchange energy originates from the interaction of electrons of parallel spin. Because of the antisymmetric property of the wave functions, the probability is small that two electrons with parallel spins will be very close neighbors. In other words, the exchange forces tend to repel electrons with parallel spins; that is, the exchange energy is negative. The exchange energy occurs in the denominator of equations 5-2.2 and 5-2.5, and hence the effect is to increase the spin susceptibility of the electron gas. The exchange energy is of great importance in ferromagnetism and is discussed further in Chapter 6. A calculation of the exchange energy for free electrons can be carried out rigorously. The details will not be given here; they may be found in a number of textbooks.[5]

The exchange energy is calculated from a wave function that is the product of two wave functions, each of which contains the coordinates of one electron only; for example, the wave function of equation 5-1.1. This is only an approximation. Actually, the wave function must be a function of the distance between the two electrons, and hence the energy will depend on the correlation between their positions. The correlation energy is important only between electrons of opposite spin. It turns out that the correlation energy is negative; that is, electrons of opposite spin also tend to be repelled. The effect of the correlation energy is to decrease the susceptibility, since the action of the exchange forces is now of less importance. The calculation of the correlation energy is very difficult; Pines[6] has given a careful treatment of the problem.

Conventional methods of measuring the susceptibility of a metal yield a value that is the sum of the paramagnetic susceptibility already considered and the diamagnetic susceptibility both of the electron gas and of the positive ion cores. Consequently, comparison of these measurements with theory is postponed until the diamagnetic properties of the electron gas have been treated in the next section. However, it is possible to isolate the paramagnetic susceptibility by magnetic resonance techniques,[7] since the paramagnetic absorption line arises solely from the spin magnetic moment of the conduction electrons.

The procedure adopted by Schumacher and Slichter[7] will now be outlined. By the Kramers-Kronig relationships (equation 3-4.7) we have

$$\chi'(\omega_0) = \frac{2}{\pi}\int_0^\infty \frac{\omega\chi''(\omega)}{\omega^2 - {\omega_0}^2}\,d\omega + \chi'(\infty);$$

[5] F. Seitz, *Modern Theory of Solids*, McGraw-Hill Book Co., New York (1940); A. H. Wilson, *Theory of Metals*, Cambridge University Press, Cambridge (1958).

[6] D. Pines, *Phys. Rev.* **95,** 1090 (1954); D. Bohm and D. Pines, *Phys. Rev.* **92,** 609 (1953); D. Pines, *Phys. Rev.* **92,** 626 (1953).

[7] R. T. Schumacher and C. P. Slichter, *Phys. Rev.* **101,** 58 (1956); R. T. Schumacher, T. R. Caver, and C. P. Slichter, *Phys. Rev.* **95,** 1089 (1954).

hence the static susceptibility $\chi'(0)$ is proportional to the integrated area under the absorption line. For a narrow line χ'' is important only close to the electron resonance frequency $\omega_L = \gamma H$, and therefore $1/\omega_L$ may be taken outside the integral. Since the field rather than the frequency is varied, it is appropriate to replace $d\omega$ by $\gamma\, dH$; we then have

$$\chi = \chi'(0) = \frac{2}{\pi\omega_L}\gamma\int_0^\infty \chi''\, dH, \tag{5-2.6}$$

where it has been assumed that $\chi'(\infty) = 0$. However, it is difficult to measure this absorption absolutely. This difficulty can be circumvented by measuring the magnetic resonance of the metal's nuclei with the identical apparatus at the same frequency ω_L, which can be achieved simply by increasing the magnetic field H to a sufficiently large value. Now, since the static nuclear susceptibility χ_n is given by the Hund formula (equation 2-7.2), we have

$$\chi_n = \chi_n'(0) = \frac{N g_n^2 I(I+1)\mu_{B_n}^2}{3kT}$$

$$= \frac{2}{\pi\omega_L}\gamma_n\int_0^\infty \chi_n''\, dH, \tag{5-2.7}$$

where

$$\gamma_n = \frac{g_n e}{2M_p c}.$$

Hence, by dividing equation 5-2.6 by (5-2.7), we obtain

$$\chi = \chi_n \frac{\gamma}{\gamma_n}\frac{\int \chi''\, dH}{\int \chi_n''\, dH}$$

$$= \chi_n \frac{\gamma}{\gamma_n}\frac{A_e}{A_n}, \tag{5-2.8}$$

where A_e and A_n are the areas under the conduction electron and nuclear absorption curves, respectively. Experimental results have been obtained only for lithium and sodium; they are shown in Table 5-2.1. Theoretical values, determined from Pauli's theory (equation 5-2.2) and from Pine's theory,[8] in which corrections for exchange and correlation have been made, are also listed for comparison. The effective mass has been used in these calculations. The best agreement is with the values of Pines;

[8] D. Pines, *Phys. Rev.* **95,** 1090 (1954).

Table 5-2.1. Experimental and Theoretical Paramagnetic Susceptibilities ($\times 10^{-6}$ emu per cm^3)

			Theoretical	
Metal	Experiment	m^*/m	Pauli	Pines
Li	2.08 ± 0.1	1.46	1.17	1.87
Na	0.95 ± 0.1	0.985	0.64	0.85

this shows the importance of including the effect of exchange and correlation. It should be mentioned also that it is possible to isolate the paramagnetic susceptibility from measurements of the Knight shift. The Knight shift is discussed later in this chapter.

3. The Diamagnetism of the Electron Gas

The translational or orbital motion of conduction electrons in the presence of a magnetic field gives rise to diamagnetic effects which will now be considered. The Hamiltonian for an electron in a magnetic field is given, either classically or quantum mechanically, by equation 2-8.4 as

$$\mathcal{H} = \frac{1}{2m}\left(\mathbf{p} - \frac{e}{c}\mathbf{A}\right)^2 + V,$$

where **A** is the vector potential. The spin is omitted, since to the order of the approximations to be used in this section the orbital and spin terms may be treated separately. For a free electron the potential V is zero. In this Hamiltonian p is the *canonical* momentum. It is the momentum that is employed in phase space and that is replaced by the operator $(\hbar/i)(\partial/\partial q)$ in quantum mechanics. The canonical momentum is related to the kinetic momentum $m\mathbf{v}$ by the equation

$$m\mathbf{v} = \mathbf{p} - \frac{e}{c}\mathbf{A}. \tag{5-3.1}$$

First, the orbital susceptibility of a free electron gas will be calculated by classical theory. Substitution for the canonical momentum **p** from equation 5-3.1 in the Hamiltonian gives for the energy E of the electron,

$$\mathcal{H} = E$$
$$= \frac{1}{2m}(m\mathbf{v})^2.$$

A similar expression is obtained for each of the particles of the electron gas. Hence the partition function $\mathfrak{Z} = \sum_i e^{-E_i/kT}$ (equation 5-1.14) may be transformed so that it is independent of the magnetic field. Therefore by equation 5-1.16 the magnetic moment vanishes. This is an example of van Leeuwen's theorem, mentioned in Section 2-3, which states that in classical theory the total magnetic susceptibility is zero. (See also problem 5-5.)

However, quantum theory does lead to a nonzero value for the orbital susceptibility of the conduction electrons, since then van Leeuwen's theorem no longer applies. The reason for the difference between the classical and quantum mechanical results may be understood from the following physical argument. The magnetic field causes each electron to move in a circular orbit (actually helical orbit) whose axis is along the field direction. However, classically the speed of each electron is unchanged, hence the energy is also unchanged. It follows by the same argument previously used that classically the susceptibility is zero. Now, the circular motion of an electron may be considered to be composed of two oscillatory components, each a linear harmonic oscillator. In quantum mechanics a harmonic oscillator may have only certain discrete values of energy. It is this quantum effect that leads to a diamagnetic susceptibility.

The quantum mechanical case will now be treated quantitatively. Suppose, as usual, that the magnetic field H is applied along the z-direction. A convenient choice of the vector potential is then

$$A_x = -Hy \qquad A_y = 0 \qquad A_z = 0, \tag{5-3.2}$$

since by equation 1-8.6 $\mathbf{H} = \nabla \times \mathbf{A}$. Other choices for $\mathbf{A}$ are possible; for example, that used to obtain equation 2-8.5. The results obtained with different choices of gauge are the same, although this is not immediately obvious at all points of the calculation; problem 5-6 is an illustration of gauge invariance. Use of equation 5-3.2 gives for the Hamiltonian of a free electron

$$\mathcal{H} = \frac{1}{2m}\left[\left(p_x - \frac{e}{c}\,yH\right)^2 + p_y^{\,2} + p_z^{\,2}\right].$$

Schrödinger's equation is then

$$\left(\frac{\partial}{\partial x} - \frac{ie}{hc}\,yH\right)^2\psi + \frac{\partial^2\psi}{\partial y^2} + \frac{\partial^2\psi}{\partial z^2} + \frac{2mE}{\hbar^2}\,\psi = 0. \tag{5-3.3}$$

This equation may be solved if we assume

$$\psi(x, y, z) = e^{i(k_x x + k_z z)}\,\Phi(y).$$

We then obtain

$$\frac{d^2\Phi}{dy^2} + \left[\frac{2mE_\perp}{h^2} + \left(\frac{eH}{\hbar c}\,y - k_x\right)^2\right]\Phi = 0, \tag{5-3.4}$$

where $E_\perp = E - \hbar^2 k_z^2/2m$ is the energy of the motion in the plane perpendicular to the direction of the field. If we substitute

$$k_x = \frac{eH}{\hbar c} y_0 \tag{5-3.5}$$

and

$$\omega = -\frac{eH}{mc}, \tag{5-3.6}$$

we have

$$\frac{d^2\Phi}{dy^2} + \frac{2m}{\hbar^2}[E_\perp + \tfrac{1}{2}m\omega^2(y - y_0)^2] = 0.$$

This is just Schrödinger's equation for the one-dimensional linear harmonic oscillator, centered at y_0, and with the potential energy $V = \frac{1}{2}fy^2 = \frac{1}{2}m\omega^2 y^2$, where f is the restoring constant. The solution to this equation may be found in any textbook on quantum mechanics. The wave functions are Hermite polynomials. The energy eigenvalues are

$$\begin{aligned} E_\perp &= (n + \tfrac{1}{2})\hbar\,|\omega| \\ &= (2n + 1)\,|\mu_B|\,H, \end{aligned} \tag{5-3.7}$$

where $n = 0, 1, 2, \ldots$, etc. Note that $E_\perp$ is independent of k_x. This result was obtained first by Landau.[9] It is usual to call ω the cyclotron resonance frequency; it is twice the Larmor frequency. The energy levels are sometimes called Landau levels.

Hence the energy levels of this two-dimensional system, which in the absence of a field were quasi-continuous, become discrete when a field is applied. These discrete energy levels are highly degenerate, since all the states that in zero field were in an energy interval $\Delta E_\perp = 2n\,|\mu|_B H - 2(n - 1)\,|\mu_B|\,H = 2\,|\mu_B|\,H$ now coincide in energy. For example, all the quasi-continuous levels with energies from 0 to $2\,|\mu_B|\,H$ now have the energy $|\mu_B|\,H$, as indicated in Fig. 5-3.1. The number of electronic states for a level with a given n and for a range dp_z will now be calculated. If we write $E_\perp = p_\perp^2/2m$, the number of states in the momentum range dp at p is given by equation 5-1.4 as

$$2\,\frac{2\pi p_\perp\, dp_\perp\, dp_z\, V}{h^3},$$

where V is the volume of the metal. Now, since

$$\begin{aligned} p_\perp\, dp_\perp &= \tfrac{1}{2}d(p_\perp^2) \\ &= m\,\Delta E_\perp \\ &= 2m\,|\mu_B|\,H, \end{aligned}$$

[9] L. Landau, *Z. Physik* **64**, 629 (1930).

we have for the number of states

$$\frac{8\pi Vm\,|\mu_B|\,H\,dp_z}{h^3} = \frac{2V\,|e|\,H}{ch^2}\,dp_z$$

$$= \frac{2V\,|e|\,H}{ch}\,dk_z. \tag{5-3.8}$$

The level degeneracy may also be calculated from the boundary conditions (problem 5-7).

It is instructive first to assume that Boltzmann statistics apply. The diamagnetism may then be easily understood qualitatively. In the absence of a magnetic field, the lower energy levels will be more heavily populated

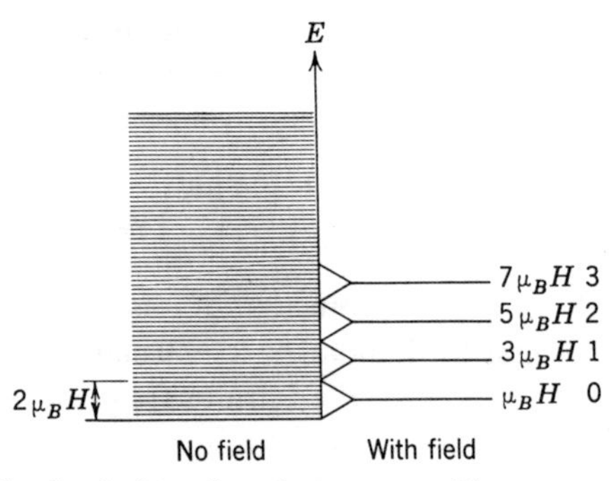

Fig. 5-3.1. The Landau levels for a free electron gas. The group of levels of the continuum combine to form highly degenerate discrete levels in the presence of a magnetic field.

than the higher ones. Consider the group of electrons that occupies a given Landau level when a field is applied (see Fig. 5-3.1). Without a field, the mean energy of these electrons will be somewhat less than the energy of the Landau level. Consequently, with a field this group of electrons will experience a net increase of energy. This is true for all the groups; hence the energy of the whole system is increased. Now, since by equation 2-8.7 the magnetic moment is given by $-\partial E/\partial H$, it follows that the susceptibility is negative, that is, diamagnetic.

In order to calculate the susceptibility for the case in which Boltzmann statistics apply, we first compute the partition function $\mathfrak{Z}$. From equations 5-1.14, 5-3.4, 5-3.7, and 5-3.8 it is

$$\mathfrak{Z} = \sum_i e^{-E_i/kT}$$

$$= \sum_{n=0}^{\infty} \frac{2eVH}{ch} \int_{-\infty}^{\infty} \left\{ \exp \left[-\frac{\hbar^2 k_z^{\,2}}{2m} - (2n+1)\,|\mu_B|\,H \right] \Big/ kT \right\} dk_z.$$

The integration over k_z is performed first. This definite integral occurs frequently in the kinetic theory of gases; its value is given in standard tables. Then we have

$$\mathfrak{Z} = \frac{2\,|e|\,VH}{ch^2}(2\pi m \mathrm{k}T)^{1/2}\sum_{n=0}^{\infty}\exp\left[-\frac{(2n+1)\,|\mu_B|\,H}{\mathrm{k}T}\right].$$

The series may be summed easily in the following way:

$$\begin{aligned}\sum &= e^{-x}(1 + e^{-2x} + e^{-4x} + \cdots)\\ &= \frac{e^{-x}}{1 - e^{-2x}}\\ &= \frac{1}{e^{x} - e^{-x}}\\ &= \frac{1}{2\sinh x},\end{aligned}$$

where

$$x = \frac{|\mu_B|\,H}{\mathrm{k}T}.$$

Therefore the result for the partition function is

$$\mathfrak{Z} = \frac{eVH}{ch^2}\,\frac{(2\pi m\mathrm{k}T)^{1/2}}{\sinh(|\mu_B|\,H/\mathrm{k}T)}. \tag{5-3.9}$$

By applying equation 5-1.16 we obtain for the magnetization

$$\begin{aligned}M &= -N\mu_B\left(\coth\frac{|\mu_B|\,H}{\mathrm{k}T} - \frac{\mathrm{k}T}{|\mu_B|\,H}\right)\\ &= -N\mu_B L\left(\frac{|\mu_B|\,H}{\mathrm{k}T}\right),\end{aligned} \tag{5-3.10}$$

where $L(x)$ is just the Langevin function of equation 2-13.2. For $|\mu_B|\,H/\mathrm{k}T \ll 1$ we have $L(x) = x/3 - x^3/45 + \cdots$ (problem 2-6). Then the susceptibility per unit volume is

$$\chi = -\frac{1}{3}\frac{N\mu_B{}^2}{\mathrm{k}T}. \tag{5-3.11}$$

Now if in the corresponding expression for the paramagnetic susceptibility of an electron gas (equation 5-2.1) we substitute $S = \frac{1}{2}$, $g = 2$, we get $\chi_{\text{para}} = N\mu_B{}^2/\mathrm{k}T$. Hence we have the remarkable result that except for sign the diamagnetism is just one third as great as the spin paramagnetism. It will turn out that this result holds for Fermi-Dirac as well as Boltzmann statistics. Equation 5-3.11 gives, as equation 5-2.1 did also, both too large a value and the wrong temperature dependence for the susceptibility. In addition, in the limit of high fields and low temperatures equation 5-3.10 predicts that the magnetic moment will approach $-N\,|\mu_B|$; this is in disagreement with experiment. These difficulties are removed when Fermi-Dirac statistics are employed.

We start with the expression for the Gibb's function of a free-electron gas that obeys Fermi-Dirac statistics, (equation 5-1.23),

$$G = \mathcal{N}E_F - kT\sum_i \log\,[1 + e^{(E_F - E_i)/kT}].$$

We assume that the summation may be replaced by an integral; that is

$$G = \mathcal{N}E_F - kT\int_0^\infty C(E)\log\,[1 + e^{(E_F - E)/kT}]\,dE, \qquad (5\text{-}3.12)$$

where $C(E)\,dE$ is the density of states for a free-electron gas in the presence of a field (see equation 5-3.8). On integrating equation 5-3.12 by parts and employing the usual limits, we have

$$\begin{aligned} G &= \mathcal{N}E_F - \int_0^\infty \frac{c(E)}{e^{(E - E_F)/kT} + 1}\,dE \\ &= \mathcal{N}E_F - \int_0^\infty c(E)\,F(E)\,dE, \qquad (5\text{-}3.13) \end{aligned}$$

where

$$c(E) = \int_0^E C(E)\,dE$$

is the number of states with energy less than E. To evaluate $c(E)$, we employ equation 5-3.8; the limits of the integration over dp_z are determined so that the energy equation $E = (2n + 1)\,|\mu_B|\,H + p_z^2/2m$ (equations 5-3.4 and 5-3.7) is satisfied. We obtain

$$\begin{aligned} c(E) &= \frac{2eHV}{ch^2}\sum_n \int_{-\{2m[E-(2n+1)\,|\mu_B|\,H]\}^{1/2}}^{\{2m[E-(2n+1)\,|\mu_B|\,H]\}^{1/2}} dp_z \\ &= \frac{4eHV}{ch^2}\sum_n \{2m[E - (2n + 1)\,|\mu_B|\,H]\}^{1/2}, \qquad (5\text{-}3.14) \end{aligned}$$

where the summation is $n = 0$ to the maximum integer for which the radicand is positive. The evaluation of equation 5-3.13 will be easier if the expression $\partial F(E)/\partial E$ instead of $F(E)$ appears in the integral. This can be arranged by integrating again by parts. By substituting equation

5-3.14 into (5-3.13) and performing this operation, we find

$$G = \mathcal{N}E_F - K\int_{-\infty}^{\infty}\left[\sum_n (\epsilon - n - \tfrac{1}{2})^{3/2}\right]\frac{d}{d\epsilon}\left[\frac{1}{e^{(\epsilon-\epsilon_F)/\vartheta}}\right]d\epsilon,$$

where

$$K = \frac{16m^{3/2}(|\mu_B|\,H)^{5/2}\,V}{3\pi^2\hbar^3},$$

$$\epsilon_F = \frac{E_F}{2\,|\mu_B|\,H}, \tag{5-3.15}$$

$$\epsilon = \frac{E}{2\,|\mu_B|\,H},$$

$$\vartheta = \frac{\mathrm{k}T}{2\,|\mu_B|\,H},$$

and the summation is $n = 0$ to $\epsilon - \frac{1}{2}$. A somewhat long and complicated calculation is required to evaluate the Gibb's function from this equation. The reader who is uninterested in the mathematical details may proceed directly to the result given in equation 5-3.21. We shall give only a brief outline of the calculation.

The sum $\sum_n (\epsilon - n - \frac{1}{2})^{3/2}$ may be computed by means of Poisson's formula, given by Courant and Hilbert as[10]

$$\sum_{n=-\infty}^{\infty}\Gamma(n) = \frac{1}{2\pi}\sum_{l=-\infty}^{\infty}\int_{-\infty}^{\infty}\Gamma\left(\frac{\tau}{2\pi}\right)e^{-il\tau}\,d\tau.$$

As adapted to our case, it becomes

$$\sum_{n=0}^{\epsilon-1/2}(\epsilon - n - \tfrac{1}{2})^{3/2} = \frac{1}{2\pi}\sum_{l=-\infty}^{\infty}\int_{-1/2}^{2\pi(\epsilon-1/2)}\left(\epsilon - \frac{\tau}{2\pi} - \frac{1}{2}\right)^{3/2}e^{-il\tau}\,d\tau.$$

On setting $x = \tau/2\pi + \frac{1}{2}$ and taking the complex conjugate, we find

$$\sum_{n=0}^{\epsilon-1/2}(\epsilon - n - \tfrac{1}{2})^{3/2} = \sum_{l=-\infty}^{\infty}(-1)^l\int_0^{\epsilon}(\epsilon - x)^{3/2}e^{+2\pi ilx}\,dx. \tag{5-3.16}$$

The term for $l = 0$ is simply

$$-\int_0^{\epsilon}(\epsilon - x)^{3/2}\,d(\epsilon - x) = \tfrac{2}{5}\epsilon^{5/2}. \tag{5-3.17}$$

[10] R. Courant and D. Hilbert, *Methoden der mathematischen Physik*, Vol. I, J. Springer, Berlin (1931), pp. 63–65. Translated edition, Interscience Publishers, New York (1953), p. 76.

We now evaluate the integral of equation 5-3.16 for $l \neq 0$. If we set $z^2 = \epsilon - x$, we have

$$\int_0^\epsilon (\epsilon - x)^{3/2} e^{2\pi i l x}\, dx = e^{2\pi i l \epsilon} \int_0^{\epsilon^{1/2}} 2z^4 e^{-2\pi i l z^2}\, dz$$

$$= -\frac{e^{2\pi i l \epsilon}}{2\pi i l} \int_0^{\epsilon^{1/2}} z^3\, d(e^{-2\pi i l z^2}).$$

Integration by parts yields

$$\int = -\frac{e^{2\pi i l \epsilon}}{2\pi i l}\left[(z^3 e^{-2\pi i l z^2})\Big|_0^{\epsilon^{1/2}} - \int_0^{\epsilon^{1/2}} 3z^2 e^{-2\pi i l z^2}\, dz\right]$$

$$\int = -\frac{\epsilon^{3/2}}{2\pi i l} - \frac{e^{2\pi i l \epsilon}}{2(2\pi i l)^2} \int_0^{\epsilon^{1/2}} 3z^2\, d(e^{-2\pi i l z^2}).$$

A further integration by parts gives

$$\int = -\frac{\epsilon^{3/2}}{2\pi i l} + \frac{3\epsilon^{1/2}}{8\pi^2 l^2} - \frac{3e^{2\pi i l \epsilon}}{8\pi^2 l^2} \int_0^{\epsilon^{1/2}} e^{-2\pi i l z^2}\, dz. \tag{5-3.18}$$

It will turn out that the last term is important only at low temperatures and high fields. The situation in which these conditions apply is discussed in the next section; for the present we omit this term. Hence, on substituting equations 5-3.17 and 5-3.18 into (5-3.16), we have

$$\sum_{n=0}^{\epsilon - 1/2} (\epsilon - n - \tfrac{1}{2})^{3/2} = \tfrac{2}{5}\epsilon^{5/2} + \sum_{\substack{l=-\infty \\ l \neq 0}}^{\infty} (-1)^l \left(-\frac{\epsilon^{3/2}}{2\pi i l} + \frac{3\epsilon^{1/2}}{8\pi^2 l^2}\right)$$

$$= \tfrac{2}{5}\epsilon^{5/2} + \frac{\epsilon^{1/2}}{16}, \tag{5-3.19}$$

since

$$\sum_{\substack{l=-\infty \\ l \neq 0}}^{\infty} (-1)^l \frac{1}{l^2} = \frac{\pi^2}{6}.$$

From equation 5-3.15 we then obtain

$$G = \mathcal{N} E_F - K \int_{-\infty}^{\infty} \left(\tfrac{2}{5}\epsilon^{5/2} + \frac{\epsilon^{1/2}}{16}\right) \frac{d}{d\epsilon}\left[\frac{1}{e^{(\epsilon - \epsilon_F)/\vartheta} + 1}\right] d\epsilon.$$

It is reasonable to assume that $E_F \gg 2\,|\mu_B|\,H$, so that ϵ_F is a fairly large number. Now, the last factor in the integrand has an appreciable value only in the region near ϵ_F (see Fig. 5-1.1), provided $\epsilon_F \gg \vartheta$, that is, $E_F \gg kT$. Hence the first factor of the integrand may be evaluated at

$\epsilon = \epsilon_F$ and then removed outside the integral. Use of equation 5-2.3 then gives

$$G = \mathcal{N}E_F + \tfrac{2}{5}K\epsilon_F^{5/2} + \frac{K}{16}\epsilon_F^{1/2}. \qquad (5\text{-}3.20)$$

Only the last term depends on H. By employing the definition of equation 5-3.15 we find

$$G = (\text{term independent of } H) + \frac{2\pi m V}{3h^2}\left(\frac{3N}{\pi}\right)^{1/3}\mu_B^2 H^2. \qquad (5\text{-}3.21)$$

The susceptibility per unit volume calculated with the aid of equation 5-1.18 is

$$\chi = -\frac{4\pi m}{3h^2}\left(\frac{3N}{\pi}\right)^{1/3}\mu_B^2. \qquad (5\text{-}3.22)$$

It is just $-\frac{1}{3}$ of the spin susceptibility given by equation 5-2.2. Thus we see that this proportionality relationship holds for both Boltzmann statistics (equation 5-3.11) and the low-temperature case of Fermi-Dirac statistics. Since Boltzmann statistics may be considered as the high-temperature limit of Fermi-Dirac statistics, it is not surprising that the minus one third factor can be shown to hold for the entire temperature range. Hence a more complete expression for the diamagnetic susceptibility is given by multiplying equation 5-2.5 for the paramagnetic susceptibility by $-\frac{1}{3}$.

The reason the susceptibility is smaller for Fermi-Dirac than for Boltzmann statistics is easily understood physically. When a field is applied, groups of the quasi-continuous levels coalesce to form Landau levels. Now, all the lower quasi-continuous levels are occupied in Fermi-Dirac statistics; hence in the presence of a field the mean energy of these electrons is unchanged. Only the few electrons close to the Fermi level will have their mean energy changed. Since there is a small increase in energy, the system will be weakly diamagnetic.

We now consider how the diamagnetism is changed if account is taken of the interaction of the conduction electrons with the lattice ions. For this case a periodic potential term V must be included in Schrödinger's equation. The calculation is difficult, for ordinary perturbation methods cannot be used. Although important progress has been made,[11] we shall not treat this problem here. Instead, if we replace the mass m with the

[11] R. E. Peierls, *Z. Physik* **80,** 763, and **81,** 186 (1933); L. Onsager, *Phil. Mag.* **43,** 1006 (1952); P. J. Harper, *Proc. Phys. Soc.* (*London*) **A-68,** 874, 879 (1955); G. E. Zil'berman, *J. Exptl. Theoret. Phys.* (*USSR*) **32,** 296 (1957) [trans. *Soviet Phys.-JEPT* **5,** 208 (1957)]; A. H. Wilson, *Proc. Cambridge Phil. Soc.* **49,** 292 (1953); I. M. Lifschitz and A. M. Kosevitch, *J. Exptl. Theoret. Phys.* (*USSR*) **29,** 730 (1955) [trans. *Soviet Phys.-JEPT* **2,** 636 (1956)]; Y. Yafet, *Phys. Rev.* **115,** 1172 (1959); W. Kohn, *Phys. Rev.* **115,** 1460 (1959).

effective mass m^*, equation 5-3.22 becomes

$$\begin{aligned}\chi_{\text{dia}} &= -\frac{4\pi m^*}{3h^2}\left(\frac{3N}{\pi}\right)^{1/3}\left(\frac{|e|\,\hbar}{2m^*c}\right)^2 \\ &= -\frac{4\pi m^*}{3h^2}\left(\frac{3N}{\pi}\right)^{1/3}\mu_B{}^2\left(\frac{m^2}{m^{*2}}\right) \\ &= -\frac{1}{3}\frac{m^2}{m^{*2}}\chi_{\text{para}}.\end{aligned} \qquad (5\text{-}3.23)$$

Note that $\mu_B = \hbar e/2mc$ is replaced by $\hbar e/2m^*c$ in the diamagnetic case because μ_B appears in the equation simply as a replacement for a certain combination of constants. In the paramagnetic formula the spin moment is equal to μ_B, regardless of whether the electrons are free, and therefore an effective mass is not used in connection with μ_B.

It should also be mentioned that the orbital diamagnetism may be calculated by a method that employs the density matrix. This method is mathematically more elegant; on the other hand, it is less physical. For details the reader should consult Wilson's book.[12]

4. Comparison of Susceptibility Theory with Experiment

We now turn to a comparison of the theory with experiment. Unfortunately, the very small susceptibilities of metals are difficult to measure

Table 5-4.1. Susceptibilities of Li and Na
($\times 10^{-6}$ emu/cm^3)

Metal	Experimental				Theoretical		
	χ_T	χ_{spin}	χ_{ion}	χ_{dia}	χ_{dia} (free)	χ_{dia} (m^*)	χ_{dia} (Kohn)
Li	1.89 ± 0.05	2.08 ± 0.1	-0.05	-0.14 ± 0.15	-0.28	-0.19	-0.07
Na	0.68 ± 0.03	0.95 ± 0.1	-0.18	-0.10 ± 0.13	-0.22	-0.22	-0.26

accurately and thus make a critical test of the theory impossible. The best test probably is for lithium and sodium, since the spin susceptibility has been isolated and measured accurately for these metals. The relevant data are given in Table 5-4.1. The second column lists the experimental values of the total susceptibility as determined by Pugh and Goldman[13] for Li and by Bowers[14] for Na. Bowers' paper contains references to

[12] A. H. Wilson, *Theory of Metals*, Cambridge University Press, Cambridge (1953), pp. 160–175.

[13] E. W. Pugh and J. E. Goldman, *Phys. Rev.* **99**, 1641 (1955).

[14] R. Bowers, *Phys. Rev.* **100**, 1141 (1955).

earlier measurements. The experimental spin susceptibilities are those in Table 5-2.1. The susceptibilities of the ion cores are the values calculated by Slater and listed in Table 2-4.3. The experimental value of the orbital susceptibilities is the second column minus the sum of the third and fourth columns. The theoretical values computed, first for free electrons from equation 5-3.22, second for electrons with effective mass m^* from equation 5-3.23, and third from a recent refined theory,[15] are listed in the last three columns. The values of effective mass employed are those in

Table 5-4.2. Susceptibilities of Some Monovalent and Diavalent Metals at Room Temperature
($\times 10^{-6}$ emu/cm^3)

	Experimental			Theoretical
Metal	χ_T	χ_{ion}	$\chi_{f.e.} = \chi_T - \chi_{ion}$	$\chi_{f.e.} = \chi_{dia} + \chi_{para}$
K	0.47	−0.31	0.76	0.35
Rb	0.33	−0.46	0.79	0.33
Cu	−0.76	−2.0	1.24	0.65
Ag	−2.1	−3.0	0.9	0.60
Au	−2.9	−4.3	1.4	0.60
Mg	0.95	−0.22	1.2	0.65
Ca	1.7	−0.43	2.1	0.5

Table 5-2.1. It is seen that both theoretical values are within the experimental error; however, this error is of the same magnitude as the orbital susceptibilities themselves.

A check of the theory for other metals will be made by comparing the sum of the spin and orbital susceptibilities for free electrons calculated from equations 5-2.2 and 5-3.22 with that deduced from experiment. This experimental quantity $\chi_{f.e.}$ is the total susceptibility observed minus the diamagnetic susceptibility of the ion cores calculated from Langevin's formula (equation 2-3.2). Values for some monovalent and diavalent metals at room temperature are given in Table 5-4.2. In general, the theoretical susceptibilities are less than those observed. Some of the discrepancy probably develops from the use of m instead of m^* and from the neglect of the exchange and correlation forces. The susceptibilities of Zn and Cd are strongly anisotropic and moreover are strongly dependent on temperature; these results are at present unexplained.

Atoms of the transition group elements have an unfilled inner shell; hence the corresponding band of the metals is believed also to be unfilled.

[15] T. Kjeldaas, Jr., and W. Kohn, *Phys. Rev.* **105,** 806 (1957).

These bands are relatively narrow in energy and thus the density of states is high. For example, a $3d$ band can accommodate 10 electrons compared to two for a $4s$ band; nevertheless, the $3d$ band is considerably narrower in energy than the $4s$ band. From equation 5-1.8, it follows that the transition metals will have values of $m^*/m \gg 1$. Now from equations

Table 5-4.3. Observed Susceptibilities of Transition Elements at Room Temperature
($\times 10^{-6}$ emu/cm^3)

Group	Element	χ_T
$3d$	Ti	14.4
	V	28.5
	Cr	22.8
	Mn	71.6
$4d$	Zr	8.4
	Nb	18.9
	Mo	8.5
	Ru	5.18
	Rh	12.4
	Pd	63.5
$5d$	Ta	13.9
	W	5.6
	Os	0.11
	Ir	4.0
	Pt	20.8

5-2.2 and 5-3.23 the susceptibility of the conduction electrons is given by

$$\chi_{\text{f.e.}} = \frac{4\pi m^*}{h^2}\left(\frac{3N}{\pi}\right)^{1/3} \mu_B^{\,2}\left(1 - \frac{m^2}{3m^{*2}}\right). \tag{5-4.1}$$

Hence we would expect the transition metals to have comparatively large paramagnetic susceptibilities. Experimental data, compiled by Kriessman and Callen[16] and shown in Table 5-4.3, in general confirm this expectation. In addition, these authors analyze the variation of the susceptibilities with temperature in terms of the density of states.

It should also be mentioned that the conduction band of the transition metals overlaps the unfilled inner band in energy. This explains the susceptibility of palladium,[17] which as a free atom has an outer electronic configuration of $4d^{10}$. According to Wohlfarth,[18] about 0.6 electron per

[16] C. J. Kriessman and H. B. Callen, *Phys. Rev.* **94,** 837 (1954).
[17] F. E. Hoare and J. C. Matthews, *Proc. Roy. Soc.* (*London*) **A-212,** 137 (1952).
[18] E. P. Wohlfarth, *Proc. Leeds. Phil. Soc.* **5,** 89 (1949).

atom is present in 5*s* states and the same number of vacant 4*d* states. These vacant states, or holes, contribute to the susceptibility in exactly the same way as electrons. Indeed, the earlier equations hold, provided n is replaced by n_h, the number of holes per unit volume.

Bismuth exhibits a large diamagnetism. On the basis of equation 5-4.1, this requires that $m^*/m \ll 1$. This situation may occur for electrons or holes near Brillouin zone boundaries.[19]

An alloy is a substitutional solid solution of one or more metals in another. In early theories it was assumed that each constituent atom contributed its valence electrons (one or more) to the Fermi distribution. The predictions of these theories were often in good agreement with the experimental measurements of the susceptibility and other properties of alloys. More recent work has shown that the situation cannot always be treated in such a simple fashion. One alternative theory assumes that some or all of the outer electrons of the solute atoms occupy localized energy states which may or may not overlap in energy the electrons of the Fermi distribution.[20]

The study of alloys formed from the solution of transition metals in noble metals has been particularly interesting. For small amounts of iron dissolved in copper or gold the susceptibility follows the Curie-Weiss law. Hence the electrons of the iron atoms apparently are localized. Silver-palladium alloys, however, are best described on the band model. It is assumed that the silver contributes its valence electrons to the holes in the 4*d* band until all the 4*d* states are filled. Then electrons go into states at the top of the Fermi distribution of 4*s* electrons. Since there is 0.6 hole per atom in 4*d* states, as already mentioned, when 60 atomic per cent of silver has been added all the 4*d* states will be filled. A similar situation exists for Cu-Pd, Au-Pd, and nickel-rich Cu-Ni alloys. Neither the band nor the localized electron models appears to fit the results for the copper-rich Cu-Ni alloys.[21]

Dilute solutions of iron in an extensive series of alloys of the 3*d* metals have also been investigated.[22] The susceptibility of some alloys is almost independent of temperature. For others χ follows a Curie-Weiss law. Magnetization appears suddenly for alloys near the electron concentration corresponding to Mo and in the vicinity of Rh and Pd. The results have

[19] H. Jones, *Proc. Roy. Soc. (London)* **A-144,** 225 (1934).

[20] J. Friedel, *Advan. Physics* **3,** 446 (1954), *Can. J. Phys.* **34,** 1190 (1956), *J. Phys. radium* **19,** 573 (1958), *Nuovo cimento Suppl.* **7,** 287 (1958); A. Blandin and J. Friedel, *J. Phys. radium* **19,** 573 (1958), **20,** 160 (1959); J. Friedel, G. Leman, and S. Olszewski, *J. Appl. Phys.* **32,** 325*S* (1961).

[21] E. W. Pugh, B. R. Coles, A. Arrott, and J. E. Goldman, *Phys. Rev.* **105,** 814 (1957).

[22] B. T. Matthias, M. Peter, H. J. Williams, A. M. Clogston, E. Corenzwit, and R. C. Sherwood, *Phys. Rev. Letters* **5,** 542 (1960).

been interpreted with a model similar to that proposed by Friedel, in which the magnetization is ascribed to a virtual level of the iron atom.[23] An iron-palladium alloy as dilute as 1.25 atomic per cent iron has been found to be ferromagnetic at low temperatures.[24] Even more surprising, a cobalt-palladium alloy with 0.1 atomic per cent Co exhibited ferromagnetism below 7°K.[25]

5. The De Haas-Van Alphen Effect

When Boltzmann statistics are employed, the orbital moment of an electron gas is given by equation 5-3.10. In high fields and at low temperatures this equation predicts that the moment will approach $-N\mu_B$. However, earlier experiments on bismuth showed that the susceptibility was an oscillatory function of the applied field. This phenomena, known as the de Haas-van Alphen effect in honor of its discoverers,[26] can be explained when Fermi-Dirac statistics are employed.[27]

We now consider the third term of equation 5-3.18,

$$-\frac{3e^{2\pi il\epsilon}}{8\pi^2 l^2}\int_0^{\epsilon^{1/2}} e^{-2\pi ilz^2}\, dz,$$

which was previously neglected. Since it is assumed that $2\mu_B H \ll E_F$, the upper limit may be taken as infinity. This term then becomes

$$-\frac{e^{-(\pi/4)i}}{2\sqrt{2}}\frac{3e^{2\pi il\epsilon}}{8\pi^2 l^{5/2}}.$$

The contribution made by this term to the Gibb's function G' is found by substituting first into equation 5-3.16 and then into equation 5-3.15. We obtain

$$G' = K\sum_{l=1}^{\infty}(-1)^l \frac{3}{8\sqrt{2}\pi^2 l^{5/2}}\int_{-\infty}^{\infty} e^{i(2\pi l\epsilon-\pi/4)}\frac{d}{d\epsilon}\left[\frac{1}{e^{(\epsilon-\epsilon_F)/\vartheta}+1}\right]d\epsilon.$$

The integral may be evaluated by contour integration. The result for the

[23] P. W. Anderson, *Phys. Rev.* **124,** 41 (1961); P. A Wolff, *Phys Rev.* **124,** 1030 (1961); P. A. Wolff, P. W. Anderson, A. M. Clogston, B. T. Matthias, M. Peter, and H. J. Williams, *J. Appl. Phys.* **33,** 1173 (1962).

[24] J. Crangle, *Phil. Mag.* **5,** 335 (1960).

[25] R. M. Bozorth, P. A. Wolff, D. D. Davis, V. B. Compton, and J. H. Wernick, *Phys. Rev.* **122,** 1157 (1961).

[26] W. J. de Haas and P. M. van Alphen, *Commun. Kamerlingh Onnes Lab. Univ. Leiden* 208*d*, 212*a* (1930), and 220*d* (1933).

[27] R. Peierls, *Z. Physik* **81,** 186 (1933); D. Shoenberg and L. D. Landau, *Proc. Roy. Soc. (London)* **A-170,** 341 (1939); M. Blackman, *Proc. Roy. Soc. (London)* **166,** 1 (1938).

Gibb's function is

$$G' = \frac{8\sqrt{2}\pi m^{3/2}(|\mu_B|\, H)^{3/2}\, kT}{h^3} \sum_{l=1}^{\infty} (-1)^l \frac{\cos\,[(\pi l E_F/|\mu_B|\, H) - (\pi/4)]}{l^{3/2}\, \sinh\,(2\pi l kT/|\mu_B|\, H)}\,. \tag{5-5.1}$$

The susceptibility may be found by taking $-(1/H)(\partial G'/\partial H)$. Usually only the term with $l = 1$ is important.

When T/H becomes appreciable, the hyperbolic sine in the denominator of the expression becomes large. Hence under these conditions the contribution of this term to the susceptibility will be unobservable.

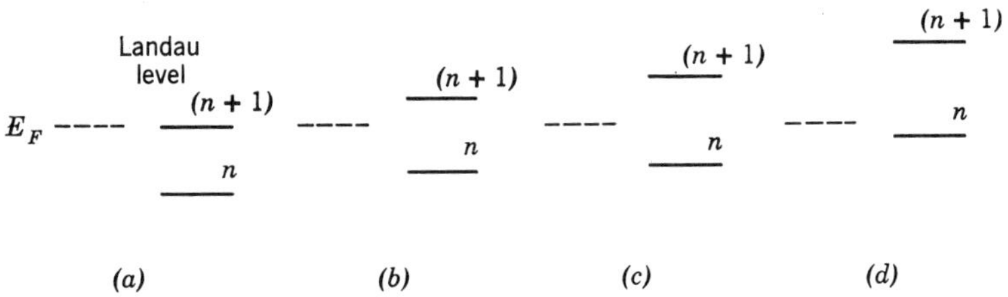

Fig. 5-5.1. Position of Landau levels (heavy lines) with respect to the Fermi energy (dashed lines) as the applied field is increased.

The oscillations in the susceptibility obviously arise from the trigonometric term in the numerator. These oscillations are a function of $1/H$; one cycle of oscillation occurs when $E_F/|\mu_B|\, H$ changes by 2. This periodicity can be understood from the following physical argument. First, consider a two-dimensional system of free electrons at absolute zero in the presence of a field. When the field H is changed, both the energy of the levels

$$\begin{aligned} E_\perp &= (n + \tfrac{1}{2})\hbar\, |\omega| \\ &= (2n + 1)\, |\mu_B|\, H \end{aligned}$$

and their degeneracy change in proportion to H. The only electrons that will contribute to a change in the total energy will be those near the Fermi level. The position of the Fermi level with respect to the nearby Landau levels is shown for four different values of H in Fig. 5-5.1. The maximum total energy results when the Fermi level coincides with a Landau level (Fig. 5-5.1*a*) and the minimum when the Fermi level lies half way between two Landau levels (Fig. 5-5.1*c*). The variation of the total energy with applied field is indicated in Fig. 5-5.2; the energies that correspond to the situations of Fig. 5-5.1 are labeled. As a result the magnetic moment $(= -\partial E/\partial H)$ will alternate in sign; hence the susceptibility will oscillate with H. In three dimensions only Landau levels with energies less than the Fermi level will be occupied by electrons; that is, the condition

$$(n + \tfrac{1}{2})\hbar\,|\omega| \leqslant E_F$$

must be fulfilled. For example, in Fig. 5-5.1*b* the electrons near the Fermi level will not go into the $(n + 1)$ Landau level but instead into states with a certain value of k_z different from zero. One cycle of the oscillation in the energy (and susceptibility) will occur each time a Landau level is removed as a possible level for occupancy. This will occur when $E_F/\hbar\,|\omega|$ changes by 1 or when $E_F/|\mu_B|\,H$ changes by 2, in agreement with our earlier result.

The question of the interaction between the electron gas and the lattice ions is a difficult theoretical problem. We may simply introduce an

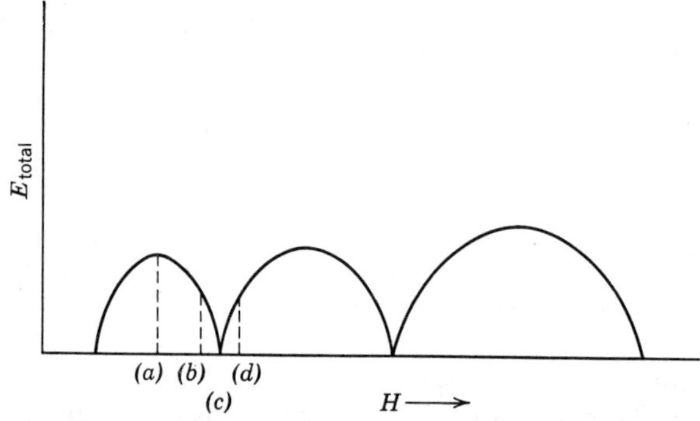

Fig. 5-5.2. Variation of the total energy with field for a free electron gas.

effective mass, but it is doubtful if this is an adequate approximation for most cases. Also, when high fields are employed, the spin-orbit coupling is no longer negligible. Further, collisions of the electrons influence the susceptibility; the effect is equivalent to an increase of temperature. The reader is referred to the literature for a discussion of these problems.[28]

Experimental results for zinc and gallium, obtained by Shoenberg,[29] are shown in Fig. 5-5.3. The ordinates are proportional to the torque on a specimen in a homogeneous field divided by H^2; this quantity is related to the susceptibility. The de Haas-van Alphen effect has also been observed for aluminum, lead, tin, indium, graphite, zinc, cadmium, thallium, beryllium, magnesium, bismuth, antimony, arsenic, mercury, molybdenum, copper, gold, and silver.

[28] I. M. Lifschitz and A. M. Kosevitch, *J. Exptl. Theoret. Phys.* (*USSR*) **29,** 730 (1955) [trans. *Soviet Phys.-JEPT* **2**, 636 (1956)]; V. Heine, *Proc. Phys. Soc.* (*London*) **A-69,** 505 (1956); A. B. Pippard, *Reports 10th Solvay Congress* (1955), p. 123; R. B. Dingle, *Proc. Roy. Soc.* (*London*) **A-211,** 517 (1952).

[29] D. Shoenberg, *Phil. Trans. Roy. Soc.* (*London*) **A-245,** 1 (1952); J. S. Dhillon and D. Shoenberg, *Phil. Trans. Roy. Soc.* (*London*) **A-248,** 1 (1955).

We have seen that the period of the oscillations is determined by $E_F/\mu_B H$. This quantity is independent of the mass, even when an effective mass is used, since then $\mu_B = e\hbar/2m^*c$ and $E_F = (h^2/2m^*)(3N/8\pi)^{2/3}$ (equation 5-1.12). By experiment it is found that N is about 10^{18} to 10^{19}/cm^3; that is, the number of conduction electrons per atom is about 10^{-3}

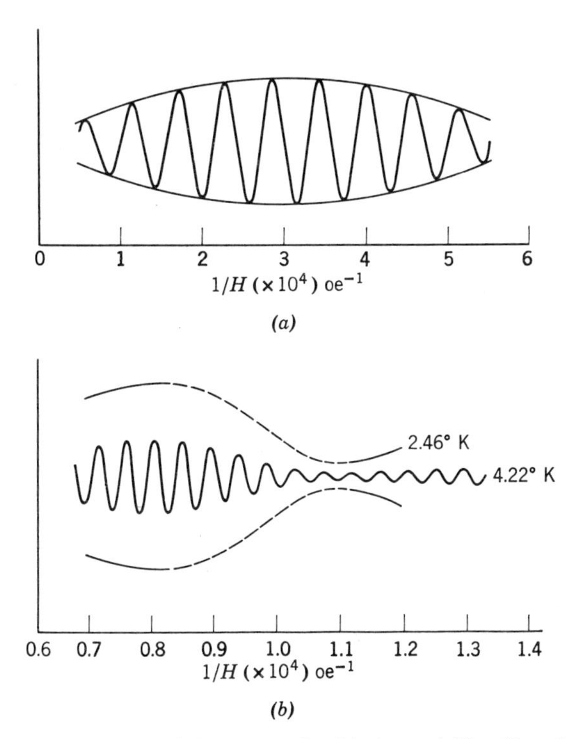

Fig. 5-5.3. The de Haas-van Alphen curves for (*a*) zinc and (*b*) gallium in the *ac* plane. Only the envelope of the curve for gallium is shown for the lower temperature. (After D. Shoenberg and J. S. Dhillon.)

to 10^{-4}. This result is difficult to reconcile with the much larger values of N obtained from transport experiments.

The period of the oscillations yields considerable information concerning the energy or Fermi surface of the conduction electrons. Indeed, the de Haas-van Alphen effect is one of the most important sources of such information.[30]

[30] D. Shoenberg, *Progress in Low-Temperature Physics*, Vol. II, Interscience Publishers, New York (1957), pp. 226–265, *The Fermi Surface*, John Wiley and Sons, New York (1960), pp. 74–83.

6. Galvanomagnetic, Thermomagnetic, and Magnetoacoustic Effects

When a metal or semiconductor that carries an electric or a phonon current is placed in a magnetic field certain effects occur, which are classified as galvanomagnetic when the current is electrical, thermomagnetic when it is thermal, and magnetoacoustic when it is acoustical. For electrical or thermal currents further classifications are made, depending on whether the magnetic field is applied parallel or perpendicular to the direction of the current and on whether the effects are measured parallel or perpendicular to its direction. In all there are twelve fundamental effects, although probably not all are independent. There is a vast literature on this subject, but it is not appropriate that it be treated in any detail here.[31] Instead, we shall be content to give a brief and elementary discussion of two topics—the Hall effect and magnetoresistance. The section will be concluded with a few remarks concerning the magnetoäcoustic effect.

Suppose that the conductor is in the form of a plane slab and that the electric current consists of electrons flowing in the x-direction with a velocity v_x. When a magnetic field H_z is applied along the z-direction, the moving electrons will experience a force that tends to move them to one side. As a result, a negative charge will develop on the lower surface and a positive charge on the upper one. Equilibrium will be established when the electric field E_y, associated with the surface charge, is large enough that the force on the electrons in the y-direction equals zero (see Fig. 5-6.1). The field E_y developed in the Hall effect is easily computed by considering equation 1-8.9 for the Lorentz force, namely,

$$\mathbf{F} = e\left(\mathbf{E} + \frac{1}{c}\,\mathbf{v} \times \mathbf{H}\right).$$

Since $F_y = 0$ at equilibrium, we get for this field

$$E_y = \frac{v_x}{c} H_z$$

$$= \frac{j_x H_z}{Nec},$$

[31] For a review, see, for example, D. K. C. MacDonald and K. Sarginson, *Rept. Progr. Phys.* **15,** 249 (1952); D. K. C. MacDonald, *Encyclopedia of Physics* **14,** 137 (1956); A. N. Gerritsen, *Encyclopedia of Physics* **19,** 183 (1956); H. Jones, *Encyclopedia of Physics* **19,** 298 (1956); J.-P. Jan, *Solid State Phys.* **5,** 1 (1957); B. Lax, *Revs. Mod. Phys.* **30,** 122 (1958); A. H. Kahn and H. P. R. Frederikse, *Solid State Phys.* **9,** 257 (1959).

where j_x is the current density, given by $j_x = Nev_x$. The Hall coefficient R_H is defined as

$$R_H = \frac{E_y}{j_x H_z} = \frac{1}{Nec}. \tag{5-6.1}$$

The same result is obtained if the electrons obey Fermi-Dirac statistics. For free electrons the Hall coefficient is negative and of the magnitude 10^{-24} for gaussian units. Clearly, the measurement of the Hall coefficient gives the density of electron carriers. For the alkali and other monovalent metals the value of n observed is in good agreement with the density of the valence electrons. For some metals and p-type semiconductors the Hall coefficient is positive. This implies that the current is carried by positive

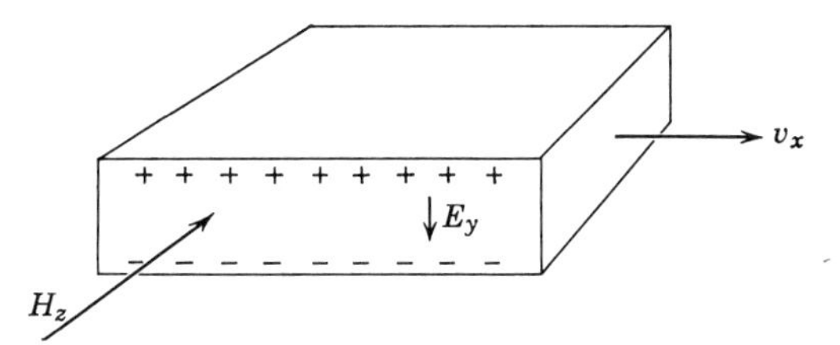

Fig. 5-6.1. The Hall effect. At equilibrium the current in the y-direction vanishes.

charges, or holes. This situation occurs when an energy band is almost filled; the vacant states at the top of the band act as positive carriers. For some metals and semiconductors the current is carried by both electrons and holes. The general situation is then much more complicated. For ferromagnetic metals there is a contribution to the Hall effect that arises from the magnetization.[32]

When a magnetic field is applied, perpendicular to the direction of current flow in a metal or semiconductor, a change of resistance is observed. The magnetoresistance is usually measured as the change of resistance ΔR relative to the resistance in zero magnetic field R. Free electron theory predicts that to first order the magnetoresistance will be zero. The reason is that since each electron has the same mean velocity the electric field of the Hall effect will cause the transverse current to vanish. Magnetoresistance arises because groups of the conduction electrons have different mean velocities. Hence, although the transverse electric field on the average balances the effect of the magnetic field, individual electrons will travel in a curved path. For most metals the magnetoresistance is positive, but there are a few cases for which it is negative, notably ferromagnetic metals and

[32] E. M. Pugh and N. Rostoker, *Revs. Mod. Phys.* **25,** 151 (1953); R. Karplus and J. M. Luttinger, *Phys. Rev.* **95,** 1154 (1954).

some dilute alloys. For small fields $\Delta R/R$ is proportional to H^2, whereas for large fields it is proportional to H. A magnetic field is considered large if the mean free path of the electrons is large compared to the radius of the electron orbits. There is also a magnetoresistance effect when the magnetic field is applied parallel to the current direction; it is smaller than the transverse effect. Hall effect and magnetoresistance measurements, particularly on single crystals, yield valuable information concerning the energy surfaces near the edges of the conduction and valence bands.[33]

At low temperatures and high magnetic fields the galvanomagnetic and thermomagnetic effects exhibit an oscillatory behavior which has the same origin as the de Haas-van Alphen effect, namely the Landau levels that result when a field is applied. Oscillations in the Hall effect have been observed in bismuth, graphite, beryllium, zinc, and antimony, in the magnetoresistance of bismuth, graphite, zinc, tin, antimony, and germanium, in the thermal conductivity of bismuth, zinc, and tin, and in the thermoelectric power of bismuth.[34]

The phonons generated in the material may be of acoustical rather than thermal origin. Typically, the source is a quartz transducer and the frequencies employed range from 10 to 200 Mc/sec. When a magnetic field is applied, oscillatory variations in the attenuation of the ultrasonic wave may be observed, provided the electron mean free path is longer than the phonon wavelength. These conditions may be fulfilled at liquid helium temperatures in very pure metals. The magnetoacoustic effect was first observed by Bömmel in tin.[35] The loss in the phonon current arises from a direct interaction between the conduction electrons and the acoustic vibrations of the lattice.[36] Theory apparently shows that the period of the oscillations in $1/H$ is related to the momentum at an extremal cross section of the Fermi surface.[37] The maxima and minima occur when the size of the electron orbits are certain multiples of the wavelength of the sound wave. Besides tin, the magnetoacoustic effect has been detected in

[33] Semiconductors: see, for example, G. L. Pearson and H. Suhl, *Phys. Rev.* **83,** 768 (1951); C. Herring and E. Vogt, *Phys. Rev.* **101,** 944 (1956); W. M. Bullis and W. E. Krag, *Phys. Rev.* **101,** 580 (1956). Metals: see, for example, B. Abeles and S. Meiboom, *Phys. Rev.* **101,** 544 (1956); E. S. Borovik, *Bull. Acad. Sci. USSR* **19,** 429 (1955).

[34] A. H. Kahn and H. P. R. Frederikse, *Solid State Phys.* **9,** 257 (1959).

[35] H. E. Bömmel, *Phys. Rev.* **100,** 758 (1955).

[36] A. B. Pippard, *Rept. Prog. Phys.* **23,** 176 (1960).

[37] A. B. Pippard, *Phil. Mag.* **2,** 1147 (1957), *Proc. Roy. Soc.* (*London*) **A-257,** 165 (1960); M. H. Cohen, M. J. Harrison, and W. A. Harrison, *Phys. Rev.* **117,** 937 (1960); V. L. Gurevich, *J. Exptl. Theoret. Phys.* (*USSR*) **37,** 71 (1959) [trans. *Soviet Phys.-JEPT* **10,** 51 (1960)]; E. A. Kaner, *J. Exptl. Theoret. Phys.* (*USSR*) **38,** 212 (1960) [trans. *Soviet Phys.-JEPT* **11,** 154 (1960)]; J. J. Quinn and S. Rodriguez, *Phys. Rev.* **128,** 2494 (1962).

copper, silver, gold, aluminum, lead, bismuth, cadmium, and zinc.[38] The observations have yielded important information concerning the shape of the Fermi surface.[39]

7. Electron Spin Resonance in Metals

Electron spin resonance from conduction electrons was observed first in sodium.[40] Schumacher and Slichter's resonance experiments on Li and Na have already been discussed in Section 5-3. Conduction electron spin resonance has also been observed[41] in the metals K,[41] Cs,[41] Be,[42] and white Sn,[43] the semimetal Bi,[44] and the semiconductors Si,[45] Ge,[46] Sb,[47] and C (graphite.)[48]

The resonance line of metals has a relatively small intensity, since the only unpaired electrons are those near the Fermi level. This line intensity is almost independent of temperature. Because of the good conductivity property of a metal, the penetration of the microwave radiation is limited. If the dimensions of the metal are large compared to the skin depth,[49] the shape of the resonance line is modified by two factors. One is variation of the phase and amplitude of the microwaves with depth and the other is the diffusion of the magnetic carriers in and out of the skin depth. The line shapes expected under different conditions have been calculated by

[38] R. W. Morse, H. V. Bohm, and J. D. Gavenda, *Phys. Rev.* **109,** 1394 (1958); R. W. Morse and J. D. Gavenda, *Phys. Rev. Letters* **2,** 250 (1959); R. W. Morse, A. Meyers, and C. T. Walker, *Phys. Rev. Letters* **4,** 605 (1960); A. A. Galkin and A. P. Korolyuk, *J. Exptl. Theoret. Phys.* (*USSR*) **37,** 310 (1959) [trans. *Soviet Phys.-JEPT* **10,** 219 (1960)]; B. W. Roberts, *Phys. Rev.* **119,** 1889 (1960); L. Mackinnon and M. R. Daniel, *Phys. Letters* **1,** 157 (1962); L. Mackinnon, M. T. Taylor, and M. R. Daniel, *Phil. Mag.* 7, 523 (1962); A. R. Mackintosh, *Phys. Rev.* **131,** 2420 (1963).

[39] See for example, W. A. Harrison and M. B. Webb, Editors, *The Fermi Surface*, John Wiley and Sons, New York (1960), pp. 213–263.

[40] T. W. Grisswold, A. F. Kip, and C. Kittel, *Phys. Rev.* **88,** 951 (1952). See also G. J. King, B. S. Miller, F. F. Carlson, and R. C. McMillan, *J. Chem. Phys.* **32,** 940 (1960).

[41] R. A. Levy, *Phys. Rev.* **102,** 31 (1956).

[42] G. Feher and A. F. Kip, *Phys. Rev.* **98,** 337 (1955).

[43] M. S. Khaikin, *J. Exptl. Theoret. Phys.* (*USSR*) **39,** 899 (1960) [trans. *Soviet Phys.-JEPT* **12,** 623 (1961)].

[44] G. E. Smith, J. K. Galt, and F. R. Merritt, *Phys. Rev. Letters* **4,** 276 (1960).

[45] A. M. Portis, A. F. Kip, C. Kittel, and W. H. Brattain, *Phys. Rev.* **90,** 988 (1953); F. K. Willenbrock and N. Bloembergen, *Phys. Rev.* **91,** 1281 (1953).

[46] K. J. Button, L. M. Roth, W. H. Kleiner, S. Zwerdling, and B. Lax, *Phys. Rev. Letters* **2,** 161 (1959).

[47] G. Bemski, *Phys. Rev. Letters* **4,** 62 (1960).

[48] G. Wagoner, *Phys. Rev.* **118,** 647 (1960).

[49] For $\omega \approx 30$ Gc the skin depth of a metal is about 10^{-4} to 10^{-5} cm.

Dyson[50] and confirmed by experiment.[51] These complications can be avoided by employing a suspension of metal particles whose size is small compared to the skin depth.

The observed g-values are close to the free-spin value. There is, however, a small difference that is believed to be the result of spin-orbit coupling.[52] Line widths are of the order of a few oersteds. Their dependence on temperature can be described approximately by the equation $\Delta H = a + bT$. Here a is a term independent of temperature. For Li, $b = 0$. Elliott[53] has proposed a mechanism for the spin-lattice relaxation. When an electron collides with a phonon, or an imperfection or impurity in the lattice, the spin-orbit coupling may cause the electron spin to flip. The value of τ_1 calculated is in order of magnitude agreement with experiment. In addition the theory predicts a linear dependence of ΔH on T, as is observed.

Electron spin resonance has been observed from the paramagnetic ion, manganese, present as an impurity in copper, silver, and magnesium.[54] The resonance line is almost isotropic, with a g-value close to the free-spin value. The line shape is that expected for fixed magnetic carriers located within the skin depth. The results are consistent with the manganese being present as Mn^{2+} ions. The interaction between the magnetic electrons of the manganese ions and the conduction electrons of the host metal might be expected to produce a shift of the g-value, but none was observed.

8. Cyclotron Resonance

As discussed in Section 5-3, Landau energy levels result when a steady magnetic field is applied to an electron gas. The energy of the Landau levels (equation 5-3.7) is given by

$$E_\perp = (n + \tfrac{1}{2})\hbar\,|\omega|,$$

where ω is the cyclotron frequency of equation 5-3.6, that is,

$$\omega = -\frac{eH}{mc}.$$

For a charge carrier with effective mass m^* the cyclotron frequency is

$$\omega = -\frac{eH}{m^*c}. \qquad (5\text{-}8.1)$$

[50] F. J. Dyson, *Phys. Rev.* **98,** 349 (1955).
[51] G. Feher and A. F. Kip, *Phys. Rev.* **98,** 337 (1955).
[52] H. Brooks, *Phys. Rev.* **94,** 144 (1954).
[53] R. J. Elliott, *Phys. Rev.* **96,** 226, 280 (1954).
[54] J. Owen, M. E. Brown, V. Arp, and A. F. Kip, *J. Phys. Chem. Solids* **2,** 85 (1957).

Note that this frequency is positive for electrons and negative for holes; the difference in sign indicates that holes and electrons rotate in opposite senses in a magnetic field. Transitions between adjacent Landau levels ($\Delta n = \pm 1$) may be induced when an oscillating field of this frequency is applied perpendicular to the steady field. The resonant absorption of energy that occurs involves electric dipole transitions rather than the magnetic dipole transitions of electron and nuclear magnetic resonance. Hence in cyclotron resonance experiments it is necessary to employ an oscillating *electric* field. Cyclotron resonance in solids was suggested and discussed theoretically[55] before the first successful experiment on germanium.[56] Although cyclotron resonance effects have been observed in some metals, the most fruitful results have been obtained in studies of the semiconductors germanium and silicon. The charge carrier density in these semiconductors is so low that Boltzmann rather than Fermi-Dirac statistics apply. Cyclotron resonance experiments are of particular importance in that they provide the only method for the direct determination of the effective mass in solids (from equation 5-8.1). Usually it is found that the effective mass is anisotropic, that is, m^* depends on the direction of the magnetic field with respect to the crystallographic axes. A knowledge of the effective mass permits the energy surfaces near the bottom of the conduction band and the top of the valence band to be determined. By this we mean the energy $E(k)$ can be determined as a function of the wave vector $\mathbf{k}$ by means of the relationship (equation 5-1.9)

$$\left(\frac{1}{m^*}\right)_{ij} = \frac{1}{\hbar^2}\frac{\partial^2 E}{\partial k_i\,\partial k_j},$$

where $i, j = x, y, z$. These energy surfaces are frequently ellipsoids oriented along certain crystallographic directions.

The energy surfaces at the bottom of the conduction band of germanium and silicon are believed to be of the form

$$E(k) = \frac{\hbar^2}{2}\left(\frac{k_x^{\,2} + k_y^{\,2}}{m_t} + \frac{k_z^{\,2}}{m_l}\right). \tag{5-8.2}$$

For germanium the energy surfaces are spheroids oriented along the [111]-directions; for silicon they are along the [100]-directions.

The energy surfaces at the top of the valence band in germanium and silicon are more complicated. Two energy bands coincide at $\mathbf{k} = 0$; this is the value of $\mathbf{k}$ at which the maximum of the energy surface occurs. Actually, we might expect three valence bands to coincide, since the bands

[55] J. Dorfmann, *Dokl. Akad. Nauk SSSR* **81,** 765 (1951); R. B. Dingle, *Proc. Roy. Soc.* (*London*) **A-212,** 38 (1952); W. Shockley, *Phys. Rev.* **90,** 491 (1953).

[56] G. Dresselhaus, A. F. Kip, and C. Kittel, *Phys. Rev.* **92,** 827 (1953).

originate from the threefold degenerate p-levels of the isolated atoms. However the spin-orbit coupling, which in the isolated atom splits the p-levels into $p_{3/2}$ and $p_{1/2}$ states, lowers the energy of one of the bands. The two remaining bands give rise to two holes with different effective masses, classified as the light and heavy hole.

An example of a cyclotron resonance spectrum observed for germanium[57] is shown in Fig. 5-8.1. The identification of a given peak as an electron or hole resonance is made either by employing a circularly

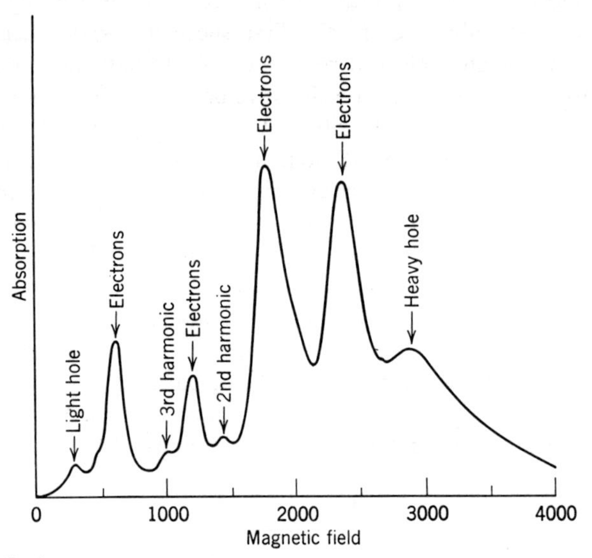

Fig. 5-8.1. Cyclotron resonance lines in germanium at 4°K at 23,000 Mc/sec. The steady magnetic field was 10° out of (110) plane and 30° from [100]-direction. The number of absorption lines for electrons and holes arise from the differently oriented energy surfaces. (After R. N. Dexter, H. J. Zeiger, and B. Lax.)

polarized oscillatory electric field or by using n- or p-type material; for the p-type only one kind of carrier will be excited. Lines that satisfy the frequency condition

$$\omega = -\Delta n \frac{eH}{m^*c},$$

where $\Delta n = 2$ and 3, as well as $\Delta n = 1$, are often observed for the heavy hole. These second and third harmonics arise from transitions between the Landau levels for which $\Delta n = \pm 2, \pm 3$.

A quantum mechanical calculation of the cyclotron resonance for holes

[57] R. N. Dexter, H. J. Zeiger, and B. Lax, *Phys. Rev.* **104,** 637 (1956).

in a semiconductor[58] shows that for low quantum numbers the spacing of the harmonic oscillators' levels is unequal. The theory predicts that a fine structure in the hole spectrum should be observable at about 1°K or below. Such a structure has been observed.[59]

We now turn to a discussion of the relaxation processes in cyclotron resonance. The electrons or holes may suffer collisions with lattice impurities, imperfections, or phonons. Therefore it is appropriate to introduce a collision relaxation time τ. If the scattering of a carrier in a collision is isotropic, then τ is equal to the mean time between collisions. The motion of a carrier in an electric field may be described classically by

$$\frac{d}{dt}(\mathsf{m}^* \cdot \mathbf{v}) + \frac{\mathsf{m}^* \cdot \mathbf{v}}{\tau} = e\left(\mathbf{E} + \frac{\mathbf{v} \times \mathbf{H}}{c}\right), \tag{5-8.3}$$

where the term $(\mathsf{m}^* \cdot \mathbf{v})/\tau$ has been introduced into the Lorentz equation (1-8.9) as a damping term to take account of the collisions. Here m^* is the effective mass tensor given for electrons by

$$\mathsf{m}^* = \begin{pmatrix} m_t & 0 & 0 \\ 0 & m_t & 0 \\ 0 & 0 & m_l \end{pmatrix}. \tag{5-8.4}$$

However, we simplify our discussion by assuming that the effective mass m^* as well as the relaxation time are isotropic, that $\mathbf{H}$ is applied along the z-direction, and that the alternating electric field, of frequency ω, is plane polarized along the x-direction. We then have

$$m^*\left(i\omega + \frac{1}{\tau}\right)v_x = eE_x + \frac{e}{c}\,v_y H,$$

$$m^*\left(i\omega + \frac{1}{\tau}\right)v_y = -\frac{e}{c}\,v_x H. \tag{5-8.5}$$

On solving for v_x, we find for the complex conductivity

$$\sigma = \frac{j_x}{E_x} = \frac{Nev_x}{E_x}$$

$$= \sigma_0\left[\frac{1 + i\omega\tau}{1 + (\omega_c^{\,2} - \omega^2)\tau^2 + 2i\omega\tau}\right] \tag{5-8.6}$$

[58] J. M. Luttinger and W. Kohn, *Phys. Rev.* **97,** 869 (1955); J. M. Luttinger, *Phys. Rev.* **102,** 1030 (1956).

[59] R. C. Fletcher, W. A. Yager, and F. R. Merritt, *Phys. Rev.* **100,** 747 (1955).

where $\sigma_0 = Ne^2\tau/m^*$ is the static conductivity and $\omega_c = -eH/m^*c$. The real part of the conductivity is

$$\text{Re}\,\sigma = \sigma_0\left\{\frac{1 + (\omega^2 + \omega_c^{\,2})\tau^2}{[1 + (\omega_c^{\,2} - \omega^2)\tau^2]^2 + 4\omega^2\tau^2}\right\}. \tag{5-8.7}$$

The absorption of microwave power is proportional to the real part of the conductivity for a skin depth that is large compared to the sample thickness, a condition easy to satisfy for semiconductors. This absorption is

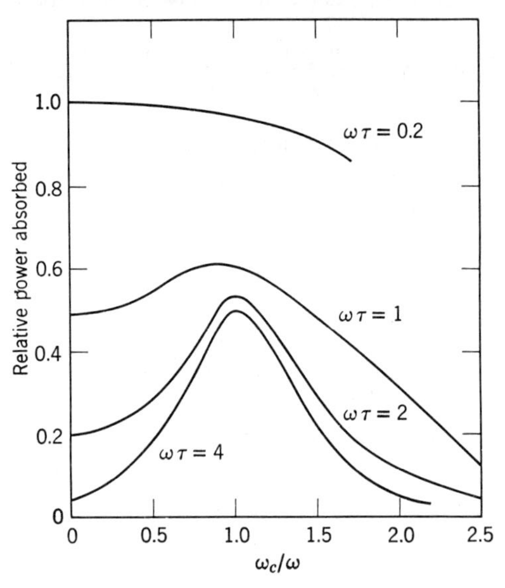

Fig. 5-8.2. Power absorption as a function of the steady magnetic field in units of ω_c/ω for different relaxation times in units of $\omega\tau$. [From B. Lax, H. J. Zeiger, and R. N. Dexter, *Physica* **20,** 818 (1954).]

shown for different values of $\omega\tau$ in Fig. 5-8.2 as a function of ω_c/ω. It is clear that in order to detect the resonant frequency we must have $\omega\tau \geqslant 1$. In other words, the average electron must be able to traverse at least one radian of its circular path between collisions. For a typical microwave frequency of $\omega \approx 10^{11}$ sec^{-1}, we must have $\tau \approx 10^{-11}$ sec at least. This condition is satisfied only for very pure semiconductor crystals at low temperatures. For an infrared frequency the experiment may be carried out at considerably higher temperatures.

The half-width at half-maximum of Reσ occurs at

$$\tau(\omega_c - \omega) = \tau\,\Delta\omega = 1. \tag{5-8.8}$$

The determination of the line width as a function of temperature yields information concerning the relaxation processes involved. In such an experiment on n-type germanium[60] it appears that the main contribution to the line width below 10°K arises from electron collisions with neutral impurities; collisions with ionized impurities are apparently not important. Above 10°K electron-phonon collisions become increasingly important as the temperature is increased. An anomoly in the relaxation time was observed at about 25°K; it is unexplained at present.

For pure samples of germanium and silicon at low temperatures the carrier density is low. When the density increases, for example, at higher temperatures, space charge or plasma effects must be considered. These effects may be taken into account by the introduction of a depolarization

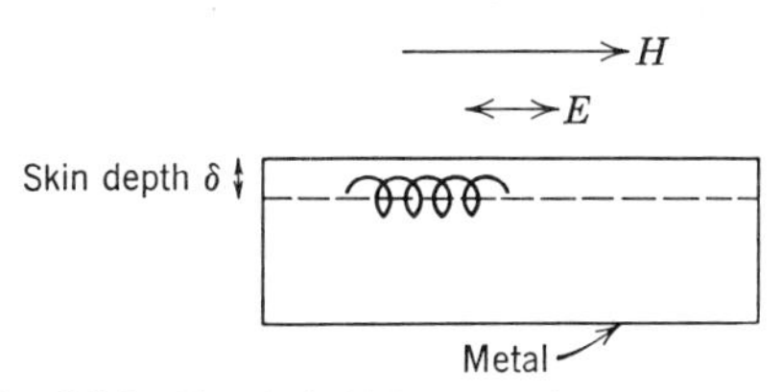

Fig. 5-8.3. The Azbel-Kaner cyclotron resonance in a metal.

term into equation 5-8.3. The result is that first the resonance frequency is reduced and then, at high enough electron densities, the resonance is independent of the effective mass. Thus plasma effects impose an upper limit to the carrier concentration possible for the study of cyclotron resonance.

For metals the carrier density is so large that the skin depth is less than the specimen's dimensions. It can then be shown that only a nonresonant absorption of microwaves is to be expected.[61] An inflection in the nonresonant absorption that is close to the cyclotron resonance frequency has been observed for bismuth.[62]

Cyclotron resonance in metals can, however, be observed by a scheme proposed by Azbel and Kaner.[63] It is necessary that anomolous skin effect conditions hold; that is, the mean free path of the electrons must be considerably larger than the skin depth. The steady magnetic field and the oscillating electric field are applied parallel to the surface of the metal. An electron, which will travel in a helical path, will spend only part of each cycle in the skin-depth region, as indicated in Fig. 5-8.3. Effectively, the

[60] D. M. S. Bagguley, R. A. Stradling, and J. S. S. Whiting, *Proc. Roy. Soc.* (*London*) **A-262,** 340 (1961), **A-262,** 365 (1961).

[61] G. Dresselhaus, A. F. Kip, and C. Kittel, *Phys. Rev.* **100,** 618 (1955).

[62] J. K. Galt, W. A. Yager, F. R. Merritt, B. B. Cetlin, and H. W. Dingle, *Phys. Rev.* **100,** 748 (1955); R. N. Dexter and B. Lax, *Phys. Rev.* **100,** 1216 (1955).

[63] M. Ya. Azbel and E. A. Kaner, *J. Exptl. Theoret. Phys.* (*USSR*) **30,** 811 (1956) [trans. *Soviet Phys.-JEPT* **3,** 772 (1956)].

electric field will interact with the electron only while it is in this region. At resonance the electron will experience the same force each time it is within the skin depth, and hence energy will be absorbed. Subharmonic resonance frequencies may also be expected, since electrons that enter the skin depth after two or three periods will also be in the correct phase to interact. Azbel and Kaner cyclotron resonance has been observed in tin, copper,[64] and bismuth.[65]

9. Nuclear Magnetic Resonance in Metals

The magnetic resonance from nuclei differs from that in nonmetallic solids because of the interaction of the conduction electrons with the nuclei. In addition, when the sample dimensions are large compared to the skin depth, the energy absorption depends on both the real and imaginary parts of the susceptibility. The shape of the resonance line is then altered[66] (see also Section 5-7), an effect that can be eliminated by using a specimen which has at least one dimension that is small compared to the skin depth.

Nuclear magnetic resonance in a metal was first observed by Pound in copper.[67] The magnetic resonance line from nuclei in a metal occurs at a higher magnetic field for a given radio frequency than that from the same nuclei in a nonmetallic salt.[68] The change in the field required for resonance, ΔH, is known as the Knight shift. The Knight shift is usually an order of magnitude larger than a typical chemical shift (Section 4-5). Experimentally it is found that the shift ΔH is proportional to the applied field H, almost independent of temperature, and progressively larger for elements with a larger atomic weight.

In order to account for the magnitude of the Knight shift, the local magnetic field at a nucleus has to be larger than the average local field in the sample. The conduction electrons with unpaired spins produce this local field at the nucleus[69]; it is large because conduction electrons in s-states have a large average probability density, $\overline{|\psi_F(0)|^2}$, at the nucleus. The contribution from the orbital motion of the conduction electrons to the local field at the nucleus is believed to be small, since it arises mainly from the probability density at larger distances from the nucleus.

The energy of the hyperfine interaction between the nuclear moment and local field at the nucleus produced by the s-state conduction electrons,

[64] E. Fawcett, *Phys. Rev.* **103,** 1582 (1956); A. F. Kip, D. N. Langenberg, B. Rosenblum, and G. Wagoner, *Phys. Rev.* **108,** 494 (1957).

[65] J. E. Aubrey and R. G. Chambers, *J. Phys. Chem. Solids* **3,** 128 (1957).

[66] N. Bloembergen, *J. Appl. Phys.* **23,** 1383 (1952).

[67] R. V. Pound, *Phys. Rev.* **73,** 112 (1948).

[68] W. D. Knight, *Phys. Rev.* **76,** 1259 (1949).

[69] C. H. Townes, C. Herring, and W. D. Knight, *Phys. Rev.* **77,** 852 (1950).

$g_n \mu_{B_n} \mathbf{I} \cdot \Delta \mathbf{H}$, is mainly given by the last term of equation 3-11.2:

$$\frac{8\pi}{3} g g_n \mu_B \mu_{B_n} \delta(r) \mathbf{s} \cdot \mathbf{I}.$$

Hence the Knight shift may be written as

$$\Delta H = \frac{\chi_{\text{para}} H}{N} \overline{|\psi_F(0)|^2} \tag{5-9.1}$$

where χ_{para} is the spin susceptibility and N is the number of atoms, both per unit volume; $\overline{|\psi_F(0)|^2}$ has been calculated for some of the alkali metals,[70] notably Li and Na. The observed Knight shift then gives values of χ_{para} that are in reasonable agreement with the Schumacher and Slichter values discussed in Section 5-2. For other metals it is usual to express $\overline{|\psi_F(0)|^2}$ in terms of the probability density for the free atom $\overline{|\psi_A(0)|^2}$. Conversely, measurement of the Knight shift and the spin susceptibility leads to information concerning the wave functions of the conduction electrons.[71]

The nuclear spin-lattice relaxation time in metals arises primarily from the hyperfine interaction between the nucleus and the s-state conduction electrons. When an electron moves in the vicinity of a nucleus, the large time-varying local field may, with a certain probability, induce transitions between the nuclear levels. The electron emits or absorbs these quanta by a change in its kinetic energy. Only those electrons near the Fermi level can participate in this energy exchange, since only they are close, in energy, to unoccupied states. Nevertheless, this mechanism leads to values of the spin-lattice relaxation time, τ_1, of about 10^{-2} sec at room temperature to about 1 sec at 1°K. These values are in good agreement with experiment.[72]

The nuclear resonance lines from metals are narrower than those from nonmetallic solids. Indeed, except at very low temperatures, the line widths are similar to those observed from liquids (see Section 4-3). Primarily this is the result of the small spin-lattice relaxation time τ_1, which limits the lifetime of the nuclear states.[73]

[70] T. Kjeldaas and W. Kohn, *Phys. Rev.* **101,** 66 (1956).

[71] R. T. Schumacher and C. P. Slichter, *Phys. Rev.* **101,** 58 (1956); Ch. Ryter, *Phys. Rev. Letters* **5,** 10 (1960); F. J. Milford and W. B. Gager, *Phys. Rev.* **121,** 716 (1961); P. L. Sagalyn and J. A. Hofmann, *Phys. Rev.* **127,** 68 (1962).

[72] N. Bloembergen, *Physica* **15,** 588 (1949); R. E. Norberg and C. P. Slichter, *Phys. Rev.* **83,** 1074 (1951); M. A. Ruderman and C. Kittel, *Phys. Rev.* **96,** 99 (1954).

[73] J. Korringa, *Physica* **16,** 601 (1950); A. G. Redfield, *Phys. Rev.* **101,** 67 (1956); A. G. Anderson and A. G. Redfield, *Phys. Rev.* **122,** 583 (1959).

Finally, it is appropriate to give a brief outline of the Overhauser effect,[74] since it also rises from the hyperfine interaction. Consider a metal in a uniform magnetic field H. Suppose that two oscillating magnetic fields are applied so that one has the frequency required for spin resonance from the conduction electrons and the other has the frequency required for nuclear resonance. When the electron spin resonance is saturated by increasing the input power to the electron system, Overhauser predicted that the nuclear resonance signal would be enhanced. This effect was confirmed in an experiment on lithium.[75] The main reason for the enhancement is the fact that the sum $m_S + m_I$ is conserved in the hyperfine interaction between a conduction electron and a nucleus. The transition $\Delta m_S = +1$ implies a corresponding transition $\Delta m_I = -1$, and vice versa. Therefore, since the difference in populations of the electrons' levels approaches zero when the electron system is saturated, the difference in populations of the nuclear levels is greatly increased. This dynamical method of polarizing nuclear spins does not require the use of low temperatures, as do the static methods described in Section 4-9.

It is not necessary to employ a metallic specimen in order to observe the dynamic polarization effect[76]; indeed, it has also been observed for semiconductors,[77] free radicals,[78] and paramagnetic crystals.[79]

10. Some Magnetic Properties of Superconductors

Certain metals and chemical compounds have zero resistance at very low temperatures; when in this state, they are called superconductors. Elements that are superconductors are generally relatively poor conductors in the normal state; examples are Al, Zn, Sn, Hg, Pb, and U. The monovalent metals are not superconductors. The transition temperature, T_c, at which the change from the normal to the superconducting state occurs ranges from a few tenths of a degree to about 10°K for metals

[74] A. W. Overhauser, *Phys. Rev.* **92,** 411 (1953); G. R. Khutsishvili, *Soviet Phys.-Uspekhi* **3,** 285 (1964); R. H. Webb, *Am. J. Phys.* **29,** 428 (1961).

[75] T. R. Carver and C. P. Slichter, *Phys. Rev.* **92,** 212 (1953), **102,** 975 (1956); R. Hecht and A. G. Redfield, *Phys. Rev.* **132,** 972 (1963).

[76] A. Abragam, *Phys. Rev.* **98,** 1729 (1955); A. Abragam and W. G. Proctor, *Compt. rend. (Paris)* **246,** 2253 (1958); A. Abragam, *Compt. rend. (Paris)* **254,** 1267 (1962); M. Goldman, *Magnetic and Electric Resonance and Relaxation*, J. Smidt, Editor, North-Holland Publishing Co., Amsterdam (1962), p. 689.

[77] A. Abragam, J. Combrisson, and I. Solomon, *Compt. rend. (Paris)* **246,** 1035 (1958); A. Abragam, A. Landesman, and J. M. Winter, *Compt. rend. (Paris)* **247,** 1849 (1958).

[78] H. G. Beljers, L. Van Der Kint, and J. S. Van Wieringen, *Phys. Rev.* **95,** 1683 (1954).

[79] C. D. Jeffries, *Phys. Rev.* **106,** 164 (1957); D. S. Liefson and C. D. Jeffries, *Phys. Rev.* **122,** 1781 (1961).

($T_c = 0.14°K$ for Ir; $T_c = 9.13°K$ for Nb). The highest transition temperature observed at present is 18.0°K for the intermetallic compound Nb_3Sn.

The application of a magnetic field that exceeds a certain critical value will destroy superconductivity. The critical field is a function of temperature of the form

$$H_{\rm cr} = H_0\left(1 - \frac{T^2}{T_c^{\,2}}\right), \tag{5-10.1}$$

so that at $T = T_c$, $H_{\rm cr} = 0$. This equation may be derived from thermodynamical considerations.[80] The magnetic field need not be of external origin; it may also be the result of the electric current flow in the conductor. Indeed, the critical current rather than the critical field is now considered to be the fundamental quantity that induces the phase transition.

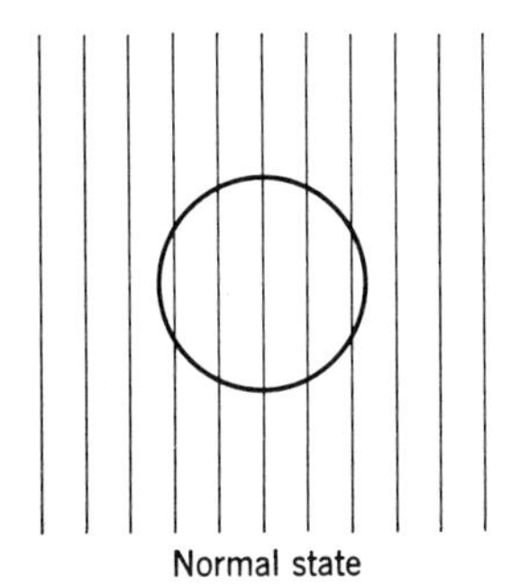

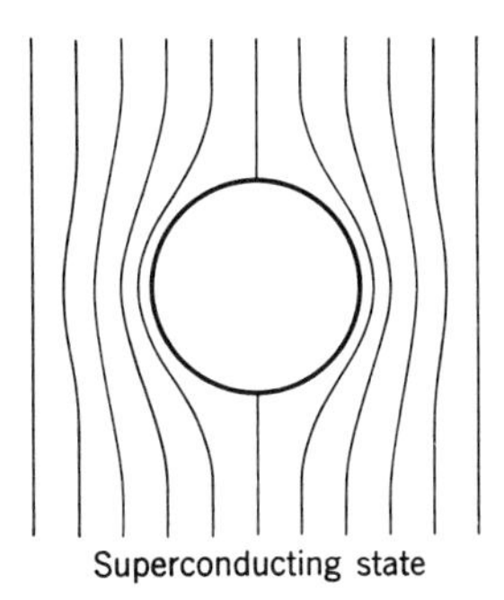

Fig. 5-10.1. The Meissner effect. The lines of induction are ejected at the transition temperature.

Any lines of induction present in a metal in the normal state are expelled in the superconducting state[81] (see Fig. 5-10.1); this important experimental result is known as the Meissner effect. Since inside the specimen $B = 0$, the susceptibility of a superconductor is given by equation 1-4.1 as

$$\chi = -\frac{1}{4\pi}. \tag{5-10.2}$$

This is the maximum negative value of susceptibility possible; consequently, a superconducting metal is said to exhibit *perfect* diamagnetism. A few elements and most alloys and chemical compounds that are superconducting show an incomplete Meissner effect and have a broad magnetic transition. They are called variously *hard*, *Type II*, or *London*

[80] C. J. Gorter, *Nuovo cimento, Suppl.* **3,** 1168 (1957).
[81] W. Meissner and R. Ochsenfeld, *Naturwiss.* **21,** 787 (1933).

superconductors. By contrast those materials with sharp transitions are known as *soft*, *Type I*, or *Pippard* superconductors. Type II materials have important applications as superconducting solenoids and are discussed later in this section.

The magnetic field H_i inside a specimen is related to the applied field by equation 1-5.1:

$$H_i = H - DM,$$

where D is the demagnetization factor. For a Type I superconductor $H_i = -4\pi M$, hence

$$M = -\frac{H}{4\pi - D}. \tag{5-10.3}$$

For a sphere $D = 4\pi/3$, and we obtain

$$M = -\frac{3}{8\pi} H$$

or

$$\chi = -\frac{3}{8\pi}. \tag{5-10.4}$$

The maximum tangential component of the field at the surface of a sphere occurs at a point on the equator. There it is equal to the internal field H_i. Substitution of equation 5-10.4 into (1-15.1) gives for this field

$$H_i = \tfrac{3}{2}H.$$

Hence this field will exceed the critical field when

$$H > \tfrac{2}{3}H_{cr}. \tag{5-10.5}$$

When the applied field exceeds $\frac{2}{3}H_{cr}$, the entire sphere does not become normal; instead, certain regions, called domains, become normal, whereas others remain superconducting. When $H = H_{cr}$, the entire sphere becomes normal. When, for a sphere,

$$\tfrac{2}{3}H_{cr} < H < H_{cr},$$

the material is said to be in the intermediate state.

The first theories of superconductivity were phenomenological; indeed a satisfactory microscopic theory was not developed until 1956. It is interesting to note that the diamagnetic rather than the zero resistivity property provided the clue that led to a successful theory.

In order to account for the Meissner effect other than by setting $\chi = -\frac{1}{4}\pi$, the London brothers[82] postulated the equation

$$\Lambda_s \nabla \times \mathbf{j}_s = -\mathbf{H}, \tag{5-10.6}$$

where $\Lambda_s = \dfrac{mc}{N_s e^2}$, N_s is the number of electrons per unit volume that are superconducting, and $\mathbf{j}_s$ is the superconducting current density. To eliminate $\mathbf{j}_s$, we start with equation 1-8.4 of Maxwell with $\mathbf{D} = 0$,

$$\nabla \times \mathbf{H} = \frac{4\pi \mathbf{j}_s}{c}$$

and take the curl of each side. Since it is now assumed that $\mu = 1$, $\nabla \cdot \mathbf{H} = 0$ and

$$\nabla^2 \mathbf{H} = -\frac{4\pi}{c} \nabla \times \mathbf{j}_s.$$

On substituting this expression into equation 5-10.6 we get

$$\lambda_L{}^2 \nabla^2 H = H, \tag{5-10.7}$$

where

$$\lambda_L = \frac{\Lambda_s c}{4\pi}.$$

The solution of this differential equation for one dimension is

$$H_i = He^{-x/\lambda_L}. \tag{5-10.8}$$

This equation states that the internal field inside the superconductor will decay exponentially and be essentially zero for $x \gg \lambda_L$. If $N_s \approx N$, the number of conduction electrons per unit volume, λ_L is of the order of 10^{-6} cm; hence equation 5-10.8 satisfactorily describes the Meissner effect.

On the basis of high frequency surface impedance measurements on superconductors, Pippard[83] proposed a modification of the London theory. If $\mathbf{H} = \nabla \times \mathbf{A}$, where $\mathbf{A}$ is the vector potential (equation 1-8.7), it follows from the London equation 5-10.6 that

$$\mathbf{j}_s = -\mathbf{A}/\Lambda_s. \tag{5-10.9}$$

This equation implies that the current density at some position depends solely on the vector potential existing at that same position. Pippard suggested instead that the current density depended on the value of **A**

[82] F. London and H. London, *Proc. Roy. Soc.* (*London*) **A-149,** 72 (1935), *Physica* **2,** 341 (1935). For pertinent work that preceded the London equations see R. Becker, G. Heller, and F. Sauter, *Z. Physik* **85,** 772 (1933).

[83] A. B. Pippard, *Proc. Roy. Soc.* (*London*) **A-203,** 210 (1950), **A-216,** 547 (1953).

over a region with dimensions of ξ, known as the coherence distance, surrounding the position. The sharp transition in type I superconductors is another indication that the electrons are behaving coherently. The terms nonlocal and local are often applied to the Pippard and London theories, respectively, for the purposes of distinction. If **A** is a slowly varying function of position, the Pippard value of $\mathbf{j}_s$ reduces to that of the London theory.

For a type I superconductor ξ is estimated to be of the order of 10^{-4} cm, which is considerably greater than the penetration distance λ_L. This result implies that energy is associated with the boundary between the normal and superconducting phases of a material in the intermediate state, as may be seen from the following simple argument. The magnetostatic energy associated with the Meissner effect is reduced because of the partial penetration of the magnetic field into the superconducting region; this decrease of energy is given approximately by $-\lambda_L H_{cr}^2/8\pi$ per unit volume. On the other hand, the energy gained in the condensation of normal to superconducting electrons is less because of the correlation effect by about $\xi H_{cr}^2/8\pi$ per unit volume. The net energy per unit area of the interphase boundary is then

$$F_b = (\xi - \lambda_L)\frac{H_{cr}^2}{8\pi} \tag{5-10.10}$$

and is positive if $\xi > \lambda_L$, as for a type I material. In spite of the positive boundary energy, domains do occur, and the reason is that the total free energy is thereby reduced.

In order to treat boundary problems more adequately, Ginzburg and Landau proposed a further extension of the London equations.[84] They introduced the order parameter Ψ which is a sort of macroscopic electronic wave function; $\Psi^*\Psi$ may be interpreted as N_s, the volume density of the superconducting electrons. Ginzburg and Landau then postulated the two differential equations

$$\nabla^2\mathbf{A} = \frac{2\pi i q\hbar}{mc}(\Psi^*\nabla\Psi - \Psi\nabla\Psi^*) + \frac{4\pi q^2}{mc^2}|\Psi|^2\mathbf{A} \tag{5-10.11}$$

and

$$\nabla^2\Psi = \frac{\kappa_L^2}{\lambda_L^2}\left[-\left(1 - \frac{A^2}{2H_{cr}^2\lambda_L^2}\right)\Psi + \beta\Psi^3\right], \tag{5-10.12}$$

where

$$\kappa_L = \frac{\sqrt{2}qH_{cr}\lambda_L^2}{\hbar c} \tag{5-10.13}$$

[84] V. L. Ginzburg and L. D. Landau, *J. Exptl. Theoret. Phys.* (*USSR*) **20,** 1064 (1950); V. L. Ginzburg, *Nuovo cimento* **2,** 1234 (1955).

and β is a constant. In the original equations q was set equal to e; however, as will become clearer when the microscopic theory is discussed later, $q = 2e$. Now, equation 1-8.4 with $D = 0$ may be written in terms of A as

$$\nabla^2 \mathbf{A} = -\frac{4\pi}{c}\,\mathbf{j}_s.$$

This expression, together with equation 5-10.11, yields for the current

$$\mathbf{j}_s = \frac{iq\hbar}{2m}(\Psi\nabla\Psi^* - \Psi^*\nabla\Psi) - \frac{q^2}{mc}\,\mathbf{A}\,|\Psi|^2. \tag{5-10.14}$$

If the applied field is small compared to the critical field H_{cr}, the spatial variation of the order parameter will be slight. Then both $\nabla\Psi^*$ and $\nabla\Psi$ approach zero and $\mathbf{j}_s$ reduces to the London value. This result establishes that the London equations hold only for $H \ll H_{cr}$ and therefore cannot be used to analyze the field induced phase transition. On the other hand, the Ginzburg-Landau equations are valid for $H \approx H_{cr}$. However, since the Ginzburg-Landau theory is a local one, its validity is restricted, except for some special circumstances, to temperatures close to T_c where $\xi < \lambda_L$.

The Ginzburg-Landau equations have provided satisfactory answers to a number of problems. Their greatest achievement is probably the prediction of type II superconductivity.[85] This result can be easily obtained by considering a one-dimensional form of equation 5-10.12, that is, with $\nabla^2\Psi = d^2\Psi/dx^2$ and A a function only of x. Now, in a type II superconductor the internal magnetic field will approach the external applied field, and it therefore follows that we may write

$$A(x) = Hx.$$

Substitution into equation 5-10.12 yields

$$\frac{d^2\Psi}{dx^2} = -\frac{\kappa_L^2}{\lambda_L^2}\left(1 - \frac{H^2x^2}{2H_{cr}^2\lambda_L^2}\right)\Psi,$$

where the term in $\beta\Psi^3$ has been omitted since it is small because of the lack of much order. This is just the Schrödinger equation for a harmonic oscillator; it has well-behaved solutions only if[86]

$$\frac{H}{H_{cr}} = \frac{\sqrt{2}\kappa_L}{2n+1},$$

[85] A. A. Abrikosov, *Dokl. Akad. Nauk SSSR* **86,** 489 (1952), *J. Exptl. Theoret. Phys. (USSR)* **32,** 1442 (1957) [trans. *Soviet Phys.-JEPT* **5,** 1174 (1957)]; N. V. Zavaritsky, *Dokl. Akad. Nauk SSSR* **86,** 501 (1952); B. B. Goodman, *Phys. Rev. Letters* **6,** 597 (1961).

[86] See, for example, V. Rojansky, *Introductory Quantum Mechanics*, Prentice-Hall, Englewood Cliffs, N.J. (1938), pp. 82–83.

where $n = 0, 1, 2, 3, \ldots$. For $n = 0$, H/H_{cr} has its maximum value given by

$$\frac{H}{H_{cr}} = \sqrt{2}\kappa_L. \tag{5-10.15}$$

Hence, if $\kappa_L > 1/\sqrt{2}$, equation 5-10.15 predicts that $H > H_{cr}$, even though $\Psi > 0$, that is, that the material will remain superconducting for fields greater than the critical one. Calculations show that when $\kappa_L > \frac{1}{2}$ the interphase boundary energy is negative. From the approximate equation 5-10.10 this implies that the coherence distance ξ is less than the penetration distance λ_L in a type II superconductor.

The only other phenomenological theory that merits a brief mention here is the two-fluid model of Gorter and Casimir.[87] In their picture a fraction of the conduction electrons are condensed into a phase with lower free energy and give rise to the superconductivity; the other electrons remain in the normal state. At $T = 0°K$ all conduction electrons are assumed to be in the superfluid phase. As the temperature is raised, the number of electrons in the condensed state decreases until at $T = T_c$ all electrons are normal. The two-fluid model was successful in accounting for some thermal properties of superconductors, including the latent and specific heats.

Two clues that were particularly important for the development of the microscopic theory are now mentioned. One is the idea of an energy gap between the superconducting and the normal electrons. Besides having been suggested in connection with the two-fluid model, some thermal experiments provided supporting evidence. The other clue came from the suggestion that the interactions that led to superconductivity were between the electrons and the lattice vibrations.[88] Confirmation of the correctness of this view was given by the observation that the critical temperature depended on the isotopic mass of an element.[89] For nontransition elements the relationship was found to be $T_c \propto M^{-1/2}$, where M is the atomic weight.

Following on the heels of some preliminary forward steps in theory,[90] Bardeen, Cooper, and Schrieffer[91] finally developed a successful microscopic theory, now popularly known as the BCS theory. We shall not

[87] C. J. Gorter and H. B. G. Casimir, *Phys. Z.* **35,** 963 (1934).

[88] H. Fröhlich, *Phys. Rev.* **79,** 845 (1950).

[89] E. Maxwell, *Phys. Rev.* **78,** 477 (1950), **79,** 173 (1950); C. A. Reynolds, B. Serin, W. H. Wright, and L. B. Nesbitt, *Phys. Rev.* **78,** 487 (1950).

[90] H. Fröhlich, *Proc. Roy. Soc.* (*London*) **A-215,** 291 (1952); J. Bardeen, *Phys. Rev.* **97,** 1724 (1955); J. M. Blatt, S. T. Butler, and M. R. Schafroth, *Phys. Rev.* **100,** 481 (1955); M. R. Schafroth, *Phys. Rev.* **100,** 502 (1955); L. N. Cooper, *Phys. Rev.* **104,** 1189 (1956); M. R. Schafroth, S. T. Butler, and J. M. Blatt, *Helv. Phys. Acta* **30,** 93 (1957).

[91] J. Bardeen, L. N. Cooper, and J. R. Schrieffer, *Phys. Rev.* **108,** 1175 (1957).

go into the quantum mechanical details here, but instead shall be content to present a qualitative outline of some of the concepts and their consequences. The BCS theory assumed that pairs of conduction electrons with opposite spin and equal and opposite momenta or wave vectors **k** and −**k** interact via the lattice. This interaction, which is attractive in nature, consists of an exchange of virtual phonons between the electrons. The two states of a pair must either be both occupied or both empty. The pairs represent the ground state of the superconductor, and, just as in the two-fluid model, more and more electrons condense into this phase as the temperature is lowered. The density of states as a function of energy for a normal metal (see Fig. 5-1.2) is modified when it becomes a superconductor. The occupancy of states with energies between $E_F + \Delta$ and

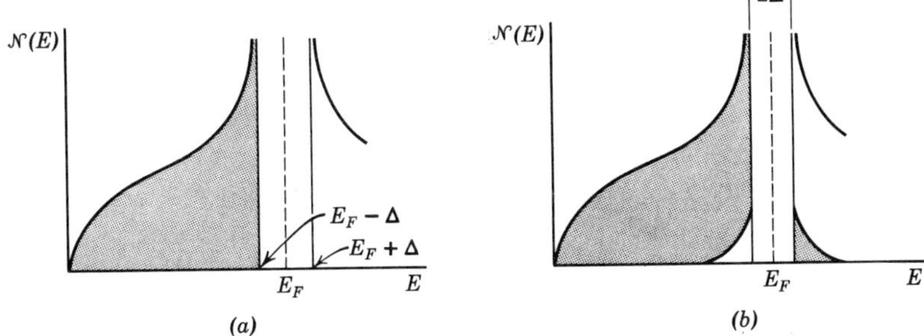

Fig. 5-10.2. The density of states in a superconductor according to the BCS theory (*a*) at $T = 0°K$ and (*b*) at $0 < T < T_c$. The shaded areas indicate the states occupied by electrons.

$E_F - \Delta$, where E_F is the Fermi energy of the normal metal, is now forbidden. For energies just above and below the forbidden gap the density of states becomes very large (see Fig. 5-10.2). In this way the total number of states below the Fermi level is kept the same. At $T = 0°K$ all the electrons are coupled as pairs and occupy the states with energy $E < E_F - \Delta$ (Fig. 5-10.2*a*). As the temperature is raised, some pairs are decoupled into normal electrons by excitation to states with energies $E > E_F - \Delta$ (Fig. 5-10.2*b*). The theory predicts that at absolute zero

$$2\Delta = 3.5kT_c \tag{5-10.16}$$

and that the energy gap decreases with increasing temperature becoming zero at the critical temperature. The appearance of kT_c in equation 5-10.16 should not be surprising, since kT_c is the thermal agitation energy required to destroy the superconductivity.

A major task of any superconductivity theory is to account for the zero resistance property. The BCS theory does this in the following way. Suppose the current flow corresponds to electrons moving with velocity **v**. Actually, the current consists of the transport of electron pairs. The total momentum of a pair is $(\hbar\mathbf{k} + m\mathbf{v}) + (-\hbar\mathbf{k} + m\mathbf{v}) = 2m\mathbf{v}$ and is the same for all pairs. In other words, the entire system with interactions is displaced in momentum space when there is a flow of current. Now, in a normal conductor the collisions of electrons with the phonons and lattice irregularities result in the transfer of kinetic energy to the lattice; this process is the origin of joule heating or, equivalently, the resistance. In a superconductor collisions still occur between the pairs and the lattice. However, if one of the electrons of a pair is to suffer a loss in velocity, hence a decrease in kinetic energy, a decoupling of the pair must occur with a consequent increase in the potential energy of 2Δ. Therefore, as long as the decrease in kinetic energy by scattering is less than the increase in potential energy, the electron pair will not be decoupled and the common momentum or velocity of the pair will remain the same. It is for this reason that the critical current rather than the critical field is the fundamental cause of the destruction of superconductivity. At absolute zero the critical current j_c can be shown to be given approximately by

$$j_c = -\frac{ne\Delta}{mv_F},$$

where v_F is the Fermi velocity defined by $E_F = \frac{1}{2}mv_F^2$.

The microscopic theory leads to a nonlocal relationship between $\mathbf{j}_s$ and $\mathbf{A}$ that is essentially the same as that proposed by Pippard; ξ is interpreted as the distance over which the attractive force between the electrons of a pair acts. The theory predicts that for a pure metal

$$\xi = 0.18\,\frac{\hbar v_F}{kT_c}.$$

Further, Gor'kov[92] has shown that the Ginzburg-Landau equations follow from the BCS theory near the transition temperature if $q = 2e$. He also succeeded in deriving a nonlocal version of the Ginzburg-Landau equations, which may be of great importance in the future. It should also be mentioned that several other authors have made significant advances in the formulation of the BCS theory.[93]

[92] L. P. Gor'kov, *J. Exptl. Theoret. Phys. (USSR)* **36,** 1918 (1959) [trans. *Soviet Phys.-JEPT* **9,** 1364 (1959)], *J. Exptl. Theoret. Phys. (USSR)* **37,** 835 (1960) [trans. *Soviet Phys.-JEPT* **10,** 593 (1960)], *J. Exptl. Theoret. Phys. (USSR)* **37,** 1407 (1959) [trans. *Soviet Phys.-JEPT* **10,** 998 (1960)].

[93] N. N. Bogoliubov, V. V. Tolmachev, and D. V. Skirkov, *A New Method in the Theory of Superconductivity*, Acad. Sci. USSR Press, Moscow (1958) [trans. Consultants Bureau, Inc., New York (1959)]; J. G. Valatin, *Nuovo cimento* **7,** 843 (1958); P. W. Anderson, *Phys. Rev.* **110,** 827 (1958), **112,** 1900 (1959); G. Rickayzen, *Phys. Rev.* **115,** 795 (1959); P. C. Martin and J. Schwinger, *Phys. Rev.* **115,** 1342 (1959); L. P. Kadanoff and P. C. Martin, *Phys. Rev.* **124,** 670 (1961); L. P. Gor'kov, *J. Exptl. Theoret. Phys. (USSR)* **34,** 735 (1958) [trans. *Soviet Phys.-JEPT* **7,** 505 (1958)]; A. A. Abrikosov and L. P. Gor'kov, *J. Exptl. Theoret. Phys. (USSR)* **35,** 1558 (1958) [trans. *Soviet Phys.-JEPT* **8,** 1090 (1959)].

Since the various phenomenological theories all follow as special cases of the microscopic theory, it is obvious that any experimental results that can be interpreted satisfactorily by the former theories are described at least as well if not better by the BCS theory. In addition, the microscopic theory can account for some experimental data that are inexplicable in terms of the phenomenological theories. An example is the nuclear relaxation time of a superconductor. The relaxation time in superconducting aluminum was determined by a field cycling technique in which the field H was changed from zero to a value larger than the critical field at periodic intervals. In this way the relaxation that occurred in the superconducting state was studied by making measurements in the normal state. The relaxation time was found first to decrease markedly as the temperature was decreased below T_c and then to increase at still lower temperatures.[94] Inasmuch as the principal relaxation mechanism is likely to be via the conduction electrons, the relaxation rate below T_c might at first sight be expected to decrease. It turns out, however, that the contributions of the two electrons of a pair to the relaxation add coherently to yield a result in agreement with the experiments. On the other hand, in the absorption of ultrasonic waves, the contributions add destructively. The theory then predicts that the ultrasonic attenuation will decrease rapidly below T_c; this behavior is indeed observed experimentally.[95] These two experiments provide strong confirmation of the pair hypothesis.

At least two other experiments give direct evidence of the existence of an energy gap. In one of these high-frequency electromagnetic radiation is employed. At low frequencies little energy is absorbed. However, when the frequency ν reaches a certain critical value, an abrupt increase in the absorption of the radiation occurs.[96] Clearly, this will take place when the photon energy equals that of the gap, that is, when

$$\hbar\nu = 2\Delta.$$

If the critical temperature of the superconductor is low enough, as for aluminum, the critical frequency will fall in the microwave region. Often, as for tin and lead, the critical frequency lies in the far infrared (see page 614), where the experimental techniques are rather difficult. The coefficient of 3.5 in equation 5-10.16 is approximately equal to that found experimentally.

In the other experiment a potential difference is applied to two metals separated by a thin insulating film. Quantum mechanics predicts that

[94] L. C. Hebel and C. P. Slichter, *Phys. Rev.* **113,** 1504 (1959).

[95] R. W. Morse and H. V. Bohm, *Phys. Rev.* **108,** 1094 (1957).

[96] R. E. Glover, *Phys. Rev.* **108,** 243 (1957); P. L. Richards and M. Tinkham, *Phys. Rev.* **119,** 575 (1960).

the electrons near the Fermi level, by virtue of their wavelike properties, may penetrate the potential barrier posed by the film and give rise to an observable current; this process is commonly called tunneling. If the two metals are in the normal state, the current is almost a linear function of the voltage; that is, the junction is ohmic. However, if one metal is in the superconducting state, electrons at the Fermi level of the normal metal cannot tunnel into the forbidden gap of the superconductor. Therefore, in order to have appreciable current flow, it is necessary to shift the relative positions of the Fermi levels by applying a critical voltage $\mathcal{E}_c$, given approximately by

$$e\mathcal{E}_c = \Delta$$

(see Figs. 5-1.2 and 5-10.2). This effect was first observed by Giaever,[97] and has since been the subject of extensive experimental and theoretical investigations.[98] The variation of the energy gap with temperature found experimentally is in good accord with that calculated from the BCS theory. If both metals are in the superconducting state, some new phenomena occur. If the two superconductors have unequal energy gaps, a part of the current versus voltage curve will have a negative slope; that is, in this region the junction will exhibit negative resistance. It has also been predicted[99] and verified[100] that a current can flow with no potential drop between the metals; this current originates from the tunneling of electron pairs. Finally, in agreement with expectations,[99] current has been observed to flow when the voltage equals $nh\nu/2e$, where n is an integer and ν is the frequency of the applied microwave radiation.[101]

It should be pointed out that to date the BCS theory has not enjoyed unlimited success. There do exist experimental data that cannot yet be considered to be completely understood, although in some areas progress is being made. An example is the Knight shift in superconductors. It will be recalled that the Knight shift in a normal metal is believed to arise from the magnetic field created at the nucleus by the spin polarization of the conduction electrons. Since the BCS theory predicts that the

[97] I. Giaever, *Phys. Rev. Letters* **5,** 147 (1960).

[98] I. Giaever, *Phys. Rev. Letters* **5,** 464 (1960); J. Nicol, S. Shapiro, and P. H. Smith, *Phys. Rev. Letters* **5,** 461 (1960); I. Giaever and K. Megerle, *Phys. Rev.* **122,** 1101 (1961); J. Bardeen, *Phys. Rev. Letters* **6,** 57 (1961); N. V. Zavaritsky, *J. Exptl. Theoret. Phys.* (*USSR*) **41,** 657 (1961) [trans. *Soviet Phys.-JEPT* **14,** 470 (1962)]; I. Giaever, H. R. Hart, and K. Megerle, *Phys. Rev.* **126,** 941 (1962); M. H. Cohen, L. M. Falicov, and J. C. Phillips, *Phys. Rev. Letters* **8,** 316 (1962); J. R. Schrieffer, *Revs. Mod. Phys.* **36,** 200 (1964).

[99] B. D. Josephson, *Phys. Letters* **1,** 251 (1962), *Revs. Mod. Phys.* **36,** 216 (1964).

[100] P. W. Anderson and J. M. Rowell, *Phys. Rev. Letters* **10,** 230 (1963); J. M. Rowell, *Phys. Rev. Letters* **11,** 200 (1963).

[101] S. Shapiro, *Phys. Rev. Letters* **11,** 80 (1963).

number of pairs of opposite spin will increase as the temperature approaches absolute zero, it is to be anticipated that the Knight shift of a superconductor will be small. Instead, the shift is found to be comparable to that observed for a normal metal.[102] Several mechanisms that attempt to account for this experimental result have been proposed.[103,104] In one of these theories[104] the size of the spin-orbit coupling and its scattering by impurities and imperfections are considered. It appears that progress is being made and that in time this particular problem will be resolved.

Whenever transition elements enter the picture, several new mysteries appear. The transition temperature of the superconducting transition elements and their compounds has a much smaller dependence on the isotopic mass than that observed for the nontransition elements. The isotope effect is particularly small for Os and is zero or almost zero for Ru.[105] These results have led to the suggestion that the attractive interaction between electrons is sometimes due to some mechanism other than the virtual phonon one of the BCS theory; this conclusion, however, has been disputed.[106] The presence of transition elements as impurities in metals or alloys appears to reduce or even erase the energy gap.[107] To illustrate, only after Mo and Ir were properly purified was it realized that these materials are superconductors.[108] It is also found that the addition of Fe to Mo, for example, reduces the transition temperature at the rate of 200° per 1 per cent Fe.[109] Finally, it should be mentioned that ferromagnetism

[102] F. Reif, *Phys. Rev.* **102,** 1417 (1956), **106,** 208 (1957); W. D. Knight, *Phys. Rev.* **104,** 852 (1956); G. M. Androes and W. D. Knight, *Phys. Rev.* **121,** 779 (1961).

[103] K. Yosida, *Phys. Rev.* **110,** 769 (1958); P. C. Martin and L. P. Kadanoff, *Phys. Rev. Letters* **3,** 322 (1959); A. B. Pippard and V. Heine, *Phil. Mag.* **3,** 1046 (1958); L. N. Cooper, *Phys. Rev. Letters* **8,** 367 (1961); J. C. Fisher, *Australian J. Phys.* **13,** 1446 (1960).

[104] R. D. Ferrell, *Phys. Rev. Letters* **3,** 262 (1959); P. W. Anderson, *Phys. Rev. Letters* **3,** 323 (1959); A. A. Abrikosov and L. P. Gor'kov, *J. Exptl. Theoret. Phys.* (*USSR*) **42,** 1088 (1962) [trans. *Soviet Phys.-JEPT* **15,** 752 (1962)]; A. M. Clogston, A. C. Gossard, V. Jaccarino, and Y. Yafet, *Revs. Mod. Phys.* **36,** 170 (1964).

[105] T. H. Geballe, B. T. Matthias, G. W. Hull, and E. Corenzwit, *Phys. Rev. Letters* **6,** 275 (1961); D. K. Finnemore and D. E. Mapother, *Phys. Rev. Letters* **9,** 288 (1962); R. A. Hein and J. W. Gibson, *Phys. Rev.* **131,** 1105 (1963).

[106] J. W. Garland, *Phys. Rev. Letters* **11,** 114 (1963).

[107] F. Reif and M. A. Woolf, *Phys. Rev. Letters* **9,** 315 (1962); J. C. Phillips, *Phys. Rev. Letters* **10,** 96 (1963); H. Suhl and D. R. Fredkin, *Phys. Rev. Letters* **10,** 131 (1963); R. Hilsch, G. v. Minnigerode, and K. Schwidtal, *Proceedings Eighth International Conference on Low Temperature Physics*, Butterworths, London (1963), p. 155.

[108] T. H. Geballe, B. T. Matthias, E. Corenzwit, and G. W. Hull, *Phys. Rev. Letters* **8,** 313 (1962); R. A. Hein, J. W. Gibson, B. T. Matthias, T. H. Geballe, and E. Corenzwit, *Phys. Rev. Letters* **8,** 408 (1962).

[109] B. T. Matthias, M. Peter, H. J. Williams, A. M. Clogston, E. Corenzwit, and R. C. Sherwood, *Phys. Rev. Letters* **5,** 542 (1960); A. M. Clogston, B. T. Matthias, M. Peter, H. J. Williams, E. Corenzwit, and R. C. Sherwood, *Phys. Rev.* **125,** 541 (1962).

and superconductivity have been observed to occur simultaneously in some metals and compounds containing rare earth elements; examples are Gd in La, Gd in YOs_2, and Gd in $CeRu_2$.[110]

We now turn to a discussion of a remarkable phenomenon discovered in 1961, namely flux quantization. That the flux within a hollow superconducting cylinder ought to be quantized had already been recognized by London[111] several years earlier. His argument ran as follows. The current produced by a charge q will be proportional to

$$m\mathbf{v} = \left(\mathbf{p} - \frac{q}{c}\mathbf{A}\right)$$

(see equation 2-8.4). Now, the current at the interior of a superconductor is zero, and therefore $\mathbf{p} = (q/c)\mathbf{A}$. The Bohr quantization rule requires that $\oint \mathbf{p} \cdot d\mathbf{s} = nh$, where n is an integer. Since

$$\frac{q}{c}\oint \mathbf{A} \cdot d\mathbf{s} = \frac{q}{c}\int \mathbf{H} \cdot \mathbf{n}\, dS = \frac{q}{c}\int \mathbf{B} \cdot \mathbf{n}\, dS = \frac{q}{c}\Phi$$

(equation 1-4.2), it follows that

$$\Phi = \frac{nhc}{q}.$$

In the BCS theory $q = 2e$, and we conclude that[112]

$$\Phi = \frac{nhc}{2e}. \tag{5-10.17}$$

Experiments[113] have established that within hollow cylinders the flux is quantized in units of $hc/2e$. Actually, in the general case a quantity called a *fluxoid* rather than the flux is quantized.[114] The canonical

[110] B. T. Matthias, H. Suhl, and E. Corenzwit, *Phys. Rev. Letters* **1**, 92 (1958); R. A. Hein, R. L. Folge, B. T. Matthias, and E. Corenzwit, *Phys. Rev. Letters* **2**, 500 (1959); B. T. Matthias, *J. Appl. Phys.* **31**, 23S (1960); R. M. Bozorth and D. D. Davis, *J. Appl. Phys.* **31**, 321S (1960); S. V. Vonsovsky and M. S. Svirskii, *Fiz. Metall. i Metalloved.* **15**, 316 (1963).

[111] F. London, *Superfluids*, Vol. I, John Wiley and Sons, New York (1950), p. 152.

[112] L. Onsager, *Phys. Rev. Letters* **7**, 50 (1961); see also N. Byers and C. N. Yang, *Phys. Rev. Letters* **7**, 46 (1961).

[113] B. S. Deaver and W. M. Fairbank, *Phys. Rev. Letters* **7**, 43 (1961); R. Doll and M. Näbauer, *Phys. Rev. Letters* **7** 51 (1961).

[114] J. Bardeen, *Phys. Rev. Letters* **7**, 162 (1961); J. B. Keller and B. Zumino, *Phys. Rev. Letters* **7**, 164 (1961).

momentum is given by

$$\mathbf{p} = m\mathbf{v} + \frac{q}{c}\mathbf{A}$$

$$= m\frac{\mathbf{j}_s}{N_s q} + \frac{q}{c}\mathbf{A}$$

$$= \frac{q}{c}\left(\frac{mc}{N_s q^2}\mathbf{j}_s + \mathbf{A}\right).$$

Hence

$$\oint \mathbf{p}\cdot d\mathbf{s} = nh = \frac{q}{c}\left(\frac{mc}{N_s q^2}\oint \mathbf{j}_s\cdot d\mathbf{s} + \int \mathbf{B}\cdot\mathbf{n}\,dS\right). \qquad (5\text{-}10.18)$$

The quantity

$$\frac{mc}{N_s q^2}\oint \mathbf{j}_s\cdot d\mathbf{s} + \int \mathbf{B}\cdot\mathbf{n}\,dS$$

is called a fluxoid. For a thick superconductor $\mathbf{j}_s = 0$ and equation 5-10.18 reduces to the condition of equation 5-10.17. It is worth emphasizing that the fluxoid or flux quantization occurs only for superconductors; normal metals will not exhibit this effect.

From the flux quantization experiments we also draw the interesting conclusion that the behavior of conduction electrons depends on the flux existing in a region that is inaccessible to the electrons. It is appropriate to note that this property had already been demonstrated in quite a different experiment suggested by Aharonov and Bohm.[115] Consider a coherent beam of electrons and suppose that it is split into two. Let a solenoid with a flux Φ be inserted between the two beams in such a way that no flux lies in the paths traversed by the electron beams. Then suppose that the electron beams are brought together again; the interference of the two beams may be detected by observing the fringe patterns. It is found that the phase difference between the two beams is proportional to the solenoid's flux![116]

In Section 1-7 it was stated that solenoids constructed with hard superconductors were important sources of large continuous magnetic fields; it is now appropriate that some additional comments concerning this topic be made. To characterize type II superconductors, two critical

[115] Y. Aharonov and D. Bohm, *Phys. Rev.* **115,** 485 (1959), **123,** 1511 (1961), **125,** 2192 (1962), **130,** 1625 (1963); see also W. Ehrenberg and R. E. Siday, *Proc. Phys. Soc. (London)* **B-62,** 8 (1949).

[116] F. G. Werner and D. R. Brill, *Phys. Rev. Letters* **4,** 349 (1960); R. G. Chambers, *Phys. Rev. Letters* **5,** 3 (1960); H. Boersch, H. Hamisch, D. Wohlleben, and K. Grohmann, *Z. Phys.* **159,** 397 (1960).

fields H_{cr1} and H_{cr2} have been introduced[117]; H_{cr1} represents the field at which flux penetration first begins, whereas H_{cr2} represents that at which the flux penetration is just completed and the superconductivity is destroyed.[118] As the field is increased between H_{cr1} and H_{cr2}, there is a gradual increase in the flux penetration. The region between H_{cr1} and H_{cr2} is known as the *mixed state*, a term which is not to be confused with the intermediate state, defined earlier in this section. It has been suggested that in the mixed state there is a lattice of flux filaments each consisting of one flux quantum or fluxon.

It has gradually emerged that an additional subdivison into ideal and nonideal type II superconductors should be made. Ideal materials are relatively pure and free of imperfections[119]; they behave reversibly in the manner corresponding to the predictions of the Ginzburg-Landau theory as deduced by Abrikosov. Nonideal materials, on the other hand, have a defect structure arising from imperfect sintering, cold working, precipitation, or some other process, and exhibit irreversible behavior. Several new theories have been developed to account for the properties of the nonideal materials.[120] The motion of the fluxons is inhibited, probably because they are pinned by the defects.

Nonideal materials have a greatly increased current-carrying capacity and are therefore employed in the construction of high field solenoids. Two groups of alloys have been used most frequently. One family is comprised of intermetallic compounds with the β-wolfram structure and includes Nb_3Sn, V_3Si, and V_3Ga. The upper critical field H_{cr2} of these materials is well in excess of 100,000 oe.[121] Unfortunately, the intermetallic compounds are brittle, and as a consequence special methods must be used for the fabrication of coils. The other group consists of

[117] At the International Conference on the Science of Superconductivity held in August, 1963 and reported in *Revs. Mod. Phys.* **36,** 2 (1964).

[118] In the literature the notation H_c, H_{c1}, and H_{c2} rather than H_{cr}, H_{cr1}, and H_{cr2} are commonly used. However, in Ch. 7 of this book the symbol H_c denotes the coercive force and appears frequently.

[119] L. V. Schubnikov, W. I. Khotkerich, J. D. Schepelev, and J. N. Riabinin, *Phys. Z. Sowjet.* **10,** 165 (1936), *J. Exptl. Theoret. Phys.* (*USSR*) **7,** 221 (1937); A. Calverley and A. C. Rose-Innes, *Proc. Roy. Soc.* (*London*) **A-255,** 267 (1960); P. S. Swartz, *Phys. Rev. Letters* **9,** 448 (1962); J. D. Livingston, *Phys. Rev.* **129,** 1943 (1963).

[120] K. Mendelssohn, *Proc. Roy. Soc.* (*London*) **A-152,** 34 (1935); C. P. Bean, *Phys. Rev. Letters* **8,** 250 (1962); P. W. Anderson and Y. B. Kim, *Revs. Mod. Phys.* **36,** 39 (1964); J. Friedel, P. -G. de Gennes, and J. Matricon, *Appl. Phys. Letters* **2,** 119 (1963).

[121] J. E. Kunzler, E. Buehler, F. S. L. Hsu, and J. H. Wernick, *Phys. Rev. Letters* **6,** 89 (1961); J. E. Kunzler, *Revs. Mod. Phys.* **33,** 501 (1961); F. J. Morin, J. P. Maita, H. J. Williams, R. C. Sherwood, J. H. Wernick, and J. E. Kunzler, *Phys. Rev. Letters* **8,** 275 (1962); B. B. Goodman, *Phys. Letters* **1,** 215 (1962); P. S. Swartz and C. H. Rosner, *J. Appl. Phys.* **33,** 2292 (1962).

alloys with the body-centered cubic lattice; important examples are the Nb-Zr alloys.[122] Although these materials have a lower value of H_{cr2}, they have the advantage of being easy to work. Finally, it is worth pointing out that some nonideal hard superconductors, because of their ability to trap fluxons, possess permanent magnet properties superior to those of conventional ferromagnetic materials (see Section 7-4).[123]

Problems

5-1. Consider a free electron gas in a cube of edge L. Suppose that periodic boundary conditions are imposed on the wave functions; for example

$$\psi(x + L, y, z) = \psi(x, y, z).$$

Find the possible states and their energy values under these conditions.

5-2. By making use of equation 5-1.21, show that the chemical potential is given by

$$\mu = -T\left(\frac{\partial S}{\partial N}\right)_{E,H}.$$

In the derivation of the Fermi-Dirac distribution function the equation

$$\delta(\log W - \alpha\mathcal{N} - \beta E) = 0$$

was employed. By recalling the famous Boltzmann relationship for entropy $S = \text{k} \log W$, show that the undetermined multiplier α is given by

$$\alpha = \frac{1}{\text{k}}\left(\frac{\partial S}{\partial \mathcal{N}}\right)_{E,H}$$

and then show that the chemical potential and Fermi energy are equal, that is, $\mu = E_F$.

5-3. Suppose that a magnetic system were approaching thermal equilibrium under the conditions of constant temperature and applied magnetic field. From the second law of thermodynamics show that when this equilibrium is reached the Gibbs function will be a minimum. With the aid of equation 5-1.21 show that this condition can be expressed by

$$\begin{aligned} dG &= \mu\, d\mathcal{N} \\ &= 0. \end{aligned}$$

[122] J. K. Hulm, M. J. Fraser, H. Riemersma, A. J. Venturino, and R. E. Wien, *High Magnetic Fields*, John Wiley and Sons, New York (1962), p. 332; S. H. Autler, *ibid.*, p. 324; R. G. Treuting, J. H. Wernick, and F. S. L. Hsu, *ibid.*, p. 597; G. D. Kneip, J. O. Betherton, D. S. Easton, and J. O. Scarbrough, *ibid.*, p. 603.

[123] C. P. Bean and M. Doyle, *J. Appl. Phys.* **33**, 3334 (1962).

Next, consider two systems, each an electron gas, which are in thermal contact. Show that when the two systems are in thermal equilibrium the Fermi energies of the two systems must be equal. (*Hint.* Use the condition on dG when $d\mathcal{N}$ electrons are transferred from one system to the other.)

5-4. By employing the reasoning used to derive equation 5-2.4 show that

$$\int_0^\infty F(E)\frac{dG(E)}{dE}\,dE = G(E_F) + \frac{\pi^2}{6}(\mathrm{k}T)^2 G''(E_F) + \cdots$$

where $G(E)$ is any function of E that is zero when $E = 0$. Then, when $G(E)$ is defined as

$$G(E) = \int_0^E C(E)\,dE,$$

show that

$$\mathcal{N} = \int_0^{E_F(T)} C(E)\,dE + \frac{\pi^2}{6}(\mathrm{k}T)^2 C'(E_F),$$

where $C(E)$ is the density of states for a free electron gas (equation 5-1.5). By subtracting

$$\mathcal{N} = \int_0^{E_F(0)} C(E)\,dE$$

from this equation show that the Fermi energy as a function of temperature is given by

$$E_F(T) \approx E_F(0)\left[1 - \frac{\pi^2}{12}\left(\frac{\mathrm{k}T}{E_F(0)}\right)^2\right]$$

to a good order of approximation.

5-5. Start with the integral form for the partition function, given by equation 5-1.15, and prove that classical theory gives zero for the orbital susceptibility of a free-electron gas. (*Hint.* Show that the volume element in phase space is invariant under a transformation from canonical to kinetic momentum by showing that the Jacobian is equal to unity.)

5-6. According to equation 2-8.7, the expectation value of magnetic moment in a state n of energy E_n is

$$\mu_n = -\frac{\partial\langle n|\,\mathcal{H}\,|n\rangle}{\partial H} = -\frac{\partial E_n}{\partial H},$$

where the Hamiltonian is given by equation 2-8.4. Show that for the field H along the z-direction the choices of vector potential

(a) $A_x = -\frac{1}{2}Hy \qquad A_y = \frac{1}{2}Hx \qquad A_z = 0,$

(b) $A_x = -Hy \qquad A_y = 0 \qquad A_z = 0,$

(c) $A_x = 0 \qquad A_y = Hx \qquad A_z = 0,$

all lead to the same value for the magnetic moment. [*Hint.* For a bounded orbit recall that $d(xy)/dt = 0$.]

5-7. By requiring that the origin of the harmonic oscillators y_0 (equation 5-3.5) lie within the volume of the metal, show that the degeneracy of the energy levels $E_{\perp} = (n + \frac{1}{2})\hbar\,|\omega|$ is $L_1L_2(eH/ch)$, where L_1 and L_2 are the lengths of the metal in the x- and y-directions, respectively.

5-8. Consider equation 5-8.3 for the motion of an electron at the bottom of the conduction band. For the situation in which H lies in the x, z plane, $\mathrm{E} = \mathrm{E}_0 e^{i\omega t}$, and m^* is given by equation 5-8.4 show that if $\omega\tau \gg 1$ the resonance condition gives

$$\left(\frac{1}{m^*}\right)^2 = \frac{\cos^2\theta}{m_t^{\,2}} + \frac{\sin^2\theta}{m_t m_l},$$

where θ is the angle H makes with the z-axis.

5-9. Van Vleck discusses the Overhauser effect in terms of three temperatures, since the Boltzmann factor may be written

$$\exp\left(-\frac{E_1}{\mathrm{k}T_k} - \frac{E_2}{\mathrm{k}T_z} - \frac{E_3}{\mathrm{k}T_I}\right),$$

where T_k is the temperature of the electron translational motion, T_z is the electron Zeeman temperature, T_I is the nuclear Zeeman temperature, and E_1, E_2, and E_3 are the corresponding energies. Show that these temperatures are related by

$$\frac{1}{T_I} = \frac{(-g\mu_B + g_I\mu_{B_n})}{g_I\mu_{B_n}T_k} + \frac{g\mu_B}{g_I\mu_{B_n}T_z}.$$

Consider the situation when complete saturation of the electron system occurs. [*Hint.* Consider transitions at equilibrium and employ the conservation of energy principle. See J. H. Van Vleck, *Nuovo cimento. Suppl.* **3,** 1081 (1957).]

Bibliography

Andrew, E. R., *Nuclear Magnetic Resonance*, Cambridge University Press, Cambridge (1956), Ch. 7.

Bagguley, D. M. S., and J. Owen, "Microwave Properties of Solids," *Rept. Prog. Phys.* **20,** 304 (1957).

Conference on the Science of Superconductivity, *Revs. Mod. Phys.* **36,** 1–331 (1964).

Dekker, A. J., *Solid State Physics*, Prentice-Hall, Englewood Cliffs, N.J. (1957), Chs. 9, 10, and 13.

Harrison, W. A., and M. B. Webb, Editors, *The Fermi Surface*, John Wiley and Sons, New York (1960).

Jan, J.-P., "Galvanomagnetic and Thermomagnetic Effects in Metals," *Solid State Phys.* **5,** 3 (1957).

Kahn, A. H., and H. P. R. Frederiske, "Oscillatory Behavior of Magnetic Susceptibility and Electronic Conductivity," *Solid State Phys.* **9,** 257 (1959).

Kittel, C., *Solid State Physics*, John Wiley and Sons, New York (1956), Chs. 9, 12, 13, and 16.

Knight, W. D., "Electron Paramagnetism and Nuclear Magnetic Resonance in Metals," *Solid State Phys.* **2,** 93 (1956).

Lax, B., *Revs. Mod. Phys.* **30,** 122 (1958).

London, F., *Superfluids*, John Wiley and Sons, New York (1950).

Lynton, E. A., *Superconductivity*, John Wiley and Sons, New York (1962).

Matthias, B. T., "Superconductivity in the Periodic System," *Progr. Low Temperature Phys.* **2,** 138 (1957).

Newhouse, V. L., *Applied Superconductivity*, John Wiley and Sons, New York (1964).

Peierls, R. E., *Quantum Theory of Solids*, Oxford University Press, Oxford (1955), Ch. 7.

Pippard, A. B., "Experimental Analysis of the Electronic Structure of Metals," *Rept. Prog. Phys.* **23,** 176 (1960).

Seitz, F., *Modern Theory of Solids*, McGraw-Hill Book Co., New York (1940), Ch. 16.

Shoenberg, D., "The de Haas-van Alphen Effect," *Progr. Low Temperature Phys.* **2,** 226 (1957).

Shoenberg, D., *Superconductivity*, 2nd ed., Cambridge University Press, Cambridge (1960).

Tanenbaum, M., and W. V. Wright, Editors, *Superconductors*, Interscience Publishers, New York (1962).

Van Vleck, J. H., "Magnetic Properties of Metals," *Nuovo cimento Suppl.* **3,** 857 (1957).

Wilson, A. H., *The Theory of Metals*, Cambridge University Press, Cambridge (1953), Chs. 6 and 8.

Yafet, Y., "g-Factors and Spin-Lattice Relaxation of Conduction Electrons," *Solid State Phys.* **14,** 1 (1963).

6

Ferromagnetism

1. Introduction

A ferromagnetic material has been described in Section 1-3 as one in which the strongly coupled atomic dipole moments tend to be aligned parallel. As a result, a *spontaneous magnetization* exists in such materials; that is, even in the absence of a magnetic field there is a magnetic moment. Above a critical temperature, T_f called the ferromagnetic Curie temperature, the spontaneous magnetization vanishes. The material is then paramagnetic with a susceptibility given approximately by the Curie-Weiss law (equation 2-6.1). Only a few materials exhibit ferromagnetism. These include the elements Fe, Ni, Co, Gd, and Dy, their alloys with one another and with some other elements, and a few substances not containing any of the ferromagnetic elements cited. The great majority of ferromagnetic materials are metals or alloys; a few, however, are ionic compounds, such as $La_{1-\delta}$ $Ca_{\delta}MnO_3$ $(0.2 < \delta < 0.4)$,[1] $CrBr_3$,[2] EuO,[3] EuS,[4] EuSe,[4] EuI_2,[4] and Eu_2SiO_4[4] (see Section 9-2). Ferrimagnetic materials also exhibit a spontaneous magnetization; they are discussed in Chapter 9. The elements Tb, Ho, Er, and Tm are also ferromagnets at low temperatures (see Table 8-8.1). These rare earth metals, together with Dy, are unusual in that they possess a nonparallel spin arrangement in the ordered state.

[1] G. H. Jonker and J. H. van Santen, *Physica* **19,** 120 (1953).
[2] I. Tsubokawa, *J. Phys. Soc. Japvn* **15,** 1664 (1960).
[3] B. T. Matthias, R. M. Bozorth, and J. H. Van Vleck, *Phys. Rev. Letters* **7,** 160 (1961).
[4] T. R. McGuire and M. W. Shafer, *J. Appl. Phys.* **35,** 984 (1964).

It is well known that in spite of the spontaneous magnetization property a ferromagnetic specimen may exhibit no magnetic moment when the applied field is zero. However, the application of even a small field usually produces a magnetic moment which is many orders of magnitude larger than that produced in a paramagnetic substance. In order to explain these results, Weiss[5] postulated the existence of small regions, called domains, each spontaneously magnetized. The magnetic moment of the entire specimen is then the vector sum of the magnetic moments of each domain. Since the direction of magnetization of each domain need not be parallel, certain domain configurations lead to a zero net moment. The application of a relatively small field changes the domain arrangement, and hence leads to an appreciable net magnetization. This hypothesis of Weiss is now well established.

The occurrence of these phenomena makes it convenient to divide the study of ferromagnetism into two main areas. The first is primarily concerned with the origin and properties of the spontaneous magnetization within a single domain.[6] This area deals more with atomic or microscopic aspects of the subject and forms the subject matter of this chapter. The second area is concerned with the behavior of the magnetization of a specimen in an applied magnetic field. The specimen will, in general, consist of many domains; however, it may, under certain conditions, consist of a single domain. This latter area then is concerned with the bulk or macroscopic properties; it is covered in Chapter 7.

The first part of this chapter outlines the classical theory of the spontaneous magnetization developed by Weiss. In this phenomenological theory the strong interaction between the atomic dipoles, called the molecular or Weiss field, is assumed to be proportional to the magnetization. The origin of this field is in the quantum mechanical exchange forces alluded to earlier in this book and discussed in some detail here. Although the Weiss theory is in satisfactory agreement with the main features observed, there are differences concerning the finer details. Other theories are developed that are in better accord with experiment; in particular, (a) a successive approximation method, (b) the Bethe-Peierls-Weiss short range order theory, and (c) spin-wave theory are discussed. All of these models assume the electronic wave functions are localized, whereas in fact almost all the ferromagnetic materials are metallic. Thus it is appropriate that the collective electron theory, which is based on a band model, is briefly considered next. Finally the anisotropy of the magnetization in a single crystal and magnetoelastic effects are treated.

[5] P. Weiss, *J. Phys.* **6,** 667 (1907).

[6] It is assumed that the material is free of imperfections, for if not the magnetization within a domain may be nonuniform. See Ch. 7.

2. The Classical Molecular Field Theory and Comparison with Experiment

The strong interaction which tends to align the atomic dipoles parallel in a ferromagnetic material may be considered as equivalent to some internal magnetic field, H_m. The thermal agitation of the atoms opposes the orienting effect of this field. The Curie point must be the temperature at which the thermal agitation energy is sufficient to destroy the spontaneous magnetization. This permits an estimate of H_m to be made. Hence for atoms with a dipole moment of one Bohr magneton

$$\mu_B H_m \approx \mathrm{k} T_f.$$

For $T_f \approx 10^{3\circ}$K, a value close to that observed for iron, we get

$$H_m \approx 10^7 \text{ oe}.$$

This is a larger field than any that have been produced in the laboratory, even by pulse methods (Section 1-7). It is also much larger than that produced by the dipole-dipole interaction, which is only of the order $\mu_B/a^3 \approx 10^3$ oe (see equation 1-3.3), where a is the lattice constant of the solid.

Although Weiss was unable to explain the origin of the internal magnetic field H_m, he proceeded to develop a phenomenological theory of ferromagnetism by assuming that H_m was proportional to the magnetization M (actually the spontaneous magnetization), so that

$$H_m = N_W M. \tag{6-2.1}$$

Here N_W is a constant called the molecular field constant. The field H_m is usually called the molecular field, although it is also sometimes called the Weiss or exchange field. The molecular field constant is of the order of magnitude $N_W = H_m/M \approx 10^7/10^3 \approx 10^4$; this may be compared with the much smaller Lorentz factor of $4\pi/3$ (equation 1-6.5). In the presence of any applied magnetic field H the actual field acting on a given dipole is

$$H_T = H + N_W M, \tag{6-2.2}$$

where the demagnetizing and Lorentz (dipole-dipole) fields are omitted, since their effect is small compared to the molecular field.

It is convenient to discuss the molecular field theory of ferromagnetic materials under the following three headings:

1. The spontaneous magnetization region
2. The paramagnetic region
3. Thermal effects.

At the same time it is pertinent to outline the principal experimental results and to compare them with the theory.

The spontaneous magnetization region. We begin with the most general case and assume that the atoms of the solid have a total angular momentum quantum number J. Then the magnetization of the specimen is given by the results of Section 2-13 as

$$M = Ng\mu_B J\, B_J(x), \tag{6-2.3}$$

where

$$x = \frac{Jg\mu_B H}{kT}, \tag{6-2.4}$$

$$B_J = \frac{2J+1}{2J} \coth \frac{2J+1}{2J} x - \frac{1}{2J} \coth \frac{x}{2J} \tag{6-2.5}$$

and N is the number of atoms per unit volume. Since we are dealing with a ferromagnetic material, the field H in equation 6-2.4 is to be replaced by the field H_T of equation 6-2.2; hence

$$x = \frac{Jg\mu_B}{kT}(H + N_W M). \tag{6-2.6}$$

In order to investigate the spontaneous magnetization, we set $H = 0$. On rewriting equations 6-2.3 and 6-2.6 we have

$$\frac{M(T)}{M(0)} = B_J(x) \tag{6-2.7}$$

and

$$\frac{M(T)}{M(0)} = \frac{kT}{NN_W g^2 \mu_B{}^2 J^2}\, x, \tag{6-2.8}$$

where

$$M(0) = Ng\mu_B J. \tag{6-2.9}$$

The dependence of the magnetization on temperature has been emphasized by writing $M(T)$. Equation 6-2.9 follows from equation 6-2.3, since $B_J(x) \to 1$ as $x \to \infty$, that is, as $T \to 0$. According to atomic theory, $M(0)$ is the maximum magnetization possible.

Since $M(T)/M(0)$ must satisfy equations 6-2.7 and 6-2.8 simultaneously, its value for a given temperature may be obtained graphically as the intersection of the two corresponding $M(T)/M(0)$ versus x curves. This method is illustrated schematically in Fig. 6-2.1, where several straight lines (equation 6-2.8) are plotted to show the nature of the results for different temperatures. The case in which the straight line is tangent to the Brillouin function at the origin corresponds to a critical temperature.

Below this temperature there are two intersections of the curves, one at a nonzero value of $M(T)/M(0)$ and the other at zero. It is easy to show (problem 6-1) that only the nonvanishing value of the relative magnetization represents a stable situation. Thus there is spontaneous magnetization in this temperature region. At and above the critical temperature, the only intersection is at $M(T)/M(0) = 0$, that is, the spontaneous magnetization vanishes. Obviously, this critical temperature is to be identified as the ferromagnetic Curie temperature T_f.

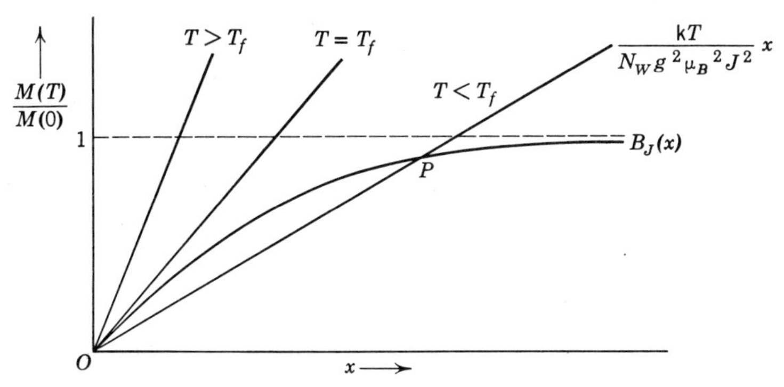

Fig. 6-2.1. Illustration of a graphical method for the determination of the spontaneous magnetization at a temperature T.

We now obtain a relationship between T_f and N_W. For small x the Brillouin function (equation 6-2.5) is given by

$$B_J(x) \doteq \frac{J+1}{3J}\,x - \frac{J+1}{3J}\,\frac{2J^2+2J+1}{30J^2}\,x^3 \qquad (x \ll 1). \tag{6-2.10}$$

Hence, for $x \to 0$, the slope of the tangent is equal to $(J+1)/3J$. When this is equated to the slope of the straight line (equation 6-2.8) for $T = T_f$, we get

$$T_f = \frac{Ng^2\mu_B^2\,J(J+1)}{3k}\,N_W. \tag{6-2.11}$$

This is to be compared with the order of magnitude relationship between T_f and N_W assumed at the beginning of this section.

By employing equation 6-2.11, equation 6-2.8 may be written

$$\frac{M(T)}{M(0)} = \frac{J+1}{3J}\left(\frac{T}{T_f}\right)x. \tag{6-2.12}$$

The value of $M(T)/M(0)$ as a function of T/T_f may be determined either with the aid of the graphical procedure already outlined or by obvious algebraic substitutions. Then, for a given value of J, the plot of $M(T)/M(0)$ versus T/T_f yields a universal curve. Such curves for $J = \frac{1}{2}$, $J = 1$, and $J = \infty$ are shown in Fig. 6-2.2. It is to be recalled that the latter corresponds to the classical case (see problem 6-1).

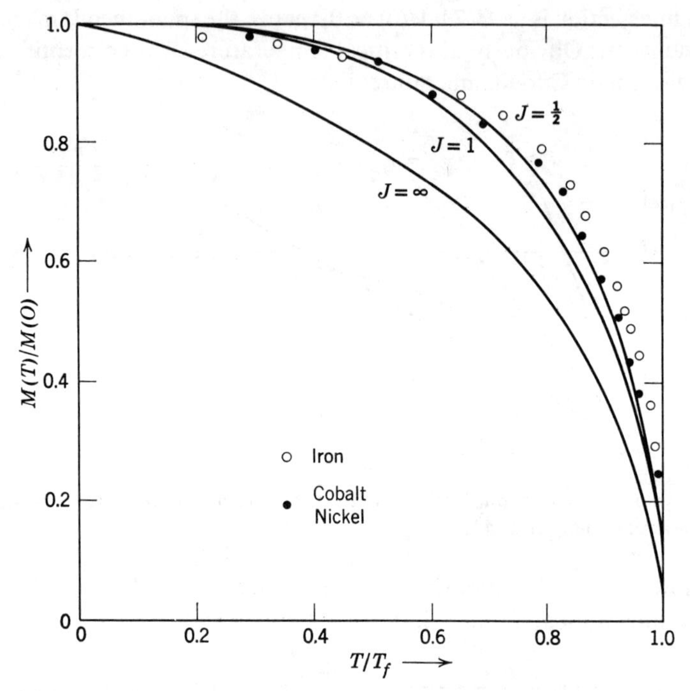

Fig. 6-2.2. The spontaneous magnetization as a function of temperature. The curves are obtained from theory; the points represent experimental data. [F. Tyler, *Phil. Mag.* **11,** 596 (1931).]

It is of interest to obtain expressions for the relative magnetization close to absolute zero and close to the Curie point. For the former limit $T/T_f \to 0$ and x is large. By recalling that $\tanh y \to 1 - 2e^{-2y}$ for $y \to \infty$, expansion of the Brillouin function yields

$$\frac{M(T)}{M(0)} = 1 - \frac{e^{-x/J}}{J} - \cdots$$

$$= 1 - \frac{e^{-[3/(J+1)](T_f/T)}}{J} \quad \text{for} \quad \frac{T}{T_f} \to 0, \qquad (6\text{-}2.13)$$

where the last equality is obtained by using equation 6-2.12 and by setting $M(T)/M(0) = 1$ in the exponent. Near the Curie point x is small since

$M(T)$ is small. Equation 6-2.12 may then be legitimately substituted in equation 6-2.10. The result is

$$\left(\frac{M(T)}{M(0)}\right)^2 = \frac{10}{3}\frac{(J+1)^2}{J^2+(J+1)^2}\left(1-\frac{T}{T_f}\right) \quad \text{for} \quad \frac{T}{T_f}\to 1. \tag{6-2.14}$$

This equation shows that the magnetization disappears continuously but has an infinite slope at the Curie point.

In turning now to a comparison with the experimental results, it is pertinent to discuss how the spontaneous magnetization is measured. It has already been mentioned that a ferromagnetic specimen consists of many domains whose magnetization vectors have a variety of orientations. It is then not feasible to measure the spontaneous magnetization (i.e., the magnetization at $H = 0$) of a single domain. Application of a magnetic field of the order of 10^3 oe, however, is usually sufficient, first, to remove the domain structure, and, second, almost to orient the net magnetization vector along the direction of the field in most ferromagnetic materials. For fields from about 10^3 to 3×10^4 oe it is then found that the empirical relationship[7]

$$M_H = M_s(T)\left(1 - \frac{a}{H} - \frac{b}{H^2}\right) + cH \tag{6-2.15}$$

is followed for many materials. Here M_H is the component of magnetization along the field direction in a field H, M_s is the magnetization when $H \to 3 \times 10^4$ oe, and a, b, and c are constants. The a/H and b/H^2 terms are probably the result of domain and imperfection effects; they are discussed briefly in Chapter 7. The term cH is small for magnetic fields in the neighborhood of 10^4 oe. It probably originates from a redistribution of the populations of the spin states, so that it represents an actual increase of the magnetization within one domain over the spontaneous value ($M_{H=0}$). $M_s(T)$ is often called the saturation magnetization for obvious reasons. It should, however, be carefully distinguished from what we will call the absolute saturation magnetization $M_s(0)$ [equal to the spontaneous magnetization, $M_{H=0}(0)$], which is obtained only at absolute zero. Of course, if H is increased beyond 3×10^4 oe, the term cH will become larger. If the applied field approached infinity, $M_s(T)$ would presumably approach the ultimate limit corresponding to complete alignment, $M_s(0)$. In recent years static fields in the vicinity of 10^5 oe and transient fields of the order of 10^6 oe have been produced (Section 1-7).

By making measurements of M_H as a function of H at a certain temperature, $M_s(T)$ can be determined. The value of the spontaneous magnetization $M_{H=0}$ presumably may then be obtained by extrapolating the strong

[7] W. Steinhaus, A. Kussmann, and E. Schoen, *Phys. Z.* **38,** 777 (1937); E. Czerlinsky, *Ann. Phys. (Leipzig)* **13,** 80 (1952).

field part of the curves to $H = 0$. A more convenient procedure, used in practice, is to plot H versus T curves for M constant and to extrapolate again to zero field. In addition, there is a method of obtaining $M(T)$ that employs the magnetocaloric effect; this method is briefly described in the thermal part of this section.

For temperatures not too close to the Curie point, that is, less than about $0.8T_f$, the difference between the saturation and spontaneous magnetization observed is negligible. Near the Curie point there is a difference, although it is only of the order of 1% except within a few degrees of T_f. This small difference is predicted by the Weiss theory. For $T \ll T_f$, H is small compared to $N_W M$ and may be neglected in equation 6-2.6. However, near T_f, $H \approx N_W M$ and may not be neglected. The former case is identical to setting $H = 0$, as in equation 6-2.8; the latter is not.[8] Since the spontaneous and saturation magnetization are almost identical except near T_f, it has become common although loose practice to use the terms interchangeably.

Experimental values[9] of the spontaneous magnetization of iron, cobalt, and nickel as a function of temperature are shown also in Fig. 6-2.2. There is a fair fit between the data and the $J = \frac{1}{2}$ curve, particularly for the higher temperatures. This suggests that the magnetization originates from the electron spin rather than from the electron orbital angular momentum, hence that the g-value is close to 2. The best evidence that $g \approx 2$ comes from gyromagnetic experiments, to be discussed in Chapter 10. We shall anticipate these results and assume henceforth that ferromagnetism is due only to the spin. This does not rule out the possibility that the orbital momentum may be incompletely quenched in some ferromagnetic materials and will make a small contribution to the magnetic moment.

For $J = S = \frac{1}{2}$, $g = 2$, equation 6-2.7 becomes

$$\frac{M(T)}{M(0)} = \tanh x, \qquad (6\text{-}2.16)$$

where

$$x = \frac{\mu_B H}{kT}.$$

For temperatures close to absolute zero equation 6-2.13 predicts

$$\frac{M(T)}{M(0)} = 1 - 2e^{-2x}.$$

[8] P. Weiss and R. Forrer, *Compt. rend. (Paris)* **178,** 1670 (1924).

[9] P. Weiss and R. Forrer, *Ann. phys. (Paris)* **5,** 153 (1926), **12,** 279 (1929); F. Tyler, *Phil. Mag.* **11,** 596 (1931); R. I. Allen and F. W. Constant, *Phys. Rev.* **44,** 232 (1932); C. Guillaud and M. Roux, *Compt. rend. (Paris)* **229,** 1062 (1949).

Instead, it is found that the best fit with the experimental data[10] is given by

$$\frac{M(T)}{M(0)} = (1 - A_{\exp} T^{3/2}), \tag{6-2.17}$$

although replacing $T^{3/2}$ by T^2 is almost as good. At somewhat higher temperatures a T^2 law gives the best agreement with the observations

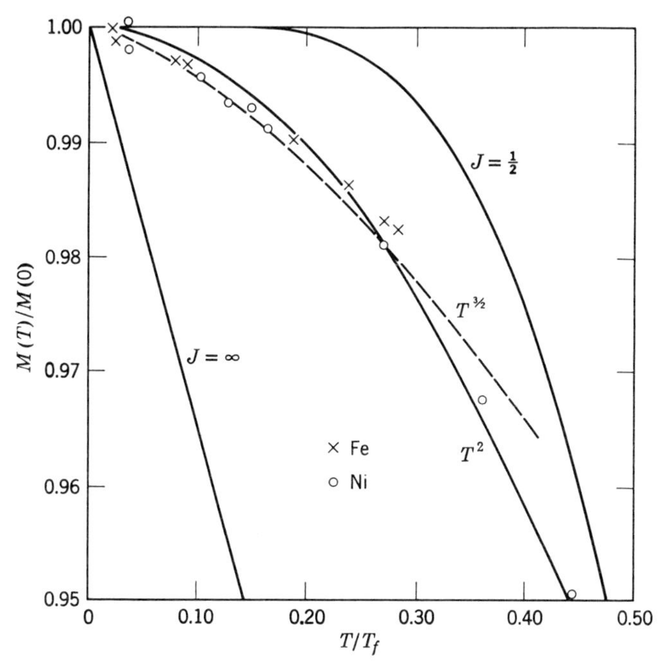

Fig. 6-2.3. The variation of the spontaneous magnetization near $T = 0°K$. The data is in best agreement with $T^{3/2}$ curve of equation 6-2.17, although the T^2 curve is also reasonably satisfactory. The predictions of the Weiss theory are unsatisfactory. (After R. M. Bozorth.)

(see Fig. 6-2.3). In any case, the Weiss theory is not in agreement with experiment. Equation 6-2.17 can be derived from spin-wave theory, a topic to be discussed in Section 6-6.

Extrapolation of these low temperature data gives the value of $M(0)$, the spontaneous magnetization at absolute zero. The effective number of Bohr magnetons per atom, n_{eff}, can then be calculated, provided the number of atoms per unit volume is known, that is, $n_{\text{eff}} = M(0)/N\mu_B$. The effective number n_{eff} refers to the maximum component of magnetization; it is not to be confused with the effective magneton number p_{eff} used

[10] P. Weiss, *Extr. Actes VII Cong. Int. Froid.* **1,** 508 (1937).

in paramagnetism (see equation 2-7.5). Representative values[11] are given in Table 6-2.1. It is to be noted that n_{eff} is nonintegral, although each atom has an integral number of electrons. This surprising feature is considered further in Section 6-8. Discussion of the investigations of several alloys is also postponed until this section.

Table 6-2.1. Magnetic Data for Ferromagnetic Metals below the Curie Point

Material	$M(290°K)$ (cgs)	$M(0)$ (cgs)	n_{eff} (0°K)
Fe	1707	1752	2.221
Co	1400	1446	1.716
Ni	485	510	0.606
Gd	1090	1980	7.10

The paramagnetic region. For temperatures above the ferromagnetic Curie point T_f the spontaneous magnetization is zero. The application of a magnetic field H will, however, produce a magnetization. Provided H is small enough that saturation effects can be neglected, the magnetization is given by substituting the first term of equation 6-2.10 into equation 6-2.3, namely

$$M = \frac{Ng\mu_B(J+1)}{3}x, \tag{6-2.18}$$

where x is given now by equation 6-2.6, that is,

$$x = \frac{Jg\mu_B}{kT}(H + N_W M).$$

After substituting for x and solving for M/H, equation 6-2.18 yields

$$\chi = \frac{M}{H} = \frac{C}{T-\theta}, \tag{6-2.19}$$

where

$$C = \frac{Ng^2\mu_B{}^2 J(J+1)}{3k}$$

and

$$\theta = N_W \frac{Ng^2\mu_B{}^2 J(J+1)}{3k} = N_W C.$$

[11] For references see R. M. Bozorth, *Ferromagnetism*, D. Van Nostrand, Princeton, N.J. (1951).

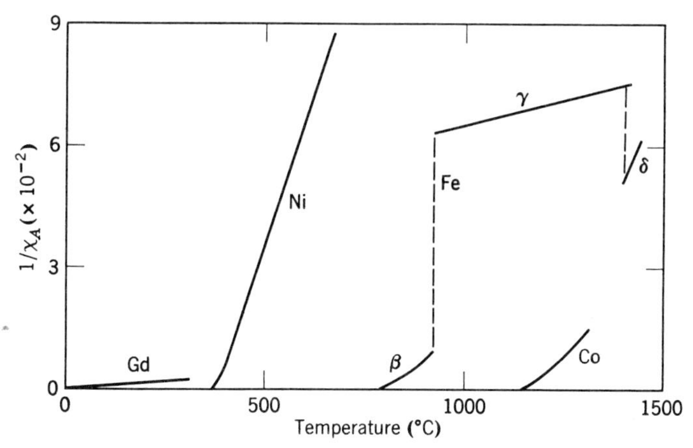

Fig. 6-2.4. Experimental susceptibility of ferromagnetic metals above the ferromagnetic Curie point. (After W. Sucksmith and R. R. Pierce.)

Equation 6-2.19 is just the Curie-Weiss law, derived earlier (equation 2-7.11) from a slightly different starting point. Comparison of this value for θ with that for T_f (equation 6-2.11) shows that the Weiss theory predicts $\theta = T_f$.

The results of experimental investigation of the susceptibility of Fe, Co, Ni, and Gd above the ferromagnetic Curie point are shown[12] in Fig. 6-2.4. For temperatures a few degrees above T_f the $1/\chi$ versus T curves are linear over a considerable temperature range. Hence in this region the Curie-Weiss law describes the data excellently. In the neighborhood of the

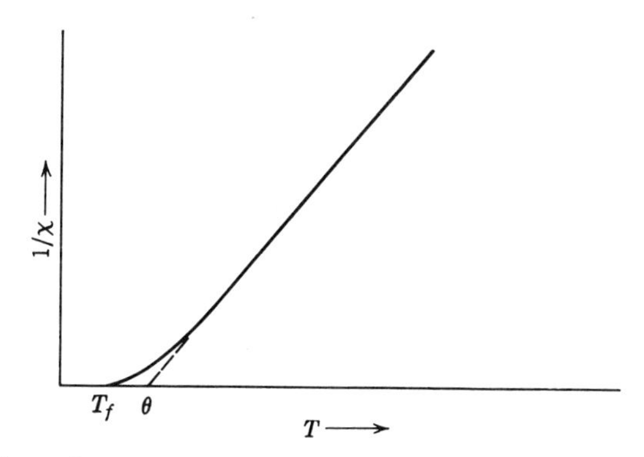

Fig. 6-2.5. Illustration of the susceptibility of a ferromagnetic metal above the Curie point. T_f is the ferromagnetic Curie point; θ is the paramagnetic Curie point.

[12] W. Sucksmith and R. R. Pierce, *Proc. Roy. Soc.* (*London*) **A-167,** 189 (1938); M. Fallot, *J. Phys. radium* **5,** 153 (1944); F. Trombe, *Ann. phys.* (*Paris*) **7,** 385 (1937).

ferromagnetic Curie point, however, all four metals show a curvature, concave upward. This curvature is presented schematically in the larger scale drawing of Fig. 6-2.5. The intercept of the linear part of the curve with the T-axis is just θ. Since θ does not coincide with T_f, as predicted by the Weiss theory, it is often called the paramagnetic Curie temperature.

Iron is, of course, complicated because of the phase changes, first from the β-(body-centered cubic) to the γ-(face-centered cubic) and then to the δ-(body-centered cubic) structure. By alloying iron with certain soluble nonmagnetic metals such as vanadium or aluminum, the β-γ transition can be raised or even entirely eliminated. By varying the amount of added metal and extrapolating to zero concentration, reasonably good data can be obtained.

Values of θ and C for the linear range of the susceptibility curves, together with T_f, are listed[13] in Table 6-2.2 for the ferromagnetic metals.

Table 6-2.2. Magnetic Data for Ferromagnetic Metals above the Curie Point

Metal	T_f (°K)	θ (°K)	C (cgs units/mole)
Fe	1043	1093	1.26
Co	1403	1428	1.22
Ni	631	650	0.32
Gd	289	302.5	7.8
Dy	105		

From the Curie constant the effective number of Bohr magnetons involved can be calculated. Except for gadolinium, these values are considerably larger than the values of n_{eff} found from the spontaneous magnetization at $T = 0$°K. The paramagnetic region will be discussed further in later sections.

Thermal effects. The change in energy that occurs when the spontaneous magnetization of a ferromagnetic specimen is changed gives rise to thermal effects. The specific heat and magnetocaloric effect are discussed here. We begin with the specific heat.

The spontaneous magnetization gives an additional contribution to the internal energy of a ferromagnetic body. In analogy with equation 1-7.7, this internal energy per unit volume, U_{sp}, is given by

$$U_{\text{sp}} = -\int H_m \, dM,$$

[13] See also L. F. Bates, R. E. Gibbs, and D. R. Pantulu, *Proc. Phys. Soc. (London)* **48,** 665 (1936); J. M. Bryant and J. S. Webb, *Rev. Sci. Instr.* **10,** 47 (1939).

where H_m is the molecular field (equation 6-2.1). On substituting, we get

$$U_{\mathrm{sp}} = -\tfrac{1}{2}N_W\, M(T)^2. \tag{6-2.20}$$

The additional specific heat due to the spontaneous magnetization is then

$$C_{\mathrm{sp}} = \frac{d}{dT}U_{\mathrm{sp}} = -\tfrac{1}{2}N_W\frac{dM^2}{dT}. \tag{6-2.21}$$

Now $|dM(T)^2/dT|$ is largest just below the Curie point and zero above. Hence the molecular field theory predicts that the specific heat rises to a maximum at the Curie temperature and then drops discontinuously to zero.

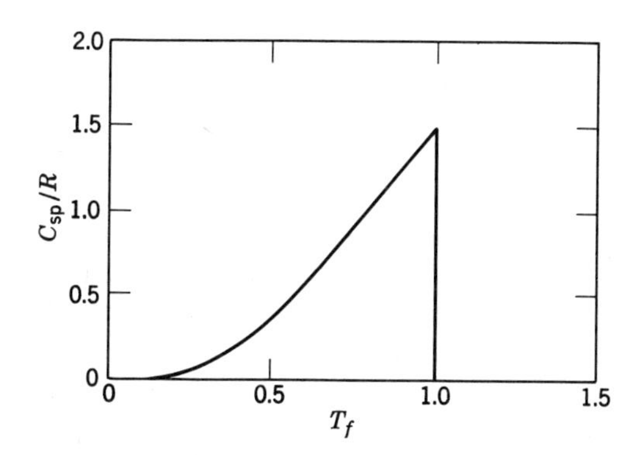

Fig. 6-2.6. Specific heat per mole for $J = \frac{1}{2}$ according to the molecular field theory.

The magnitude of this discontinuity will now be calculated. Substitution from equations 6-2.9 and 6-2.11 into equation 6-2.21 yields

$$C_{\mathrm{sp}} = -\frac{3JNk}{2(J+1)}\frac{d[M(T)/M(0)]^2}{d(T/T_f)}.$$

Near the Curie point the value of $\dfrac{d[M(T)/M(0)]^2}{d(T/T_f)}$ can be easily evaluated from equation 6-2.14. Hence the discontinuity in C_{sp} at $T = T_f$ is

$$\Delta C_{\mathrm{sp}} = \frac{5J(J+1)}{J^2 + (J+1)^2}Nk. \tag{6-2.22}$$

For $J = \frac{1}{2}$, $\Delta C_{\mathrm{sp}} = \frac{3}{2}Nk$; for a mole $\Delta C_{\mathrm{sp}} = \frac{3}{2}R$. The variation of C_{sp}/R with temperature for this case is given by the result of problem 6-2 and is shown in Fig. 6-2.6.

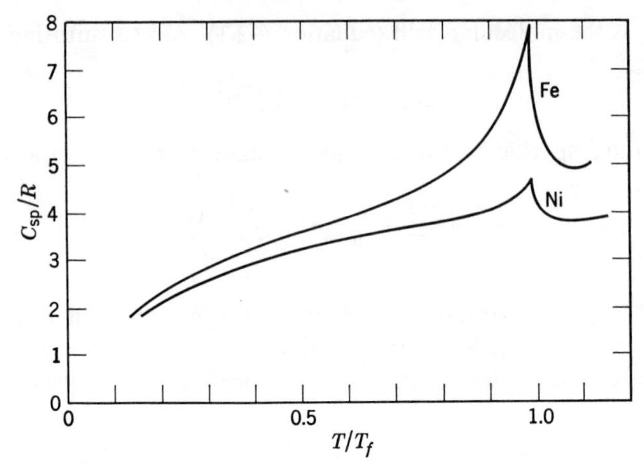

Fig. 6-2.7. Specific heat, in calories per mole, for nickel and iron.

Experimental results for the specific heat of nickel[14] and iron[15] are shown in Fig. 6-2.7. In order to compare the results with the preceding theory, it is necessary to subtract the nonmagnetic contributions to the specific heat. These are the lattice vibrations, the thermal expansion, and the electronic specific heats. The most detailed analyses have been made for nickel.[16] The theoretical and experimental curves are somewhat similar. For example, the values of ΔC_{sp} (at T_f) and the areas under the curves are of the same order of magnitude. However, there are important differences. First, the experimental curves rise more rapidly than the theoretical just below the Curie point. This feature is associated with the difference between dM^2/dT from experiment and theory. Second, the change in C_{sp} near the Curie point is not discontinuous but occurs over a temperature range of 7 to 50°K, depending on the sample. Third, there is a magnetic specific heat above the Curie point. There is some question about the amount of C_{sp} in this range because of uncertainty concerning the electronic specific heat. It is known, however, that the electronic specific heat is much larger than that for a normal metal, such as copper. These last two features may be understood at least qualitatively on the basis of the short-range order theory, to be discussed in Section 6-5.

We turn now to a discussion of the magnetocaloric effect, which has already been discussed in connection with paramagnetic materials. There it was shown that the change of magnetization that results from an

[14] C. Sykes and H. Wilkinson, *Proc. Phys. Soc.* (*London*) **50,** 834 (1938). See also E. Lapp, *Ann. phys.* (*Paris*) **12,** 442 (1929); E. Ahrens, *Ann. Phys.* (*Leipzig*) **21,** 169 (1934).

[15] J. H. Awbery and E. Griffiths, *Proc. Roy. Soc.* (*London*) **A-174,** 1 (1940).

[16] E. C. Stoner, *Phil. Trans. Roy. Soc.* (*London*) **A-235,** 165 (1936).

adiabatic change in the applied field produces a reversible temperature change given by equation 3-3.1,

$$\Delta T = -\frac{T}{C_H}\left(\frac{\partial M}{\partial T}\right)_H \Delta H. \qquad (6\text{-}2.23)$$

For ferromagnetic materials the same equation presumably applies when there is a true change in the magnetization, that is, when there is a change in the net number of spins oriented favorably with respect to the applied field within one domain. This effect has already been mentioned in connection with the cH term of equation 6-2.15. This equation does not apply to irreversible domain effects, which produce a much smaller temperature change. Since $(\partial M/\partial T)_H < 0$, there will be an increase in temperature when H is increased; moreover, the increase will be expected to be largest for $T = T_f$ (from Fig. 6-2.2).

It is also useful to have an expression for ΔT in terms of ΔM. For an adiabatic change equation 3-2.19 gives

$$\Delta T = \frac{T}{C_M}\left(\frac{\partial H}{\partial T}\right)_M \Delta M. \qquad (6\text{-}2.24)$$

The value of $(\partial H/\partial T)_M$ may be obtained from equation 6-2.3,

$$\frac{M(T)}{M(0)} = B_J\left\{\frac{Jg\mu_B[H + N_W M(T)]}{kT}\right\}$$

by differentiating with respect to H and T and keeping M constant. We get

$$\left(\frac{\partial H}{\partial T}\right)_M = \frac{H + N_W M}{T}.$$

On substituting into equation 6-2.24 we obt-in

$$\Delta T = \frac{H + N_W M}{C_M}\Delta M.$$

For $T < T_f$, $H < N_W M$, and we have

$$\Delta T = \frac{N_W}{2C_M}\Delta M^2$$

$$= \frac{N_W}{2C_M}(M_H{}^2 - M_{H=0}^2). \qquad (6\text{-}2.25)$$

Experimental values of the magnetocaloric temperature change as a function of temperature are shown for iron[17] in Fig. 6-2.8; this change is

[17] H. H. Potter, *Proc. Roy. Soc. (London)* **A-146,** 362 (1934). For Ni see P. Weiss and R. Forrer, *Ann. Phys. (Leipzig)* **5,** 153 (1926).

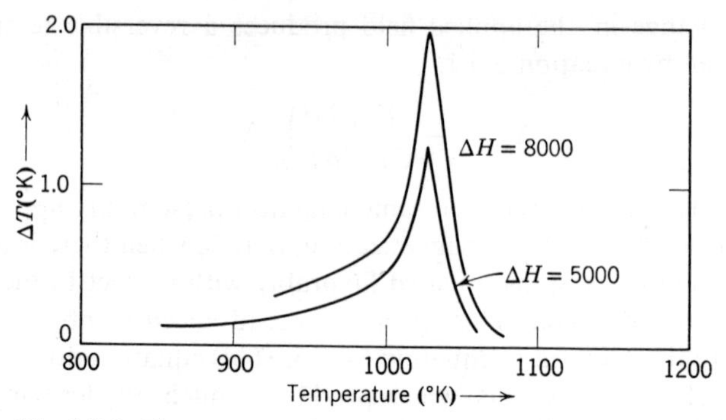

Fig. 6-2.8. The magnetocaloric effect in iron. (After H. H. Potter.)

of the order of 1°K near $T = T_f$. These results can be used to evaluate $\Delta T/\Delta H$ for $H \rightarrow 0$; further $(\partial M/\partial T)_{H=0}$ can be determined from the spontaneous magnetization curves. Then, by substituting into equation 6-2.23, $C_{H=0}$ can be calculated. The values of specific heat so calculated are in reasonable agreement with those obtained from direct measurements.

Determination of the magnetization M_H in a magnetocaloric experiment permits a check of equation 6-2.25. Some of Potter's results for iron are shown in Fig. 6-2.9 in which ΔT is plotted versus $M_H{}^2$. For high fields the curves are linear and parallel, as predicted by theory. The departure from a linear relationship at lower fields is the result of domain effects. The intercept with the $\Delta T = 0$-axis, obtained by extrapolating the linear

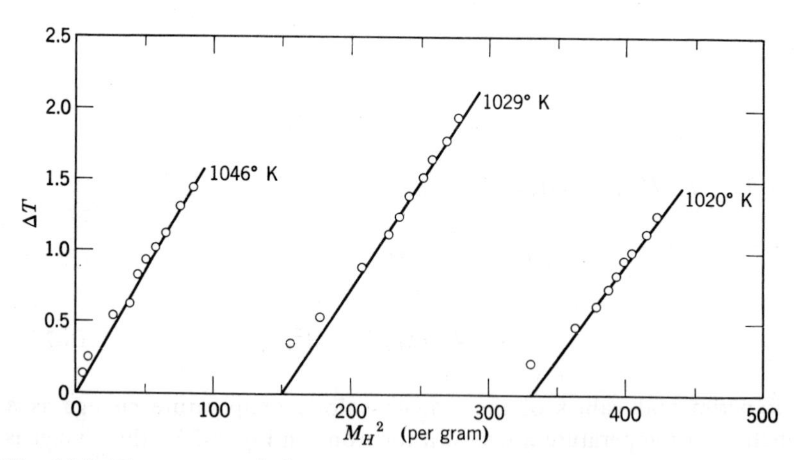

Fig. 6-2.9. The magnetocaloric temperature change as a function of M^2 (for iron). (After H. H. Potter.)

part of the curves, yields the spontaneous magnetization $M_{H=0}(T)$ for a particular temperature. Indeed, this is probably the best way to determine the spontaneous magnetization experimentally, particularly for temperatures near the Curie point.

3. The Exchange Interaction

The origin of the Weiss molecular field remained obscure until the advent of quantum mechanics. Then Heisenberg[18] showed that this field is the result of the quantum mechanical exchange interaction. This exchange interaction has no classical analog. Heisenberg's theory is based on the Heitler-London method developed for the hydrogen molecule. The main features of the theory, together with some pertinent background material, are presented here; further details may be found in several textbooks on quantum mechanics.

First consider two electrons, moving in similar potential fields; common examples are the helium atom and the hydrogen molecule. If at first the interaction between the electrons is neglected, the Schrödinger equation is

$$\left[-\frac{\hbar^2}{2m}(\nabla_1^{\ 2} + \nabla_2^{\ 2}) + V(q_1) + V(q_2)\right]\psi = E\psi,$$

where the subscripts 1 and 2 refer to the two electrons, considered distinguishable. It is easily seen by substitution that possible solutions are

$$\psi_a(1)\,\psi_b(2) \quad \text{and} \quad \psi_a(2)\,\psi_b(1), \tag{6-3.1}$$

with $E = E_a + E_b$ for both cases. Here $\psi_a(1)$ is the one-electron wave function when electron 1 is in state a and is a solution of the one-electron Schrödinger equation,

$$\left[-\frac{\hbar^2}{2m}\nabla_1^{\ 2} + V(q_1)\right]\psi = E_a\,\psi$$

and $\psi_b(2)$ is that when electron 2 is in state b. On the other hand, $\psi_a(2)$ and $\psi_b(1)$ are the wave functions when electrons 1 and 2 are interchanged, as in comparison to the first case. The electrons, however, are indistinguishable, and as a consequence it is necessary that

$$|\psi(1, 2)|^2\, dq_1\, dq_2 = |\psi(2, 1)|^2\, dq_1\, dq_2, \tag{6-3.2}$$

where $\psi(1, 2)$ is the wave function that describes the two electron systems

[18] W. Heisenberg, *Z. Physik* **49,** 619 (1928).

and $\psi(2, 1)$ is that when the two electrons are interchanged; q_1 and q_2 represent the coordinates for the two electrons. Hence either

$$\psi(1, 2) = +\psi(2, 1) \tag{6-3.3}$$

or

$$\psi(1, 2) = -\psi(2, 1). \tag{6-3.4}$$

The wave functions of equation 6-3.3 are called symmetrical; those of equation 6-3.4 are called antisymmetrical. Neither of the wave functions of equation 6-3.1 has these properties and therefore neither is an acceptable solution. However, the linear combinations

$$\psi_{\text{sym}}(1, 2) = \frac{1}{\sqrt{2}} [\psi_a(1)\,\psi_b(2) + \psi_a(2)\,\psi_b(1)] \tag{6-3.5}$$

$$\psi_{\text{anti}}(1, 2) = \frac{1}{\sqrt{2}} [\psi_a(1)\,\psi_b(2) - \psi_a(2)\,\psi_b(1)] \tag{6-3.6}$$

do satisfy equations 6-3.3 and 6-3.4, respectively, and are the only two that do so. The factors $1/\sqrt{2}$ are the normalizing constants if it is assumed that $\psi_a(1)$ and $\psi_b(2)$ are already normalized. Note that the antisymmetric solution may also be written as a determinant, namely,

$$\psi_{\text{anti}} = \frac{1}{\sqrt{2}} \begin{vmatrix} \psi_a(1) & \psi_a(2) \\ \psi_b(1) & \psi_b(2) \end{vmatrix}. \tag{6-3.7}$$

Experiment shows that electrons have wave functions that are always antisymmetric (see problem 6-3). Indeed, this is another way of stating the Pauli exclusion principle, since for antisymmetric wave functions the probability that two electrons will be in the same state is zero ($\psi_{\text{anti}} = 0$). However, it must be understood that the one-electron functions are not only functions of the spatial coordinates but also of the spin.

If it is assumed that the orbital motion is quenched, the one-electron wave functions may be written

$$\psi = (\phi r)\chi,$$

where $\phi(r)$ is the solution of a Schrödinger equation for an electron without spin and χ is a function of the spin coordinates only. Now, the component of the spin angular momentum along the axis of quantization (the z-axis) can assume only two values, namely $\pm\hbar/2$. In other words, if a magnetic field H is applied in the z-direction, there will be a magnetic moment μ_z with magnitude of one Bohr magneton lying either parallel or antiparallel to the field. We denote the wave functions describing the spin states by χ_α and χ_β, where

$$\chi_\alpha = 1, \qquad \chi_\beta = 0 \quad \text{for} \quad \mu_z \text{ parallel to } H$$

and (6-3.8)

$$\chi_\alpha = 0, \qquad \chi_\beta = 1 \quad \text{for} \quad \mu_z \text{ antiparallel to } H.$$

For example,

$$|\phi(r)\chi_\alpha|^2 \, d\tau$$

is to be interpreted as the probability that the electron is in the volume element $d\tau$ and has a spin such that the associated magnetic moment would be parallel to an applied magnetic field.

At this point we digress briefly in order to remind the reader that it is often convenient to represent the spin angular momentum operator with the aid of the Pauli matrices. In units of $\hbar$ the spin operator components are

$$s_x = \tfrac{1}{2}\begin{pmatrix} 0 & 1 \\ 1 & 0 \end{pmatrix}, \qquad s_y = \tfrac{1}{2}\begin{pmatrix} 0 & -i \\ i & 0 \end{pmatrix}, \qquad s_z = \tfrac{1}{2}\begin{pmatrix} 1 & 0 \\ 0 & -1 \end{pmatrix}. \tag{6-3.9}$$

The spin wave functions in matrix form are then

$$\chi_\alpha = \begin{pmatrix} 1 \\ 0 \end{pmatrix}, \qquad \chi_\beta = \begin{pmatrix} 0 \\ 1 \end{pmatrix}, \tag{6-3.10}$$

provided it is understood that, for electrons, if a magnetic field is applied it is set up along the negative z-direction.[19] It is left to the reader (problem 6-4) to verify certain commutation and other relationships.

Let us return now to a discussion of the two-electron case. The antisymmetrical wave functions of the electrons may have either of two forms:

$$\phi_{\text{sym}}(1, 2)\chi_{\text{anti}}(1, 2) \tag{6-3.11}$$

or

$$\phi_{\text{anti}}(1, 2)\chi_{\text{sym}}(1, 2), \tag{6-3.12}$$

where $\phi_{\text{sym}}(1, 2)$ and $\phi_{\text{anti}}(1, 2)$ are symmetrical and antisymmetrical wave functions that involve only the spatial coordinates, respectively, whereas $\chi_{\text{anti}}(1, 2)$ and $\chi_{\text{sym}}(1, 2)$ are antisymmetrical and symmetrical spin wave functions, respectively. Explicitly, these wave functions of equations

[19] This is necessary, since (a) $s_z\chi_\alpha = \frac{1}{2}\chi_\alpha$ and $s_z\chi_\beta = -\frac{1}{2}\chi_\beta$ and (b) for an electron the spin angular momentum vector is opposite in direction to the magnetic-moment vector.

6-3.11 and 6-3.12, in terms of one-electron wave functions, are, respectively,

$$\psi_{\mathrm{I}} = A[\phi_a(1)\,\phi_b(2) + \phi_a(2)\,\phi_b(1)][\chi_\alpha(1)\,\chi_\beta(2) - \chi_\alpha(2)\,\chi_\beta(1)] \qquad (6\text{-}3.13)$$

and

$$\psi_{\mathrm{II}} = B[\phi_a(1)\,\phi_b(2) - \phi_a(2)\,\phi_b(1)]\begin{bmatrix} \chi_\alpha(1)\quad \chi_\alpha(2) \\ \chi_\alpha(1)\chi_\beta(2) + \chi_\alpha(2)\,\chi_\beta(1) \\ \chi_\beta(1)\quad \chi_\beta(2) \end{bmatrix}, \qquad (6\text{-}3.14)$$

where A and B are normalizing factors. If the two electrons have a common z-axis, equation 6-3.13 represents the situation in which the electron spins are antiparallel ($S = 0$, singlet state), whereas equation 6-3.14 represents the situation in which the spins are parallel ($S = 1$, triplet states) with components $M_S = 1, 0, -1$, respectively.

Next, suppose that there is an interaction between the two electron systems, given by the Hamiltonian $\mathcal{H}_{12}$. For example, for the hydrogen molecule, considered as two hydrogen atoms, whose nuclei are designated by a and b, the interaction potential of the two atoms is given by

$$\mathcal{H}_{12} = \frac{e^2}{r_{ab}} + \frac{e^2}{r_{12}} - \frac{e^2}{r_{1b}} - \frac{e^2}{r_{2a}}, \qquad (6\text{-}3.15)$$

where r_{ab} is the distance between nuclei, r_{12} is the distance between electrons and r_{1b} and r_{2a} are the distances between a given nucleus and the electron on the other atom. The values of the additional energies of the two-electron system are then given[20] by applying equation 2-8.3: $E = \int \psi^* \mathcal{H}_{12} \psi \, d\tau$. These energies for the states of equations 6-3.13 and 6-3.14 are, respectively,

$$E_{\mathrm{I}} = A^2(K_{12} + J_{12}) \qquad (6\text{-}3.16)$$

and

$$E_{\mathrm{II}} = B^2(K_{12} - J_{12}), \qquad (6\text{-}3.17)$$

where

$$K_{12} = \int \phi_a^*(1)\,\phi_b^*(2)\mathcal{H}_{12}\,\phi_a(1)\,\phi_b(2)\,d\tau_1\,d\tau_2 \qquad (6\text{-}3.18)$$

and

$$J_{12} = \int \phi_a^*(1)\,\phi_b^*(2)\mathcal{H}_{12}\,\phi_a(2)\,\phi_b(1)\,d\tau_1\,d\tau_2. \qquad (6\text{-}3.19)$$

Here K_{12} is the average Coulomb interaction energy, whereas J_{12} is called an exchange integral and occurs as a direct result of the requirement that the electrons be indistinguishable.

[20] Actually, for the hydrogen molecule it is not necessary to assume that equations 6-3.13 and 6-3.14 represent the correct zero-order wave functions. Application of the variation method of approximation leads directly to these wave functions.

For the hydrogen molecule J_{12} is negative; hence in the ground state the spins are antiparallel (equation 6-3.13). In order to have ferromagnetism, that is, spins parallel, J_{12} has to be positive. The conditions that favor a positive J_{12} may be examined in the following qualitative way.[21] Suppose that the wave functions ϕ_a and ϕ_b have no nodes in the region of appreciable overlap, so that $\phi_a^*(1)\,\phi_b^*(2)\,\phi_a(2)\,\phi_b(1)$ is positive. Then the sign of the exchange integral will be positive if the contributions from the positive terms of the Hamiltonian exceed those from the negative terms. For example, consider the Hamiltonian of equation 6-3.15. A large value for the term e^2/r_{12} occurs if the wave functions are large only midway between the nuclei, since only in this region is r_{12} small. Further, the terms $-e^2/r_{1b}$ $-e^2/r_{2a}$ are smallest for wave functions that are small near the nuclei.

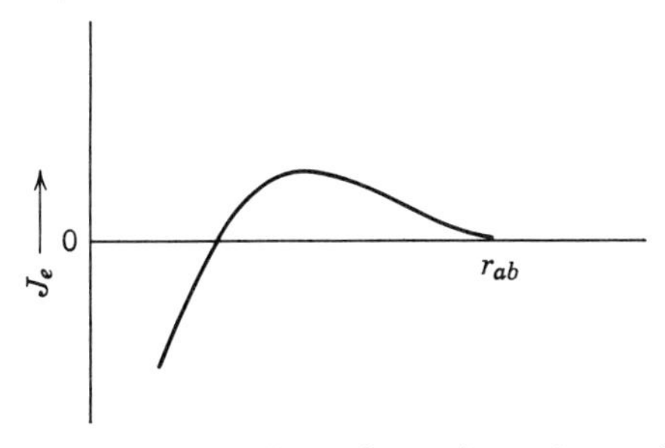

Fig. 6-3.1. Schematic variation of the exchange integral as a function of the interatomic distance r_{ab}.

These conditions are best met if the interatomic spacing r_{ab} is large compared to the radii of the orbitals, in the manner indicated in Fig. 6-3.1. Hence J_{12} is most likely to be positive for d and f wave functions for the atoms of some iron and rare earth metals; obviously it is positive for iron, cobalt, nickel, and gadolinium. Although some metals of the iron group elements such as Mn and Cr are not ferromagnetic, some of their compounds are. The probable explanation is that there is a small change in the interatomic spacing of the alloys which favors ferromagnetism. For example, MnAs and MnSb are both ferromagnetic, with a lattice spacing of 2.85 and 2.89 Å, respectively; the lattice spacing of pure Mn is 2.58 Å.

It is useful to express the energy of the two states of the two-electron system (equations 6-3.13 and 6-3.14) in terms of the electron spins. Recall that equations 6-3.16 and 6-3.17 give the energy of the states with symmetric and antisymmetric orbital wave functions, respectively. However,

[21] J. C. Slater, *Phys. Rev.* **35,** 509 (1930), **36,** 57 (1930); E. C. Stoner, *Proc. Leeds Phil. Soc.* **2,** 391 (1933); A. Sommerfeld and H. A. Bethe, *Handbuch der Physik*, 2nd Ed., Vol. XXIV/2, J. Springer, Berlin (1933), p. 596.

the orbital symmetry and the spin alignment are correlated. Hence the two electrons may be considered to be coupled via their spins. This coupling must be proportional to the eigenvalues of the scalar product $\mathbf{s}_1 \cdot \mathbf{s}_2$. To calculate the constant of proportionality, we will assume that the z-axis for the two electrons coincide; in general, however, it is not necessary that this be so. We have

$$\mathbf{s}_{12}{}^2 = (\mathbf{s}_1 + \mathbf{s}_2)^2$$
$$= \mathbf{s}_1{}^2 + \mathbf{s}_2{}^2 + 2\mathbf{s}_1 \cdot \mathbf{s}_2,$$

where $\mathbf{s}_{12}$ is the vector sum of $\mathbf{s}_1$ and $\mathbf{s}_2$. The eigenvalues in terms of $\hbar$, when $\mathbf{s}_1{}^2$, $\mathbf{s}_2{}^2$, $\mathbf{s}_{12}{}^2$, $\mathbf{s}_1 \cdot \mathbf{s}_2$ and $-\frac{1}{2} - 2\mathbf{s}_1 \cdot \mathbf{s}_2$ operate on the wave

Table 6-3.1

Operator	Eigenvalue
$\mathbf{s}_1{}^2 = \mathbf{s}_2{}^2$	$s_1(s_1 + 1) = s_2(s_2 + 1) = \frac{3}{4}$
$\mathbf{s}_{12}{}^2$	$s_{12}(s_{12} + 1) = 0$ for $s_{12} = 0$
	$= 2$ for $s_{12} = 1$
$\mathbf{s}_1 \cdot \mathbf{s}_2$	$\frac{1}{2}[s_{12}(s_{12} + 1) - 2s(s + 1)] = -\frac{3}{4}$ for $s_{12} = 0$
	$= \frac{1}{4}$ for $s_{12} = 1$
$-\frac{1}{2} - 2\mathbf{s}_1 \cdot \mathbf{s}_2$	$+1$ for $s_{12} = 0$
	-1 for $s_{12} = 1$

functions of equations 6-3.13 and 6-3.14, are given in Table 6-3.1. Hence equations 6-3.16 and 6-3.17 may be written in the operator form

$$E = K - \tfrac{1}{2}J_{12} - 2J_{12}\mathbf{s}_1 \cdot \mathbf{s}_2. \tag{6-3.20}$$

Although this equation makes it appear that there is a direct coupling between the spins, it is important to remember that this interaction energy is actually of electrostatic origin. It is difficult to calculate the actual magnitude of J_{12}. A recent machine computation appears to show that a strict Heitler–London model does not yield a large enough value for J_{12} to account for the size of the Weiss molecular field.[22] It is therefore suggested that departures from this model are important for the origin of ferromagnetism.

In problems concerning ferromagnetism usually only the spin dependent term of equation 6-3.20 is of interest. This term we will call the two-electron exchange Hamiltonian. For two atoms, i and j, that have one electron each the exchange Hamiltonian is

$$\mathcal{H} = -2J_{ij}\mathbf{s}_i \cdot \mathbf{s}_j, \tag{6-3.21}$$

[22] R. Stuart and W. Marshall, *Phys. Rev.* **120,** 353 (1960).

where J_{ij} is the exchange integral for the two electrons. Next, consider two atoms, each of which has more than one electron with an unpaired spin. Then, the exchange Hamiltonian is

$$\begin{aligned}\mathcal{H} &= -2J_{ij} \sum_{i,j}{}' \mathbf{s}_i \cdot \mathbf{s}_j \\ &= -2J_{ij} \sum_i \mathbf{s}_i \cdot \sum_j \mathbf{s}_j \\ &= -2J_{ij} \mathbf{S}_i \cdot \mathbf{S}_j, \end{aligned} \tag{6-3.22}$$

where $\mathbf{S}_i = \Sigma\, \mathbf{s}_i$ and $\mathbf{S}_j = \Sigma\, \mathbf{s}_j$ are the total spins of atoms i and j, respectively. We have assumed here (a) that all the electrons have the same exchange integral J_{ij} and (b) that the exchange energy between electrons of the same atom is constant, hence can be omitted. The first assumption is reasonable if the orbital angular momentum is quenched, say by crystalline fields, whereas the second is valid if the spin angular momentum is not quenched. Further, consider a solid and let us assume that the exchange integral is negligible except for nearest neighbors. Then the exchange Hamiltonian of a given atom i with its neighbors is

$$\mathcal{H} = -2 \sum_j J_{ij} \mathbf{S}_i \cdot \mathbf{S}_j, \tag{6-3.23}$$

where the summation is over the nearest neighbors of atom i. If the exchange integral is isotropic, equal to J_e, we get

$$\mathcal{H} = -2J_e \sum_j \mathbf{S}_i \cdot \mathbf{S}_j. \tag{6-3.24}$$

Finally, the Hamiltonian of an entire crystal is found by summing equation 6-3.24 over all the atoms in the crystal; that is,

$$\mathcal{H} = -2J_e \sum_{ij}{}' \mathbf{S}_i \cdot \mathbf{S}_j. \tag{6-3.25}$$

It is of interest to note in passing that recent experiments indicate that a biquadratic exchange term of the form $\mathcal{H}_b = 2J_b \Sigma\, (\mathbf{S}_i \cdot \mathbf{S}_j)^2$ is important for some ionic compounds.[23] No theoretical derivation of this term has been given yet.

By solving for the properties of a system with such a Hamiltonian we should arrive at a theory of ferromagnetism. However, it has not been found possible to solve this problem, except at very low or very high temperatures, without making further approximations. One approach to this problem is discussed in the next section. Another approach may be

[23] E. A. Harris and J. Owen, *Phys. Rev. Letters* **11**, 9 (1963).

illustrated by considering the Hamiltonian of equation 6-3.24. Expansion of the scalar product yields

$$\mathcal{H} = -2J_e \sum_j (S_{xi}S_{xj} + S_{yi}S_{yj} + S_{zi}S_{zj}).$$

The assumption is then made that the interaction $-2J_e \sum_j (S_{xi}S_{xj} + S_{yi}S_{yj})$ may be neglected, and thus the Hamiltonian is just

$$\mathcal{H} = -2J_e \sum S_{zi}S_{zj}. \tag{6-3.26}$$

This is called the *Ising* model.[24] The approximation amounts to assuming that the instantaneous values of the neighboring spins may be replaced by their time averages. Thus for z nearest neighbors

$$\mathcal{H} = -2zJ_e(\bar{S}_{xi}\bar{S}_{xj} + \bar{S}_{yi}\bar{S}_{yj} + \bar{S}_{zi}\bar{S}_{zj}).$$

If the z-axis is the axis of quantization,

$$\bar{S}_{xj} = \bar{S}_{yj} = 0;$$

hence

$$\begin{aligned}\mathcal{H} &= -2zJ_e S_{zi}\bar{S}_{zj}\\ &= -2J_e \sum S_{zi}S_{zj},\end{aligned}$$

which is just equation 6-3.26. There is an extensive literature on calculations made with the Ising model.[25] These calculations have the advantage that they can often be carried out with rigor, at least for one- and two-dimensional cases. Among the results obtained are (*a*) a one-dimensional lattice cannot show ferromagnetism, (*b*) a square two-dimensional lattice does exhibit ferromagnetism, for which case (*c*) $J_e/kT_f = 0.441$. Neglect of the x- and y-components of the spin however is not justifiable, particularly for low temperatures. The results obtained with the Ising model, although instructive, must be treated with reserve.

A relationship between the exchange integral J_e and the Weiss molecular field constant N_W may easily be obtained if equation 6-3.26 is assumed.[26] Since the magnetization, assumed along the z-direction of the specimen, is given by $M = Ng\mu_B\bar{S}_{zj}$, equation 6-3.26 becomes

$$\mathcal{H} = -2z\frac{S_{zi}J_eM}{Ng\mu_B}.$$

Alternatively, the Hamiltonian in terms of the Weiss field N_WM is given by

$$\mathcal{H} = -gS_{zi}\mu_BN_WM.$$

[24] E. Ising, *Z. Physik* **31,** 253 (1925).

[25] For a review see G. E. Newell and E. W. Montroll, *Revs. Mod. Phys.* **25,** 353 (1953).

[26] E. C. Stoner, *Magnetism and Matter*, Methuen, London (1934), p. 358.

By equating these last two equations we get the desired relationship, namely,

$$N_W = \frac{2zJ_e}{Ng^2\mu_B^{\ 2}}. \tag{6-3.27}$$

Further, by using equation 6-2.11, we find that

$$\frac{J_e}{\mathrm{k}T_f} = \frac{3}{2zS(S+1)}. \tag{6-3.28}$$

For example, with $z = 6$ and $S = \frac{1}{2}$, $J_e/\mathrm{k}T_f = 0.333$.

In certain problems, particularly in domain theory, to be discussed in Chapter 7, it is both appropriate and convenient to consider the spin matrix operators of equation 6-3.24 as classical vectors. This approximation is reasonably valid when the neighboring spins make small angles with one another. Equation 6-3.24 may then be rewritten as

$$\mathcal{H} = E = -2J_eS^2 \sum_j \cos\phi_{ij}, \tag{6-3.29}$$

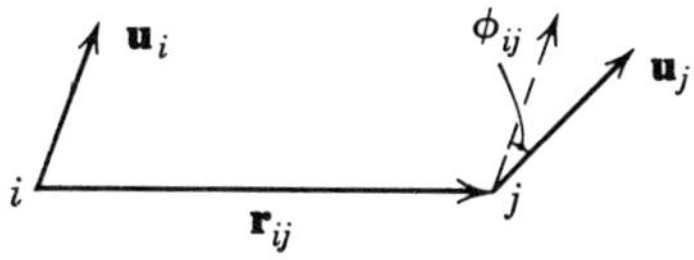

Fig. 6-3.2. Unit vectors $\mathbf{u}_i$ and $\mathbf{u}_j$ connected by a displacement vector $\mathbf{r}_{ij}$.

where ϕ_{ij} is the angle between the directions of the classical spin angular momentum vectors. If the angle between neighboring spins is very small, we may make the approximation that $\cos\phi_{ij} \doteqdot 1 - \phi_{ij}^{\ 2}/2$. The variable part of the exchange energy is then given by

$$E = J_eS^2 \sum \phi_{ij}^{\ 2}. \tag{6-3.30}$$

Equation 6-3.29 is often expressed in another form. Let $\mathbf{u}_i$ and $\mathbf{u}_j$ be unit vectors parallel to the spin vectors of adjacent atoms i and j, and let $\mathbf{r}_{ij} = x_{ij}\mathbf{i} + y_{ij}\mathbf{j} + z_{ij}\mathbf{k}$ be the displacement vector between the atoms, as indicated in Fig. 6-3.2. Then

$$\begin{aligned}\cos\phi_{ij} &= \mathbf{u}_i \cdot \mathbf{u}_j \\ &= \alpha_{1i}\alpha_{1j} + \alpha_{2i}\alpha_{2j} + \alpha_{3i}\alpha_{3j},\end{aligned} \tag{6-3.31}$$

where α_1, α_2, and α_3 are the direction cosines of a unit vector relative to the x-, y-, and z-axes, respectively. Since the angle between the unit vectors $\mathbf{u}_i$ and $\mathbf{u}_j$ is small, the direction cosines of $\mathbf{u}_j$ may be expanded in a Taylor series in the direction cosines of $\mathbf{u}_i$. For example, the first term of equation 6-3.31 is

$$\alpha_{1i}\alpha_{1j} = \alpha_{1i}(\alpha_{1i} + \mathbf{r}_{ij} \cdot \nabla\alpha_{1i} + \tfrac{1}{2}(\mathbf{r}_{ij} \cdot \nabla)^2\alpha_{1i} + \cdots).$$

Next, we sum this expression over the nearest neighbors j. For cubic crystals, terms like $\sum_j \mathbf{r}_{ij} \cdot \nabla\alpha_{1i}$ and cross terms from $\sum_j \frac{1}{2}(\mathbf{r}_{ij} \cdot \nabla)^2\alpha_{1j}$ like $\sum x_{ij}y_{ij}(\partial^2\alpha_{1j}/\partial x_{ij}\partial y_{ij})$ are zero because of the symmetry. We have

$$\sum_j \cos\phi_{ij} = z + \tfrac{1}{2}\alpha_{1i}\frac{\partial^2\alpha_{1i}}{\partial x_{ij}^2}\sum_j x_{ij}^2 + \tfrac{1}{2}\alpha_{1i}\frac{\partial^2\alpha_{1i}}{\partial y_{ij}^2}\sum_j y_{ij}^2 + \tfrac{1}{2}\alpha_{1i}\frac{\partial^2\alpha_{1i}}{\partial z_{ij}^2}\sum_j z_{ij}^2$$
$$+ \tfrac{1}{2}\alpha_{2i}\frac{\partial^2\alpha_{2i}}{\partial x_{ij}^2}\sum_j x_{ij}^2 + \cdots.$$

Now since $\sum_j x_{ij}^2 = \sum_j y_{ij}^2 = \sum_j z_{ij}^2 = \frac{1}{3}\sum_j r_{ij}^2$, we obtain

$$\sum_j \cos\phi_{ij} = z + \tfrac{1}{6}\sum_j r_{ij}^2\mathbf{u}\cdot\nabla^2\mathbf{u}.$$

where we have omitted the subscript i on $\mathbf{u}_i$. By considering only the variable part of the energy, the exchange energy is therefore given by

$$\mathcal{H} = E = -\frac{J_eS^2}{3}\sum_j r_{ij}^2\,\mathbf{u}\cdot\nabla^2\mathbf{u}.$$

With the aid of the vector identity,

$$\nabla^2(\mathbf{u}\cdot\mathbf{u}) = 2[(\nabla\alpha_1)^2 + (\nabla\alpha_2)^2 + (\nabla\alpha_3)^2] + 2(\mathbf{u}\cdot\nabla^2\mathbf{u}),$$
$$= 0$$

the last equation may be written

$$\mathcal{H} = E = \frac{J_eS^2}{3}\sum_j r_{ij}^2[(\nabla\alpha_1)^2 + (\nabla\alpha_1)^2 + (\nabla\alpha_3)^2]. \qquad (6\text{-}3.32)$$

For a cubic structure, either simple, body-centered, or face-centered, the expression $\sum_j r_{ij}^2$ is equal to $6a^2$, where a is the lattice spacing. Hence we have finally

$$\mathcal{H} = E = 2J_eS^2a^2[(\nabla\alpha_1)^2 + (\nabla\alpha_2)^2 + (\nabla\alpha_3)^2] \qquad (6\text{-}3.33)$$

as the exchange energy for one cube of edge a.

4. The Series Expansion Method

In Section 6-2 it was noted that some of the predictions of the molecular field theory were in disagreement with experiment. Various theories which attempt to clear up these discrepancies have been developed. The early theories were valid only over a restricted temperature range, namely, (1) temperatures well above the Curie point, (2) temperatures in the vicinity of the Curie point, and (3) temperatures well below the Curie point. More recent theories that employ either the random phase approximation[27] or the time-dependent Green's function[28] apply to the entire temperature range. They reduce to the restricted results of the older theories with a good degree of approximation. However, the mathematics employed in the present form of the new theories is relatively advanced. We shall therefore be content in the first edition of this book to present a resumé of the older theories.

In this section the theory valid for $T > T_f$ is treated; the two subsequent sections consider theories valid for $T \approx T_f$ and $T \ll T_f$, respectively.

A quantum mechanical theory of ferromagnetism should presumably employ a Heisenberg exchange Hamiltonian. Consequently, let us consider the Hamiltonian

$$\mathcal{H} = -2J_e \sum_{\mathrm{nn}} \mathbf{S}_i \cdot \mathbf{S}_j - g\mu_B \mathbf{H} \cdot \sum \mathbf{S}_k, \tag{6-4.1}$$

where the first summation is over all nearest-neighbor pairs of the crystal, and the second summation is over all atoms. The first term of this equation is the exchange term (equation 6-3.25), whereas the second term represents the interaction energy with the applied field. The problem is to evaluate the quantum mechanical partition function, given in analogy with equation 5-1.14 by

$$\mathfrak{Z} = \operatorname{tr} e^{-\mathcal{H}/kT}, \tag{6-4.2}$$

where tr means the trace of the matrix. Once $\mathfrak{Z}$ is known, the magnetization, hence the susceptibility, is obtained immediately with the aid of equation 5-1.16. The partition function is computed by expansion[29] into a power series in $1/T$. For $T \gg T_f$ the series converges rapidly and the coefficients for only a few terms need to be evaluated. The calculation essentially involves the evaluation of terms like $\overline{(\Sigma\, \mathbf{S}_i \cdot \mathbf{S}_j)^n}$ where $\mathbf{S}_i$, etc., is given by equation 6-3.9; the reader is referred to the literature[29] for

[27] R. Brout, *Phys. Rev.* **115,** 824 (1959), **118,** 1009 (1960); F. Englert, *Phys. Rev. Letters* **5,** 102 (1960); G. Horwitz and H. B. Callen, *Phys. Rev.* **124,** 1757 (1961).

[28] S. V. Tyablikov, *Ukrain. Math. Zhur.* **11,** 287 (1959); D. N. Zubarev, *Soviet Phys.-Uspekhi* **3,** 320 (1960); Yu. A. Izyumov and E. N. Yakovlev, *Fiz. Metal. Metalloved.* **9,** 667 (1960); K. Kawasaki and H. Mori, *Progr. Theor. Phys.* (*Kyoto*) **25,** 1045 (1961); R. A. Tahir-Kheli and D. ter Haar, *Phys. Rev.* **127,** 88, 95 (1962), **130,** 108 (1963); R. A. Tahir-Kheli, *Phys. Rev.* **132,** 689 (1963); H. B. Callen, *Phys. Rev.* **130,** 890 (1963).

[29] W. Opechowski, *Physica* **3,** 181 (1937); H. A. Kramers, *Comm. Kamerlingh Onnes Lab. Leiden* No. 83; H. A. Brown and J. M. Luttinger, *Phys. Rev.* **100,** 685 (1955); H. A. Brown, *Phys. Rev.* **104,** 624 (1956); G. S. Rushbrook and P. J. Wood, *Proc. Phys. Soc.* (*London*) **A-68,** 1161 (1955), **A-70,** 765 (1957); C. Domb and M. J. Sykes, *Proc. Phys. Soc.* (*London*) **A-240,** 214 (1957).

details. The result is that the susceptibility may be written as the series (in the notation of Brown and Luttinger[29])

$$\chi = \frac{Ng^2\mu_B{}^2S(S+1)}{3kT}\sum_{n=0}^{\infty}\frac{a_n}{t^n}, \tag{6-4.3}$$

where $t = kT/J_e$. The coefficients have been calculated through to a_5 for arbitrary spin S for various crystal lattices; for $S = \frac{1}{2}$ the coefficient a_6 has been calculated.

Termination of equation 6-4.3 at the zeroth-order term and the first-order term yields, respectively, Curie's law and the Curie-Weiss law. Inclusion of further terms gives a deviation from the linearity predicted by the molecular field theory and thus represents at least a qualitative improvement in the description of the $1/\chi$ versus T curves. The terms through a_2 depend on the number of nearest neighbors and not on their arrangement in the crystal lattice; higher terms do depend on the kind of crystal lattice.

Equation 6-4.3 to the term in a_2 had been derived earlier by Heisenberg[30] by another method. Heisenberg attempted to evaluate the partition function of equation 5-1.14, $\mathfrak{Z} = \Sigma\, e^{-E/kT}$, from a knowledge of the energy levels of the system. The evaluation of these energy levels, however, is a virtually impossible task. Suppose a crystal has a certain total spin. The maximum spin this crystal may have is $\mathcal{N}S$, where $\mathcal{N}$ is the number of atoms, each with a spin S, in the crystal; further, there is only one such state and it has the lowest exchange energy. For any other spin configuration there are many states with widely differing energies. For example, suppose $S = \frac{1}{2}$ and that a certain number of spins are inverted, compared to the lowest energy state. Then, when the inverted spins are located within a small volume so that many of them are neighbors, the exchange energy is much less than when these spins are distributed more or less uniformly throughout the crystal. As a first approximation, Heisenberg assumed that all states of the crystal with the same total spin had the same energy; this assumption led to the Curie-Weiss law. As a second approximation, he assumed that the energy levels for a given total spin had a Gaussian distribution. He was able to compute the mean and standard deviation of this distribution. This second approximation leads to the second-order approximation of equation 6-4.3 alluded to earlier.

The Curie temperature may be obtained from the series expansion method provided the questionable assumption is made that it is valid to extrapolate from the high temperature region. The Curie temperature is calculated either from the value of t for which $1/\chi = 0$ or from the value

[30] W. Heisenberg, *Z. Physik* **49**, 619 (1928).

of t for which $a_n/a_{n-1} = t$. This latter criterion is obtained by calculation of the radius of convergence of the series. Some values calculated by Brown[31] with the first criterion are given in Table 6-4.1. It is evident that there is rapid convergence of the successive approximations.

Table 6-4.1. Values of J_e/kT_f for Various Spins by the Series Expansion Method

Spin	Lattice	$n = 1$	$n = 2$	$n = 3$	$n = 4$	$n = 5$
$\frac{1}{2}$	*s c*	0.333	0.578	0.500	0.518	0.546
1	*s c*	0.125	0.194	0.169	0.186	0.176
$\frac{3}{2}$	*s c*	0.0667	0.0920	0.0862	0.0939	0.0903
$\frac{1}{2}$	*b c c*	0.250	0.500	0.338	0.418	0.358
1	*b c c*	0.0938	0.121	0.115	0.128	0.122
$\frac{3}{2}$	*b c c*	0.0500	0.0612	0.0598	0.0648	0.0628

Mention should be made of an attempt by Néel[32] to extend the Weiss theory by considering that the instantaneous value of the molecular field fluctuates (compare this with the assumption made in equation 6-3.26). He was able with this theory to predict a difference between the paramagnetic and ferromagnetic Curie points.

5. The Bethe-Peierls-Weiss Method

The ordering of the dipoles in a ferromagnetic material and the change from an ordered to a disordered state at the Curie temperature are examples of what are often called *cooperative phenomena*. It is reasonable to expect that some methods developed for other order-disorder phenomena may be usefully adapted to the ferromagnetic case. In particular, Weiss[33] and others have applied the Bethe[34]–Peierls[35] short-range-order theory of alloys to ferromagnetism. The principle of the method, as applied to the magnetic case, is as follows. Consider a cluster of atoms that consists of a certain atom i, called the central atom, and its z nearest neighbors. The interaction of the neighbor atoms with each other and with the central atom is treated by employing a Heisenberg exchange interaction. The interactions of the atoms outside the cluster is represented by an effective

[31] H. A. Brown, *Phys. Rev.* **104,** 624 (1956).

[32] L. Néel, *Ann. phys. (Paris)* **18,** 5 (1932), **8,** 237 (1937), *Le magnétisme* **II,** 67 (1940).

[33] P. R. Weiss, *Phys. Rev.* **74,** 1493 (1948); U. Firgau, *Ann. Phys. (Leipzig)* **40,** 295 (1941); H. A. Brown and J. M. Luttinger, *Phys. Rev.* **100,** 685 (1955).

[34] H. Bethe, *Proc. Roy. Soc. (London)* **A-150,** 552 (1935).

[35] R. E. Peierls, *Proc. Roy. Soc. (London)* **A-154,** 207 (1936).

magnetic field, H_e, which acts on the z neighbor atoms of the cluster. The effective field H_e is then determined by a consistency condition. The Curie point is recognized as the temperature below which H_e is nonzero. The Bethe-Peierls-Weiss method is concerned only with the local or short-range ordering within a cluster and neglects long-range order[36] and spin-wave effects that are important at low temperatures. The validity of this method therefore may be expected to be confined to temperatures in the vicinity of the Curie temperature.

The Hamiltonian of the cluster is then assumed to be given by

$$\mathcal{H} = -2J_e\mathbf{S}_i \cdot \sum_{j=1}^{z} \mathbf{S}_j - g\mu_B\mathbf{H}_e \cdot \sum_{j=1}^{z} \mathbf{S}_j - g\mu_B\mathbf{H} \cdot \mathbf{S}_i, \qquad (6\text{-}5.1)$$

where $\mathbf{S}_i$ is the spin of the central atom, $\mathbf{H}_e$ is the effective field acting on the z neighbor atoms of spin $\mathbf{S}_j$, and $\mathbf{H}$ is the applied field, assumed to interact only with the central atom. The summations are over the nearest neighbors. In this equation it is assumed that the interaction of the neighboring atoms is absent. A little consideration will show that this assumption is valid for some lattices, such as the simple cubic and body-centered cubic. It is not valid for the face-centered and hexagonal structure, in which case an additional term of the form $-2J_e\sum \mathbf{S}_j \cdot \mathbf{S}_k$ must be added to the Hamiltonian.

We shall illustrate the Bethe-Peierls-Weiss method for the simplest case, namely, that in which the Ising model is employed. This treatment is essentially equivalent to the binary alloy case. The Hamiltonian (equation 6-5.1) then becomes

$$\mathcal{H} = -2J_eS_{zi}\sum S_{zj} - g\mu_BH_e\sum S_{zj} - g\mu_BHS_{zi}, \qquad (6\text{-}5.2)$$

where H_e and H are assumed to be the fields in the z-direction. Further, let us assume that $S_{zi} = \pm\frac{1}{2}$ and $S_{zj} = \pm\frac{1}{2}$. Then the total spin of the neighbor atoms $S_{zn} = \Sigma S_{zj}$ has the values $-z/2, -(z/2 - 1), \ldots, +(z/2)$. For a given value of S_{zn} the number of states is the number of ways $(S_{zn} + z/2)$ spins can be arranged on the z sites, that is,

$$\frac{z!}{(S_{zn} + z/2)!\,(z/2 - S_{zn})!}.$$

The partition function (equation 5-1.14) is then

$$\mathfrak{Z} = \sum \frac{z!}{(S_{zn} + z/2)!\,(z/2 - S_{zn})!}\, e^{-E/kT},$$

[36] For an elementary discussion of long- and short-range ordering in alloys see A. J. Dekker, *Solid State Physics*, Prentice-Hall, Englewood Cliffs, N.J. (1956), Ch. 4.

where E is the eigenvalue of the foregoing Hamiltonian and is easily computed; for example, the eigenvalue of S_{zi} is just M_{si}. It is convenient to introduce the notation

$$x = e^{-g\mu_B H/kT}; \qquad x_1 = e^{-g\mu_B H_e/kT}; \qquad y = e^{-J_e/kT}. \tag{6-5.3}$$

Then the partition function becomes

$$\mathfrak{Z} = \sum \frac{z!}{(S_{zn} + z/2)!(z/2 - S_{zn})!}\left[\frac{1}{\sqrt{x}}\left(\frac{1}{x_1 y}\right)^{S_{zn}} + \sqrt{x}\left(\frac{y}{x_1}\right)^{S_{zn}}\right], \tag{6-5.4}$$

where the summation is from $S_{zn} = -z/2$ to $S_{zn} = +z/2$. In order to carry out the summation, consider the expression

$$\sum_{S_{zn}=-z/2}^{z/2} \frac{z!}{(S_{zn} + z/2)!\,(z/2 - S_{zn})!}\, w^{S_{zn}}. \tag{6-5.5}$$

Let

$$t = \frac{z}{2} + S_{zn},$$

so that

$$S_{zn} = t - \frac{z}{2}; \qquad \frac{z}{2} - S_{zn} = z - t.$$

The limits $S_{zn} = -z/2$ and $S_{zn} = +z/2$ then correspond to $t = 0$ and $t = z$, respectively. Equation 6-5.5 becomes

$$\sum_{t=0}^{z} \frac{z!}{t!\,(z-t)!}\, w^{t(z/2)} = w^{-z/2}(1 + w)^z$$
$$= \left(\frac{1 + w}{\sqrt{w}}\right)^z \tag{6-5.6}$$

by the binomial theorem. Application of this result to equation 6-5.4 then yields

$$\mathfrak{Z} = \frac{1}{\sqrt{x}}\left(\sqrt{x_1 y} + \frac{1}{\sqrt{x_1 y}}\right)^z + \sqrt{x}\left(\sqrt{\frac{x_1}{y}} + \sqrt{\frac{y}{x_1}}\right)^z. \tag{6-5.7}$$

The magnetic moment (actually average) of the central atom i and of one of the neighbor atoms is given by an adaption of equation 5-1.16, namely,

$$\bar{\mu}_i = g\mu_B x \frac{\partial \log \mathfrak{Z}}{\partial x} \quad \text{and} \quad \bar{\mu}_j = g\mu_B x_1 \frac{\partial \log \mathfrak{Z}}{\partial x_1}. \tag{6-5.8}$$

Hence we obtain

$$\bar{\mu}_i \frac{2\mathfrak{Z}}{g\mu_B} = \sqrt{x}\left(\sqrt{\frac{x_1}{y}} + \sqrt{\frac{y}{x_1}}\right)^z - \frac{1}{\sqrt{x}}\left(\frac{1}{\sqrt{x_1 y}} + \sqrt{x_1 y}\right)^z \tag{6-5.9}$$

and

$$\bar{\mu}_j \frac{2\mathfrak{Z}}{g\mu_B} = \sqrt{x}\left(\sqrt{\frac{x_1}{y}} + \sqrt{\frac{y}{x_1}}\right)^{z-1}\left(\sqrt{\frac{x_1}{y}} - \sqrt{\frac{y}{x_1}}\right)$$

$$+ \frac{1}{\sqrt{x}}\left(\frac{1}{\sqrt{x_1 y}} + \sqrt{x_1 y}\right)^{z-1}\left(\sqrt{x_1 y} - \frac{1}{\sqrt{x_1 y}}\right). \quad (6\text{-}5.10)$$

Now, physically, all atoms are equivalent, so we apply the condition

$$\bar{\mu}_i = \bar{\mu}_j. \quad (6\text{-}5.11)$$

This is the consistency condition, mentioned earlier, that determines H_e. For the limiting case of zero applied field $x = 1$, and equation 6-5.11 gives

$$(1 + x_1 y)^{z-1} x_1 = (x_1 + y)^{z-1}. \quad (6\text{-}5.12)$$

Obviously $x_1 = 1$ is always a solution; this corresponds to $H_e = 0$. We shall now show that below a certain temperature there is another root with $x_1 < 1$, corresponding to $H_e > 0$. Let us look for a solution $x_1 = 1 - \alpha$, where α is small. From equation 6-5.12 we have

$$(1 + y - y\alpha)^{z-1}(1 - \alpha) = (y + 1 - \alpha)^{z-1}.$$

Then, on working only to terms that are first order in α, we get first

$$(1 + y)^{z-1}\left(1 - \frac{y\alpha}{1 + y}\right)^{z-1}(1 - \alpha) = (1 + y)^{z-1}\left(1 - \frac{\alpha}{1 + y}\right)^{z-1}$$

and

$$\left[\frac{(z - 1)y}{1 + y} + 1\right]\alpha = \left(\frac{z - 1}{1 + y}\right)\alpha.$$

If $\alpha \neq 0$, then

$$y = \frac{z - 2}{z}$$

or, using equation 6-5.3,

$$\frac{J_e}{kT_c} = \log\left(\frac{z}{z - 2}\right), \quad (6\text{-}5.13)$$

an equation which determines a critical temperature. For temperatures below T_c, $x_1 < 0$, $H_e \neq 0$, and there is ferromagnetism[37]; hence the critical temperature T_c is to be identified as the Curie temperature T_f. It is left to the reader to show (problem 6-6) that just below the Curie point the internal field is given by

$$\left(\frac{g\mu_B H_e}{2kT}\right)^2 \doteqdot \frac{3}{z}(z - 1)\left(\frac{z}{2} - 1\right)\left(e^{J_e/kT} - e^{J_e/kT_f}\right) \quad (6\text{-}5.14)$$

[37] We can show that for $T < T_f$ the root $x_1 < 1$ has lower energy than that for $x_1 = 1$.

and the average atomic magnetic moment is given by

$$\bar{\mu}_i \doteq -\mu_B \frac{z}{z-1}\left(\frac{g\mu_B H_e}{2kT}\right). \tag{6-5.15}$$

The calculations have also been carried out[38] for the Hamiltonian of equation 6-5.1. One merit of the calculations is that closed-form expressions are obtained. We shall be content here merely to quote a few of the results. Values of J_e/kT_f for various lattices and spins are given in Table 6-5.1. The good agreement with the results of the series expansion method

Table 6-5.1. Values of J_e/kT_f Obtained from the Bethe-Peierls-Weiss Method

	Spin		
Lattice Type	$\frac{1}{2}$	1	$\frac{3}{2}$
s c	0.540	0.169	0.0864
b c c	0.343	0.148	0.0597

(Table 6-4.1) should be noted. Above the Curie temperature the susceptibility does not follow the Curie-Weiss law; instead the $1/\chi$ versus T curve is convex, and there is a paramagnetic and ferromagnetic Curie temperature, as observed experimentally. For example, for a spin $\frac{1}{2}$ and a body-centered cubic lattice the theory gives $\theta = 1.05T_f$. The reason for the curvature is essentially because the effective field H_e, in the presence of an applied field, is not quite zero because of the exchange interaction.

The discontinuity in the specific heat per mole at the Curie temperature for the body-centered cubic structure is calculated to be

$$\frac{\Delta C_{\text{sp}}}{R} = 2.05 \quad \text{for} \quad S = \tfrac{1}{2}$$

and

$$\frac{\Delta C_{\text{sp}}}{R} = 3.40 \quad \text{for} \quad S = 1.$$

Although the value for $S = \frac{1}{2}$ is somewhat larger than that observed for Ni, the value for $S = 1$ is in reasonable agreement with that observed for Fe (see Fig. 6-2.7). The variation of the specific heat with temperature according to the Ising model is shown schematically in Fig. 6-5.1. The

[38] P. R. Weiss, *Phys. Rev.* **74,** 1493 (1948); H. A. Brown and J. M. Luttinger, *Phys. Rev.* **100,** 685 (1955).

theory predicts a nonzero value of the specific heat above the Curie temperature as observed; this behavior is the result of the persistence of short-range ordering.

Finally, mention should be made of calculations based on a simpler model, called the constant-coupling approximation.[39] In this model the exchange interaction between two atoms only is considered; the coupling

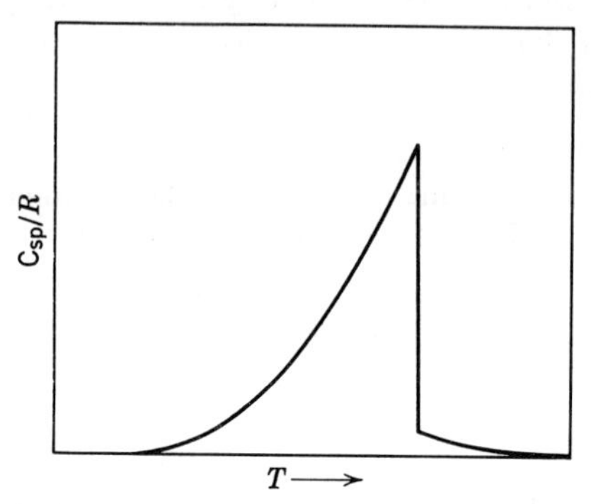

Fig. 6-5.1. Schematic illustration of the variation of specific heat with temperature according to the Bethe-Peierls-Weiss method for the Ising model.

with the other atoms is represented by an effective field. The values of J_e/kT_f obtained are close to those obtained with the Bethe-Peierls-Weiss method.

6. Spin Waves

Consider a ferromagnetic specimen of unit volume at absolute zero. Assume that an axis of quantization is established, say by a small magnetic field applied along the negative z-direction. The third law of thermodynamics requires that the spin system be completely ordered. Since the system must also be in its ground state, it follows that the spin quantum number of each atom will have its maximum value. For $S = \frac{1}{2}$ and $g = 2$, the only one we shall consider in detail, the magnetic moment is then given by $M = N\,|\mu_B|$, where N is the number of atoms per unit volume. Next, suppose that the temperature is raised slightly so that one spin is reversed; this presumably is the lowest excited state of the system. Now, each atom has an equal probability of being the one whose spin is reversed. This suggests that the reversed spin will not remain localized at one atom. However, for the moment consider that the reversed spin is located at a

[39] P. W. Kasteleijn and J. Van Kranendonk, *Physica* **22,** 317 (1956); T. Oguchi, *Prog. Theor. Phys. (Kyoto)* **13,** 148 (1955).

particular atom. The exchange forces will tend to invert the reversed spin. One possibility is a transition back to the ground state; this, however, is relatively unlikely. Instead, it turns out that the reversed spin travels from one atom to another, the exchange always occurring between neighbors. The propagation of the reversed spin through the crystal is called a *spin wave*.[40] Because of the boundary conditions only certain wavelengths are possible. As will be shown, the quantum mechanical problem of one reversed spin can be solved exactly.[41]

Now suppose that as a result of a further increase in temperature the crystal has two reversed spins. Two additional complications occur. First, because in general the two reversed spins, or spin waves, will be traveling with different velocities, they will meet at some time. The result is a scattering. Second, there is a possibility that the reversed spins will be bound together on adjacent atoms. This state, sometimes called a *spin complex*,[42] has a lower exchange energy than when the two reversed spins are separated. If more than two spins are reversed, the same types of complication occur, although now there will be more collisions and also the spin complexes may consist of more than two reversed spins. The usual approximation in spin-wave theory is to neglect these complications and to assume that the spin waves are independent of each other. Then the total energy of a number of spin waves is just the sum of the energies of the individual spin waves. This superposition can be expected to be valid only as long as the number of reversed spins is small, that is, for temperatures well below the Curie temperature. According to Dyson,[43] the error in the calculation of the magnetization when the spin-wave interactions are neglected is less than 5% even at $T = 0.5T_f$.

We now turn to a quantitative quantum mechanical treatment of spin waves. For simplicity, we shall assume that the Hamiltonian consists of only the exchange term given by equation 6-3.5, namely,

$$\mathcal{H} = -2J_e \sum_{ij}' \mathbf{S}_i \cdot \mathbf{S}_j,$$

where the operators **S** are the Pauli matrices of equation 6-3.9, since we assume $S = \frac{1}{2}$. It is convenient to introduce the matrices

$$S_i^+ = S_{xi} + iS_{yi}$$

and

$$S_i^- = S_{xi} - iS_{yi}. \tag{6-6.1}$$

[40] F. Bloch, *Z. Physik* **61,** 206 (1930).

[41] J. C. Slater, *Phys. Rev.* **35,** 509 (1930).

[42] H. A. Bethe and A. Sommerfeld, *Handbuch der Physik*, XXIV/2, J. Springer, Berlin (1933), p. 333.

[43] F. J. Dyson, *Phys. Rev.* **102,** 1217, 1230 (1956); T. Oguchi, *Phys. Rev.* **117,** 117 (1960); F. Keffer and R. Loudon, *J. Appl. Phys.* **32,** 2S (1961).

Then the exchange Hamiltonian may be written

$$\begin{aligned}\mathcal{H} &= -2J_e \sum \mathbf{S}_i \cdot \mathbf{S}_j \\ &= -2J_e \sum (S_{xi}S_{xj} + S_{yi}S_{yj} + S_{zi}S_{zj}) \\ &= -2J_e \sum [\tfrac{1}{2}(S_i^+S_j^- + S_i^-S_j^+) + S_{zi}S_{zj}].\end{aligned} \qquad (6\text{-}6.2)$$

We now develop certain relationships that will be useful in performing operations with this Hamiltonian. By using the Pauli matrices, it follows immediately that

$$S^+ = \begin{pmatrix} 0 & 1 \\ 0 & 0 \end{pmatrix} \quad \text{and} \quad S^- = \begin{pmatrix} 0 & 0 \\ 1 & 0 \end{pmatrix}. \qquad (6\text{-}6.3)$$

Further, recalling that the spin eigenfunctions of an atom are given by equations 6-3.10, it is easy to show that

$$\begin{aligned} S^+ \,|\chi_\beta\rangle &= 1\,|\chi_\alpha\rangle, \\ S^- \,|\chi_\alpha\rangle &= 1\,|\chi_\beta\rangle, \\ S_z \,|\chi_\alpha\rangle &= \tfrac{1}{2}\,|\chi_\alpha\rangle, \\ S_z \,|\chi_\beta\rangle &= -\tfrac{1}{2}\,|\chi_\beta\rangle. \end{aligned} \qquad (6\text{-}6.4)$$

For a system of N spins the ground-state eigenfunction (equation 6-6.2) is

$$\chi_0 = \chi_{\alpha 1}\chi_{\alpha 2}\chi_{\alpha 3}\chi_{\alpha 4} \cdots \chi_{\alpha N}. \qquad (6\text{-}6.5)$$

To show this, consider

$$\mathcal{H}\chi_0 = -2J_e \sum [\tfrac{1}{2}(S_i^+S_j^- + S_i^-S_j^+) + S_{zi}S_{zj}]\chi_0.$$

The first two terms of the operator act to invert a reversed spin; since, however, there are no reversed spins, these terms have no effect. Explicit mathematical evaluation leads to the null matrix. Only the third term leads to a nonzero result. Hence we get

$$\begin{aligned}\mathcal{H}\chi_0 &= -\tfrac{1}{4}NzJ_e\chi_0 \\ &= E_0\chi_0,\end{aligned} \qquad (6\text{-}6.6)$$

where z is the number of nearest neighbors. The extra factor of $\frac{1}{2}$ arises because interactions between pairs of atoms should be counted only once.

Next, suppose the system has one reversed spin, located at the wth atom. Let us see if

$$\chi_w = \chi_{\alpha 1}\chi_{\alpha 2} \cdots \chi_{\alpha(w-1)}\chi_{\beta w}\chi_{\alpha(w+1)} \cdots \chi_{\alpha N} \qquad (6\text{-}6.7)$$

is an eigenfunction of the Hamiltonian. Consider

$$\mathcal{H}\chi_w = -2J_e \sum [\tfrac{1}{2}(S_i^+S_j^- + S_i^-S_j^+) + S_{zi}S_{zj}]\chi_w.$$

First consider all terms except those in which there is an operation on the atomic function $\chi_{\beta w}$. We obtain

$$-2J_e \sum [\tfrac{1}{2}(S_i^+ S_j^- + S_i^- S_j^+)]\chi_w = 0\chi_w$$

as before and

$$-2J_e \sum S_{zi} S_{zj} \chi_w = -\tfrac{1}{4}J_e z(N-2)\chi_w. \qquad (6\text{-}6.8)$$

Next, consider the remaining terms. Then

$$-2J_e \sum_{i=1}^{z} \tfrac{1}{2}(S_w^+ S_{w+i}^- + S_w^- S_{w+i}^+)\chi_w = \sum_{i=1}^{z} - J_e \chi_{w+i},$$

since the second operator yields a null matrix and

$$-2J_e \sum_{i=1}^{z} S_{zw} S_{z(w+i)} \chi_w = \tfrac{1}{2} J_e z \chi_w, \qquad (6\text{-}6.9)$$

where the summations are over the nearest neighbors of atom w. On collecting terms, we have

$$\mathcal{H}\chi_w = (-\tfrac{1}{4}J_e N z + z J_e)\chi_w - \sum_{i=1}^{z} J_e \chi_{w+i}. \qquad (6\text{-}6.10)$$

Obviously χ_w is not an eigenfunction. Further, it is clear that the reversed spin does not stay at the position w, but instead propagates. Hence an eigenfunction may be formed by a linear combination of N functions like χ_w, each of which represents a reversed spin located at one of the N atoms, that is,

$$X = \sum c_w \chi_w. \qquad (6\text{-}6.11)$$

The recursion relationship between the coefficients and its solution is left as problem 6-7. The result is (apart from a normalizing factor)

$$c_w = e^{i\mathbf{k}\cdot\mathbf{R}_w},$$

where $\mathbf{R}_w$ is a vector from the origin to the atom w. For periodic boundary conditions[44] $c_w = c_{w+N}$, and therefore the components of $\mathbf{k}$ may have only the values

$$\frac{k_x x_{ij} N_x}{2\pi} = 0; \pm 1; \pm 2; \cdots; \pm \frac{N_x}{2} \qquad (6\text{-}6.12)$$

and similar equations for k_y and k_z. Here N_x is the number of atoms and x_{ij}, the distance between nearest neighbors, both along the x-direction. Equation 6-6.11 then becomes

$$X = \sum_w e^{i\mathbf{k}\cdot\mathbf{R}_w}\chi_w; \qquad (6\text{-}6.13)$$

that is, the eigenfunctions have wavelike properties.

[44] For the one-dimensional case this requirement may be pictured as bending a row of N_x atoms into a ring so that the first atom is also the $N_x + 1$ atom.

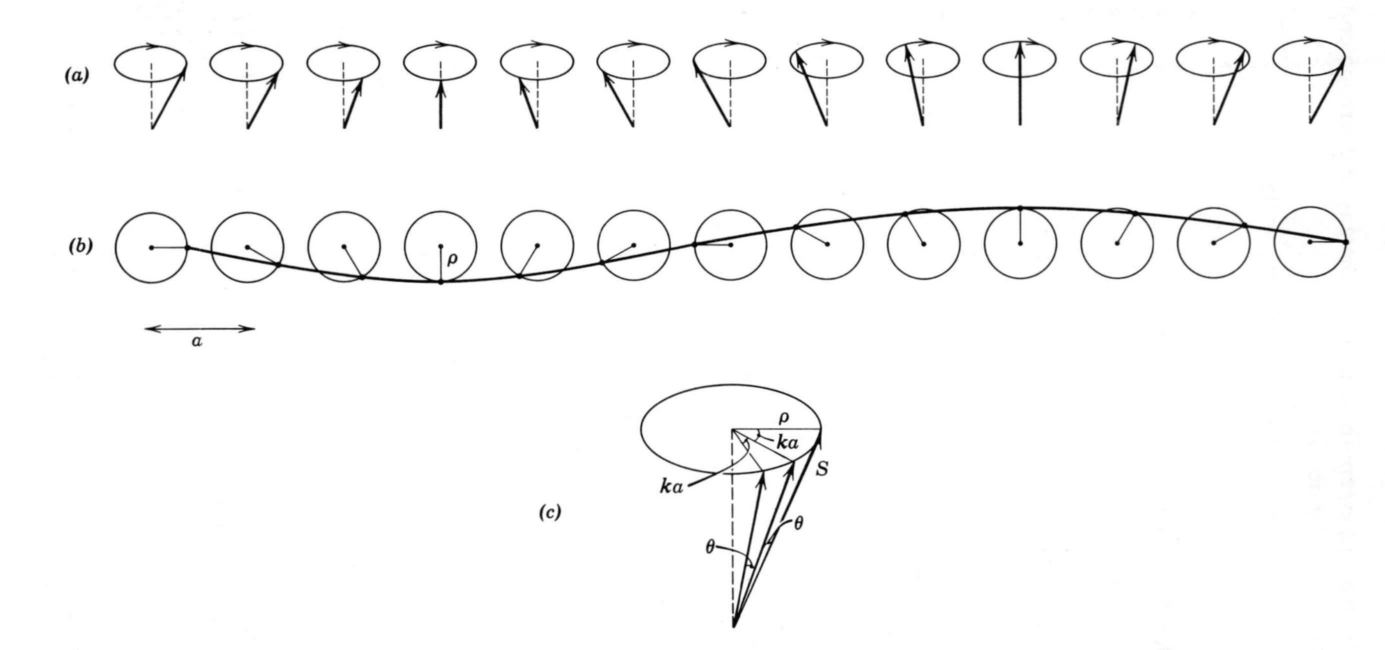

Fig. 6-6.1. To illustrate a spin wave with wave vector k. The precessing spins at a given time are shown in perspective in (a) and from above in (b). The angular relationships between the spins of three successive atoms are shown in (c).

We therefore multiply equation 6-6.10 by $e^{i\mathbf{k}\cdot\mathbf{R}_w}$ and sum over w from 1 to N. This gives

$$\begin{aligned}\mathcal{H}\mathrm{X} &= (E_0 + zJ_e)\mathrm{X} - J_e \sum e^{i\mathbf{k}\cdot\mathbf{r}}\mathrm{X} \\ &= E\mathrm{X}, \end{aligned} \tag{6-6.14}$$

since

$$\begin{aligned}\sum_w e^{i\mathbf{k}\cdot\mathbf{R}_w}\chi_{w+i} &= \sum_{w+i} e^{i\mathbf{k}\cdot\mathbf{R}_{w+i}}\chi_{w+i}e^{i\mathbf{k}\cdot(\mathbf{R}_w-\mathbf{R}_{w+i})} \\ &= \sum_w e^{i\mathbf{k}\cdot\mathbf{R}_w}\chi_w e^{i\mathbf{k}\cdot\mathbf{r}},\end{aligned}$$

where $\mathbf{r}$ is the vector from a given atom to one of its nearest neighbors. The summation in equation 6-6.14 is, of course, over the z nearest neighbors. Hence the energy of a spin wave is

$$E - E_0 = J_e(z - \sum e^{i\mathbf{k}\cdot\mathbf{r}}). \tag{6-6.15}$$

For small values of k this reduces to

$$E - E_0 = \tfrac{1}{2}J_e \sum (\mathbf{k}\cdot\mathbf{r})^2,$$

which for the case of cubic symmetry becomes

$$E - E_0 = \tfrac{1}{6}J_e k^2 \sum r^2,$$

where r is the distance between nearest neighbors. For a simple, body-centered or face-centered cubic lattice this becomes

$$E - E_0 = J_e k^2 a^2, \tag{6-6.16}$$

where a is the lattice spacing. Since a wave may be considered as a harmonic oscillator of the same frequency, we may apply a second quantization to equation 6-6.16. We then obtain the important dispersion relationship

$$\hbar\omega = J_e k^2 a^2. \tag{6-6.17}$$

Spin waves may also be treated semiclassically[45]; then the spins are considered as classical vectors of length $[S(S+1)]^{1/2}$ precessing about the axis of quantization. In the ground state all the N spins have a z-component $M_S = S$ and are precessing in phase. In the first excited state the reversed spin is considered as averaged over all the N spins, that is, the z-component of each spin is $M_S = S - 1/N$. Further, there is a phase difference between the precessing spins, corresponding to the different possible functions χ_w. The situation for a linear chain of atoms is shown in Fig. 6-6.1. If we assume that the length $[S(S+1)]^{1/2}$ can be replaced by S,

[45] G. Heller and H. A. Kramers, *Proc. Roy. Acad. Sci. (Amsterdam)* **37,** 378 (1934); C. Herring and C. Kittel, *Phys. Rev.* **81,** 869 (1951); F. Keffer, H. Kaplan, and Y. Yafet, *Am. J. Phys.* **21,** 250 (1953); J. Van Kranendonk and J. H. Van Vleck, *Revs. Mod. Phys.* **30,** 1 (1958).

the radius of the circle of precession, ρ, is the amplitude of the spin wave. It is left to the reader as problem 6-8 to show that this semiclassical approach leads to the same dispersion relationship as that given by equation 6-6.17.

Since the energy of a wave for a particular vibrational mode of wave vector k is the same as that of a harmonic oscillator of the same frequency, we have

$$E = n_k E_k,$$

where

$$E_k = \hbar\omega_k$$

and

$$n_k = 1, 2, 3, \ldots,$$

where n_k is the vibrational quantum number. The zero-point energy has been neglected for simplicity.[46] The average value of the quantum number $\bar{n}_k$ for a particular temperature is

$$\bar{n}_k = \frac{\sum n_k e^{-n_k E_k/kT}}{\sum e^{-n_k E_k/kT}},$$

where the sum is over all states. This gives

$$\bar{n}_k = \frac{-d/[d(E_k/kT)] \sum e^{-n_k E_k/kT}}{(1 - e^{-E_k/kT})^{-1}}$$

$$= \frac{1}{e^{E_k/kT} - 1}, \qquad (6\text{-}6.18)$$

which is just the well-known Planck distribution formula.

A spin wave may also be considered as a particle in that an energy E_k and a momentum p_k can be associated with it. Since interchange of any two reversed spins represents the same state of the system, spin-wave particles obey Bose-Einstein statistics. Further, since the total number of spin waves is not fixed, the spin-wave gas is completely degenerate. Hence the average number $\bar{n}_k$ of spin waves for a given k, corresponding to an average of $\bar{n}_k$ reversed spins, at a certain temperature T is just given by

$$\bar{n}_k = \frac{1}{e^{E_k/kT} - 1}.$$

Since this result is identical to that of equation 6-6.18, n_k may be identified either as a vibrational quantum number or as the number of quanta.

[46] It may be shown that the zero-point energy arises because of the x- and y-components of the spin. See M. J. Klein and R. S. Smith, *Phys. Rev.* **80,** 1111 (1950).

Spin waves have properties like those possessed by photons and phonons; indeed they are often referred to as *magnons.*

The total number of spin waves $\bar{n}$ is given by the sum of $\bar{n}_k$ over all the k states, that is,

$$\bar{n} = \sum_k \bar{n}_k.$$

Now the number of states between k and $k + dk$ is

$$C(k)\,dk = \frac{4\pi k^2\,dkV}{(2\pi)^3}, \tag{6-6.19}$$

as may be seen by substituting $p = \hbar k$ into equation 5-1.3*a*. Since $E_k = J_e k^2 a^2$ for a spin wave, equation 6-6.19 becomes

$$C(E)\,dE = \frac{V}{4\pi^2}\left(\frac{1}{a^2 J_e}\right)^{3/2} E^{1/2}\,dE. \tag{6-6.20}$$

Hence

$$\bar{n} = \frac{V}{4\pi^2}\left(\frac{1}{a^2 J_e}\right)^{3/2} \int_0^\infty \frac{E^{1/2}\,dE}{e^{E/\mathrm{k}T} - 1}$$

$$= \frac{V}{4\pi^2}\left(\frac{\mathrm{k}T}{a^2 J_e}\right)^{3/2} \int_0^\infty \frac{x^{1/2}\,dx}{e^x - 1},$$

where $x = E/\mathrm{k}T$. The integration can be performed by making a series expansion. The result is

$$\bar{n} = 0.1174V\left(\frac{\mathrm{k}T}{a^2 J_e}\right)^{3/2}. \tag{6-6.21}$$

The change in the spontaneous magnetization because of the excitation of spin waves is then

$$\frac{\Delta M}{M(0)} = \frac{M(0) - M(T)}{M(0)} = \frac{\bar{n}}{N}.$$

For a simple cubic, body-centered cubic, and face-centered cubic the volume V is equal to Na^3, $Na^3/2$, and $Na^3/4$, respectively. Hence

$$\frac{\Delta M}{M(0)} = \frac{0.1174}{f}\left(\frac{\mathrm{k}T}{J_e}\right)^{3/2}, \tag{6-6.22}$$

where $f = 1$, 2, and 4 for the simple, body-centered, and face-centered cubic lattice, respectively. This equation is known as the Bloch $T^{3/2}$ law. It is identical with equation 6-2.17, which describes the experimental data, provided that

$$A_{\text{exp}} = \frac{0.1174}{f}\left(\frac{\mathrm{k}}{J_e}\right)^{3/2}. \tag{6-6.23}$$

According to Marshall,[47] the same result holds for $S = 1$. Experimental determinations[48] of A_{exp} lead to the values $J_e = 205\text{k}$ for iron and $J_e = 230\text{k}$ for nickel.

In a like manner the internal energy per unit volume is given by

$$U = \int_0^\infty \bar{n}_k E_k \, C(E_k) \, dE_k$$
$$= \frac{V}{4\pi^2}\left(\frac{1}{a^2 J_e}\right)^{3/2} \int_0^\infty \frac{E^{3/2} \, dE}{e^{E/\text{k}T} - 1}$$
$$= \frac{V}{4\pi^2}\left(\frac{\text{k}T}{a^2 J_e}\right)^{3/2} \text{k}T \int_0^\infty \frac{x^{3/2} \, dx}{e^x - 1}.$$

The important thing to note is that the specific heat $C_v = (dU/dT)_v$ will depend on temperature according to the relation

$$C_v \propto T^{3/2}. \qquad (6\text{-}6.24)$$

For ferromagnetic metals and alloys the conduction electrons contribute a term to the specific heat that is proportional to T. This latter term is the prominent one. However, some insulators have a spontaneous magnetization; examples are the ferrites to be discussed in Chapter 9. The behavior predicted by equation 6-6.24 has been observed for magnetite and yttrium iron garnet.[49]

The effect of applying large magnetic fields has been considered in detail by Holstein and Primakoff[50]; there is then an additional term, $-g\mu_B \Sigma \, \mathbf{S} \cdot \mathbf{H}$, in the Hamiltonian. These authors find that to obtain even a small increase in the spontaneous magnetization an extremely large field is required. This increase in magnetization is the origin of the cH term of equation 6-2.15.

Spin waves also play an important role in resonance phenomena in strongly coupled dipole systems. This topic is treated in Chapter 10.

7. Band Model Theories of Ferromagnetism

The preceding theories of ferromagnetism have all been based on the Heisenberg model in which it is assumed that the electrons are localized at the atoms. Since the ferromagnetic materials are either metals or

[47] W. Marshall, *Proc. Phys. Soc. (London)* **A-67,** 85 (1954).

[48] M. Fallot, *Ann. phys. (Paris)* **6,** 305 (1936); E. Kondorsky and L. N. Fedotor, *Akad. Nauk. SSSR*, ser. *Fiz.* **16,** 432 (1952).

[49] J. S. Kouvel, *Phys. Rev.* **102,** 1489 (1956); D. T. Edmonds and R. G. Petersen, *Phys. Rev. Letters* **2,** 499 (1959), **4,** 92 (1960).

[50] T. Holstein and H. Primakoff, *Phys. Rev.* **58,** 1098 (1940).

alloys, it is obvious that this assumption is invalid. Theories that consider mobile electrons or holes in unfilled bands have been developed; two in particular are briefly discussed here.

Calculations in which the interactions between the electrons of an electron gas are considered have become known as *collective electron* theories. The earliest theory considered the free electron gas.[51] It was shown that because of correlation effects it was very unlikely that ferromagnetism would result. Subsequent theories consider the interaction between the electrons and the ion cores; that is, they employ Bloch-type wave functions. Since the main interest has concerned the iron group metals or alloys, the calculations have been made for the electrons or holes of the 3*d* band; the 4*s*-band electrons are assumed not to contribute to the ferromagnetism. The first calculations based on the band model were made by Slater;[52] he obtained results for nickel that were in fair agreement with experiment. However, the discussion here is limited to the theory initiated by Stoner[53] and known as *collective electron ferromagnetism*.

Stoner's theory is based on the following three well-defined assumptions:

1. The 3*d* band is parabolic in the neighborhood of the Fermi level; that is, the density of states has the form $C(E)\,dE = KE^{1/2}\,dE$, as given by equation 5-1.5 in which m is replaced by m^*. The kinetic energy of the electrons or holes is related to the wave vector by

$$E(k) = \frac{\hbar^2 k^2}{2m^*}$$

(equation 5-1.7).

2. The exchange interaction between the electrons may be represented by a molecular field. Thus the energy of an electron with spin parallel or antiparallel to the magnetization is (equation 1-6.1)

$$E_1 = \pm N_W M \mu_B,$$

where N_W is the molecular field constant and M is the magnetization at a temperature T. Stoner finds it convenient to introduce the notation

$$\zeta = \frac{M}{N\mu_B} \quad \text{and} \quad \mathrm{k}\theta' = N_W N \mu_B{}^2, \tag{6-7.1}$$

[51] F. Bloch, *Z. Physik* **57**, 545 (1929); L. Brillouin, *J. Phys. radium* **3**, 565 (1932); E. P. Wigner, *Phys. Rev.* **46**, 1002 (1934), *Trans. Faraday Soc.* **34**, 678 (1938).

[52] J. C. Slater, *Phys. Rev.* **49**, 537 (1936), **49**, 931 (1936), **52**, 198 (1937), *Revs. Mod. Phys.* **25**, 199 (1953).

[53] E. C. Stoner, *Proc. Roy. Soc.* (*London*) **A-165**, 372 (1938), **A-169**, 339 (1939), *Phil. Mag.* **25**, 899 (1938).

where N is the number of $3d$ electrons or holes per unit volume. Then

$$E_1 = \pm k\theta'\zeta. \tag{6-7.2}$$

The total exchange energy per unit volume is $\frac{1}{2}Nk\theta'\zeta^2$ (from equation 1-7.6).

3. The electrons or holes obey Fermi-Dirac statistics. The quantum mechanical bases for these assumptions has been examined.[54]

It is important to note that N is not necessarily an integral multiple of the number of atoms per unit volume in the band picture. This comes about because it is believed that the $3d$ band is overlapped in energy by a much wider $4s$ band in the manner shown in Fig. 6-7.1.[55] Since the bands

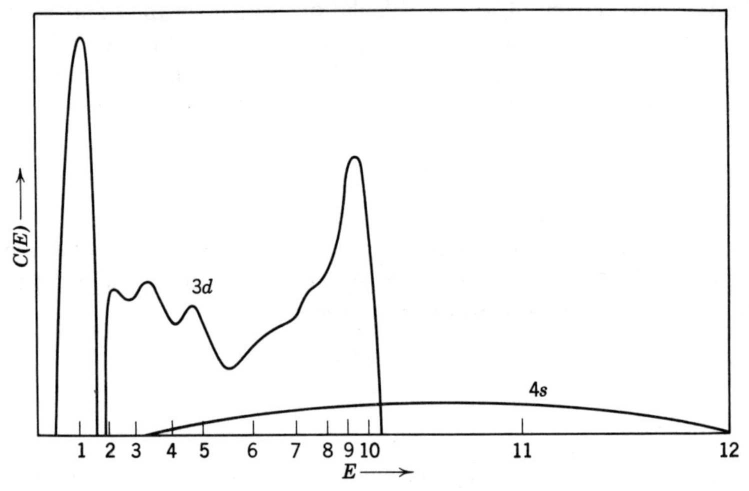

Fig. 6-7.1. The density of states as a function of the energy for the $4s$ and $3d$ bands of copper. The numbers indicate the portion of the band filled by 1 to 12 electrons. [J. C. Slater, *Phys. Rev.* **49,** 537 (1936).]

will be filled to the same energy level, the number of electrons per atom in each band will usually not be integral. For example, in the case of nickel in which the sum of the $3d$ and $4s$ electrons is 10, there are approximately 9.4 electrons in the $3d$ band and 0.6 electrons in the $4s$ band. Hence $N = 0.6$, the number of holes in the $3d$ band. This important feature provides an immediate explanation for the nonintegral value of the number of magnetic carries, n_{eff}, as observed from measurements of the spontaneous magnetization at 0°K (see Table 6-2.1).

[54] E. P. Wohlfarth, *Revs. Mod. Phys.* **25,** 211 (1953); A. B. Lidiard, *Proc. Phys. Soc.* (*London*) **A-64,** 814 (1951), **A-65,** 885 (1952).

[55] J. C. Slater, *J. Appl. Phys.* **8,** 385 (1937); H. M. Krutter, *Phys. Rev.* **48,** 664 (1935).

The number of electrons with magnetic moments either parallel or antiparallel to the magnetization is then just given by the integral of the Fermi-Dirac distribution function of equation 5-1.11 where the energy is

$$E = \frac{\hbar^2 k^2}{2m^*} \pm N_W M \mu_B.$$

If there is an applied field H, an additional energy term of $\pm\mu_B H$ must be included. The magnetization is equal to μ_B multiplied by the difference between the number of electrons with spins parallel and antiparallel, that is,

$$M = \mu_B \int \{F[E(k) - N_W N \mu_B{}^2] - F[E(k) + N_W N \mu_B{}^2]\} \frac{C(E)}{2V}\, dE \tag{6-7.3}$$

(compare with the derivation of equation 5-2.4). At $T = 0°K$ it is easy to show that the magnetization is given by

$$\begin{aligned} M &= \mu_B \frac{1}{(2\pi)^2} \\ &\quad \times \left[\int_0^{E_F + N_W N \mu_B{}^2} \left(\frac{2m^*}{\hbar^2}\right)^{3/2} E^{1/2}\, dE - \int_0^{E_F - N_W N \mu_B{}^2} \left(\frac{2m^*}{\hbar^2}\right)^{3/2} E^{1/2}\, dE\right] \\ &= \frac{\mu_B}{6\pi^2}\left(\frac{2m^*}{\hbar^2}\right)^{3/2} [(E_F + N_W N \mu_B{}^2)^{3/2} - (E_F - N_W N \mu_B{}^2)^{3/2}]. \end{aligned} \tag{6-7.4}$$

It turns out that the magnetization depends on the magnitude of the exchange forces compared to the Fermi energy. For $k\theta'/E_F > 2^{-1/3}(0.794)$, $\zeta(0)[= M(0)/N\mu_B] = 1$; for $k\theta'/E_F < \frac{2}{3}$, $\zeta(0) = 0$. For intermediate values,

$$\frac{2}{3} < \frac{k\theta'}{E_F} < 2^{-1/3}, \tag{6-7.5}$$

the relative magnetization $\zeta(0)$ lies between 0 and 1; that is, symbolically, $0 < \zeta(0) < 1$ (see problem 6-11). Thus the collective electron ferromagnetism theory provides another mechanism for the occurrence of nonintegral values of the effective number of magnetic carriers, n_{eff}.

The calculation of the magnetization as a function of temperature, although straightforward in principle, is not easy and must be done numerically; the reader is referred to Stoner's papers for details. Some of the results are shown in Fig. 6-7.2, plotted as a function of T/T_f. The ordinate on the left is the spontaneous magnetization relative to that at absolute zero, that is, $\zeta(T)/\zeta(0)$. The ordinate at the right is proportional to the reciprocal of the susceptibility, that is,

$$\frac{\zeta(0)}{\zeta(T)}\frac{\mu_B H}{kT_f} = \frac{\mu_B M(0)}{kT_f}\frac{1}{\chi}. \tag{6-7.6}$$

The spontaneous magnetization curves, of course, apply for $T < T_f$, whereas the inverse susceptibility curves apply for $T > T_f$. Different curves arise, depending on $k\theta'/E_F(0)$, where $E_F(0)$ is the Fermi energy at

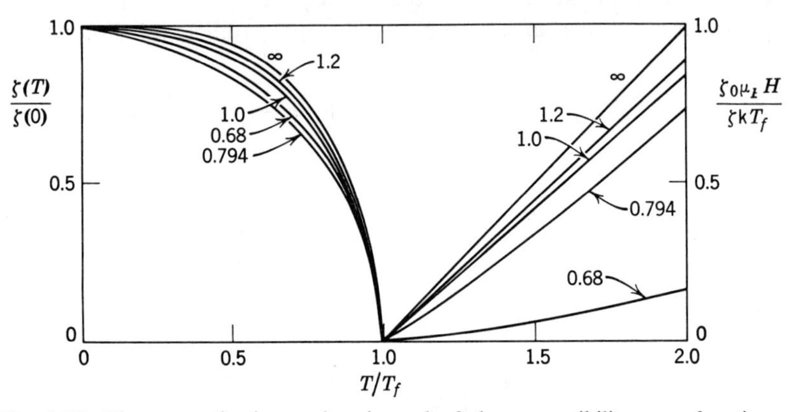

Fig. 6-7.2. The magnetization and reciprocal of the susceptibility as a function of temperature calculated from the theory of collective electron ferromagnetism. The curve $k\theta'/E_F = \infty$ corresponds to the use of Boltzmann statistics. [E. C. Stoner, *Proc. Roy. Soc.* (*London*) **A-165,** 372 (1938).]

absolute zero, as given by equation 5-1.12, with m replaced by m^*. The limiting case $k\theta'/E_F = \infty$ corresponds to the classical molecular field theory, the curve for this case being identical to that for $S = \frac{1}{2}$ in Fig. 6-2.2.

It is striking that the curves for both the magnetization and susceptibility closely resemble those derived assuming the electrons to be localized. There are two features of the susceptibility curves that represent improvements over the molecular field theory. First, there is some curvature, concave upward, in the susceptibility curves, particularly near the Curie point. Second, comparison of the experimental values of the susceptibility of Fe, Co, and Ni with the collective electron theory leads to values of the effective number of carriers, n_{eff}, in good agreement with those deduced from measurements of the magnetization at $T = 0°$K. The values obtained when the susceptibility measurements are compared to the molecular field theory are 3.39, 3.22, and 0.86 for Fe, Co, and Ni, respectively; these values are widely different from those obtained from the magnetizations at $T = 0°$K, namely, 2.22, 1.71, and 0.60 (see Table 6-2.1).

Stoner has also calculated the specific heat as a function of temperature. Other authors have extended the theory of collective electron ferromagnetism to include a number of refinements.[56] Friedel[57] has proposed a model that is intermediate to Heisenberg's localized model and Stoner's band model.

To summarize: one of the main achievements of the theory of collective electron ferromagnetism is the prediction of nonintegral, consistent values for n_{eff}, the effective number of magnetic carriers. The theory is in better agreement with experiment than the simple molecular field treatment described in Section 6-2; it is not necessarily in better agreement than the more refined theories that employ the Heisenberg model, such as those discussed in Sections 6-4, 6-5, and 6-6. Detailed comparison with experimental measurements of magnetic and thermal properties and with neutron diffraction studies show that for Ni the collective electron theory of ferromagnetism is favored, whereas for Fe and Gd the Heisenberg theory is better.

We now turn to the second theory involving mobile electrons to be discussed, namely, that proposed by Vonsovsky[58] and Zener.[59] In this model the 3*d* electrons are assumed to be localized at the atoms. Mobile electrons enter the picture in that the exchange interaction between the 3*d* electrons and the 4*s* conduction electrons is considered (see Fig. 6-7.3).

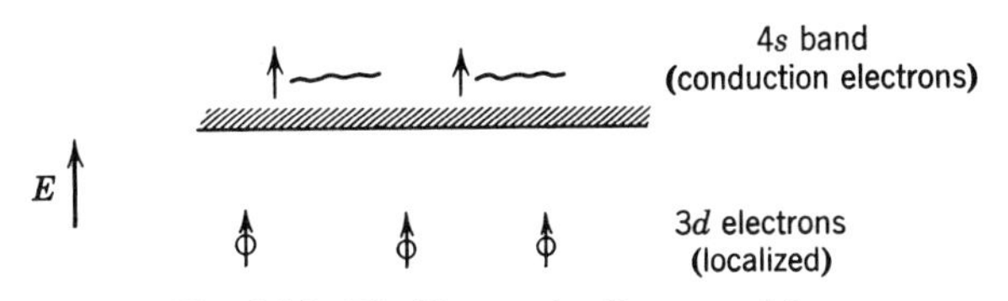

Fig. 6-7.3. The Vonsovsky-Zener model.

It is not unreasonable to consider this interaction, as its existence and importance in an isolated atom is well established from spectroscopic data. In a sense this model represents a situation intermediate between the Heisenberg and collective electron models.

In Vonsovsky's theory Heisenberg exchange interaction between the 3*d* shells of neighboring atoms aligns the spins of the 3*d* shells. The result is

[56] W. Band, *Proc. Camb. Phil. Soc.* **42,** 139, 144 (1946); E. P. Wohlfarth, *Proc. Roy. Soc. (London)* **A-195,** 434 (1949), *Phil. Mag.* **40,** 1095 (1945), **42,** 374 (1951); G. M. Bell, *Phil. Mag.* **43,** 127 (1952).

[57] J. Friedel, G. Leman, and S. Olszewski, *J. Appl. Phys.* **32,** 3255 (1961).

[58] S. V. Vonsovsky, *J. Phys. (USSR)* **10,** 468 (1946); S. V. Vonsovsky and E. A. Turov, *J. Exptl. Theoret. Phys. (USSR)* **24,** 419 (1953).

[59] C. Zener, *Phys. Rev.* **81,** 440 (1951), **83,** 299 (1951); C. Zener and R. R. Heikes, *Revs. Mod. Phys.* **25,** 191 (1953).

equivalent to the production of an internal field that interacts with the conduction electrons. As a consequence, more conduction electrons will have their spins parallel than antiparallel to the spin direction of the ion cores; that is, the conduction electrons become polarized. According to a rough calculation, the polarization corresponds to an effective number of about 0.15 Bohr magnetons per atom. For example, for Fe where $n_{\text{eff}} = 2.22$ the $3d$ electrons contribute 2.07 Bohr magnetons per atom and the $4s$ electrons contribute the remainder.

Zener has given a simple formal theory. For the spin energy per atom, E_{spin}, he writes

$$E_{\text{spin}} = \tfrac{1}{2}\alpha S_d^{\,2} - \kappa S_d S_c + \tfrac{1}{2}\gamma_c S_c^{\,2}, \tag{6-7.7}$$

where S_d and S_c are the average values per atom of the spin components along the magnetization direction of the $3d$ and $4s$ electrons, respectively. The first term represents the exchange energy of the *d-d* shell interaction, the second term, the exchange energy of the *s-d* interaction, and the last term, the increase in kinetic and correlation energy of the conduction electrons because of their polarization. When this spin energy is minimized with respect to S_c, we get

$$S_c = \frac{\kappa}{\gamma_c} S_d,$$

which leads to a minimum energy

$$E_{\text{spin}}(\text{min}) = \frac{1}{2}\left(\alpha - \frac{\kappa^2}{\gamma_c}\right) S_d^{2}. \tag{6-7.8}$$

Comparison with the Heisenberg exchange energy (equation 6-3.26) shows

$$J_e = \frac{1}{4}\left(\alpha - \frac{\kappa^2}{\gamma_c}\right). \tag{6-7.9}$$

It is not necessary that there be any interaction between adjacent $3d$ shells in order to have ferromagnetism. The exchange interaction between a conduction electron and the incomplete $3d$ shell of one atom will favor the alignment of the spin of the $4s$ electron parallel to that of the $3d$ shell. The traveling electron will interact with the $3d$ shells of all the other atoms and will tend to align the $3d$ shell spins parallel to its own spin (see problem 6-12). By assuming $\alpha = 0$ and estimating the value of κ from spectroscopic data and the value of γ_c from the theory of the free electron gas, Zener obtains a value of the Curie temperature which is of the correct order of magnitude.

The foregoing models by no means exhaust all the possibilities. Van Vleck[60] has considered a model in which $3d$ electrons are permitted to

[60] J. H. Van Vleck, *Revs. Mod. Phys.* **25**, 220 (1953).

move through the crystal subject to the constraints that states of high ionization are forbidden. As an example, consider the case of Ni which is believed to have on the average 9.4 electrons per atom in the $3d$ state. In Van Vleck's model 40% of the atoms have a configuration of $3d^{10}$, the other 60% have a configuration $3d^9$. The sites of the $3d^9$ and $3d^{10}$ configurations are constantly being shuffled, but the percentage of each remains constant. On the other hand, in the collective electron model there is a mixture of $3d^{10}$, $3d^9$, $3d^8$, $3d^7$, . . . configurations.

8. Ferromagnetic Metals and Alloys

The ferromagnetic metals of the $3d$ transition group are body-centered cubic Fe, face-centered cubic Ni, and close-packed hexagonal or face-centered cubic Co.[61] The magnetic and thermal properties of these metals have been described and compared with various theories earlier in this chapter. It was not possible to decide on the basis of the experimental data which theory was the correct one. Even the nonintegral values of n_{eff} are consistent with the Heisenberg model if it is assumed that different atoms have different electronic configurations. For example, if for nickel 60% of the atoms have the configuration $3d^{10}$ and 40%, $3d^9$, there will be an average of 0.6 holes per atom. Probably the main conclusion to be drawn is that all the theories are wrong, at least to some extent. Besides gadolinium, some of the other rare earth metals are ferromagnetic provided the temperature is low enough (see Section 8-8).

Extensive investigations of the magnetic properties of ferromagnetic alloys have been made. The motivation has been both scientific and commercial. This section considers only those magnetic properties of the type treated earlier in this chapter; other properties, many of commercial importance, are discussed in the next chapter. Further, the present discussion will be limited to a few examples that will serve to illustrate how some of the data may be understood at least superficially.

We consider first some alloys that are solid solutions of nonferromagnetic metals in ferromagnetic metals. For nickel alloys there is a decrease in the number of Bohr magnetons per solute atom approximately equal to the number of valence electrons of this atom. This behavior can be understood if it is assumed that the valence electrons go into the $3d$ band to fill up the 0.6 holes per atom of nickel. For example, suppose copper, which has one more electron than nickel, is added to nickel. The magnetization of the alloy is observed to decrease with the concentration of Cu, and to be zero when 60% of the alloy is Cu. The results for Cu

[61] At room temperature both hexagonal and cubic structures are possible for cobalt.

and other elements[62] are shown in Fig. 6-8.1. The proposed model fits the data very well for at least small concentrations of the solute atoms.

Quite another result is observed for similar alloys of Fe. For all alloys except those with transition elements with 8, 9, or 10*d* electrons there is a decrease in the spontaneous magnetization of approximately 2.2 Bohr

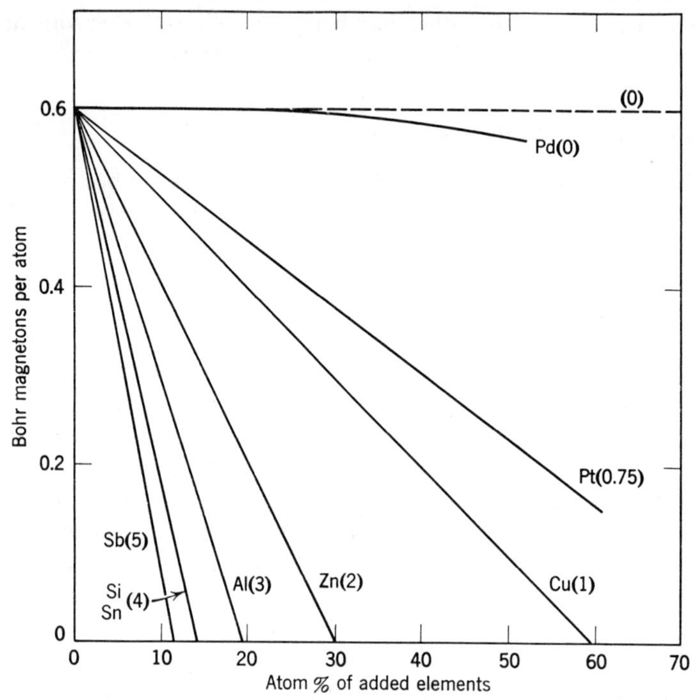

Fig. 6-8.1. Spontaneous magnetization of some nickel alloys. The number of valence electrons of the added element is shown in parentheses. Some of the curves represent extrapolation of the data. (After R. M. Bozorth.)

magnetons per solute atom. Since this is the same as the value of n_{eff} for Fe, the situation is as if the Fe atoms were replaced by atoms that do not contribute to the magnetization. One possibility is that the solute atoms do not interact with the Fe, but remain localized. Another possibility, allowed for by the collective electron theory, is that pure Fe is not completely saturated at $T = 0°\text{K}$, hence that there are more than 2.2 holes per atom in the 3*d* band.

Next, we consider alloys of the ferromagnetic metals with one another

[62] C. Sadron, *Ann. phys.* (*Paris*) **17,** 371 (1932); V. Marian, *Ann. phys.* (*Paris*) **7,** 459 (1937); G. T. Rado and A. R. Kaufmann, *Phys. Rev.* **60,** 336 (1941).

and with neighboring transition series metals.[63] For some of these alloys the magnetization is given by the average of the number of Bohr magnetons contributed by each constituent. For example, in a Ni-Co alloy with an atomic fraction of f nickel atoms we find $n_{\text{eff}} = 0.6f + 1.7(1 - f)$, since there are 0.6 and 1.7 $3d$ holes per Ni and Co atom, respectively. Alloys that show this behavior generally have $n_{\text{eff}} < 2.5$, as indicated in Fig. 6-8.2. Alloys with $n_{\text{eff}} > 2.5$ generally exhibit a decreased magnetization, possibly because neither $3d$ subband[64] is completely filled. Perhaps the

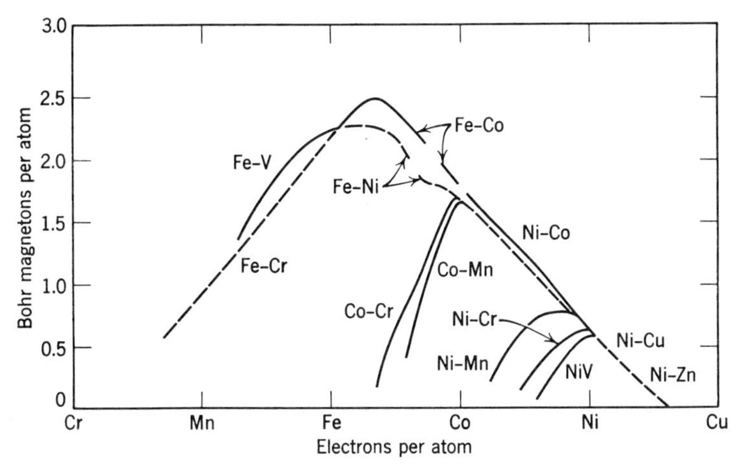

Fig. 6-8.2. The spontaneous magnetization as a function of the electron concentration. (The Slater-Pauling curve, after R. M. Bozorth.) [J. C. Slater, *J. Appl. Phys.* **8,** 385 (1937); L. Pauling, *Phys. Rev.* **54,** 899 (1938).]

most puzzling result is the increase in spontaneous magnetization when Ni or Co is added to Fe. The explanation is probably to be sought by considering the exchange interaction between unlike as well as like nearest neighbors and the band structure of alloys. An Fe-Co alloy with an approximate composition of 30% Co has a moment of almost 2.5 Bohr magnetons per atom; this is the largest moment observed in any material. Ferromagnetic alloys that do not contain a ferromagnetic element include (*a*) binary alloys of Mn with N, P, As, Sb, and Bi, (*b*) binary alloys of Cr with S, Te, and Pt, and (*c*) ternary alloys such as the Heusler alloys (for example, $MnAlCu_2$).

[63] P. Weiss, R. Forrer, and F. Birch, *Compt. rend.* (*Paris*) **189,** 789 (1929); M. Fallot, *Ann. phys.* (*Paris*) **6,** 305 (1936), **10,** 291 (1938); T. Farcas, *Ann. phys.* (*Paris*) **8,** 146 (1937); V. Marian, *Ann. phys.* (*Paris*) **7,** 459 (1937); C. Guillaud, *Compt. rend.* (*Paris*) **219,** 614 (1944).

[64] The $3d$ band may be considered as two subbands, one with parallel and the other with antiparallel spin orientations.

The addition of a nonferromagnetic metal to a ferromagnetic metal generally decreases the Curie temperature.[65] The paramagnetic susceptibility of ferromagnetic alloys above their Curie temperature usually cannot be fitted to the Curie-Weiss equation. The difference between the ferromagnetic and paramagnetic Curie temperatures frequently exceeds 100°K; in some cases there is no linear region in the $1/\chi$ versus T curves at all.[66] A reasonable fit with the data can often be obtained with a relationship between $1/\chi$ and T that involves three parameters. The theoretical significance of these parameters is not understood at present.

Alloys at or near the composition AB or A_3B usually exhibit an order-disorder transformation at some temperature. Further, the degree of order often depends on the heat treatment given the alloy. For example, an alloy quenched from a relatively high-to-low temperature may have a lower degree of order than an annealed specimen. The degree of order affects the magnetic properties.[67]

A transformation, in which there is a change of crystal structure, commonly occurs in both metals and alloys. The two phases generally have quite different magnetic properties; for example, one may be ferromagnetic, the other antiferromagnetic or even paramagnetic. Spontaneous magnetization or susceptibility versus temperature curves then have discontinuities.[68]

In the foregoing discussion it has been assumed that the alloys are single-phase. The presence of more than one phase complicates the magnetic properties. Magnetic measurements have been found to be fruitful for the study of polyphase systems.[69]

9. Crystalline Anisotropy

The Heisenberg exchange energy depends on the scalar product $\mathbf{S}_i \cdot \mathbf{S}_j$, which is an invariant with respect to the choice of coordinate system. Thus, until now, the magnetization of a ferromagnetic specimen has been considered isotropic. Experimentally, however, it is found that the magnetization tends to lie along certain crystallographic axes; this effect is known as crystalline anisotropy. It is additional to the directional

[65] For a summary see R. M. Bozorth, *Ferromagnetism*, D. Van Nostrand, Princeton, N.J. (1951), Chs. 3–8 and pp. 720–723.

[66] M. A. Wheeler, *Phys. Rev.* **56**, 1137 (1939); A. R. Kaufmann and C. Starr, *Phys. Rev.* **58**, 1098 (1940); C. Manders, *Ann. phys. (Paris)* **5**, 167 (1936); A. R. Kaufmann, S. T. Pan, and J. R. Clark, *Revs. Mod. Phys.* **17**, 87 (1945).

[67] W. Sucksmith, *Proc. Roy. Soc. (London)* **A-171**, 525 (1939).

[68] M. Fallot, *Ann. phys. (Paris)* **10**, 291 (1938).

[69] W. Sucksmith and K. Hozelitz, *Proc. Roy. Soc. (London)* **A-181**, 303 (1943); K. M. Guggenheimer, H. Heitler, and K. Hoselitz, *J. Iron Steel Inst.* **158**, 192 (1948).

effects that occur when the specimen's shape lacks spherical or cubic symmetry; these enter because of the demagnetization energy, a topic that has already been treated in Chapter 1.

The existence of crystalline anisotropy may be demonstrated by the magnetization curves of single-crystal specimens. By magnetization curve we mean the component of magnetization in the direction of the applied field M_H, plotted as a function of the applied field. Magnetization curves for single crystals of iron, nickel, and cobalt for various orientations of the applied field with respect to the crystal axes for room temperature are shown[70] in Fig. 6-9.1. It is clear that much smaller fields are required to magnetize the crystals to saturation along certain directions than along others. The crystallographic axes along which the magnetization tends to lie are called easy directions; the axes along which it is most difficult to produce saturation are called hard directions. It will turn out that the magnetization curves depend on the domain configurations, so that in a sense it would be appropriate to postpone the present discussion until the next chapter. However, crystalline anisotropy itself is a property of a single domain; hence it is pertinent to discuss it here.

Crystalline anisotropy energy, sometimes called magnetocrystalline energy, is defined as the work required to make the magnetization lie along a certain direction compared to an easy direction. If the work is performed at constant temperature, the crystalline anisotropy energy is actually a free energy. To see this, we first write the first law of thermodynamics (equation 3-2.3) in the form

$$dU = dQ - dW.$$

Then, from equation 3-2.13,

$$dF = -dW - S\,dT$$

and it follows that at constant temperature

$$F_{\text{final}} - F_{\text{initial}} = \int -dW$$

or, if F_{initial} is taken as the zero of energy,

$$F = \int -dW. \tag{6-9.1}$$

Here $+dW$ is the work done *by* the system; the sign is reversed for work done on the system.

[70] K. Honda and S. Kaya, *Sci. Rept. Tôhoku Univ.* **15,** 721 (1926). See also R. G. Piety, *Phys. Rev.* **50,** 1173 (1936) and K. Honda, H. Masumoto, and Y. Shirakawa, *Sci. Rept. Tôhoku Univ.* **24,** 391 (1935).

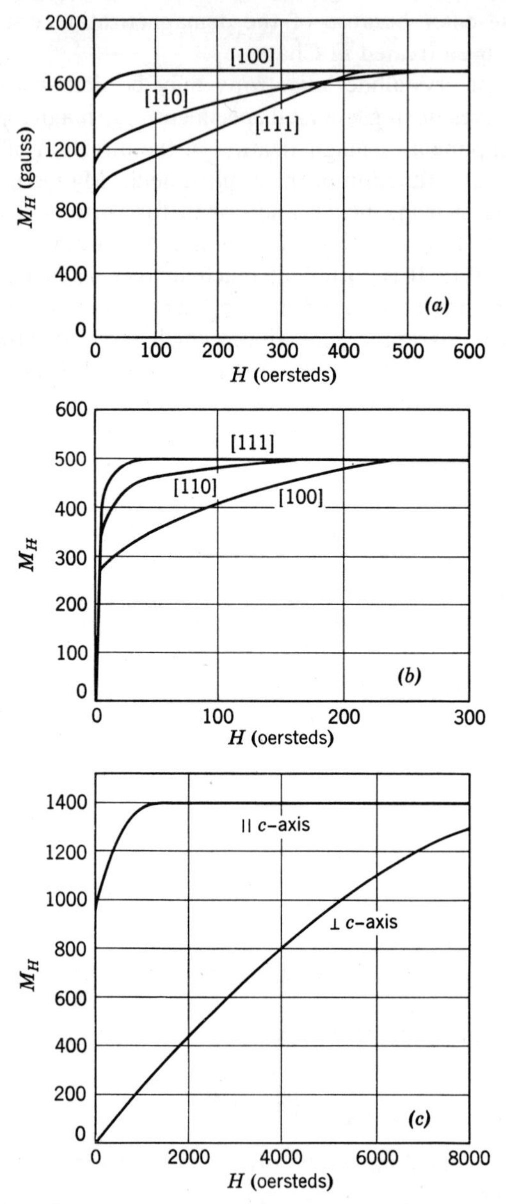

Fig. 6-9.1. Magnetization curves of single crystals of (*a*) iron, (*b*) nickel, and (*c*) cobalt. The direction of the applied field with respect to the crystallographic axes is indicated for each curve. (After Honda and Kaya.)

It is usual to express the anisotropy energy in a power series of trigonometric functions of the angles the magnetization vector makes with the principal axes of the crystal.[71] For example, consider the case of iron. As illustrated in Fig. 6-9.1, the easy directions are the cube edges [100], whereas the hard directions are the body diagonals. Let α_1, α_2, and α_3 be the direction cosines between the direction of magnetization and the cube edges. Because of the cubic symmetry, many terms will not appear in the anisotropy energy expression. This expression must be independent of a change in sign of any of the α's and will therefore have neither odd power terms nor cross terms such as $\alpha_1\alpha_2$. Further, the expression must be independent of an interchange of any two α's. Hence, in the expression $c_1\alpha_1^2 + c_2\alpha_2^2 + c_3\alpha_3^2$ it is required that $c_1 = c_2 = c_3$. Then, since $\alpha_1^2 + \alpha_2^2 + \alpha_3^2 = 1$, no anisotropy can result from such a term. A term of the fourth degree is the lowest that will occur. It is not necessary to include both the term $\alpha_1^2\alpha_2^2 + \alpha_1^2\alpha_3^2 + \alpha_3^2\alpha_2^2$ and the term $\alpha_1^4 + \alpha_2^4 + \alpha_3^4$ because of the relation

$$\alpha_1^2\alpha_2^2 + \alpha_2^2\alpha_3^2 + \alpha_3^2\alpha_1^2 = \tfrac{1}{2} - \tfrac{1}{2}(\alpha_1^4 + \alpha_2^4 + \alpha_3^4).$$

The next term is one of sixth degree: $\alpha_1^2\alpha_2^2\alpha_3^2$. Although in principle higher order terms will occur, it is often not necessary in practice to include them in order to describe the experimental results. The expression for the anisotropy free energy F_K for iron is then

$$F_K = K_1(\alpha_1^2\alpha_2^2 + \alpha_2^2\alpha_3^2 + \alpha_3^2\alpha_1^2) + K_2\alpha_1^2\alpha_2^2\alpha_3^2. \qquad (6\text{-}9.2)$$

The anisotropy energy is usually a sensitive function of temperature; this is described by permitting the coefficients to vary with temperature.

Nickel is also cubic, but the body diagonals are the easy directions, whereas the cube edges are the hard directions. Thus equation 6-9.2 may be used to describe the anisotropy energy of nickel, except that now K_1 is negative.

Cobalt is hexagonal, and the easy direction is the hexagonal c-axis at room temperature. Any direction in the basal plane (the plane perpendicular to the c-axis) is a hard direction, since the magnetization is almost isotropic for a given angle θ between the direction of magnetization and the symmetry axis. The anisotropy energy may then be expressed by

$$F_K = K_1' \sin^2\theta + K_2' \sin^4\theta. \qquad (6\text{-}9.3)$$

It has been found unnecessary to include higher order terms such as $\sin^6\theta$.

The magnetization curves of Fig. 6-9.1 may be deduced on the basis of some simple assumptions. Consider the case of iron and suppose first

[71] N. S. Akulov, *Z. Physik* **57,** 249 (1929), **69,** 78 (1931).

that H is applied along the [100]-direction. For $H = 0$ presumably the magnetization vector of a domain[72] will lie along one of the six easy directions ([100], [010], [001], [$\bar{1}$00], [0$\bar{1}$0], and [00$\bar{1}$]), and, further, the domains will be equally distributed over these directions so that the net moment is zero. The application of even an infinitesimal field should cause the specimen to become saturated along the [100]-direction. In fact, a finite field is required to obtain a particular value of M_H, although for a field of about 50 oersteds saturation is almost reached. This behavior is said to be due to "internal stresses"; possible origins of these stresses include dislocations, inclusions, and inhomogeneities (see Chapter 7 for further details).

Consider now that H is applied along the [110]-direction. By symmetry, the magnetization vectors of the domains will lie in the (001) plane. When H is very small, the domain magnetizations will lie nearly along the [100]- and [010]-directions, and there will be an equal distribution of the two types of domain, as shown in Fig. 6-9.2a. When the field H is

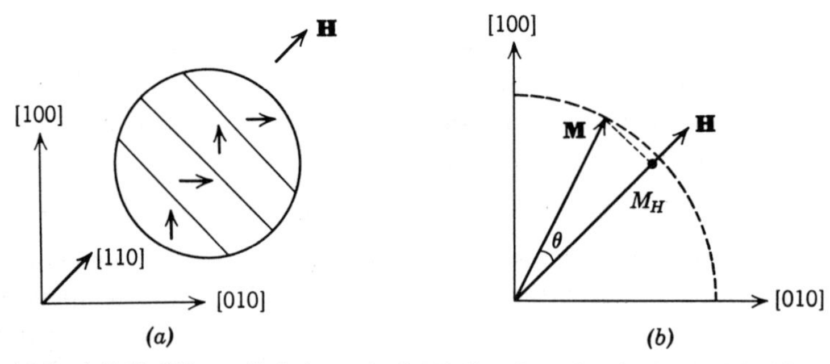

Fig. 6-9.2. A field H is applied along the [110]-direction of a single-crystal of iron. In (a) the proposed domain distribution between the [100]- and [010]-directions for a small applied field H is shown. In (b) the direction of the magnetization vector of a domain is shown for some larger field H.

increased, the magnetization vectors will tend to rotate toward the field direction ([110]). Clearly, the component of magnetization perpendicular to the field direction of the single crystal will be zero for all fields; this is observed experimentally. The angle θ the magnetization $\mathbf{M}$ of a domain will make with $\mathbf{H}$ will be that which makes the free energy a minimum. The two energy terms that depend on this angle are, one, the anisotropy energy F_K, and, two, the magnetization energy per unit volume that arises from the interaction with the applied field, $F_H = -\mathbf{M} \cdot \mathbf{H}$ (equations

[72] The term "direction of a domain" is sometimes used; it means the direction of the domain's magnetization.

1-6.2 and 6-9.1). The other energy terms need not be considered. In terms of θ, equation 6-9.1 becomes

$$F_K = K_1(\tfrac{1}{2} - \cos^2\theta)^2, \tag{6-9.4}$$

since

$$\alpha_1 = \cos\left(\frac{\pi}{4} - \theta\right) = \frac{1}{\sqrt{2}}(\cos\theta + \sin\theta),$$

$$\alpha_2 = \cos\left(\frac{\pi}{4} + \theta\right) = \frac{1}{\sqrt{2}}(\cos\theta - \sin\theta),$$

and

$$\alpha_3 = 0,$$

as may be seen from Fig. 6-9.2*b*. Hence the condition for a minimum is

$$\frac{\partial[K_1(\frac{1}{2} - \cos^2\theta)^2 - MH\cos\theta]}{\partial\theta} = 0,$$

which leads to

$$\frac{HM}{4K_1} = \frac{M_H}{M}\left[\frac{1}{2} - \left(\frac{M_H}{M}\right)^2\right], \tag{6-9.5}$$

since $M_H = M\cos\theta$. This equation fits the experimental data well, provided a suitable value for K_1 is chosen. In a similar manner the magnetization curve when H is applied along the [111]-direction may be deduced, and again a good fit with the data is obtained. When the field is applied along directions other than symmetry directions ([100], [110], and [111]), the component of magnetization perpendicular to the applied field is no longer zero because of a nonsymmetric distribution of the domain directions. For fields below a certain value the demagnetizing field must be taken into account; for details we refer the reader to the literature.[73] It suffices to say that it is believed that the shape of magnetization curves of single crystals is well understood macroscopically.

There are four main methods for the experimental determination of the anisotropy coefficients. One simply involves finding the best fit of the theoretical with the experimental magnetization curves. For example, fitting equation 6-9.4 to the [110]-magnetization curve of iron provides a value of K_1. In a second method the work required to saturate the specimen in different directions is compared. The work required to change the magnetization M_H from zero to saturation is $W = \int_0^M H\,dM_H$, which is just the area between the magnetization curve and the M_H-axis.

[73] L. Néel, *J. Phys. radium* **5,** 241 (1944); H. Lawton and K. H. Stewart, *Proc. Roy. Soc.* (*London*) **A-193,** 72 (1948); H. Lawton, *Proc. Camb. Phil. Soc.* **45,** 145 (1949).

In an attempt to allow for internal stresses, etc., differences between the work W required for two different directions is taken. This value is equal to the difference between the anisotropy energy for these same two directions. For the symmetry directions this gives

$$W_{[110]} - W_{[100]} = \frac{K_1}{4}$$

and

$$W_{[111]} - W_{[100]} = \frac{K_1}{3} + \frac{K_2}{27}$$

or

$$\begin{aligned} K_1 &= 4(W_{[110]} - W_{[100]}) \\ K_2 &= 9\{3(W_{[111]} - W_{[100]}) - 4(W_{[110]} - W_{[100]})\}, \end{aligned} \qquad (6\text{-}9.6)$$

where $W_{[100]}$, $W_{[110]}$, and $W_{[111]}$ denote the work required to saturate the magnetization along the [100]-, [110]-, and [111]-directions, respectively.

In a third method the torque produced by the action of the applied field is measured. This torque is given by

$$\mathbf{L} = \mathbf{M} \times \mathbf{H},$$

that is,

$$L = HM_{\perp}, \qquad (6\text{-}9.7)$$

where $M_{\perp}$ is the component of magnetization perpendicular to the field direction. This torque is equal to the torque required to rotate the magnetization from an easy direction to its equilibrium direction. In order to simplify the analysis, large fields are applied so that it may be assumed that the magnetization lies along the field direction. The torque on the specimen is then given by

$$-\frac{\partial(F_K + F_H)}{\partial\theta} = -\frac{\partial F_K}{\partial\theta},$$

since

$$\frac{\partial F_H}{\partial\theta} \doteqdot 0.$$

Here θ is the angle between the field and the crystal axes. The anisotropy coefficients are determined by fitting a theoretical torque curve to the measured values; a typical torque curve is shown[74] in Fig. 6-9.3 (see problem 6-14). The fourth method employs the technique of ferromagnetic resonance and is discussed in Chapter 10.

[74] H. J. Williams, *Phys. Rev.* **52**, 747 (1937); see also R. Gans, *Ann. Phys. (Leipzig)* **15**, 28 (1932); R. M. Bozorth and H. J. Williams, *Phys. Rev.* **59**, 827 (1941).

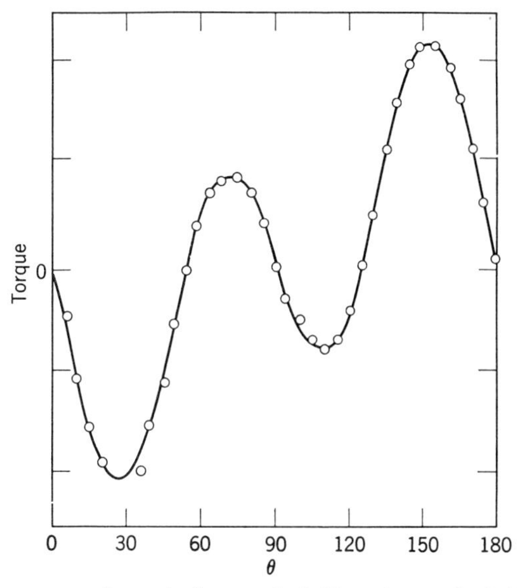

Fig. 6-9.3. Torque curve for a single crystal of silicon iron. The field is applied in the (110) plane and θ is measured with respect to a [100]-direction. The circles represent the experimental data, whereas the curve is theoretical. (After H. J. Williams.)

Representative values of the anisotropy constants for the ferromagnetic metals at room temperature are given in Table 6-9.1. There is a considerable variation between the values obtained by different observers. Little weight can be given to the values of K_2, and even for K_1 the limits of error are rather large ($\sim$10%). The anisotropy constants of a number of alloys have also been measured. The values of K_1 for the nickel-iron (Permalloy) and cobalt-iron alloy systems are shown in Fig. 6-9.4*a*, *b*. One interesting feature of these results is the change in sign of K_1 for a nickel-iron alloy at a composition of about 70% nickel and for a cobalt-iron alloy at about 60% iron. A summary of the data on alloys may be found in Bozorth's book.

Table 6-9.1. Anisotropy Constants at $T = 293°$K

Metal	K_1 or K_1'		K_2 or K_2'	
	ergs/cm^3	cm^{-1}	ergs/cm^3	cm^{-1}
Fe	4.6×10^5	0.027	1.5×10^5	0.01
Ni	-5×10^4	0.003		
Co	4.1×10^6	0.23	1.0×10^6	0.06

Microscopic theories of crystalline anisotropy have been more of a qualitative than a quantitative success. These theories are rather involved, and only a brief outline is given here. It is reasonable to expect that directional effects develop because of the spin-orbit coupling. The orbital wave functions will reflect the symmetry of the lattice because of the interactions between the lattice atoms that arise from the crystalline electric fields and the overlap of the wave functions. The spins are then made aware of this anisotropy via the spin-orbit coupling.

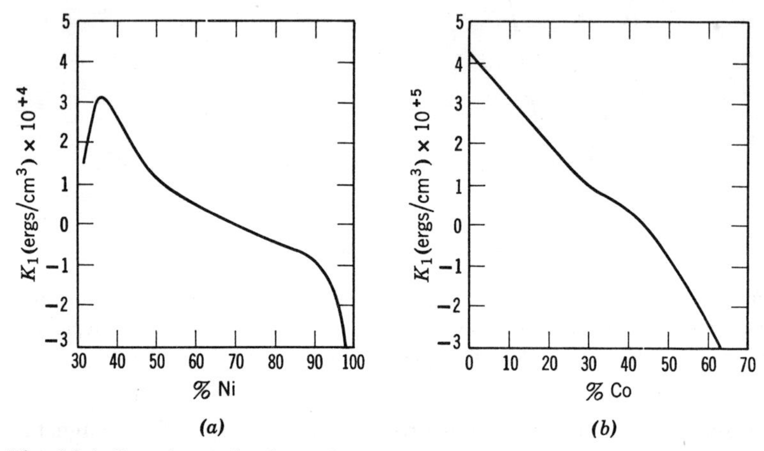

Fig. 6-9.4. Experimental values of the anisotropy constant K_1 for (*a*) Ni-Fe and (*b*) Co-Fe alloys at room temperature.

One way of viewing the anisotropy, due mainly to Van Vleck,[75] is to consider that the exchange coupling between the metal's ions has an anisotropic contribution. In the usual calculation of the exchange energy the spin-orbit coupling is neglected. When the spin-orbit interaction is considered as a perturbation, the second-order term

$$J_e\left(\frac{\lambda}{\Delta E}\right)^2\left[\mathbf{S}_i\cdot\mathbf{S}_j-\frac{3(\mathbf{S}_i\cdot\mathbf{r}_{ij})(\mathbf{S}_j\cdot\mathbf{r}_{ij})}{r_{ij}^{\ 2}}\right] \tag{6-9.8}$$

is obtained, where λ is the spin-orbit coupling constant and ΔE is the energy difference between the ground and excited energy states of the orbital levels. Except for the factor $J_e(\lambda/\Delta E)^2$, this expression is the same as that for the classical dipole-dipole interaction energy. As a result, this type of coupling is sometimes called pseudo-dipolar; it is also sometimes

[75] J. H. Van Vleck, *Phys. Rev.* **52,** 1178 (1937); F. Bloch and G. Gentile, *Z. Physik* **70,** 395 (1931); H. Brooks, *Phys. Rev.* **58,** 909 (1940); J. R. Tessman, *Phys. Rev.* **96,** 1192 (1954); W. J. Carr, *Phys. Rev.* **108,** 1158 (1950).

called anisotropic exchange. When the perturbation is calculated out to the fourth order, the quadrupolelike term

$$J_e(\lambda/\Delta E)^4(\mathbf{S}_i \cdot \mathbf{r}_{ij})^2(\mathbf{S}_j \cdot \mathbf{r}_{ij})^2 \tag{6-9.9}$$

is obtained. This type of coupling, however, requires that the spin be 1 or greater.

Now, classically, when the pseudo-dipole term is applied to a crystal with cubic symmetry, it averages to zero. This follows because the coupling is proportional to $(1 - 3\cos^2\theta_{ij})$ and the average of $\cos^2\theta_{ij}$ is $\frac{1}{3}$. Quantum mechanically the spin vectors are not parallel. The result is that in second order the pseudo-dipolar interaction does affect the energy levels; the constant of proportionality is then $J_e\left(\dfrac{\lambda}{\Delta E}\right)^4$.

For nickel the spin of an atom is unlikely to be greater than $\frac{1}{2}$; this appears to rule out the quadrupole expression. For iron this term may be the origin of the anisotropic coupling. Whether the anisotropy originates from the pseudo-dipole or quadrupole term, it is predicted that

$$K_1 \approx J_e\left(\frac{\lambda}{\Delta E}\right)^4.$$

From equation 3-9.3, this may be written

$$K_1 \approx J_e(g - 2)^4. \tag{6-9.10}$$

The term $(g - 2)$ is a measure of the orbital angular momentum which is present because of incomplete quenching. This factor enters in the fourth power because it arises from a fourth-order effect. Now, since $(g - 2) \approx 0.1$ for nickel and iron and $J_e \approx kT_f \approx 10^{-13}$ erg/atom or 10^9 ergs/cm^3, it is predicted that K_1 is of the order of 10^6 ergs/cm^3, as is indeed observed.

Another way of viewing the anisotropy is similar to that developed for paramagnetic salts, as discussed in Sections 2-10 and 3-9. In this picture[76] the magnetic electrons are subjected to a crystalline electric field of a certain symmetry, usually cubic. If it is assumed that the orbital momentum is zero, or is quenched, the spin-Hamiltonian is that given by equation 3-10.1, namely,

$$\mathcal{H} = g\mu_B\mathbf{H}\cdot\mathbf{S} + \frac{a}{6}\left(S_x^4 + S_y^4 + S_z^4 - \frac{707}{16}\right) + D\left(S_z^2 - \frac{35}{12}\right), \tag{6-9.11}$$

where the field **H** is now interpreted as the molecular field. In order that this Hamiltonian may give anisotropy, it is necessary that $S > \frac{5}{2}$. In this

[76] K. Yosida and M. Tachiki, *Prog. Theor. Phys.* (*Kyoto*) **17**, 331 (1957); W. P. Wolf, *Phys. Rev.* **108,** 1152 (1957).

model, then, the anisotropy arises from the action of the crystalline field on a single atom as well as on the spin-orbit coupling. This model may be expected to be particularly appropriate for the nonmetals with exchange coupling, such as some of the ferrimagnetic and antiferromagnetic materials. In these cases the magnetic ions usually have $S > 2$ and are surrounded by diamagnetic neighbors so that the magnetic electrons certainly can be assumed to be localized. We will return to this subject in Chapters 8 and 9.

It has been stated that the anisotropy constants are sensitive functions of temperature; this is illustrated by Fig. 6-9.5. It may be shown that theory

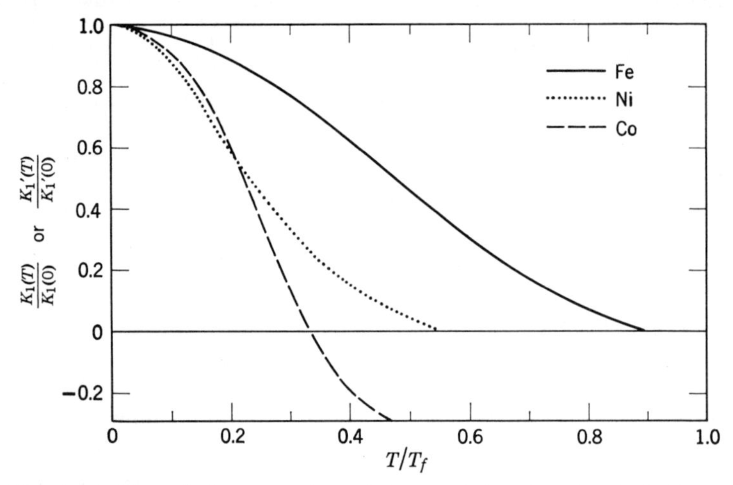

Fig. 6-9.5. The anisotropy constants K_1 (for nickel and iron) and K_1' (for cobalt) as a function of temperature (in reduced units). [Data from R. M. Bozorth, *J. Appl. Phys.* **8,** 575 (1937); H. J. Williams and R. M. Bozorth, *Phys. Rev.* **56,** 837 (1939); K. Honda and H. Masumoto, *Sci. Rept. Tôhoku Univ.* **20,** 323 (1931).]

predicts[77] that the anisotropy constant K_1 is related to the spontaneous magnetization via a tenth power law, namely

$$\frac{K_1(T)}{K_1(0)} = \left[\frac{M(T)}{M(0)}\right]^{10} \tag{6-9.12}$$

where $K_1(0)$ and $M(0)$ are the anisotropy constant and magnetization at absolute zero, respectively. This relationship can be derived, starting with

[77] N. Akulov, *Z. Physik* **100,** 197 (1936); C. Zener, *Phys. Rev.* **96,** 1335 (1954); F. Keffer, *Phys. Rev.* **100,** 1692 (1955); R. Bruner, *Phys. Rev.* **107,** 1539 (1957); W. J. Carr, *J. Appl. Phys.* **29,** 436 (1958); J. H. Van Vleck, *J. Phys. radium* **20,** 124 (1959).

any one of equations 6-9.8, 6-9.9, and 6-9.11, or it may be derived on completely classical grounds. This result is in good agreement with early experimental data for iron. However, more recent experimental results show a third or fourth power dependence below 200°K and a ninth power above.[78] These deviations from the tenth power law may be caused by thermal expansion[79] and magnetoelastic coupling.[80] Nickel is observed to obey a fiftieth power law. The theory predicts that K_2 will depend on the twenty-first power of the magnetization. Because of the lack of accurate experimental results, it has not been possible to check this prediction.

10. Magnetoelastic Effects

Until now, the various results of this chapter have been derived under the assumption that the ferromagnetic specimen was unstrained. However, experimentally it is found that there is a change in the dimensions of a specimen when its magnetization is varied. This phenomena, first observed by Joule in 1842, is called *magnetostriction.* Its existence means that there is an interaction between magnetization and strain. This magnetoelastic interaction also implies that the application of a stress will change the magnetization; this effect is indeed observed. It is sometimes called the Villari effect.

The most extensive investigations have been of the linear magnetostriction, λ, defined as the change of length δl per unit length, that is, $\delta l/l$, measured in a particular direction. The magnetostriction of single crystals of iron along certain crystallographic axes is shown[81] in Fig. 6-10.1 as a function of the magnetization M_H, measured along the same axes. The smallness of the change of length requires the use of sensitive measuring equipment; optical, mechanical, and electrical methods have been employed.[82] A recent technique has utilized strain gages.[83]

When the specimen is magnetized to saturation, the magnetostriction approaches a limiting value λ_s, called the saturation magnetostriction. The saturation magnetostriction along the [100]- and [111]-directions are denoted by $\lambda_{[100]}$ and $\lambda_{[111]}$, respectively. Values[84] for iron and nickel are

[78] C. D. Graham, Jr., *J. Appl. Phys.* **31,** 1505 (1960).

[79] W. J. Carr, *J. Appl. Phys.* **31,** 69 (1960).

[80] E. R. Callen and H. B. Callen, *Phys. Rev.* **129,** 578 (1963).

[81] S. Kaya and H. Takaki, *Sci. Rept. Tôhoku Univ.* (Honda anniv. vol.) 314 (1936).

[82] S. J. Lochner, *Phil. Mag.* **36,** 498 (1893); B. H. C. Mathews, *J. Sci. Instr.* **16,** 124 (1939); H. Lloyd, *J. Sci. Instr.* **6,** 81 (1929); R. Whiddington, *Phil. Mag.* **40,** 634 (1920); K. Honda and Y. Masiyama, *Sci. Rept. Tôhoku Univ.* **20,** 574 (1931).

[83] J. E. Goldman and R. Smoluchowsky, *Phys. Rev.* **75,** 140 (1949).

[84] R. Becker and W. Döring, *Ferromagnetismus*, J. Springer, Berlin (1939), pp. 277–280.

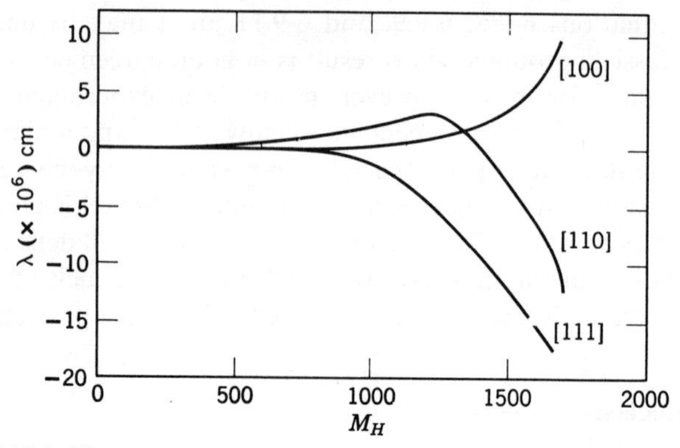

Fig. 6-10.1. The magnetostriction of single crystals of iron at room temperature.

given in Table 6-10.1. The magnetostriction is positive or negative, depending on whether there is an expansion or a contraction. For some alloys the magnetostriction is almost zero, as, for example, in the Permalloy with a composition of 82% Ni and 18% Fe.

Table 6-10.1. Saturation Magnetostriction Values ($\times 10^{-6}$)

	$\lambda_{[100]}$	$\lambda_{[111]}$	λ_s
Fe	19.5	−18.8	−7
Ni	−46	−25	−34

Magnetostriction phenomena are usually treated on the basis of a phenomological macroscopic theory developed by Akulov[85] and Becker.[86] Although this theory has been criticized,[87] it has provided an interpretation of most of the experimental results. An outline of this formal theory will now be given.

Consider some point in an unstrained crystal and let its location be described by a vector from some fixed origin. Let the length of this vector be r_0, and its components with respect to the crystal axes be x_0, y_0, and z_0. If the crystal is deformed, the new location of the point is described by a vector of length r with components x, y, and z given by

$$\begin{aligned} x &= x_0 + A_{11}x_0 + A_{12}y_0 + A_{13}z_0, \\ y &= y_0 + A_{21}x_0 + A_{22}y_0 + A_{23}z_0, \\ z &= z_0 + A_{31}x_0 + A_{32}y_0 + A_{33}z_0, \end{aligned} \tag{6-10.1}$$

[85] N. S. Akulov, *Z. Physik* **69,** 78 (1931).

[86] R. Becker, *Z. Physik* **62,** 253 (1930).

[87] W. F. Brown, Jr., *Revs. Mod. Phys.* **25,** 131 (1953).

where the A_{ij}'s are the components of the strain tensor. Let the direction cosines of the vector in the unstrained crystal be specified by β_1, β_2, and β_3, where

$$\beta_1 = \frac{x_0}{r_0}, \qquad \beta_2 = \frac{y_0}{r_0}, \quad \text{and} \quad \beta_3 = \frac{z_0}{r_0}.$$

Then equation 6-10.1 may be written

$$x = r_0\left(\beta_1 + \sum_j A_{1j}\beta_j\right),$$

$$y = r_0\left(\beta_2 + \sum_j A_{2j}\beta_j\right),$$

$$z = r_0\left(\beta_3 + \sum_j A_{3j}\beta_j\right),$$

from which it follows that the length of the vectors are related by

$$r = r_0\left(1 + \sum_{ij} 2A_{ij}\beta_i\beta_j\right)^{1/2},$$

where the higher order terms that involve products of the A_{ij}'s are neglected. Hence the change of length in a certain direction whose direction cosines are β_1, β_2, and β_3 is given by

$$\frac{\delta l}{l} = \sum_{ij} A_{ij}\beta_i\beta_j, \tag{6-10.2}$$

since

$$\frac{\delta l}{l} = \frac{r - r_0}{r_0}.$$

The assumption is made that the crystal strain depends on the direction of the spontaneous magnetization with respect to the crystal axes, that is, on the direction cosines α_1, α_2, and α_3. Thus the A_{ij}'s can be expressed as a power series in the α's. Becker and Döring find this expression by considering the free energy of the crystal. It is assumed that this energy is composed of three parts, the anisotropy energy F_K, the elastic energy F_{El}, and the magnetoelastic energy F_{ME}, that is

$$F = F_K + F_{El} + F_{ME}.$$

For a cubic crystal F_K is given by equation 6-9.1; it is a function only of the α's. F_{El} is a function only of the components A_{ij} of the strain tensor; it is given by standard texts on elasticity theory. F_{ME} is assumed to be a function of both α_i and A_{ij}. Because of the crystal symmetry, taken to be cubic in this case, only certain terms may appear in the free-energy expression. The equilibrium condition is found by minimizing the free energy with respect to A_{ij}; this leads to the desired relationship between the strain components A_{ij} and the direction cosines α_i. For details the reader is referred to Becker and Döring (1939). For our purpose it will suffice to note that the lowest order terms are given by

$$A_{ii} = a + b\alpha_i^2$$

and (6-10.3)

$$A_{ij} = c\alpha_i\alpha_j,$$

since A_{ii} must be independent of a change of sign of α_i, and A_{ij} must change sign with α_i or α_j. Substitution of these values of A_{ii} and A_{ij} into equation 6-10.2 yields

$$\frac{\delta l}{l} = 3a + b(\alpha_1^2\beta_1^2 + \alpha_2^2\beta_2^2 + \alpha_3^2\beta_3^2) + c(\alpha_1\alpha_2\beta_1\beta_2 + \alpha_2\alpha_3\beta_2\beta_3 + \alpha_3\beta_1\beta_3\beta_1). \quad (6\text{-}10.4)$$

For a demagnetized specimen, that is, when there is a random distribution of domains so that the net magnetic moment is zero, it is usual to take $\delta l/l = 0$. Then since $\overline{\alpha_i^2} = \frac{1}{3}$ and $\overline{\alpha_i\alpha_j} = 0$, equation 6-10.5 may be written

$$\frac{\delta l}{l} = h_1(\alpha_1^2\beta_1^2 + \alpha_2^2\beta_2^2 + \alpha_3^2\beta_3^2 - \tfrac{1}{3}) + 2h_2(\alpha_1\alpha_2\beta_1\beta_2 + \alpha_2\alpha_3\beta_2\beta_3 + \alpha_3\alpha_1\beta_3\beta_1). \quad (6\text{-}10.5)$$

If the magnetostriction is measured along the direction of magnetization, that is, if $\alpha_i = \beta_i$, the change in length is given by

$$\frac{\delta l}{l} = \tfrac{2}{3}h_1 + 2(h_2 - h_1)(\alpha_1^2\alpha_2^2 + \alpha_2^2\alpha_3^2 + \alpha_3^2\alpha_1^2).$$

The saturation magnetostriction measured along the [100]- and [111]-directions is then given by

$$\lambda_{[100]} = \tfrac{2}{3}h_1 \quad \text{and} \quad \lambda_{[111]} = \tfrac{2}{3}h_2.$$

Substitution of this result into equation 6-10.5 leads to the expression usually quoted for the magnetostriction, namely,

$$\frac{\delta l}{l} = \tfrac{3}{2}\lambda_{[100]}(\alpha_1^2\beta_1^2 + \alpha_2^2\beta_2^2 + \alpha_3^2\beta_3^2 - \tfrac{1}{3}) + 3\lambda_{[111]}(\alpha_1\alpha_2\beta_1\beta_2 + \alpha_2\alpha_3\beta_2\beta_3 + \alpha_3\alpha_1\beta_3\beta_1). \quad (6\text{-}10.6)$$

This expression is adequate for the description of the experimental data for most cubic crystals; sometimes, however, higher order terms must be considered.

If the magnetostriction is assumed to be isotropic, that is, $\lambda_{[100]} = \lambda_{[111]} = \lambda_s$, equation 6-10.6 reduces to

$$\frac{\delta l}{l} = \tfrac{3}{2}\lambda_s(\cos^2\theta - \tfrac{1}{3}), \tag{6-10.7}$$

where $\theta = \cos^{-1}\Sigma\,\alpha_i\beta_i$ is the angle between the spontaneous magnetization and the direction in which $\delta l/l$ is measured. This equation may also be derived in a simple direct way. Since we assume that the magnetostriction depends on this angle θ, we start with $\delta l/l = f(\theta)$. Now, whether $\theta = 0$ or $\theta = \pi$, $\delta l/l$ has the same value. The simplest function with this property is $f(\theta) = A\cos^2\theta$. Now, for a demagnetized specimen the magnetization vectors of the domains are randomly directed. Since $\overline{\cos^2\theta} = \frac{1}{3}$ and $\delta l/l = 0$ for this state by definition, we must have $\delta l/l = A(\cos^2\theta - \frac{1}{3})$. At saturation $\theta = 0$, and therefore

$$\frac{\delta l}{l} = \tfrac{3}{2}\lambda_s(\cos^2\theta - \tfrac{1}{3})$$

for a single domain, as given above. This equation finds some use for polycrystalline specimens. Values of λ_s frequently used for Fe and Ni are given in Table 6-10.1.

The application of a stress to a ferromagnetic specimen changes the direction of the spontaneous magnetization; this is the inverse of the magnetostriction effect. The usual type of stress applied is a simple tension, σ. It is easy to show from thermodynamics (problem 6-15) that the stress and magnetostriction effects are related by the equation

$$\frac{1}{l}\left(\frac{\partial l}{\partial H}\right)_\sigma = \left(\frac{\partial M}{\partial \sigma}\right)_H. \tag{6-10.8}$$

It is of interest to have an expression for the contribution to the free energy F_σ per unit volume that arises when a tension is applied. This free energy will be equal to the work done when the magnetostriction change of length occurs in the presence of the tension (equation 6-9.1). We shall consider only the simplest case, namely that in which the magnetostriction is isotropic. Then we have

$$F_\sigma = -\sigma\int d\left(\frac{\delta l}{l}\right).$$

Substitution of equation 6-10.7 into this expression then leads immediately to

$$F_\sigma = -\tfrac{3}{2}\lambda_s\sigma\cos^2\theta, \tag{6-10.9}$$

where a constant energy term has been omitted. Hence for $\lambda_s > 0$ the minimum in F_σ occurs when the direction of magnetization is parallel to the stress axis, whereas for $\lambda_s < 0$ the minimum occurs when **M** is perpendicular to the stress axis. More complicated cases are treated in Becker and Döring (1939).

Another approach to the study of magnetostriction and stress is the measurement of the elastic moduli. In particular, consider Young's modulus E_Y, defined as

$$E_Y = \frac{\sigma l}{\delta l},$$

where $\delta l/l$ is measured along the direction of the applied tension. For a specimen whose magnetization is saturated by the application of a large magnetic field, Young's modulus depends only on the elastic properties of the sort that occur in any nonferromagnetic solid. When the specimen is not saturated, the applied tension will cause alterations in the direction of magnetization of the domains. Then, as a result of the magnetostriction, there will be an additional change in length, hence in E_Y. The change in Young's modulus due to magnetostriction is known as the ΔE_Y *effect*. It has been investigated by a number of workers.[88]

The foregoing equations developed for magnetostriction and stress apply to one domain. Now, except at saturation, both single-crystal and polycrystalline specimens usually contain several domains. Thus in the analysis of the experimental data account must be taken of the domain configuration. Since domain configurations will be discussed in Chapter 7, it is appropriate to postpone an interpretation of magnetostrictive and stress experiments until then.

Finally, it is necessary to consider the microscopic origin of magnetostriction. It has been shown[89] that the dipole-dipole interaction cannot account for the magnitude of the observed magnetostriction. It is believed that the magnetostriction has the same origin as the crystalline anisotropy, namely in the spin-orbit coupling. Indeed, some quantum mechanical calculations yield the correct order of magnitude.[90] The details of the calculations, however, are in as unsatisfactory a state as they are for the crystalline anisotropy.

The crystalline anisotropy, of course, depends on the interatomic distances, that is, on the amount of strain. If the crystalline anisotropy is considered as the fundamental quantity, magnetostriction can be considered to occur because the spontaneous straining of the lattice lowers the

[88] K. Honda and T. Terada, *Phil. Mag*. **13,** 36 (1907); W. F. Brown, Jr., *Phys. Rev.* **52,** 325 (1937); W. Döring, *Z. Physik* **114,** 579 (1939); R. Kimura, *Proc. Phys.-Math. Soc. (Japan)* **22,** 45 (1940); R. Street, *Proc. Phys. Soc. (London)* **60,** 236 (1948).

[89] N. S. Akulov, *Z. Physik* **52,** 389 (1928); R. Becker, *Z. Physik* **62,** 253 (1930).

[90] S. V. Vonsovsky, *J. Phys. (USSR)* **3,** 181 (1940); J. H. Van Vleck, *Phys. Rev.* **52,** 1178 (1937); T. Katayama, *Sci. Rept. Tôhoku Univ.* **A-3,** 341 (1951).

anisotropy energy more than it raises the elastic energy. This is the point of view adopted by Kittel.[91]

Problems

6-1. Derive the molecular field theory of ferromagnetism for the case in which the Brillouin function is replaced by the Langevin function. Show that for $T/T_f < 1$, $M(T)/M(0) \neq 0$ is the stable state. In the original theory Weiss employed the Langevin rather than the Brillouin function.

6-2. For $J = \frac{1}{2}$, $g = 2$, show that the specific heat that arises from the spontaneous magnetization is given by

$$C_{sp} = \frac{[M(T)/M(0)]^2 \{1 - [M(T)/M(0)]^2\}}{(T/T_f)\,((T/T_f) - \{1 - [M(T)/M(0)]^2\})} Nk$$

for $0 < T < T_f$. From this show that near the Curie point

$$C_{sp} = \frac{3}{2}\left(3 - \frac{2T_f}{T}\right) Nk.$$

6-3. Show that if at a certain instant a system of particles is described by a symmetric wave function it is always described by a symmetric wave function and that a similar statement holds for antisymmetric wave functions.

6-4. By employing equations 6-3.9 and 6-3.10 show that the following relationships are satisfied:

$$S_xS_y - S_yS_x = iS_z \qquad S_zS_x + S_xS_z = 0$$

$$S_yS_z - S_zS_y = iS_x \qquad S_xS_y = \frac{i}{2} S_z$$

$$S_zS_x - S_xS_z = iS_y \qquad S_yS_z = \frac{i}{2} S_x$$

$$S_xS_y + S_yS_x = 0 \qquad S_zS_x = \frac{i}{2} S_y$$

$$S_yS_z + S_zS_y = 0$$

$$S_x\chi_\alpha = \frac{1}{2}\chi_\beta \qquad S_x\chi_\beta = \frac{1}{2}\chi_\alpha$$

$$S_y\chi_\alpha = \frac{i}{2}\chi_\beta \qquad S_y\chi_\beta = -\frac{i}{2}\chi_\alpha$$

$$S_z\chi_\alpha = \frac{1}{2}\chi_\alpha \qquad S_z\chi_\beta = -\frac{1}{2}\chi_\beta.$$

[91] C. Kittel, *Revs. Mod. Phys.* **21,** 541 (1949).

6-5. For n atoms in a crystal, each with $S = \frac{1}{2}$, show that

$$\overline{(\Sigma \mathbf{S}_i \cdot \mathbf{S}_j)^2} = \frac{nz}{2}\overline{(\mathbf{S}_i \cdot \mathbf{S}_j)^2} + nz(z-1)\,\overline{(\mathbf{S}_i \cdot \mathbf{S}_j)(\mathbf{S}_i \cdot \mathbf{S}_k)}$$

$$+ \frac{zn}{2}\left(\frac{zn}{2} - 2z + 1\right)\overline{(\mathbf{S}_i \cdot \mathbf{S}_j)(\mathbf{S}_k \cdot \mathbf{S}_l)},$$

where z is the number of nearest neighbors. Evaluate each of the terms on the right-hand side in terms of the total spin of the crystal.

6-6. Show that equation 6-5.12 can be manipulated into the form

$$e^{J_e/kT} = \frac{\sinh\left(\dfrac{z/2}{z-1}\dfrac{g\mu_B H}{kT}\right)}{\sinh\left(\dfrac{z/2 - 1}{z-1}\dfrac{g\mu_B H_1}{kT}\right)}$$

From this derive the result of equation 6-5.13. Further, derive equations 6-5.14 and 6-5.15.

6-7. Consider a one-dimensional chain of N atoms, each with spin $S = \frac{1}{2}$. Show that the wave function of equation 6-6.11 is an eigenfunction of the Hamiltonian of equation 6-6.2, provided

$$[E + \tfrac{1}{2}(N-4)J_e]c_w + J_e(c_{w+1} + c_{w-1}) = 0.$$

Find the solutions of this equation.

6-8. Adapt the classical equation for gyroscopic motion with no damping (equation 3-7.2) for the case in which only exchange forces are acting to show that

$$\frac{d\hbar\mathbf{S}_i}{dt} = 2J_e\mathbf{S}_i \times \sum_j \mathbf{S}_j.$$

Consider a linear chain of atoms, each with a spin $S = \frac{1}{2}$, and make the approximation $S \doteqdot [S(S+1)]^{1/2}$. Show, with the help of Fig. 6-6.1, that for small values of k the foregoing equation leads to the dispersion relationship

$$\hbar\omega = J_e k^2 a^2$$

for a spin wave.

6-9. Show that spin-wave theory predicts that a two-dimensional lattice will not be ferromagnetic. [*Hint.* Use equation 6-6.19 adapted to the two-dimensional case.]

6-10. The orbital part of the atomic wave functions were not considered in the analysis of spin waves in Section 6-6. By generalization of equation 6-3.7 the ground state is

$$\psi_0 = \begin{vmatrix} \phi_a(1)\,\chi_\alpha(1) & \phi_a(2)\,\chi_\alpha(2) & \cdots \\ \phi_b(1)\,\chi_\alpha(1) & \phi_b(2)\,\chi_\alpha(2) & \cdots \\ \cdots & \cdots & \cdots \\ \cdots & \cdots & \cdots \end{vmatrix}$$

in an obvious notation. When the spin at atom w is reversed, the wave function is

$$\psi_w = \begin{vmatrix} \phi_a(1)\,\chi_\alpha(1) & \phi_a(2)\,\chi_\alpha(2) & \cdots \\ \cdots & \cdots & \cdots \\ \phi_w(1)\,\chi_\beta(1) & \phi_w(2)\,\chi_\beta(2) & \cdots \\ \cdots & \cdots & \cdots \\ \phi_N(1)\,\chi_\alpha(1) & \phi_N(2)\,\chi_\alpha(2) & \cdots \end{vmatrix}$$

The wave function for the first excited state will be of the form $\psi_I = \sum_w c_w \psi_w$. Assume that the lattice is a linear chain. Find the recursion relationship between the coefficients c_w. Then find the energy of a spin wave.

6-11. Derive equation 6-7.4 from 6-7.3. Show that the necessary condition for $M \neq 0$ at $T = 0°K$ is that $k\theta'/E_F > \frac{2}{3}$ and for $\zeta = 1$ is $k\theta'/E_F > 2^{-1/3}$.

6-12. In the Vonsovsky-Zener theory of ferromagnetism it is assumed that the localized unpaired $3d$ electrons of one ion polarize the $4s$ conduction electrons and these electrons in turn polarize the other ions. Assume that the $3d$-$4s$ interaction can be represented by a molecular field and neglect the interaction between $3d$ shells and $4s$ electrons. If the magnetization of the ion cores obeys Curie's law, show that the Curie temperature is given by $T_f = \chi_{\text{para}} C N_W{}^2$, where χ_{para} is the Pauli susceptibility of the conduction electrons.

6-13. A system consists of two particles at fixed positions. Each particle has a magnetic moment of fixed magnitude μ, capable of orientation along the positive or negative z-axis. The two particles are coupled by exchange forces in such a way that the internal energy of the system is $+c$ when the two moments are parallel and $-c$ when the two moments are antiparallel (c = constant). The system is subject to a magnetic field H directed along the z-axis.

To interpret the magnetic behavior of a certain solid material, we adopt as our theoretical model an assemblage of systems of the foregoing type, N per unit volume. The individual systems of the assemblage are assumed not to interact with one another or with their environment, except for small interactions sufficient to establish thermodynamic equilibrium. (a) Derive an exact formula for the magnetization M of the solid at an arbitrary absolute temperature T and field intensity H in thermodynamic equilibrium. (b) Derive and discuss approximate formulas valid under the following conditions:

(i) $|\mu H| \ll kT, \quad |c| \ll kT;$

(ii) $|\mu H| \ll kT, \quad |c| \gg kT,\ c < 0;$

(iii) $|\mu H| \ll kT, \quad |c| \gg kT,\ c > 0;$

(iv) $|c| \ll kT \ll |\mu H|;$

(v) $kT \ll |\mu H| \ll |c|,\ c < 0;$

(vi) $kT \ll |\mu H| \ll |c|,\ c > 0.$

6-14. (a) Consider a single crystal of iron, cut in the form of a disk with surfaces that are (100) planes. If θ and ϕ are the angles an applied field H and the magnetization M make with the [100]-direction, show that the condition for equilibrium is

$$HM \sin (\theta - \phi) = \frac{K_1}{2} \sin 4\phi,$$

provided H is large enough that all the domain vectors lie along the same direction. (b) For the case described, show that the torque in an infinite field is

$$L_{(100)} = -\frac{K_1 \sin 4\theta}{2}.$$

Also, for a disk cut in the (110) plane show that the torque is given by

$$L_{(110)} = -K_1 \frac{(2 \sin 2\theta + 3 \sin 4\theta)}{8} - K_2 \frac{(\sin 2\theta + 4 \sin \theta - 3 \sin 6\theta)}{64}.$$

6-15. When a stress is applied to a system, an expression for the work done by the stress must be included in the thermodynamic expressions of Section 3-2. Consider the expression for Gibb's function and show that when a tension σ is applied to a ferromagnetic body,

$$\frac{1}{l}\left(\frac{\partial l}{\partial H}\right)_\sigma = \left(\frac{\partial M}{\partial \sigma}\right)_H,$$

where l is the length of the specimen.

Bibliography

Bates, L. F., *Modern Magnetism*, 4th ed., Cambridge University Press, Cambridge (1962).

Becker, R., and W. Döring, *Ferromagnetismus*, J. Springer, Berlin (1939).

Belov, K. P., *Magnetic Transitions*, Consultants Bureau, New York (1961).

Birss, R. R., "Saturation Magnetostriction of Ferromagnetics," *Advan. Phys.* **8,** 252 (1959).

Bitter, F., *Introduction to Ferromagnetism*, McGraw-Hill Book Co., New York (1937).

Bonch-Bruevich, V. L., and S. V. Tyablikov, *Green Function Methods in Statistical Mechanics*, North-Holland Publishing Co., Amsterdam (1962).

Bozorth, R. M., *Ferromagnetism*, D. Van Nostrand, Princeton, N.J. (1951).

Carr, W. J., Jr., "Magnetostriction," *Magnetic Properties of Metals and Alloys*, American Society for Metals, Cleveland, Ohio (1959), pp. 200–251.

DeBarr, A. E., *Soft Magnetic Materials Used in Industry*, Chapman and Hall, London (1953).

Domb, C., "Theory of Cooperative Phenomena in Crystals," *Advan. Phys.* **9,** 149, 245 (1960).

Hellwege, K. H., and A. M. Hellwege, *Magnetic Properties I*, J. Springer, Berlin (1962).

Hoselitz, K., *Ferromagnetic Properties of Metals and Alloys*, Oxford University Press, Oxford (1952).

Kneller, E., *Ferromagnetismus*, J. Springer, Berlin (1962).

Kittel, C., "Ferromagnetism," *Nuovo cimento Suppl.* **3,** 895 (1957).

Lee, E. W., "Magnetostriction and Magnetomechanical Effects," *Rept. Progr. Phys.* **18,** 184 (1955).

Mott, N. F., and H. Jones, *Theory of the Properties of Metals and Alloys*, Oxford University Press, Oxford (1936).

Nix, F. C., and W. Shockley, "Order-disorder Transformations in Alloys," *Revs. Mod. Phys.* **10,** 1 (1938).

Peierls, R. E., *Quantum Theory of Solids*, Oxford University Press, Oxford (1955), Ch. 8.

Seitz, F., *Modern Theory of Solids*, McGraw-Hill Book Co., New York (1940), Ch. 16.

Stewart, K. H., *Ferromagnetic Domains*, Cambridge University Press, Cambridge (1954).

Stoner, E. C., *Magnetism and Matter*, Methuen, London (1934).

Stoner, E. C., *Magnetism*, Methuen, London (1947).

Stoner, E. C., "Ferromagnetism," *Rept. Progr. Phys.* **11,** 43 (1948).

Stoner, E. C., "Collective Electron Ferromagnetism," *J. Phys. radium* **12,** 372 (1951).

Wannier, G. H., *Elements of Solid State Theory*, Cambridge University Press, Cambridge (1959), Ch. 4.

Wilson, A. H., *Theory of Metals*, Cambridge University Press, Cambridge (1958), Ch. 7.

Van Vleck, J. H., "Electric and Magnetic Susceptibilities," Oxford University Press, Oxford (1932), Ch. 12.

Van Vleck, J. H., "Survey of Theory of Ferromagnetism," *Revs. Mod. Phys.* **17,** 27 (1945).

Van Kranendonk, J., and J. H. Van Vleck, "Spin Waves," *Revs. Mod. Phys.* **30,** 1 (1958).

Vonsovsky, S. V., and Y. S. Shur, *Ferromagnetism*, State Publishing House for Technical-Theoretical Literature, Moscow (1948).

7

The Magnetization of Ferromagnetic Materials

1. Introduction

When a magnetic field is applied to a ferromagnetic specimen, the magnetization may vary from zero to the saturation value. As mentioned in Section 6-1, Weiss, in order to explain this behavior, postulated the existence of domains. According to Weiss, although each domain is spontaneously magnetized, the direction of magnetization may vary from one domain to another. In general, a specimen consists of many domains, with a domain configuration that is a function of the applied field. Since the magnetic moment of the specimen is the vector sum of the magnetic moments of each domain, the magnetization, or average magnetic moment per unit volume, may take on any value between zero and saturation.

Experimentally, the component of the magnetization along the direction of the applied field, M_H, rather than the magnetization of the ferromagnetic material, is always measured. Frequently these two quantities are the same, but not always. It has become the practice to refer to M_H as the magnetization and to drop the subscript H. However, in some equations that will be developed in this chapter the spontaneous magnetization $\mathbf{M}$ occurs. We therefore retain the symbols M_H and $\mathbf{M}$. The reader is reminded that the symbol $\mathbf{M}$ has also been used for the (average) magnetization of a ferromagnetic material, for example, in Chapter 1. Usually the context makes the meaning of the symbol clear.

The variation of M_H with the applied field depends, of course, on the material, but frequently the curves have the general appearance shown in

Fig. 7-1.1. It is assumed that the material is initially in a demagnetized state, indicated by O. When a field is applied, the magnetization M_H increases until the saturation value is reached: this behavior is indicated by the curve $OABC$. This part of the plot of M_H versus H is known as the magnetization curve; magnetization curves of single crystals have already been discussed in Section 6-9. By performing the sequence of operations of

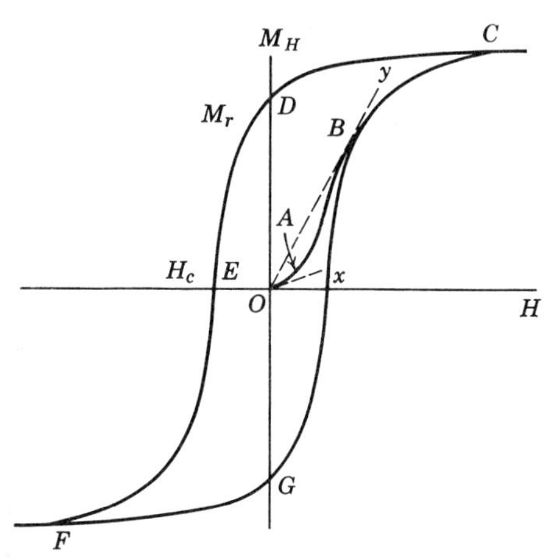

Fig. 7-1.1. Magnetization curve ($OABC$) and hysteresis loop ($CDEFGC$) of a typical ferromagnetic material.

reducing the field to zero, increasing it in the reverse direction, decreasing it to zero, and then increasing it to the original value, the curve $CDEFGC$, known as a hysteresis curve or loop, is obtained. This curve is symmetric about the origin and is reproducible, at least after the first few cycles have been completed. Since the magnetization lags the applied field, work is performed on taking the material through a cycle. This work is given by the line integral

$$W = \oint H \, dM_H \tag{7-1.1}$$

taken along the hysteresis loop and is obviously equal to the area enclosed by the loop.

Certain numerical quantities are employed to describe the salient features of magnetization curves and hysteresis loops. Since the susceptibility $\chi\,(=M_H/H)$ of a ferromagnetic material depends both on the applied field and on the previous magnetic history, several susceptibilities may be defined. The initial susceptibility χ_0 is the slope of the tangent of the magnetization curve at the origin O (for Fig. 7-1.1, the slope of OX). The maximum susceptibility χ_m normally occurs near the knee of the magnetization curve (point B): it is given by the slope of OY for Fig. 7-1.1. The differential susceptibility is the slope of the M_H versus H curve, or dM_H/dH. On the hysteresis loop the value of the magnetization for $H = 0$ is called the residual magnetization or remanence, M_r (point D in Fig. 7-1.1). The value of the field for which $M_H = 0$ is called the coercive force H_c (point E in Fig. 7-1.1). The magnetization at point C is, of course, the saturation value.

Sometimes the scale of the ordinate of Fig. 7-1.1 is changed in that $4\pi M_H$ $(=B - H)$ is plotted against H. However, the commonest way of displaying the data is to plot the component of the magnetic induction along the field direction B_H against the applied field H. Then the initial and maximum permeabilities μ_0 and μ_m are the quantities that correspond to the initial and maximum susceptibilities, respectively, whereas the residual induction B_r corresponds to the remanence M_r. The symbols ${}_MH_c$ and ${}_BH_c$ are sometimes used to distinguish the coercive force of an M_H versus H curve from that of a B_H versus H curve, respectively; in this book the additional subscript will be employed only when there is a danger of ambiguity.

In theoretical calculations it is usually more direct to compute the M_H versus H curves; this is the type of curve that will mainly be considered in this chapter. Many experimental arrangements, however, yield the B_H versus H curves directly. For example, consider a toroid of the ferromagnetic material, with both a primary and a secondary winding. If a current i flows in the primary, the average field parallel to the axis of the toroid is given by $H = 4\pi ni$ (equation 1-7.2), where n is the number of turns per unit length. If the radius R of the toroid is large, this field is almost uniform over the cross-sectional area A of the toroid. When the current is changed, a voltage will be induced in the secondary by Faraday's law. From equation 1-7.5 the average value of B_H is given by

$$B_H = -\frac{1}{NA}\int \mathcal{E}\,dt,$$

where $\mathcal{E}$ is the induced emf and N is the total number of turns on the secondary. When i is changed quickly, the change in B_H may be determined by the deflection of a ballistic galvanometer connected to the secondary. If ac current is applied to the primary and an integrator inserted in the secondary circuit, the hysteresis loop may be displayed on an oscilloscope screen. A solenoid may be employed instead of a toroid; the secondary often consists of a small coil, called a pick-up coil, inserted inside the solenoid (primary). A complication now arises, for unless the sample is in the form of a very long needle the demagnetization field must be taken into account.

In the solenoidal arrangement the sample may be moved while the field is held constant; the resulting deflection of a ballistic galvanometer is then proportional to the magnetization M_H. The magnetization may also be measured directly if the secondary consists of two series-connected coils wound in opposition, for then the change in H does not induce a voltage in the secondary. Sometimes electromagnets are employed instead of solenoids when large H fields are desired[1]; a correction for the image of the sample in the pole tips must then be made. Magnetization curves may also be determined by the magnetometer method. A magnetometer consists of a magnetic needle of known moment, suspended so that it may rotate in a horizontal plane. The deflection of this needle when the specimen is brought into its vicinity permits the magnetization to be calculated. Other methods for the investigation of magnetization curves are described in the literature.[2]

Domains, originally postulated by Weiss in order to account for the observed magnetization and hysteresis curves, have been the subject of extensive investigations. Although important work was done in the 1930's, the greatest advances have been made since 1945. In addition to the magnetization curve, there is considerable experimental evidence for the existence of domains, some of it of a very direct nature: this evidence will be reviewed later in this chapter. On the theoretical side a thermodynamic approach is always employed. The stable equilibrium state of the specimen's magnetization is sought by minimizing an expression for the free energy.

The total free energy of a ferromagnetic specimen in an applied magnetic field may be written as the sum of several free energy terms (equation 6-9.1):

$$F_T = F_H + F_D + F_K + F_\sigma + F_e + F_0, \tag{7-1.2}$$

where the symbols on the right-hand side of the equation have the following meaning. The first term, F_H, is the energy of the specimen's magnetization in the applied field H and is given by (equation 1-6.2),

$$F_H = -\int \mathbf{M} \cdot \mathbf{H} \, dv. \tag{7-1.3}$$

For $\mathbf{M}$ and $\mathbf{H}$ uniform $F_H = -\mathbf{M} \cdot \mathbf{H}$ per unit volume. The second term, F_D, is the self-energy of the magnetization in its own field. By equation 1-6.6 it is equal to

$$F_D = -\tfrac{1}{2}\int \mathbf{M} \cdot \mathbf{H}_s \, dv, \tag{7-1.4}$$

[1] P. Weiss and R. Forrer, *Ann. phys.* (*Paris*) **12,** 297 (1929).

[2] For a summary see R. M. Bozorth, *Ferromagnetism*, D. Van Nostrand, Princeton N.J. (1951), Ch. 19.

where $\mathbf{H}_s$ is the field of the surface and volume pole densities. For a specimen of ellipsoidal shape (equation 1-6.7)

$$F_D = \tfrac{1}{2} DM^2 \tag{7-1.5}$$

per unit volume. The next term, F_K, is the crystalline anisotropy energy. For cubic crystals (equation 6-9.2)

$$F_K = K_1({\alpha_1}^2{\alpha_2}^2 + {\alpha_2}^2{\alpha_3}^2 + {\alpha_3}^2{\alpha_1}^2) + K_2{\alpha_1}^2{\alpha_2}^2{\alpha_3}^2 \tag{7-1.6}$$

per unit volume, whereas for uniaxial crystals (equation 6-9.3)

$$F_K = K_1' \sin^2 \theta + K_2' \sin^4 \theta \tag{7-1.7}$$

per unit volume. F_σ is the magnetostrictive energy when a tension σ is present; for a specimen with isotropic magnetization (equation 6-10.9)

$$F_\sigma = -\tfrac{3}{2}\lambda_s \, \sigma \cos^2 \theta \tag{7-1.8}$$

per unit volume. F_e represents the exchange free energy; the expression used for this energy requires some discussion. Strictly, the exchange free energy should be calculated from the partition function (see Section 5-1, especially equation 5-1.19, and recall $F = G + HM$). However, in the summation over all states, a knowledge of the energy levels of the system is required. As discussed in Section 6-4, the determination of these energies is not feasible. Now the free energy at absolute zero is clear enough, for then by equation 6-3.25

$$\begin{aligned} F_e &= U \\ &= -2J_e \sum_{ij}{}' \mathbf{S}_i \cdot \mathbf{S}_j. \end{aligned} \tag{7-1.9}$$

The problem is then to arrive at an expression that will approximately describe the system for $T \neq 0$. It is assumed that the exchange free energy for this case is given by equation 7-1.9, provided the spin operators are considered as classical vectors. This leads to (equation 6-3.29)

$$F_e = -2J_e \, S^2 \, \Sigma \cos \phi_{ij}. \tag{7-1.10}$$

Thus for $T \neq 0$ the energy differs from that at $T = 0°$ inasmuch as some spins make nonzero angles with one another. Hence the energy, and therefore the free energy, is increased by an amount proportional to the cosine of the angle between adjacent spins. An alternative form for equation 7-1.10 is (equation 6-3.33)

$$F_e = \frac{2J_eS^2}{a} \int [(\nabla\alpha_1)^2 + (\nabla\alpha_2)^2 + (\nabla\alpha_3)^2] \, dv, \tag{7-1.11}$$

where α_1, α_2 and α_3 are the direction cosines between the direction of magnetization and the axes of a rectangular coordinate system. The integral is justified if it is assumed that the direction of magnetization is a continuous function of position. Finally, the term F_0 represents any other contributions to the free energy that may be present. For example, if the specimen were inhomogenous, the resulting perturbations would give rise to such a term.

The minimization of the total free energy leads to nonlinear partial differential equations,[3] together with certain boundary conditions. For example, let us assume that $F_\sigma = F_0 = 0$. Then equation 7-1.2 becomes

$$F_T = \int \{-\mathbf{M}\cdot\mathbf{H} - \tfrac{1}{2}\mathbf{M}\cdot\mathbf{H}_s + F_K + A[(\nabla\alpha_1)^2 + (\nabla\alpha_2)^2 + (\nabla\alpha_3)^2]\}\, dv, \tag{7-1.12}$$

where $A = 2J_eS^2/a$ and equations 7-1.3, 7-1.4, and 7-1.11 have been employed. On assuming that the magnetization has a constant magnitude and a direction that is a continuous function of position, it can be shown[3] that minimization of this equation, without constraints and for a finite specimen, yields the partial differential vector equation

$$\mathbf{u}\times\left[M(\mathbf{H}+\mathbf{H}_s) - \frac{\partial F_K}{\partial \mathbf{u}} + 2A\nabla^2\mathbf{u}\right] = 0 \tag{7-1.13}$$

and the boundary condition

$$\mathbf{u}\times\frac{\partial \mathbf{u}}{\partial n} = 0, \tag{7-1.14}$$

where $\partial/\partial n$ indicates differentiation along the outward normal to the specimen's surface and $\mathbf{u}$ is a unit vector

$$\mathbf{u} = \alpha_1\mathbf{i} + \alpha_2\mathbf{j} + \alpha_3\mathbf{k}$$

directed along the magnetization such that $\mathbf{M} = M\mathbf{u}$.

Several methods of solution have been attempted. One employs the Ritz variational method.[4] A second uses numerical integration[5] in which fast digital computers are employed. In a third method the differential equations are linearized: such a procedure is appropriate for cases in which the vector $\mathbf{u}$ is almost uniform. The problem of solution is difficult and progress has been made in only a small number of cases. Probably the most significant progress to date has been made with the third method, for which rigorous results in a few cases and approximate results in several more have been obtained. Some of these results are described subsequently in this chapter. The foregoing general theoretical approach belongs to an area that is sometimes called *micromagnetics*.

[3] W. F. Brown, Jr., *Phys. Rev.* **58**, 736 (1940).
[4] E. Kondorsky, *Izvest. Akad. Nauk. SSSR, Ser. Fiz.* **16**, 398 (1952).
[5] W. F. Brown, Jr., *Phys. Rev.* **105**, 1479 (1957), *J. Appl. Phys.* **29**, 470 (1958).

An alternative theoretical approach is to employ a model that appears to be physically reasonable for a certain situation and then to minimize the sum of the free-energy terms that are most important for the particular model. This type of theory depends on physical intuition and lacks the rigor of the micromagnetic approach. It has the great merit, however, of providing a qualitative explanation of most of the experimental results on ferromagnetic materials. For some cases there is quantitative agreement with experiment. Many examples of this approach are given in subsequent sections of this chapter. Now, however, let us see from a simple qualitative argument why domains may be expected to occur.

Consider a single crystal of ferromagnetic material that is a parallelepiped. Suppose that the applied field is zero, that there are no stresses, internal or external, and that the specimen has no imperfections. Then $F_H = F_\sigma = F_0 = 0$. Further, suppose that the long edge of the crystal lies along an easy direction of magnetization. Then F_K will be a minimum if all the atomic magnetic moments lie parallel to this long edge. Such a single-domain configuration is shown in Fig. 7-1.2*a*. In addition, the

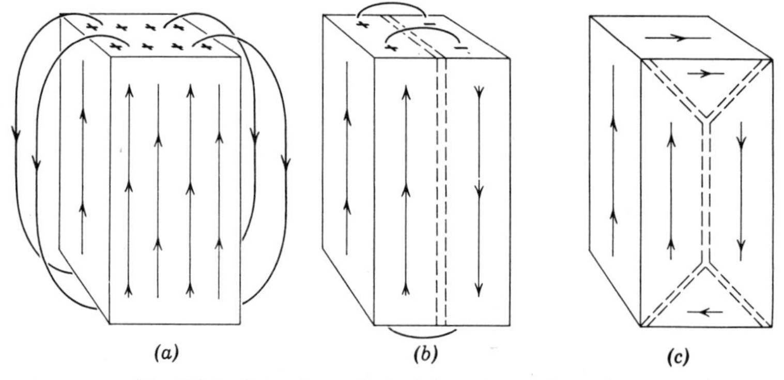

Fig. 7-1.2. Some hypothetical domain configurations.

exchange energy F_e will also be a minimum. However, the remaining term, F_D, the demagnetization energy, will not have its minimum value. Indeed, for iron F_D is of the order of $M^2 \approx 10^6$ ergs/cm³. This large magnetic energy may be decreased by introducing a domain structure, although it may be accomplished only at the expense of increasing the other energy terms. For example, for the configuration of two antiparallel domains, shown in Fig. 7-1.2*b*, the demagnetization energy is reduced by a factor of about one half. There is now, however, a boundary or transition region between the two domains in which the atomic moments are not parallel to each other and moreover do not lie along easy directions. Hence

F_e and F_K are increased. Nevertheless, it may be shown that the total free energy for the two domain arrangement is usually less than that for the single domain, primarily because a comparatively small number of atomic moments are involved in increasing F_e and F_K. Subdivision into additional linear domains magnetized alternately parallel and antiparallel to the long edge is likely to occur as long as it will reduce F_T.

The boundary region between two domains is often called a *domain wall*, although the term Bloch wall is also used, particularly with reference to the type of wall that occurs in this example. Although the width of this wall is probably not clearly defined, the wall is frequently considered an entity in domain configuration calculations. Examples of the use of this concept are given later in the chapter.

Certain domain arrangements make the demagnetization energy zero: an example is given in Fig. 7-1.2*c*. The triangular or prismatic domains, called *closure domains*, are most likely to occur if their magnetization lies along an easy direction. Such may be the case for cubic crystals like iron. For uniaxial crystals closure domains would have to lie along a hard direction.

Although the examples considered here are oversimplified, they do show clearly that it is the demagnetization energy term that is primarily responsible for the occurrence of domains. Now, it is to be recalled that the origin of the demagnetization energy is the classical dipole-dipole interaction, which is very much smaller than the exchange interaction between adjacent atoms. Hence it may at first appear anomalous that a domain structure rather than a state of uniform magnetization is usually favored. The reason is that the dipole-dipole forces are long range, dropping off slowly with distance, whereas the exchange forces are short range being limited almost to the nearest neighbors. Thus, overall, a domain structure is to be expected, whereas over short distances, that is, within one domain, the magnetization is expected to be uniform or almost uniform.

Therefore, as a general rule, a domain configuration is expected that will reduce, if not remove, the uncompensated poles on the surface of the specimen. Moreover, it is reasonable to expect that uncompensated poles will not occur along domain walls, since they would create a magnetic field, and hence magnetic energy (see Section 1-3). The condition for the avoidance of poles on domain walls is that the component of the magnetization normal to the wall be continuous across the wall. This condition is fulfilled for the example of Fig. 7-1.2*c*. Here the domain walls that separate the vertical and closure domains bisect the 90° angle between the magnetization vectors of these two domains.

The next question to consider is how the magnetization of the ferromagnetic specimen changes when a magnetic field is present. It is believed that two main processes may occur. In one the volume of domains favorably oriented with respect to the applied field grow at the expense of those unfavorably oriented; that is, there is a displacement of the domain walls. In the other process the magnetization of the domain rotates toward the field direction. The two processes may occur either reversibly or irreversibly, depending on the strength of the applied field and on the nature of the specimen. Further details are given in later sections.

Ferromagnetic materials are often classified as hard or soft, depending on whether their coercive force is large or small. No definition of the dividing line between the two cases has ever been made. However, if $H_c > 100$ oe, the material would certainly be called hard, whereas if $H_c < 5$ oe it would certainly be called soft. Coercive forces as large as 10^4 oe and as small as 10^{-3} oe in order of magnitude have been observed. Various theories have been developed in an attempt to understand the physical basis for the range of values observed. The theories fall roughly into two main classifications. One class is concerned with bulk materials. These theories depend on various structure-sensitive properties of the material, such as the internal strains and the nonmagnetic inclusions. As a result, it is difficult to set up realistic models. These theories are mainly relevant for soft magnetic materials. The other class is concerned with small particles. A particularly simple model applies, provided the particles are sufficiently small. The most important of the magnetically hard materials are essentially composed of small particles. Small particles will be discussed first in Sections 7-2, 7-3, and 7-4; bulk material is discussed in later sections.

2. Single-Domain Particles

As has been stated already, a ferromagnetic specimen in general consists of many domains. Under certain circumstances, however, the specimen may consist of a single-domain. One example has been met before, in Sections 6-2 and 6-3, in which the specimen became a single-domain when it was magnetized to saturation by the application of a sufficiently large field. Another example is presented when the specimen is a very small particle. Then the exchange forces may dominate, so that in spite of the presence of the demagnetization energy, and the absence of any applied field, the particle is uniformly magnetized; that is, it is a single domain. Such single-domain particles are the subject matter of this section.

Critical size. First, let us make an estimate of the critical size below which the particle will be a single domain in zero applied field. Presumably, the single-domain configuration will occur if its energy is less than the energy of other configurations which reduce or remove the uncompensated poles. For the sake of simplicity, we assume that the particle is a prolate ellipsoid of revolution, with polar (long) semiaxes a and equatorial (short) semiaxes b, as indicated in Fig. 7-2.1. Let D_a denote the longitudinal demagnetization factor and D_b, the transverse demagnetization factor. We also assume that $F_K = 0$ when the magnetization **M** lies along the polar axis, that is, that this is an easy direction and $K_1 > 0$. Further, we assume $F_\sigma = 0$. Since $D_a < D_b$, the demagnetization energy F_D will also be a minimum when **M** lies along the polar axis. The energy of the single-domain configuration is then

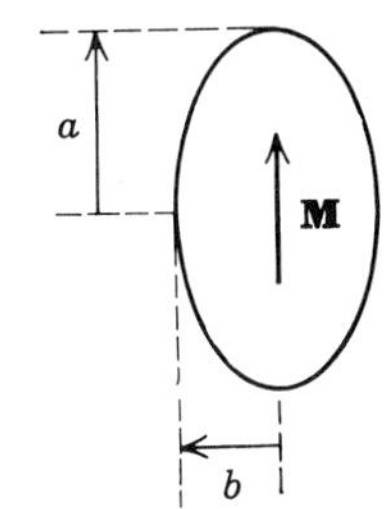

Fig. 7-2.1. A prolate ellipsoidal particle with uniform magnetization (single-domain particle).

$$F_T = F_D$$

$$= \tfrac{1}{2} D_a M^2 \left[\left(\frac{4\pi}{3} \right) a b^2 \right], \qquad (7\text{-}2.1)$$

since the volume of the particle is $(4\pi/3)ab^2$.

The next step is to calculate the energy of other configurations. Strictly speaking, all other configurations should be considered, as well as whether it is possible to pass from these configurations to the single-domain state. It is not feasible to make such an exhaustive study. Instead, only the configuration that appears likely for a particular crystalline anisotropy is considered. The stress, and hence the magnetostrictive energy, is assumed to be zero. If the ferromagnetic material has a negligibly small crystalline anisotropy, the atomic moments may be expected to point along closed rings about the polar axis.[6] This arrangement is illustrated in Fig. 7-2.2*a*,

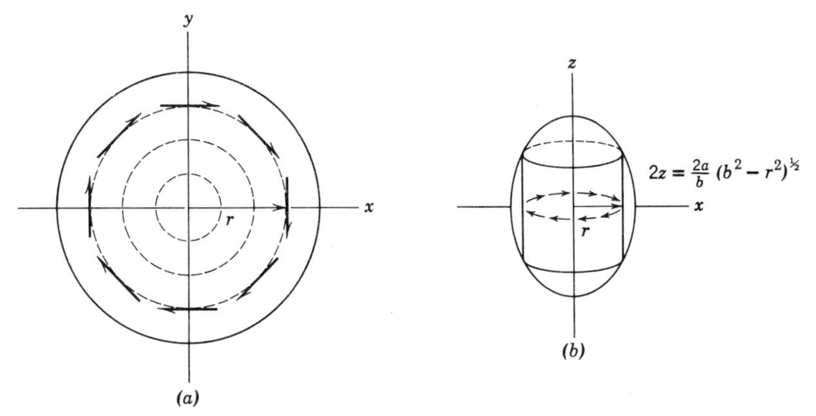

Fig. 7-2.2. Illustration of flux closure produced by a ring configuration in a prolate ellipsoidal particle with negligible crystalline anisotropy. The rings in the equatorial plane are shown in (*a*), the shells of circular cylinders in (*b*).

[6] L. Néel, *Compt. rend.* (*Paris*) **224,** 1488 (1947).

which shows the rings in the equatorial plane. If the crystalline anisotropy is relatively large, most of the atomic moments may be expected to lie along easy directions. The arrangement that may be expected for a cubic crystal[7] is shown in Fig. 7-2.3*a*, whereas that for a uniaxial crystal is shown in Fig. 7-2.3*b*. The regions where the directions of the moments change rapidly become essentially domain walls.

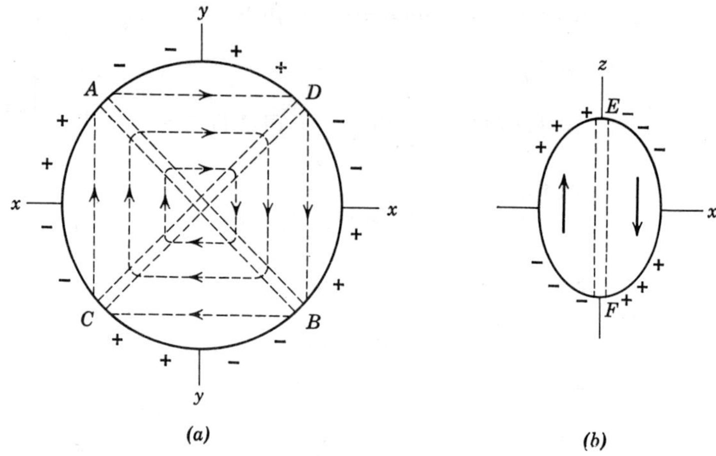

Fig. 7-2.3. Proposed simple domain arrangements in a small prolate ellipsoidal particle with large crystalline anisotropy: (*a*) is for a cubic crystal with easy directions along the x- and y-axes: the equatorial plane is shown; (*b*) is for a uniaxial crystal with easy direction along the z-axis. The surface poles are indicated. *AB*, *CD*, and *EF* are domain walls.

We proceed to calculate the energy for the situation in Fig. 7-2.2. Under the assumed conditions, all energies except the exchange energy F_e are zero. The increase in the exchange energy per atom is given by equation 6-3.30 as

$$F_{\text{atom}} = J_e S^2 \Sigma \phi_{ij}^2.$$

Next we calculate the exchange energy of the atoms in one ring of radius r, as indicated in Fig. 7-2.2*a*. The total change in angle of the spins in going around the ring is 2π and the number of atoms in such a ring is $N = 2\pi r/a_l$. Here a_l is the lattice spacing if the crystal structure is simple cubic; otherwise it is the distance between nearest neighbors. Hence the angle ϕ_{ij} between adjacent spins is

$$\phi_{ij} = \frac{a_l}{r}.$$

[7] C. Kittel, *Phys. Rev.* **70**, 965 (1946).

The exchange energy per ring is then

$$F_{\text{ring}} = (F_{\text{atom}})\left(\frac{2\pi r}{a_l}\right)$$

$$= \frac{2\pi J_e S^2 a_l}{r}.$$

Now the number of rings in a circular cylindrical shell, as shown in Fig. 2-2.2*b*, is $2z/a_l$ where $z = (a/b)(b^2 - r^2)^{1/2}$. Therefore

$$F_{\text{cyl}} = F_{\text{ring}} \cdot \frac{2a(b^2 - r^2)^{1/2}}{a_l b}$$

and the exchange energy for the prolate ellipsoid is

$$F_{\text{ellip}} = \frac{4\pi a J_e S^2}{a_l b} \int_{a_l}^{b} \frac{(b^2 - r^2)^{1/2}}{r}\, dr.$$

The lower limit is taken as a_l rather than zero in order to avoid a mathematical singularity. This arises because the direction in which this center atom may point is indeterminate. On physical grounds this problem need not worry us, since we may, for example, assume that there is a vacancy at this lattice site. The integral is a standard one and is equal to

$$(b^2 - r^2)^{1/2} - b \log \frac{b + (b^2 - r^2)^{1/2}}{r}$$

For $b \gg a_l$ we have finally

$$F_{\text{ellip}} = \frac{4\pi a J_e S^2}{a_l}\left[\log\left(\frac{2b}{a_l}\right) - 1\right]. \tag{7-2.2}$$

The critical size is found by equating equation 7-2.2 to equation 7-2.1, since for particles below this size the single-domain configuration has the lowest energy. This may be verified, for example, by numerical substitution. The critical size is then given by the transcendental equation

$$\frac{b^2}{\log(2b/a_l) - 1} = \frac{6J_e S^2}{a_l D_a M^2}. \tag{7-2.3}$$

Although the semiaxis a does not appear explicitly, it enters implicitly through the demagnetization factor D_a. Of the other factors the magnetization M is the one that is likely to have the widest range of values for different materials. For a given ratio a/b, b is approximately inversely proportional to M.

A numerical estimate of the critical size may be obtained by substituting some suitable expression for J_e from Chapter 6, such as equation 6-3.28. Then insertion of the numerical values for iron gives for the critical values of $2b$ approximately 1.5×10^{-6} cm (150 Å) and 6×10^{-6} cm (600 Å) for $a/b = 1$ (the sphere) and $a/b = 10$, respectively. The calculation of other critical values is left as problem 7-1*a*.

Of course, complete neglect of the crystalline anisotropy is not valid for iron. The calculation of the critical size when the models of Fig. 7-2.3 apply is left as problem 7-1*b*. The results give about the same critical size for iron as that already obtained. However, it may be shown that this size is less than the thickness of a domain wall for iron. Hence the crystalline anisotropy of iron is not large enough for these models to be applicable. They may, however, apply for materials with anisotropy energies of 10^6 ergs/cm³ or greater.

In summary: although there are several approximations and assumptions involved, nevertheless the foregoing calculations provide an estimate of the critical size below which a particle will be a single domain in a zero applied field.

Hysteresis loops. Next, we shall investigate the behavior of the magnetization of a single-domain particle when a magnetic field is applied. The critical single-domain size of a particle in a zero applied field has just been calculated. It does not necessarily follow that such a particle will remain a single domain when a field is applied. However, we shall assume for the present that the magnitude of the particle's magnetization (M) does remain constant for all values of this applied field, since this greatly simplifies the analysis. The validity of this assumption is discussed later in this section.

Consider a prolate ellipsoidal particle with negligible crystalline and strain energy, that is, $F_K = F_\sigma = 0$, and suppose a magnetic field is applied. At equilibrium, the magnetization $\mathbf{M}$ will lie in the plane defined by the directions of the field and the polar axis, as indicated in Fig. 7-2.4. This follows since there are no forces present that will act to move $\mathbf{M}$ out of this plane.

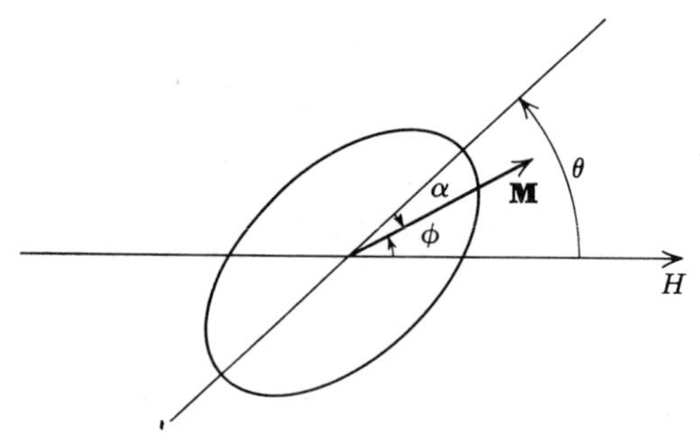

Fig. 7-2.4. The prolate ellipsoidal particle in a magnetic field.

The demagnetization energy is given by

$$\begin{aligned} F_D &= \tfrac{1}{2}M^2(D_a \cos^2 \alpha + D_b \sin^2 \alpha)V \\ &= \tfrac{1}{4}(D_a + D_b)M^2V - \tfrac{1}{4}(D_b - D_a)M^2V \cos 2\alpha, \qquad (7\text{-}2.4) \end{aligned}$$

where α is the angle between **M** and the polar axis. The particle may be said to possess *shape* anisotropy, for this energy term is a minimum if **M** lies along the polar axis (the easy direction) and a maximum if **M** lies along an equatorial axis (the hard direction). The energy in the applied field is

$$F_H = -HMV \cos \phi,$$

where ϕ is the angle between **H** and **M**. Hence the total energy is

$$F_T = \text{constant} - \tfrac{1}{4}(D_b - D_a)M^2V \cos 2\alpha - HMV \cos \phi. \qquad (7\text{-}2.5)$$

For stable equilibrium in a given field H, the magnetization **M** will point along a direction that makes F_T a minimum. We then obtain[8]

$$\frac{\partial F_T}{\partial \phi} = \tfrac{1}{2}(D_b - D_a)M^2 \sin 2\alpha + HM \sin \phi = 0, \qquad (7\text{-}2.6)$$

since $\phi = \theta + \alpha$ ($\alpha < 0$ in Fig. 7-2.4). This corresponds to a minimum, provided that

$$\frac{\partial^2 F_T}{\partial \phi^2} = (D_b - D_a)M^2 \cos 2\alpha + HM \cos \phi > 0.$$

Equation 7-2.6 which is actually a torque equation states that equilibrium is achieved when the torques (or couples) on the magnetization from the external and demagnetization fields are equal. It is convenient to rewrite this equation as

$$\tfrac{1}{2} \sin 2(\phi - \theta) + h \sin \phi = 0, \qquad (7\text{-}2.7)$$

where h is a field in reduced units given by

$$h = \frac{H}{(D_b - D_a)M} \,. \qquad (7\text{-}2.8)$$

We wish to solve this equation for ϕ as a function of h and θ. It turns out that this solution is difficult to obtain directly, and instead it is best to use a method of inverse interpolation. Details are given by Stoner and

[8] E. C. Stoner and E. P. Wohlfarth, *Phil. Trans. Roy. Soc. (London)* **A-240,** 599 (1948).

Wohlfarth.[9] Nowadays, fast digital computers are usually employed to reduce the labor of solution. Only the general nature of the results are given here.

When the applied field is zero, the magnetization **M** lies along the polar axis of the particle. Whether this is the positive or negative direction of the axis depends on the immediate history. Since the magnitude of the magnetization $|M|$ is constrained to be constant, a magnetic field acts merely to rotate the magnetization **M** toward the direction of **H** (or **h**). This rotation may be reversible or irreversible. Let us consider some examples.

Suppose the field **h** is applied perpendicular to the polar axis ($\theta = \pi/2$). The orientation of **M**, as **h** is successively increased and decreased through positive and negative values, is shown in Fig. 7-2.5*a*. The component of magnetization M_H is proportional to the field until $h = 1$; then the magnetization **M** lies along the field direction. In plotting magnetization or hysteresis curves, it is convenient to use reduced units. Thus the value of the reduced magnetization $M_H/|M|$ ($= \cos \phi$) is shown plotted against h in Fig. 7-2.5*b*. The curves, of course, may be reinterpreted as a plot of M_H versus H by a simple change of scale. It is to be noted that for $\theta = \pi/2$ the rotation of **M** is reversible and therefore the hysteresis is zero.

The situation for the field applied along the polar axis is illustrated in Fig. 7-2.6. There are no torques acting on the magnetization either from the demagnetization or the applied field. In practice, however, there will always be some perturbation present that will tend to cause a very small rotation of the magnetization. Suppose the field is directed oppositely to the magnetization. When small rotations occur, the shape anisotropy will tend to maintain the magnetization in its old orientation, whereas the field will tend to rotate the magnetization along the field direction. For small fields the shape effect wins out. At a certain critical field, actually for $h = 1$, the magnetization jumps 180°. Thus there is a discontinuity in M_H, and hysteresis occurs (see Fig. 7-2.6*b*). This energy is dissipated as heat.

The foregoing two cases represent limiting ones, inasmuch as in the first only a reversible rotation occurs, whereas in the second only an irreversible rotation (jump or flip) occurs. For all other cases part of the rotation is reversible and part is irreversible. The situation for $\theta = 10°$, 45°, and 80° is illustrated in Figs. 7-2.7, 7-2.8, and 7-2.9, respectively. It may be noted that when $\theta < 45°$ the critical field at which a jump occurs is equal to the coercive force. For $135° > \theta > 45°$ the critical field is greater than the coercive force. For $103° > \theta > 77°$ it is interesting to note that M_H is larger before the jump than after. One jump occurs each half cycle,

[9] E. C. Stoner and E. P. Wohlfarth, *Phil. Trans. Roy. Soc.* (*London*) **A-240,** 599 (1948).

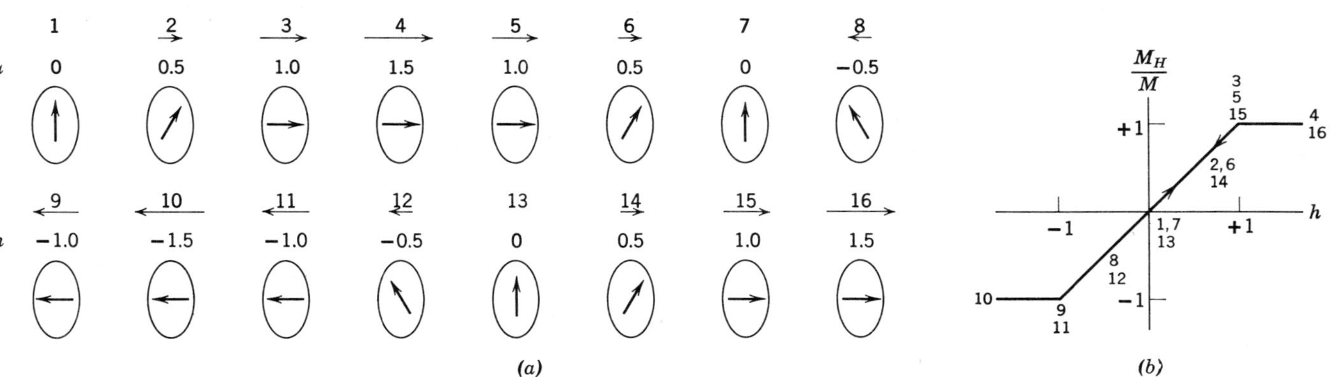

Fig. 7-2.5. The orientation of the magnetization **M** of a single-domain ellipsoidal particle as a function of a field applied perpendicular to the particle's polar axis ($\theta = \pi/2$) is illustrated. In (*a*) the vectors above the ellipses represent the field, whose magnitude is also indicated in units of h. The arrows within the ellipse represent the magnetization. The magnetization curve, plotted in the reduced units of M_H/M versus h, is shown at (*b*). The numbers 1, 2, 3, . . . , 16 correspond to successive stages in the application of the field. The hysteresis is zero.

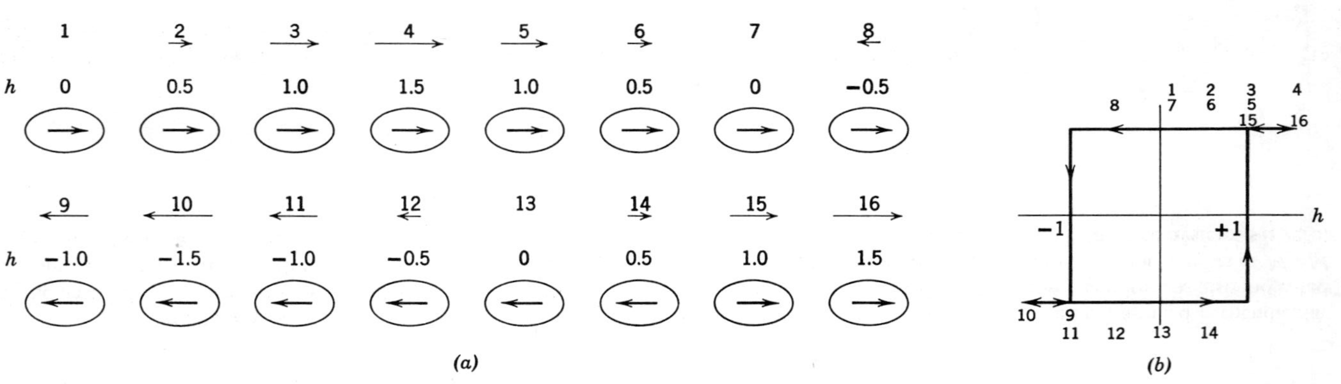

Fig. 7-2.6. Similar to Fig. 7-2.5, except that the field is applied along the polar axis ($\theta = 0°$). The jump in the magnetization occurs at $h = 1$. No reversible rotation occurs.

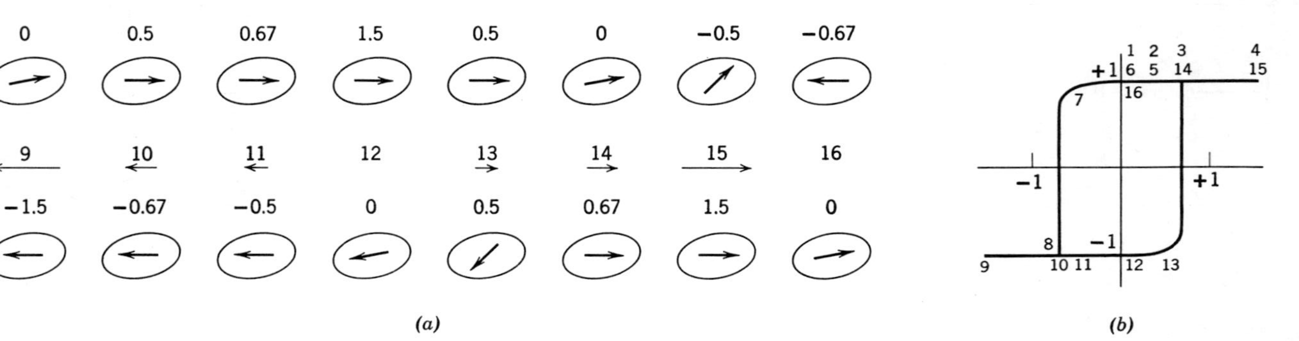

Fig. 7-2.7. The field is applied at $\theta = 10°$ (to the polar axis). The rotation of **M** is reversible until $h = 0.67$; then the magnetization jumps. The coercive force h_c is also equal to 0.67.

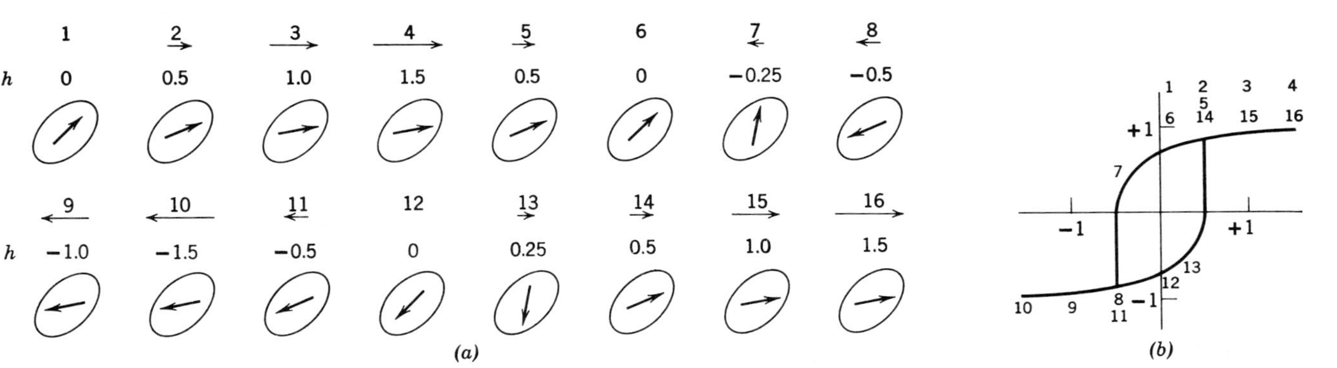

Fig. 7-2.8. $\theta = 45°$. The jump occurs at $h = 0.5$, which is the smallest field at which a jump takes place for any value of θ. The magnetization is double valued for $-0.5 < h < 0.5$.

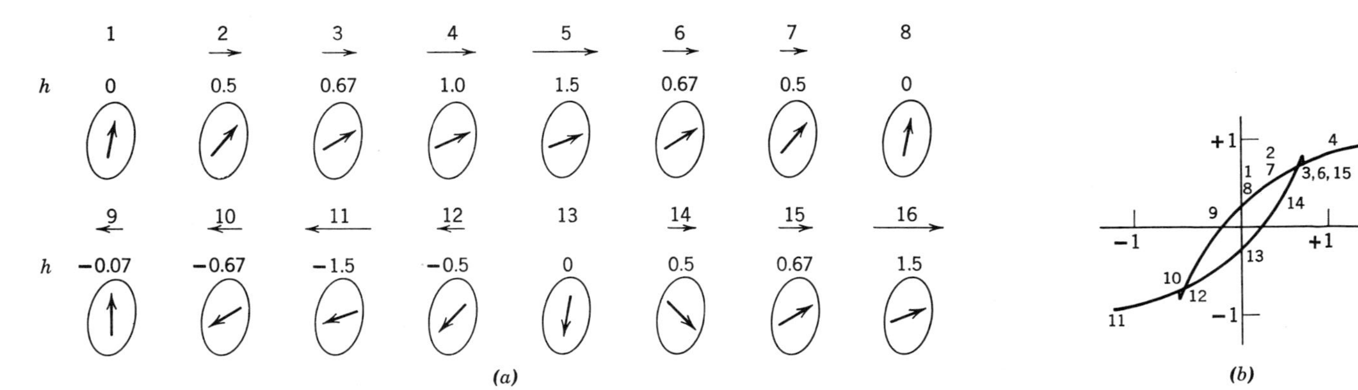

Fig. 7-2.9. $\theta = 80°$. The critical field for a jump of the magnetization **M** occurs at $h = 0.67$: then there is a *decrease* in M_H. The magnetization **M** does not lie along **h** until $h \to \infty$.

except perhaps during the first, depending on the initial orientation of the magnetization (see Figs. 7-2.6, 7-2.7, 7-2.8, and 7-2.9).

Only for $\theta = 0°$ and $\pi/2$ does the magnetization **M** ever lie along the field direction for a finite field **h**: for all others an infinite field is required. The reason is as follows. The torque arising from the field $MH \sin \phi$ approaches zero for $\phi \to 0$ and H finite. When the opposing torque, arising from the particles's shape, is finite, alignment is possible only for $\mathbf{H} \to \infty$. For $\theta = 0°$ and $\pi/2$ the torque from the shape is zero, and alignment is possible with **H** finite; in fact it occurs at $h = 1$.

The critical field for a jump in the magnetization occurs when the free energy changes from a minimum to a maximum. This condition is

$$\frac{\partial F_T}{\partial \phi} = \frac{\partial^2 F_T}{\partial \phi^2} = 0.$$

It may be shown (problem 7-2) that the critical field for a given θ, as calculated from this condition, is

$$h_{\text{crit}} = -(1 - w^2 + w^4)^{1/2}/(1 + w^2),$$

where (7-2.9)

$$w = \tan^{1/3} \theta.$$

The maximum value of $|h_{\text{crit}}|$ occurs for $\theta = 0$ or π; then $|h_{\text{crit}}| = 1$.

Experimentally, an assembly of particles is usually studied. Suppose the polar axes of the prolate ellipsoids are orientated at random. Since the coercive force of the individual particles ranges from $h = 0$ to $h = 1$ (see the foregoing figures), a reasonable guess for the coercive force of the powder would be 0.5. By actual calculation Stoner and Wohlfarth[10] find $h_c = 0.479$. The remanence is expected and is found to be $\overline{\cos \phi} = 0.5$. The magnetization and hysteresis curves for ellipsoidal particles with random orientation are shown in Fig. 7-2.10.

These results on hysteresis curves of particles with shape anisotropy may also apply to spherical particles with crystalline or strain anisotropy. For a material with uniaxial anisotropy, such as cobalt, the crystalline anisotropy energy, to first order, is given by (equation 7-1.7)

$$F_K = K_1' \sin^2 \alpha,$$

where α is the angle between **M** and the easy direction. In an applied field the total energy is then

$$F_T = (K_1' \sin^2 \alpha - HM \cos \phi)V$$

and the equilibrium equation is

$$\tfrac{1}{2} \sin 2(\phi - \theta) + h \sin \phi = 0,$$

[10] E. C. Stoner and E. P. Wohlfarth, *Phil. Trans. Roy. Soc.* (*London*) **A-240,** 599 (1948).

where

$$h = \frac{HM}{2K_1'}. \tag{7-2.10}$$

Equation 7-2.10 is identical to equation 7-2.7 except for the reinterpretation of the reduced field. Hence the maximum coercive force occurs for $h = 1$, that is, for $H_c = 2K_1'/M$. The cubic anisotropy is more complicated.[11] It is easy to show, however, that a field applied along the [100]-direction yields the maximum coercive force, equal to $H_c = 2K_1/M$. For

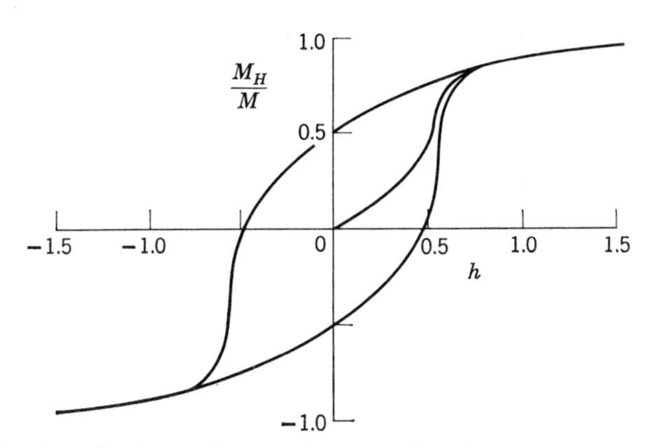

Fig. 7-2.10. Magnetization and hysteresis curves of prolate ellipsoidal particles orientated at random. The interaction between particles is neglected. [E. C. Stoner and E. P. Wohlfarth, *Phil. Trans. Roy. Soc. (London)* **A-240**, 599 (1948).]

a powder of spherical particles with cubic crystalline anisotropy, oriented at random, the coercive force is

$$H_c = \frac{0.64K_1}{M} \tag{7-2.11}$$

according to a calculation by Néel.[12] There may, however, be some question concerning this calculation.[13]

For a spherical particle with $F_K = 0$ and $H = 0$, but with a nonzero tension, the energy is just the result of the magnetostriction and is given by (equation 7-1.8)

$$F_\sigma = -\tfrac{3}{2}\lambda_s\sigma\cos^2\alpha.$$

For $\lambda_s\sigma > 0$, that is, positive magnetostriction and tension or negative magnetostriction and compression, the energy is a minimum if **M** lies

[11] C. E. Johnson and W. F. Brown, *J. Appl. Phys.* **32,** 243S (1961).

[12] L. Néel, *Compt. rend. (Paris)* **224,** 1488 (1947).

[13] C. E. Johnson and W. F. Brown, *loc. cit.*

Table 7-2.1. Maximum Coercive Forces of Single-Domain Particles Assumed to Possess Only One Type of Anisotropy
(room temperature)

Material	Coercive Force (oersteds)		
	Shape $2\pi M$	Crystalline $\frac{2K_1}{M}$ or $\frac{2K_1'}{M}$	Strain $\frac{3\lambda_s\sigma}{M}$
Fe	10,700	540	600
Ni	3,050	5850	4000
Co	8,800	200	600

along the strain axis ($\alpha = 0$) and a maximum if **M** lies in the plane perpendicular to the strain axis. Therefore it is customary to say that such a particle possesses *strain* anisotropy. Then, if $H \neq 0$, the total energy is

$$F_T = (-\tfrac{3}{2}\lambda_s\sigma \cos^2 \alpha - HM \cos \phi)V.$$

The equilibrium equation is

$$\tfrac{1}{2} \sin 2 (\phi - \theta) + h \sin \phi = 0,$$

provided

$$h = \frac{HM}{3\lambda_s\sigma}. \tag{7-2.12}$$

This is again identical to the prolate ellipsoidal case, provided, of course, $\lambda_s\sigma > 0$. The maximum coercive force is given by $H_c = 3\lambda_s\sigma/M$.

Some examples in which more than one type of anisotropy is present have been investigated. The reader is referred to the literature for details.[14]

Numerical values of the maximum coercive force, under the assumption that only one anisotropy is operative, may be readily estimated. When shape anisotropy is present, the coercive force is a maximum for a long circular cylinder. From equation 7-2.8 it is given by

$$H_c = 2\pi M, \tag{7-2.13}$$

since then $h = 1$ and $(D_b - D_a)$ approaches 2π, its maximum value. When crystalline anisotropy is present, the maximum coercive force is given by $2K_1'/M$ or $2K_1/M$, depending on whether the anisotropy is uniaxial or cubic. When strain anisotropy is present, the maximum coercive force is given by $3\lambda_s\sigma/M$, provided $\lambda_s\sigma > 0$. When the stresses are internal, an upper limit for σ is probably about 2×10^{10} dynes/cm^2. Maximum coercive forces for iron, nickel, and cobalt, as predicted by these equations, are listed in Table 7-2.1. It is these large values that make single-domain

[14] E. P. Wohlfarth and D. G. Tonge, *Phil. Mag.* **2,** 1333 (1957); C. E. Johnson and W. F. Brown, *J. Appl. Phys.* **30,** 320 (1959).

particles important materials for permanent magnets (see Section 7-4).

It is of interest to consider the oblate ellipsoid of revolution, that is, when $b > a$. Then, because of the shape anisotropy, any direction in the equatorial plane is an easy one, whereas the polar axis is a hard direction. Suppose a field is applied at some angle to the polar axis. When the field is reduced from a positive value to zero, the magnetization **M** rotates in the plane defined by the field and the polar axis to an axis in the equatorial plane. When H is made infinitesimally negative, the magnetization rotates in the equatorial plane until it again lies in the $H - a$ plane; that is, it rotates 180°. Thus a discontinuity occurs only for H (or h) = 0, and there is no hysteresis. The magnetization curves for oblate ellipsoids for various angles between the applied field and the polar axis are shown in Fig. 7-2.11. These curves may also apply to spherical particles with either

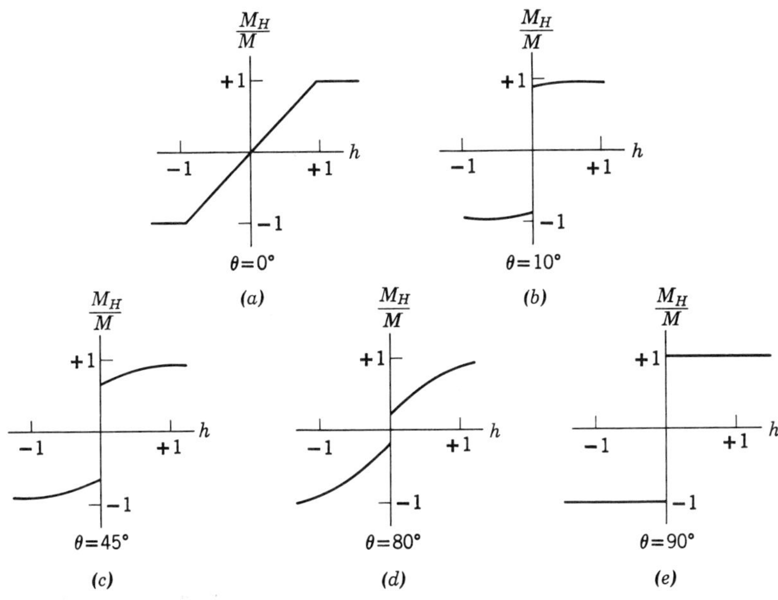

Fig. 7-2.11. Magnetization curves for an oblate ellipsoid of revolution plotted in reduced units. The successive curves are for various angles α between the applied magnetic field and the polar (a) axis, as indicated. In no case is there any hysteresis. [Based on calculations by E. C. Stoner and E. P. Wohlfarth, *Phil. Trans. Roy. Soc. (London)* **A-240,** 599 (1948).]

crystalline or strain anisotropy. This would occur in crystalline anisotropy for a material with uniaxial anisotropy if $K_1' < 0$ and in the strain anisotropy if $\lambda_s\sigma < 0$. An example of $K_1' < 0$ is cobalt at higher temperatures, whereas $\lambda_s\sigma < 0$ either for positive magnetostriction and compression or for negative magnetostriction and tension.

The case of the general ellipsoidal particle, that is, one with three unequal semiaxes, $a > b > c$, is much more complicated. The magnetization then may no longer lie in the plane defined by the directions of the field and the principal axis a. Fortunately, it appears[15] that the results for the prolate and oblate ellipsoids cover the main possibilities that are physically significant.

Actual particles are seldom ellipsoids. However, it can be shown[16] that a single-domain particle of any arbitrary shape is precisely equivalent to a suitably chosen ellipsoid as far as the behavior of its magnetization in an applied (uniform) field is concerned. Hence no more general case than the ellipsoid need be considered.

Incoherent rotations. In the foregoing analysis it has been assumed that the magnitude of the magnetization $|\mathbf{M}|$ remains constant (uniform) both without a field and during reversible and irreversible changes with a field. This assumption has been investigated,[17] starting with equations 7-1.13 and 7-1.14 of micromagnetic theory. Suppose both the easy direction of the crystalline anisotropy and any applied field lie along the z-axis. If the anisotropy and field are large, the magnetization will be at least almost uniform, and the direction cosine α_3 of the unit vector $\mathbf{u}$ will be nearly equal to 1, whereas the direction cosines α_1 and α_2 will be small. The micromagnetic equations, when expressed to first order in α_1 and α_2, become linear. Then, starting with a state of uniform magnetization, the possibility that small deviations from uniformity will occur on reduction of the field may be examined rigorously. If there are small nonuniformities, they will become larger for an infinitesimal decrease of the field: the analysis will then require the use of the nonlinear equations. To summarize, the nonlinear equations provide the following information: (*a*) whether the magnitude of the magnetization will become nonuniform or remain uniform when the magnetic field is changed, (*b*) if the magnetization becomes nonuniform, the field, often called the *nucleation* field, at which it will occur, and (*c*) the form of the nonuniform magnetization. However, if nonuniform magnetization begins, the magnetization curve can be followed only by employing the nonlinear equations.

It turns out[18] that the reversible magnetization changes continue to occur coherently, whereas the irreversible changes may be incoherent. Further, the nucleation field for an incoherent reversal is lower than the

[15] E. C. Stoner and E. P. Wohlfarth, *Phil. Trans. Roy. Soc. (London)* **A-240,** 599 (1948); C. E. Johnson, *J. Appl. Phys.* **33,** 2515 (1962).

[16] W. F. Brown and A. H. Morrish, *Phys. Rev.* **105,** 1198 (1957).

[17] W. F. Brown, *Phys. Rev.* **105,** 1479 (1957); E. H. Frei, S. Shtrikman, and D. Treves, *Phys. Rev.* **106,** 446 (1957).

[18] For a summary see W. F. Brown, *J. Appl. Phys.* **30,** 62S (1959) and A. Aharoni, *J. Appl. Phys.* **30,** 70S (1959).

corresponding critical field of the Stoner-Wohlfarth theory. As a result, a lower coercive force is predicted. For example, consider a prolate ellipsoidal particle. For values of the equatorial semiaxis b, less than the value

$$b_c = \frac{\alpha}{M} \frac{A^{1/2}}{D_b},$$

the irreversible changes continue to take place by rotation in unison of the atomic magnetic moments, as in the Stoner-Wohlfarth model. Here α is a constant about equal to 2, and A is the exchange constant of equation 7-1.12. This critical value is smaller than that given by equation 7-2.3. For larger values of b the irreversible change takes place by a mode called *curling*, in which there is a circumferential component of magnetization that is a function of the distance from the polar axis. These two modes are illustrated in Fig. 7-2.12. Note that the magnetization curling

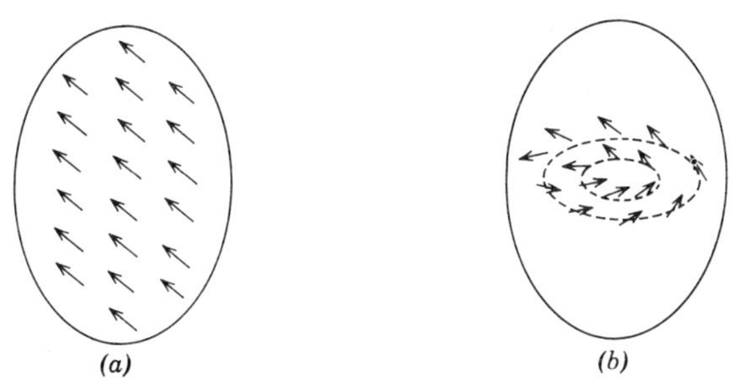

Fig. 7-2.12. Irreversible magnetization changes in a prolate ellipsoid by (*a*) coherent rotation and by (*b*) the curling mode. [E. H. Frei, S. Shtrikman, and D. Treves, *Phys. Rev.* **106,** 446 (1957).]

mode passes through the assumed stable state shown in Fig. 7-2.2. Curling again occurs in the infinite circular cylinder. For smaller radii, there is another mode, called *buckling*, instead of the rotation in unison. This buckling mode consists of a rotation whose amplitude varies sinusoidally along the cylinder's axis (see Fig. 7-2.13). It can be shown, nevertheless, that rotation in unison is a good approximation to the buckling mode; the rotation in unison mode is frequently employed in calculations for the circular cylinder.

It is not certain whether incoherent rotations actually take place in particles. On the theoretical side some energy terms, such as the magnetostriction F_σ and other F_0 energy terms, have been omitted from the micromagnetic differential equations. On the experimental side there exists some indirect but no direct evidence of incoherent reversal modes. In any case, it should be emphasized that the Stoner-Wohlfarth model still applies to sufficiently small particles as well as to the reversible magnetization changes that occur up to the nucleation fields in larger particles.

Incoherent magnetization reversal may be expected in a particle that is not a single unit. For example, Jacobs and Bean[19] consider that an elongated particle may actually consist of several spheres attached together to form a linear chain, as indicated in Fig. 7-2.14. In their idealized

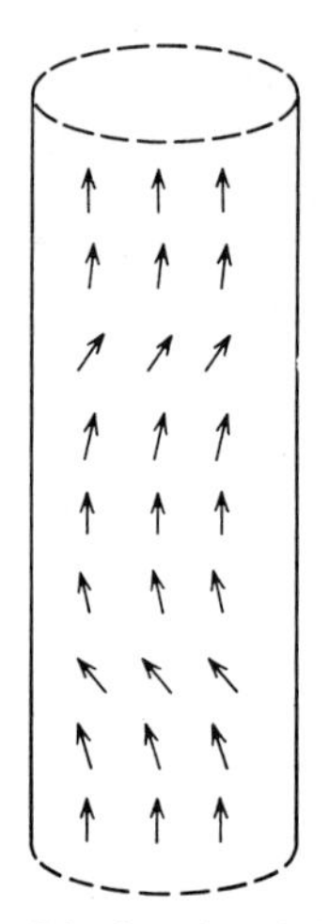

Fig. 7-2.13. The change of magnetization in an infinite circular cylinder by the buckling mode. [E. H. Frei, S. Shtrikman, and D. Treves, *Phys. Rev.* **106,** 446 (1957).]

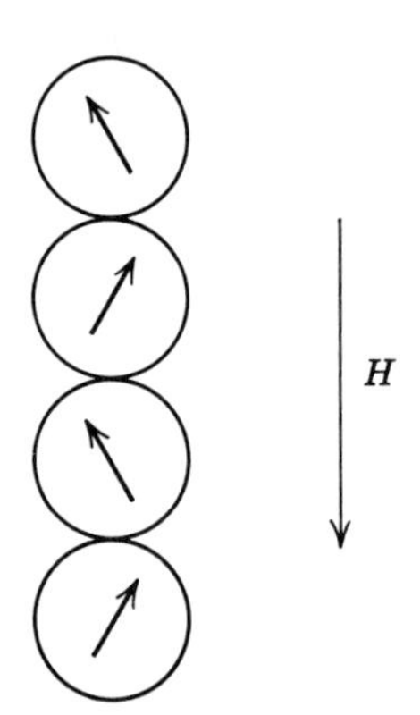

Fig. 7-2.14. Magnetization reversal by "fanning" in a chain of spheres. (After I. S. Jacobs and C. P. Bean.)

model the magnetization of each sphere is considered to be uniform in magnitude, and only a dipole-dipole interaction is assumed between the spheres. Then, when a field is applied parallel to the axis of the chain, the magnetization reverses by rotations of the sphere's magnetization in opposite directions (fanning), as illustrated in Fig. 7-2.14. This incoherent magnetization reversal results in a lower coercive force than would arise for a simultaneous parallel rotation of the moments (see problem 7-3).

Some experimental results. The experimental evidence for the existence of single-domain particles will now be reviewed. The coercive force of powders containing small particles is observed to increase as the particle

[19] I. S. Jacobs and C. P. Bean, *Phys. Rev.* **100,** 1060 (1955).

size decreases, at least until extremely small sizes are reached (the superparamagnetic particles to be discussed in Section 7-3). Typical results[20] for iron and cobalt are shown in Fig. 7-2.15. Although large coercive forces are observed, they are still almost always considerably less than the values predicted by theory. It is worth noting that no sudden discontinuity in the coercive force is seen at the critical size and further that the coercive force continues to increase for particles smaller than the critical size. Part of the explanation probably lies in the fact that actual

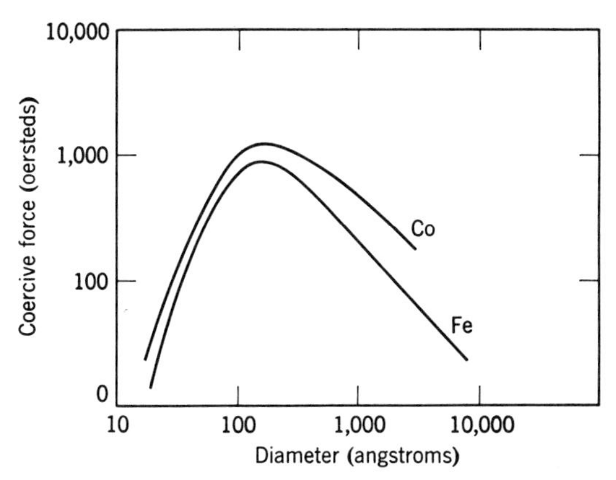

Fig. 7-2.15. Coercive force as a function of particle size. The coercive force of the powders develops from both crystalline and shape anisotropy. $T = 76°K$.

powders always contain a distribution of particles; this would lead to a smearing out of the curves. However, incoherent reversals, such as curling, would also tend to produce these results; this is indirect evidence for their existence.

Elmore[21] measured the magnetization curves of colloidal particles. If the particles were permanent magnets, they would be expected to behave as the molecules of a classical paramagnetic gas and have a magnetization curve given by the Langevin equation (see problem 2-6)

$$M_H = N\mu L\left(\frac{\mu H}{kT}\right).$$

Experiment showed that this was so. Of course, this experiment does not show whether the particles were single domain or merely had a nonzero remanent magnetization.

[20] W. H. Meiklejohn, *Revs. Mod. Phys.* **25,** 302 (1953); F. E. Luborsky, *J. Appl. Phys.* **32,** 171S (1961).

[21] W. C. Elmore, *Phys. Rev.* **54,** 1092 (1938).

Other experiments[22] involving powders have been performed. These experiments always suffer from a number of complications which make interpretation of the results more difficult. Chief among them are the interaction effects between the particles, the distribution of particle sizes and shapes, and the averaging effect due to the different orientations of the particles.

Thus an experiment on one individual particle has several advantages. Such an experiment[23] has been performed with an ultrasensitive quartz-fiber torsion balance that had a 0.5 micron suspension fiber. The remanent

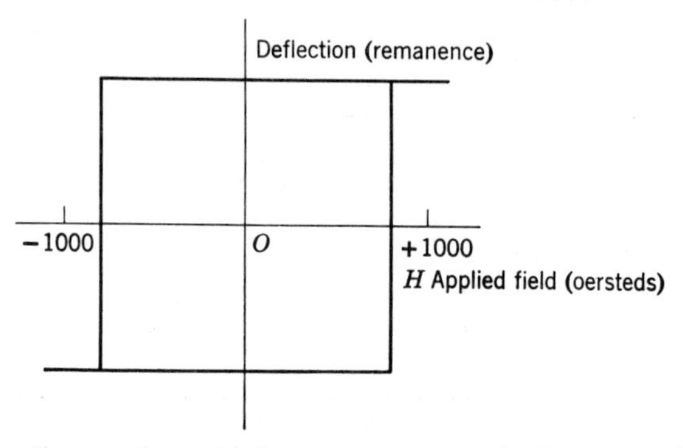

Fig. 7-2.16. The discontinuous change in the remanent magnetization expected for a single-domain particle after saturation by a large field and subsequent application and removal of a reverse field H. For the particles studied the discontinuity occurred for $H \approx 800$ oe.

magnetization of a micron-sized particle[24] was determined by noting the deflection of the balance in a small inhomogeneous magnetic field. Now, the remanence of a single-domain particle with some sort of uniaxial anisotropy is expected to have only two possible values, since in a zero applied field the magnetization will lie along either the negative or positive easy direction. Further, one particular value of remanence will be changed to the other value by the application and subsequent removal of a reverse magnetic field that is large enough to cause an irreversible rotation of the magnetization. The deflection, or remanence, expected as a function of an applied reverse field H is shown in Fig. 7-2.16. Such discontinuous changes in the remanence were indeed observed for individual small

[22] C. Kittel, J. K. Galt, and W. E. Campbell, *Phys. Rev.* **25,** 302 (1950); L. Weil, *J. Phys. radium* **12,** 437 (1951); A. H. Morrish and S. P. Yu, *J. Appl. Phys.* **26,** 1049 (1955).

[23] A. H. Morrish and S. P. Yu, *Phys. Rev.* **102,** 670 (1956).

[24] The particles studied were actually ferrimagnetic rather than ferromagnetic. This in no way restricts the validity of the general results.

particles. On the other hand, the remanence was observed to change continuously for larger particles, a result to be expected if the particles were multidomain in character. This experiment then provides direct evidence for the existence of single-domain particles. It leaves unanswered, however, the question whether the irreversible magnetization change occurred by a coherent or an incoherent mode.

Other effects. In most powders the distance between particles is rather small, and consequently the magnetic interaction between particles is important. The calculation of this interaction is a complicated many-body problem. To date, the theories have been based on somewhat specialized models.[25] For single-domain particles whose coercive force is primarily determined by shape anisotropy, several of the theories predict the relationship

$$H_c(p) = H_c(0)(1 - p) \quad (7\text{-}2.14)$$

where p is the packing fraction defined as the ratio of the density of the powder to the density of the bulk material. This result is in good agreement with most, though not all, of the experimental results.[26] If the coercivity of particles is determined mainly by crystalline anisotropy, it is predicted that there will be little or no effect on H_c from the interaction; this is observed experimentally.

Particle interactions may be expected to influence the critical single-domain size. According to theory,[27] the critical size increases when the packing fraction increases. There is experimental evidence for this effect.[28]

Mention should also be made of interesting work on the remanence and rotational hysteresis of powders containing single-domain particles. Measurement of the reduced remanence (M_R/M) as a function of a previously applied field H permits, in principle, the determination of the shape distribution of the powders.[29] This follows, since the remanence changes occur when the applied field equals a critical field for an irreversible change of magnetization. By rotational hysteresis we mean the work done on rotating a specimen 360° in a magnetic field. This work is

[25] L. Néel, *Compt. rend. (Paris)* **224,** 1550 (1947), *Appl. Sci. Res.* **B-4,** 13 (1954); C. Kittel, *Revs. Mod. Phys.* **21,** 541 (1949); E. Kondorsky, *Dokl. Akad. Nauk SSSR* **68,** 37 (1951); E. P. Wohlfarth, *Proc. Roy. Soc. (London)* **A-232,** 208 (1955); A. Aharoni, *J. Appl. Phys.* **30,** 70S (1959).

[26] E. H. Carman, *Brit. J. Appl. Phys.* **6,** 426 (1955); F. E. Luborsky, L. I. Mendelsohn, and T. O. Paine, *J. Appl. Phys.* **28,** 344 (1957).

[27] E. Kondorsky, *Izvest. Akad. Nauk SSSR, Ser. Fiz.* **16,** 398 (1952); A. Aharoni, *J. Appl. Phys.* **30,** 70S (1959).

[28] A. H. Morrish and L. A. K. Watt, *Phys. Rev.* **105,** 1476 (1957).

[29] E. P. Wohlfarth, *J. Appl. Phys.* **29,** 595 (1958); C. E. Johnson and W. F. Brown, *J. Appl. Phys.* **29,** 313 (1958).

the result of the irreversible changes of magnetization. Hence the rotational hysteresis, measured as a function of the applied field, also gives information on the distribution of the critical fields for the powder.[30] In addition, such measurements provide evidence for incoherent rotations of the magnetization.

3. Superparamagnetic Particles

In the preceding section it has been assumed that when the applied field is zero the magnetization is stable; that is, the direction of magnetization remains unchanged with time. To be specific, consider a single-domain particle with uniaxial anisotropy. The variable part of the energy is then given by an expression of the type

$$F_T = \tfrac{1}{2}CV \sin^2 \alpha, \qquad (7\text{-}3.1)$$

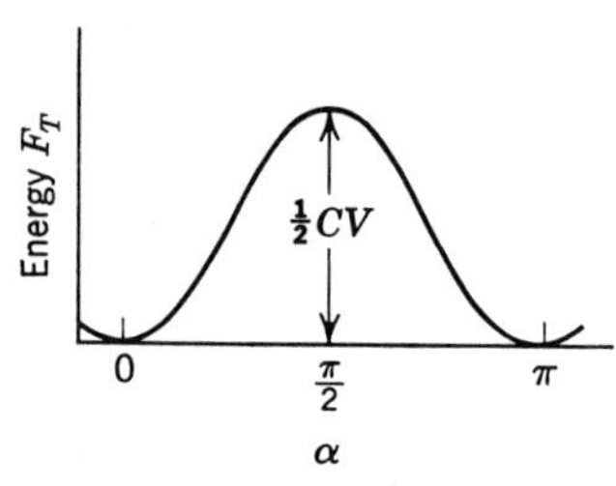

Fig. 7-3.1. The energy of a single-domain particle with uniaxial anisotropy as a function of the angle α between the magnetization and the easy direction. The height of the energy barrier is $\frac{1}{2}CV$. $H = 0$.

where V is the volume of the particle, α is the angle between the magnetization $\mathbf{M}$ and the easy direction, and C depends on the anisotropy present but is a constant for a given particle. From equations 7-2.4, 7-1.7 and 7-1.8 it follows that for shape anisotropy, $C = (D_b - D_a)M^2$, for crystalline anisotropy $C = 2K_1'$, whereas for strain anisotropy $C = 3\lambda_s\sigma$. The dependence of the energy F_T on α is shown graphically in Fig. 7-3.1. Either $\alpha = 0$ or π is obviously a direction of minimum energy, and these directions are separated by an energy barrier of height $\frac{1}{2}CV$. The magnetization will remain stable and lie along the direction defined by $\alpha = 0$ or by $\alpha = \pi$ unless some perturbing force exists that can take the magnetization over the energy barrier. Thermal agitation may provide such a perturbation. This process is most likely to occur if the volume V of the particle is small, so that the height of the energy barrier is lowered, or if the temperature T is high. If the process does occur, the time average of the remanence will be zero. This process is presumably the reason for the decrease in the coercive force shown in Fig. 7-2.15 for the small particle sizes. Particles whose magnetization changes spontaneously are analogous to paramagnetic atoms, except that their magnetic moment is

[30] I. S. Jacobs and F. E. Luborsky, *J. Appl. Phys.* **28,** 467 (1957); R. B. Campbell, H. Amar, A. E. Berkowitz, and P. J. Flanders, *Proc. Conf. Mag. and Mag. Materials, Boston* **T-91,** 118 (1957).

much larger. Such particles are said to exhibit *superparamagnetism*[31]; their existence was first predicted by Néel.[32]

The process may be characterized by a relaxation time τ. For example, consider a powder of uniaxial particles whose easy directions all lie along the z-axis. Suppose a field is applied along the positive z-axis and that the powder is magnetized to saturation M by the application of a field. When the field is removed, the magnetization will decay according to the equation

$$M_H = Me^{-t/\tau}. \qquad (7\text{-}3.2)$$

For τ very large $M_H = M$ for all times of interest, and the system is stable. For τ small enough $M_H \to 0$ in time.

Now $1/\tau$ is the sum of the probability per second for a single transition of the magnetization from a $+z$- to a $-z$-direction, plus that for the transition from $-z$ to $+z$; the analogous problem for a system of paramagnetic atoms has been treated in Section 3-5. Thus $1/\tau$ must be proportional to the Boltzmann factor $e^{-CV/2kT}$, since $\frac{1}{2}CV$ is just the activation energy for a transition. The constant of proportionality may be estimated from the following argument. The upper limit for $1/\tau$ must be given by the frequency of precession from a $+z$- to a $-z$-direction, namely $\gamma H_{\text{crit}}/2\pi$, where H_{crit} is the equivalent applied field necessary to produce a magnetization reversal in a stable particle with this geometry. Thus H_{crit} is given by C/M, regardless of the type of anisotropy present (see Section 7-2). Since $\gamma \doteqdot 10^7$, the order of magnitude of $\gamma C/2\pi M$ is 10^9 or 10^{10}. More sophisticated arguments[33] give the same order of magnitude. Thus we have

$$\frac{1}{\tau} \doteqdot 10^9 e^{-CV/2kT}. \qquad (7\text{-}3.3)$$

If the anisotropy is cubic, this equation is modified to

$$\frac{1}{\tau} = 10^9 e^{-CV/8kT},$$

provided the cube edges are the easy directions. If the magnetization reversal is incoherent, the energy barrier $\frac{1}{2}CV$ will be reduced. The application of a field also modifies the foregoing result (see problem 7-4).

For a given temperature and particle shape the relaxation time τ changes from less than a second to several years when the volume V changes by a relatively small amount. For example, a spherical particle

[31] C. P. Bean, *J. Appl. Phys.* **26,** 1381 (1955).
[32] L. Néel, *Compt. rend.* (*Paris*) **228,** 604 (1949).
[33] L. Néel, *Ann. geophys.* **5,** 99 (1949); W. F. Brown, *J. Appl. Phys.* **30,** 1308 (1959).

with $C = 10^6$ ergs/cm^3 at $T = 300$°K has $\tau = 10^{-1}$ sec for a radius of 17 Å and $\tau = 10^8$ sec for a radius of 22 Å. Only over a very narrow range of sizes is τ the order of the time of an experiment. Thus, to a good degree of approximation, there exists a critical particle size above which the particles are stable and below which the particles are always in thermodynamic equilibrium and exhibit superparamagnetism.

To be definite, the critical size may be defined as that which corresponds to $\tau = 10^2$ sec. Then $\frac{1}{2}CV \doteq 25\mathrm{k}T$. For a given particle shape and volume the temperature that satisfies this equality is often called the *blocking* temperature T_B. For spherical particles with crystalline anisotropy, and with $T_B = 300$°K, the critical radius is 125 Å for Fe and 40 Å for Co.

One experimental test for the existence of superparamagnetism is to measure the remanence of a system of single-domain particles as a function of temperature. Of course, in all such systems there is a distribution of particle sizes. Hence, as the temperature decreases, the remanence is expected to increase in a manner dependent on the portion of particles that are superparamagnetic. This effect has been observed experimentally.[34] The size distribution of the particles may also be inferred from such measurements.

Another experimental test is to measure the magnetization curve at different temperatures. If all the particles are superparamagnetic, the magnetization should be given by the Langevin equation

$$M_H = NL(\mu H/\mathrm{k}T).$$

Experimentally, it is indeed found[35] that when the magnetization is plotted as a function of H/T the curves taken at different temperatures are superimposed. The curves do not coincide with the Langevin curve calculated under the assumption that all particles have the same magnetic moment. This is presumably because of the presence of a distribution of particle sizes.

The spontaneous magnetization can be determined from the initial slope of the magnetization curve. It is found that for particles as small as about 10 Å the spontaneous magnetization is independent of particle size[36] and of temperature[37] in the range from 4.2 to 300°K.

[34] L. Weil, *J. chim. phys.* **51,** 715 (1954); L. Weil and L. Gruner, *Compt. rend.* (*Paris*) **243,** 1629 (1956); W. Henning and E. Vogt, *Z. Naturforsh.* **12a,** 754 (1957).

[35] C. P. Bean and I. S. Jacobs, *J. Appl. Phys.* **27,** 1448 (1956); J. J. Becker, *Trans. Am. Inst. Mining, Met. Petrol. Engrs.* **209,** 59 (1957); A. E. Berkowitz and P. J. Flanders, *J. Appl. Phys.* **30,** 111S (1959).

[36] F. E. Luborsky, *J. Appl. Phys.* **29,** 309 (1958).

[37] J. W. Cahn, I. S. Jacobs, and P. E. Lawrence, quoted in *J. Appl. Phys.* **30,** 124S (1959).

Finally, it should be mentioned that in addition to being in the form of powder the particles may be suspended in a liquid or a solid. In a solid the particles are usually produced by precipitation out of solid solution in an alloy. Application of the principles of superparamagnetism permits studies of the precipitated particles to be made as a function of the metallurgical treatment of the alloy.[38]

4. Permanent Magnet Materials

The two main features a material should possess in order to serve as a superior permanent magnet are the ability to produce a large magnetic field and to continue to produce this field for an extended period of time. A large value of the remanence is necessary for the first requirement, whereas a large value of the coercive force is necessary for the second. In the analysis of the performance of permanent magnets it is appropriate to consider the B_H versus H hysteresis loop. Thus a figure of merit for a permanent magnet material could be defined as the product $B_r H_c$ where it is understood that the reference is to ${}_BH_c$. Actually, a better figure of merit is the maximum value of the product of B_H and H, denoted by $(B_H H)_{\text{max}}$, where B_H and H lie on that part of the hysteresis curve in the second quadrant (often called the demagnetization curve). The reason for this is as follows. Consider the integral $\int \mathbf{B} \cdot \mathbf{H}\, dv$ taken over all space. Now we have

$$\int \mathbf{B} \cdot \mathbf{H}\, dv = -\int \mathbf{B} \cdot \nabla\varphi\, dv = \int \varphi \nabla \cdot \mathbf{B}\, dv - \int \nabla \cdot (\varphi \mathbf{B})\, dv.$$

Now $\nabla \cdot \mathbf{B} = 0$ by equation 1-8.2 and $\int \nabla \cdot (\varphi \mathbf{B})\, dv = 0$ by equations 1-6.8 and 1-6.9. Hence

$$\int B_H H\, dv = 0.$$

A permanent magnet is usually in the shape of a horseshoe or sometimes a bar, so that there is always an air gap of some geometry. Thus we may write

$$\int_{\text{magnet}} B_H H\, dv = -\int_{\text{outside}} H^2\, dv,$$

where the integral on the left-hand side is over the volume of the magnet and the integral on the right-hand side is over the space outside the magnet. Therefore

$$(B_H H)V = -qH_a^{\,2}V_a, \qquad (7\text{-}4.1)$$

[38] For a summary see C. P. Bean and J. D. Livingston, *J. Appl. Phys.* **30,** 124S (1959).

where V is the volume of the magnet, V_a is the useful volume of the air gap, H_a is the field in this gap, and q is a leakage factor, greater than unity, that takes account of the flux that lies outside the gap (see Section 1-9). Thus for a given volume of material V the largest magnetic field is produced when $(B_H H)$ is a maximum. Two design principles immediately become apparent. One, the magnet should have such a geometry that in the presence of its own demagnetizing field it operates at the point on the demagnetization curve for which the energy product $B_H H$ is a maximum. Two, a material with a large value of $(B_H H)_{\max}$ should be chosen. In addition, there are other design factors such as cost and mechanical properties of the material that need to be considered.

Single-domain particles possess properties that make them almost ideal permanent magnet materials. If a large anisotropy is present, they have a large coercive force (see Table 7-2.1). If they are aligned and of a material with a large spontaneous magnetization, they have both a large remanence and a large value of $(B_H H)_{\max}$. There are two general methods of manufacturing small-particle magnets. In one method the single-domain particles of a powder are first aligned and then compressed; sometimes a nonmagnetic binder is added. In the other method an alloy is treated metallurgically so that small ferromagnetic particles are precipitated out into a relatively nonmagnetic matrix; that is, at least two phases are produced. Historically, the second method came first and indeed preceded the development of the theory of single-domain particles. Also, to date, the highest value obtained for $(B_H H)_{\max}$ has been for a material produced by this second method. The demagnetization and the $B_H H$ curves of some typical permanent magnet materials are shown in Fig. 7-4.1.[39,40] The quantity $(B_H H)$ is often called the energy product and has the units of gauss-oersteds or ergs per cubic centimeter.

Permanent magnets have been fabricated from the powders of several different materials. One of the earliest was made from manganese-bismuth particles, whose predominant anisotropy is crystalline.[41] Values of $(B_H H)_{\max}$ of 5.4×10^6 gauss-oe have been achieved. A more promising approach has been the use of elongated iron or iron-cobalt particles, whose anisotropy is primarily the result of their shape. Values of $(B_H H)_{\max}$ as high as 6.5×10^6 gauss-oe have been obtained[42] for Fe-Co particles. This is still very much less than the maximum energy product of about

[39] F. E. Luborsky, L. I. Mendelsohn, and T. O. Paine, *J. Appl. Phys.* **28,** 344 (1957).

[40] A. E. Luteijn and K. J. de Voss, *Philips Res. Rept.* **11,** 489 (1956); A. J. J. Koch, M. G. van der Steeg, and K. J. de Voss, *Berichte der Arbeitsgemeinschaft Ferromagnetismus*, Rieder Verlag, Stuttgart (1959).

[41] C. Guillaud, *J. Rech. Centre Natl. Rech. Sci.* **2,** 267 (1949); E. Adams, M. W. Hubbard, and A. M. Syeles, *J. Appl. Phys.* **23,** 1207 (1952).

50×10^6 gauss-oe that appears to be possible. A list[42] of the properties of permanent magnets made from various powders is given in Table 7-4.1.

Of the permanent magnet materials fabricated by treating an alloy, the Alnico family is by far the most important at present. The first Alnico-type material was discovered by Mishima[43] in Japan. The early Alnicos had a $(B_H H)_{\max} \approx 1.6 \times 10^6$ gauss-oe. It was found that if the specimen

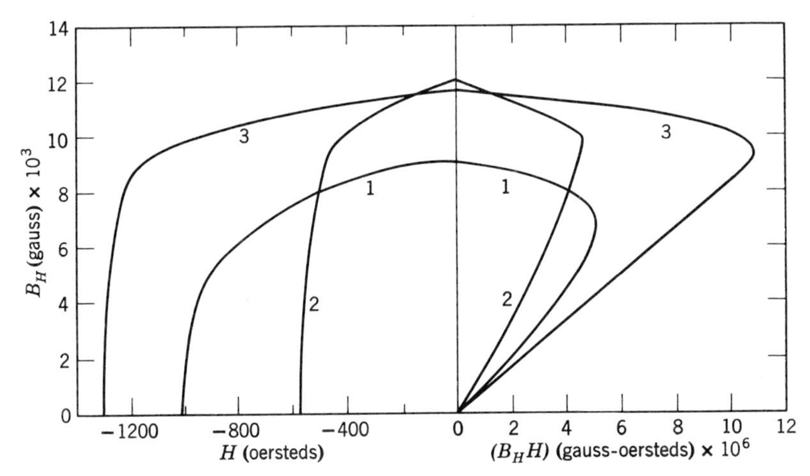

Fig. 7-4.1. The demagnetization and $B_H H$ curves of some typical permanent magnet materials. The numbers refer to the following materials: (1) a compressed powder of aligned elongated Fe-Co particles, (2) Alnico V, and (3) Ticonal XX (a type of Alnico V). [(1) F. E. Luborsky, L. I. Mendelsohn, and T. O. Paine, *J. Appl. Phys.* **28,** 344 (1957). (3) A. E. Luteijn and K. J. de Voss, *Philips. Res. Rept.* **11,** 489 (1956).]

was cooled in the presence of a magnetic field the maximum energy product was greatly increased. Materials produced in this manner are known as Alnico V. Electron-microscope and other observations show that the materials have rodlike precipitates, which have almost the composition of iron, imbedded in a matrix that is rich in Ni and Al. Thus it appears that the materials consist of single-domain particles with a large shape anisotropy distributed in an relatively nonmagnetic matrix.

An Alnico V material with the addition of Ti, known as Ticonal XX, has been produced with $(B_H H)_{\max} \fallingdotseq 11 \times 10^6$ gauss-oe (see Fig. 7-4.1). The properties of permanent magnets[44] made with other alloys, treated in such a manner that small particle precipitates occur, are listed in Section

[42] F. E. Luborsky, *J. Appl. Phys.* **30,** 174S (1961).

[43] T. Mishima, *Ohm* **19,** 353 (1932).

[44] For a general review see E. P. Wohlfarth, *Advan. Phys.* **8,** 87 (1959).

Table 7-4.1. Data on Permanent Magnets Made from Powders

Material	Predominant Anisotropy	B_r (gauss)	${}_BH_c$ (oersted)	$(B_HH)_{max} \times 10^{+6}$ (gauss-oersteds)
Cobalt ferrite				
$CoFe_2O_4$	crystalline	4000	500	0.95
Barium ferrite				
$BaO \cdot 6Fe_2O_3$	crystalline	4000	1930	3.7
Manganese bismuth				
(MnBi)	crystalline	4800	3650	5.4
Iron	crystalline	6000	475	1.0
Iron-cobalt	crystalline	8600	410	1.6
Manganese aluminum	crystalline	4280	2750	3.5
Iron	shape	9150	765	4.3
Iron-cobalt	shape	10800	980	6.5
Iron-cobalt ferrite	shape	6000	1600	4.0

(a) of Table 7-4.2. Fine particle magnets have also been made by electrodeposition.[45]

It is worth emphasizing that the magnetization processes that occur in permanent magnets of the alloy fine-particle type may not be limited to the simple rotations in single-domain particles. For one thing, the particles sometimes appear to be joined by small fibers or islands of

Table 7-4.2. Data on Permanent Magnets Made from Alloys

Name	Percent Composition (Remainder Iron)	B_r (gauss)	${}_BH_c$ (oersteds)	$(B_HH)_{max} \times 10^{+6}$ (gauss-oersteds)
(a)				
Alnico II	12Al, 18Ni, 13Co, 6Cu	7300	560	1.6
Alnico V	14Ni, 8Al, 24Co, 3Cu	12700	650	5.5
Ticonal XX	15Ni, 7Al, 34Co, 4Cu, 5Ti	11800	1315	11.0
Vicalloy II	52Co, 13V	10400	570	4.2
Cunife I	60Cu, 20Ni	5400	550	1.5
Remalloy, Comalloy				
or Indalloy	12Co, 17Mo	10500	250	1.2
Platinum-iron	40Pt	6000	1900	3.3
Platinax II	50Pt, 50Co	6400	4800	9.2
(b)				
Carbon steel	1Mn, 0.9C	10000	50	0.2
Tungsten steel	5W, 0.7C, 0.3Mn	10500	65	0.3
Chromium steel	3.5Cr, 1C, 0.4Mn	9500	65	0.3
Cobalt steel	36Co, 7W, 3.5Cr, 0.9C	10000	230	1.0

[45] J. S. Sallo and K. H. Olson, *J. Appl. Phys.* **32,** 203S (1961).

materials; for another, it is not clear what role the weakly ferromagnetic matrix may play. The problem is likely to be complicated.

Before the introduction of fine-particle magnets various types of quench-hardened steels (iron with about 1% C) were used as permanent magnet materials. These inhomogeneous materials probably derived their properties by impeding the movement of domain walls, perhaps by one of the mechanisms to be discussed in Section 7-7. A list of these materials is tabulated in Section (b) of Table 7-4.2.

5. Domain Walls

For particles larger than the critical single-domain size (for $H = 0$) some sort of nonuniform magnetization is expected. One possibility is the circular spin arrangement illustrated in Fig. 7-2.2. This arrangement would appear to be likely if the particle size is less than the thickness of a domain wall. It is questionable if the concept of spontaneous magnetization applies to such a particle, but if it does the spontaneous magnetization is zero. The behavior of the magnetization when a field is applied could be calculated. However, since particles with a circular spin arrangement have never been reported, it does not appear to be appropriate to pursue the matter very far. For particles somewhat larger than the thickness of a domain wall simple domain configurations such as those illustrated in Fig. 7-2.3 may be expected. Even here rather different properties from those of bulk material may be expected. These differences probably arise from magnetostatic and wall nucleation energies. Some calculations on small multidomain particles have been made,[46] but it is fair to say that to date such particles are not understood. Small multidomain particles will not be considered further here. Another intermediate case, that of thin films, has been the subject of extensive investigation since early in the 1950's. Because a number of advances have been made, this topic will be discussed further in Section 7-11.

At this point we turn our attention to the study of bulk soft magnetic materials. Various properties are discussed in the succeeding sections (7-6 to 7-10). In this section domain walls are treated.

Domain walls are also often called Bloch walls, for their properties were first considered by Bloch.[47] More extensive and refined treatments, however, have been given by others.[48] As already mentioned in connection

[46] C. Kittel, *Phys. Rev.* **73,** 810 (1948); L. Néel, *J. Phys. radium* **17,** 250 (1956); H. Amar, *J. Appl. Phys.* **29,** 542 (1958).

[47] F. Bloch, *Z. Physik* **74,** 295 (1932).

[48] L. Landau and E. Lifshitz, *Phys. Z. Sowjet.* **8,** 153 (1935); L. Néel, *Cahiers Phys.* **25,** 1 (1944).

with Fig. 7-1.2, energy is associated with the domain wall. The atomic moments are not parallel to one another, and therefore there is an increase in the exchange energy F_e. Also, the moments in the wall are not lying along easy directions of magnetization, hence the anisotropy energy F_K is increased. Moreover, if there are any uncompensated poles in the wall, that is, if there is a divergence of **M**, there will also be an increase in the magnetostatic energy F_D. To be specific, suppose that the domain wall is a plane, lying in the OXY plane of Fig. 7-5.1 so that OZ is normal to the wall. There is experimental evidence that the domain walls are frequently plane. Then the direction of an atomic moment or spin may

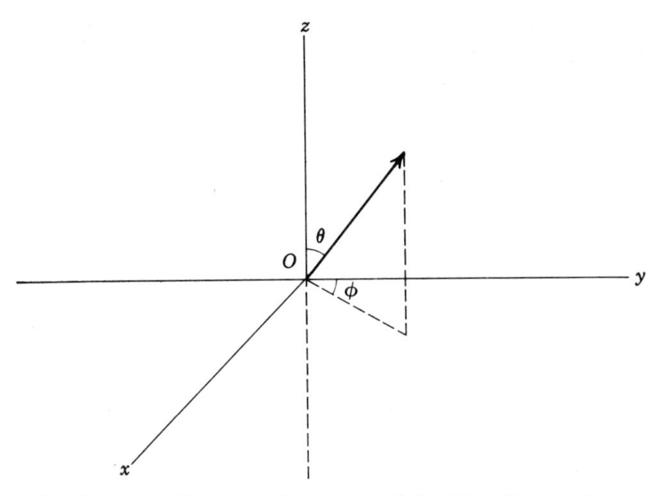

Fig. 7-5.1. The direction of an atomic moment defined by the angular coordinates (θ, ϕ). The domain wall, assumed to be a plane, lies parallel to the OXY plane. Then OZ is normal to the wall.

be defined by the angles θ and ϕ; θ is the angle between the moment and the normal to the wall and ϕ is that between the projection of the moment on the wall and a fixed axis, in this case the OY-axis. Further, under the assumptions made, the problem is one-dimensional, and θ and ϕ are functions of the location of the moment in the wall; that is, they are functions of the coordinate z.

Now, if θ does vary through the wall, div **M** will not be zero, and there will be uncompensated poles. In calculations made in the 1930's variations in θ through the wall were permitted.[49] However, since this leads to a large magnetostatic energy contribution, it has usually been assumed in

[49] See for example, F. Bitter, *Introduction to Ferromagnetism*, McGraw-Hill Book Co., New York (1937).

recent years that θ, and thus the normal component of the magnetization, remains constant throughout the wall. This constraint probably goes too far, since the total energy may be minimized by allowing a small magnetostatic contribution. Nevertheless, it is convenient, and probably not greatly in error, to assume θ equals a constant, and this is done in the subsequent calculations in this section. Thus the magnetization of two adjacent domains will change direction in going through a wall by θ remaining constant and ϕ changing from its initial to its final value. For the common case in which the magnetization of adjacent domains is antiparallel $\theta = \pi/2$ and ϕ goes from 0 to 180°, as shown in Fig. 7-5.2.

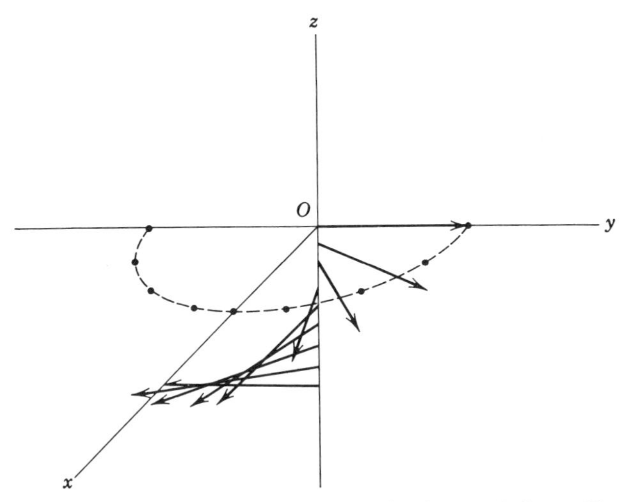

Fig. 7-5.2. To illustrate a 180° domain wall lying in the OXY plane. Here $\theta = \pi/2$ and ϕ goes from 0 to 180°.

The transition region for this case is often called a 180° wall. Other types of wall also occur, and the angles take different values, as will be discussed later.

Besides possessing an energy, domain walls have a certain width, which is the order of many atomic spacings. The reason is that the exchange energy is lower when the change of the spin direction takes place gradually over several atoms rather than abruptly over a couple of atoms, as may be seen from the following argument. The variable part of the exchange energy between two atoms is given by equation 6-3.30 as $F_e = J_e S^2 \phi^2$, where ϕ is the angle between the two spins. For an abrupt change in the magnetization of π radians the energy is $F_{\text{abrupt}} = J_e S^2 \pi^2$. On the other hand, for a gradual change over a row of N atoms, with equal angles of

π/N between adjacent spins, the energy is $F_{\text{gradual}} = J_e S^2(\pi/N)^2 N = J_e S^2(\pi^2/N)$. Clearly F_{gradual} is less than F_{abrupt}. However, the width of the walls is limited by the presence of the crystalline energy, inasmuch as the wider the wall, the greater this energy becomes. The actual width of the wall is determined by the competition between these two energies, F_e and F_K (assuming, of course, $F_D = 0$).

Because a variety of uncertainties is involved, a crude estimate of the energy and width of a domain wall is often sufficient for many purposes. Consider a 180° wall parallel to a (100) plane of a cubic lattice with $K_1 > 0$. The increase in the exchange energy for one row of atoms in the wall has been given as $J_e S^2(\pi^2/N)$. The number of rows of atoms per unit area of wall depends on the crystal structure, but for a simple cubic lattice it is equal to $1/a^2$, and in any event it is of this order of magnitude. The exchange energy per unit area of wall is then $J_e S^2\pi^2/Na^2$, or $J_e S^2\pi^2/\delta a$, where $\delta(= Na)$ is the width of the wall. The crystalline energy per unit area of wall is proportional to $K_1\delta$ and is of this order of magnitude. Then, if $F_D = 0$, the total energy per unit area associated with a wall is

$$F_T \doteqdot F_{\text{wall}} = \frac{J_e S^2 \pi^2}{\delta a} + K_1 \delta. \tag{7-5.1}$$

This energy is a minimum with respect to δ when

$$\frac{\partial F_{\text{wall}}}{\partial \delta} = 0 = -\frac{J_e S^2 \pi^2}{\delta^2 a} + K_1$$

or

$$\delta = \left(\frac{J_e S^2 \pi^2}{K_1 a}\right)^{1/2}. \tag{7-5.2}$$

The wall energy per unit area for this value of δ is

$$F_{\text{wall}} = 2\pi\left(\frac{J_e S^2 K_1}{a}\right)^{1/2}. \tag{7-5.3}$$

Values roughly appropriate for iron are $K_1 = 4.6 \times 10^5$ erg/cm³, $a = 2.9 \times 10^{-8}$ cm, $J_e \approx kT_f/5$, $kT_f = 1.4 \times 10^{-13}$, and $S = 1$. Substitution of these values gives approximately

$$\delta = 0.5 \times 10^{-5} \text{ cm},$$

or about 200 lattice spacings, and

$$F_{\text{wall}} = 4 \text{ ergs/cm}^2.$$

If the magnetostriction energy is important, K_1 should be replaced by $K_1 + \lambda_s \sigma$.

It is worthwhile to calculate the energy and width of a domain wall more carefully. In addition to the merit of providing a somewhat better estimate of these quantities, such a calculation makes clear their dependence on various factors, such as the wall orientation, the type of wall, and the presence of an applied field. From equation 7-1.11 the exchange energy is given by

$$F_e = A \int [(\nabla\alpha_1)^2 + (\nabla\alpha_2)^2 + (\nabla\alpha_3)^2] \, dv.$$

For the coordinate system of Fig. 7-5.1, $\alpha_1 = \sin\phi \sin\theta$, $\alpha_2 = \cos\phi \sin\theta$, and $\alpha_3 = \cos\theta$. Thus the exchange energy reduces to

$$F_e = A \sin^2\theta \int \left(\frac{d\phi}{dz}\right)^2 dv.$$

The value of the crystalline anisotropy energy depends on the direction of the normal to the wall (OZ) with respect to the crystallographic axes; some of the more important cases are given by the results of problem 7-6. In any case, since θ = constant, the crystalline anisotropy energy is some function of ϕ. Thus the increase in the crystalline anisotropy energy, as compared to that in the adjacent domain, may be written

$$F_K - F_K\,(\text{domain}) = \int f(\phi)\, dv.$$

For $K_1 > 0$, F_K (domain) = 0, but in general F_K (domain) is not zero. The total increase of energy per unit area of the domain wall is then given by

$$F_{\text{wall}} = \int_{-\infty}^{\infty} \left[B\left(\frac{d\phi}{dz}\right)^2 + f(\phi)\right] dz, \qquad (7\text{-}5.4)$$

where

$$B = A \sin^2\theta.$$

Now, ϕ must depend on z in such a way that the integral is a minimum. From the calculus of variations[50] an integral $\int g\, dx$ will have its minimum value if the integrand is a solution of the differential equation, known as the Euler equation, $\partial g/\partial y - (d/dx)\left(\frac{\partial g}{\partial y'}\right) = 0$ where $y' = dy/dx$. Application of this condition to equation 7-5.4 yields

$$\frac{d\,f(\phi)}{d\phi} - 2B\,\frac{d^2\phi}{dz^2} = 0.$$

[50] For proof, see, for example, E. S. Woods, *Advanced Calculus*, Ginn and Co., Boston (1934), pp. 317–319.

Multiplication by $d\phi/dz$ and integration over z then gives

$$f(\phi) = B\left(\frac{d\phi}{dz}\right)^2 + C$$

for the dependence of ϕ on z. For $|z| \to \infty$, that is, within a domain, $f(\phi) = 0$ and $d\phi/dz = 0$ and therefore $C = 0$. Hence we have finally

$$f(\phi) = B\left(\frac{d\phi}{dz}\right)^2. \tag{7-5.5}$$

This equation states that at each point in the wall the crystalline anisotropy energy is equal to the exchange energy. In other words, in regions of the wall in which the crystalline anisotropy energy is large, the angle between adjacent spins is greater than in regions in which the crystalline anisotropy energy is small. Substitution of equation 7-5.5 into (7-5.4) then gives

$$F_{\text{wall}} = 2\int_{-\infty}^{\infty} f(\phi)\, dz;$$

that is, the wall energy is just twice the crystalline energy. Since, from equation 7-5.5

$$dz = \sqrt{B}\,\frac{d\phi}{[f(\phi)]^{1/2}}, \tag{7-5.6}$$

the wall energy can be written as

$$F_{\text{wall}} = 2\sqrt{B}\int_{\phi_1}^{\phi_2} [f(\phi)]^{1/2}\, d\phi$$

or

$$F_{\text{wall}} = 2\sqrt{A}\sin\theta\int_{\phi_1}^{\phi_2} [f(\phi)]^{1/2}\, d\phi, \tag{7-5.7}$$

where ϕ_1 and ϕ_2 are the initial and final values of ϕ in the wall.

The quantities θ, ϕ_1, ϕ_2, and $f(\phi)$ depend on the type of wall, that is, on the change of angle through the wall, and on the wall orientation, but if these quantities are known the wall energy can be calculated. Generally, when there is no applied field and negligible strain, the direction of magnetization of a domain may be expected to lie along an easy direction. For cubic crystals with $K_1 > 0$ the directions of magnetization of adjacent domains are expected to be at angles of 90 or 180° with each other. The corresponding transition regions are called 90 or 180° domain walls. For cubic crystals with $K_1 < 0$, 70.5, 109.5, and 180° walls are expected, whereas for uniaxial crystals only 180° walls are likely.

As an example, the value of $f(\phi)$ for walls with normals along the [100]-, [111]-, and [110]-directions are given by the results of (*a*), (*b*), and (*c*) in problem 7-6. For 90° walls with these orientations the angles take the values of (a) $\theta = \pi/2$, $\phi_1 = 0$, $\phi_2 = \pi/2$, (b) $\cos\theta = 1/\sqrt{3}$, $\phi_1 = 0$, $\phi_2 = 2\pi/3$, and (c) $\theta = \pi/4$, $\phi_1 = 0$, $\phi_2 = \pi$, respectively. It is left as problem 7-7 to show that the wall energies per unit area are

$$(a) \qquad F_{\text{wall}[100]} = \sqrt{AK_1}, \tag{7-5.8}$$

$$(b) \qquad F_{\text{wall}[111]} = \tfrac{32}{27}\sqrt{AK_1}, \tag{7-5.9}$$

and

$$(c) \qquad F_{\text{wall}[110]} = 1.73\sqrt{AK_1}, \tag{7-5.10}$$

respectively. The wall does not necessarily assume the orientation with the least surface energy, as other factors enter; this is illustrated by the structure shown in Fig. 7-1.2c.

For a 180° wall the wall normal may be expected to lie along the [100]-direction. Then since

$$\theta = \frac{\pi}{2}, \qquad \phi_1 = 0, \qquad \phi_2 = \pi, \quad \text{and} \quad f(\phi) = K_1 \sin^2\phi \cos^2\phi, \tag{7-5.11}$$

the wall surface energy density is

$$F_{\text{wall}} = 2\sqrt{AK_1}. \tag{7-5.12}$$

Numerical values of the wall energy per unit area depend on the value selected for the exchange constant A. Values of A may be obtained from any of several methods outlined in Chapter 6, such as (a) from the Curie temperature, (b) from a fit to the Bloch $T^{3/2}$ law for the magnetization at low temperatures, or (c) from the Curie constant above the Curie point. For a 180° wall in iron, as given by equation 7-5.12, values of F_{wall} quoted in the literature vary from 1.4 to 3.0 ergs/cm².

Wall energies may also depend on the applied field. For example, suppose the field is applied along the [110]-direction in a cubic crystal such as iron. The normal to the domain walls will also lie along the [110]-direction. The domain arrangement when H is small is shown in Fig. 6-9.2a. For very small fields the angle between the magnetizations of adjacent domains is close to 90°. As the field increases, this angle will decrease. For the purpose of calculation, $\phi_1 = 0$, $\phi_2 = \pi$, whereas θ is a function of H and may be determined from equation 6-9.4.[51] According to a calculation by Néel,[52] when H varies from 0 to 385 oe, F_{wall} for iron varies from 1.21 to 0.32 ergs/cm² for this case.

[51] Note that there is a change of notation of the angle compared to the equations of this section.

[52] L. Néel, *J. Phys. radium* **5,** 241 (1944).

The wall width may be calculated by integrating equation 7-5.6. For a 90° wall, with the value of $f(\phi)$ as given by equation 7-5.11, the width is given by

$$z = \sqrt{\frac{A}{K_1}} \int \frac{d\phi}{\sin\phi \cos\phi}$$

or

$$z = \sqrt{\frac{A}{K_1}} \log \tan \phi, \tag{7-5.13}$$

where the origin is chosen at $\phi = \pi/4$ in order to make the constant of integration zero. This equation shows that a 90° wall has no well-defined width, inasmuch as ϕ approaches its extreme values of 0 and 90° only asymptotically. The wall width may be defined as $\Delta z \approx \pi\sqrt{A/K_1}$, since about 75% of the change in ϕ takes place over this distance. For a 180° wall there is an additional difficulty in that the wall apparently will separate into two 90° walls an infinite distance apart. However, this argument neglects the presence of magnetostriction, which will require that the separation between the two 90° walls be as small as possible in order to reduce the state of strain in this region. It can be shown that the magnetostriction does not change the wall energy of iron appreciably.[53] The calculation of wall energy and width for other cases may be found in the literature.[54]

6. Domain Structure

Some domain structures have been proposed earlier (Section 7-1, Fig. 7-1.2). It is fortunate that experimental methods for the direct observation of domain structures permit a check of such proposals. The two most widely used are the magnetic-powder technique and the Kerr or Faraday magneto-optical technique.

In the magnetic-powder technique[55] a suspension or colloidal solution of small magnetic particles,[56] usually magnetite (Fe_3O_4), is placed on the surface of the ferromagnetic specimen. When surface poles are present, the particles tend to concentrate in certain regions to form a pattern. The pattern is readily observed with an optical microscope. Patterns

[53] C. Kittel, *Revs. Mod. Phys.* **21,** 541 (1949).

[54] B. A. Lilley, *Phil. Mag.* **41,** 792 (1950).

[55] First suggested by F. Bitter, *Phys. Rev.* **38,** 1903 (1931) and L. von Hámos and P. A. Thiessen, *Z. Physik* **77,** 442 (1931).

[56] Recipes for the preparation of a colloidal suspension of magnetite are given by W. C. Elmore, *Phys. Rev.* **54,** 309 (1938) and C. Kittel and J. K. Galt, *Solid State Phys.* **3,** 439 (1956), p. 493.

with some regularity are observed only on single crystals. When the surfaces of these crystals are finely polished mechanically, a mazelike pattern which is a result of the surface strain conditions that are produced by the polishing is often seen. On electrolytically polishing the surface, significant patterns are detected. Finer details in the pattern may be investigated by making the pattern set into a dry film and then viewing this film with the electron microscope.[57]

A number of discussions of the formation of powder patterns have been presented.[58] These patterns depend on whether the particles are single-domains. In any case, the particles may be expected to collect in regions in which there is a large change in the magnetic field at the surface of the specimen. One such region will be where a domain wall intersects the surface. Then, because of the change of direction of the atomic moments within a wall, there is a component of magnetization parallel to the wall (see Fig. 7-5.2) which will give rise to surface poles. Therefore surface domain boundaries will be observed as lines in the pattern. Lines may also be expected at boundaries between regions with different surface density of poles. Some concentration of the colloid also occurs over regions with a surface pole density. It is usually assumed that the magnetic moment of the particles does not affect the domain structure. Although this assumption is probably valid for large specimens, it is not clear that it holds for very small specimens, such as thin films. Some powder patterns are illustrated in Figs. 7-11.6 and 7-11.8.

In the magneto-optical technique a beam of polarized light is either transmitted through the ferromagnetic specimen or reflected from its surface. Provided there is a component of the magnetization along the direction of propagation of the light, the plane of polarization will be rotated by an amount dependent on the direction of magnetization. Rotation of the plane of polarization of transmitted light is called the Faraday effect, whereas rotation of reflected light is called the Kerr effect.

In the experimental arrangement[59] a light beam is first passed through a Nicol or some other suitable prism (the polarizer) to polarize it. Domains whose direction of magnetization differ rotate the plane of polarization by different amounts. The light beam is then passed through another prism (the analyzer) and observed with a microscope. When the analyzer is adjusted to extinguish the light from some domains, some light from other domains will still be transmitted. Consequently, a domain pattern

[57] D. J. Craik and P. M. Griffiths, *Brit. J. Appl. Phys.* **9,** 279 (1958).

[58] C. Kittel, *Phys. Rev.* **76,** 1527 (1949); E. C. Stoner, *Rept. Progr. Phys.* **12,** 83 (1950).

[59] H. J. Williams, F. G. Foster, and E. A. Wood, *Phys. Rev.* **82,** 119 (1951); C. A. Fowler and E. M. Fryer, *Phys. Rev.* **86,** 426 (1952).

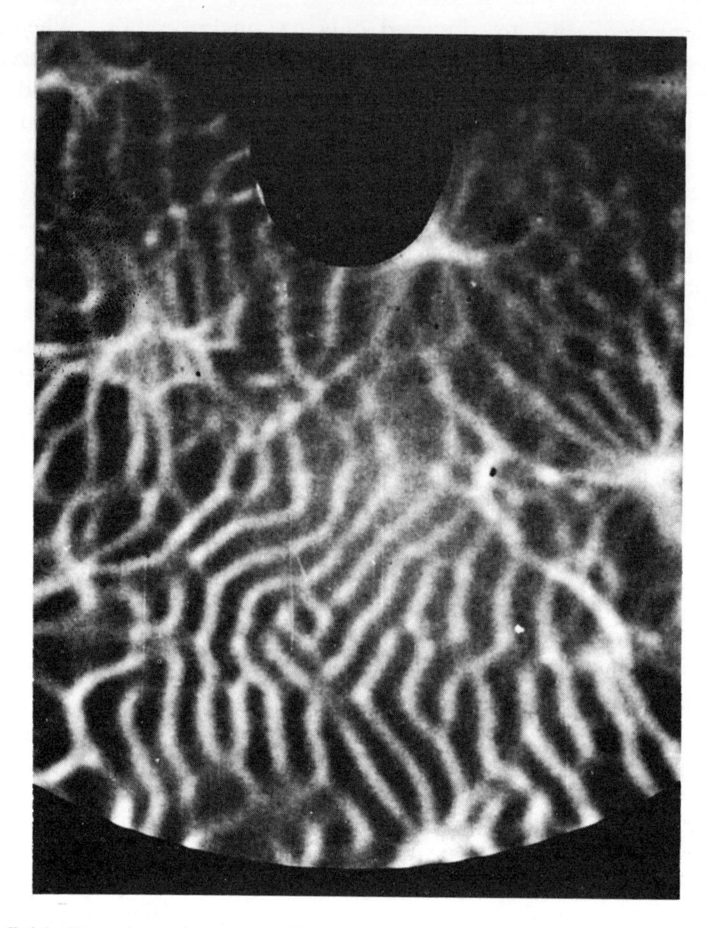

Fig. 7-6.1. Domain pattern on a silicon-iron specimen as revealed by electron mirror microscopy. (Courtesy of the late L. Mayer, General Mills Inc., Minneapolis, Minn., now Litton Industries.)

consisting of regions with different light intensities will be observed. In contrast to the powder pattern technique, domains are observed when the surface pole density is small or zero. On the other hand, when the difference in direction of magnetization of adjacent domains is small, the light contrast is so small that observation is difficult. Further, domains with antiparallel directions of magnetization are not differentiated when their magnetization is perpendicular to the direction of light propagation.

Domains have been observed by other methods. When an electron beam is deflected at a surface of a ferromagnetic sample, the interaction with the magnetic fields associated with the surface pole distribution causes a change in the trajectories. As a result, a pattern may be observed on a photographic plate or on a screen.[60] In this method it is important that the surface be smooth, for any inhomogeneities will affect the electron deflection and increase the complexity of the pattern. An example of a domain pattern obtained with electron mirror microscopy is shown in Fig. 7-6.1. Domain patterns have also been observed by the transmission of electrons through ferromagnetic thin films,[61] by X-ray diffraction,[62] and by the depolarization of polarized neutrons.[63]

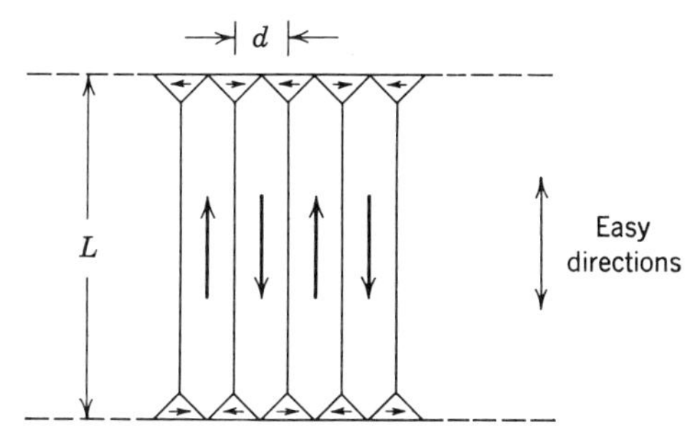

Fig. 7-6.2. Proposed domain structure of a uniaxial crystal. ($H = 0$.) The direction of magnetization is indicated by the arrows.

In order to obtain domain patterns with some regularity and therefore to have some hope of being able to interpret the pattern quantitatively, it is usually necessary to use single crystals. Domain structures of the type shown in Fig. 7-1.2*b* have been observed, for example, on a surface parallel to the *c*-axis of single crystals of cobalt,[64] whereas structures of the type shown in Fig. 7-1.2*c* have been observed on a surface parallel to the (100) plane of single crystals of iron.[65] In general, however, there is a larger number of domains, as indicated in Fig. 7-6.2. For a uniaxial crystal of length L the domain spacing d may be estimated by the following argument.[66] The area of domain wall per unit area of the crystal surface is

[60] L. H. Germer, *Phys. Rev.* **62,** 295 (1942); L. Marton, *Phys. Rev.* **73,** 1475 (1948); L. Mayer, *J. Appl. Phys.* **28,** 975 (1957); M. E. Hale, H. W. Fuller, and H. Rubinstein, *J. Appl. Phys.* **30,** 789 (1959).

[61] H. W. Fuller and M. E. Hale, *J. Appl. Phys.* **31,** 238 (1960); J. T. Michalak and R. C. Glenn, *J. Appl. Phys.* **32,** 1261 (1961).

[62] K. M. Merz, *J. Appl. Phys.* **31,** 147 (1960).

[63] M. T. Burgy, D. J. Hughes, and J. R. Wallace, *Phys. Rev.* **74,** 1207 (1948).

[64] W. C. Elmore, *Phys. Rev.* **53,** 757 (1938).

[65] H. J. Williams and R. C. Sherwood, *Magnetic Properties of Metals and Alloys,* Am. Soc. Metals, Cleveland, Ohio (1959), p. 38.

[66] L. Landau and E. Lifshitz, *Phys. Z. Sowjet.* **8,** 153 (1935).

L/d, per unit volume of crystal, $1/d$. Hence the wall energy per unit volume of crystal is F_w/d. The magnetization of the closure domains points along the hard direction, and therefore there is crystalline anisotropy energy. The volume of the domains of closure per unit area of crystal surface is $d/2$ and per unit volume of crystal, $d/2L$. The anisotropy energy is therefore $K_1'd/2L$ where K_1' is the uniaxial anisotropy constant. If the wall energy of the closure domains is neglected, the total energy is

$$F_T = \frac{F_w}{d} + \frac{K_1'd}{2L}. \tag{7-6.1}$$

This energy is a minimum with respect to the domain spacing when

$$\frac{\partial F_T}{\partial d} = -\frac{F_w}{d^2} + \frac{K_1'}{2L} = 0;$$

that is, the spacing is given by

$$d = \left(\frac{2LF_w}{K_1'}\right)^{1/2} \tag{7-6.2}$$

and the minimum energy per unit volume is

$$F_T\,(\text{min}) = \left(\frac{2F_wK_1'}{L}\right)^{1/2}. \tag{7-6.3}$$

For $L = 1$ cm, $F_w = 2$ ergs/cm^2 and $K_1' = 10^5$ ergs/cm^3, $d = 0.6 \times 10^{-3}$ cm and F_T (min) $= 0.6 \times 10^3$ ergs/cm^3. Domain spacing of this order of magnitude has been observed (for example, for cobalt).

Of course, a uniaxial material with large crystalline anisotropy would not be so likely to have closure domains because there is magnetostatic energy to consider in the calculation of the domain spacing (problem 7-8a). For a cubic crystal with $K_1 > 0$ the magnetization of the closure domains of the type indicated in Fig. 7-6.2 also lies along easy directions. However, because of the magnetostriction change of length of the domains, there is an elastic energy that must be considered. The same expressions as those given by equations 7-6.2 and 7-6.3 are obtained, except that the anistropy energy constant K_1' is replaced by the elastic energy (problem 7-8b).

This simple example illustrates the approach employed for the analysis of domain structures. Some domain structure that appears to be physically reasonable is assumed. The domain wall is treated as an entity and its energy is explicitly included in the total energy of the structure. The total energy is then minimized with respect to one or more of the parameters involved in the structure. Of course, more than one domain structure may appear to be possible, in which case the total energy of each structure is minimized and the one with the least energy may be expected to be the most likely. This approach has led to the prediction of domain patterns that have been subsequently observed. The reverse procedure, in which an observed pattern is analyzed, has also been employed. A much more satisfactory approach to the whole problem would be to predict the domain structure by minimizing the total free energy as given by equation 7-1.2; this problem has not been solved.

We now proceed to discuss briefly some of the principal domain patterns observed and their interpretation. In the analysis of the patterns it is helpful if the direction of magnetization within the domains is known; there are several tricks available for this determination. In the powder-pattern technique a scratch may be made on the surface of the specimen. If the magnetization is not parallel to the scratch, surface poles will be formed. Therefore there will be a deposit of the colloid along the scratch except when the magnetization is parallel to the scratch. In the magneto-optical technique use is made of the fact that there is no effect when the magnetization is perpendicular to the direction of propagation of the light. Thus the direction of magnetization may be found by rotating the sample. In either case the application of a magnetic field will increase the area of domains magnetized along the field direction at the expense of the other domains. The application of a tension will serve the same purpose. Thin ferromagnetic probes may also be used; they tend to introduce domains of reverse magnetization except when their magnetization is parallel to that of the domain.

When the surface of a silicon-iron (3.8% Si) is inclined at a small angle θ to the (100) plane, the pattern of Fig. 7-6.2 is modified in that the main domain now possesses additional spikelike domains.[67] This so-called "tree" pattern is illustrated by Fig. 7-6.3*a*. Since the magnetization of the main domains lies along the [100]-direction, there is a surface pole density of $M \sin \theta$. The branches of the tree thus act as partial closure domains by reducing the magnetostatic energy of these surface poles, as illustrated by Fig. 7-6.3*b*.

When an iron crystal is cut with its top and bottom surfaces parallel to the (100) plane and its length along a [110]-direction, the domain structure of the type shown in Fig. 6-9.2*a* may be expected. By assuming the closure domain structure of Fig. 7-6.4, Néel has calculated[68] the domain spacing as a function of a magnetic field applied along the [110]-direction. Both the change in the anisotropy energy of the main domains and the change in the wall energy with the applied field are taken into

[67] H. J. Williams, R. M. Bozorth, and W. Shockley, *Phys. Rev.* **75,** 155 (1949).
[68] L. Néel, *J. Phys. radium* **5,** 265 (1944).

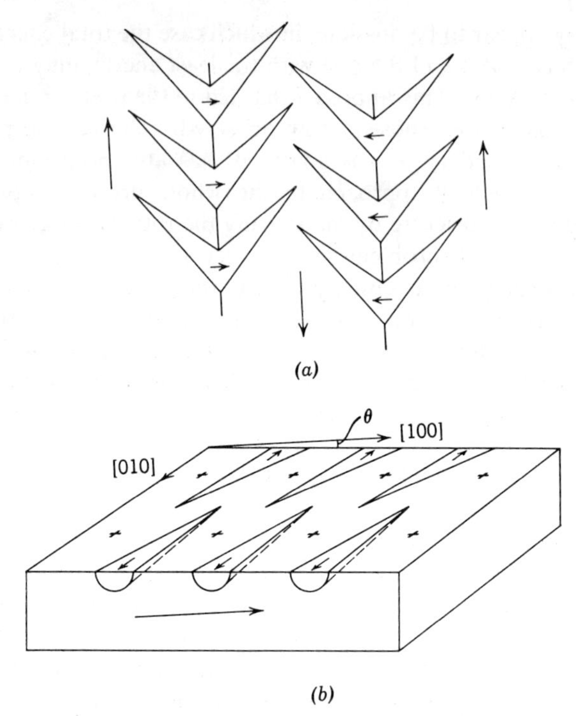

Fig. 7-6.3. To illustrate "tree" patterns observed on silicon-iron crystals on a surface inclined at a small angle θ with the (100) plane. In (*a*) a typical powder pattern is shown. In (*b*), the interpretation of the pattern is indicated. [H. J. Williams, R. M. Bozorth, and W. Shockley, *Phys. Rev.* **75,** 155 (1949).]

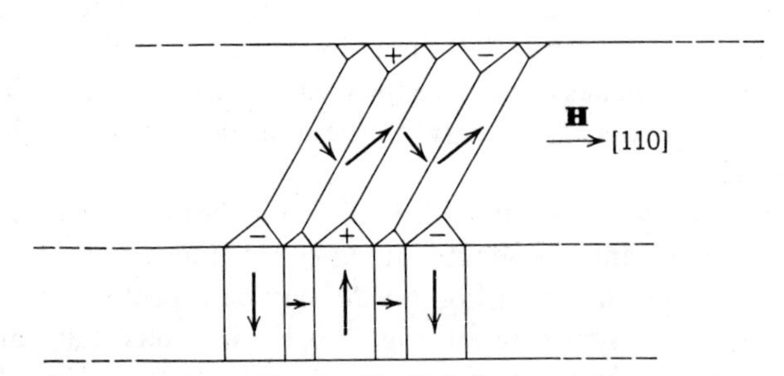

Fig. 7-6.4. Proposed domain structure in cubic specimen with length along the [110]-direction and upper and lower surfaces parallel to the (100) plane. [L. Néel, *J. Phys. radium* **5,** 265 (1944).]

account. He predicts that the spacing will be large in small fields, becoming first smaller and then almost constant as the field is increased. This behavior has been observed experimentally.[69]

Domain patterns on picture-frame and whisker specimens of iron crystals are discussed in Section 7-7. Lozenge patterns on iron crystals have been reported.[70] The 180, 109, and 71° domain walls expected for nickel single crystals have been observed.[71] On the basal plane of single crystals of Co a lace-like structure has been seen.[72] Although the details are not understood, the structure undoubtedly originates because of the

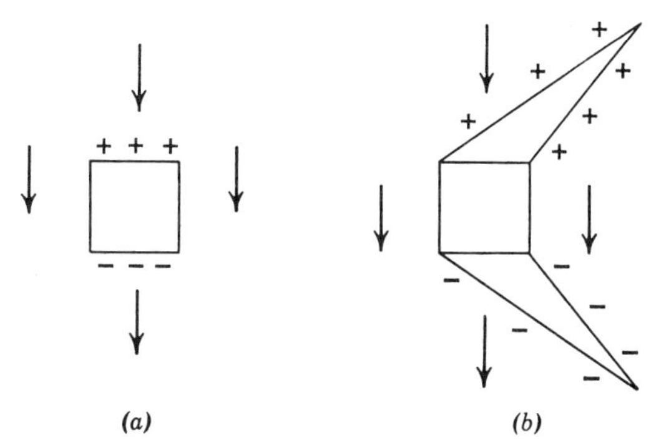

Fig. 7-6.5. In (*a*) the surface poles associated with a cavity lying within a domain are indicated. In (*b*) the reduction of the surface pole density by a secondary domain structure is shown. [H. J. Williams, R. M. Bozorth, and W. Shockley, *Phys. Rev.* **75,** 155 (1949).]

otherwise large magnetostatic energy of the surface poles on this plane. Only a few studies of domain patterns on alloys have been reported; these include NiCo (40% Ni)[73] and MnBi.[74]

When a specimen possesses an imperfection such as a cavity or inclusion within a domain, surface poles will be present (see Fig. 7-6.5*a*). The magnetostatic energy associated with these poles is greatly reduced by a secondary domain structure (see Fig. 7-6.5*b*), for then the poles are distributed over a larger area. In spite of the additional wall energy,

[69] L. F. Bates and F. E. Neal, *Proc. Phys. Soc.* (*London*) **A-63,** 374 (1950).

[70] W. S. Paxton and T. G. Nilon, *J. Appl. Phys.* **26,** 494 (1955); J. B. Goodenough, *Phys. Rev.* **103,** 356 (1956); C. D. Graham, *Proc. Conf. Mag. and Mag. Materials, Boston* **T-91,** 410 (1957).

[71] M. Yamamoto and T. Iwata, *Phys. Rev.* **81,** 887 (1951); H. J. Williams and J. G. Walker, *Phys. Rev.* **83,** 634 (1951); L. F. Bates and G. W. Wilson, *Proc. Phys. Soc.* (*London*) **A-66,** 819 (1953).

[72] W. C. Elmore, *Phys. Rev.* **53,** 757 (1938).

[73] R. M. Bozorth and J. G. Walker, *Phys. Rev.* **79,** 888 (1950).

[74] B. W. Roberts and C. P. Bean, *Phys. Rev.* **96,** 1494 (1954).

Néel[75] has shown that the total energy is then reduced. Such a structure has been observed.[76]

More complicated patterns are observed on polycrystalline materials. Spikelike domains occur often at grain boundaries, presumably to minimize the effect of surface poles. It is not clear whether the exchange interaction is important at grain boundaries. Patterns on polycrystalline thin films are discussed in Section 7-11.

Ferrimagnetic materials also possess a spontaneous magnetization, and hence may be expected to exhibit a domain structure determined by the same principles that govern ferromagnetic materials. However, it is more difficult to obtain a suitable surface. The surface must be prepared by etching, since electropolishing cannot be employed. Thin specimens of garnets and orthoferrites (about 1 mil thick) are transparent to visible light. Therefore the Faraday effect may be used to observe domain patterns.[77,78] In this way it has been established that the domain structure within the body of the specimen is the same as the surface structure. Mention should also be made of interesting patterns observed on the plane perpendicular to the easy axis of hexagonal ferrimagnetic materials with a large uniaxial anisotropy.[78,79]

7. The Analysis of the Magnetization Curves of Bulk Materials

It is now appropriate to consider the magnetization processes that occur in bulk materials. By the term bulk materials we exclude those made up essentially of small particles. Most bulk materials are magnetically soft, although some have coercive forces as high as about 100 oe. It has already been mentioned (Section 7-1) that changes in the magnetization are believed to be the result of domain-wall movements or of rotations of the domain magnetizations. The magnetization curves of single crystals in the high field region in which the rotational processes predominate are considered to be well understood. This subject has already been treated in Section 6-9 and will not be discussed further. Rotational processes in polycrystalline materials, as well as wall movements in both single crystals and polycrystalline materials, remain to be discussed. Magnetization changes by these mechanisms depend strongly on the imperfections of the material. Various models have been proposed and elaborate calculations of such

[75] L. Néel, *Cahiers Phys.* **25,** 21 (1944).

[76] H. J. Williams, *Phys. Rev.* **71,** 646 (1947).

[77] J. F. Dillon, *J. Appl. Phys.* **29,** 539 (1958).

[78] H. J. Williams and R. C. Sherwood, *Magnetic Properties of Metals and Alloys,* Am. Soc. Metals, Cleveland, Ohio (1959), p. 35.

[79] J. Kaczer and R. Gemperle, *Czech. J. Phys.* **B-10,** 505 (1960).

quantities as the coercive force, initial permeability, and remanence have been made. The assumptions made in these theories are questionable, however. For most materials the best that can be hoped for is that theory will provide a qualitative, or perhaps a semiquantitative, description of the actual experimental situation. Moreover, it is unlikely that any one theory can be developed that will apply to all materials. A detailed account of the existing theoretical calculations will not be given here; instead, emphasis is placed on the physical principles involved and the method of analysis appropriate to the various regions of the magnetization and hysteresis curves.

Other aspects of the magnetization curves require consideration. One of the most important is the question of the nucleation of domains when a large field applied to a material is removed. This problem is at present largely unsolved, although it is certain that imperfections play a major role. Other topics to be considered are minor loops, Barkhausen effect, and, in Section 7-8, magnetothermal effects. It is convenient to discuss these topics under various subheadings.

Domain wall movements. Suppose a single crystal of a ferromagnetic material is perfect in the sense that it has no imperfections. For simplicity, assume that there are 180° domain walls in the crystal and then consider one such wall. When a magnetic field H is applied parallel to the direction of magnetization of a domain on one side of the wall, the wall will tend to be displaced, as indicated in Fig. 7-7.1*a*. The field required to produce this displacement will be very small, provided the domains in question do not give rise to a change in the surface pole distribution. A perfect single crystal therefore may be expected to possess a very large initial permeability and a very small coercive force.

However, single crystals have imperfections of one kind or another. As a result, the movement of a domain wall is impeded, and this leads to smaller initial permeabilities and larger coercive forces. The effect is expected to be even greater in polycrystalline materials, since the presence of grain boundaries introduces an additional imperfection.

Because of the presence of the imperfections, the total energy of a specimen generally will depend on the location of the domain walls. Various energy terms may be affected, such as the wall energy, F_w, itself, and the magnetostatic energy. It is convenient to let F_{wt} denote the energy arising from a unit area of one wall located at a certain position, regardless of the origin of this energy. Again, as in Fig. 7-7.1*a*, let us consider a 180° domain wall that lies parallel to the y, z plane and moves in the x-direction when a field is applied. Suppose F_{wt} for this wall depends on x in the manner shown in Fig. 7-7.1*b*. When the field is zero, the wall will lie at an energy minimum, assumed in this example to be at

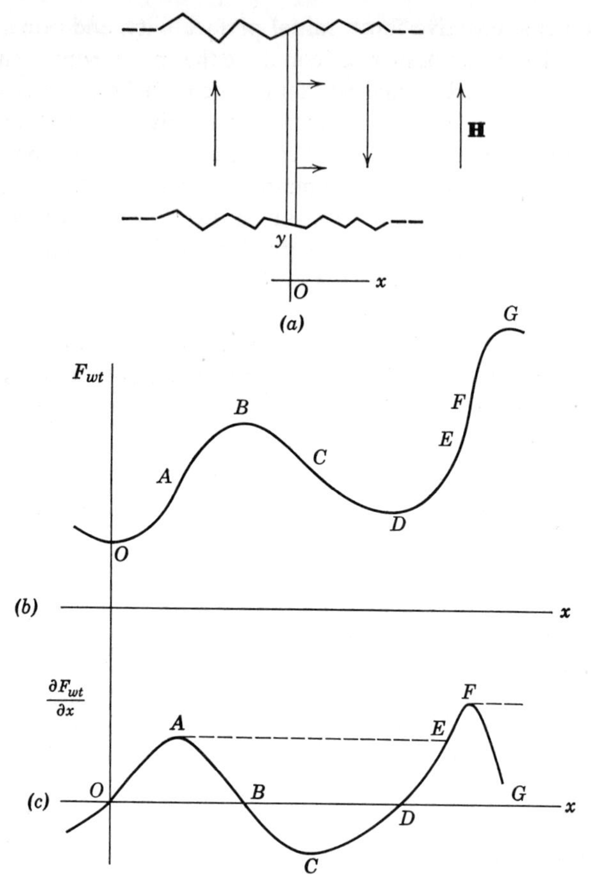

Fig. 7-7.1. To illustrate domain wall movement in soft materials possessing imperfections. In (*a*) the direction of motion of a 180° wall under the action of a field H is indicated. In (*b*) a possible form for the dependence of F_{wt} on x is shown; in (*c*) the slope of F_{wt} is plotted as a function of x.

$x = 0$. When a field is applied, the total energy per unit area of wall is given by

$$F_T = F_{wt} - 2HMx$$

and the wall will lie at a position determined by the equilibrium condition

$$\frac{d}{dx} F_T = 0;$$

that is,

$$2MH = \frac{dF_{wt}}{dx}. \tag{7-7.1}$$

Hence the action of the field will be to displace the wall to the right by an amount depending on the slope of the F_{wt} versus x curve. For positions of the wall between O and A in Fig. 7-7.1c the motion is reversible, since on decreasing the field the wall will return to the value of x appropriate to that field. Once the wall has reached position A, however, it will move spontaneously to position E. This motion is irreversible, since on decreasing the field the wall moves toward position D and then, on reversing the field direction, toward C. On a further increase of field to the range EF the motion is again reversible, whereas at F another spontaneous motion occurs.

This simple model shows how various parts of the hysteresis curve may be related to domain wall movements. The initial permeability is determined by the magnitude of dF_{wt}/dx near the origin. The coercive force is a measure of the field required to move a wall past the energy barriers and therefore depends on the maximum value of dF_{wt}/dx. The hysteresis arises from the irreversible energy changes. The remanence is the result of a wall being taken from one energy minimum to another by the application and subsequent removal of a field; in the foregoing example the wall's position may be changed from the minimum at O to that at D via the sequence $OAEFED$. Of course, an actual specimen will have many walls instead of just one. The behavior of the material may then be found either by superposing the effects of each wall or by considering an average situation for one wall.

Coercive force. In order to calculate a quantity such as the coercive force, it is necessary to have an expression for the energy F_{wt}. For this a knowledge of two things is required: one, the nature of the imperfections and how they contribute to the energy and, two, the distribution of the imperfections throughout the material. Several models have been considered, which are not necessarily mutually exclusive. Hence they may apply singly or in combination to some materials, although probably not to all. A brief outline of the various theories of the coercive force will now be given.

The strain theory was historically the first one to be developed. The imperfections in this case are assumed to be inhomogeneous internal stresses. These stresses may affect two energy terms, namely the magnetostrictive energy of the domains and the domain wall energy. Now, the coercive force must be determined primarily by the movement of 180° domain walls, since reversal of the magnetization is involved. The movement of such domain walls leaves the volume magnetostrictive energy unchanged, since F_σ depends on $\cos^2\theta$ (equation 7-1.8) and $\cos^2\theta = \cos^2(\pi + \theta)$. Therefore it is necessary to consider only the domain wall energy.[80] This energy for the case in which $\lambda_s\sigma > K_1$ is easy to calculate.

[80] E. Kondorsky, *Physik. Z. Sowjet.* **11,** 597 (1937); M. Kersten, *Probleme der technischen Magnetisierungskurve*, J. Springer, Berlin (1938), p. 42.

Then, since K_1 may be neglected, $F_\sigma = -\frac{3}{2}\lambda_s\sigma \cos^2 \theta$ is substituted for the function $f(\phi)$ with $\phi = \theta$ in equation 7-5.7. A simple integration then gives

$$F_w = 2(3A\lambda_s\sigma)^{1/2}. \tag{7-7.2}$$

When the crystalline anisotropy cannot be neglected, the integration is more difficult.[81] The wall energy may then be written as

$$F_w = \sqrt{2A}(K_1 + c\lambda_s\sigma)^{1/2}, \tag{7-7.3}$$

where c is a numerical factor, of order of unity, and depends on the direction of σ with respect to the wall (problem 7-9).

In order to calculate the coercive force, it is necessary to know the spatial variation of the stresses. For simplicity, we assume that this variation is sinusoidal, that is, that

$$\sigma = \sigma_0 + \Delta\sigma \sin \frac{2\pi x}{l}, \tag{7-7.4}$$

where σ_0 is a uniform stress, $\Delta\sigma$ is the amplitude, and l is the wavelength of the stress fluctuation. If l is large compared to the wall width δ, the stress σ may be taken as constant within the wall. Then, from equation 7-7.3, we have

$$\frac{\partial F_w}{\partial x} = c\lambda_s\left[\frac{A}{2(K_1 + c\lambda_s\sigma)}\right]^{1/2}\frac{d\sigma}{dx}$$

$$= \frac{2\pi^2 c\lambda_s\,\Delta\sigma\delta}{l}\cos\frac{2\pi x}{l},$$

where the definition of the wall thickness has been employed (from equation 7-5.13). Now the coercive force depends on the maximum value of $\partial F_{wt}/\partial x$, and here $F_{wt} = F_w$. Thus

$$\left(\frac{\partial F_{wt}}{\partial x}\right)_{\max} = \frac{2\pi^2 c\lambda_s\,\Delta\sigma\delta}{l}$$

and (from equation 7-7.1),

$$H_c = \frac{\pi^2 c\lambda_s\Delta\sigma}{M}\left(\frac{\delta}{l}\right). \tag{7-7.5}$$

This calculation applies to only one wall. When many walls are present, an average may be obtained by taking into account the variation in c and $\Delta\sigma$ throughout the material.

When $\delta > l$, the wall energy may be written as the sum of two terms: one, the usual wall energy, and two, a surface magnetostrictive energy

[81] B. A. Lilley, *Phil. Mag.* **41,** 792 (1950).

arising from the stress variation. It may then be shown (problem 7-9) that

$$H_c \approx \frac{\lambda_s \Delta\sigma}{M}\left(\frac{l}{\delta}\right). \tag{7-7.6}$$

Now, since for $l > \delta$, H_c increases as l decreases, whereas for $l < \delta$, H_c increases as l increases, it follows that the coercive force is a maximum for $l \approx \delta$ given approximately by

$$H_c \approx \frac{\lambda_s \Delta\sigma}{M}. \tag{7-7.7}$$

The internal stresses present in a ferromagnetic specimen may be changed by some metallurgical process, such as cold working, plastic deformation, or heat treatment. Then, if the internal stresses are known, the theory can be compared with experiment. Although no direct way of accurately measuring internal stresses has been devised, they may be estimated by some other independent magnetic measurement. The values obtained for the internal stresses lead to the prediction of coercive forces that are generally within an order of magnitude of the experimental values of H_c.

An experiment, first performed by Sixtus and Tonks,[82] which involves the motion of only one 180° domain wall, is worthy of special mention. Such a simple situation is likely to provide the best test of the theory. A tension is applied to a wire, as indicated in Fig. 7-7.2*a*. The wire is made of some material that possesses a positive magnetostriction ($\lambda_s > 0$) and a small crystalline anisotropy ($K_1 \approx 0$), such as Permalloy with 68% Ni. When the applied field is zero, there will be an equal volume of domains lying parallel or antiparallel to the tension axis, for then the magnetostriction energy F_σ is a minimum and $F_K \approx 0$. When a sufficiently large field produced by coil M is applied along the tension axis, the material becomes a single domain whose magnetization lies along the field direction. The wire retains this single-domain configuration when first the field is reduced to zero and then when a small field is applied in the reverse direction. The application of a pulse of magnetic field, produced by coil P, nucleates a region of reverse magnetization. Because this pulse is of very short duration, this region does not propagate down the wire, but remains stationary in the form of an "island" in the wire, as indicated in Fig. 7-7.2*b*. The shape of such an island may be determined by sliding a

[82] K. J. Sixtus and L. Tonks, *Phys. Rev.* **37,** 930 (1931), **42,** 419 (1932), **43,** 70, 931 (1933); F. Preisach, *Phys. Z.* **33,** 913 (1932); K. J. Sixtus, *Phys. Rev.* **48,** 425 (1935); W. Döring, *Z. Physik.* **108,** 137 (1938); W. Döring and H. Haake, *Phys. Z.* **39,** 865 (1938) and L. J. Dijkstra and J. L. Snoek, *Philips Res. Rept.* **4,** 334 (1949).

search coil along the wire. Then, when the field of coil M is increased, a critical value causes the island to grow suddenly, and the magnetization of the wire quickly reverses by the propagation of a domain wall (see Fig. 7-7.2*c*), along the wire. The velocity of the wall is determined by measuring the time between the voltage pulses produced in search coils S_1 and S_2.

A knowledge of the properties of the islands of reverse magnetization permits an estimate of the wall energy F_w to be made. The results appear

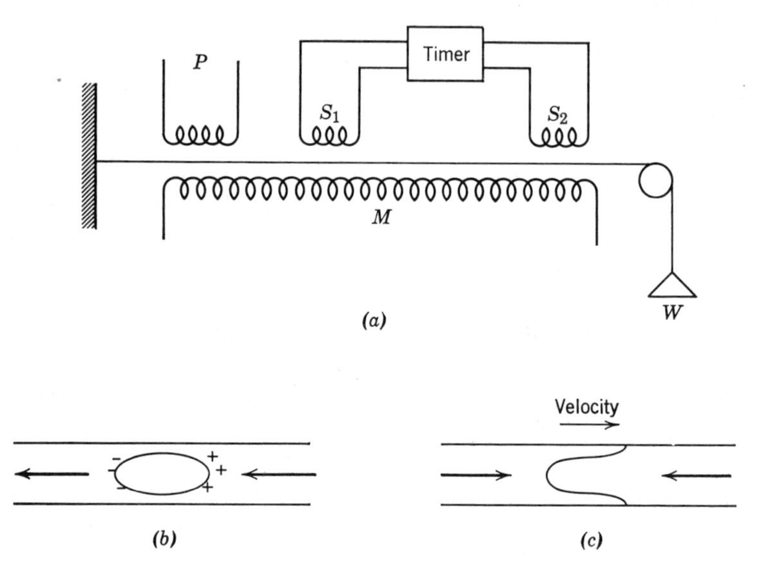

Fig. 7-7.2. To illustrate the Sixtus-Tonks experiment a schematic of the experimental arrangement is shown in (*a*), an "island" of reverse magnetization in (*b*), and the profile of the propagating domain wall in (*c*). (After K. J. Sixtus and L. Tonks.)

to confirm equation 7-7.2, in which σ is the externally applied tension.[83,84] The critical field, below which the domain wall will not propagate, presumably corresponds to the coercive force. This field leads to values of the internal stresses that are in agreement with those estimated from other independent magnetic measurements.[84] The velocity of the wall is also of interest, since it depends on such factors as eddy currents and relaxation effects.[84]

It should be mentioned that the strain theory is not able to account for large coercive forces. For one thing, the maximum value $\Delta\sigma$ may have must be less than the breaking stress. For another, the assumption has

[83] S. Ogawa, *Sci. Rept. Tôhoku Univ.* **1,** 53 (1949).
[84] R. Becker and W. Döring, *Ferromagnetismus*, J. Springer, Berlin (1939), pp. 184–217.

been made that the stress distribution is uniform in the plane of the domain wall, whereas, in fact, the distribution usually will be irregular and probably random. Consideration of these points leads to the conclusion that the maximum coercive force that can arise from internal stresses is about 10 oe.[85]

The *inclusion theory* considers nonferromagnetic inclusions and the change in wall energy when the wall intersects one or more of these inclusions. Nonmagnetic inclusions can occur as the result of the presence

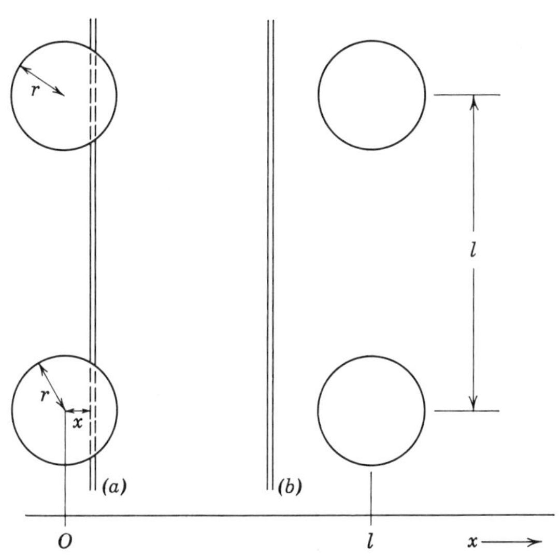

Fig. 7-7.3. Simplified model to illustrate the inclusion theory. In (*a*) the wall intersects some nonmagnetic inclusions and the wall energy is reduced. In (*b*) no intersections occur; 180° walls are assumed.

of elements such as C, N, or S in amounts larger than their solubility limit or as the result of the presence of gas bubbles. When a domain wall intersects an inclusion, the area, hence the energy of the wall, is reduced. The wall energy is the only contribution to F_{wt} considered in this theory.[86]

For purposes of illustration, assume that the inclusions are spheres of radius r located at the corners of a simple cubic lattice with lattice spacing l, as indicated in Fig. 7-7.3. For an intersection with one inclusion the

[85] L. Néel, *Ann. Univ. Grenoble* **22,** 299 (1946).

[86] M. Kersten, *Grundlagen einer Theorie der ferromagnetischen Hysterese und Koerzitivkraft*, Hirzel, Leipzig (1943) [reproduced by Edwards, Ann Arbor, Mich. (1943)], *Z. Physik* **124,** 714 (1948).

wall area is reduced an amount $\pi(r^2 - x^2)$. If there are N inclusions per unit volume, that is, $N = 1/l^3$, the wall energy per unit area is

$$F_{wt} = F_w[1 - N^{2/3}\,\pi(r^2 - x^2)] \quad \text{for} \quad x < |r|,$$

and

$$F_{wt} = F_w \quad \text{for} \quad |l| - |r| > x > |r|.$$

Similar relationships hold for x outside this range. For $x < |r|$ we have

$$\frac{\partial F_{wt}}{\partial x} = 2F_w N^{2/3}\pi x,$$

which has a maximum at $x = r$, that is

$$\left(\frac{\partial F_{wt}}{\partial x}\right)_{\max} = 2F_w N^{2/3}\pi a.$$

Then, from equation 7-7.1, the coercive force is given by

$$H_c = \frac{F_w N^{2/3}\pi a}{M} \tag{7-7.8}$$

or, since $F_w \approx \delta K_1$ (equations 7-5.2 and 7-5.3),

$$H_c \approx N^{2/3}\pi a \delta \frac{K_1}{M}. \tag{7-7.9}$$

Other cases, for example, in which the inclusions have cube or rodlike shapes or in which 90° domain walls are present, have been calculated by Kersten. Néel[87] has considered a random distribution of inclusions. The effect is to reduce the size of the hills and valleys of the F_{wt} versus x function. He concludes that the maximum coercive force possible is less than 1 oe if the domain walls remain rigid planes and less than 10 oe if the wall becomes deformed so as to intersect the maximum number of inclusions possible.

The *variable internal field theory*, first proposed by Néel,[87] takes into account magnetostatic energy omitted from the two theories just given. That there is magnetostatic energy present when there are inclusions is clear from Figs. 7-6.5 and 7-7.4*a*. For a spherical inclusion this energy is (equation 7-1.6)

$$F_D = \frac{1}{2} DM^2V$$

$$= \frac{1}{2}\left(\frac{4\pi}{3}\right)\left(\frac{4\pi r^3}{3}\right)M^2.$$

[87] L. Néel, *Ann. Univ. Grenoble* **22,** 299 (1947).

When such an inclusion is bisected by a 180° wall, a pole distribution like that shown in Fig. 7-7.4*b* reduces this magnetostatic energy by about 50%. The difference in F_D for these two cases is considerably larger than the change in the wall energy F_w calculated from the inclusion theory for all but very small inclusions. The situation is probably not exactly as pictured in the drawing; Néel assumes that the magnetization will be nonuniform in the vicinity of the inclusion. Uncompensated poles may also occur when internal stresses are present. If these stresses vary in direction within one domain, the divergence of the magnetization will tend to be

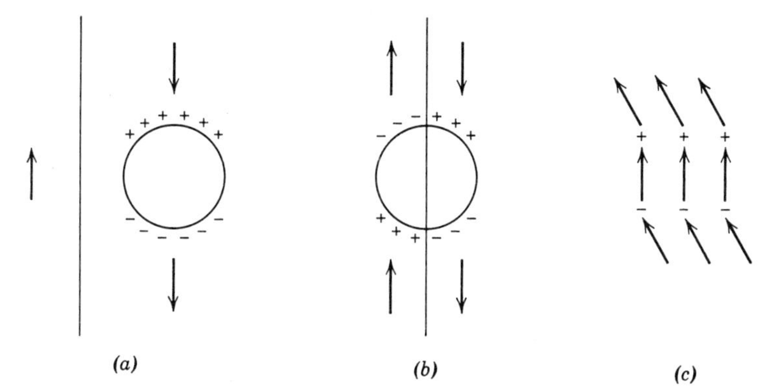

Fig. 7-7.4. To illustrate the variable field theory. The magnetostatic energy associated with the uncompensated poles varies with the position of the domain wall.

nonzero; a possible situation is illustrated in Fig. 7-7.4*c*. When this stress system is intersected by a domain wall, the magnetostatic energy is decreased.

Néel[88] has made an involved calculation on the basis of this model and finds for iron

$$H_c = 2.1v_s + 360v_i$$

and for nickel

$$H_c = 330v_s + 97v_i,$$

where v_s is the fractional volume of the internal stresses and v_i is the fractional volume of the nonmagnetic inclusions. It was assumed that the stresses were tensions of magnitude 30 kg/mm² and that $\delta \approx r$ for the inclusions. For polycrystalline materials, it appears that the uncompensated poles at grain boundaries are particularly important.[89] It has already been pointed out that secondary domain structures are likely in the vicinity of a large inclusion (see Fig. 7-6.5*b*); the same remark is

[88] L. Néel, *Ann. Univ. Grenoble* **22,** 299 (1947).

[89] J. B. Goodenough, *Phys. Rev.* **95,** 917 (1954).

valid for stress centers. When such domain structures occur, they will influence the value of the coercive force.[90] Other theoretical contributions to the variable internal field theory may be found in the literature.[91]

Regardless of the uncertainty concerning the details, the variable field theory appears to be able to account for coercive forces larger than 10 oe. This mechanism may be the origin of the permanent magnet properties of the quench-hardened steels mentioned in Section 7-4.

On the experimental side the coercive force arising from the presence of inclusions has been studied by a number of workers. It has been found, for example,[92] that for spherical particles of Fe_3C in Fe the maximum coercive force occurs when $2r \approx \delta$, where r is the radius of an inclusion.

All of these theories of the coercive force fail to take account of the changes in the over-all domain configuration when individual walls move. As emphasized in Section 7-5, the occurrence of uncompensated poles on domain walls and on crystal surfaces will tend to be avoided. The neglect of such considerations is likely to be particularly important for polycrystalline materials.

Initial permeability. According to the picture already given, the initial permeability (and initial susceptibility) depends on the value of $\partial F_{wt}/\partial x$ near the origin. The actual values of $\partial F_{wt}/\partial x$ depend on the details of the specific model; problems 7-10 and 7-11 serve to illustrate two models. Further, the initial permeability is reversible in this picture, that is, the same magnetization curve is traced out no matter whether the field is increased or decreased, provided only that the magnitude of the field remains small. In actual fact it is found that the magnetization curve is not completely reversible; that is, there is a small amount of hysteresis. This should not be a complete surprise, since it is reasonable to expect that some walls will undergo irreversible movements even in small fields. From experimental studies Rayleigh[93] found first that

$$M_H = \alpha H + \beta H^2$$

and second that

$$M_H = \alpha(H_1 - H) + \tfrac{1}{2}\beta(H_1 - H)^2,$$

provided the applied field is small. Here H_1 is the maximum field applied and α and β are constants that characterize a particular material. The second relationship is valid only for $-H_1 < H < H_1$. Many materials

[90] E. Kondorsky, *Dokl. Akad. Nauk SSSR* **68,** 37 (1949).

[91] F. Vicena, *Czech. J. Phys.* **4,** 419 (1954), **5,** 11, 480 (1955); M. Kersten, *Z. angew. Phys.* **7,** 397 (1955), **8,** 496 (1956).

[92] L. J. Dijkstra and C. Wert, *Phys. Rev.* **79,** 979 (1950).

[93] Lord Rayleigh, *Phil. Mag.* **23,** 225 (1887).

are found to follow these empirical relationships.[94] Rayleigh's relationships have been derived from theories based on certain domain models.[95]

Picture-frame specimens. Because of the complexity of the problem, the best hope for the achievement of significant information concerning the magnetization process lies in the study of specimens with unusually simple domain arrangements. One simple case, the Sixtus-Tonks experiment, has already been discussed. Another results when a single crystal is cut in the form of a hollow rectangle with edges lying along the crystalline easy directions. "Picture-frame" specimens were first made by Williams and Shockley[96] out of single crystals of silicon iron; similar specimens have been made out of other single crystals, including the ferrimagnetic materials magnetite[97] and YIG.[98]

The simplest domain arrangement observed is shown in Fig. 7-7.5*a*; the specimen then consists of four domains, one lying along each leg,

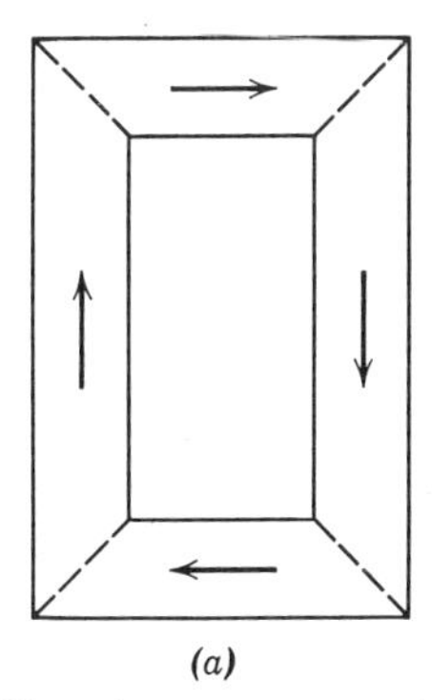

(*a*)

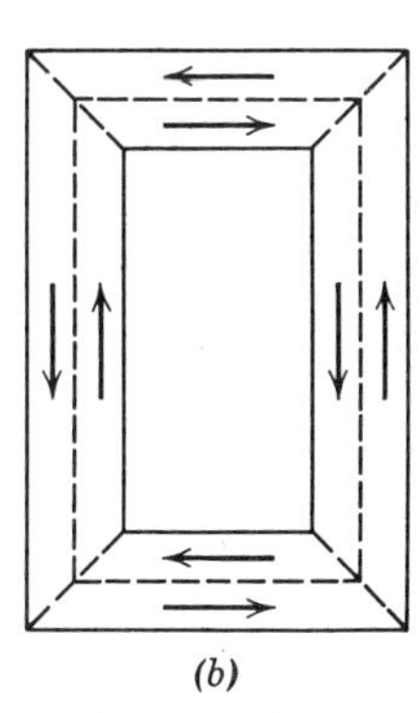

(*b*)

Fig. 7-7.5. Domain arrangements in picture-frame specimens. The domain walls are indicated by the dashed lines; the arrows represent the direction of magnetization of the domains.

separated by 90° domain walls oriented at 45° to each leg. Although in this state the specimen as a whole is demagnetized, it may in a sense be considered saturated in a clockwise direction about the crystal. The application of a reverse field along one leg serves to produce the domain pattern shown in Fig. 7-7.5*b*. By increasing the field, the 180° walls can be made to move across each leg and to leave the specimen magnetized

[94] W. B. Ellwood, *Physics* **6,** 295 (1935).

[95] F. Preisach, *Z. Physik* **94,** 277 (1935); E. Kondorsky, *J. Phys.* (*USSR*) **6,** 93 (1942); L. Néel, *Cahiers Phys.* **12,** 1 (1942), **13,** 18 (1943).

[96] H. J. Williams and W. Shockley, *Phys. Rev.* **75,** 178 (1949).

[97] J. K. Galt, *Phys. Rev.* **85,** 664 (1952).

[98] F. B. Hagedorn and E. M. Gyorgy, *J. Appl. Phys.* **32,** 282S (1961).

in the counterclockwise direction. Measurements[99] show that the net magnetization (or induction) of one leg varies linearly with the position of the 180° wall, and hence provides a direct proof of the correlation between magnetization and domain structure.

On the basis of the model outlined on page 385 and illustrated in Fig. 7-7.1*b*, *c*, hysteresis occurs because the applied field is held constant during the sudden wall movement from position A to position E. If, however, the field were quickly reduced when the wall reaches position A, the wall would move to B and then to C, etc., and hysteresis would be zero. Such an experiment was attempted for the 180° wall of a picture-frame specimen by Stewart.[100] He was able to reduce the hysteresis by about 12%. With a more refined experimental arrangement, it is likely that the hysteresis could have been reduced further, although it probably could not have been reduced to zero. This residual hysteresis may imply that there are discontinuities in the F_{wt} versus x curve.

The approach to saturation. When the applied field is increased to larger values, domain-wall movement becomes relatively unimportant, and magnetization changes occur primarily by domain rotations. Only polycrystalline materials will be the subject of discussion here, since rotational processes in single crystals have already been treated in Section 6-9. It has been stated that the approach to the saturated state may be described by the empirical relationship (equation 6-2.15)

$$M_H = M_s\left(1 - \frac{a}{H} - \frac{b}{H^2}\right) + cH,$$

where M_s is the saturation magnetization and M_H is the magnetization (along the field direction) in the field H. The term cH is small; it has already been discussed in Chapter 6 and need not concern us further here. The problem is then to interpret the $1/H$ and $1/H^2$ terms.

Suppose each crystallite has become a single domain whose magnetization vector lies almost along the field direction. The torque arising from the action of the field, $HM_s \sin\theta$ (equation 1-10.1) is opposed by the torque resulting from the crystalline anisotropy L_K. Actually, other torques may be present, but they will be assumed to be negligible here. At equilibrium we have

$$\begin{aligned} L_K &= HM_s \sin\theta \\ &= HM_s\,\theta, \end{aligned}$$

[99] H. J. Williams W. and Shockley, *Phys. Rev.* **75,** 178 (1949).
[100] K. H. Stewart, *J. Phys. radium* **12,** 325 (1951).

since θ is presumably very small. Then we get

$$\begin{aligned} M_H &= M_s \cos\theta \\ &= M_s\left(1 - \frac{\theta^2}{2}\right) \\ &= M_s\left(1 - \frac{L_K^{\,2}}{2M_s^{\,2}H^2}\right) \end{aligned}$$

for the approach law for one crystallite.[101] Then, on assuming that the crystallites are oriented at random, an averaging[102,103] leads to the expression

$$M_H = M_s\left(1 - \frac{b}{H^2}\right),$$

where b depends on the anisotropy constant K_1. When interactions between grains are considered,[104,105] it is found that b depends on the value of the applied field for smaller fields and becomes a constant when $H \gg 4\pi M_s$. Experimental measurements on nickel wire[105] in the range $150 < H < 750$ appear to confirm the calculations that take the interactions into account. A value for K_1 is obtained that is in good agreement with that found for single crystals.

The origin of the $1/H$ term is less well understood. Calculation shows that when imperfections such as dislocations,[106] or nonmagnetic inclusions[107] are present, a $1/H$ term predominates for intermediate fields, whereas a $1/H^2$ term predominates for very large fields (when $H \gg 4\pi M_s$). For iron an intermediate field is one from, say, 1000 to 50,000 oe. Experimentally, the $1/H$ term is found to be the important one for fields above a few thousand oersteds. The maximum field normally employed in the laboratory, however, is only about 25,000 oe, so that no real test of the term that is ultimately most important appears to have been made.

Remanence. Suppose that when a polycrystalline ferromagnetic specimen goes from the saturated to the remanent state, only reversible rotations of the magnetization vectors of the crystallites occur. A straightforward calculation of the ratio M_r/M, where M is the spontaneous or saturation magnetization (assumed to be equal), may then be made. For an assembly

[101] P. Weiss, *J. Phys. radium* **9,** 373 (1910).
[102] N. S. Akulov, *Z. Physik* **69,** 822 (1931).
[103] R. Gans, *Ann. Phys. (Leipzig)* **15,** 28 (1932).
[104] T. Holstein and H. Primakoff, *Phys. Rev.* **59,** 1098 (1941).
[105] L. Néel, *J. Phys. radium* **9,** 193 (1948).
[106] W. F. Brown, Jr., *Phys. Rev.* **60,** 139 (1941), **82,** 94 (1951).
[107] Néel, 184 *op. cit.*, 184.

of crystallites oriented at random, whose predominant anisotropy is crystalline, it may be shown[108] that (*a*) $M_r/M = 0.83$ for cubic crystals with [100]-axes as easy directions, (*b*) $M_r/M = 0.86$ for cubic crystals with [111]-axes as easy directions, and (*c*) $M_r/M = 0.50$ for uniaxial crystals. The last is, of course, identical to the Stoner-Wolhfarth calculation, the result of which is illustrated in Fig. 7-2.10. Experimentally, it is found[109] that M_r/M for iron and nickel is almost always less than 0.85 and usually less than 0.5.

The foregoing calculations are inadequate for two main reasons. One, no account is taken of interactions between the crystallites and, two, the assumption that no domain wall movements occur is probably incorrect for most materials. The general problem in which both rotations and domain wall movements take place in an increasing or decreasing applied field is a difficult one. Nevertheless, attempts have been made to distinguish between the contributions made by the two processes to the change in magnetization in sintered ferrites.[110]

Nucleation of domains: whiskers. Suppose a ferromagnetic specimen is saturated by the application of a large magnetic field so that the specimen has become one large domain. Then, when the field is reduced, there is the question how and when domains of reverse magnetization are started. The problem of the nucleation of domains is one of fundamental importance to the understanding of the magnetization process.

If the specimen were a perfect single crystal, the moments presumably would have to be rotated against the crystalline anisotropy. For iron saturated along a [100]-direction this rotation would require a reverse field of $H = 2K_1/M$, which is about 540 oe (see Table 7-2.1). From experiment, however, it is found that most soft materials possess coercive forces from a few oersteds to a fraction of an oersted. Hence some other nucleation mechanism must be sought. It is not unreasonable to suspect that imperfections are involved, since large demagnetization fields may be present in their vicinity. For example, for a spherical nonmagnetic inclusion the reverse field in the cavity is $H = -(4\pi/3)M$, which for iron is about 7000 oe. The presence of secondary domain structures near cavities has already been illustrated in Fig. 7-6.5.

Support for these ideas has come from the study of iron whiskers. Whiskers contain a relatively small number of imperfections and are, in fact, the nearest thing to perfect crystals known. They may be grown by the reduction of a ferrous halide in hydrogen. A typical whisker is about 1 cm long and a few microns thick. It also has gleaming strain-free

[108] R. Gans, *Ann. Phys. (Leipzig)* **15,** 28 (1932).
[109] R. M. Bozorth, *Z. Physik* **124,** 519 (1948).
[110] D. M. Grimes, *J. Phys. Chem. Solids* **3,** 141 (1957).

surfaces that are usually parallel to the principal crystallographic planes; for example, the (100) planes.

In experiments on certain selected (100) iron whiskers regions were found[111] in which the reversal of the magnetization required fields of almost 500 oe, a value close to the theoretical limit of $2K_1/M$ expected for a perfect specimen. It appears likely that these reversals took place via the curling mode. For whiskers that possessed growth defects on the surface, or damaged ends, the field required for magnetization reversal was much less, sometimes of the order of a few oersteds.

The domain patterns observed on whiskers is also of interest. A demagnetized whisker with four (100) faces often has two main domains, separated by 180° walls, together with two closure domains,[112] as shown in Fig. 7-7.6*a*. If, however, a field is applied perpendicular to the whisker's

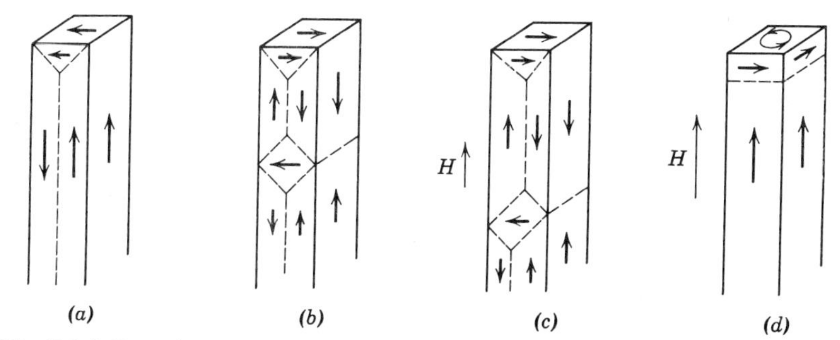

Fig. 7-7.6. Domain structures of iron whiskers, as inferred from powder pattern and Kerr effect studies. The surfaces of the whisker are (100) planes. (After G. G. Scott and R. V. Coleman and C. A. Fowler, E. M. Fryer, and D. Treves.)

axis and then removed, the structure shown in Fig. 7-7.6*b* is obtained. Then, when a field is applied along the whisker's axis, the structure is modified as shown in Fig. 7-7.6*c*. When the applied field is increased to a few hundred oersteds, most of the whisker becomes saturated. However, near the ends of the whisker small reverse domains persist.[113] On increasing the field, spikelike transverse domains are observed up to 2000 oe, whereas the structure shown in Fig. 7-7.6*d* is found up to about 5000 oe. Now, the tips of whiskers often possess sharp corners. In a saturated specimen there would be a large local demagnetizing field in the vicinity of these corners. It therefore appears likely that the domain structures

[111] R. W. DeBlois and C. P. Bean, *J. Appl. Phys.* **30,** 225S (1959); R. W. DeBlois, *J. Appl. Phys.* **32,** 1561 (1961).

[112] G. G. Scott and R. V. Coleman, *J. Appl. Phys.* **28,** 1512 (1957).

[113] C. A. Fowler, E. M. Fryer, and D. Treves, *J. Appl. Phys.* **31,** 2267 (1960).

occur in order to reduce the large fields that would otherwise be present at the ends of the whiskers.[114] It is reasonable to suppose that a similar situation exists when irregularly shaped imperfections lie within the body of ferromagnetic specimens. Since extremely large applied fields are not normally employed for the determination of hysteresis loops, the implication is that true saturation is not achieved in many materials. The retention of small domains means that the nucleation of domains may not be involved in these hysteresis loops. Additional evidence for the persistence of small domains has been obtained from studies on single crystals of cobalt. It is found that there is little change in the domain pattern on the basal plane until fields of several thousand oersteds are applied.[115] Finally, mention should be made of domain studies on barium ferrite crystals. After these crystals have been saturated along the easy direction, it is found that domains always reappear at the same sites.[116] This result supports the idea that nucleation occurs at certain localized imperfections.

For most materials the applied field required for domain nucleation is probably less than that required for domain wall movement. However, if the nucleation field is larger, a rectangular-shaped hysteresis loop is to be expected. This follows, for after nucleation has been achieved the walls will move rapidly across the specimen and reverse the magnetization.

Barkhausen effect. Discontinuous changes in the magnetization of a ferromagnetic specimen were first observed by Barkhausen.[117] In his experimental arrangement the applied field, produced by a solenoid, is changed continuously and the signals induced in a search coil are amplified and detected by earphones or a loudspeaker. Subsequent investigators have improved on the method of detection and oscillographs,[118] galvanometers,[119] and pulse counting techniques[120] have been employed. In this way information on the size of the magnetic moment involved in the magnetization change, on the size distribution, and on the reversal time has been obtained. For soft materials the sudden magnetization changes presumably arise from irreversible wall movements; for example, the spontaneous motion of the wall from position A to position E in Fig. 7-7.1. For a picture-frame specimen (Fig. 7-7.5*b*) the Barkhausen effect

[114] S. Shtrikman and D. Treves, *J. Appl. Phys.* **31,** 725 (1960).

[115] D. J. Craik and R. S. Tebble, *Rept. Prog. Phys.* **24,** 116 (1961).

[116] C. Kooy and U. Enz, *Philips Res. Rept.* **15,** 7 (1960).

[117] H. Barkhausen, *Physik Z.* **20,** 401 (1919).

[118] R. M. Bozorth, *Phys. Rev.* **34,** 772 (1929); K. Murakawa, *Proc. Phys.-Math. Soc. Japan* **19,** 715 (1937).

[119] R. M. Bozorth and J. Dillinger, *Phys. Rev.* **35,** 733 (1930).

[120] R. S. Tebble, I. C. Skidmore, and W. D. Corner, *Proc. Phys. Soc.* (*London*) **A-63,** 739 (1950).

was observed when the wall changed its position suddenly.[121] Discontinuous magnetization changes may also occur when irreversible domain rotations take place in single-domain particles. It is possible that some of the Barkhausen effect in polycrystalline materials has this origin.

Preisach-type models. Brief mention is now made of a method of analyzing magnetization curves by what we will call Preisach-type models. Consider a specimen consisting of many domains and assume that each possesses a rectangular hysteresis loop.[122] Further suppose the interaction between domains can be represented by a local field acting on each domain. Let h denote this local field for a particular domain with coercive force c. Then, when a field is applied, this domain is characterized by the critical fields $-c - h$ and $+c - h$ shown in Fig. 7-7.7*a*. The ensemble

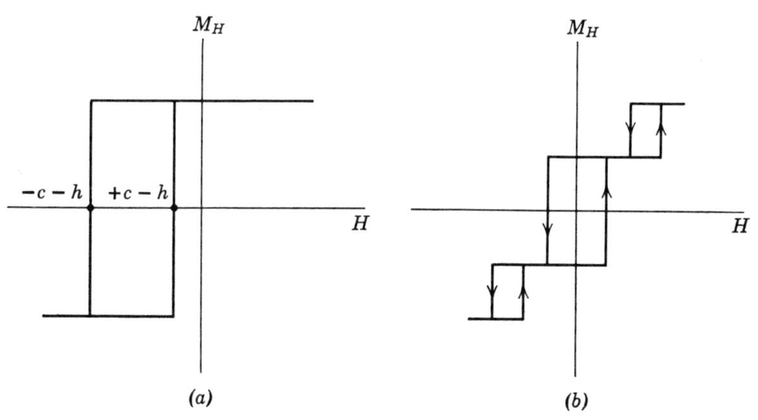

Fig. 7-7.7. Hysteresis loops assumed in Preisach type models. In (*a*) the rectangular hysteresis loop of one grain (domain) is displaced because of the presence of a local field h ($h > c$). In (*b*) a possible hysteresis loop for two coupled grains is shown. [L. Néel, *J. Phys. radium* **20,** 215 (1959).]

of domains is then described by the distribution in the values of c and h. Calculations based on this model lead to results concerning the magnetization curve that are in accord with experiment; for example, the Rayleigh relationships have been obtained.[123]

The assumption that each domain possesses a rectangular hysteresis loop in general appears to be artificial. This situation may develop if each domain is associated with one grain of a polycrystalline specimen.

[121] H. J. Williams and W. Shockley, *Phys. Rev.* **75,** 178 (1949).

[122] F. Preisach, *Z. Physik* **94,** 277 (1935); see also P. Weiss and J. de Freundenreich, *Arch. Sc. Nat.* (*Geneva*) **42,** 449 (1916).

[123] E. Kondorsky, *J. Phys.* (*USSR*) **6,** 93 (1942); L. Néel, *Cahiers Phys.* **12,** 1 (1942), **13,** 18 (1943).

Then, if each grain is a single domain, we have already seen (Section 7-2) that a rectangular hysteresis loop is a possibility. If each grain possesses more than one domain, a rectangular loop is still possible. This would occur, for example, if the nucleation field were larger than the field required to cause domain wall movement or if the F_{wt} versus x function had the form of a square wave. Even granted rectangular hysteresis loops, however, it has not been shown that it is valid to represent the interaction effects by a local field.

Néel[124] has extended this model by considering two interacting grains, called a *couple*. Each grain is assumed to possess a rectangular hysteresis loop and a coercive force and saturation magnetic moment that differ from each other. Six possible hysteresis loops are envisioned for the couple; one is shown in Fig. 7-7.7*b*. An ensemble of these couples, assumed not to interact with one another, is then considered. Two new effects are predicted. One, called *bascule*,[125] is the change in magnetization in a certain applied field H_1 after this field has been taken from H_1 to another field H_2 and back to H_1 a number of times. The other, called *reptation*,[126] is the gradual displacement of the hysteresis loop as the number of cycles is increased. Both effects appear to have been observed.[127]

Although Preisach-type models are based on doubtful assumptions, they do in some cases give interesting results that are in agreement with experimental data. The precise reason for this is not clear. It may well be that the models provide a set of parameters that is suitable for magnetization problems.

External stresses. The magnetization curves of most ferromagnetic materials are greatly changed when external stresses are applied. Probably the simplest situation to treat is one in which the crystalline anisotropy may be neglected because it is small compared to the magnetostrictive energy. Suppose in addition that the magnetostriction can be considered isotropic. Then there are two possibilities, depending on whether the product $\lambda_s\sigma$ is positive or negative. For $\lambda_s\sigma > 0$, that is, positive magnetostriction and a tension or negative magnetostriction and a compressive stress, F_σ is a minimum when $\theta = 0$ or π. Thus, in the absence of other factors, some domains will lie along one and some along the other direction parallel to the tension axis. Then, when a field is applied along this axis, the specimen will tend to become a single domain whose magnetization lies along the field direction. This applies reasonably well to Permalloy with slightly less than 70% Ni (see Fig. 6-9.4*a*). The hysteresis loop of a

[124] L. Néel, *Compt. rend. (Paris)* **246,** 2313 (1958), *J. Phys. radium* **20,** 215 (1959).

[125] Perhaps translated as "tilting."

[126] Perhaps translated as "creep."

[127] Nguyen-Van-Dang, *J. Phys. radium* **20,** 222 (1959).

specimen of this material with and without a tension applied is shown in Fig. 7-7.8. When $\sigma = 0$, as usual a certain field is required to produce saturation. When $\sigma \neq 0$, the magnetization is constrained to lie along one or another of two possible directions, as expected, the actual direction being determined by the strength of the applied field. This effect is employed in the Sixtus-Tonks experiment.

When $\lambda_s\sigma < 0$, F_σ is a minimum when $\theta = \pi/2$ or $3\pi/2$, that is, the domains tend to lie at right angles to the tension axis. A field applied

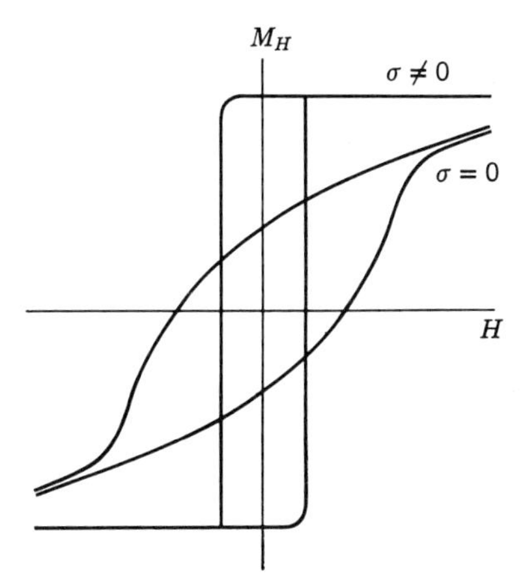

Fig. 7-7.8. Typical hysteresis loop of Permalloy with 68% Ni with and without applied tension. (After R. M. Bozorth.)

along the tension axis will cause the magnetization of the domains to rotate toward the direction of the field. If there are no complicating factors, the total energy for one domain is

$$F_T = (-\tfrac{3}{2}\lambda_s\sigma \cos^2 \theta - HM \cos \theta)V$$

where θ is the angle between the direction of the domain's magnetization and that of the field and tension and V is the volume of the domain. Application of the condition for a minimum, $\partial F_T/\partial\theta = 0$, then leads to the equation

$$M_H = -\frac{M^2}{3\lambda_s\sigma} H \tag{7-7.10}$$

for the magnetization at equilibrium. For nickel $\lambda_s < 0$ and K_1 is relatively small, so that if a sufficiently large tension is applied the conditions are reasonably well met. The effect of tension on the hysteresis loop of Ni is indicated in Fig. 7-7.9. It is clear that the application of a tension reduces the remanence and increases the field necessary to produce saturation, as expected. In the ideal case the remanence would be reduced to zero; it is not zero here, probably because the crystalline anisotropy energy F_K and the magnetostatic energy F_D are not zero. Experimentally, it is found that increasing the tension reduces the remanence further. Also, the linear relationship between M_H and H predicted by equation 7-7.10 is observed[128] provided the tension is large enough.

For materials with a large crystalline anisotropy, the term F_K must be retained in the analysis. If, in addition, the magnetostriction can no

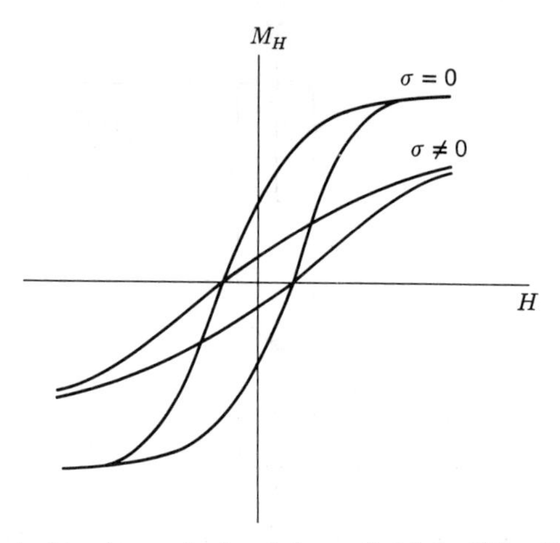

Fig. 7-7.9. Effect of tension on hysteresis loop of nickel. (After R. M. Bozorth.)

longer be assumed to be isotropic, the problem in general becomes much more difficult. However, the case in which $F_\sigma \ll F_K$ is relatively simple. Then the direction along which the magnetization lies will be determined primarily by F_K, the action of F_σ being to make some easy directions preferred over others. The effect a tension has on the magnetization curve can be predicted with the aid of the equation (problem 6-15)

$$\left(\frac{\partial M}{\partial \sigma}\right)_H = \frac{1}{l}\left(\frac{\partial l}{\partial H}\right)_\sigma.$$

Now, if the magnetization changes are occurring by coherent rotations from the easy directions to a harder one, such as that presumably occurring above the knee of the magnetization curve, $(\partial l/\partial H)_\sigma$ will be approximately independent of σ and will be given by its value for $\sigma = 0$. For iron, since the rotation will be away from a [100]-direction, $(\partial l/\partial H)_\sigma$ is negative (see Table 6-10.1). Hence this argument predicts that the magnetization of

[128] R. Becker and M. Kersten, *Z. Physik* **64**, 660 (1930).

iron between the knee and saturation will be decreased when a tension is applied, a result that is observed experimentally.

Minor hysteresis loops. If an alternating field $H_1 \cos \omega t$ is applied to a ferromagnetic specimen and H_1 is insufficient to produce saturation, a typical B_H versus H curve will have the form of one of the curves shown in Fig. 7-7.10. Such curves are called minor hysteresis loops. When saturation is achieved, the curve obtained is sometimes called a major hysteresis loop, and frequently just the hysteresis loop, as in this book,

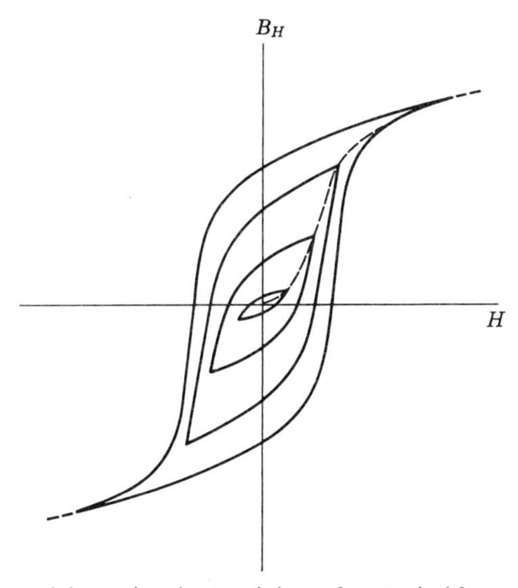

Fig. 7-7.10. Major and three minor hysteresis loops for a typical ferromagnetic material.

for example. The tips of the minor hysteresis loops lie on the magnetization curve. Hence the maximum permeability, as measured at the field H_1, occurs when the tip of the minor loop lies in the vicinity of the knee of the magnetization curve. The analysis of minor loops is based on the same principles already described in this chapter. For most soft materials the changes of magnetization during the traversal of minor loops takes place primarily by reversible and irreversible wall motion.

A knowledge of minor hysteresis loops is important in the design of transformers for use in electrical engineering applications. Power transformers are operated on a minor loop that provides close to the maximum permeability. The effective transfer of power is aided if the material has a large saturation magnetization and if the induction at the knee of the magnetization curve approaches this saturation value. The latter feature may be achieved if the material is anisotropic, with the field applied along the easy direction, and if the magnetization changes take place by 180° domain wall movements. It is also desirable that the material have a small hysteresis loop area in order to reduce the energy losses.

Transformers employed in communication circuits are usually operated with a constant field superposed on the alternating field. The constant field is present because the transformer is in the plate circuit of a conducting electronic tube. It is then more convenient to tolerate than to eliminate the steady field. Typical minor hysteresis loops superposed on the magnetization curve are shown in Fig. 7-7.11. The ratio $\Delta B_H/\Delta H$ is called

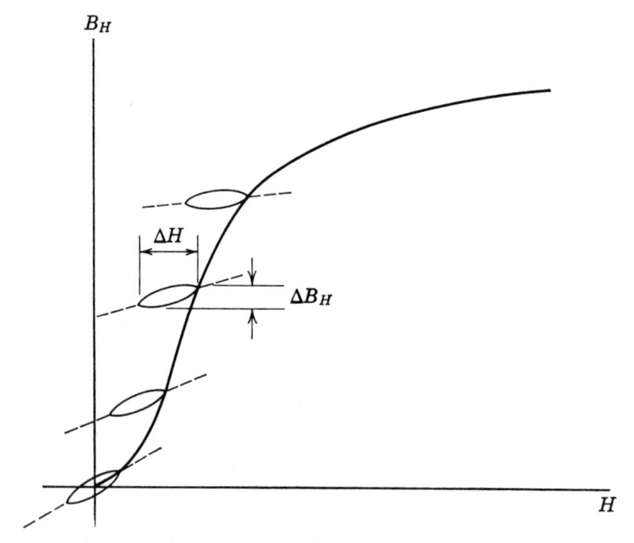

Fig. 7-7.11. Minor hysteresis loops superposed on the magnetization curve.

the incremental permeability and the limit when the area of the minor loop approaches zero, dB_H/dH, is called the reversible permeability for obvious reasons. It is desirable to reduce the area of the minor loops in order to make the induction versus H relationship as linear as possible. This reduces the introduction of harmonics into the signal and hence improves the fidelity. Also, the greater the permeability of the minor loop, the greater the sensitivity of the amplifier. The factors that affect these biased minor loops are the same as those that are important for the initial permeability.

8. Thermal Effects Associated with the Hysteresis Loop

The work required to take a ferromagnetic material once around the hysteresis loop is given by the area enclosed by the loop (equation 7-1.1).

This work corresponds to the sum of the irreversible changes in magnetization that occur in one cycle. If the traversal of the loop is carried out under adiabatic conditions, the temperature of the material will rise, the amount depending on the specific heat. Although this temperature rise is small, being of the order of 10^{-4}°C for most soft magnetic materials, it can be measured accurately without undue difficulty. Such measurements, however, do not provide additional information concerning the irreversible magnetization changes that occur. Of much greater interest are the temperature changes that take place after each small step of a step-by-step traversal of the hysteresis loop. These temperature changes are only of the order of 10^{-6}°C, and hence require very sensitive apparatus. In the early experiments a number of thermocouples were connected in series to a galvanometer.[129] Bates and his collaborators,[130] in an extensive series of measurements carried out since 1940, employ a different thermocouple arrangement. In their apparatus about 20 thermocouples are used. One junction of each thermocouple is attached to the specimen; the other junction, although nearby, is thermally insulated from the specimen. Each thermocouple has its own primary winding on a transformer core. The temperature changes are detected by a sensitive galvanometer connected to the secondary.

The data for a CoNi alloy with 59% Co, shown[131] in Fig. 7-8.1, will serve to illustrate the general nature of the experimental results and the method of analysis. Instead of the change of temperature, ΔT, it is convenient to consider the corresponding change in heat per cubic centimeter, $\Delta Q'$, given by $\Delta Q' = C\,\Delta T$, where C is the specific heat. The curve labeled $Q_T'(=\Sigma\,\Delta Q_T')$ gives the total observed heat changes in the traversal of the ascending branch of the hysteresis loop; that is, when a large magnetic field is first applied in one (the negative) direction, decreased to zero, and then increased to the same large value in the opposite (positive) direction. The Q_T' curve for the descending branch of the hysteresis loop is symmetrical to the one shown and is not usually plotted. Now the Q_T' curve includes both the irreversible and the reversible thermal changes. The separation of reversible and irreversible changes is accomplished by making small decreases in the magnetic field at several particular points on the hysteresis curve. If the minor loop is sufficiently small so that the area enclosed is essentially zero, the observed heat changes may reasonably be

[129] U. Adelsberger, *Ann. Phys. (Leipzig)* **83,** 184 (1927); T. C. Hardy and L. Quimby, *Phys. Rev.* **54,** 217 (1938).

[130] L. F. Bates and J. C. Weston, *Proc. Phys. Soc. (London)* **53,** 5 (1941); L. F. Bates and G. Marshall, *Revs. Mod. Phys.* **25,** 17 (1953); L. F. Bates and N. P. R. Sherry, *Proc. Phys. Soc. (London)* **B-68,** 642 (1955).

[131] L. F. Bates and H. Clow, *J. Phys. radium* **20,** 93 (1959).

assumed to be the result of reversible processes only. The curve labeled Q_R' gives the reversible heat changes found by the backward step method.

It is now pertinent to consider the processes that contribute to the reversible heat changes. It may be shown[132] that the thermal change arising from changes in the spontaneous magnetization M is given by

$$\Delta Q_{MC}' = a \int d(HM_H), \tag{7-8.1}$$

where $a = -\dfrac{T}{M}\left(\dfrac{\partial M}{\partial T}\right)$. This is just the magnetocaloric effect, already discussed in Section 6-2 (see equation 6-2.23), and may be calculated with a

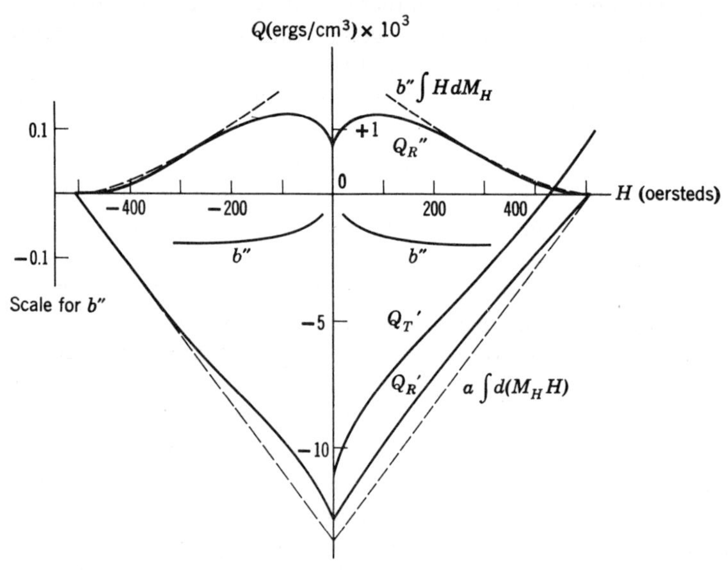

Fig. 7-8.1. Changes in heat as a function of applied field arising from various processes in a CoNi alloy with 59% Co.

good degree of accuracy. It also may be shown[132] that the thermal change arising from the reversible rotation of the magnetization against the crystalline anisotropy is given by

$$\Delta Q_{Ro}' = b \int H \, dM_H \tag{7-8.2}$$

where $b = (T/K_1)/(\partial K_1/\partial T)$, K_1 being the anisotropy constant. Since K_1 and $\partial K_1/\partial T$ are usually not known with a high degree of precision and since stresses may be present that modify the value of K_1, $\Delta Q_{Ro}'$ cannot be

[132] E. C. Stoner and P. Rhodes, *Phil. Mag.* **40,** 481 (1949).

Table 7-8.1. Values of Thermal Constants

Metal	b	b''
Iron	−0.48	−0.29
Nickel	−4.5	−4.35
Cobalt	−1.27	−1.1

calculated so accurately as $\Delta Q_{MC}'$. If the calculated value of $\Delta Q_{MC}'$ is subtracted from the observed thermal change Q_R', we get

$$Q_R'' = Q_R' - a\int d(HM_H)$$

$$= \int b''H\,dM_H \tag{7-8.3}$$

and

$$b'' = \frac{\Delta Q_R''}{\Delta\int H\,dM_H} \tag{7-8.4}$$

Then b'' may be compared with b. Curves of $a\int d(M_H H)$ and Q_R'' as a function of H are plotted in Fig. 7-8.1. It is interesting to note that the contribution from the magnetocaloric effect is considerable, even though only small or moderate fields are applied. Also the curves $b''\int H\,dM_H$ for $b'' = -0.09$ are indicated by a dashed line. The agreement between the $b''\int H\,dM_H$ and Q_R'' curves for moderate fields (in this case fields greater than 200 oe) leads to the conclusion that in this range the reversible heat changes are primarily the result of spontaneous magnetization changes and rotations. A similar conclusion has been reached for several other materials. The values of b and b'' for iron,[133] nickel, and cobalt[134] are compared in Table 7-8.1. For small applied fields $b'' \neq b$, and it is reasonable to suppose that domain wall motions are important. Attempts to estimate the contributions to b'' from other processes have been made,[135] but at present no definitive conclusions have been reached.

9. Soft Magnetic Materials

An analysis of the properties of a particular soft magnetic material may be given in terms of the principles described earlier in this chapter. Conversely, a knowledge of these principles should help in the design of a material with certain specific properties. Properties that are usually desirable in a soft magnetic material include large saturation magnetization, small coercive force, large initial and maximum permeabilities, and small hysteresis loss. A large saturation magnetization is achieved, of

[133] L. F. Bates and D. J. Sansom, *Proc. Phys. Soc.* (*London*) **B-74,** 53 (1959).

[134] L. F. Bates and H. Clow, *J. Phys. radium* **20,** 93 (1959).

[135] R. W. Teale and G. Rowlands, *Proc. Phys. Soc.* (*London*) **B-70,** 1123 (1957).

course, by choosing a metal or alloy with a large atomic magnetic moment, such as iron or an iron-cobalt alloy. The other properties are improved by increasing the freedom of motion of the domain walls. This may be accomplished by reducing or removing the inclusions, cavities, grain boundaries, and internal stresses. The inclusions can be decreased by removing the impurities, whereas the internal stresses of a mechanical origin can be reduced by suitable annealing. The choice of materials with small values of crystalline anisotropy K_1 and magnetostriction λ_s also helps. A small K_1 reduces the effect of inclusions on wall motions, and a small λ_s reduces that of stresses. In addition, magnetic softness requires that the field for domain nucleation be small; this is favored by small values of K_1 and λ_s.

The qualities of soft magnetic materials can also be improved by various other metallurgical techniques. Two important techniques will now be described. In the first, called *grain orientation*, a suitable alloy is subjected to a sequence of cold-rolling and annealing processes. As a result of this action, the crystal grains become oriented, with the easy crystalline direction tending to lie parallel to the rolling direction. For fields applied parallel to this rolling direction the hysteresis loop is approximately rectangular; hence the maximum permeability is increased and the coercive force and losses are decreased compared to the isotropic material. By changing the metallurgical treatment it has been found possible to make the directions parallel and perpendicular to the rolling direction both easy directions; the alloy is then said to possess *cube texture*.

In the other technique, called magnetic annealing, a magnetic field is applied to the alloy as it cools through a certain temperature range. The material is then found to possess, in addition to the usual crystalline anisotropy, a uniaxial anisotropy that lies along the direction of the annealing field. This again gives rise to an approximately rectangular loop in the uniaxial direction, with consequently higher μ_{max} and lower losses. The origin of the uniaxial anisotropy is not certain. It has been suggested that neighboring pairs of like atoms tend to become aligned along the direction of the applied field when the temperature is high enough to permit diffusion. Then, at lower temperatures, the aligned pairs are immobile and produce the uniaxial anisotropy. This is known as the directional-order theory.[136]

Iron possesses a relatively large saturation magnetization and is comparatively inexpensive. Because of these properties, iron is widely used as a soft magnetic material. Although an average commercial grade of

[136] L. Néel, *J. Phys. radium* **15,** 225 (1954); S. Taniguchi and M. Yamamoto, *Sci. Rept. Tôhoku Univ.* **A-6,** 330 (1954); S. Chikazumi and T. Oomura, *J. Phys. Soc. Japan* **10,** 842 (1955).

iron has a maximum permeability of about 5×10^3 and a coercive force of about 0.9 oersteds, proper purification procedures can make it one of the softest materials known. By starting with electrolytically pure iron and then by heating in a moist hydrogen atmosphere, a single crystal of iron has been made with $\mu_{max} \approx 1.4 \times 10^6$. The disadvantage to the use of pure iron, besides cost, is the low electrical resistance and mechanical softness. If a commercial grade of iron is not good enough for some application, an alloy is usually substituted. Pure cobalt and nickel are too costly for most uses.

Alloys of silicon iron have been in widespread use since 1900. The alloy starts to become less ductile when more than 4% Si is present, and as a result most of the alloys in use contain less than 6%, the commonest amount being 3 to 4%. The addition of silicon to iron reduces the saturation magnetization by only a small amount, reduces the crystalline anisotropy and magnetostriction by an appreciable amount, and greatly increases the electrical resistivity. The presence of silicon apparently makes it easier to reduce the effect of inclusions and to increase the grain size by metallurgical treatment. A 3% Si-Fe alloy can be grain oriented and is in common use commercially (Hypersil). Cube texture has also been produced in a 3% alloy. The maximum permeability of the oriented alloy may exceed 4×10^4. However, when a very large μ_{max} is desired and a large saturation magnetization is not important, it is usually cheaper to use some other alloy.

Some nickel-iron alloys (Permalloys) possess very large permeabilities. The alloys have a face-centered cubic structure for 35 to 100% Ni content. In this range the 50 Ni-50 Fe alloy (Hypernik) has the largest saturation magnetization ($M_s = 1230$). This 50% alloy can be easily purified. In addition, it has a relatively large electrical resistivity, it is ductile, and it can be annealed by a magnetic field and grain oriented. As a result it has found considerable use.

Of greater interest, however, is the Permalloy with about 78% nickel. This alloy possesses both a small magnetostriction and a small crystalline anisotropy. Actually, $\lambda_s = 0$ for 82% Ni, whereas $K_1 = 0$ at about 75% Ni, so that the 78% Ni represents a compromise composition. With proper heat treatment, a maximum permeability of about 10^5 and a coercive force of about 0.05 oe can be obtained.

The 78 alloy is near the composition Ni_3Fe and thus may have a tendency to order. The largest magnetic annealing effects occur near a 70% Ni alloy, and it may well be that the ordering is involved. Ordering can be suppressed by adding a third element. In addition, the presence of a third element can increase the electrical resistivity and the ductility. One example is Mumetal, in which there is 76 Ni, 18 Fe, 5 Cu, and 2 Cr.

Table 7-9.1. Data for Typical Soft Magnetic Materials

Material	Remarks	Per Cent Composition	Initial Permea-bility	Maximum Permea-bility	Coercive Force (oe)	Saturation Induction $T = 300°K$
Iron	Commercial	99 Fe	200	6,000	0.9	21,600
Iron	Pure	99.9 Fe	25,000	350,000	0.01	21,600
Si—Fe		96 Fe, 4 Si	1,200	6,500	0.5	19,500
Si—Fe	Grain oriented (Hypersil)	97 Fe, 3 Si	9,000	40,000	0.15	20,100
50 Permalloy	(Hypernik)	50 Fe, 50 Ni		100,000	0.05	16,000
78 Permalloy		78 Ni, 22 Fe	4,000	100,000	0.05	10,500
Mumetal		18 Fe, 75 Ni, 5 Cu, 2 Cr	20,000	100,000	0.05	7,500
Supermalloy		15 Fe, 79 Ni, 5 Mo, 0.5 Mn	90,000	10^6	0.004	8,000
Permendur		50 Fe, 50 Co	500	6,000	0.2	24,600
Fe—Co—V		49 Fe, 49 Co, 2 V		100,000	0.2	23,000
Perminvar	Magnetic annealed	34 Fe, 43 Ni, 23 Co		400,000	0.03	15,000
Fe—Si—Al	(Sendust in powder)	85 Fe, 9.5 Si, 5.5 Al	35,000	120,000	0.02	12,000

Another, Supermalloy, with 79 Ni, 15 Fe, 5 Mo, and 0.5 Mn, possesses a maximum permeability of about 10^6, which is the largest value of any commercially available material. Supermalloy also has a coercive force of about 0.002 oe and a hysteresis loss of about 5 ergs/cm³ cycle.

Iron-cobalt alloys have large saturation magnetizations; the maximum occurs in a 50 Co-50 Fe alloy (Permendur). The addition of 2% vanadium reduces the coercive force and the hysteresis losses and increases the electrical resistivity and ductility. This material can also be magnetically annealed.

The ternary alloys of Fe-Co-Ni are also of interest. Perminvar, with 43 Ni, 34 Fe, and 23 Co, has hysteresis loops with a wasp waist near the origin. This alloy can be magnetically annealed and then has a rectangular loop. A ternary iron-silicon-aluminum alloy (9.5 Si, 5.5 Al) is the only one known for which $\lambda_s = 0$ and $K_1 = 0$. It has a large permeability of about 10^5.

The composition and main properties of some of the more important soft magnetic materials[137] are listed in Table 7-9.1. Hysteresis loops of a

[137] For additional data and extensive references the following books are suggested: R. M. Bozorth, *Ferromagnetism*, D. Van Nostrand and Co., Princeton, N.J. (1951); *Magnetische Werkstoffe*, **59**, Gmelins Handbuch der Anorganischen chemie, Verlag chemie, G. m. b. H. Weinheim/Bergstrasse (1959); F. Brailsford, *Magnetic Materials*, Methuen, London (1948).

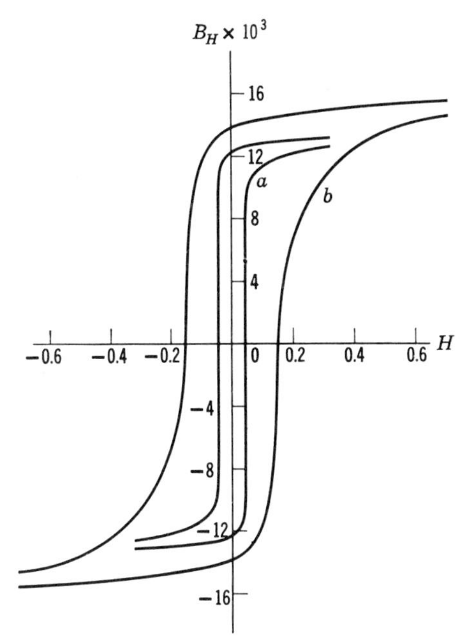

Fig. 7-9.1. Hysteresis loops of (*a*) pure iron and (*b*) grain-oriented 97 Fe 3 Si (Hypersil). [A. Arrott and J. E. Goldman, *Electrical Manufacturing*, March (1959), Copyright C-M Technical Publications Corporation.]

few of these materials are shown in Figs. 7-9.1, 7-9.2, and 7-9.3. The properties of ferrimagnetic materials are discussed in Chapter 9.

10. Time Effects

When a magnetic field that varies with time is applied to a ferromagnetic material, the magnetization changes do not take place instantaneously, but instead lag the field changes. This behavior is described, as usual, by one or more relaxation times τ. Although there are several processes that may cause the time lag, one of the most important is the following. The changing magnetic field will, by Faraday's law, induce an electromotive force in the material (given by equations 1-7.4 or 1-7.5). Now, since ferromagnetic metals and alloys are relatively good conductors, the currents, called eddy currents, that flow are appreciable. These eddy currents will have a direction given by Lenz's law. Hence the magnetic field produced by the eddy currents will oppose the change in the applied field. The result is that it takes time for the applied field to penetrate the material. Consequently, the change in magnetization is slower than it would otherwise be. Indeed, if the applied field varies sinusoidally with time, the

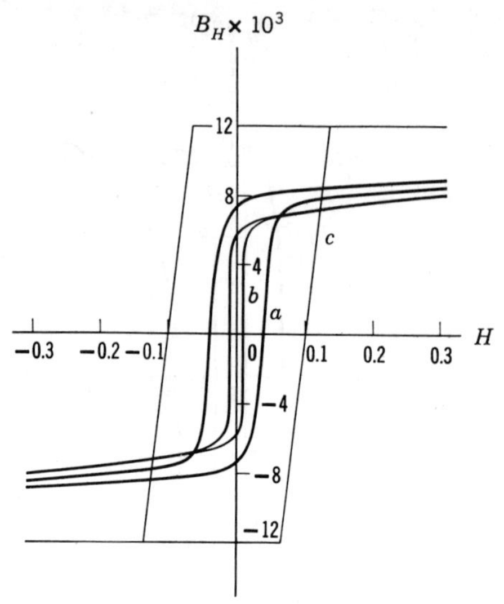

Fig. 7-9.2. Hysteresis loops of (*a*) Hypernik (after R. M. Bozorth); (*b*) Supermalloy and (*c*) cube-textured 50 Ni 50 Fe. [A. Arrott and J. E. Goldman, *Electrical Manufacturing*, March (1959), Copyright C-M Technical Publications Corporation.]

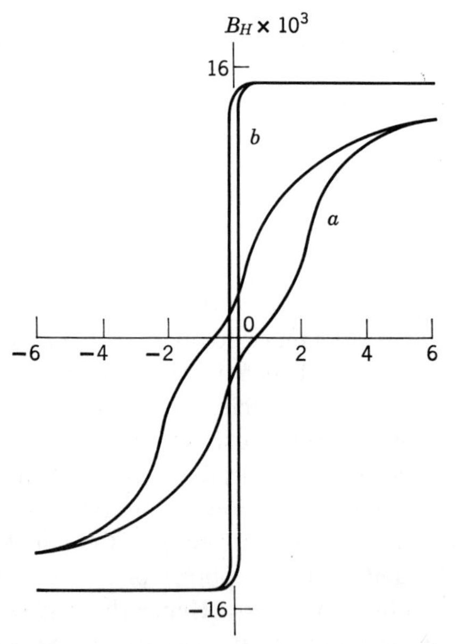

Fig. 7-9.3. Hysteresis loops of Perminvar (*a*) after 1000°C anneal and (*b*) after magnetic anneal. Note the waspwaist for curve (*a*). [H. J. Williams and M. Goertz, *J. Appl. Phys.* **23,** 316 (1952).]

amplitude of the field at the interior may be considerably less than that at the surface. Only when the applied field remains constant long enough for all the eddy currents to decay to zero does the field at the interior equal that at the surface.

An exact calculation of the effects of eddy currents is feasible only if the permeability μ is constant, that is, if there is a linear relationship between B_H and H. For small minor hysteresis loops, this condition is almost met. Although the calculations are straightforward, they are by no means trivial; the reader is referred to the literature for details.[138] Only a few results will be quoted here. Suppose an alternating field $H_1 \cos \omega t$ is applied to an infinite sheet of thickness $2d$. Then the amplitude of the field, H_x, at a distance x measured from the center of the sheet is given by

$$H_x = H_1\left(\frac{\cosh 2x/\delta + \cos 2x/\delta}{\cosh 2d/\delta + \cos 2d/\delta}\right). \tag{7-10.1}$$

Here δ is the skin depth, defined as the depth in the material at which $H_{x-d}/H_1 = 1/e$; it is given by

$$\delta = \frac{1}{2\pi}\sqrt{\frac{\rho}{\mu\nu}}, \tag{7-10.2}$$

where $2\pi\nu = \omega$. The phase angle by which the field at the distance $x - d$ below the surface lags that at the surface is proportional to $(x - d)/\delta$. The energy loss from the eddy currents is computed by integrating $i^2\rho$ over the sheet. If $\delta > d$, a condition that is likely to be satisfied at low frequencies, it is found that the energy loss per unit volume per cycle from eddy currents is

$$W_e = eB_1^{\,2}\nu, \tag{7-10.3}$$

where $e = 2\pi^2 d^2/3\rho$ and $B_1 = \mu H_1$. An equation of the same form holds for wires (cylinders) and spheres.

If the minor hysteresis loop is small enough, it is reasonable to assume that the hysteresis losses are independent of the eddy current losses. The hysteresis losses may then be calculated by using Rayleigh's relationship (Section 7-7). It is found that

$$W_h = aB_1^{\,3}, \tag{7-10.4}$$

where W_h is the hysteresis loss per cubic centimeter per cycle and a is a constant. The total loss, W_T, can be measured with an electrical bridge. In order to account for the total losses observed, it is necessary to add a term to the sum of equations 7-10.3 and 7-10.4. We then have

$$\frac{W_T}{B_1^{\,2}} = aB_1 + e\nu + c, \tag{7-10.5}$$

[138] J. J. Thomson, *Electrician* **28,** 559 (1892); W. Cauer, *Arch. Elecktrotech* **15,** 308 (1925); K. L. Scott, *Proc. IRE* **18,** 1750 (1930).

where c is independent of B_1 and ν. This equation provides a good description of the data obtained from a number of materials.[139] It is not certain, however, that the constants have the meaning of the theory.

For larger minor hysteresis loops the situation is not nearly so simple. The permeability can no longer be assumed constant and indeed will vary both with position and time in the material. Also the hysteresis losses and the eddy current losses cannot be considered as independent of each other. The interdependence of these two losses may be easily demonstrated by observing the hysteresis loop as a function of frequency on an oscilloscope screen. The area enclosed by the loop is found to increase with frequency. No empirical relationship that describes this effect appears to have been given.

The foregoing analysis has not taken any account of the domain structure. For illustrative purposes consider a 180° domain wall. When the wall moves with a velocity v, the magnetic moment of the two domains adjacent to the wall will change at the rate $2Mv$ per unit area of wall. In general this velocity will depend on the frequency of the applied field and will increase with frequency. However, it is worth pointing out that even when the applied field is changed very slowly the wall will sometimes travel with a nonzero velocity. This will occur when the field is large enough to cause an irreversible wall movement; an example is the spontaneous motion of a 180° wall from position A to E in Fig. 7-7.1c. As a result of the wall motion, eddy currents will be induced. These eddy currents, in turn, will produce a magnetic field that will act to oppose the motion. A straightforward calculation[140] shows that this opposing field, H_{op}, is given by

$$H_{op} = \beta_H v, \tag{7-10.6}$$

where β_H is a constant. The loss per cubic centimeter in moving the wall against this field is therefore $W = 2M\beta_H\bar{v}$, where $\bar{v}$ is the average wall velocity. If the wall motion is reversible for zero frequency, that is, if the wall remains in the vicinity of a minimum in the energy F_{wt}, the field acting on the wall is equal to $k_H x/\chi_r$. Here $k_H = 2M$, χ_r is the reversible susceptibility, that is, the susceptibility for $\nu = 0$, and x is measured from the energy minimum. If the applied field is sinusoidal and eddy current damping is present, the equation of motion is

$$\beta_H \frac{dx}{dt} + \frac{k_H x}{\chi_r} = H_1 \cos \omega t. \tag{7-10.7}$$

[139] V. E. Legg, *Bell System Tech. J.* **16,** 39 (1936); W. B. Ellwood and V. E. Legg, *J. Appl. Phys.* **8,** 351 (1937).

[140] R. Becker, *Physik Z.* **39,** 856 (1938).

The relaxation time and energy losses may then be found from this equation (problem 7-15). If the wall motion is irreversible, the eddy current damping will act to reduce the wall's velocity. The contour of the propagating domain wall in the Sixtus-Tonks experiment (Fig. 7-7.2*c*) is probably the result of eddy currents producing an opposing field that is larger near the center than near the surface of the wire. Calculation of the effect of eddy currents based on domain models may be found in the literature.[141]

In engineering applications it is important to reduce the losses that arise because of the presence of eddy currents. This may be accomplished by choosing materials with large resistivity and by making at least one dimension of the material small.

The additional losses described by the third term on the right-hand side of equation 7-10.5 may be caused by one or more of several processes. The time lag corresponding to this loss term may be demonstrated by suddenly applying a field H to a specimen and noting the response of the magnetization. The approach of the magnetization to its equilibrium value can be approximately described by the equation $M_H = M(1 - e^{-t/\tau})$. A better fit with the data can be made if more than one relaxation time τ is assumed. It is found that the time taken to reach equilibrium is much too long to be caused by eddy currents. A number of names have been used to describe these phenomena, including magnetic viscosity, magnetic aftereffect, and magnetic *traînage*.

The magnetic aftereffect can be caused by the diffusion of interstitial atoms. Carbon and nitrogen atoms occupy interstitital positions in the iron lattice. When these atoms are removed by some purification process, the magnetic aftereffect disappears.[142] Further, this aftereffect depends on temperature, as is to be expected for a diffusion process. Presumably, the interstitial atoms, which probably lie on a cube edge midway between the iron atoms, diffuse so that one particular cube edge direction is favored. This preferential alignment may occur because of the relaxation of strains caused by the magnetostriction.[143] However, crystalline anisotropy energy is more likely to be involved, at least according to Néel.[144,145] A similar magnetic aftereffect occurs in substitutional alloys. A pair of neighboring substituted atoms define an axis, which may tend to be aligned in some preferred direction.[146]

[141] H. J. Williams, W. Shockley, and C. Kittel, *Phys. Rev.* **80,** 1090 (1950); R. H. Pry and C. P. Bean, *J. Appl. Phys.* **29,** 532 (1958).

[142] G. Richter, *Ann. Phys. (Leipzig)* **32,** 683 (1938); J. L. Snoek, *Physica* **6,** 161 (1939).

[143] J. L. Snoek, *Physica* **8,** 711 (1941); D. Polder, *Philips Res. Rept.* **1,** 5 (1945).

[144] L. Néel, *J. Phys. radium* **12,** 339 (1951), **13,** 249 (1952).

[145] For a recent survey of this subject see G. W. Rathenau, *Magnetic Properties of Metals and Alloys*, Am. Soc. Metals, Cleveland, Ohio (1959), p. 168.

[146] S. Chikazumi, *J. Phys. Soc. Japan* **5,** 327, 333 (1950).

The initial permeability of some demagnetized specimens is observed to decrease with time. This phenomena, sometimes called disaccommodation, is also the result of diffusion of interstitial atoms.[147]

The magnetic properties of a material may also change over long periods of time as the result of the slow diffusion of some atoms. These atoms may diffuse in order to relieve strains or because some chemical change is occurring, such as the precipitation of a new phase, perhaps a carbide or a nitride.

Thermal fluctuations can also produce a magnetic aftereffect.[148] In soft magnetic materials these fluctuations may enable a domain wall to pass an energy barrier it would not otherwise overcome.[149] When a certain magnetic field is applied to a single-domain particle, the thermal fluctuations may supply the additional energy required for an irreversible rotation.[150] The effects of thermal agitation in superparamagnetic particles have been discussed earlier in Section 7-3.

Some materials exhibit a magnetic aftereffect that does not appear to originate from any of the foregoing causes. In particular, a time lag, sometimes called a Jordan lag, which is independent of both temperature and frequency, has been observed. Although it is possible to suggest additional mechanisms, such as short-range ordering and dislocation movements, the actual source of these other time effects is unknown at present.

When a domain wall moves, a precessional motion of the moments associated with the wall occurs. As a result, the wall behaves as though it possessed a mass. This inertial effect is best studied in materials that have a large electrical resistivity, for then no eddy currents will be present to complicate matters. Many ferrimagnetic materials are either insulators or semiconductors. When a steady magnetic field is applied in addition to the alternating field, ferromagnetic resonance, a phenomenon analogous to paramagnetic resonance, may be observed. Domain wall resonance and ferromagnetic resonance are discussed in Chapter 10.

11. Thin Films

Although thin films have been investigated for almost a century, an extensive study of their magnetic properties dates only from the early

[147] J. L. Snoek, *Physica* **5,** 663 (1938).

[148] L. Néel, *J. Phys. radium* **11,** 49 (1950), **12,** 339 (1951); P. Brissonneau, *J. Phys. radium* **19,** 490 (1958).

[149] J. C. Barbier, *Ann. phys.* (*Paris*) **9,** 84 (1954).

[150] R. Street and J. C. Woolley, *Proc. Phys. Soc.* (*London*) **A-62,** 562 (1949), **B-63,** 509 (1950); R. Street, J. C. Woolley, and P. B. Smith, *Proc. Phys. Soc.* (*London*) **B-65,** 679 (1952).

1950's. The impetus for this increased interest stems largely from the potential use of thin films of ferromagnetic materials as the memory elements of digital computers. However, the magnetic properties of thin films are also of considerable interest from a purely scientific point of view.

Thin films can be produced by vacuum evaporation, by sputtering, by electrodeposition, and by chemical deposition. The first of these methods, vacuum evaporation, has been employed the most often to date. Thin films of the ferromagnetic elements Fe, Ni or Co and of various alloys such as MnBi and NiFeMo have all received attention. However, by far the largest number of investigations have been made on thin films of Permalloy with 79 to 82% Ni. Permalloy with this composition has a magnetostriction λ_s close to zero, hence the magnetic properties are changed little if at all when a stress is applied. This feature is a great advantage in engineering applications. Further, when the evaporation is made in the presence of a magnetic field, the Permalloy film is found to possess a uniaxial anisotropy[151] that lies along the direction of the field. This anisotropy cannot be the result of the crystalline anisotropy, since for this Permalloy composition K_1, although not zero, is small. The uniaxial anisotropy can be described by

$$F_K = K_1' \sin^2 \theta, \tag{7-11.1}$$

where θ is the angle between the magnetization **M** and the easy or longitudinal axis. Some remarks concerning the possible origin of this anisotropy will be made later.

Experiments show that a film can consist of a single domain when no field is applied. If the magnetization changes occur by coherent rotations only, the hysteresis curve can be calculated by using the same methods employed for single-domain particles (Section 7-2). Since the thickness of a film is always very small compared to the other dimensions, the demagnetization field is large unless the magnetization lies in the plane of the film. Therefore, in the absence of any energy that opposes the action of the shape anisotropy, the magnetization **M** will always lie in the plane of the film. The magnetostatic energy F_D may then be set equal to zero. Further, because of the single-domain assumption, the exchange energy F_e can be set equal to zero.

It is common practice to apply two independent fields to the film. One of these, the longitudinal field H_z, is applied parallel to the easy axis of the uniaxial anisotropy, whereas the other, the transverse field H_x, is applied perpendicular to this easy axis. The energy per unit volume in the applied field is then $F_H = -H_zM \cos \theta - H_xM \sin \theta$. Therefore the total energy

[151] M. S. Blois, Jr., *J. Appl. Phys.* **26**, 975 (1955).

of the film is given by

$$F_T = (K_1' \sin^2 \theta - H_z M \cos \theta - H_x M \sin \theta)V. \qquad (7\text{-}11.2)$$

This energy is a minimum when

$$\frac{\partial F_T}{\partial \theta} = 0 = 2K_1' \sin \theta \cos \theta + H_z M \sin \theta - H_x M \cos \theta, \qquad (7\text{-}11.3)$$

provided $\partial^2 F_T/\partial \theta^2 > 0$. In reduced units equation 7-11.3 may be written

$$\tfrac{1}{2} \sin 2\theta + h_z \sin \theta - h_x \cos \theta = 0, \qquad (7\text{-}11.4)$$

where

$$h_z = \frac{H_z}{H_K}, \qquad h_x = \frac{H_x}{H_K} \quad \text{and} \quad H_K = \frac{2K_1'}{M}. \qquad (7\text{-}11.5)$$

Equation 7-11.4 is similar to equation 7-2.7 employed in the theory of single-domain particles. There is, however, the following difference.

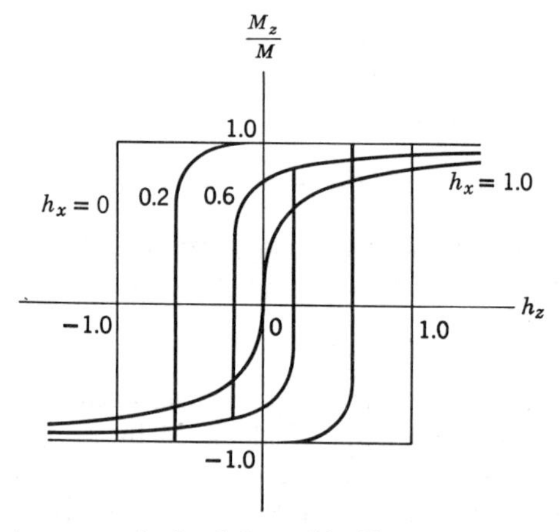

Fig. 7-11.1. Hysteresis curves calculated for a thin film. The component of the magnetization along the easy direction M_z is plotted as a function of a field H_z, applied along this same direction for a given applied transverse field H_x. Reduced units are used. (After J. B. Goodenough and D. O. Smith.)

For the single-domain particle the angle between the applied field H and the easy axis remains constant as H is varied. On the other hand, for the thin film, the angle between the resultant field $H[=(H_x^2 + H_z^2)^{1/2}]$ and the easy axis in general does not remain constant. Indeed, the usual procedure is to vary H_z, whereas H_x is kept constant. Hysteresis curves calculated from equation 7-11.4 under these conditions are shown[152] in Fig. 7-11.1.

[152] J. B. Goodenough and D. O. Smith, *Magnetic Properties of Metals and Alloys*, Am. Soc. Metals, Cleveland, Ohio (1959), Chapter 7.

Note that the component of the magnetization along the easy direction rather than that along the resultant field direction is plotted; this is the customary procedure. For $h_x = 0$, the hysteresis curve is, of course, identical to that shown in Fig. 7-2.6*b*. The critical fields at which an irreversible rotation in the magnetization occurs can be calculated by application of the condition $\partial F_T/\partial\theta = \partial^2 F_T/\partial\theta^2 = 0$. It is easy to show (problem 6-16) that the critical fields satisfy the equation

$$h_x^{2/3} + h_z^{2/3} = 1. \tag{7-11.6}$$

Hence, as h_x is increased, the longitudinal field required for an irreversible rotation is reduced. For $h_x > 0$ no hysteresis occurs. This behavior is clearly shown in Fig. 7-11.1.

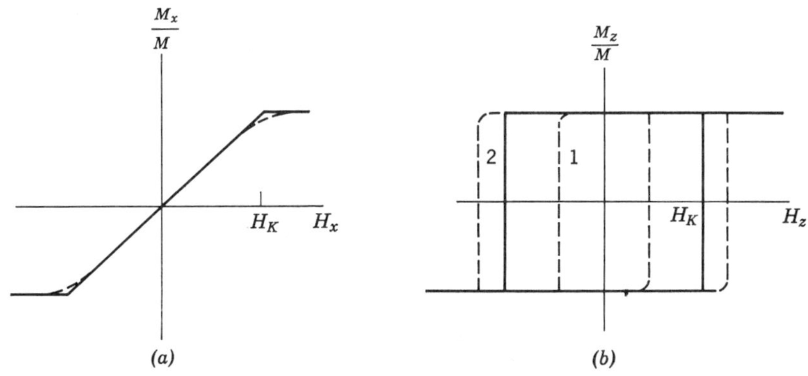

Fig. 7-11.2. Hysteresis curves for thin films (*a*) in the hard direction and (*b*) in the easy direction. Theoretical curves are shown by the solid lines, experimental curves by the dashed lines. In (*b*) curve 1 is typical of most films, where $H_c < H_K$. Curve 2 is sometimes observed ($H_c > H_K$); the film is then said to be inverted.

When $H_z = 0$, the calculated hysteresis curve in the transverse (hard) direction is as shown by the solid line in Fig. 7-11.2*a*; this curve is identical to that of Fig. 7-2.5*b*. Experimentally, it is indeed found that the transverse hysteresis curve of most films is similar to the predicted one. Often, however, there are some small differences. The hysteresis curve is usually slightly curved in the vicinity of H_K, as indicated by the dashed curve in Fig. 7-11.2*a*. In addition, sometimes the curve has a slight opening; that is, there is a small amount of hysteresis. The value of H_K for a particular film is assumed to be given by the extrapolation of the linear part of the experimental curve to the saturation magnetization line. The anisotropy field $H_K(=2K_1'/M)$ is by no means a constant for all films. The majority of films have an anisotropy field of a few oersteds.

The hysteresis curve measured in the easy direction with $H_x = 0$ usually has a coercive force H_c that is less than the theoretical value H_K. The reason is that the magnetization reversal occurs by the movement of domain walls rather than by a coherent rotation. The longitudinal hysteresis loop of most films, however, is rectangularly shaped, presumably because the field required for domain nucleation is larger than that required for wall movement. A typical hysteresis loop for the easy direction is shown as curve 1 of Fig. 7-11.2; it may be compared with the theoretical loop for coherent rotation shown by the solid line. Some films have a coercive force that is larger than the anisotropy field (curve 2 of Fig. 7-11.2*b*). This unusual property is present when the magnetization rotates clockwise in some parts of the film and counterclockwise in other parts. This behavior is expected if the anisotropy direction varies from place to place in the film. Domains then appear whose walls tend to be perpendicular to the easy direction. This is a relatively stable state inasmuch as further rotation or wall motion is resisted. The domains remain locked until a field large enough to cause reversal is applied.

When a steady transverse or biasing field H_x is applied, a switch of the magnetization by a coherent rotation will occur if the critical field H_z for an irreversible rotation is less than the domain nucleation field. The M_z versus H_z hysteresis loops for $H_x \neq 0$ shown in Fig. 7-11.1 can, in fact, be observed. Another experimental approach employs a steady biasing field H_x together with a pulsed field H_z. If the rise time of the pulse is short ($\approx 10^{-9}$ sec) compared to the time taken for wall motion, a switch by rotation is expected. It is found[153] that the critical fields, h_x and h_z, for a switch (irreversible rotation) are indeed equal to the values predicted by equation 7-11.6. Provided $H_x \neq 0$, the time taken for a film to switch by rotation can be as short as a few millimicroseconds.[153,154] It is this fast switching property that makes thin films attractive as potential memory elements in the high-speed computers planned for the future. Since the rapid switching takes place by a precessional motion of the magnetization, it is necessary to take the relaxation processes into account in the analysis of the switching time.[154]

The origin of the induced uniaxial anisotropy in thin films has not been established. It is very possible that some, although not all, of the anisotropy is the result of the directional ordering of pairs of iron atoms.[155] According to this theory, neighboring iron atoms define an axis that tends to become oriented along the direction of the magnetization. The magnetic field applied during the evaporation aligns the magnetization along the

[153] C. D. Olson and A. V. Pohm, *J. Appl. Phys.* **29**, 274 (1958).
[154] D. O. Smith and G. P. Weiss, *J. Appl. Phys.* **29**, 290 (1958).
[155] S. Chikazumi, *J. Phys. Soc. Japan* **5**, 327, 333 (1950), **11**, 551 (1956).

field direction. Consequently, the atom pairs also tend to diffuse so that they lie parallel to the field direction. Then, after evaporation, the temperature of the film drops, the diffusion stops and the atom pairs are frozen into place, and hence give rise to the uniaxial anisotropy.

It has been reported that a uniaxial anisotropy develops which depends on the angle of incidence the evaporation beam makes with the substrate.[156] For Permalloy the easy axis lies at right angles to the line of incidence of the beam. The origin of this anisotropy appears to be in doubt.[157]

In some Permalloy films it has been found possible to select the easy direction by applying a large enough magnetic field along this direction.[158] It has been suggested that this rotatable anisotropy property is associated with the presence of minor constituents, in particular oxygen. Indeed, it may well be that oxygen plays a role in the formation of the induced uniaxial anisotropy itself.[159]

Small crystalline and stress anisotropies in general are also present in a film. Since the film is polycrystalline, the direction of these anisotropies varies from place to place in the film. Consequently, these small local anisotropies produce a dispersion in the magnitude and direction of the uniaxial anisotropy, and it is appropriate to talk about an average easy direction. This dispersion of the easy axis has an important influence on the domain pattern observed.[160]

The magnetization of a thin film at low temperatures can be calculated from spin-wave theory. The calculation for a three-dimensional (bulk) material has been given in Section 6-6. It has also been pointed out (problem 6-9) that a two-dimensional lattice does not exhibit ferromagnetism. A thin film is a three-dimensional material but with one dimension reduced. Hence it is still appropriate to use equation 6-6.19, except that a summation instead of an integration is made in the direction perpendicular to the plane of the film (y-axis). The calculations[161] predict that for films less than 100 Å thick the spontaneous magnetization decreases with film thickness. It is easy to see the physical reason for this behavior. Atoms lying in the surface layers have a smaller number of nearby neighbors than the other layers. Hence the exchange forces are unable to produce the same amount of alignment of the atomic moments found in bulk

[156] T. G. Knorr and R. W. Hoffman, *Phys. Rev.* **113,** 1039 (1959).

[157] D. O. Smith, *J. Appl. Phys.* **32,** 70S (1961).

[158] R. J. Prosen, J. O. Holmen, and B. E. Gran, *J. Appl. Phys.* **32,** 91S (1961).

[159] R. J. Prosen, J. O. Holmen, T. Cebulla, and B. E. Gran, *J. Phys. Soc. Japan, Suppl. B-1* **17,** 580 (1962).

[160] S. Middelhoek, Thesis, University of Amsterdam (1961); R. W. Olmen and S. M. Rubens, *J. Appl. Phys.* **34,** 1107 (1962).

[161] M. J. Klein and R. S. Smith, *Phys. Rev.* **81,** 378 (1951); S. J. Glass and M. J. Klein, *Phys. Rev.* **109,** 288 (1958); L. Valenta, *Czech. J. Phys.* **7,** 127 (1957).

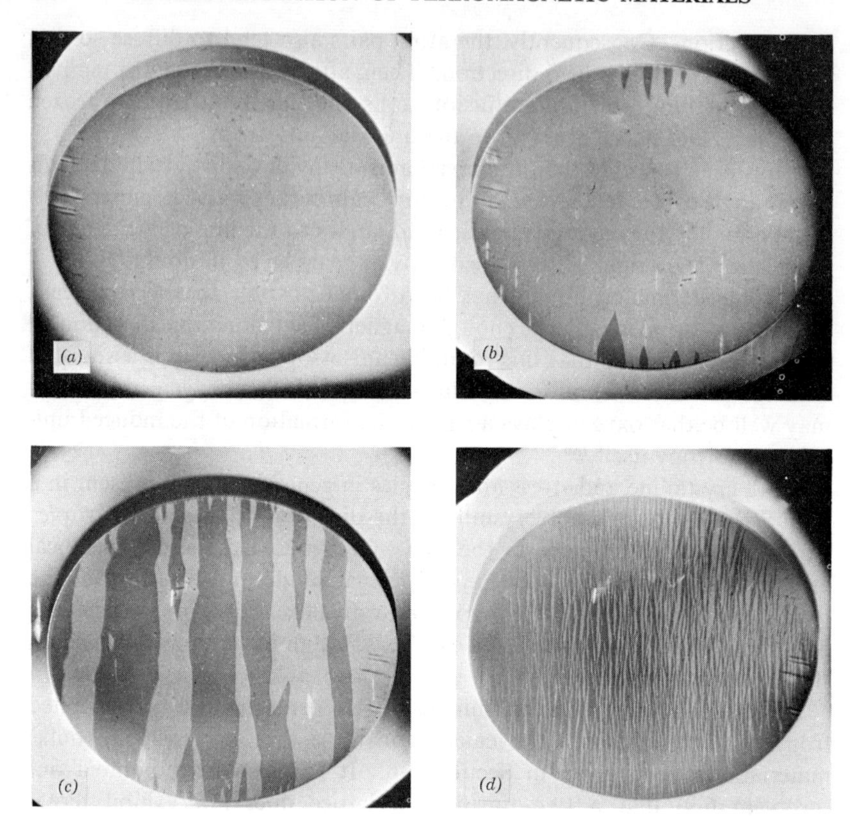

Fig. 7-11.3. Domain patterns on a Permalloy film with 81 % Ni observed with the Kerr technique. The film is 8 mm in diameter and approximately 2000 Å thick. (*a*) A single-domain film ($H = 0$). (*b*) Nucleation spikes in a reverse field of $0.2H_c$. (*c*) Domain pattern when demagnetized with an alternating field of decreasing amplitude (frequency 60 cps). (*d*) Domain pattern when demagnetized by means of a transverse field $\geqslant 1.5H_K$. (Courtesy R. W. Olmen-Remington Rand Univac, St. Paul, Minnesota.)

material. This effect is important only if the surface layers constitute a large percentage of the total material, that is, for very thin films. Experiments on thin films prepared in vacuums of 10^{-5} mm of Hg do indeed show[162] that the saturation magnetization decreases with film thicknesses below 100 Å. On the other hand, the saturation magnetization of films evaporated at 10^{-9} mm of Hg remains constant and equal to the bulk value[163]

[162] A. Drigo, *Nuovo cimento* **8,** 498 (1951); E. C. Crittenden, Jr. and R. W. Hoffman, *Revs. Mod. Phys.* **25,** 310 (1953); M. H. Seavey, Jr. and P. E. Tannenwald, *J. Appl. Phys.* **29,** 292 (1958); W. Ruske, *Ann. Phys. (Leipzig)* **2,** 274 (1958); H. J. Bauer, *Z. Physik* **153,** 484 (1959).

[163] C. A. Neugebauer, *Structure and Properties of Thin Films*, John Wiley and Sons, New York (1959), p. 358.

down to thicknesses of about 20 Å. The difference in the films that leads to such experimental results has not yet been established. On the theoretical side, it is to be noted that the spontaneous and not the saturation magnetization was calculated. It is conceivable that the inclusion of an applied field and possibly other factors in the calculation would lead to a different result.

Interesting domain patterns are observed in thin films. Some patterns obtained by means of the Kerr effect are shown in Fig. 7-11.3. The easy axis in each picture is along the vertical. A film in its remanent state can be a single domain, and an example is shown in Fig. 7-11.3*a*. When a field is applied in the direction opposite to the magnetization, small spike-like domains are nucleated at the film edge (Fig. 7-11.3*b*). These spikes propagate across the film when the reverse field is increased. The pattern on a film demagnetized by an alternating field applied along the easy direction is illustrated in Fig. 7-11.3*c*. Quite a different pattern is observed when the single-domain film is demagnetized by applying a field along the hard direction (Fig. 7-11.3*d*). This fine domain structure is believed to be due to the dispersion of the easy axis. If a field larger than H_K is applied in the hard direction, the magnetization will then also be along this direction. Suppose there are small variations in the anisotropy. Then, when the field is switched off, part of the magnetization will rotate clockwise and part counterclockwise, and fine domains whose magnetizations are parallel to the easy direction will result. The spacing of these domains depends on the wavelength of the anisotropy variation.

An unusual domain pattern is shown in Fig. 7-11.4*a*. There is considerable magnetostatic energy associated with the uncompensated poles on the domain walls, as indicated in Fig. 7-11.4*b*. The zigzag domain boundary reduces this magnetostatic energy to some extent. However, it is clear that some other anisotropy that favors this domain arrangement must be present.

Domain walls in thin films differ from those in bulk materials. Consider a 180° wall. Suppose, as in Section 7-5, that the normal component of the magnetization remains constant throughout the wall, that is, that there are no uncompensated poles. In terms of the notation of Fig. 7-5.2, θ remains constant throughout the wall and ϕ changes from 0 to 180°; that is, the magnetization changes direction on going through the wall by a rotation about an axis perpendicular to the wall. At the intersection of the wall with the surface of the material, uncompensated poles, however, do occur; this is illustrated in Fig. 7-11.5*a*. In bulk material the magnetostatic energy arising from these surface poles is small and may be neglected. In thin films, however, the distance between the poles is small and the magnetostatic energy is considerable.

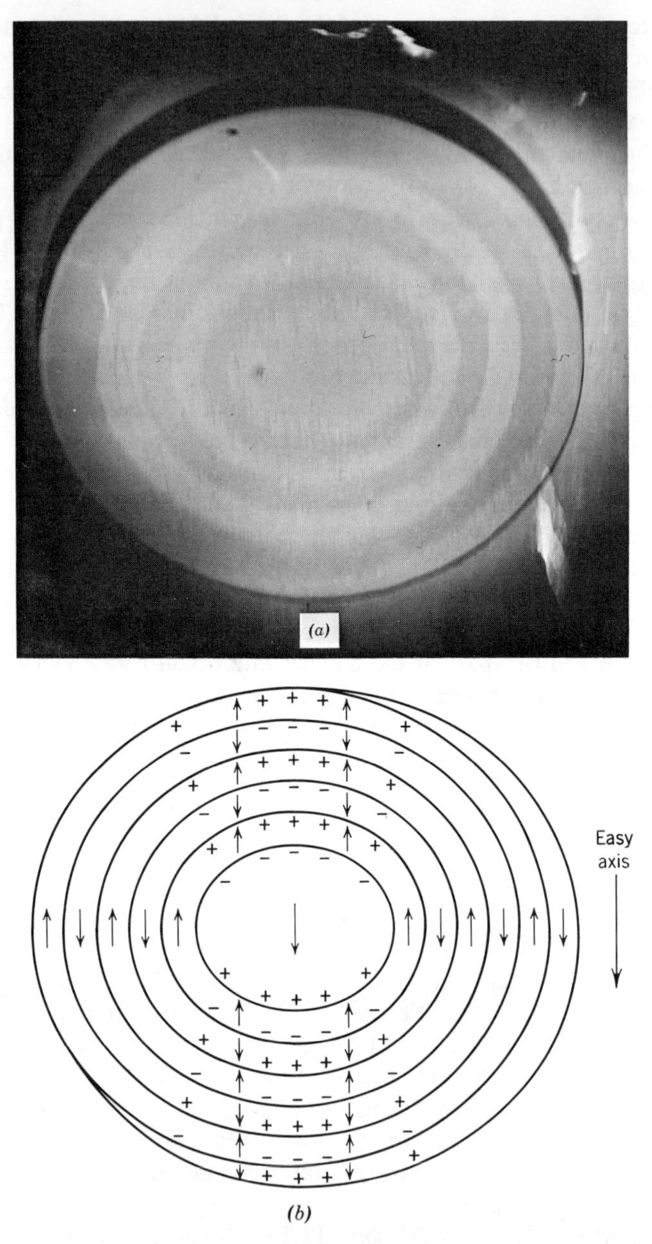

Fig. 7-11.4. Annular uniaxial domain pattern on a Permalloy film. The film possesses uniaxial anisotropy with the easy direction along the vertical. (*a*) Pattern observed by means of Kerr effect. (*b*) Sketch showing the magnetization directions (the zigzag spikes are not shown). (University of Minnesota magnetism group.)

A simple approximate calculation of this magnetostatic energy may be made by assuming first that the angles between adjacent spins are equal and second that the wall is a cylinder with an elliptical cross section.[164] It is found that in a thin film the energy of a wall, F_w, is larger and the width of a wall is smaller than in bulk material.

For very thin films the magnetostatic energy term is so large that it is reasonable to suspect that the assumption that θ remains constant in the wall is no longer valid. The extreme case in which ϕ remains constant and θ changes by 180° is illustrated in Fig. 7-11.5*b*. In the literature this type of wall is often called a Néel wall; the other (Fig. 7-11.5*a*) is often called a

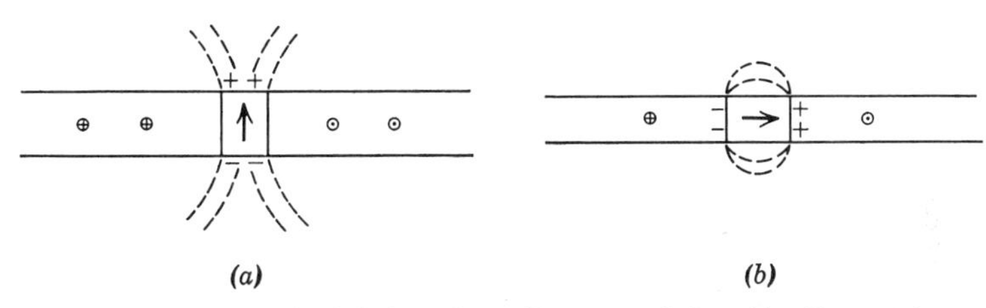

Fig. 7-11.5. Two types of 180° domain wall proposed for thin films. The atomic moments in the wall in (*a*) turn about an axis perpendicular to the wall and in (*b*) turn about an axis perpendicular to the plane of the film.

Bloch wall. A calculation made with the assumptions already given shows that in films less than about 400 Å thickness the energy of a Néel wall is less than that of a Bloch wall. More sophisticated calculations lead to essentially the same conclusion.[165] It would appear that walls in which both θ and ϕ vary are possible, but no calculations have been reported.

Powder patterns have provided evidence for the existence of Néel walls in very thin films. The idea is the following. The colloidal particles tend to form chains lying along the flux paths indicated in Fig. 7-11.5. When dark field illumination is used, the light is incident almost parallel to the plane of the film. Hence more light is expected to be reflected from the chains above a Néel wall than those above a Bloch wall. Indeed, the most distinct powder patterns of walls are observed for films of 100 to 300 Å thickness,[166] as expected. In films less than 100 Å thickness double Néel walls, that is, two Néel walls with opposite polarity, are frequently observed.[167]

[164] L. Néel, *Compt. rend.* (*Paris*) **241,** 533 (1955).

[165] H. Stephani, *Wiss. Z. Friedrich-Schiller-Univ. Jena* **7,** 373 (1957/58); H. D. Dietze and H. Thomas, *Z. Physik* **163,** 523 (1961).

[166] S. Middelhoek, Thesis, University of Amsterdam (1961); Y. Gardo and Z. F. Unatogawa, *J. Phys. Soc. Japan* **15,** 1126 (1960).

[167] H. J. Williams and R. C. Sherwood, *J. Appl. Phys.* **28,** 548 (1957).

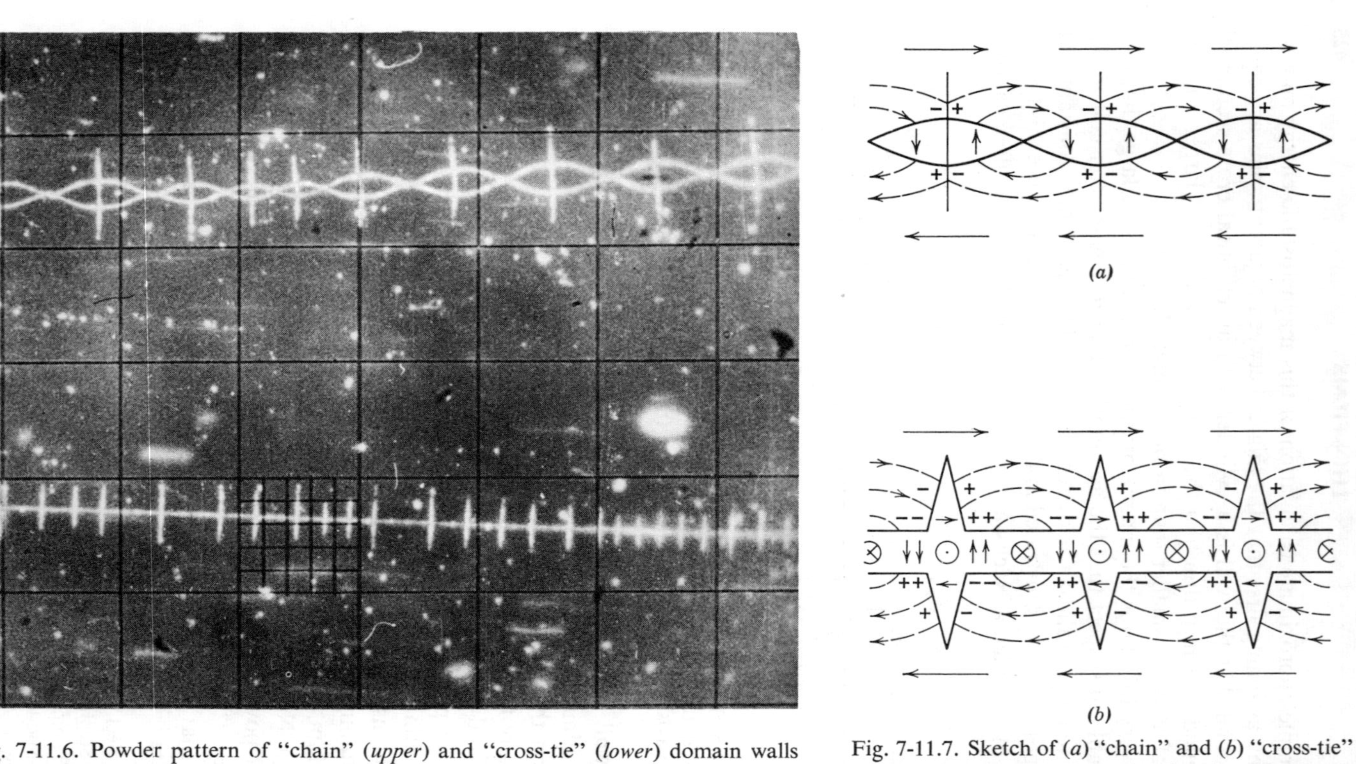

Fig. 7-11.6. Powder pattern of "chain" (*upper*) and "cross-tie" (*lower*) domain walls in a Permalloy film. (Courtesy of R. M. Moon-Lincoln Laboratory, M.I.T., Lexington, Mass.)

Fig. 7-11.7. Sketch of (*a*) "chain" and (*b*) "cross-tie" walls. The arrows indicate the direction of magnetization. Note the polarity of the uncompensated poles at the walls.

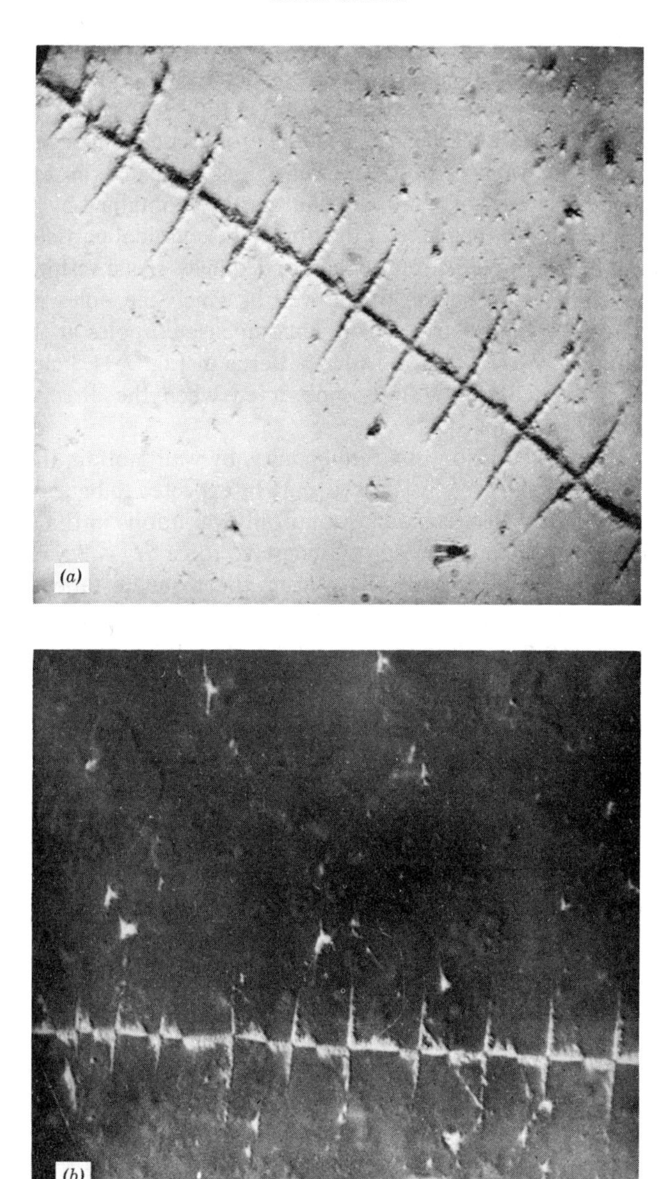

Fig. 7-11.8. Powder patterns of a cross-tie wall in a Permalloy film (*a*) without and (*b*) with a magnetic field applied perpendicular to the plane of the film. Note the shift of the colloidal particles toward the part of the domain wall with opposite polarity. (Courtesy of R. M. Moon-Lincoln Laboratory, M.I.T., Lexington, Mass.)

Other types of domain wall have been observed in thin films. "Chain" and "crosstie" walls observed by the powder technique[168] are shown in Fig. 7-11.6. These unusual walls are found in films whose thickness lies in a range of about 300 to 1000 Å. A model proposed for the magnetization directions in the vicinity of these walls[169] is indicated in Fig. 7-11.7. Support for the model of a crosstie wall has been obtained by applying a field perpendicular to the plane of the film. The colloidal particles become magnetized along the direction of the applied field. As a result, the particles should shift to the region of the wall in which the poles provide an attractive force. The polarity of the uncompensated poles at the walls is indicated in Fig. 7-11.7. The powder patterns of Fig. 7-11.8 clearly show this effect.[168] An opposite shift is observed when the direction of the applied field is reversed.

When the magnetization of a film changes by wall motion, the function $F_{wt}(x)$, hence also the hysteresis curve, may be expected to be a function of the film thickness. Néel[170] assumes that only variations in the film thickness are important and predicts the coercive force H_c will vary as $d^{-4/3}$ where d is the film thickness. However, the magnetostatic energy F_D arising from uncompensated poles at the domain walls and on the film surfaces can also have an important influence on the coercive force.[171]

Problems

7-1. (a) By employing equation 7-2.3 plot the critical length ($2a$) for single-domain behavior of a prolate ellipsoidal particle of iron as a function of b/a. (b) Calculate the critical single-domain size for the cases in which the models of Fig. 7-2.3 apply. Neglect the demagnetization energy for these models. Calculate the critical size for a sphere of (a) iron, (b) nickel, (c) cobalt, and (d) MnBi.

7-2. Show that the jump in the magnetization of a single-domain particle occurs at the critical field given by equation 7-2.9. Assume the Stoner-Wohlfarth model. Calculate the critical field for $\theta = 10°, 45°, 80°$, and $90°$ and compare with the values given in the text.

7-3. Consider a chain of four spheres of radius a, dipole moment μ. Assume that the magnetization of each sphere is uniform and that the only interaction between the spheres is dipole-dipole. When a magnetic field is applied parallel to the axis of the chain, what is the expression for the total energy (neglect magnetostriction, crystalline, and exchange energies). Calculate and compare

[168] R. M. Moon, *J. Appl. Phys.* **30**, 82S (1959).

[169] E. E. Huber, D. O. Smith, and J. B. Goodenough, *J. Appl. Phys.* **29**, 294 (1958); S. Methfessel, S. Middelhoek, and H. Thomas, *IBM J. Res. Develop* **4**, 96 (1960).

[170] L. Néel, *J. Phys. radium* **17**, 250 (1956).

[171] S. Middelhoek, Thesis, University of Amsterdam (1961).

the coercive force for rotation of the spheres' magnetization M in unison and by fanning (rotations in alternative directions). Consider the situation in which the angle between the field direction and the axis of the chain increases from 0 to 90°.

7-4. Show that when a field is applied to a system of aligned uniaxial superparamagnetic particles, the energy barrier is given by

$$\tfrac{1}{2}CV - MVH + \frac{M^2V^2H^2}{2CV}$$

Derive an expression for the magnetization curve and compare it with the Langevin curve.

7-5. Show that an upper limit to the maximum energy product is

$$(B_H H)_{\max} = \left(\frac{B_r}{2}\right)^2.$$

7-6. Suppose the direction of magnetization is described by the angles (θ, ϕ) of Fig. 7-5.1. Show that the crystalline anisotropy energy per unit volume of a cubic crystal is given by (neglect K_2)

(a) $$F_K = K_1(\sin^2\theta - \tfrac{7}{8}\sin^4\theta - \tfrac{1}{8}\sin^4\theta\cos 4\phi)$$

when the normal to the wall is along a cube edge,

(b) $$F_K = K_1\left(\frac{\cos^4\theta}{3} + \frac{\sin^4\theta}{4} - \frac{\sqrt{2}}{3}\cos\theta\sin^3\theta\cos 3\phi\right)$$

when the normal to the wall is along a cube diagonal, and

(c) $$F_K = \frac{K_1}{4}[1 - 4\sin^2\theta + 4\sin^4\theta + (6\sin^2\theta - 4\sin^4\theta)\sin^2\phi - 3\sin^4\theta\sin^4\phi]$$

when the normal to the wall is along a face diagonal. For a uniaxial crystal show that

(d) $$F_K = K_1'\sin^2\phi$$

when the easy direction is along the OY axis.

7-7. Show that the surface energy of a 90° wall in a cubic crystal with $K_1 > 0$ is given by the results of equations 7-5.5, 7-5.9, or 7-5.10, depending on whether the orientation of the normal to the wall lies along the [100]-, [111]- or [110]-direction, respectively.

7-8. (a) Consider a uniaxial material with slablike domains of length L and spacing d. Assume that the direction of magnetization of the domains lies along the easy axis. Calculate the domain spacing. [*Hint*. The magnetostatic energy per unit volume is given approximately by M^2d/L.] (b) A cubic crystal with $K_1 > 0$ has the domain structure on a surface parallel to a (100) plane shown in Fig. 7-6.2. Calculate the domain spacing.

7-9. (a) Calculate the domain wall energy for the case in which crystalline anisotropy and magnetostrictive energies are present. Assume that the stress acts along the easy direction of the crystalline anisotropy. (b) Suppose that an internal stress inhomogeneity of magnitude $\Delta\sigma$ extends over a length l. If $l < \delta$, show that the coercive force (for one wall) is given approximately by equation 7-7.6.

7-10. Suppose that the internal stress in a material varies sinusoidally between tension and compression along the x-axis, so that domains separated by 90° walls may be expected. Show that the initial susceptibility is given by

$$\chi_0 \approx \frac{M^2}{\lambda \Delta\sigma}$$

where $\Delta\sigma$ is the maximum tension or compression present. Neglect any variation of F_w with x.

7-11. Calculate the initial susceptibility on the basis of the inclusion model illustrated in Fig. 7-7.3. (*Ans.* $\chi_0 \approx \dfrac{2M^2}{\pi N^{2/3} K_1 \delta}$ for one wall per centimeter.)

7-12. Consider a material that possesses internal stresses σ_i with $\lambda_s \sigma_i > 0$. Suppose an external stress σ and a magnetic field H are applied at an angle ϕ to the internal stress σ_i, assumed uniform, in one domain. If the magnetization changes by coherent rotations only and the other energy terms may be neglected, calculate the change in remanence with the stress σ for this domain. If the internal stresses are distributed at random over the material, show that

$$\frac{\partial M_r}{\partial \sigma} = \frac{M}{4\sigma_i}.$$

7-13. Consider an ensemble of single-domain particles, shaped as prolate ellipsoids of revolution, with polar semiaxis a and equatorial semiaxis b. Suppose that the polar axis of each particle is parallel to a certain direction, say the z-axis, and that the magnetization of half the particles lies along the positive direction and the other half along the negative z-direction. Neglect particle interactions and crystalline anisotropy and assume that there are no stresses or imperfections present. (a) If a magnetic field is applied along a direction perpendicular to the z-axis, calculate the initial susceptibility in terms of the demagnetizing factors D_a and D_b appropriate to the a- and b-axis, respectively. (b) Give numerical values for $a/b = 10$, 5, and 1 for iron at $T = 300$°K. (c) Calculate the initial susceptibility if the field makes an angle θ with the z-axis. (d) Suggest a situation for which the foregoing ideas could apply to a bulk soft magnetic material.

7-14. Derive equations 7-8.1 and 7-8.2.

7-15. If the motion of a 180° domain wall is described by equation 7-10.7, what is the relaxation time? Solve this equation for x and show that the susceptibility is given by

$$\chi = \chi_r \frac{(k_H/\chi_r)^2}{(k_H/\chi_r)^2 + \omega^2 \beta_H^{\,2}}$$

and the energy lost per cubic centimeter by

$$W = \frac{\omega k_H \beta_H H_1^2}{2(k_H/\chi_r)^2 + \omega^2 \beta_H^2}.$$

7-16. Consider a thin film with uniaxial anisotropy. If both a longitudinal and a transverse field are applied, show that the critical fields for irreversible rotation satisfy equation 7-11.6. Show that the tangent to the asteroid, $h_x^{2/3} + h_z^{2/3} = 1$, is given by equation 7-11.4. Illustrate with a graphical plot of these equations for various values of the parameter θ.

Bibliography

Arrott, A. and J. E. Goldman, "Fundamentals of Ferromagnetism," *Elec. Mfg.* March, 109 (1959).

Bates, L. F., *Modern Magnetism*, 4th ed., Cambridge University Press, Cambridge (1962).

Becker, R. and W. Döring, *Ferromagnetismus*, J. Springer, Berlin (1939).

Bitter, F., *Introduction to Ferromagnetism*, McGraw-Hill Book Co., New York (1937).

Bozorth, R. M., *Ferromagnetism*, D. Van Nostrand, Princeton, N.J. (1951).

Brailsford, F., *Magnetic Materials*, Methuen, London (1948).

Brown, Jr., W. F. *Micromagnetics*, Interscience Publishers, New York (1963).

Craik, D. J. and R. S. Tebble, "Magnetic Domains," *Rept. Prog. Phys.* **24,** 116 (1961).

Elochner, B. and W. Andia, *Fortschr. Physik* **3,** 163 (1955).

Hatfield, D., *Permanent Magnets and Magnetism*, John Wiley and Sons, New York (1962).

Hoselitz, K., *Ferromagnetic Properties of Metals and Alloys*, Oxford University Press, Oxford (1950).

"International Conference on Magnetism," *J. Phys. radium* **12,** 149–508 (1951), **20,** 65–442 (1959), *J. Phys. Soc. Japan, Suppl. B-1* **17,** 1–718 (1962).

Kersten, M., *Grundlagen einer Theorie der ferromagnetischen Hysterese und Koerzitivkraft*, Hirzel, Leipzig (1943). Reproduced Edwards Brothers, Ann Arbor, Mich. (1946).

Kittel, C. and J. K. Galt, "Ferromagnetic Domain Theory," *Solid State Phys.* **3,** 439 (1956).

Magnetic Properties of Metals and Alloys, Am. Soc. Metals, Cleveland, Ohio (1959).

Parker, R. J., and R. J. Studdes, *Permanent Magnets and Their Application*, John Wiley and Sons, New York (1962).

Rado, G. T. and H. Suhl, Editors, *Magnetism*, Vol. III, Academic Press, New York (1963).

Snoek, J. L., *New Developments in Ferromagnetic Materials*, Elsevier Publishing Co., Amsterdam (1947).

Stewart, K. H., *Ferromagnetic Domains*, Cambridge University Press, Cambridge (1954).

Stoner, E. C., "Ferromagnetism: Magnetization Curves," *Rept. Progr. Phys.* **13,** 83 (1950).

"Symposiums on Magnetism and Magnetic Materials," *AIEE*, **T-91** (Feb. 1957); *J. Appl. Phys. Suppl.*, **29,** 237–548 (1958), **30,** 1S–322S (1959), **31,** 1S–419S (1960), **32,** 1S–399S (1961), **33,** 1019–1394 (1962), **34,** 1005–1390 (1963), **35,** 737–1106 (1964).

"Washington Conference on Magnetism," *Revs. Mod. Phys.* **25,** 1–351 (1953).

Wohlfarth, E. P., "Hard Magnetic Materials," *Advan. Phys.* **8,** 87 (1959).

8

Antiferromagnetism

1. Introduction

An antiferromagnetic material has been defined as one (Section 1-3) in which an antiparallel arrangement of the strongly coupled atomic dipoles is favored. Néel[1] originally envisaged an antiferromagnetic substance as composed of two sublattices, one of whose spins tends to be antiparallel to those of the other. He assumed the magnetic moment of the two sublattices to be equal, so that the net moment of the materials was zero. Since Néel's original hypothesis the term antiferromagnetism has been extended to include materials with more than two sublattices and those with triangular, spiral, or canted spin arrangements; the latter may have a small nonzero magnetic moment.

Now magnetism is an experimental science. Therefore, if the above definition of antiferromagnetism is to be meaningful, some experimental technique should be available that will provide direct evidence for the presence of an antiparallel dipole arrangement. Although such a spin arrangement can be inferred from susceptibility and other measurements, this evidence can hardly be called direct since use is made of a chain of arguments that involves certain assumptions. The neutron diffraction method, developed essentially since 1945, fortunately does provide a direct test of the dipole arrangement in a material. A short account of the principles and achievements of neutron diffraction is given in Section 8-2.

[1] L. Néel, *Ann. phys. (Paris)* **17**, 64 (1932).

In ferromagnetism the exchange integral J_e is positive, whereas in antiferromagnetism J_e is negative. The Curie temperature T_f of ferromagnetism has its counterpart, the Néel temperature T_N, in antiferromagnetism; that is, T_N is the temperature above which the ordered antiparallel arrangement of the dipoles disappears. Indeed, the subject of antiferromagnetism may be said to be similar to that of ferromagnetism with, however, certain important differences. Use of this similarity is made in the development of theories of antiferromagnetism. The molecular field theory of antiferromagnetism for a two-sublattice system is outlined in Section 8-3. The predictions of this theory are in accord with the main features of the experimental results. Although the molecular field has its origin in an exchange interaction, as in ferromagnetism, there is the following difference. Most antiferromagnetic materials are compounds in which the cations are separated by much larger distances than those in ferromagnetic materials. It is concluded that the exchange interaction must occur indirectly via nonmagnetic anions such as O^{2-}. Several theories of the indirect exchange interaction are described. The application of the series expansion method, the Bethe-Peierls-Weiss method, and the spin-wave theory to antiferromagnetism is considered briefly. Other topics discussed in this chapter include the crystalline anisotropy, antiferromagnetic metals, canted dipole arrangements, antiferromagnetic domains, and the antiferromagnetic-ferromagnetic exchange interaction. Antiferromagnetic resonance is treated in Chapter 10.

2. Neutron Diffraction Studies

Neutrons, by virtue of their wave properties, are diffracted by crystalline materials. In order to provide useful information concerning crystal structure, it is necessary that neutrons have a wavelength comparable to the interatomic spacing in crystals, that is, a wavelength λ of the order of 1 Å. According to the de Broglie equation, this wavelength is related to the neutron's momentum p by $\lambda = h/p$. If the neutrons are in thermal equilibrium at a temperature T, their root mean square momentum is given by $p^2/2M_n = \frac{3}{2}\,kT$, where M_n is the mass of the neutron. Hence it follows that $\lambda^2 = h^2/3mkT$. Now the large flux of neutrons necessary for diffraction experiments is available only from atomic piles. These piles are usually cooled with water and thus are operated at temperatures between 0 and 100°C. The neutrons produced in the pile lose energy in collisions with the moderator of heavy water or graphite and tend to come into thermal equilibrium at the pile temperature. For $T = 20$°C the last equation gives $\lambda = 1.49$ Å, whereas for $T = 50$°C, $\lambda = 1.42$ Å. Fortunately, these are just the wavelengths required for diffraction studies of crystals.

A beam of neutrons can be obtained from the pile with the aid of a collimator. These neutrons have a Maxwellian distribution of velocities, that is, various wavelengths. In the study of crystal structures it is highly desirable to use a monochromatic beam of neutrons, then better resolution is obtained. A monochromatic beam can be produced with a single crystal as indicated in Fig. 8-2.1. Suitable single crystals include LiF, NaCl, CaF_2, Pb, and Cu. When the neutron beam makes an angle of incidence θ with the crystal, those neutrons that have a wavelength which satisfies the Bragg equation

$$n\lambda = 2d \sin \theta \tag{8-2.1}$$

will be reflected. Here d is the spacing of successive atomic planes that lie parallel to the crystal surface and n is an integer, called the order of the

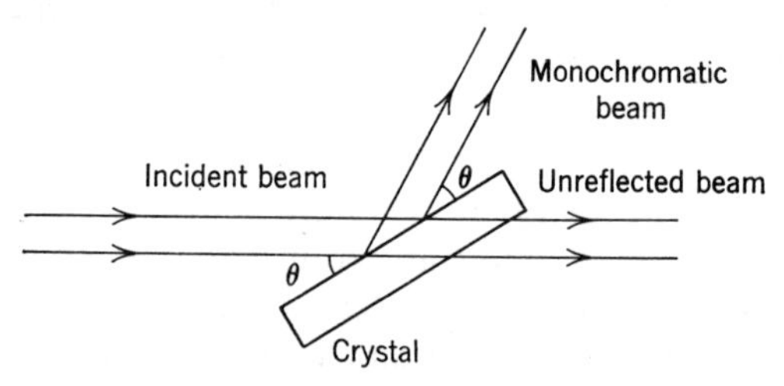

Fig. 8-2.1. Production of a monochromatic beam of neutrons. Only neutrons with $\lambda = 2d \sin \theta$ are reflected by the crystal.

reflection. It is usually arranged that only the first-order reflection ($n = 1$) is present. In actual practice the reflected neutrons have a small spread in wavelength of the order of 0.05 Å which develops mainly from a small divergence in the incident beam. This spread of wavelengths is much larger than that in an X-ray beam. In addition, the beam intensity is considerably less for neutrons than for X rays. Neutron diffraction nevertheless does possess some advantages over X-ray diffraction, and these will now be made clear.

Neutrons are scattered by the atomic nuclei of the crystalline material. The amplitude of this scattering is about the same order of magnitude for all atoms. X rays, on the other hand, are scattered only by the atomic electrons, and the amplitude of their scattering is proportional to the atomic number. Hence for the determination of the positions of the light elements, such as hydrogen and carbon, in the crystal lattice, neutrons have an advantage over X rays.

Further, the neutron-nuclear scattering amplitude varies in an irregular manner with atomic number, so that the scattering from neighboring elements in the periodic table may be appreciably different. Thus it is possible to distinguish between elements such as iron and cobalt when both occur in a material, for example, a ferrite. With X rays it is virtually impossible to make such a distinction, for the scattering amplitudes are almost the same.

Finally, neutrons are also scattered appreciably by atoms that have a permanent electronic dipole moment. This occurs because neutrons, although electrically neutral, possess a nuclear spin, hence a dipole moment. The scattering therefore arises from the interaction of the dipole moment of the neutron with the electronic dipole moment of the atoms. The phenomena of magnetic scattering has led to important results concerning magnetic materials. It is this application of neutron diffraction that is of primary concern here.

It is pertinent to compare the scattering of neutrons by nuclei with that by magnetic atoms. It will be recalled that for X rays the beam interacts with the atomic electrons, and, as a consequence, the electrons emit radiation of the same wavelength in all directions. Now, the wavelength of the incident radiation is comparable to the atomic dimensions. Hence the emitted radiation is in general not in phase, and there is some cancellation because of interference. It is convenient to introduce an atomic scattering factor f_X, defined as the ratio of the amplitude of the wave scattered by the atom to that of the wave scattered by a free electron. The variation of f_X with $(\sin \theta)/\lambda$, where θ is the angle of incidence of the X-ray beam, is illustrated in Fig. 8-2.2. For $\theta = 0, f_X = Z$, the total number of electrons in the atom. For all other values of θ the scattering from the atom is less than that for Z free electrons. The situation is entirely different for the scattering of neutrons by nuclei because the dimensions of the nucleus are only of the order of 10^{-13} cm, which is very small in comparison with the neutron wavelength of 10^{-8} cm. Consequently, the atomic scattering factor for neutron-nuclear scattering f_N is almost independent of the angle θ, as indicated in Fig. 8-2.2. In magnetic scattering, on the other hand, atomic electrons are involved. However, in contrast to the X-ray case, only a few electrons in outer shells contribute to the scattering. Hence the atomic scattering factor for neutron-magnetic scattering f_M decreases more rapidly with the angle θ than does the X-ray factor f_X; this behavior is also indicated in Fig. 8-2.2.

The diffraction patterns obtained from magnetic materials in the paramagnetic state differs from those in an ordered state in the following important way. In a paramagnetic substance, such as paramagnetic salts, and ferromagnetic and antiferromagnetic materials above their Curie and Néel temperatures, respectively, the atomic moments are oriented at random in space. Consequently, the magnetic scattering is entirely incoherent and contributes to the diffuse background. The magnetic scattering is obtained by subtracting from the background a calculation of the diffuse scattering arising from other effects; for example, the thermal motion of the atoms. In ferromagnetic and ferrimagnetic materials below their Curie point, and in antiferromagnetic materials below their Néel point, the atomic moments are ordered in a parallel or antiparallel arrangement. Hence the magnetic scattering is coherent, and this leads to lines in the diffraction pattern that are superposed on the lines from the coherent

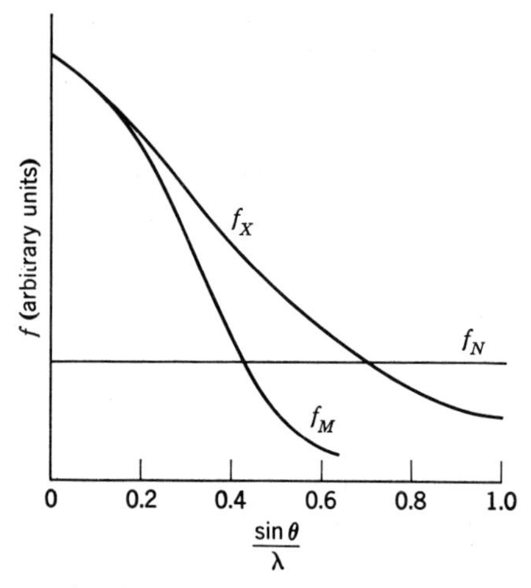

Fig. 8-2.2. Atomic scattering factors as a function of (sin θ)/λ. The curves labeled f_X, f_N, and f_M are for X-ray, neutron-nuclear, and neutron-magnetic scattering, respectively. The factors f_X and f_M are normalized with respect to their values at $\theta = 0$. The units of λ are Å.

nuclear scattering. The difference between the factors f_N and f_M permits the magnetic scattering to be separated from the nuclear scattering.

The magnetic scattering cross section per unit solid angle, measured as a function of (sin θ)/λ for either an ordered or disordered magnetic system, may then be compared with a curve derived from theory.[2] In this comparison it is necessary to employ a value for the angular momentum quantum number, and this value can then be checked against that obtained from magnetic measurements. In general, good agreement is found. For example, for Mn^{2+} in the paramagnetic state of MnF_2, a good fit between the experimental and theoretical scattering cross-section curves is obtained[3] if it is assumed that $S = \frac{5}{2}$. This is the value obtained from susceptibility measurements (see Table 2-11.1). For polycrystalline iron below the Curie

[2] J. Steinberger and G. C. Wick, *Phys. Rev.* **76**, 994 (1949).

[3] C. G. Shull, W. A. Strauser, and E. O. Wollan, *Phys. Rev.* **83**, 333 (1951).

temperature the curves are in good agreement if it is assumed[4] that the effective spin quantum number S is 1.11 (Fig. 8-2.3); this is just the value given by saturation magnetization experiments. Other examples may be found in the literature.[5]

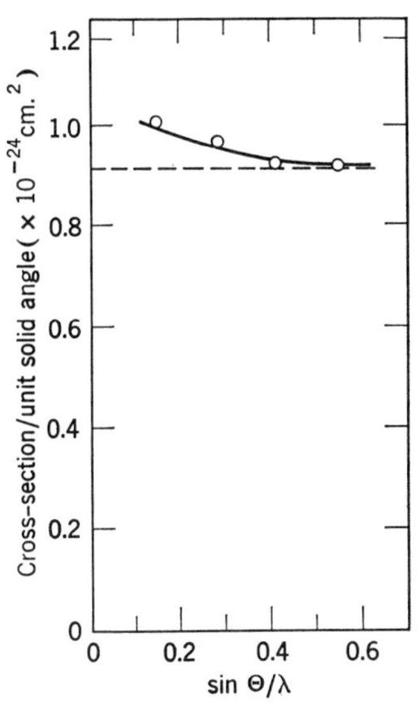

Fig. 8-2.3. The differential scattering cross section of polycrystalline iron. The circles represent the experimental data. The magnetic scattering is obtained by subtracting the isotropic nuclear scattering indicated by the dashed line. The full curve is obtained from the theory of magnetic scattering with the assumption that $S = 1.11$. [C. G. Shull, E. O. Wollan, and W. C. Koehler, *Phys. Rev.* **84,** 912 (1951).]

For a discussion of the relative intensities of the coherent diffraction lines it is convenient to introduce the structure factor F. For X rays the structure factor F_X is defined as the ratio of the amplitude of the wave scattered by all the atoms in a unit cell to that scattered by a free electron. In general, the atoms in the unit cell emit waves that are not in phase. Let a certain atom at the corner of the unit cell be chosen as origin. If ϕ_j is the phase difference between the wave scattered by the jth atom and that scattered by the atom at the origin, the value of F_X is given by

$$F_X = \sum f_{Xj} e^{i\phi_j}, \qquad (8\text{-}2.2)$$

where f_{Xj} is the atomic scattering factor of the jth atom. The summation extends over all the atoms in the unit cell. An expression will now be obtained for ϕ_j with the aid of Fig. 8-2.4. Consider the waves scattered by two atoms P_1 and P_2, the latter being at the origin. Let $\mathbf{s}$ and $\mathbf{s}'$ represent unit vectors lying along the directions of the incident and scattered beams, respectively, and let $\boldsymbol{\rho}$ be the vector from the origin to P_2. The path difference between the two scattered waves is

$$\begin{aligned} AP_2 - P_1B &= \boldsymbol{\rho} \cdot \mathbf{s} - \boldsymbol{\rho} \cdot \mathbf{s}' \\ &= \boldsymbol{\rho} \cdot (\mathbf{s} - \mathbf{s}'). \end{aligned}$$

It is worth noting that $\mathbf{s} - \mathbf{s}'$ is a vector perpendicular to the reflecting plane, that is, the plane that would reflect the incident beam. The phase

[4] C. G. Shull, E. O. Wollan, and W. C. Koehler, *Phys. Rev.* **84,** 912 (1951).

[5] W. C. Koehler and E. O. Wollan, *Revs. Mod. Phys.* **25,** 128 (1953), *Phys. Rev.* **92,** 1380 (1953).

difference between the two beams is given by

$$\phi = \frac{2\pi}{\lambda} \boldsymbol{\rho} \cdot (\mathbf{s} - \mathbf{s}');$$

for example, if the path difference is one wavelength λ, the phase difference is 2π. This equation may be written

$$\begin{aligned} \phi &= \boldsymbol{\rho} \cdot (\mathbf{k} - \mathbf{k}') \\ &= \boldsymbol{\rho} \cdot \boldsymbol{\varkappa}, \end{aligned} \tag{8-2.3}$$

where $\mathbf{k}$ and $\mathbf{k}'$ are the propagation vectors of the incident and reflected beams, respectively, and $\boldsymbol{\varkappa}$ is a vector perpendicular to the reflecting plane.

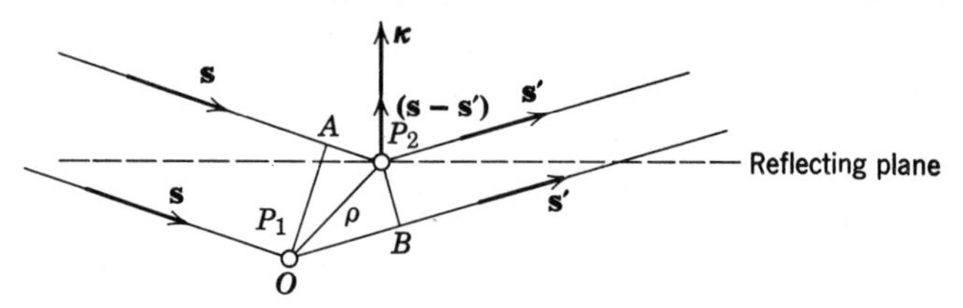

Fig. 8-2.4. Calculation of the structure factor F_X.

Now $\boldsymbol{\rho}$ may be expressed in terms of the primitive translation vectors $\mathbf{a}$, $\mathbf{b}$, and $\mathbf{c}$ of the unit cell, namely

$$\boldsymbol{\rho} = u_j\mathbf{a} + v_j\mathbf{b} + w_j\mathbf{c}. \tag{8-2.4}$$

Also, a diffraction line is observed only if the von Laue equations are satisfied; that is, if

$$\begin{aligned} \mathbf{a} \cdot \boldsymbol{\varkappa} &= 2\pi h, \\ \mathbf{b} \cdot \boldsymbol{\varkappa} &= 2\pi k, \\ \mathbf{c} \cdot \boldsymbol{\varkappa} &= 2\pi l, \end{aligned} \tag{8-2.5}$$

where h, k, and l are integers. These equations merely state that the phase difference between the beams scattered from the lattice points must be an integral multiple of 2π if reinforcement of the amplitude of the beams is to occur. It is easy to show that the integers h/n, k/n, and l/n may be identified as the Miller indices of the reflecting plane, where n is the largest common integral factor of h, k, and l. Substitution of equations 8-2.4 and 8-2.5 into 8-2.3 yields

$$\phi_j = 2\pi(hu_j + kv_j + lw_j)$$

and therefore

$$F_X = \sum f_{Xj} e^{2\pi i(hu_j + kv_j + lw_j)}. \tag{8-2.6}$$

If each atom is identical

$$F_X = f_X \sum e^{2\pi i(hu_j+kv_j+lw_j)}. \tag{8-2.7}$$

The intensity of the reflected wave is proportional to $|F_X|^2$.

The structure factors for neutrons are defined in analogy with that for X rays. For the scattering of neutrons by the atomic nuclei the structure factor F_N is given by

$$F_N(hkl) = \sum b_j e^{2\pi i(hu_j+kv_j+lw_j)}, \tag{8-2.8}$$

where b_j is the amplitude of the scattered wave from the jth nucleus, and $F_N(hkl)$ is proportional to the amplitude of the diffracted beam for the reflection from the planes whose Miller indices are (hkl). The length b is proportional to the atomic scattering factor f_N. For the scattering of neutrons by the magnetic atoms the structure factor F_M is calculated from[6]

$$F_M(hkl) = \sum p_j e^{2\pi i(hu_j+kv_j+lw_j)}, \tag{8-2.9}$$

where p_j is the amplitude of the scattered wave from the jth magnetic atom and $F_M(hkl)$ is proportional to the amplitude of the diffracted beam from the (hkl) reflection. Here p_j is proportional to f_M. For an antiferromagnetic material p is positive or negative, depending on whether the magnetic moments lie parallel or antiparallel to the preferred direction. If the beam of neutrons is unpolarized, the intensities of the nuclear and magnetic scattering, proportional to $|F_N|^2$ and $|F_M|^2$, respectively, are additive. The resultant intensity is proportional to $|F|^2$, which is given by[6]

$$|F|^2 = |F_N|^2 + q^2|F_M|^2, \tag{8-2.10}$$

where $q^2 = \sin^2\theta$, and θ is the angle between $\boldsymbol{\varkappa}$ and the directions of magnetization, so that $0 \le q^2 \le 1$. If the beam is polarized, there is coherence between the amplitude of the nuclear and magnetic reflected waves.

It is instructive to consider an example. For a body-centered cubic lattice of identical atoms, the atoms may be considered to have $(u_j v_j w_j)$-values of (000) and $(\frac{1}{2}\frac{1}{2}\frac{1}{2})$. Only one corner atom is included, since each corner atom is shared by eight unit cells. Thus we have

$$F_N = b(1 + e^{\pi i(h+k+l)}).$$

Therefore, if $(h + k + l)$ is odd, $F_N = 0$ and the line for this (hkl) reflection is absent. In particular, the (100) line is missing, whereas the (200) is present in the diffraction spectrum of the nuclear scattering. The physical

[6] O. Halpern and M. H. Johnson, *Phys. Rev.* **55**, 898 (1939).

reason for this result can be easily seen by referring to Fig. 8-2.5. The first-order (100) reflection requires that the path difference between beams (1) and (3) be one wavelength. In the body-centered cubic lattice there is also a reflection from the plane halfway between the (100) planes. The path difference between beam (2) reflected from this midplane and beams (1) or (3) is $\lambda/2$. Since there are equal numbers of identical atoms on each plane, the amplitude of the reflected beams is identical. Therefore there is cancellation and the (100) line does not appear. If the path difference between beams (1) and (2) is λ, a line does appear. Because of the choice of the unit cell this is the (200) line.

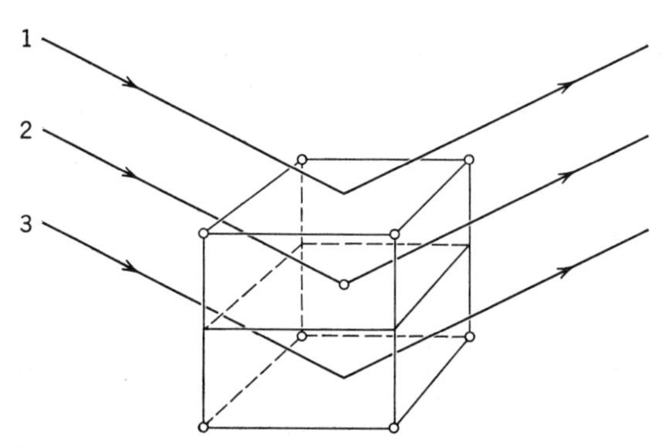

Fig. 8-2.5. To illustrate the absence of the (100) line in the diffraction spectrum of a body-centered cubic lattice of identical atoms, the phase difference between beams (1) and (3) is 2π, that between beams (1) and (2), π.

Suppose the atoms of the body-centered cubic structure under consideration possess identical magnetic moments. If the material is ferromagnetic, the magnetic scattering from each atom is alike, since the magnetic moment of an atom at the center of the cube is parallel to that of one of the corner atoms. An example of such a material is iron. It is convenient to choose a magnetic unit cell that is the same size as the chemical unit cell employed earlier. Then we get

$$F_M = p(1 + e^{\pi i(h+k+l)}).$$

Hence the lines from the magnetic scattering appear at the same angle θ as those from the nuclear scattering. With reference to the specific discussion, the (100) line is absent, whereas the (200) one is present.

If, however, the material is antiferromagnetic, the situation is quite different. Suppose the magnetic moments of the body-centered atoms are antiparallel to the moments of the corner atoms. It is still convenient to employ a magnetic unit cell of the same dimensions as the chemical cell, but now

$$F_M = p_1 + p_2 e^{\pi i(h+k+l)},$$

where p_1 is the amplitude of the wave scattered by the corner atom and p_2 that scattered by the body-centered atom; $p_1 \neq p_2$. In terms of Fig. 8-2.5, there is no longer complete cancellation between beams (1) and (2), and a (100) magnetic diffraction line appears. As an example, a neutron diffraction spectrum of chromium, which has a body-centered cubic structure, is shown[7] in Fig. 8-2.6. The presence of the (100) line leads to

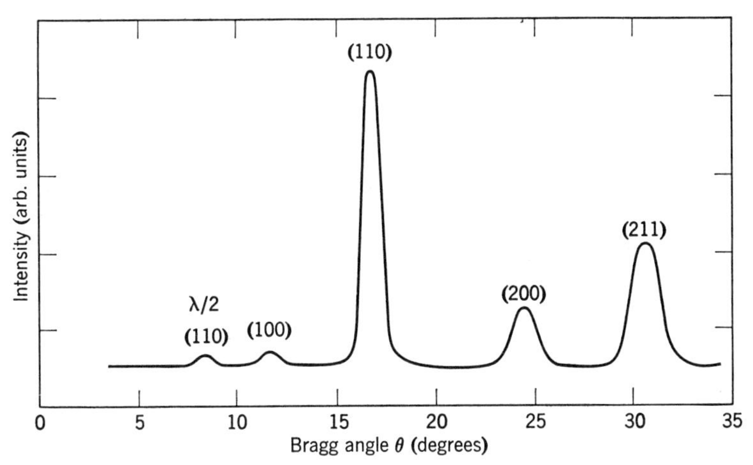

Fig. 8-2.6. Neutron diffraction pattern of chromium at $T = 20°K$. [C. G. Shull and M. K. Wilkinson, *Revs. Mod. Phys.* **25,** 100 (1953).]

the conclusion that chromium is antiferromagnetic with the magnetic moment arrangement previously assumed. Antiferromagnetic metals and alloys are discussed further in a later section.

A rather similar situation exists for the rutile structure. This is a body-centered tetragonal lattice with $c < a$, as illustrated in Fig. 8-2.7. If the body-centered atoms have magnetic moments that are antiparallel to those of the corner atoms, a (100) line would again appear. This is indeed observed[8] for MnF_2, FeF_2, CoF_2, and NiF_2. A neutron diffraction pattern for MnF_2 is shown in Fig. 8-2.8. Since the Néel temperature of MnF_2 is

[7] C. G. Shull and M. K. Wilkinson, *Revs. Mod. Phys.* **25,** 100 (1953); see also R. J. Weiss, L. M. Corliss, and J. M. Hastings, *Phys. Rev. Letters* **3,** 211 (1959); V. N. Bykov, V. S. Golovkin, N. V. Ageev, V. A. Levdik, and S. I. Vinogradov, *Dokl, Akad. Nauk SSSR* **128,** 1153 (1959) [trans. *Soviet Phys.-Doklady*. **4,** 1070 (1959)].

[8] R. A. Erickson, *Phys. Rev.* **90,** 779 (1953).

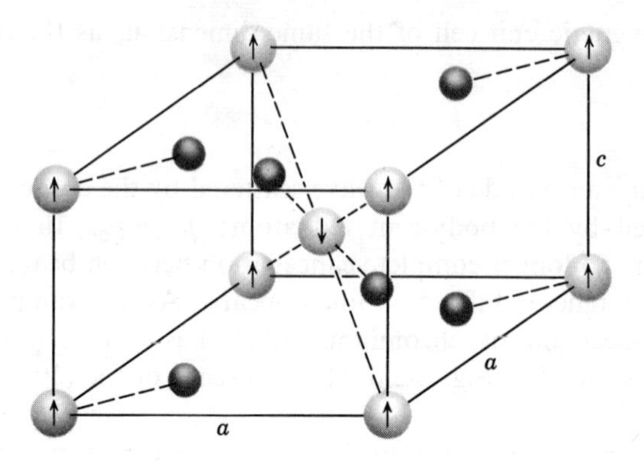

Fig. 8-2.7. The rutile lattice. The antiferromagnetic arrangement of the magnetic ions for MnF_2, FeF_2, and CoF_2 is indicated by the arrows. The dark spheres represent the fluorine ions. [R. A. Erickson, *Phys. Rev.* **90,** 779 (1953).]

about 75°K, a (100) line is present at 23°K and is absent at 300°K. For MnF_2, FeF_2, and CoF_2 no (001) line is observed. From this result it is deduced that the magnetic moments lie parallel to the c-axis, since q^2 of equation 8-2.10 is zero for this reflection. This arrangement of magnetic moments is indicated in Fig. 8-2.7. For NiF_2 a small (001) line does appear, and it is concluded that the moments are perpendicular to the c-axis.[9] The

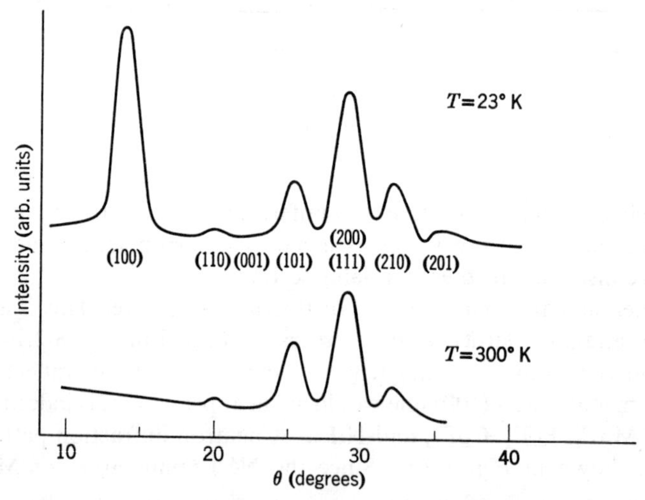

Fig. 8-2.8. Neutron diffraction pattern for MnF_2 below and above the Néel temperature. Note the antiferromagnetic (100) line at $T = 23°K$. [R. A. Erickson, *Phys. Rev.* **90,** 779 (1953).]

[9] T. Moriya, *Phys. Rev.* **117,** 635 (1960).

locations of the fluorine ions for these materials are also shown in Fig. 8-2.7. MnO_2 has the rutile lattice and is antiferromagnetic, but it appears that there are four magnetic sublattices.[10]

A consideration of the structure factor F_N for a face-centered cubic lattice shows that (hkl) lines appear only if the indices are all odd or all even integers (problem 8-1). For some compounds lines are observed in addition to those expected from the nuclear scattering. For example, diffraction patterns of MnO below and above the Néel temperature of

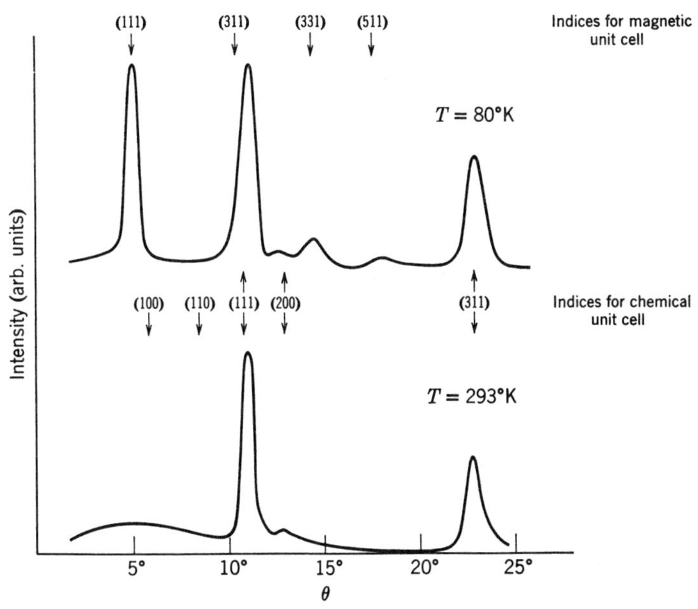

Fig. 8-2.9. Neutron diffraction spectrum of MnO below and above the Néel temperature.

122°K are shown[11] in Fig. 8-2.9. From the diffraction data it is concluded that the magnetic moments are arranged as indicated in Fig. 8-2.10. The sides of the magnetic unit cell are twice as large as those of the chemical cell. The additional lines can be indexed on the basis of the magnetic cell, as indicated in Fig. 8-2.9. (See also problem 8-1.) The magnetic moments of the Mn^{2+} ions are parallel on a given (111) plane, but the moments on adjacent (111) planes are antiparallel. In the earlier work[12] it appeared that the magnetic moments of the Mn^{2+} ions were oriented parallel to the [100]-axes.

[10] R. A. Erickson, *Phys. Rev.* **85,** 745 (1952).

[11] C. G. Shull, W. A. Strauser, and E. O. Wollan, *Phys. Rev.* **83,** 333 (1951).

[12] *Ibid.*

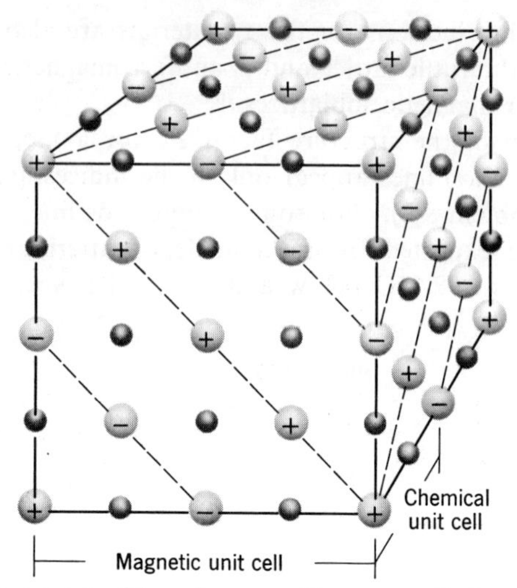

Fig. 8-2.10. Arrangement of ions for an antiferromagnetic face-centered cubic lattice such as MnO. The circles with plus and minus signs represent the magnetic ions (Mn^{2+}) with moments parallel and antiparallel to the preferred direction, respectively. The dark spheres represent the oxygen ions. One magnetic unit cell is shown. (After C. G. Shull, W. A. Strauser, and E. O. Wollan).

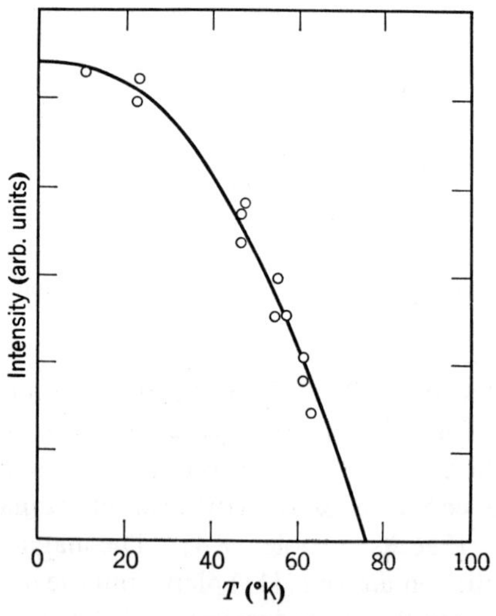

Fig. 8-2.11. The intensity of the (100) line of MnF_2 as a function of temperature. The circles represent the experimental data; the curve is a plot of the Brillouin function for $S = \frac{5}{2}$. [R. A. Erickson, *Phys. Rev.* **90**, 779 (1953).]

More recent results obtained with neutron spectrometers possessing higher resolution suggest that the moments lie in the (111) ferromagnetic plane.[13] NiO, CoO, FeO, MnS, and MnSe also have the magnetic structure shown in Fig. 8-2.10. For NiO the moments appear to lie in the (111) plane, whereas for FeO they lie along the [111]-direction perpendicular to the ferromagnetic sheets.

In addition to structural studies, the neutron diffraction technique provides information concerning the variation of the magnetization of a sublattice with temperature. This is obtained by measuring the intensity of one of the magnetic lines as a function of temperature. The data on intensities also leads to an estimate of the Néel temperature. For example, the intensity of the (100) line of MnF_2 at various temperatures is shown[14] in Fig. 8-2.11. The Brillouin function with $S = \frac{5}{2}$ and $T_N = 75°K$ fits the data well.

In the last decade neutron diffraction studies have established the antiferromagnetism of a number of other compounds. These include $CoMnO_2$ which has the same magnetic structure[15] as MnO, CrF_2, and $CrCl_2$ which have rutile-type structures,[16] α-Fe_2O_3[17] and Cr_2O_3[18] which have a rhombohedral unit cell, CrF_3, FeF_3, CoF_3, MnF_3[19] and MoF_3[20] which have a modified cubic structure, and FeS[21] which is hexagonal with the NiAs structure. The compounds $FeBr_2$, $CoBr_2$, $FeCl_2$, and $CoCl_2$, which have a hexagonal-layer-type structure, possess some unusual properties.[22] When an applied field is made large enough a transition from the antiferromagnetic to the ferromagnetic state occurs. The term *metamagnetism*, considered to be a special branch of antiferromagnetism, is sometimes applied to such materials.[23] The orthoferrites have the chemical formula $MFeO_3$ where M is yttrium or a rare earth metal ion. These compounds have the perovskite structure. Although they are

[13] W. L. Roth, *Phys. Rev.* **110,** 1333 (1958).

[14] R. A. Erickson, *Phys. Rev.* **90,** 779 (1953).

[15] G. E. Bacon, R. Street, and R. H. Tredgold, *Proc. Roy. Soc.* (*London*) **A-217,** 252 (1953).

[16] J. W. Cable, M. K. Wilkinson, and E. O. Wollan, *Phys. Rev.* **118,** 950 (1960).

[17] C. G. Shull, W. A. Strauser, and E. O. Wollan, *Phys. Rev.* **83,** 333 (1951).

[18] B. N. Brockhouse, *J. Chem. Phys.* **21,** 961 (1953).

[19] E. O. Wollan, H. R. Child, W. C. Koehler, and M. K. Wilkinson, *Phys. Rev.* **112,** 1132 (1958).

[20] M. K. Wilkinson, E. O. Wollan, H. R. Child, and J. W. Cable, *Phys. Rev.* **121,** 74 (1961).

[21] S. S. Sidhu and D. Meneghetti, *Phys. Rev.* **91,** 435 (1953); J. T. Sparks, W. Mead, A. J. Kirshbaum, and W. Marshall, *J. Appl. Phys.* **31,** 356S (1960).

[22] M. K. Wilkinson, J. W. Cable, E. O. Wollan, and W. C. Koehler, *Phys. Rev.* **113,** 497 (1959).

[23] J. Becquerel and J. van den Handel, *J. Phys. radium* **10,** 10 (1939).

basically antiferromagnetic,[24] they possess a weak ferromagnetism, probably because of a canted spin arrangement (see Section 8-9). It is interesting to note in passing that the antiferromagnetic structure of $CuCl_2 \cdot 2H_2O$ has been deduced from the nuclear resonance spectra of the protons.[25]

The results of some neutron diffraction studies of ferrimagnetic materials are reported in Chapter 9. The discussion of neutron diffraction would be incomplete without a brief mention of some of the significant contributions that have been made to the subject of ferromagnetism. The lack of any superlattice lines or diffuse magnetic scattering in iron leads to the conclusion that the iron atoms have identical magnetic moments.[26] The diffraction pattern of the face-centered cubic form of cobalt is consistent[26] with the assumption of an effective spin of $S = 0.87$; this is just the value obtained from saturation magnetization measurements. Experiments on nickel are difficult because the magnetic scattering is small. Some rare earth metals and their compounds have been shown to be ferromagnetic at low temperatures. For example, it has been established[27] that dysprosium undergoes a change from an antiferromagnetic state to a ferromagnetic state at 87°K. Also the rare earth nitrides, HoN and TbN are found to be ferromagnetic with Curie temperatures of 18 and 43°K, respectively.[28] The inelastic scattering of neutrons by spin waves in iron has been observed.[29] A study of the scattering by spin waves in face-centered cubic cobalt has confirmed[30] the quadratic form of the dispersion relationship (equation 6-6.17). A beam of neutrons becomes polarized[31] on passing through a piece of iron magnetized to saturation. When the polarized beam passes through a sheet of iron containing many domains the neutrons are depolarized. This effect may be detected by employing another piece of magnetized iron as an analyzer. In this way information on domain size and the refraction of the beam at domain walls is obtained.[32]

[24] W. C. Koehler and E. O. Wollan, *J. Phys. Chem. Solids* **2,** 100 (1957); W. C. Koehler, E. O. Wollan, and M. K. Wilkinson, *Phys. Rev.* **118,** 58 (1960).

[25] N. J. Poulis and G. E. G. Hardeman, *Physica* **18,** 201, 315 (1952).

[26] C. G. Shull and M. K. Wilkinson, *Revs. Mod. Phys.* **25,** 100 (1953); C. G. Shull, E. O. Wollan, and W. C. Koehler, *Phys. Rev.* **84,** 912 (1951).

[27] M. K. Wilkinson, W. C. Koehler, E. O. Wollan, and J. W. Cable, *J. Appl. Phys.* **32,** 48S (1961).

[28] M. K. Wilkinson, H. R. Child, J. W. Cable, E. O. Wollan, and W. C. Koehler, *J. Appl. Phys.* **31,** 358S (1960).

[29] R. D. Lowde, *Phys. Rev. Letters* **4,** 452 (1960), *Proc. Roy. Soc.* (*London*) **A-235,** 305 (1956).

[30] R. N. Sinclair and B. N. Brockhouse, *Phys. Rev.* **120,** 1638 (1960).

[31] J. Fleeman, D. B. Nicodemus, and H. H. Staub, *Phys. Rev.* **76,** 1774 (1949).

[32] M. T. Burgy, D. J. Hughes, J. R. Wallace, R. B. Heller, and W. Woolf, *Phys. Rev.* **80,** 953 (1950).

3. Molecular Field Theory of Antiferromagnetism

The molecular field theory for the simplest case, namely an antiferromagnetic material with two sublattices A and B will be developed first.[33] A specific example is the body-centered cubic lattice, with the A lattice consisting of the corner positions and the B lattice, the body positions. An atom at an A site then has nearest neighbors that all lie on B sites and next nearest neighbors that all lie on A sites. A similar situation holds for an atom on a B site. The molecular field $\mathbf{H}_{mA}$ acting on an atom at an A site then may be written in analogy with equation 6-2.1 as

$$\mathbf{H}_{mA} = -N_{AA}\mathbf{M}_A - N_{AB}\mathbf{M}_B,$$

where $\mathbf{M}_A$ and $\mathbf{M}_B$ are the magnetizations of the A and B sublattices, respectively, N_{AB} is a molecular field constant for the nearest neighbor interaction, and N_{AA} is a molecular field constant for the next nearest neighbor interaction. Similarly, the molecular field $\mathbf{H}_{mB}$ acting on an atom at a B site may be written

$$\mathbf{H}_{mB} = -N_{BA}\mathbf{M}_A - N_{BB}\mathbf{M}_B.$$

Since the same type of atoms occupy the A and B lattice sites, $N_{AA} = N_{BB} = N_{ii}$ and $N_{AB} = N_{BA}$. Then, if a field $\mathbf{H}$ is also applied, the fields $\mathbf{H}_A$ and $\mathbf{H}_B$ at an atom on the A and B lattices, respectively, are given by

$$\mathbf{H}_A = \mathbf{H} - N_{ii}\mathbf{M}_A - N_{AB}\mathbf{M}_B$$

and

$$\mathbf{H}_B = \mathbf{H} - N_{AB}\mathbf{M}_A - N_{ii}\mathbf{M}_B. \tag{8-3.1}$$

The interaction between nearest neighbors is antiferromagnetic and therefore the molecular field constant N_{AB} must be positive. On the other hand, it is conceivable that N_{ii} may be positive, negative, or even zero, depending on the particular material.

At thermal equilibrium the magnetizations of the sublattices (see Sections 2-13 and 6-2) are given by

$$M_A = \tfrac{1}{2}Ng\mu_B S\, B_S(x_A),$$

where

$$x_A = \frac{Sg\mu_B}{kT}H_A \tag{8-3.2}$$

[33] L. Néel, *Ann. phys. (Paris)* **18,** 5 (1932) **5,** 232 (1936); F. Bitter, *Phys. Rev.* **54,** 79 (1938); J. H. Van Vleck, *J. Chem. Phys.* **9,** 85 (1941).

and

$$B_S(x_A) = \frac{2S+1}{2S} \coth \frac{2S+1}{2S} x_A - \frac{1}{2S} \coth \frac{x_A}{2S},$$

and

$$M_B = \tfrac{1}{2} N g \mu_B S\, B_S(x_B),$$

where

$$x_B = \frac{S g \mu_B}{kT} H_B$$

and

$$B_S(x_B) = \frac{2S+1}{2S} \coth \frac{2S+1}{2S} x_B - \frac{1}{2S} \coth \frac{x_B}{2S}. \qquad (8\text{-}3.3)$$

Here N is the total number of atoms (or ions) with a permanent dipole moment per unit volume, and J has been set equal to S. It is convenient to continue the development of the molecular field theory under various subheadings.

Behavior above the Néel temperature. Although there is no antiferromagnetic ordering above the Néel temperature, a small magnetization is induced by the applied field. For the usual values of applied field, saturation effects are negligible and the Brillouin function can be replaced by the first term of the series expansion in x (equation 6-2.10), namely, $B_S(x) = [(S+1)/3S]x$. Then equations 8-3.2 and 8-3.3 become

$$M_A = \frac{N g^2 \mu_B{}^2 S(S+1)}{6kT} H_A$$

and

$$M_B = \frac{N g^2 \mu_B{}^2 S(S+1)}{6kT} H_B.$$

Now

$$\begin{aligned} H_A &= |\mathbf{H} - N_{ii}\mathbf{M}_A - N_{AB}\mathbf{M}_B| \\ &= H - N_{ii}M_A - N_{AB}M_B, \end{aligned}$$

since H, M_A, and M_B are parallel in the paramagnetic region. Similarly

$$H_B = H - N_{AB}M_A - N_{ii}M_B.$$

Substitution of these values of H_A and H_B into the foregoing equations for M_A and M_B gives

$$M_A = \frac{N g^2 \mu_B{}^2 S(S+1)}{6kT} (H - N_{ii}M_A - N_{AB}M_B) \qquad (8\text{-}3.4)$$

and

$$M_B = \frac{N g^2 \mu_B{}^2 S(S+1)}{6kT} (H - N_{AB}M_A - N_{ii}M_B). \qquad (8\text{-}3.5)$$

Addition of equations 8-3.6 and 8-3.7 then yields

$$M = M_A + M_B = \frac{Ng^2\mu_B^2S(S+1)}{6kT}[2H - (N_{ii} + N_{AB})M].$$

Hence the susceptibility $\chi = M/H$ is given by

$$\chi = \frac{C}{T+\theta}, \qquad (8\text{-}3.6)$$

where

$$C = \frac{Ng^2\mu_B^2S(S+1)}{3k}$$

and

$$\theta = \tfrac{1}{2}C(N_{ii} + N_{AB}).$$

Since, generally, $N_{AB} > N_{ii}$, θ is positive. It is interesting to compare equation 8-3.6 with equation 6-2.9, $\chi = C/(T - \theta)$, the Curie-Weiss law. These equations are plotted in Fig. 8-3.1 for illustrative purposes.

The Néel temperature. Below the Néel temperature both sublattices possess a spontaneous magnetization ($H = 0$) of equal magnitude. The critical temperature at which the spontaneous magnetization of one of these sublattices vanishes can be found from equation 8-3.2 or 8-3.3 by employing the method described in Section 6-2. Alternatively, the critical temperatures can be found by approaching from the high temperature side ($T > T_N$). Equations 8-3.4 and 8-3.5 are valid in the vicinity of T_N, for saturation effects are still unimportant. With $H = 0$, these equations become

$$M_A = \frac{C}{2T}(-N_{ii}M_A - N_{AB}M_B)$$

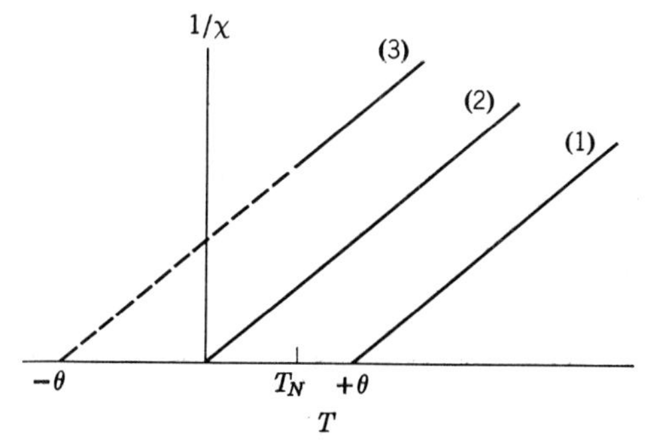

Fig. 8-3.1. Plot of $1/\chi$ versus temperature. Curve 1, $\chi = C/(T - \theta)$ (Curie-Weiss law); curve 2, $\chi = C/T$ (Curie's law), and curve 3, $\chi = C/(T + \theta)$.

and

$$M_B = \frac{C}{2T}(-N_{AB}M_A - N_{ii}M_B).$$

For nonzero values of the magnetizations M_A and M_B the determinant of the coefficients of M_A and M_B must be zero. Application of this condition yields the Néel temperature, namely,

$$T_N = \tfrac{1}{2}C(N_{AB} - N_{ii}). \tag{8-3.7}$$

Hence the Néel temperature is higher, the stronger the AB interaction and the weaker the AA or BB interaction, as is to be expected on physical grounds. Substitution for C from equation 8-3.6 then gives

$$\frac{\theta}{T_N} = \frac{N_{AB} + N_{ii}}{N_{AB} - N_{ii}} \tag{8-3.8}$$

for the ratio of the "paramagnetic" temperature θ to the Néel temperature. If $N_{ii} = 0$, $T_N = \theta$, whereas if $N_{ii} > 0$, $\theta > T_N$. Further, there is an upper limit to the ratio θ/T_N. If N_{ii} becomes too large in comparison to N_{AB}, the two sublattice arrangement assumed is unstable; this point is discussed further in a later subsection.

Susceptibility below the Néel temperature. In the absence of any applied field the sublattice magnetizations $\mathbf{M}_A$ and $\mathbf{M}_B$ are antiparallel to one another below the Néel temperature. The crystalline anisotropy, assumed for simplicity to be uniaxial, determines the common direction of $\mathbf{M}_A$ and $\mathbf{M}_B$, since these sublattice magnetizations will be collinear with the easy direction. Except to provide a preferred direction, the crystalline anisotropy will be neglected in the present calculation. This is justified if the crystalline anisotropy field is small compared to the molecular field and if the applied field is not too large. The crystalline anisotropy will be explicitly introduced into the equations developed in Section 8-7.

In general, the applied field may make any arbitrary angle with the easy axis. First, however, we calculate the susceptibility of a single crystal when the magnetic field is applied parallel to the easy axis. Suppose the applied field $\mathbf{H}$ is parallel to $\mathbf{H}_{mA}$ and antiparallel to $\mathbf{H}_{mB}$. Then x_A and x_B of equations 8-3.2 and 8-3.3 may be written

$$x_A = \frac{g\mu_B S}{kT}(H - N_{ii}M_A + N_{AB}M_B) \tag{8-3.9}$$

and

$$x_B = \frac{g\mu_B S}{kT}(-H + N_{AB}M_A - N_{ii}M_B). \tag{8-3.10}$$

On the other hand, when $H = 0$, $M_A = -M_B = M_0$, it follows from equations 8-3.2 and 8-3.3 that $x_A = -x_B = x_0$, where

$$x_0 = \frac{g\mu_B S}{kT}\,[(N_{AB} - N_{ii})M_0]. \qquad (8\text{-}3.11)$$

Provided saturation effects are unimportant, the Brillouin function may be expanded as a Taylor series in H. If only the first-order term is retained, we have

$$B_S(x_A) = B_S(x_0) + \frac{g\mu_B S}{kT}\,[H + N_{ii}(M_0 - M_A) + N_{AB}(M_B - M_0)] \times B_S'(x_0) \qquad (8\text{-}3.12)$$

and

$$B_S(x_B) = B_S(x_0) - \frac{g\mu_B S}{kT}\,[H + N_{AB}(M_0 - M_A) + N_{ii}(M_B - M_0)] \times B_S'(x_0) \qquad (8\text{-}3.13)$$

where $B_S'(x_0)$ is the derivative of the Brillouin function with respect to its argument. The coefficient of $B_S'(x_0)$ in equation 8-3.12 is the difference of equations 8-3.9 and 8-3.11, whereas that in equation 8-3.13 is the difference of equation 8-3.11 and 8-3.10. Multiplication of equation 8-3.12 and 8-3.13 by $\frac{1}{2}g\mu_B SN$ yields the sublattice magnetizations M_A and M_B, respectively (equations 8-3.2 and 8-3.3). The magnetization M induced by the field is then $M = M_A - M_B$. From this equation it is easy to show that the susceptibility $\chi_{\|} = M/H$ is given by

$$\chi_{\|}(T) = \frac{N\mu_B{}^2 g^2 S^2\, B_S'(x_0)}{kT + \frac{1}{2}(N_{ii} + N_{AB})\mu_B{}^2 g^2 S^2 N\, B_S'(x_0)}. \qquad (8\text{-}3.14)$$

At absolute zero, this equation predicts $\chi_{\|} = 0$. The physical reason is that in the approximation of the molecular field theory all the atomic moments are either parallel or antiparallel to the applied field at $T = 0°K$. Hence the field does not exert any torque on the moments and the induced magnetization is zero. As the temperature T increases, the $\chi_{\|}$ also increases until at the Néel temperature $\chi_{\|}(T_N)$ becomes equal to the susceptibility $\chi(T)$ given by equation 8-3.6. For $T > T_N$, equation 8-3.14 reduces to equation 8-3.6 (problem 8-3). For $N_{ii} = 0$, $\chi_{\|}$ can be plotted as a function of T/T_N without the necessity of assuming any explicit value for N_{AB}. Such curves for various values of S are shown in Fig. 8-3.2.

We now calculate the susceptibility of a single crystal when the magnetic field is applied perpendicular to the easy axis. The applied field produces a torque that tends to rotate the sublattice magnetizations, as indicated

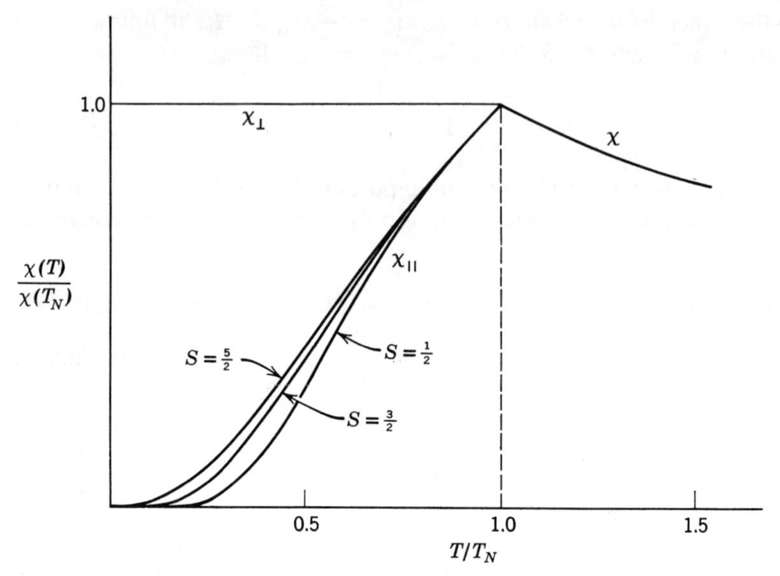

Fig. 8-3.2. The susceptibility of an antiferromagnetic material as a function of temperature in reduced units. $N_{ii} = 0$. [A. B. Lidiard, *Rept. Prog. Phys.* **25,** 441 (1962).]

in Fig. 8-3.3. This rotation is opposed by the molecular field. The magnitude of the sublattice magnetizations also increases when the field is applied. However since the angle of rotation ϕ is normally small, this effect may be neglected to first order. At equilibrium the total torque on each sublattice magnetization must equal zero, that is, for M_A, say,

$$|\mathbf{M}_A \times (\mathbf{H} + \mathbf{H}_{mA})| = 0.$$

By employing equation 8-3.1 it follows that

$$|\mathbf{M}_A \times \mathbf{H} - N_{AB}\mathbf{M}_A \times \mathbf{M}_B| = 0$$

or

$$M_A H \cos\phi - N_{AB} M_A M_B \sin 2\phi = 0.$$

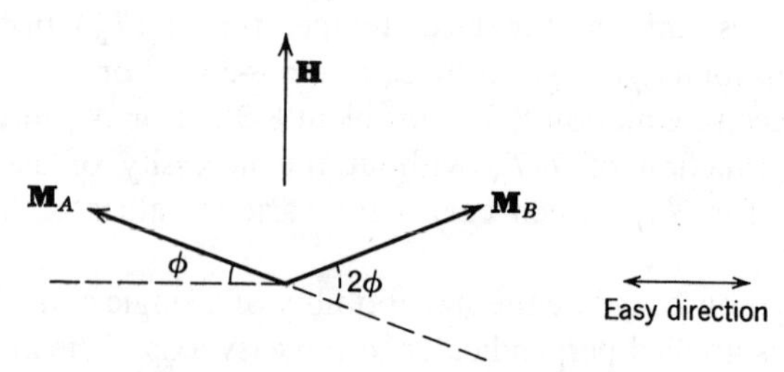

Fig. 8-3.3. To illustrate the calculation of the susceptibility $\chi_\perp$.

Hence

$$2M_B \sin \phi = \frac{H}{N_{AB}}. \tag{8-3.15}$$

Now the net magnetization M along the field direction is just

$$M = (M_A + M_B) \sin \phi = 2M_B \sin \phi$$

since $M_A = M_B$. Therefore the susceptibility $\chi_\perp (= M/H)$ is given by

$$\chi_\perp = \frac{1}{N_{AB}}; \tag{8-3.16}$$

that is, $\chi_\perp$ is a constant and is independent of N_{ii}. The susceptibility $\chi = C/(T + \theta)$ for $T \geqslant T_N$ (equation 8-3.6) also reduces to $\chi = 1/N_{AB}$ at $T = T_N$. Curves of $\chi_\perp$ and χ are shown in Fig. 8-3.2.

Materials that are polycrystalline or in the form of powders are often studied in experiments. It is reasonable to assume that the easy directions in the specimen are distributed at random. In general, the applied field will make some angle θ with the easy direction of one crystallite or particle. The susceptibility of the specimen can be calculated by considering the components of field parallel and perpendicular to the easy direction, namely $H \cos \theta$ and $H \sin \theta$, respectively. If $M_\parallel$ and $M_\perp$ denote the induced magnetization parallel and perpendicular to the easy direction, the susceptibility of one crystallite χ_c is given by

$$\chi_c = \frac{\mathbf{M} \cdot \mathbf{H}}{H^2} = \frac{M_\parallel \cos \theta + M_\perp \sin \theta}{H}.$$

Now, since $\chi_\parallel = M_\parallel/(H \cos \theta)$ and $\chi_\perp = M_\perp/(H \sin \theta)$, it follows that

$$\chi_c = \chi_\parallel \cos^2 \theta + \chi_\perp \sin^2 \theta.$$

An average taken over a unit sphere then yields the susceptibility χ_p of the powder or polycrystalline material. Thus

$$\chi_p = \chi_\parallel \overline{\cos^2 \theta} + \chi_\perp \overline{\sin^2 \theta}$$

or

$$\chi_p = \tfrac{1}{3}\chi_\parallel + \tfrac{2}{3}\chi_\perp. \tag{8-3.17}$$

In the comparison of experimental results with theory the ratio of the susceptibility of a powder at absolute zero to that at the Néel temperature $\chi_p(0)/\chi_p(T_N)$ is often considered. The molecular field then predicts that

$$\frac{\chi_p(0)}{\chi_p(T_N)} = \frac{2}{3},$$

since $\chi_\parallel = 0$ at $T = 0°\text{K}$, $\chi_\parallel = \chi_\perp$ at $T = T_N$, and $\chi_\perp$ is constant.

Sublattice arrangements. The arrangement of the atomic moments assumed for the body-centered cubic structure in the foregoing discussion is shown in Fig. 8-3.4*a*, in which the plus and minus signs indicate moments aligned parallel and antiparallel to the easy direction, respectively. It should be remembered that it is only at absolute zero that the plus and minus signs represent the atomic moments of each magnetic atom (or ion).

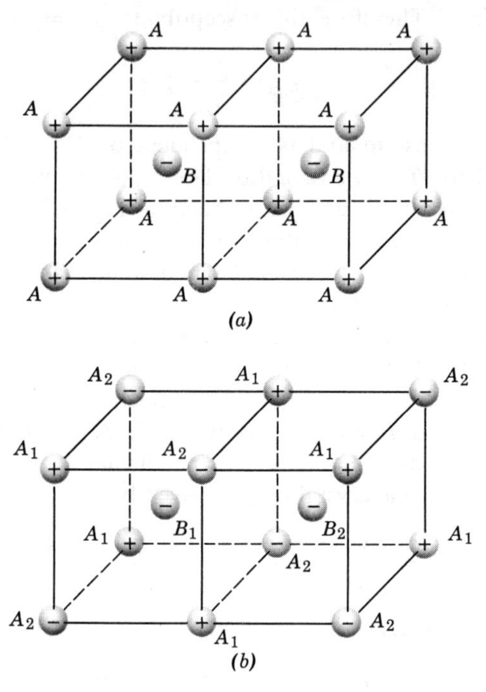

Fig. 8-3.4. Antiferromagnetic arrangement of atomic moments in a body-centered cubic lattice. (*a*) The first kind of order. (*b*) The second kind of order.

At any other temperature below the Néel temperature these signs represent the mean atomic moment per atom predicted by equations 8-3.2 and 8-3.3 for the sublattice magnetizations.

When N_{ii} becomes larger, compared to N_{AB}, an arrangement in which the next nearest neighbors are antiparallel is favored; this second kind of ordering is indicated in Fig. 8-3.4*b*. This dipole arrangement can be analyzed by considering four sublattices, A_1, A_2, B_1, and B_2. As an example the Néel temperature will now be calculated.[34] Any given corner or

[34] P. W. Anderson, *Phys. Rev.* **79,** 350, 750 (1950); J. S. Smart, *Phys. Rev.* **86,** 968 (1953); J. H. Van Vleck, *J. Phys. radium* **12,** 262 (1951).

body-centered atom has as many parallel as antiparallel nearest neighbors; hence this interaction term is zero. Then, for temperatures in the vicinity of the Néel temperature the sublattice magnetizations in analogy with equation 8-3.4 and 8-3.5 are given by

$$M_{A_1} = \frac{C}{2T}(H - N_{ii}M_{A_2}); \qquad M_{A_2} = \frac{C}{2T}(H - N_{ii}M_{A_1}) \tag{8-3.18}$$

and

$$M_{B_1} = \frac{C}{2T}(H - N_{ii}M_{B_2}); \qquad M_{B_2} = \frac{C}{2T}(H - N_{ii}M_{B_1}). \tag{8-3.19}$$

The Néel temperature is found by setting $H = 0$ and requiring that the sublattice magnetizations be nonzero. It follows that

$$T_N = \frac{N_{ii}C}{2} \tag{8-3.20}$$

or, on substituting for C from equation 8-3.6,

$$\frac{\theta}{T_N} = \frac{N_{AB} + N_{ii}}{N_{ii}}. \tag{8-3.21}$$

Also, it may be noted that $M_{A_1} = -M_{A_2}$ and $M_{B_1} = -M_{B_2}$.

The antiferromagnetic arrangement with the largest Néel temperature is the one that will occur, since this type of ordering is the state of lower energy. For $N_{ii} < \frac{1}{2}N_{AB}$ equation 8-3.7 yields a larger value than equation 8-3.20 for T_N, and therefore the first kind of ordering is stable. For $N_{ii} > \frac{1}{2}N_{AB}$ the situation is reversed, and the second kind of ordering is stable. For $N_{ii} = \frac{1}{2}N_{AB}$ equation 8-3.8 or 8-3.21 gives $\theta/T_N = 3$. Thus for the first kind of ordering the molecular field theory predicts that $1 \leqslant \theta/T_N \leqslant 3$.

In the rutile structure the nearest neighbors of the body-centered atom are the two atoms that lie on either side along the c-axis; the corner atoms are the next nearest neighbors. The molecular field theory just developed applies to the rutile structure provided N_{AB} is assumed to represent the molecular field constant for the interaction between the body-centered and corner atoms. This is not an unreasonable assumption, since the antiferromagnetic structure inferred from neutron diffraction studies (Fig. 8-2.7) indicates that the next nearest neighbor interactions are the important ones.

The face-centered cubic structure represents a more complicated situation, since some of the nearest neighbors of one atom are nearest neighbors. The antiferromagnetic structure of MnO, as revealed by neutron diffraction studies, is shown in Fig. 8-2.9. Any particular Mn^{2+}

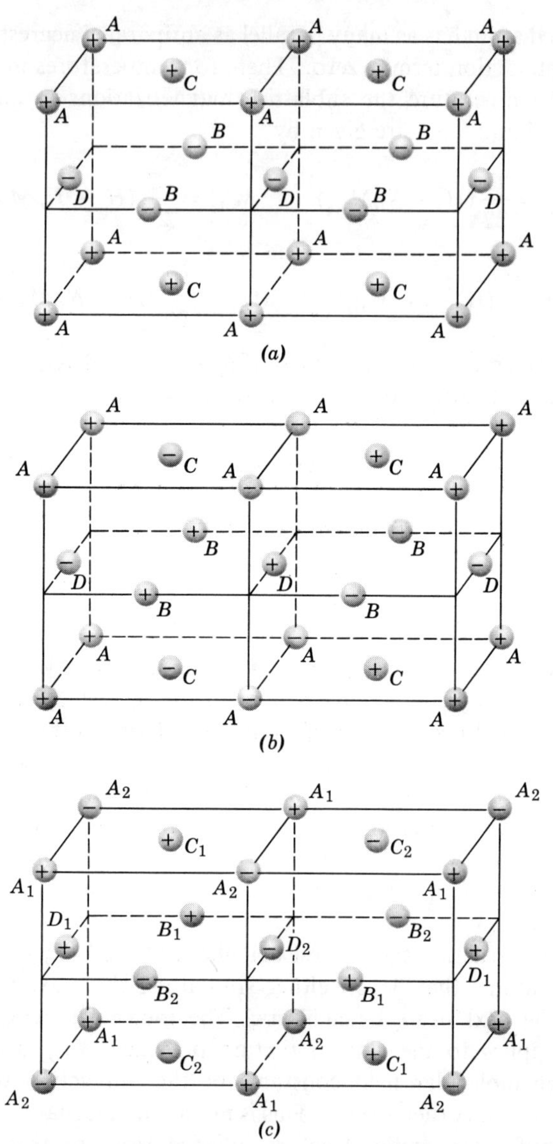

Fig. 8-3.5. Antiferromagnetic arrangement of atomic moments in a face-centered cubic structure. (*a*) The first kind of order. (*b*) Improved first kind of order. (*c*) Second kind of order (MnO structure).

ion has six parallel and six antiparallel nearest neighbors, whereas the six next nearest neighbors are antiparallel. Hence the important interaction that produces the antiferromagnetism must be between the next nearest neighbors, that is, between two Mn^{2+} ions lying along a cube edge and separated by an O^{2-} ion. This interaction, called indirect exchange, is discussed in Section 8-5. For the face-centered cubic structure it is therefore appropriate to let N_{AB} denote the molecular field constant for the coupling between next nearest neighbors and to let N_{ii} denote that for the coupling between nearest neighbors.

The face-centered cubic lattice can be conveniently analyzed by decomposing the structure into four sublattices, A, B, C, and D, as indicated in Fig. 8-3.5. A given atom on, say, the A sublattice then has four nearest neighbors on the B, C, and D sublattices and six nearest neighbors on its own (A). It is easy to show (problem 8-4) that the susceptibility above the Néel temperature is given by $\chi = C/(T + \theta)$ (equation 8-3.6), provided

$$\theta = \frac{C}{4}(N_{AB} + 3N_{ii}). \tag{8-3.22}$$

The MnO antiferromagnetic arrangement can be analyzed by subdividing each sublattice into two, as shown in Fig. 8-3.5c. Then, in analogy with equations 8-3.18 and 8-3.19, the A_1 and A_2 sublattice magnetizations are

$$M_{A_1} = -\frac{C}{4T}N_{AB}M_{A_2} \quad \text{and} \quad M_{A_2} = -\frac{C}{4T}N_{AB}M_{A_1}. \tag{8-3.23}$$

It follows that the Néel temperature is given by

$$T_N = \frac{N_{AB}C}{4}. \tag{8-3.24}$$

This result together with equation 8-3.22, yields

$$\frac{\theta}{T_N} = \frac{N_{AB} + 3N_{ii}}{N_{AB}}. \tag{8-3.25}$$

By comparing this Néel temperature with those for the first kind of ordering (Fig. 8-3.5a) and for the improved first kind of ordering (Fig. 8-3.5b), it can be shown (problem 8-4) that the second kind of ordering is stable only if $N_{AB} > \frac{3}{4}N_{ii}$. The improved kind of ordering occurs if $N_{AB} < \frac{3}{4}N_{ii}$, but the first kind of ordering occurs only if $N_{AB} < 0$. For $N_{AB} = \frac{3}{4}N_{ii}$, we find $\theta/T_N = 5$.

The paramagnetic-antiferromagnetic transition in the presence of an applied magnetic field. The Néel temperature (equation 8-3.7) was calculated under the assumption that no magnetic field had been applied.

The application of a field shifts the Néel temperature to a lower value. The physical reason for this effect is easily understood. Consider an antiferromagnetic material at a temperature T just below the Néel temperature when $H = 0$. Then suppose a field $\mathbf{H}$ is applied parallel to $\mathbf{M}_A$ and antiparallel to $\mathbf{M}_B$. As H is increased, M_B first decreases in magnitude until it becomes zero, then becomes parallel to $\mathbf{M}_A$, and finally coincides with M_A. A larger field is required for this coincidence if the sample is at a lower temperature.

By expanding the Brillouin function in powers of its argument, it can be shown[35] that the Néel temperature $T_N(H)$ for a field H applied along the easy direction is given by

$$\frac{T_N(H) - T_N(0)}{T_N(0)} = \left(\frac{3S}{S+1}\right)^3 \frac{B_S'''(0)}{2(Ng\mu_B S)^2} \chi_\perp{}^2 H^2, \qquad (8\text{-}3.26)$$

where

$$B_S'''(0) = -\frac{(S+1)[(S+1)^2 + S^2]}{90S^3}$$

and $T_N(0)$ is the Néel temperature given by equation 8-3.7. This equation may be rewritten in terms of the sublattice magnetizations for $H = 0$, $M_0 = (Ng\mu_B S/2)B_S[(N_{AB} - N_{ii})M_0(g\mu_B S/\mathrm{k}T)]$ (see equation 8-3.11)

$$\frac{T_N(H) - T_N(0)}{T_N(0)} = -3\left[\frac{(T_N(0) - T)}{TM_0{}^2}\right]\chi_\perp{}^2 H^2, \qquad (8\text{-}3.27)$$

provided that T is close to $T_N(0)$. The variation of T_N with H appears to have been observed in some antiferromagnetic materials with a low Néel temperature; for example, in $CuCl_2 \cdot 2H_2O$[36] and in $Co(NH_4)_2(SO_4)_2 \cdot 6H_2O$.[37]

Thermal effects. The specific heat of an antiferromagnet can be calculated by the same method as that used for a ferromagnet (Section 6-2). For $H = 0$ we can, as usual, write $M_A = -M_B = M_0$ and $H_m = H_{mA} = -H_{mB} = (N_{AB} - N_{ii})M_0$. Hence the internal energy per unit volume, U_{sp}, arising from the antiferromagnetism is just

$$\begin{aligned} U_{sp} &= -2\int H_m\, dM \\ &= -(N_{AB} - N_{ii})M_0{}^2. \end{aligned}$$

[35] J. A. Sauer and H. N. V. Temperley, *Proc. Roy. Soc.* (*London*) **A-176,** 203 (1940); C. G. B. Garrett, *J. Chem. Phys.* **19,** 1154 (1951); J. M. Ziman, *Proc. Phys. Soc.* (*London*) **64,** 1108 (1951); J. Kanamori, K. Motizuki, and K. Yosida, *Besseiron-Kenkyu* **63,** 28 (1953).

[36] N. J. Poulis and G. E. G. Hardeman, *Physica* **18,** 201, 315, 429 (1952).

[37] C. G. B. Garrett, *Proc. Roy. Soc.* (*London*) **A-206,** 242 (1951).

The specific heat is then given by $C_{sp} = dU_{sp}/dT$. By carrying out the same type of computation used in Section 6-2 it is easy to show that the discontinuity in the specific heat ΔC_{sp} is given by

$$\Delta C_{sp} = \frac{5S(S+1)}{S^2 + (S+1)^2} N\mathrm{k}, \qquad (8\text{-}3.28)$$

an expression that is the same as equation 6-2.22 provided that J is replaced by S. For example, for $S = \frac{1}{2}$ the discontinuity in the specific heat either of an antiferromagnet at $T = T_N$ or of a ferromagnet at $T = T_f$ is equal to 3/2 $N\mathrm{k}$ according to the molecular field theory.

Although a magnetocaloric effect is, of course, to be expected for antiferromagnetic materials, it does not appear to have been studied in detail. It has been assumed that the molecular field constants N_{AB} and N_{ii} are independent of temperature. However, with the thermal expansion of the lattice, it is reasonable to anticipate that N_{AB} and N_{ii} would indeed change. If this change does occur, the magnetic internal energy will make a contribution to the coefficient of expansion α. In fact, α is then proportional[38] to C_{sp}, hence to dM_0/dT. Consequently, a discontinuity in α is expected at the Néel temperature.

4. Some Experimental Results for Antiferromagnetic Compounds

It is now appropriate to compare the predictions of the molecular field theory with the experimental data. It is reasonable to expect the best agreement for antiferromagnetic materials that are certain to possess magnetic electrons localized at the atoms. Consequently, only the experimental results for compounds are considered here; the data for metals and alloys are discussed in Section 8-8.

Although susceptibility measurements were made on antiferromagnetic materials as early as the 1910's,[39] it was not until 1938 that Bizette, Squire, and Tsai[40] clearly recognized that their data inferred the presence of antiferromagnetism. The results they[40] obtained for a powdered sample of MnO are shown in Fig. 8-4.1. The experimental curves are of the general shape expected on the basis of the theoretical curves of Fig. 8-3.2. For the susceptibility measured in the smaller field (of 5000 oe), the ratio $\chi_p(0)/\chi_p(T_N)$ is found to be 0.69, which is very close to the predicted value

[38] See for example, C. Kittel, *Introduction to Solid State Physics*, John Wiley and Sons, New York (1956), p. 154.

[39] K. Honda and T. Soné, *Sci. Rept. Tôhoku Univ.* **3**, 223 (1914); T. Ishiwara, *Sci. Rept. Tôhoku Univ.* **3**, 303 (1914); K. Honda and T. Ishiwara, *Sci. Rept. Tôhoku Univ.* **4**, 215 (1915); P. Weiss and G. Foëx, *J. Phys.* **1**, 274, 744, 805 (1911).

[40] H. Bizette, C. Squire, and B. Tsai, *Compt. rend.* (*Paris*) **207**, 449 (1938).

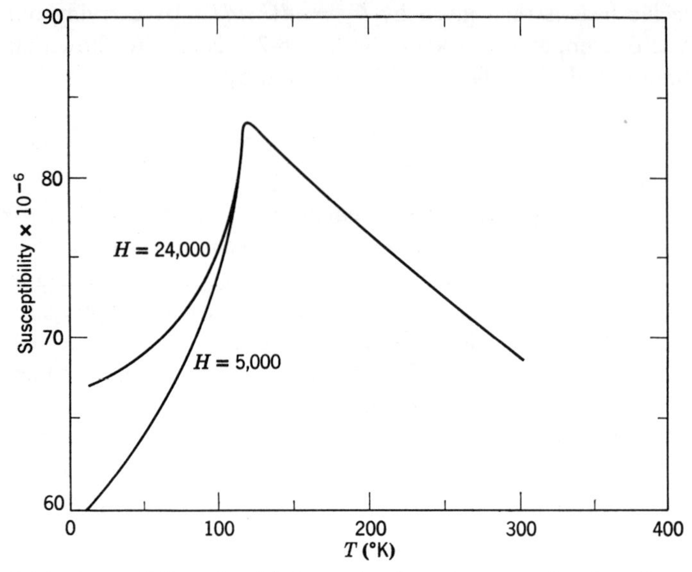

Fig. 8-4.1. Susceptibility of an MnO powdered specimen as a function of temperature for $H = 5000$ and $H = 24{,}000$ oe. [H. Bizette, C. Squire, and B. Tsai, *Compt. rend.* (*Paris*) **207,** 449 (1938).]

of 0.67 (equation 8-3.17). Below the Néel temperature the susceptibility depends on the applied field; a possible explanation is obtained when the crystalline anisotropy is explicitly introduced into the theory (see Section 8-7).

Most of the susceptibility measurements have been made on powders or polycrystalline materials, but a growing amount of data on single crystals have been reported in the last few years. Results[41] obtained for a single crystal of MnF_2 are shown in Fig. 8-4.2. The values of $\chi_\perp - \chi_\parallel$ are in accord with those obtained in another experiment.[42] The $\chi_\parallel$ and $\chi_\perp$ curves agree remarkably well with the theoretical ones (Fig. 8-3.2). For other fluorides, namely, FeF_2, CoF_2, and NiF_2, the agreement is not so good.[43]

The specific heat curves for MnF_2[44] and MnO[45] are shown in Fig. 8-4.3. As predicted in Section 8-3, the curves have a pronounced maximum in the vicinity of the Néel temperature. However, the peaks of the curves

[41] H. Bizette and B. Tsai, *Compt. rend.* (*Paris*) **238,** 1575 (1954).

[42] M. Griffel and J. W. Stout, *J. Chem. Phys.* **18,** 1455 (1950).

[43] J. W. Stout and L. M. Martarresse, *Revs. Mod. Phys.* **25,** 338 (1953); L. M. Martarresse and J. W. Stout, *Phys. Rev.* **94,** 1792 (1954).

[44] J. W. Stout and H. E. Adams, *J. Am. Chem. Soc.* **64,** 1535 (1942).

[45] R. W. Millar, *J. Am. Chem. Soc.* **50,** 1875 (1928).

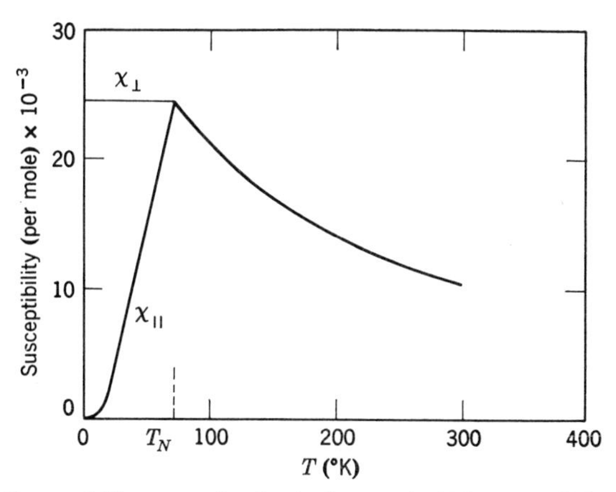

Fig. 8-4.2. Susceptibility per mole of a single crystal of MnF_2 parallel and perpendicular to the *c*-axis. [H. Bizette and B. Tsai, *Compt. rend.* (*Paris*) **238,** 1575 (1954).]

are sharper than expected from the theory. Also, there is still a substantial specific heat above T_N, indicating, as in the ferromagnetic case, the persistence of short-range ordering. The thermal expansion coefficient possesses an anomaly at T_N as anticipated; curves[46] for some oxides are illustrated in Fig. 8-4.4.

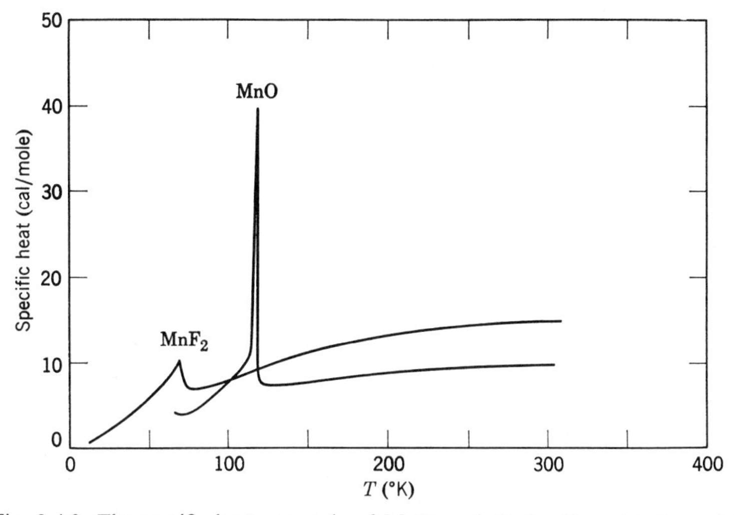

Fig. 8-4.3. The specific heat per mole of MnF_2 and MnO. Recent but unpublished data obtained by J. W. Stout on a large crystal of MnF_2 puts the peak at a slightly higher temperature. [MnF_2: J. W. Stout and H. E. Adams, *J. Am. Chem. Soc.* **64,** 1535 (1942); MnO: R. W. Millar, *J. Am. Chem. Soc.* **50,** 1875 (1928).]

[46] M. Foëx, *Compt. rend.* (*Paris*) **227,** 193 (1948).

Pertinent data for a selection of antiferromagnetic compounds are listed in Table 8-4.1. The Néel temperature given is a weighted mean of values obtained from susceptibility, specific heat, and neutron diffraction measurements. It is worth mentioning that T_N can be determined by other methods; these include conductivity, optical, elastic constant, proton resonance, dilatometric, and solubility measurements. The constants C and θ of the Curie-Weiss law, deduced from the susceptibility above the Néel temperature, and the ratio $\chi_p(0)/\chi_p(T_N)$ are also tabulated if available. The results reported by different investigators for the same

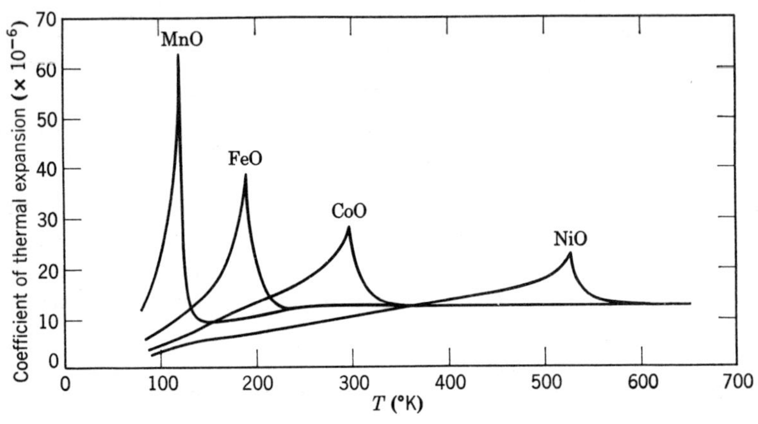

Fig. 8-4.4. Coefficient of thermal expansion for MnO, FeO, CoO, and NiO. [M. Foëx, *Compt. rend.* (*Paris*) **227**, 193 (1948).]

material sometimes disagree. The reason probably lies in the purity of the sample; small deviations from stoichiometry are likely to have a profound effect on antiferromagnetic properties.

Although the ratio of the powder susceptibilities, $\chi_p(0)/\chi_p(T_N)$, is close to the predicted value of 0.67 (equation 8-3.17) for many materials, there is a large discrepancy for some compounds. The majority of the ratios of θ/T_N listed lie between 1 and 2, the largest value being 5.0 (for MnO). These are consistent with the predicted range of $1 \leqslant \theta/T_N \leqslant 3$ for the b.c.c. and rutile structures and with that of $1 \leqslant \theta/T_N \leqslant 5$ for the face-centered cubic structure (see Section 8-3). The substitution of the observed values of θ/T_N into equation 8-3.5 or equation 8-3.25 provides an estimate of the ratio of the molecular field constants, N_{AB}/N_{ii}. For the metamagnetic materials, $FeCl_2$, $CoCl_2$, and $NiCo_2$, θ is negative; equation 8-3.6 then implies that N_{ii} is negative and larger in magnitude than N_{AB}. However, from equation 8-3.7 it follows that $T_N > |\theta|$, whereas in fact it is found from experiment that $T_N < |\theta|$. The values of the effective number of Bohr magnetons per atom calculated from the Curie constant C are usually in good agreement with those expected from theory. For example, for MnO, experiment yields $p_{\text{eff}} = 5.95$; the theoretical value for $S = 5/2$ is $p_{\text{eff}} = 5.92$.

Table 8-4.1. Data for Some Antiferromagnetic Compounds[a]

Compound	Crystal Structure	T_N(°K)	θ(°K)	$\dfrac{\theta}{T_N}$	C_{mole}	$\dfrac{\chi_p(0)}{\chi_p(T_N)}$
MnO	f.c.c.	122	610	5.0	4.40	0.69
FeO	f.c.c.	185	570	3.1	6.24	0.77
CoO	f.c.c.	291	280	0.96	3.0	—
NiO	f.c.c.	515	—	—	—	0.67
MnS	f.c.c.	165	528	3.2	4.30	0.82
MnF_2	rutile	74	113	1.5	4.08	0.75
FeF_2	rutile	85	117	1.4	3.9	0.72
CoF_2	rutile	40	53	1.3	3.3	—
NiF_2	rutile	78	116	1.5	1.5	—
MnO_2	rutile	86	—	—	—	0.93
Cr_2O_3	rhombohedral	307	1070	3.5	2.56	0.76
α-Fe_2O_3	rhombohedral	950	2000	2.1	4.4	—
FeS	hex. layer	613	857	1.4	3.44	—
$FeCl_2$	hex. layer	24	−48	−2.0	3.59	<0.2
$CoCl_2$	hex. layer	25	−38.1	−1.5	3.46	~ 0.6
$NiCl_2$	hex. layer	50	−68.2	−1.4	1.36	—
$FeCO_3$	complex	57	—	—	—	0.25
$CuCl_2 \cdot 2H_2O$	rhombohedral	4.3	5	1.16	—	—
$FeCl_2 \cdot 4H_2O$		1.6	2	1.2	3.61	—
$NiCl_2 \cdot 6H_2O$		5.3	—	—	—	—

[a]For a list of references see T. Nagamiya, K. Yosida, and R. Kubo, *Advan. Phys.* **4,** 1 (1955). See also R. D. Pierce and S. A. Friedberg, *J. Appl. Phys.* **32,** 66S (1961); T. Haseda and E. Kanda, *J. Phys. Soc. Japan* **12,** 1051 (1957).

It has often been concluded in the past that a material was antiferromagnetic because a sharp peak in the susceptibility versus temperature curve of the polycrystalline substance was observed. This conclusion is not always valid. For example, there is a discontinuity in the susceptibility of polycrystalline Ti_2O_3[47] and V_2O_3.[48] However, neutron diffraction

[47] M. Foëx and J. Wucher, *Compt. rend.* (*Paris*) **241,** 184 (1955); A. D. Person, *J. Phys. Chem. Solids* **5,** 316 (1958).

[48] V. Hoschek and W. Klemm, *Z. anorg. u. allgem. Chem.* **242,** 63 (1939); M. Foëx, S. Goldsztaub and R. Wey, *J. Rech. Centre Nat. Rech. Sci. Lab. Bellevue* (*Paris*) **21,** 237 (1952).

experiments indicate that appreciable long range order is absent[49] below the transition temperature. Further, the single crystal susceptibilities[50] $\chi_{\parallel}$ and $\chi_{\perp}$ are almost independent of temperature below the discontinuity. It is suggested[50] that the low temperature susceptibility originates from the temperature-independent paramagnetism described in Section 2-7. The discontinuity in the susceptibility may arise from a phase transformation. A maximum in the susceptibility versus temperature curve may also occur if at some temperature kT is sufficient to populate an upper energy level. This may be the explanation of the data obtained for cupric acetate monhydrate for example.[51]

Finally it should be mentioned that X-ray studies have established that the crystal lattice of some antiferromagnetic materials undergoes a small distortion on cooling through the Néel temperature. For example, NiO, MnO, and FeO, which are face-centered cubic above T_N, expand or contract in a direction perpendicular to the (111)-ferromagnetic sheets and become slightly rhombohedral below T_N.[52] Also, α-Fe_2O_3[53] and Cr_2O_3,[54] which are rhombohedral, undergo an expansion and contraction along the [111]-direction, respectively. The origin of this lattice distortion is not fully understood at present.

5. The Indirect Exchange Interaction

It is reasonable to anticipate that the molecular field in antiferromagnetism has the same basic origin as in ferromagnetism, that is, in a quantum mechanical exchange interaction. If it is assumed that the Heisenberg theory applies, the exchange Hamiltonian is given by equation 6-3.24 as

$$\mathcal{H} = -2J_e \sum_j \mathbf{S}_i \cdot \mathbf{S}_j, \tag{8-5.1}$$

except here $J_e < 0$, that is,

$$\mathcal{H} = 2\,|J_e| \sum_j \mathbf{S}_i \cdot \mathbf{S}_j. \tag{8-5.2}$$

On this basis, the molecular field theory represents an approximation in which the instantaneous values of the spins in the exchange Hamiltonian are replaced by their time averages (see the derivation of equation 6-3.27).

[49] G. Shirane, S. J. Pickart, and R. Newnham, *J. Phys. Chem. Solids* **13,** 166 (1960); A. Paloetti and S. J. Pickart, *J. Chem. Phys.* **32,** 308 (1960).

[50] P. H. Carr and S. Foner, *J. Appl. Phys.* **31,** 344S (1960).

[51] G. Foëx, T. Karantassis, and N. Perakis, *Compt. rend. (Paris)* **236,** 982 (1953).

[52] H. P. Rooksby, *Acta Cryst.* **1,** 226 (1948); N. C. Tombs and H. P. Rooksby, *Nature (London)* **165,** 442 (1950); B. T. M. Willis and H. P. Rooksby, *Acta Cryst.* **6,** 827 (1953).

[53] B. T. M. Willis and H. P. Rooksby, *Proc. Phys. Soc. (London)* **B-65,** 950 (1952).

[54] S. Greenwald, *Nature (London)* **168,** 379 (1951).

For the antiferromagnetic compounds it takes but a little consideration to come to the conclusion that the exchange interactions must be quite different from those that occur in ferromagnetism. Consider, for example, MnO. The neutron diffraction experiments show that the strong antiferromagnetic interaction is between the next nearest Mn^{2+} ions that lie along the cube edges and that are separated by the large O^{2-} ions. Now, the next nearest neighboring Mn^{2+} ions are 4.43 Å apart. Hence the direct exchange interaction between these Mn^{2+} ions, according to Fig.

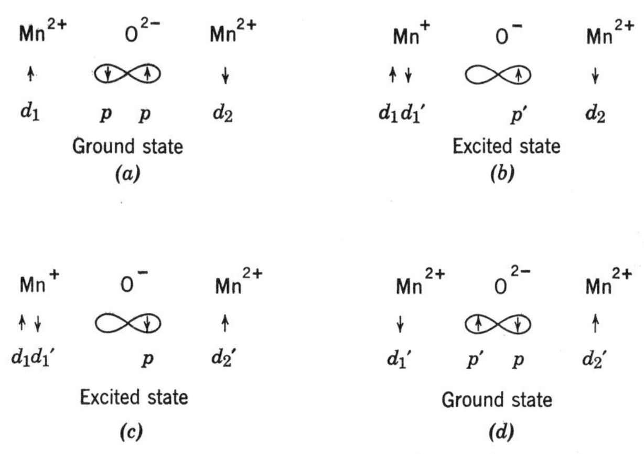

Fig. 8-5.1. The superexchange interaction.

6-3.1, is expected to be ferromagnetic and very small in magnitude. We conclude, therefore, that the direct exchange interaction cannot be the source of the strong antiferromagnetic coupling. Instead, an indirect exchange interaction that operates via the intermediate O^{2-} ions would appear to be involved. For some of the other antiferromagnetic compounds, such as MnF_2, the situation is not so clear-cut; nevertheless, the important role of the anion is again apparent. The physical principles of some of the mechanisms that have been proposed for indirect exchange will now be outlined.

The superexchange mechanism was first introduced by Kramers[55] in an attempt to account for the exchange interaction in paramagnetic salts (see Section 3-6). This theory has since been developed further by Anderson and others.[56] The idea can be illustrated by two Mn^{2+} and one O^{2-} ions arranged collinearly, as indicated in Fig. 8-5.1. The simplest model requires the consideration of four electrons. The ground state

[55] H. A. Kramers, *Physica* **1,** 182 (1934).

[56] P. W. Anderson, *Phys. Rev.* **79,** 350 (1950); J. H. Van Vleck, *J. Phys. radium* **12,** 262 (1951).

consists of one electron on each Mn^{2+} in the states d_1 and d_2 and two electrons on the O^{2-} ion in identical p orbitals (Fig. 8-5.1*a*). The p orbitals have a dumbbell shape that coincides with the axis joining the two Mn ions. There is then a certain probability that one of the oxygen's p electrons will be transferred to a manganese ion, as indicated in Fig. 8-5.1*b*. In this excited state there will be a coupling between the electrons in the d_1 and d_1' states of the Mn ion and also between the remaining oxygen's electron and the one d_2 electron of the other Mn ion. Moreover, this coupling is spin-dependent. The complete exchange of the electrons on the magnetic ions is accomplished by the electron transfers illustrated in Figs. 8-5.1*c* and 8-5.1*d*.

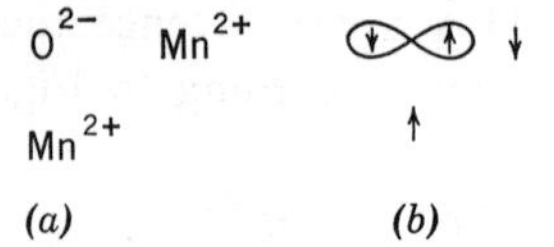

O^{2-} Mn^{2+}

Mn^{2+}

(*a*) (*b*)

Fig. 8-5.2. The nearest neighbor interaction in MnO.

This model makes it clear why the next nearest neighbor interactions are stronger than those of the nearest neighbor in MnO.[57] The bond joining two neighboring Mn ions with the intervening oxygen ion are at right angles, as indicated in Fig. 8-5.2. The dumbbell-shaped p wave function of the oxygen then appreciably overlaps the wave function of only one of the neighboring magnetic ions, hence the interaction is small (Fig. 8-5.2*b*). Indeed, the superexchange interaction that does occur between nearest neighbors probably primarily originates either from a hybridization of $2s$ and $2p$ wave functions or from the excitation of an electron to the $3s$ state of the oxygen ion. These states are probably the most important in the superexchange interaction in compounds such as MnF_2. The other p wave functions of the oxygen ion that lie along the other axes are not involved, since these states remain filled during the particular electron transfers considered here.

It is conceivable that other excited states than that in Fig. 8-5.1 play a role; for example, there may be a simultaneous transfer of the oxygen's two electrons, one each to the adjacent Mn ions.[58] In any case, the quantum mechanical calculation of superexchange requires third-order perturbation theory. It is then found that the superexchange Hamiltonian has the same form as that of the direct exchange of ferromagnetism, namely that of equation 8-5.1.

A somewhat different indirect exchange interaction has been proposed[59] for compounds that are only partly ionic in character. In such materials

[57] For an alternative model see T. N. Casselman and F. Keffer, *Phys. Rev. Letters* **4,** 498 (1960).

[58] G. W. Pratt, *Phys. Rev.* **97,** 926 (1955).

[59] J. B. Goodenough, *Phys. Rev.* **100,** 564 (1955); J. B. Goodenough and A. Loeb, *Phys. Rev.* **98,** 391 (1955); see also J. Kanamori, *J. Phys. Chem. Solids* **10,** 87 (1959).

covalent bonds are important. In this model each of the oxygen's two *p* electrons are pictured as being shared by an adjacent magnetic ion. This can come about if the energy of the 4*s* or 4*p* states of the magnetic ion are not much greater than that of the 3*d* state, or if the unfilled *d*, *s*, and *p* orbitals are hybridized. This type of indirect exchange is called semicovalent,[59] since only one electron is shared between the anion and each cation. Semicovalent exchange gives rise to an antiferromagnetic coupling between the cations because the magnetic moment of a particular cation tends to be parallel to that of the shared *p* electron and the two *p* electrons have antiparallel moments.

Finally, a mechanism called double exchange has also been discussed.[60] In this model two cations of different valency trade identities. For example, consider the system $Mn^{3+}O^{2-}Mn^{4+}$. If two electrons are transferred, one from the oxygen to the Mn^{4+} ion and the other from the Mn^{3+} to the oxygen ion, the configuration becomes $Mn^{4+}O^{2-}Mn^{3+}$. Double exchange is expected to occur only if the moments of the two cations are parallel; that is, the coupling is ferromagnetic. Moreover, the Hamiltonian cannot be expressed by equation 8-5.1.

6. More Advanced Theories of Antiferromagnetism

The molecular field theory of antiferromagnetism successfully predicts many of the main features of the experimental data. It is clear from Section 8-4, however, that there are discrepancies. It is not surprising that theories have been developed that are analogous to those in ferromagnetism. The theories that are briefly discussed here are (*a*) the series expansion method (Section 6-4), (*b*) the Bethe-Peierls-Weiss method (Section 6-5), and (*c*) spin waves (Section 6-6). These theories are valid over only a restricted temperature range. They are generally more difficult for antiferromagnetism than for ferromagnetism. Most calculations are restricted to a two-sublattice model in which only the AB interactions are considered.

The series expansion method. A problem is presented because in contrast to the ferromagnetic case the sublattice magnetizations are not constants of the motion. Nevertheless, a way of expressing the partition function has been found.[61] As before, the partition function is calculated by expansion into a power series in $1/T$; this requires the computation of terms that involve products of the spin operators. Also, the Néel temperature can be evaluated by setting $1/\chi = 0$. The values obtained for J_e/kT_N for different orders of approximation, n, are listed[62] in Table

[60] C. Zener, *Phys. Rev.* **82,** 403 (1951).
[61] R. Kubo, *Phys. Rev.* **87,** 568 (1952).
[62] R. Kubo, Y. Obata, and A. Ohno, *Busseiron-Kenkyu* **47,** 35, 46 (1952), **57,** 45 (1952).

8-6.1. The first-order approximation ($n = 1$) is identical to that obtained from the molecular field theory. The convergence of the series is poor compared to the ferromagnetic case. Thus for $n = 3$ there is no transition temperature, whereas for $n = 4$ the values of J_e/kT_N are not reasonable. An expansion in terms of the parameter $x = 4$ [tanh (J_e/kT)](2 − tanh J_e/kT) did not improve the convergence.[63]

Table 8-6.1. Values of J_e/kT_N for $S = \frac{1}{2}$ by the Series Expansion Method

Lattice	$n = 1$	$n = 2$	$n = 3$	$n = 4$
Linear chain	1.00	1.00	none	3.19
s. c.	0.333	0.333	none	1.53
b. c. c.	0.250	0.250	none	1.63

The Bethe-Peierls-Weiss method. It will be recalled that the Bethe-Peierls-Weiss method in ferromagnetism considers a cluster of $z + 1$ atoms; that is, a central atom and its z nearest neighbors. The interactions in the cluster are treated exactly by using the Heisenberg exchange Hamiltonian (equation 8-5.1). The effect of the atoms outside the cluster is represented by an effective field H_e that acts on the shell of z neighbor atoms of the cluster. In the antiferromagnetic case it is necessary to consider two clusters, one in which the central atom lies on an A site and one in which it lies on a B site. Also, the effective field H_e must then be replaced by two different effective fields, H_A and H_B, one of which may be assumed to be parallel and the other antiparallel to the applied field. The usual consistency condition requires that the average magnetic moment of the central atom on one sublattice be equal to that of a neighbor atom on the same sublattice. In this way two equations are obtained for H_A and H_B. The Néel temperature is recognized as the temperature below which H_A and H_B have nonzero values when the applied field is zero. The values of J_e/kT_N for different lattices calculated by Li[64] are listed in Table 8-6.2. It is to be noted that no transition temperature is predicted for the face-centered cubic lattice. Here only nearest neighbor interactions are considered. It is the next nearest neighbor interactions that produce the ordering observed experimentally in such materials as MnO. The values of J_e/kT_N for the simple and body-centered cubic lattice are slightly less than those obtained for the ferromagnetic case (Table 6-5.1).

[63] R. Kubo, Y. Obata, and A. Ohno, *Busseiron-Kenkyu* **43,** 22 (1951).
[64] Y. Y. Li, *Phys. Rev.* **84,** 721 (1951).

Above the Néel temperature the calculations yield a $1/\chi$ versus T curve that is convex. By extrapolating from high temperatures Li[64] finds $\theta = 1.5T_N$ and $\theta = 1.25T_N$ for the simple and body-centered cubic lattices, respectively. It will be recalled that the molecular field theory predicts $\theta = T_N$ when $N_{ii} = 0$.

Other calculations have been made that employ the Bethe-Peierls-Weiss method. In one computation[65] a slight modification in the Hamiltonian of the cluster leads to values of J_e/kT_N that are somewhat lower than those of Table 8-6.2. The constant coupling approximation, in which clusters of only two atoms are considered, yields results[66] that are almost the same as the tabulations.

Table 8-6.2. Values of J_e/kT_N Obtained for $S = \frac{1}{2}$ from the Bethe-Peierls-Weiss Method[a]

Lattice	J_e/kT_N
Linear chain	none
s.c.	0.492
b.c.c.	0.314
f.c.c.	none

[a] Y. Y. Li, *Phys. Rev.* **84**, 721 (1951).

Spin waves. Spin wave calculations for antiferromagnetism are analogous to those described in Section 6-6 for ferromagnetism. There are certain important differences in the results. For example, on the basis of equation 6-6.5, and also from the molecular field theory, it might be expected that the ground-state eigenfunction and energy for an antiferromagnet would be

$$\chi_0 = \chi_{\alpha 1}\chi_{\beta 2}\chi_{\alpha 3}\chi_{\beta 4}\chi_{\alpha 5}\cdots\chi_{\beta N} \tag{8-6.1}$$

and

$$E_0 = -Nz\,|J_e|\,S^2, \tag{8-6.2}$$

respectively. However, it may easily be shown that equation 8-6.1 is not an eigenfunction of the Hamiltonian of equation 8-5.1. The exact ground-state energy is not known, but it has been shown[67] that

$$-Nz\,|J_e|\,S^2\left(1 + \frac{1}{zS}\right) < E_0 < -Nz\,|J_e|\,S^2. \tag{8-6.3}$$

[65] T. Oguchi and Y. Obata, *Prog. Theor. Phys.* (*Kyoto*) **9**, 359 (1952).
[66] T. Nakamura, *Busseiron-Kenkyu* **63**, 12 (1953).
[67] P. W. Anderson, *Phys. Rev.* **83**, 1260 (1951).

Actually, the lower bound is not greatly different from the upper, so that the simple picture in which all the spins point up on one sublattice and down on the other at $T = 0$ is not far from being correct.

Detailed calculations,[68] either quantum mechanical or semiclassical, show that the dispersion relationship for an antiferromagnetic material is

$$\omega = \text{constant } k \tag{8-6.4}$$

where k is the wave number. This is quite a different result from that obtained for a ferromagnet, in which a quadratic dependence was found (equation 6-6.17). The antiferromagnetic case may be understood physically by considering the precession of moments of atoms on the two sublattices[69] (problem 8-6).

The susceptibility at low temperatures can be calculated in a manner similar to that employed in the calculation of the spontaneous magnetization of a ferromagnet (Section 6-6). Provided $\mathrm{k}T$ is not too small compared to the crystalline anisotropy energy, the parallel susceptibility[70] is given by

$$\chi_{\parallel} = \frac{Ng^2\mu_B^2\mathrm{k}^2T^2}{6z^3\,|J_e|^3\,S^3} \tag{8-6.5}$$

for a body-centered cubic lattice. Kubo[70] also finds

$$\chi_{\perp} = \frac{Ng^2\mu_B^2}{4z\,|J_e|}. \tag{8-6.6}$$

It is easy to show (problem 8-7) that this value of $\chi_\perp$ is identical to that given by the molecular field theory (equation 8-3.15).

7. Crystalline Anisotropy: Spin Flopping

The crystalline anisotropy energy for a uniaxial antiferromagnetic crystal may be written in analogy with the ferromagnetic case as

$$F_K = \tfrac{1}{2}K(\sin^2\theta_A + \sin^2\theta_B) \tag{8-7.1}$$

or as

$$F_K = -\tfrac{1}{2}K(\cos^2\theta_A + \cos^2\theta_B), \tag{8-7.2}$$

[68] H. A. Bethe, *Z. Physik* **71**, 205 (1931); G. Heller and H. A. Kramers, *Proc. Roy. Acad. Sci. (Amsterdam)* **37**, 378 (1934); L. Hulthen, *Proc. Roy. Acad. Sci. (Amsterdam)* **39**, 190 (1936); T. Holstein and H. Primakoff, *Phys. Rev.* **58**, 1098 (1940); J. M. Ziman, *Proc. Phys. Soc. (London)* **A-65**, 540, 548 (1952); T. Nakamura, *Prog. Theor. Phys. (Kyoto)* **7**, 539 (1952); J. R. Tessman, *Phys. Rev.* **88**, 1132 (1952); P. W. Kasteleijn, *Physica* **18**, 104 (1952).

[69] F. Keffer, H. Kaplan, and Y. Yafet, *Am. J. Phys.* **25**, 250 (1953).

[70] R. Kubo, *Phys. Rev.* **87**, 568 (1952).

where θ_A and θ_B are the angles between the sublattice magnetizations $\mathbf{M}_A$ and $\mathbf{M}_B$ and the easy direction; the subscript and superscript have been deleted from the coefficient K for convenience. Other crystal symmetries, such as cubic and orthorhombic, are also of interest but are not considered here. Many of the results for the uniaxial case apply for these crystal structures, provided the angles between the sublattice magnetizations and one of the easy axes are small.

It is useful to describe the action of the crystalline anisotropy on the sublattice magnetizations by the equivalent magnetic fields H_{KA} and H_{KB}, defined by the equation

$$\begin{aligned} F_K &= -H_{KA}M_A \cos\theta_A - H_{KB}M_B \cos\theta_B \\ &= -H_{KA}M_0 \cos\theta_A - H_{KB}M_0 \cos\theta_B, \end{aligned}$$

where it has been assumed that $M_A = M_B = M_0$, and M_0 is, as before, the sublattice magnetization for $H = 0$. For an arbitrary variation of the directions of the sublattice magnetizations, we have

$$\delta F_K = H_{KA}M_0 \sin\theta_A\,\delta\theta_A + H_{KB}M_0 \sin\theta_B\,\delta\theta_B$$

together with

$$\delta F_K = K(\sin\theta_A \cos\theta_A\,\delta\theta_A + \sin\theta_B \cos\theta_B\,\delta\theta_B).$$

It then follows that

$$H_{KA} = \frac{K}{M_0}\cos\theta_A \quad \text{and} \quad H_{KB} = \frac{K}{M_0}\cos\theta_B. \tag{8-7.3}$$

If θ_A and θ_B are small, the anisotropy fields reduce to K/M_0; this result is to be compared with those of equations 7-2.10 and 7-11.4 for ferromagnetism.

The calculation of the susceptibility when the crystalline anisotropy is explicitly introduced will now be considered. If the magnetic field $\mathbf{H}$ is applied along the easy direction, and provided it is not too large (see later), $\mathbf{M}_A$ and $\mathbf{M}_B$ will remain collinear with the easy direction. The anisotropy fields then produce no torque on the sublattice magnetizations, and equation 8-3.14 for $\chi_\parallel$ still holds.

When $\mathbf{H}$ is applied perpendicular to the easy direction, the anisotropy fields produce a restoring torque on the sublattice magnetizations (see Fig. 8-3.3). At equilibrium the torque acting on each sublattice magnetization must be zero; thus for $\mathbf{M}_A$ we have

$$|\mathbf{M}_A \times (\mathbf{H} + \mathbf{H}_{mA} + \mathbf{H}_{KA})| = 0$$

or

$$M_A H \cos\theta - N_{AB}M_A M_B \sin 2\theta - M_A \frac{K}{M_0}\sin\theta\cos\theta = 0,$$

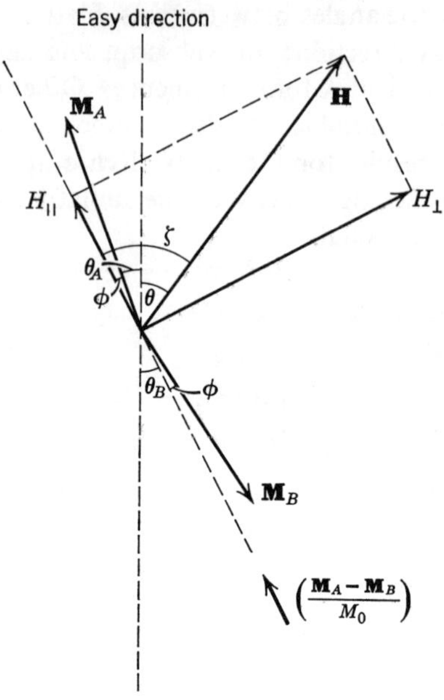

Fig. 8-7.1. The calculation of the susceptibility when **H** is applied at an angle θ to the easy direction.

where θ is the angle between M_A and the easy direction. Hence

$$2M_B \sin\theta = \frac{H}{N_{AB} + K/2M_0^2}$$

if it is assumed that only the direction and not the magnitude of M_A changes when H is applied. Finally, since the induced magnetic moment of the specimen is $2M_B \sin\theta$, we find

$$\chi_{\perp} = \frac{1}{N_{AB} + K/2M_0^2}. \tag{8-7.4}$$

If the anisotropy field is small compared to the molecular field, $\chi_{\perp} \approx 1/N_{AB}$ as before (equation 8-3.15).

The general case, in which **H** is applied at some angle, say θ, with the easy direction, is rather more complicated.[71] Then the sublattice magnetizations $\mathbf{M}_A$ and $\mathbf{M}_B$ will make angles θ_A and θ_B, respectively, with the easy direction, as indicated in Fig. 8-7.1. It is convenient to define a direction $(\mathbf{M}_A - \mathbf{M}_B)/M_0$ that makes an angle ζ with **H**. Actually, the direction

$$\frac{\mathbf{M}_A}{M_A} - \frac{\mathbf{M}_B}{M_B}$$

[71] T. Nagamiya, K. Yosida, and R. Kubo, *Advan. Phys.* **4**, 1 (1955).

should be considered, but since $|M_A| \approx |M_B|$ these two directions are almost coincident. The component of **H** parallel to $(\mathbf{M}_A - \mathbf{M}_B)/M_0$, $H_\parallel = H \cos \zeta$, causes M_A and M_B to change from M_0 to $M_0 + \frac{1}{2}\chi_\parallel H_\parallel$ and $M_0 - \frac{1}{2}\chi_\parallel H_\parallel$, respectively. The action of the perpendicular component, $H_\perp \sin \zeta$, rotates M_A and M_B through the angle ϕ. At equilibrium, the torque equations for the two sublattice magnetizations may then be written

$$(M_0 + \tfrac{1}{2}\chi_\parallel H_\parallel)H \sin(\zeta - \phi) - N_{AB}M_0^2 \sin 2\phi + K \cos\theta_A \sin\theta_A = 0$$

and (8-7.5)

$$(M_0 - \tfrac{1}{2}\chi_\parallel H_\parallel)H \sin(\zeta + \phi) - N_{AB}M_0^2 \sin 2\phi - K \cos\theta_B \sin\theta_B = 0, \qquad (8\text{-}7.6)$$

where it has been assumed that $M_0^2 \gg (\frac{1}{2}\chi_\parallel H_\parallel)^2$ and $M_0^2 \gg (\frac{1}{2}\chi_\perp H_\perp)^2$, which is justified, provided ϕ is small. On substituting for

$$\theta_A = (\zeta - \theta) - \phi \quad \text{and} \quad \theta_B = (\zeta - \theta) + \phi$$

and expansion of the trigometric functions, equations 8-7.5 and 8-7.6 become

$$\begin{aligned}(M_0 + \tfrac{1}{2}\chi_\parallel H_\parallel)(H_\perp \cos\phi - H_\parallel \sin\phi) - 2N_{AB}M_0^2 \sin\phi\cos\phi \\ + K[\cos(\zeta - \theta)\cos\phi + \sin(\zeta - \theta)\sin\phi] \\ \times [\sin(\zeta - \theta)\cos\phi - \cos(\zeta - \theta)\sin\phi] = 0 \qquad (8\text{-}7.7)\end{aligned}$$

and

$$\begin{aligned}(M_0 - \tfrac{1}{2}\chi_\parallel H_\parallel)(H_\parallel \sin\phi + H_\perp \cos\phi) - 2N_{AB}M_0^2 \sin\phi\cos\phi \\ - K[\cos(\zeta - \theta)\cos\phi - \sin(\zeta - \theta)\sin\phi] \\ \times [\sin(\zeta - \theta)\cos\phi + \cos(\zeta - \theta)\sin\phi] = 0. \qquad (8\text{-}7.8)\end{aligned}$$

Addition of these two equations then gives

$$\chi_\perp = \frac{1}{N_{AB} + (K/2M_0^2)\cos 2(\zeta - \theta)}, \qquad (8\text{-}7.9)$$

since $\chi_\parallel\chi_\perp H_\parallel^2/4 \ll M_0^2$ and $\chi_\perp = (2M_0 \sin\phi)/H$. Similarly, subtraction of these two equations yields

$$(\chi_\perp - \chi_\parallel)H^2 \sin\zeta\cos\zeta = K \sin 2(\zeta - \theta). \qquad (8\text{-}7.10)$$

By considering the components of the induced magnetization parallel and perpendicular to $(\mathbf{M}_A - \mathbf{M}_B)/M_0$, it can be shown (problem 8-8) the susceptibility of a powder is given by[72]

$$\chi_p = \tfrac{1}{3}\chi_\parallel + \tfrac{2}{3}\chi_\perp + \frac{2}{15K}(\chi_\perp - \chi_\parallel)^2 H^2. \qquad (8\text{-}7.11)$$

[72] T. Nagamiya, *Prog. Theor. Phys.* (*Kyoto*) **4,** 342 (1951); K. Yosida, *Prog. Theor. Phys.* (*Kyoto*) **6,** 691 (1951).

The susceptibility of powders below the Néel temperature is often found to depend on the strength of the applied field; this is illustrated by the data for MnO given in Fig. 8-4.2. The application of equation 8-7.11 leads to a value of $K = 6 \times 10^4$ ergs/cm³ at $T = 0$°K for MnO. From data[73] for MnF_2 the value $K = 6 \times 10^5$ ergs/cm³ is obtained. These anisotropy constants are of the same order of magnitude as those found for ferromagnetic materials. It should be mentioned that it has not been established that crystalline anisotropy is the only origin of the field

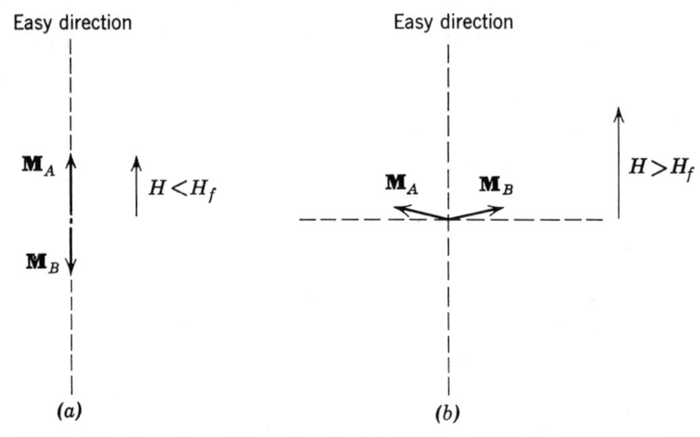

Fig. 8-7.2. Spin flopping. Note that in (*a*) $|M_A| > |M_B|$, in (*b*) $|M_A| = |M_B|$.

dependence of the susceptibility. For example Néel[74] has suggested that domain structure may be involved.

Earlier in this section the susceptibility when the magnetic field is applied parallel to the easy direction was discussed. It was stated that the analysis was valid only if the field was not too large; this point will now be examined. If, on application of H, the sublattice magnetizations remain collinear with the easy direction, as indicated in Fig. 8-7.2*a*, the change in free energy per unit volume is, according to equation 1-7.8, $\Delta F_T = -\chi_{\|}(H^2/2)$. Suppose, however, the sublattice magnetizations [actually $(\mathbf{M}_A - \mathbf{M}_B)/M_0$] lie perpendicular to the easy direction (Fig. 8-7.2*b*). If $K \approx 0$, this change in free energy is $\Delta F_T = -\chi_{\perp}(H^2/2)$. Now, for an antiferromagnetic material $\chi_{\perp} > \chi_{\|}$ for $T < T_N$. Therefore, under these conditions, the perpendicular orientation would be the ground state. For $K = 0$, $\Delta F_T = K - \chi_{\perp}(H^2/2)$ for the perpendicular orientation; the parallel orientation is then the state of lowest energy provided H is not

[73] H. Bizette and B. Tsai, *Compt. rend.* (*Paris*) **209,** 205 (1939).
[74] L. Néel, *Proc. Intl. Conf. on Theor. Phys.* (*Kyoto*) 701 (1953).

too large. At a certain critical field H_f, defined by the equation[75]

$$-\tfrac{1}{2}\chi_{\|}H_f^2 = K - \tfrac{1}{2}\chi_{\perp}H^2$$

or

$$H_f = \left[\frac{2K}{(\chi_{\perp} - \chi_{\|})}\right]^{1/2}, \qquad (8\text{-}7.12)$$

the energy of the two states are equal. Hence for $H < H_f$ the sublattice magnetizations will be parallel to the easy direction, whereas for $H > H_f$ they will be perpendicular. The change from the parallel to the perpendicular orientation is usually called *spin flopping*.

Spin flopping can be produced most easily if the Néel temperature is low, for then H_f is relatively small (problem 8-8). Indeed, this phenomena was first observed[76] in $CuCl_2{\cdot}2H_2O$ ($T_N = 4.3°K$) with a steady applied field. Since then spin flopping has been observed with large pulsed fields in other materials; for example, at $T = 4°K$, $H_f = 93$ koe in MnF_2[77] and $H_f = 59$ koe in Cr_2O_3.[78] The measurement of H_f, of course, permits K to be evaluated by equation 8-8.12. Anisotropy constants can also be determined from susceptibility measurements above the Néel temperature,[79] and from antiferromagnetic resonance experiments (Chapter 10).

Microscopic theories of the origin of the crystalline anisotropy in antiferromagnetic materials have been based on the theories developed for paramagnetic and ferromagnetic substances. Three sources have been considered: (*a*) the crystalline field action on a single ion, (*b*) the anisotropic exchange interaction, and (*c*) the dipole-dipole interaction.

As discussed for paramagnetic salts (Sections 2-10 and 3-9), the action of the crystalline field on a single atom gives rise to anisotropy. The crystalline field quenches the angular momentum of the atom. The spin-orbit coupling, considered as a perturbation, then leads to an anisotropic energy term in the spin-Hamiltonian (equations 3-9.2, 3-9.5, and 3-10.1).

Anisotropic exchange interaction occurs when, in addition to the direct or indirect exchange interaction, spin-orbit coupling is considered. The pseudo-dipole and the quadrupole interaction energies of equations 6-9.7 and 6-9.8, respectively, are then obtained. In antiferromagnetic compounds the spin-orbit coupling of the anions is also involved, since $L \neq 0$ for the excited states that give rise to superexchange.

[75] L. Néel, *Ann. phys.* (*Paris*) **5**, 232 (1936).

[76] N. J. Poulis, J. van den Handel, J. Ubbink, J. A. Poulis, and C. J. Gorter, *Phys. Rev.* **82**, 552 (1951); C. J. Gorter, *Revs. Mod. Phys.* **25**, 277, 332 (1953).

[77] I. S. Jacobs, *J. Appl. Phys.* **32**, 61S (1961).

[78] S. Foner and Shou-Ling Hou, *J. Appl. Phys.* **33**, 1289 (1962).

[79] K. Yosida, *Prog. Theor. Phys.* (*Kyoto*) **6**, 691 (1951); F. Keffer, *Phys. Rev.* **87**, 608 (1952).

Finally, the anisotropic dipole-dipole interaction is given by

$$\frac{\boldsymbol{\mu}_i \cdot \boldsymbol{\mu}_j}{r_{ij}^{\ 3}} - 3\frac{(\boldsymbol{\mu}_i \cdot \mathbf{r}_{ij})(\boldsymbol{\mu}_j \cdot \mathbf{r}_{ij})}{r_{ij}^{\ 5}}$$

(see equations 1-3.3 and 3-11.2), where $\boldsymbol{\mu}_i = \mu_B g \mathbf{S}_i$ and $\boldsymbol{\mu}_j = \mu_B g \mathbf{S}_j$ if g is isotropic. The expression is somewhat more complicated if the g-values are anisotropic.

Calculations of the various contributions to the anisotropy have been made for several compounds. For MnF_2 Keffer[80] concludes that the dipole-dipole energy is the most important. For MnO the dipole-dipole interaction apparently[81] produces a large anisotropy which constrains the atomic moments to lie in (111) planes. Within these planes there is a weaker anisotropy that determines the easy direction.

8. Metals and Alloys

Some metals of transition elements have been observed to possess antiferromagnetism. Their antiferromagnetic properties, however, generally differ considerably from those observed for compounds. It is therefore appropriate to devote a separate section to the discussion of the transition metals.

As a result of the direct exchange interaction that occurs in the Heisenberg model, that is, for localized magnetic electrons, it is reasonable to expect that all the 3*d* transition metals other than the ferromagnetic ones, Fe, Co, and Ni, will exhibit antiferromagnetism. In fact only Cr and α-Mn have been found to be antiferromagnetic.[82] For chromium, which is body-centered cubic, neutron diffraction experiments on single crystals indicate[83] that the Néel temperature is about 308°K; curiously, it has been reported that $T_N = 475°\text{K}$ for powdered specimens.[82,83] The sublattice magnetizations, as determined from the intensity of the (100) line at absolute zero, are only $0.4\mu_B$ per atom. The antiferromagnetism of α-Mn ($T_N = 100°\text{K}$) is also very weak,[82] being about $0.5\mu_B$ per atom. Neutron diffraction experiments on vanadium[82] show no measurable antiferromagnetism. Although apparently scandium and titanium have not been studied by neutron diffraction, there is no indication from other measurements that they are antiferromagnetic.

[80] F. Keffer, *Phys. Rev.* **87,** 608 (1952).

[81] J. Kaplan, *J. Chem. Phys.* **22,** 1709 (1954); F. Keffer and W. O'Sullivan, *Phys. Rev.* **108,** 637 (1957).

[82] C. G. Shull and M. K. Wilkinson, *Revs. Mod. Phys.* **25,** 100 (1953).

[83] L. M. Corliss, J. M. Hastings, and R. J. Weiss, *Phys. Rev. Letters* **3,** 211 (1959).

It is difficult to account for these results on the basis of the Heisenberg or Zener-Vonsovsky models, although such an attempt has been made.[84] However, calculations based on the collective electron theory[85] (Section 6-7) are in reasonable agreement with the experimental data. Small values of the magnetic moment per atom can arise in this theory if the Fermi energy is comparable to the exchange energy; then $\zeta(0)$ $(= M(0)/N\mu_B)$ can be small (equation 6-7.5) or even zero. Under these circumstances the collective electron theory predicts that the discontinuity in the specific heat at the Néel temperature would be too small to be observed; this is in agreement with experiment.[86]

Neutron diffraction experiments[87] also have shown that antiferromagnetism is absent in Nb($4d^4\,5s$), Mo($4d^5\,5s$), and W($5d^4\,6s^2$). The magnetic properties of the nonantiferromagnetic (and ferromagnetic) transition metals appear to be satisfactorily explained by the theory of the electron gas (equation 5-2.2) provided a large value of the effective mass is assumed (Table 5-4.3).

Only a few neutron diffraction investigations of rare earth metals with $4f$ shells less than half filled have been made. It has been determined[88] that the hexagonal close-packed form of cerium ($4f\,5d\,6s^2$) becomes antiferromagnetic below 12.4°K. Anomalous behavior in the susceptibility[89] and specific heat[90] had previously been observed in cerium. Neodymium ($4f^3\,5d\,6s^2$) is antiferromagnetic[91] at 4.2°K, and has a complex structure. Europium ($4f^7\,6s^2$), with $T_N = 87$°K, has a helical spin structure[92] (see later).

The rare earth metals with $4f$ shells more than half filled possess rather peculiar magnetic properties. They are ferromagnetic at low temperatures and antiferromagnetic at intermediate temperatures, $T_f < T < T_N$; their Curie and Néel temperatures, as determined by susceptibility[93] and

[84] C. Zener and R. R. Heikes, *Revs. Mod. Phys.* **25,** 191 (1953).

[85] A. B. Lidiard, *Proc. Phys. Soc.* (*London*) **66,** 1188 (1953).

[86] L. D. Armstrong and H. Grayson-Smith, *Can. J. Phys.* **A-28,** 51 (1950).

[87] C. G. Shull and M. K. Wilkinson, *Revs. Mod. Phys.* **25,** 100 (1953).

[88] M. K. Wilkinson, H. R. Child, C. J. McHague, W. C. Koehler, and E. O. Wollan, *Phys. Rev.* **122,** 1409 (1961).

[89] J. M. Lock, *Proc. Phys. Soc.* (*London*) **B-70,** 566 (1957).

[90] D. H. Parkinson and L. M. Roberts, *Proc. Phys. Soc.* (*London*) **B-70,** 471 (1957).

[91] M. K. Wilkinson, H. R. Child, W. C. Koehler, J. W. Cable, and E. O. Wollan, *J. Phys. Soc. Japan, Suppl. B-III* **17,** 27 (1962).

[92] C. E. Olsen, N. G. Nereson, and G. P. Arnold, *J. Appl. Phys.* **33,** 1135 (1962).

[93] D. R. Behrendt, S. Legvold, and F. H. Spedding, *Phys. Rev.* **109,** 1544 (1958); B. D. Rhodes, S. Legvold, and F. H. Spedding, *Phys. Rev.* **109,** 1547 (1958); J. F. Elliott, S. Legvold, and F. H. Spedding, *Phys. Rev.* **100,** 1595 (1955); R. W. Green, S. Legvold, and F. H. Spedding, *Phys. Rev.* **122,** 122 (1961); D. D. Davis and R. M. Bozorth, *Phys. Rev.* **118,** 1543 (1960).

neutron diffraction[94] experiments, are listed in Table 8-8.1. All these metals have the hexagonal close-packed structure. In the antiferromagnetic state the moments are aligned ferromagnetically within the hexagonal layers. In Tb, Dy, and Ho the moments in these layers lie perpendicular to the hexagonal c-axis and are arranged in a helical or screw structure about this axis. In Er and Tm the moments lie parallel to the c-axis and have an amplitude that varies sinusoidally. The rare earth metals have an appreciable magnetic moment per atom.

Table 8-8.1. Data for the Heavy Rare Earth Metals

Metal	Electron Configuration	T_f(°K)	T_N(°K)	Spin Arrangement
Tb	$4f^8\,5d\,6s^2$	218	230	Helical
Dy	$4f^9\,5d\,6s^2$	87	179	Helical
Ho	$4f^{10}\,5d\,6s^2$	20	133	Helical
Er	$4f^{11}\,5d\,6s^2$	20	80	Sinusoidal
Tm	$4f^{13}\,6s^2$	22	56	Sinusoidal

A helical spin arrangement was actually first predicted theoretically by the following classical argument. Consider a set of ferromagnetic sheets and suppose that neighboring sheets are coupled by the exchange integral[95] J_1, next nearest layers by J_2, etc. The coupling of the spins within the sheets will be designated by J_0. The exchange energy then has the form

$$F_e = -2S^2 \sum [J_0 + 2J_1 \cos(\theta_1 - \theta_0) + 2J_2 \cos(\theta_2 - \theta_0) + \cdots],$$

where θ_0, θ_1, θ_2, . . . are the angles made by the moments in each layer with respect to some arbitrary axis lying in the layer; the summation is over all layers. If $\theta_n = n\phi$, this equation becomes

$$F_e = -2S^2 N[J_0 + 2J_1 \cos\phi + 2J_2 \cos 2\phi + \cdots],$$

where N is the number of layers. This energy may be a minimum for ϕ equal to some angle other than 0 or π; if so a helical structure will be the

[94] M. K. Wilkinson, W. C. Koehler, E. O. Wollan, and J. W. Cable, *J. Appl. Phys.* **32,** 48S (1961); W. C. Koehler, *J. Appl. Phys.* **32,** 20S (1961); J. W. Cable, E. O. Wollan, W. C. Koehler, and M. K. Wilkinson, *J. Appl. Phys.* **32,** 49S (1961); W. C. Koehler, J. W. Cable, E. O. Wollan, and M. K. Wilkinson, *J. Phys. Soc. Japan, Suppl. B-III* **17,** 32 (1962); W. C. Koehler, J. W. Cable, E. O. Wollan, and M. K. Wilkinson, *J. Appl. Phys.* **33,** 1124 (1962).

[95] The additional subscript e is unnecessary here and is omitted.

stable state of the system.[96] The presence of crystalline anisotropy can modify this structure into one with a sinusoidal spin arrangement.[97]

Studies of alloys of the transition metals are clearly of interest. The antiferromagnetism of alloys of α-Mn with Fe, N, and H, for example, has been investigated.[98] Most of the neutron diffraction experiments to date, however, have been concerned with those alloys that are intermetallic compounds. Examples include CrSb,[99] which has the NiAs structure, $MnAu_2$, which has a helical spin arrangement,[100] and FeRh, which is antiferromagnetic at low temperatures[101] and becomes ferromagnetic above 333°K. Several rare earth intermetallic compounds with the MnO structure, including TbAs, TbSb, ErP, and ErSb, are antiferromagnetic with Néel temperatures in the vicinity of 10° or lower.[102]

9. Canted Spin Arrangements

Some antiferromagnetic materials also possess weak ferromagnetism. The historical example is α-Fe_2O_3, which has the rhombohedral crystal structure shown in Fig. 8-9.1*a*. According to neutron diffraction experiments,[103] from 945°K (the Néel temperature) to 263°K the magnetic moments lie in the (111) plane (see Fig. 8-9.1*a*), whereas below 263°K they lie close to the [111]-(trigonal)-direction. The weak ferromagnetism is observed for temperatures in the range $263° < T < 945°K$. The experimental situation below 250°K is not clear, but it appears that in this temperature region there may be no weak ferromagnetism in a pure specimen.[104]

[96] A. Yoshimori, *J. Phys. Soc. Japan* **14,** 807 (1959); J. Villian, *J. Phys. Chem. Solids* **11,** 303 (1959); T. A. Kaplan, *Phys. Rev.* **116,** 888 (1959).

[97] K. Yosida and H. Miwa, *J. Phys. Soc. Japan, Suppl. B-I* **17,** 5 (1962); R. J. Elliott, *Phys. Rev.* **124,** 346 (1961); T. A. Kaplan, *Phys. Rev.* **124,** 329 (1961); T. Nagamiya, *J. Appl. Phys.* **33,** 1029 (1962).

[98] A. Arrott and B. R. Coles, *J. Appl. Phys.* **32,** 51S (1961).

[99] A. I. Snow, *Phys. Rev.* **85,** 365 (1952).

[100] A. Herpin, P. Mériel, and J. Villain, *Compt. rend. (Paris)* **249,** 1334 (1959); A. Herpin and P. Mériel, *Compt. rend. (Paris)* **250,** 1450 (1960).

[101] F. de Bergevin and L. Muldawer, *Compt. rend. (Paris)* **252,** 1347 (1961); E. F. Bertaut, A. Delapalme, F. Forrat, G. Roult, E. de Bergevin, and R. Pauthenet, *J. Appl. Phys.* **33,** 1123 (1962).

[102] M. K. Wilkinson, H. R. Child, W. C. Koehler, J. W. Cable, and E. O. Wollan, *J. Phys. Soc. Japan, Suppl. B-III* **17,** 22 (1962).

[103] C. G. Shull, W. A. Strauser, and E. O. Wollan, *Phys. Rev.* **83,** 333 (1951); A. H. Morrish, G. B. Johnston, and N. A. Curry, *Phys. Letters* **2,** 177 (1963).

[104] L. Néel and R. Pauthenet, *Compt. rend. (Paris)* **234,** 2172 (1952); S. T. Lin, *Phys. Rev.* **116,** 1447 (1959); S. T. Lin, *J. Appl. Phys.* **31,** 273S (1960), **32,** 394S (1961); A. Tasaki and S. Iida, *J. Phys. Soc. Japan* **18,** 1148 (1963).

The origin of the weak ferromagnetism has been a subject of much speculation. Néel has suggested that magnetite impurities[105] or imperfections[106] may give rise to the spontaneous magnetization. Li[107] has suggested that the uncompensated moments in the walls that separate antiferromagnetic domains may produce the effect. The formation of domains probably requires the presence of imperfections (Section 8-10). Experimentally, however, it is found that the weak ferromagnetism is present even in very pure crystals. Also, not all antiferromagnetic

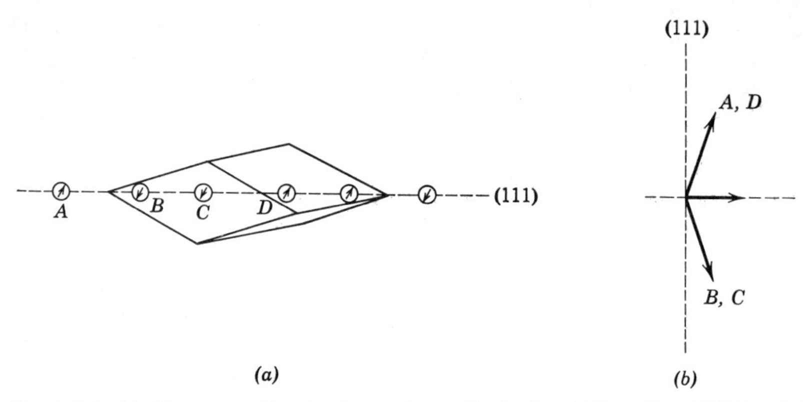

Fig. 8-9.1. (*a*) The magnetic structure of $\alpha - Fe_2O_3$ for $945° > T > 263°K$. Only the Fe^{3+} ions in the rhombohedral unit cell are shown. (*b*) Weak ferromagnetism arising from canting of the atomic magnetic moments.

materials exhibit this phenomena. These results appear to rule out the foregoing theories.

Quite another mechanism has been proposed by Dzialoshinsky.[108] A tilting of the magnetic moments toward one another, as indicated in Fig. 8-9.1, will result in a net magnetic moment. However, a canted spin arrangement is possible only if the magnetic crystal symmetry is the same as that when the spins are antiparallel. For α-Fe_2O_3 it can be shown[108] that this condition is satisfied when the moments lie in the basal (111) plane, and is not satisfied if the moments lie along the ternary [111]-axis. Then, by considering a general expression for the free energy in terms of spin variables, it turns out[108] that the term

$$\mathcal{H}_{\mathfrak{D}} = \mathfrak{D} \cdot (\mathbf{S}_i \times \mathbf{S}_j) \tag{8-9.1}$$

is allowed under the crystal symmetry. This term, which is proportional

[105] L. Néel, *Ann. phys. (Paris)* **4,** 249 (1949).

[106] L. Néel, *Revs. Mod. Phys.* **25,** 58 (1953).

[107] Y. Y. Li, *Phys. Rev.* **101,** 1450 (1956).

[108] I. Dzialoshinski, *J. Phys. Chem. Solids* **4,** 241 (1958), *J. Exptl. Theoret. Phys. (USSR)* **33,** 1454 (1959) [trans. *Soviet Phys.-JEPT* **6,** 1120 (1958)].

to sin θ, or θ if $\theta \to \pi$, where θ is the angle between $\mathbf{S}_i$ and $\mathbf{S}_j$, is responsible for the canting. For α-Fe_2O_3 the vector $\mathfrak{D}$ lies along the [111]-direction.

This theory is purely phenomenological, so that the question of the microscopic origin of the canting of the spins remains to be considered. Two mechanisms that can produce canted spins have been found. The first requires the presence in the crystal of two nonequivalent sites for the magnetic ions. The crystalline field acting on the ions in the two sites is then different. This, together with the spin-orbit coupling leads to non-parallel easy directions of crystalline anisotropy for the nonequivalent ions. This mechanism[109] is believed to be responsible for the weak ferromagnetism of NiF_2.

In the second mechanism the combined effects of the spin-orbit and the superexchange interaction are considered. For crystals with low symmetries calculation[110] shows that a term with the antisymmetric form of equation 8-9.1 can arise. This antisymmetric superexchange interaction appears to be the origin of the weak ferromagnetism of α-Fe_2O_3 and of some of the orthoferrites.[111] It is likely to be the more important of the two mechanisms for antiferromagnetic materials with high Néel temperatures, for then the exchange interaction is large.

Other antiferromagnetic materials that exhibit a weak ferromagnetism include $MnCO_3$,[112] $CoCO_3$,[112] $Mn(C_2H_3O_2)_2 \cdot 4H_2O$,[113] $KMnF_3$,[114] and CrF_3.[115]

10. Domains in Antiferromagnetic Materials

The assumption that domains are present in antiferromagnetic materials was made in several papers that appeared in the literature many years ago. The basis for this assumption is often not clear. Obviously, domains are not to be expected for the same reason as for ferromagnetic materials, that is, to reduce the magnetostatic energy (Section 7-1). The presence of domains would, however, increase the entropy because of the additional disorder, and hence lower the free energy. Consider an antiferromagnetic material as it cools below its Néel temperature. It seems reasonable to expect that at T_N the antiferromagnetic ordering nucleated in various regions of the crystal will vary in direction. There will then be transition regions, or walls, separating the domains. If the theory developed in

[109] T. Moriya, *Phys. Rev.* **117,** 635 (1960).

[110] T. Moriya, *Phys. Rev.* **120,** 91 (1960); F. Keffer, *Phys. Rev.* **126,** 896 (1962).

[111] D. Treves, *Phys. Rev.* **125,** 1843 (1962).

[112] A. S. Borovik-Romanov and M. P. Orlova, *J. Exptl. Theoret. Phys.* (*USSR*) **31,** 579 (1956).

[113] R. B. Flippen and S. A. Friedberg, *Phys. Rev.* **121,** 1591 (1961).

[114] A. J. Heeger, O. Beckman, and A. M. Portis, *Phys. Rev.* **123,** 1652 (1961).

[115] W. N. Hansen and M. Griffel, *J. Chem. Phys.* **30,** 913 (1959).

Section 7-5 for domain walls in ferromagnetic materials applies here, it follows that the wall thickness and energy depend only on the exchange and crystalline anisotropy energies (provided the magnetostrictive energy F_σ is zero). Since the latter energies are of the same order of magnitude in antiferromagnetic as in ferromagnetic materials, the wall thicknesses and energies will also be about the same. On purely energetic grounds, however, it would appear to be highly unlikely that domain walls in antiferromagnetic materials would be about 500 Å thick and have energy densities of about 1 erg per cm^2. Instead, it is more likely that these walls have a thickness of just a few lattice spacings and perhaps of only one. Even then the presence of imperfections, such as vacancies or dislocations, is probably required to stabilize the wall.

Recently, domains in antiferromagnetic specimens have been observed directly[116,117] with neutron diffraction and polarized light techniques. Single crystals of NiO, which has the magnetic structure of MnO (Fig. 8-2.9), have been studied in some detail.[117] It will be recalled that in NiO the magnetic moments are arranged in ferromagnetic sheets that lie in (111) planes. Further, on cooling below the Néel temperature, a contraction occurs along the [111]-direction. There are, however, four equivalent [111]-directions. The result is that different regions of the crystal have different contraction axes and the crystal becomes twinned. The optic axis in these regions differs, and hence the domains can be observed with polarized light. The walls between these twinned regions can be made to move by applying a suitable stress. Another type of domain wall which arises from a rotation of the moments within the (111) sheets has also been observed. This wall can be displaced by applying a magnetic field.

Néel[118] has predicted that small particles of antiferromagnetic materials will possess some novel properties. The idea is the following. Consider a two-sublattice antiferromagnet composed of ferromagnetic sheets whose moments lie alternatively parallel and antiparallel to a certain direction; an example is MnO. When the total number of sheets is even, there will be complete compensation of the moments. However, when the number is odd, there will be one plane of uncompensated moments and therefore a small net magnetization. Although in a large crystal the effect is negligible, in a small particle an appreciable percentage of the total number of atomic moments may be uncompensated. A weak ferromagnetism that increases as the particle size is decreased should then be

[116] M. K. Wilkinson, *J. Appl. Phys.* **30,** 278S (1959).

[117] W. L. Roth, *J. Appl. Phys.* **31,** 2000 (1960); W. L. Roth and G. A. Slack, *J. Appl. Phys.* **31,** 352S (1960).

[118] L. Néel, *Compt. rend.* (*Paris*) **252,** 4075 (1961), **253,** 9, 203 (1961).

observed. Also, small enough particles should exhibit a superparamagnetism analogous to that found in ferromagnetic particles (Section 7-3). Both effects have been observed experimentally in fine particles of NiO.[119]

11. Interfacial Exchange Anisotropy

When an antiferromagnetic and ferromagnetic phase coexist in a single specimen, unusual effects can arise. For example, as first observed in the Co-CoO system[120] the hysteresis loop of the ferromagnetic material may

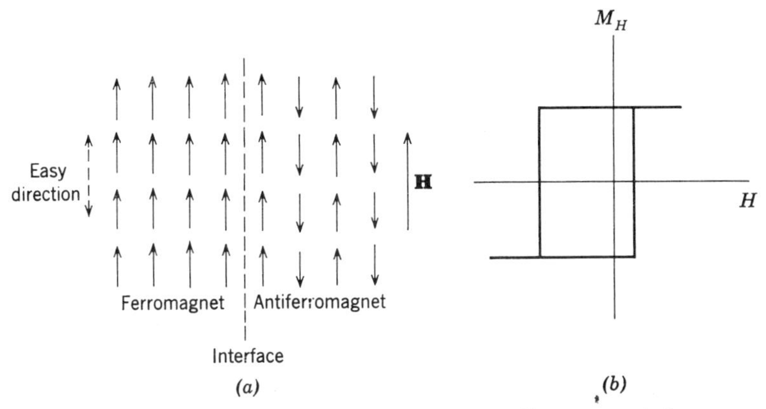

Fig. 8-11.1. (*a*) The atomic dipole arrangement of an antiferromagnetic-ferromagnetic system when cooled in a magnetic field applied along the easy direction of the crystalline anisotropy. (*b*) The shifted hysteresis loop that results from the interfacial antiferromagnetic-ferromagnetic exchange interaction.

be shifted to the left along the H-axis (Fig. 8-11.1*b*). This and other related phenomena are believed to be the result of the exchange interaction that acts across the interface between the antiferromagnetic and ferromagnetic materials. A simple model will serve to illustrate the principles involved. Consider a single domain of antiferromagnetic material and a contiguous single domain of ferromagnetic material with a plane interface (Fig. 8-11.1*a*). Suppose that the two materials have collinear uniaxial anisotropies and further that the Curie temperature T_f of the ferromagnet is larger than the Néel temperature T_N of the antiferromagnet. If a sizable magnetic field is applied along the easy direction at a temperature $T_N < T < T_f$, the ferromagnet's magnetization will become aligned with the field. Then, if the specimen is cooled through the Néel temperature, the antiferromagnet will become ordered with the dipoles in the first sheet

[119] W. J. Schuele and V. D. Deetscreek, *J. Appl. Phys.* **33,** 1136 (1962).
[120] W. H. Meiklejohn and C. P. Bean, *Phys. Rev.* **102,** 1413 (1956), **105,** 904 (1957).

lying parallel to those in the adjacent sheet of the ferromagnet (Fig. 8-11.1*a*). Consequently, on cooling to room temperature and removing the applied field, the antiferromagnet, provided it has a large crystalline anisotropy, will tend to hold the magnetization of the ferromagnet along the direction of the previously applied field. When a reverse field is applied, the rotation of the ferromagnet's magnetization is opposed by the exchange interaction at the interface of the two phases and the result is a shifted hysteresis loop with a coercive force in the forward direction that may be considerably larger than that in the reverse direction (Fig. 8-11.1*b*). Such two-phase single-domain particles obviously have great potentialities as permanent magnet materials.

If the antiferromagnetic material has a small crystalline anisotropy, its dipoles will rotate with the ferromagnet's magnetization when a reverse field is applied. The ferromagnet's hysteresis loop will then not be shifted. It can be shown,[121] however, that the material will possess a nonzero rotational hysteresis in large magnetic fields. NiO-Ni, which may be present in most Permalloy films, is believed to be an example of such a system.

Other antiferromagnetic-ferromagnetic systems in which the effects of interfacial exchange anisotropy have been observed include[121] $(Ni, Fe)_3Mn$ and other alloys containing Mn, FeO-Fe, FeS-Fe, Fe-Al, and non-stoichiometric $LaMnO_3$. Interfacial exchange interactions have also been found to be present in antiferromagnetic-ferrimagnetic and ferrimagnetic-ferromagnetic systems.[121]

Problems

8-1. Show that for a face-centered cubic lattice of identical atoms reflections from planes with indices (*hkl*) are not observed if some of the indices are even and some are odd. Interpret the neutron diffraction pattern observed for MnO (Fig. 8-2.8) below the Néel temperature. The wavelength of the incident neutrons is $\lambda = 1.06$ Å and the lattice spacing of the chemical unit cell is $a = 4.43$ Å.

8-2. Develop the molecular field theory of antiferromagnetism assuming only an *AB* interaction. Show that the equations in the text with $N_{ii} = 0$ reduce to these results.

8-3. Evaluate equation 8-3.14 for the susceptibility for a magnetic field applied parallel to the easy direction (a) at $T = 0$°K, (b) for $T > T_N$, (c) at $T = T_N$, (d) for $S = \frac{1}{2}$, and (e) for $S = \infty$ (the classical limit).

8-4. Consider the face-centered cubic lattice as composed of the four sublattices shown in Fig. 8-3.5 and derive equation 8-3.22. Calculate the Néel temperature for the first kind of ordering and for the improved first kind of

[121] For a review see W. H. Meiklejohn, *J. Appl. Phys.* **33**, 1328 (1962).

ordering illustrated in Figs. 8-3.5*a*, and *b*. Deduce the conditions that must be fulfilled for the occurrence of these types of antiferromagnetic arrangement.

8-5. Derive equations 8-3.26 and 8-3.27. Also, establish that the discontinuity in the specific heat at $T = T_N$ is given by equation 8-3.28.

8-6. Consider a linear chain consisting of two sublattices, with odd-numbered atoms belonging to one sublattice A and even-numbered atoms belonging to the other sublattice B. Suppose the linear chain is antiferromagnetic and that the temperature is close to absolute zero. For a normal mode, the spins will all precess in the same manner, that is, clockwise or anticlockwise, as viewed from, say, above (see Fig. 6-6.1). The radius of the precession circles for the spins pointing upward will, however, be different from that for the spins pointing downward. Show, from the torque equations, that the dispersion relationship is given by $\omega \propto k$ (equation 8-6.4). [*Hint*. See F. Keffer, H. Kaplan, and Y. Yafet, *Am. J. Phys*. **25**, 250 (1953).]

8-7. The perpendicular susceptibility is given by

$$\chi_\perp = \frac{1}{N_{AB}} \qquad \text{(equation 8-3.15)}$$

according to the molecular field theory, and by

$$\chi_\perp = \frac{Ng^2\mu_B{}^2}{4z\,|J_e|} \qquad \text{(equation 8-6.6)}$$

according to the spin-wave theory. Establish the equivalence of these two expressions.

8-8. (a) Show that when the crystalline anisotropy energy is explicitly considered the susceptibility of a powder is given by equation 8-7.11.

(b) Start from equation 8-7.10 and show that the critical field is given by

$$H_f = \left(\frac{2K}{\chi_\perp - \chi_\parallel}\right)^{1/2} \qquad \text{(equation 8-7.12).}$$

(c) Show that the critical field may also be expressed by

$$H_f = \left[\frac{(2H_m + H_K)H_K}{(1 - \alpha)}\right]^{1/2},$$

where $\alpha = \chi_\parallel/\chi_\perp$ and H_m is the molecular field. Hence show

$$H_f \approx (2H_mH_K)^{1/2}.$$

Bibliography

Bacon, G. E., *Neutron Diffraction*, 2nd ed. Oxford University Press, Oxford (1962).

Goodenough, J. B., *Magnetism and the Chemical Bond*, John Wiley and Sons, New York (1963).

Lidiard, A. B., "Antiferromagnetism," *Rept. Prog. Phys.* **17**, 201 (1954).

Nagamiya, T., K. Yosida, and R. Kubo, "Antiferromagnetism," *Advan. Phys.* **4**, 1 (1955).

9

Ferrimagnetism

1. Introduction

A ferrimagnetic material may be defined as one which below a certain temperature possesses a spontaneous magnetization that arises from a nonparallel arrangement of the strongly coupled atomic dipoles. The example originally considered[1] was that of a substance composed of two sublattices with the magnetic moments of one sublattice tending to be antiparallel to those of the other. When the sublattice magnetizations are not equal there will be a net magnetic moment. Just as for antiferromagnetism, the concept of ferrimagnetism has been broadened to include materials with more than two sublattices and with other spin configurations, such as triangular or spiral arrangements. There is then the question of the dividing point between ferrimagnetic materials and those antiferromagnetic materials with a small magnetic moment. It is usually assumed that a ferrimagnetic specimen has an appreciable net magnetization, although no precise definition of the term appreciable has been given.

Even for a two sublattice system, there are several schemes that can lead to ferrimagnetism. As for the antiferromagnetic case, the two sublattices are denoted by A and B. Suppose first that all the magnetic ions, N per unit volume, have identical moments, regardless of whether they are situated on A or B sites. If a fraction λ and $\mu(= 1 - \lambda)$ occupy A and B sites, respectively, the material will possess a net moment if $\lambda \neq \mu$ (Fig. 9-1.1a). This case was treated in detail by Néel.[1] An unequal

[1] L. Néel, *Ann. phys. (Paris)* **3**, 137 (1948).

partitioning of the ions can occur either if there are unequal numbers of sites on the two sublattices or if the ions prefer the sites on one sublattice over those on the other. Factors that may give rise to site preference include (a) ion size re site size, (b) electron configuration of the ion, and (c) symmetry and strength of the crystalline field at a site.

Even when $\lambda = \mu$ there will be ferrimagnetism if the magnetic moments of the ions on the *A* and *B* sites are unequal (Fig. 9-1.1*b*). This inequality may develop (a) from elements in different ionic states, for example,

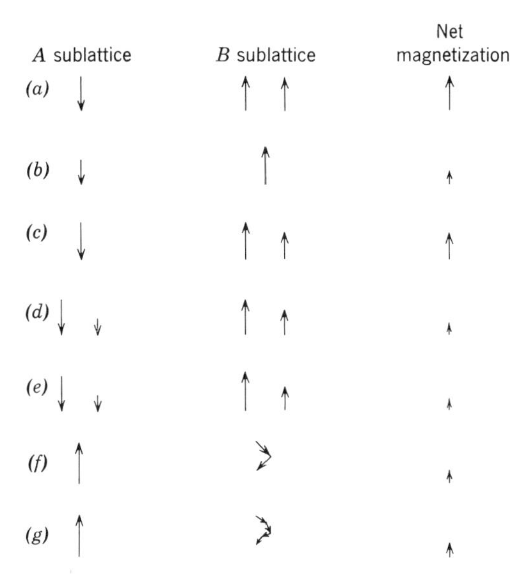

Fig. 9-1.1. Various possible ferrimagnetic arrangements for two sublattices. The magnitudes of the vectors may represent either the magnetic moments of the individual ions or the moment per chemical formula unit for each type of ion.

Fe^{2+} and Fe^{3+}, (b) from different elements in the same or different ionic states, for example, Fe^{3+} and Co^{2+}, or (c) from different crystalline fields acting at the two sites. Frequently there is a combination of the two, that is, different numbers of ions with dissimilar moments on the sublattices. A particular example of importance is shown in Fig. 9-1.1*c*. Here there are two types of ion, with all the ions of one type on *B* sites and those of the other type equally divided between the *A* and *B* sites. Other examples of a more general nature in which there are different types of ions in both sites are shown in Figs. 9-1.1*d*, *e*.

Nonantiparallel arrangements, such as those in Figs. 9-1.1*f*, *g*, are also conceivable. They are possible if the antiferromagnetic interaction within one sublattice (here *B*) is comparable to that between the two sublattices.

The competition between the forces on the B spins may then lead to a triangular configuration. The presence of a strong crystalline anisotropy of, say, cubic symmetry at one type of site may also play a role in these examples.

Systems with more than two sublattices may occur. Included are those with an odd number of sublattices; indeed, ferrimagnetic materials with three sublattices of considerable importance are known. The presence of a number of sublattices gives rise to myriad possible spin arrangements. Nevertheless, the molecular field theory for a two sublattice system predicts the main features observed for the majority of known ferrimagnetic materials. This theory is outlined in Section 9-2 and compared with experimental results for various classes of ferrimagnetic materials in subsequent sections.

Ferrimagnetism was first found in materials with the spinel crystal structure. Their chemical formula is PQ_2X_4, in which often $X = O^{2-}$, $Q = Fe^{3+}$, and P is some divalent cation. Other more general combinations of cations also are possible. In the early days of the development of the subject such materials were called *ferrites*, even though this word lacks the generic meaning usually employed in chemical nomenclature. More recently it has sometimes been used in a broader sense. Furthermore, this use has varied from author to author, with the result that the term has become ill defined. In order to be specific, a ferrite is defined here as any ferrimagnetic material with a low electrical conductivity; however, the word is used rather sparingly. The different ferrimagnetic materials are classified by a term descriptive of their crystal structures. Thes pinels are discussed in Section 9-3. The garnets, three sublattice systems that can frequently be described by the two-sublattice theory, are treated in Section 9-4. Other ferrimagnetic materials, such as the various hexagonal oxides, are considered in Section 9-5.

Almost all of the known ferrimagnetic materials are poor conductors of electricity. Exceptions include some intermetallic compounds such as Mn_2Ge_5,[2] $Mn_{2-x}Cr_xSb$ with $0.02 < x < 0.2$,[3] and Mn_4N.[4] By far the best conductor among those we call the ferrites is magnetite (Fe_3O_4), which is a spinel. Its resistivity at room temperature[5] is approximately 7×10^{-3} ohm cm.

[2] K. Yasucochi, K. Kanematsu, and T. Ohoyama, *J. Phys. Soc. Japan* **15,** 932 (1960).

[3] T. J. Swoboda, W. H. Cloud, T. A. Bither, M. S. Sadler, and H. S. Jarrett, *Phys Rev. Letters* **4,** 509 (1960); W. H. Cloud, T. A. Bither, and T. J. Swoboda, *J. Appl. Phys.* **32,** 55S (1961); H. S. Jarrett, P. E. Bierstedt, F. J. Darnell, and M. Sparks, *J. Appl. Phys.* **32,** 57S (1961).

[4] R. Juza and H. Puff, *Z. Electrochem.* **61,** 810 (1957); W. J. Takei, G. Shirane, and B. C. Frazer, *Phys. Rev.* **119,** 122 (1960).

[5] E. J. W. Verwey and P. W. Haaijman, *Physica* **8,** 979 (1941).

The property of high resistivity is of great importance both scientifically and technologically. For one thing, the magnetic electrons are certain to be localized at the ions, hence band theories need not be considered in the quantum mechanical theories. The sublattice structure does make these theories more complicated than for the corresponding ferromagnetic case. A brief outline of the assumptions and results of some quantum mechanical calculations is given in Section 9-6. In addition, because of the high resistivity, eddy currents are virtually absent. As a result, experiments can be conducted at high frequencies without the complicating presence of eddy current losses. This has led to a number of important studies, some of which are reported in Chapter 10. On the technological side, this feature has made possible a large number of applications, ranging from low-loss transformer cores to some esoteric microwave devices; some of the latter are mentioned briefly in Chapter 10.

Apart from eddy current effects, the magnetization processes in ferrimagnets are governed generally by the same principles as those developed for ferromagnets in Chapter 7. In the analysis no account is taken of the sublattice structure; rather, only the net magnetization is considered. This leads to domain effects and small particle properties similar to those described earlier. There are some differences in detail. For instance, since most polycrystalline ferrimagnetic materials are ceramics, the samples tend to be somewhat porous. The uncompensated poles on the interior surfaces have a marked effect on the soft magnetic properties. Certain aspects of soft ferrimagnetic materials are discussed in Section 9-7. Small ferrimagnetic particles usually have larger critical sizes than ferromagnetic particles because of the smaller magnetic moment per unit volume. Data on some permanent magnets fabricated from ferrite powders are listed in Table 7-4.1. The ultimate limit of the energy product $(B_H H)_{\max}$ of ferrites is smaller than for iron and iron alloys essentially because of the lower saturation magnetizations. Small particles of ferrites are important in several other connections. A common form of tape used in recording consists of fine $\gamma - Fe_2O_3$ particles suspended in a binder. Other tapes, some with a higher storage capacity, are made by electrodeposition. There is a large literature on recording tapes.[6] Ore that consists of small particles of iron oxides dispersed in quartz is a leading source of iron. Known as taconite in Minnesota, it is often processed along the lines of techniques developed at the University of Minnesota. The process involves grinding the ore into small particles, extracting the iron oxide particles magnetically, and sintering them into pellets.[7] In

[6] See for example C. D. Mee, *The Physics of Magnetic Recording*, North-Holland Publishing Co., Amsterdam (1964).

[7] Mines Experiment Station Reports, University of Minnesota, Minneapolis.

geophysics the study of rock magnetism provides information concerning the earth's magnetism. The rocks, either igneous or sedimentary, may have fine particles of some ferrite, usually magnetite, embedded in them. Various mechanisms of considerable interest have emerged from these geophysical investigations. Rock magnetism is discussed briefly in Section 9-8; the other topics concerning small ferrite particles are given no further consideration.

2. The Molecular Field Theory of Ferrimagnetism

For two sublattices the molecular fields for a ferrimagnet are formally the same as those for an antiferromagnet, namely (see Section 8-3)

$$\mathbf{H}_{mA} = -N_{AA}\mathbf{M}_A - N_{AB}\mathbf{M}_B$$

and

$$\mathbf{H}_{mB} = -N_{BA}\mathbf{M}_A - N_{BB}\mathbf{M}_B.$$

At equilibrium $N_{AB} = N_{BA}$ as before. However, now $N_{AA} \neq N_{BB}$, since the sublattices are crystallographically inequivalent. Moreover, $M_A \neq M_B$ for any one of the reasons stated in Section 9-1. Then, if a field $\mathbf{H}$ is also applied, the fields $\mathbf{H}_A$ and $\mathbf{H}_B$ at an atom on the A and B sublattices, respectively, are

$$\mathbf{H}_A = \mathbf{H} - N_{AA}\mathbf{M}_A - N_{AB}\mathbf{M}_B$$

and

$$\mathbf{H}_B = \mathbf{H} - N_{AB}\mathbf{M}_A - N_{BB}\mathbf{M}_B. \tag{9-2.1}$$

Here $N_{AB} > 0$, since the interaction between the two sublattices is antiferromagnetic. The other molecular field constants, N_{AA} and N_{BB}, may in principle be positive or negative but apparently are positive for the great majority of ferrimagnetic materials. Also, they are usually small compared to N_{AB}. The situation in which their magnitudes are comparable is considered later in this section. It is worth mentioning that in the literature it is common to express the molecular field constants N_{AA} and N_{BB} in terms of N_{AB} by the equations

$$N_{AA} = \alpha N_{AB} \quad \text{and} \quad N_{BB} = \beta N_{AB}. \tag{9-2.2}$$

This notation is of some value for the temperature region in which the system is ordered.

The magnetizations of the sublattices at thermal equilibrium are given by

$$M_A = \sum_i N_i g_i \mu_B S_i \, B_{S_i}(x_A),$$

where, as usual,

$$x_A = \frac{S_i g_i \mu_B}{kT} H_A, \tag{9-2.3}$$

$$B_{S_i}(x_A) = \frac{2S_i + 1}{2S_i} \coth \frac{2S_i + 1}{2S_i} x_A - \frac{1}{2S_i} \coth \frac{x_A}{2S_i},$$

and

$$M_B = \sum_j N_j g \mu_B S_j B_{S_j}(x_B), \tag{9-2.4}$$

together with similar expressions for x_B and $B_{S_j}(x_B)$. Here N_i is the number of atoms per unit volume with the spin quantum number S_i. If all the atoms or ions on the A (or B) sites are identical, there is, of course, no summation necessary. If the orbital angular momentum is not completely quenched, the value of g will differ from 2; if there is no quenching at all, it will be necessary to replace S with J. As in previous discussions, the subject is developed under various subheadings.

Paramagnetic region. The critical temperature above which the spontaneous magnetization of a ferrimagnet vanishes has been called variously the Curie or Néel temperature. Apart from historical connections, either term obviously has some logical merit: Curie temperature because there is spontaneous magnetization and Néel temperature because the moment ordering is nonparallel (usually antiparallel). However, it is a pity that another quite different term is not used. After all, some materials may be ferromagnetic in one temperature region and ferrimagnetic in another; other materials may have both antiferromagnetic and ferrimagnetic phases. In an effort to reduce the danger of ambiguity, we will denote the critical temperature by T_{FN} and call it either the ferrimagnetic Néel temperature, or just the Néel temperature.

Above T_{FN} the usual approximation for the Brillouin function may be made (equation 6-2.10), provided there are no saturation effects. The sublattice magnetizations are then given by

$$M_A = \frac{C_A}{T} H_A \quad \text{and} \quad M_B = \frac{C_B}{T} H_B, \tag{9-2.5}$$

where

$$C_A = \sum_i \frac{N_i g^2 \mu_B^2 S_i(S_i + 1)}{3k}$$

and

$$C_B = \sum_j \frac{N_j g^2 \mu_B^2 S_j(S_j + 1)}{3k}.$$

The notation is the same as that employed in equations 9-2.3 and 9-2.4. The magnetic fields may be written in the scalar form

$$H_A = H - N_{AA}M_A - N_{AB}M_B$$

and

$$H_B = H - N_{AB}M_A - N_{BB}M_B.$$

Substitution for the fields in equation 9-2.5 then leads to

$$\begin{aligned} (T + C_A N_{AA})M_A + C_A N_{AB} M_B &= C_A H \\ C_B N_{AB} M_A + (T + C_B N_{BB})M_B &= C_B H. \end{aligned} \tag{9-2.6}$$

These equations then yield for the sublattice magnetizations

$$M_A = \frac{C_A(T + C_B N_{BB}) - C_A C_B N_{AB}}{(T + C_A N_{AA})(T + C_B N_{BB}) - C_A C_B N_{AB}^2} H \tag{9-2.7}$$

and

$$M_B = \frac{C_B(T + C_A N_{AA}) - C_A C_B N_{AB}}{(T + C_A N_{AA})(T + C_B N_{BB}) - C_A C_B N_{AB}^2} H. \tag{9-2.8}$$

Addition of these two equations gives immediately an expression for the susceptibility $\chi[=(M_A + M_B)/H]$. Its inverse can be written in a more convenient form. By performing the indicated long division, it is easy to show that

$$\frac{1}{\chi} = \frac{T}{C} - \frac{1}{\chi_0} - \frac{\sigma}{T - \theta'}, \tag{9-2.9}$$

where

$$C = C_A + C_B$$

and

$$\frac{1}{\chi_0} = -\frac{1}{C^2}(C_A^2 N_{AA} + C_B^2 N_{BB} + 2C_A C_B N_{AB}),$$

$$\sigma = \frac{C_A C_B}{C^3}\{C_A^2(N_{AA} - N_{AB})^2 + C_B^2(N_{BB} - N_{AB})^2$$

$$- 2C_A C_B[N_{AB}^2 - (N_{AA} + N_{BB})N_{AB} + N_{AA}N_{BB}]\},$$

$$\theta' = -\frac{C_A C_B}{C}(N_{AA} + N_{BB} - 2N_{AB}).$$

The graphical representation of equation 9-2.9 is a hyperbola. Its asymptote ($T \to \infty$) is given by

$$\frac{1}{\chi} = \frac{T}{C} - \frac{1}{\chi_0}. \tag{9-2.10}$$

Since usually $N_{AB} > |N_{AA}|, |N_{BB}|$, and in addition for most cases N_{AB}, N_{AA}, and N_{BB} are all positive, the $1/\chi$-intercept ($= -1/\chi_0$) is positive. If θ is defined as the T-intercept found by extrapolation of the line, then

$$\theta = -\frac{C}{\chi_0} \tag{9-2.11}$$

and equation 9-2.10 may be written

$$\chi = \frac{C}{T + \theta}, \tag{9-2.12}$$

where $\theta > 0$. Equation 9-2.12 is formally the same as that for the susceptibility of an antiferromagnet above T_N (equation 8-3.6), wherein θ is the

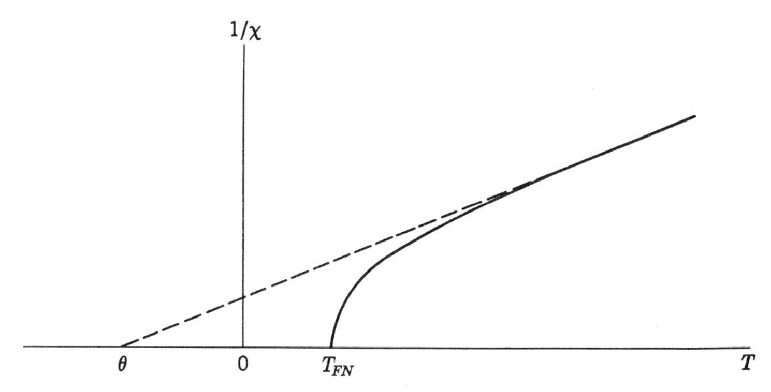

Fig. 9-2.1. Inverse susceptibility of a ferrimagnet above the Néel temperature T_{FN} according to the molecular field theory. The asymptote (equation 9-2.12) of the hyperbola is also plotted.

"paramagnetic" temperature. Therefore for a temperature sufficiently above T_{FN} the $1/\chi$ versus T plot has essentially the same linear relationship predicted by the molecular field theory for antiferromagnetic materials.

Now since, presumably, $T_{FN} > 0$ and the susceptibility is infinite at this point, the $1/\chi$ versus T curve must have considerable curvature just above the ferrimagnetic Néel temperature. This behavior is illustrated in Fig. 9-2.1. Here only that part of the hyperbola of equation 9-2.9 that is of physical interest is drawn.

The rapid change of $1/\chi$ close to T_{FN} is in marked contrast to the ferromagnetic and antiferromagnetic cases. This phenomena has its physical origin in the relative magnitudes of the sublattice magnetizations, which differ not only because $C_A \neq C_B$ but also because of an antiparallel effect (problem 9-1c).

The ferrimagnetic Nèel temperature. The ferrimagnetic Néel temperature can be found from equations 9-2.6 with $H = 0$ by setting the determinant of the coefficients equal to zero. This yields

$$T_{FN} = -\tfrac{1}{2}(C_A N_{AA} + C_B N_{BB}) + \tfrac{1}{2}[(C_A N_{AA} - C_B N_{BB})^2 + 4C_A C_B N_{AB}{}^2]^{1/2}. \quad (9\text{-}2.13)$$

Another temperature is also obtained, namely

$$T_{FN}' = -\tfrac{1}{2}(C_A N_{AA} + C_B N_{BB}) - \tfrac{1}{2}[(C_A N_{AA} - C_B N_{BB})^2 + 4C_A C_B N_{AB}{}^2]^{1/2}.$$

Geometrically this is the temperature at which the other branch of the hyperbola intersects the T-axis (problem 9-1*b*).

Spontaneous magnetization. In this subsection it is assumed that the magnetization of the A and B sublattices are aligned antiparallel. Obviously this assumption is valid if N_{AB} is the only nonzero molecular field constant; it is also reasonable if N_{AB} is large compared to N_{AA} and N_{BB}. This assumption will be examined more carefully later.

The sublattice magnetizations are then given by equations 9-2.3 and 9-2.4, namely

$$M_A(T) = M_A(0)\, B_{S_A}(x_A) \quad \text{and} \quad M_B(T) = M_B(0)\, B_{S_B}(x_B), \quad (9\text{-}2.14)$$

where

$$M_A(0) = N_A g \mu_B S_A, \qquad M_B(0) = N_B g \mu_B S_B,$$

and

$$x_A = \frac{S_A g \mu_B}{kT}(-N_{AA} M_A + N_{AB} M_B),$$

$$x_B = \frac{S_B g \mu_B}{kT}(N_{AB} M_A - N_{BB} M_B).$$

Here the summation sign formerly employed has been omitted in the interests of simplicity and the applied field H has been set equal to zero because it is the spontaneous magnetization that is of interest. Equations 9-2.14 are thus two transcendental expressions that can be solved for M_A and M_B. The magnitude of the resultant spontaneous magnetization is then found from the equation

$$M(T) = |M_A(T) - M_B(T)|. \quad (9\text{-}2.15)$$

For purposes of illustration, it is assumed that $M_B(0) > M_A(0)$.

Clearly, the variation of the sublattice magnetizations with temperature depends on the ratios $M_A(0)/M_B(0)$, $N_{AA}/N_{AB}(=\alpha)$, and $N_{BB}/N_{AB}(=\beta)$. For the special case in which all the magnetic ions have the same moment $M_A(0)/M_B(0) = \lambda/(1 - \lambda)$; in much of the literature λ is taken as one of the parameters. The various M versus T curves may be conveniently discussed either as a function of α and β with $M_A(0)/M_B(0)$ (or λ) fixed[8] or as a function of $M_A(0)$ and $M_B(0)$ with α and β kept constant.[9]

The actual computation of M_A and M_B as a function of temperature from equations 9-2.14 is, in general, laborious. A graphical method of solution has been devised by Néel (in appendix I of his paper[8]). In this procedure a number β is first arbitrarily chosen in which

$$\beta = \frac{M_A(T)}{M_B(T)} = \frac{M_A(0)}{M_B(0)} \frac{B_{S_A}(x_A)}{B_{S_B}(x_B)}.$$

Then, by eliminating $M_A(T)$, $M_B(T)$, $B_{S_A}(x_A)$ and $B_{S_B}(x_B)$, the ratio $x_A/x_B = \alpha$ is obtained in which α is a function of β. By taking logarithms these equations are written

$$\log B_{S_A}(x_A) - \log B_{S_B}(x_B) = \log \frac{M_B\beta}{M_A}$$

and

$$\log x_A - \log x_B = \log \alpha.$$

Two transparent sheets of drawing paper on which have been plotted, respectively, the curves $y = \log B_{S_A}(x_A)$, $x = \log x_A$, and $y = \log B_{S_B}(x_B)$, $y = \log x_B$ are next employed. These curves are superposed in such a fashion that the ordinate difference is equal to $\log M_B\beta/M_A$ and abscissae is equal to $\log \alpha$. This operation yields x_B and $B_{S_B}(x_B)$, hence the temperature T and the resultant magnetization M. By choosing different values of β in succession the sought-for curve is constructed point by point.

Expansion of the Brillouin functions by the series of equation 6-2.10 is valid close to the Néel temperature, that is, for approximately $0.95 < T/T_{FN} < 1$. $M_A(T)$ and $M_B(T)$ can then be evaluated numerically by procedures outlined by Néel[10] and Smart.[11] Extrapolation can be employed to yield solutions for $T < 0.95\ T_{FN}$.

By all odds, however, the best method of computing the sublattice magnetizations is to use one of the fast modern digital computers now available. Results for a particular set of parameters can be obtained very quickly with a program that utilizes a suitable iterative method.

For the region of interest the magnetizations of the A and B sublattices have the maximum value possible at absolute zero; that is, they are equal

[8] L. Néel, *Ann. phys. (Paris)* **3,** 137 (1948); J. S. Smart, *Am. J. Phys.* **23,** 356 (1955).

[9] E. W. Gorter, *Proc. IRE* **43,** 1945 (1955).

[10] Néel, *loc. cit.*

[11] Smart, *loc. cit.*

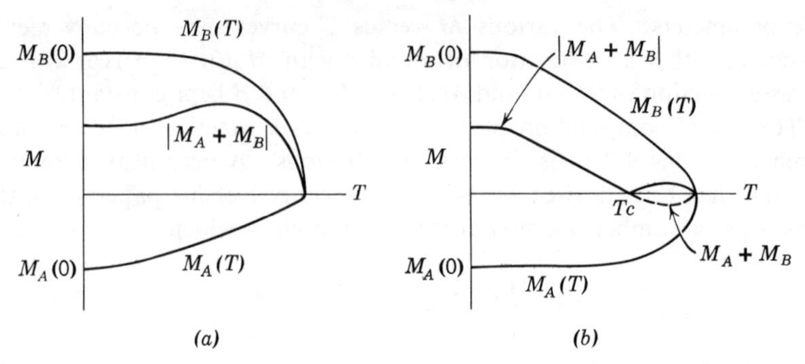

Fig. 9-2.2. Two unusual spontaneous magnetization-temperature curves predicted by the molecular field theory. For (*a*) the net molecular field on an *A*-site ion is less than that on a *B*-site ion; for (*b*) this situation is reversed.

to $M_A(0)$ and $M_B(0)$, respectively (see problem 9-3). Then, as the temperature is raised, their magnitudes decrease monotonically until they both become zero at the Néel temperature T_{FN}. Often the resultant magnetization varies with the temperature in a manner that is reminiscent of the standard ferromagnetic curve shown in Fig. 6-2.2. This is the case if the intersublattice interactions N_{AA} and N_{BB} are roughly comparable. Some unusually shaped curves are also possible; two are illustrated in

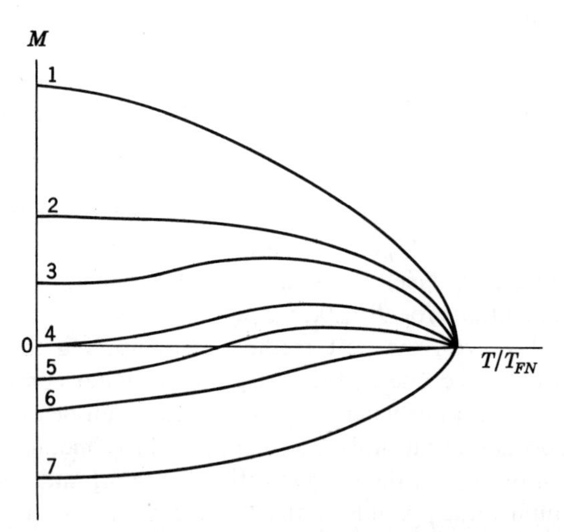

Fig. 9-2.3. Resultant magnetization as a function of temperature for various ratios of $M_A(0)/M_B(0)$. For the curves labeled 1, 2, and 3, $M_A(0)/M_B(0) < 1$; for 5, 6, and 7, $M_A(0)/M_B(0) > 1$; for curve 4, $M_A(0)/M_B(0) = 1$. For curve 2, $M_A(0)/M_B(0) = (N_{AB} + N_{AA})/(N_{AB} + N_{BB})$; for curve 6, $M_A(0)/M_B(0) = (N_{AB} - N_{AA})/(N_{AB} - N_{BB})$. (After E. W. Gorter.)

Fig. 9-2.2. The one of Fig. 9-2.2*a* occurs if, for example, the interaction within the A sublattice is antiferromagnetic, whereas that within the B sublattice is either less antiferromagnetic or ferromagnetic. Then the effective molecular field on A-site ions is less than on B-site ions. As a result $|M_A(T)|$ decreases more rapidly with temperature than $M_B(T)$ (in the low temperature range). The reverse situation produces the curve of Fig. 9-2.2*b*.

In Fig. 9-2.2*a* it is to be noted that even though the resultant magnetization has a maximum at $T > 0°K$ the slope is zero at $T = 0°K$. This is required by the third law of thermodynamics, as may readily be seen by considering equation 3-2.18, namely $(\partial M/\partial T)_H = (\partial S/\partial H)_T$. According to the third law, $(\partial S/\partial H)_T = 0$ at $T = 0°K$. Therefore in the limit at absolute zero it is necessary that $(\partial M/\partial T)_H = 0$.

For the case illustrated in Fig. 9-2.2*b* the spontaneous magnetization undergoes a reversal in direction; the temperature range in which this takes place is indicated by the dashed line. However, since the saturation magnetization is the quantity measured in practice, it is $|M_A(T) + M_B(T)|$, plotted as the full line, that is actually observed. The temperature T_C at which the resultant magnetization is zero is commonly called the *compensation point*. Some interesting resonance phenomena occur in the vicinity of the compensation temperature (see Chapter 10).

Various $(M - T)$ curves possible when the chemical composition is varied are illustrated schematically[12] in Fig. 9-2.3. For the curve labeled 4, $M_A(0) = M_B(0)$ and therefore $M(0) = 0$. However, at higher temperatures the spontaneous magnetization is, in general, nonzero. It indeed would be zero for all temperatures only if $H_{mA} = H_{mB}$, that is, if $N_{AA} = N_{BB}$. Curve 2 is a limiting case that separates curves with roughly a Brillouin form from those with a maximum above absolute zero (the type of Fig. 9-2.2*a*). This curve corresponds to the composition $M_A(0)/M_B(0) = (N_{AB} + N_{AA})/(N_{AB} + N_{BB})$. It can be shown (problem 9-4) that then $|H_{mA}| = |H_{mB}|$. Curve 6 separates the "Brillouin-type" from those with a compensation temperature. For curve 6, $M_A(0)/M_B(0) = (N_{AB} - N_{AA})/(N_{AB} - N_{BB})$; this is actually just the condition that $dM/dT = 0$ as the Néel temperature is approached (problem 9-4).

In addition to the measurement of the spontaneous magnetization as a function of temperature, the determination of the magnitude of the spontaneous magnetization at absolute zero provides a check on the theory. If the magnetic moments are due only to the spins, the sublattice magnetizations at $T = 0°K$ are

$$M_A(0) = \sum N_i g \mu_B S_i \quad \text{and} \quad M_B(0) = \sum_j N_j g \mu_B S_j.$$

[12] E. W. Gorter, *Proc. IRE* **43,** 1945 (1955).

Hence, if the number of magnetic ions on the sublattice is known, the magnitude of the moment at absolute zero can easily be calculated. Experimentally the spontaneous magnetization at $T = 0°K$ can be determined by measurements at low enough temperature together with suitable extrapolation procedures (the ferromagnetic case is discussed in Section 6-2).

When there are discrepancies between the theoretical and experimental values of $M(0)$, one or more of the following effects may be the cause.

1. The orbital angular momentum may be only partly quenched by the crystalline field and therefore may make some contribution to the magnetic moments of the ions.

2. The effective spins may be different from the free ion values. This can be caused either by the crystalline field or by covalent coupling.

3. The distribution of the magnetic ions over the two sublattices may be a variable quantity that depends on the previous history, particularly the heat treatment. The result is it is almost impossible to obtain a precise knowledge of N_i and N_j.

4. The valency of the ions may be variable. This can develop in an interchange of an outer electron between the magnetic ions. For example, when manganese and iron are present in the material, besides Mn^{2+} and Fe^{3+} ions, the expected states, there may be Mn^{3+} and Fe^{2+} ions present.

5. The sublattice magnetizations may not be antiparallel. Triangular and other spin arrangements are discussed in a later subsection.

Extension to include additional molecular fields. It has already been mentioned that more than one type of ion may occupy the A or B sites. It is not unreasonable that more than three molecular fields are then required. For example, suppose there are magnetic ions $M(1)$ and $M(2)$ on A sites and ions $M(3)$ and $M(4)$ on B sites, then a total of 10 interactions result (see Table 9-2.1) that require in general 10 molecular field

Table 9-2.1. Interactions in a Ferrite with Four Different Magnetic Ions

Within A Sites	Within B Sites	Between A and B Sites
$M(1) - M(1)$	$M(3) - M(3)$	$M(1) - M(3)$
$M(1) - M(2)$	$M(3) - M(4)$	$M(1) - M(4)$
$M(2) - M(2)$	$M(4) - M(4)$	$M(2) - M(3)$
		$M(2) - M(4)$

constants. Because there is such a large number of adjustable parameters, a fit with the experimental data does little to promote confidence in the validity of the molecular field theory. It is fortunate that the simple theory which employs only three molecular field constants predicts curves that are in reasonable agreement with most of the experimental results.

Triangular and other spin arrangements. If the interaction within a sublattice in addition to being antiferromagnetic is very large, the magnetization of this sublattice at absolute zero will be less than the saturation value of $Ng\mu_B S$. The condition for which this occurs is given in problem 9-3. Further calculation[13] then shows that the net magnetization either increases or decreases as absolute zero is approached, that is, $dM/dT \neq 0$ at $T = 0°K$. This behavior, however, is forbidden by the third law of

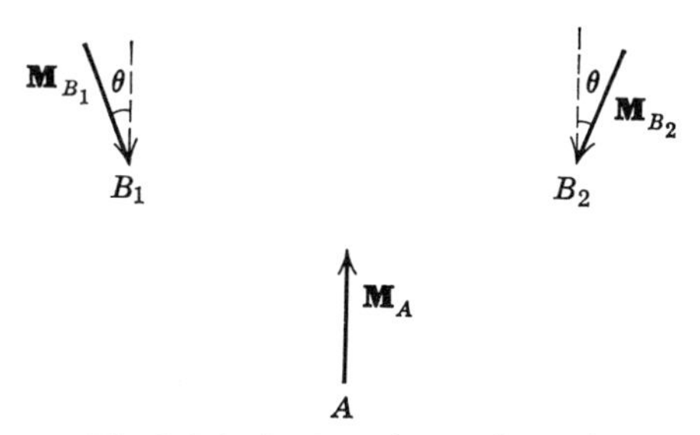

Fig. 9-2.4. A triangular configuration.

thermodynamics (see page 497). This problem can be resolved by removing the requirement that the sublattice magnetizations be antiparallel.

In principle, the spin configuration can be found from the condition that the free energy be a minimum. The free energy may be expressed by the classical form of the Heisenberg exchange Hamiltonian (equation 6-3.24), as done, for example, in Section 8-8. Alternatively, the free energy may be written as $-\mathbf{M} \cdot \mathbf{H}_m$ where $\mathbf{H}_m$ is the molecular field. The two forms, of course, are equivalent.

One possible spin configuration is a triangular arrangement, such as that illustrated in Fig. 9-1.1*f*. The way this arrangement can develop may be seen from the following highly simplified example. Suppose that the B sublattice is subdivided into two, with magnetizations $\mathbf{M}_{B_1}$ and $\mathbf{M}_{B_2}$ canted at an angle θ with respect to $\mathbf{M}_A$ (Fig. 9-2.4). Further suppose $M_{B_1} = M_{B_2} = M_B$, $M_A = M_B$, N_{BB} is the molecular field constant for the interaction between the B_1 and B_2 sublattices and $N_{AA} = 0$. The interaction free energy is then given by

$$F = 2N_{AB}M_A M_B \cos\theta + N_{BB}M_B^2 \cos^2\theta.$$

[13] L. Néel, *Ann. phys. (Paris)* **3,** 137 (1948).

This energy is a minimum with respect to θ when

$$(\sin\theta)(N_{AB} + 2N_{BB}\cos\theta) = 0.$$

Hence the B_1 and B_2 sublattice magnetizations are canted at an angle given by

$$\cos\theta = -\frac{N_{AB}}{2N_{BB}},$$

provided, of course, that $N_{AB} < 2N_{BB}$. If, however, $N_{AB} > 2N_{BB}$, it follows that $\sin\theta$, hence θ, is equal to zero; that is the A and B sublattice magnetizations are antiparallel.

In general, the sublattice magnetizations are unequal in magnitude, and moreover they depend on the temperature. Therefore there is the possibility of triangular-antiparallel transitions in the spin configuration. Examples of the conditions necessary for such transitions in some particular cases have been calculated by Lotgering.[14]

When the details of the crystal lattice are considered, it has been shown that other spin configurations can occur. For cubic spinels with sufficiently large interactions within the B sublattice the ground state appears to consist of a spiral or helical spin arrangement that is ferrimagnetic.[15] There is evidence that a spiral configuration occurs in a hexagonal ferrimagnet.[16] Other more complex spin configurations are also possible.

Three sublattice systems. A molecular field theory for three sublattice systems may be developed along the lines employed for two sublattices. The main application to date has been to the garnet system. The garnet sublattices have commonly been denoted by a, d, and c, respectively, and consequently this notation is used here. In analogy with equations 9-2.1 the fields acting at a, d, and c sites, respectively, are

$$\begin{aligned}
\mathbf{H}_a &= \mathbf{H} - N_{aa}\mathbf{M}_a - N_{ad}\mathbf{M}_d - N_{ac}\mathbf{M}_c,\\
\mathbf{H}_d &= \mathbf{H} - N_{da}\mathbf{M}_a - N_{dd}\mathbf{M}_d - N_{dc}\mathbf{M}_c,\\
\mathbf{H}_c &= \mathbf{H} - N_{ca}\mathbf{M}_a - N_{cd}\mathbf{M}_d - N_{cc}\mathbf{M}_c,
\end{aligned} \qquad (9\text{-}2.16)$$

where $N_{ad} = N_{da}$, $N_{ac} = N_{ca}$, $N_{cd} = N_{dc}$, and, in general, $M_a \neq M_d \neq M_c$. It is assumed that the magnetizations of the a and c sublattices are

[14] F. K. Lotgering, *Philips Res. Rept.* **11,** 190 (1956).

[15] T. A. Kaplan, *Phys. Rev.* **116,** 888 (1959); F. Bertaut, *Compt. rend. (Paris)* **250,** 85 (1960); T. A. Kaplan, K. Dwight, D. Lyons, and N. Menyuk, *J. Appl. Phys.* **32S,** 13 (1961); D. H. Lyons, T. A. Kaplan, K. Dwight, and N. Menyuk, *Phys. Rev.* **126,** 540 (1962).

[16] U. Enz, *J. Appl. Phys.* **32S,** 22 (1961).

antiparallel to that of the d sublattice. The sublattice magnetizations are then given by

$$M_a = N_a g\mu_B J_a B_{J_a}(x_a),$$

$$M_d = N_d g\mu_B J_d B_{J_d}(x_d), \qquad (9\text{-}2.17)$$

and

$$M_c = N_c g\mu_B J_c B_{J_c}(x_c),$$

where

$$x_a = \frac{J_a g\mu_B}{kT} H_a,$$

together with similar expressions for x_d and x_c. These equations may be evaluated by graphical or computer techniques provided that N_a, N_d, N_c, J_a, J_d, J_c and the molecular field constants are known. For the garnet system certain tricks can be employed to provide information about the molecular field constants. These procedures are discussed to some extent in Section 9-4.

Ferromagnetic interaction between sublattices. A ferromagnetic interaction between the A and B sublattices of a two-sublattice system can be treated formally simply by changing the sign before the constant N_{AB} in equations 9-2.1, *et seq.*, from minus to plus. The resultant magnetization, instead of being given by equation 9-2.1, is now the sum of the magnetizations of the two sublattices. Calculation shows that the resultant magnetization-temperature curves have shapes that are generally similar to the standard Brillouin type (Fig. 6-2.2). Of course, if the interactions within the sublattices are strong enough, a triangular antiferromagnetic or paramagnetic arrangement of the spins will occur (problem 9-5).

A few poorly conducting materials with parallel sublattice magnetizations are actually known. Mixed crystals called manganites and cobaltides are examples.[17] Their respective chemical formulas are

$$La^{3+}Mn^{3+}O_3—M^{2+}Mn^{4+}O_3,$$

in which M^{2+} is one of the alkaline earths Ca^{2+}, Ba^{2+}, or Sr^{2+} and $La^{3+}Co^{3+}O_3—Sr^{2+}Co^{4+}O_3$. These materials have the perovskite crystal structure, in which the magnetic manganese or cobalt ions occupy the body-centered positions of the cubic lattice. Complete ferromagnetism occurs only for a small range of compositions; for $La_{1-\delta}Ca_\delta MnO_3$ this range is approximately $0.2 < \delta < 0.4$. Other examples of ionic crystals that are ferromagnetic include CrO_2,[18] $CrBr_3$ (hexagonal close packed),[19]

[17] G. H. Jonker and J. H. van Santen, *Physica* **16,** 337 (1950), **19,** 120 (1953).

[18] A. Michel, G. Chaudron, and J. Bénard, *J. Phys. radium* **12,** 189 (1951).

[19] W. N. Hansen, *J. Appl. Phys.* **30,** 304S (1959); I. Tsubokawa, *J. Phys. Soc. Japan* **15,** 1664 (1960); J. F. Dillon, *J. Appl. Phys.* **33,** 1191 (1962).

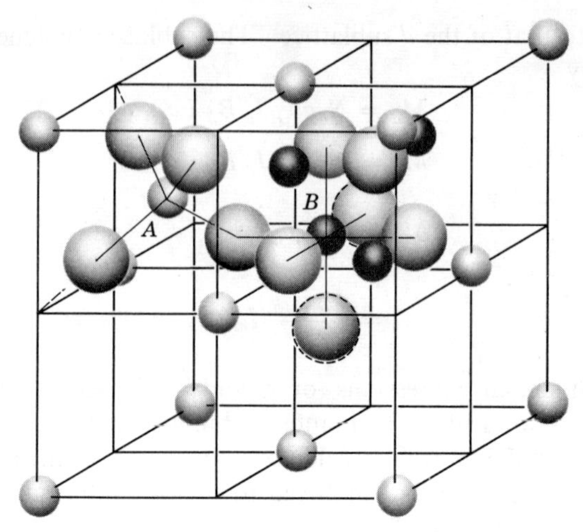

Fig. 9-3.1. The unit cell of the spinel lattice. The large spheres represent oxygen ions, the small light spheres ions in tetrahedral (A) sites and the small dark spheres ions in octahedral (B) sites. For only two octants are the positions of all the ions shown. The other octants have one or the other of these two structures and are arranged so that no two adjacent octants have the same configuration. [Adapted from E. W. Gorter, *Philips Res. Rept.* **9,** 295 (1954).]

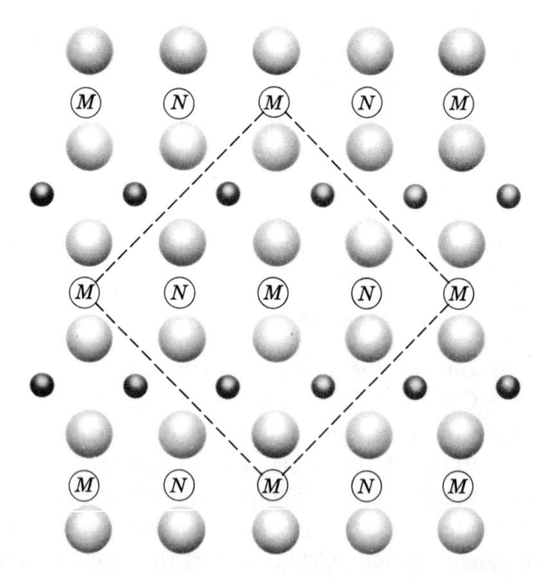

Fig. 9-3.2. The ions in a (001) plane of the spinel lattice. The small unshaded spheres represent A sites that are situated either slightly above (M) or below (N), the plane of the diagram. The dotted lines indicate a face of the unit cell. Note the f.c.c. stacking of the oxygen ions (large spheres).

EuO,[20] EuS,[21] and Eu_2SiO_4.[22] Although suggestions concerning the indirect exchange mechanism operating have been made,[23] the origin of the coupling in these ferromagnetic insulators is not known at present.

3. Spinels

Spinel was originally the name of a naturally occurring mineral with the chemical formula $MgAl_2O_4$. Ferrimagnetic spinels have the same crystal structure as the mineral but the cations are replaced by ions of the transition elements. As mentioned before, the general chemical formula is PQ_2X_4. In the simple spinels the Q ions are trivalent ions of the same element and the P ions are all divalent ions of the same type. Solid solutions of simple spinels also occur. In addition, diamagnetic ions can be substituted for some of the magnetic ions. For the last two cases cations with a valency of one, four, and possibly six may be present in the lattice.

The oxygen ions, which are large compared to the cations, form essentially a face-centered cubic lattice. The smallest cubic unit cell consists of eight molecules of PQ_2X_4, that is, 32 oxygen ions. The cations occupy interstitial positions, of which there are two distinctly different types. In one the magnetic ion is surrounded by four oxygen ions located at the corners of a tetrahedron. Such an interstice is called a tetrahedral, or *A*, site. In the other the magnetic ion is surrounded by six oxygen ions placed at the vertices of an octahedron; this is called an octahedral, or *B*, site. Not all interstitials are occupied; those that are have certain symmetry properties. Eight *A* sites and 16 *B* sites are occupied per unit cell. Tetrahedral and octahedral sites are indicated in the diagram of a unit cell of the spinel lattice shown in Fig. 9-3.1. The length of an edge of the unit cell is approximately 8 Å. The atomic arrangement in a plane of the lattice is illustrated in Fig. 9-3.2. In this sketch it can be seen that the cations in *B* sites lie in linear chains.

If the eight divalent (P) ions occupy the *A* sites and the 16 trivalent (Q) ions occupy the *B* sites, the structure is said to be a *normal* spinel. If, on

[20] B. T. Matthias, R. M. Bozorth, and J. H. Van Vleck, *Phys. Rev. Letters* **7,** 160 (1961).

[21] T. R. McGuire, B. E. Argyle, M. W. Shafer, and J. S. Smart, *Appl. Phys. Letters* **1,** 17 (1962).

[22] M. W. Shafer, T. R. McGuire, and J. C. Suits, *Phys. Rev. Letters* **11,** 251 (1963).

[23] C. Zener *Phys. Rev.* **82,** 403 (1951); P. W. Anderson, *Phys. Rev.* **79,** 350 (1950), **79,** 705 (1950); D. Polder, *J. Phys. radium* **12,** 274 (1951); J. B. Goodenough and A. L. Loeb, *Proc. Conf. Mag. and Mag. Materials, Pittsburgh* **T-78,** 14 (1955); J. Callaway and D. C. McCollum, *Phys. Rev.* **130,** 1741 (1963); S. H. Charap and E. L. Boyd, *Phys. Rev.* **133,** A811 (1964).

the other hand, the B sites are occupied half by divalent and half by trivalent ions, generally distributed at random, and the A sites by trivalent ions, the structure is said to be an *inverse* spinel. Almost all of the simple spinels that are ferrimagnetic have the inverse arrangement. Partly inverted structures are also possible.

The superexchange interaction between two cations via an intermediate oxygen ion is greatest if the three ions are collinear and if their separations are not too great (Section 8-5). The ion arrangements in the spinel that are likely to be most important are shown[24] in Fig. 9-3.3. For the situation

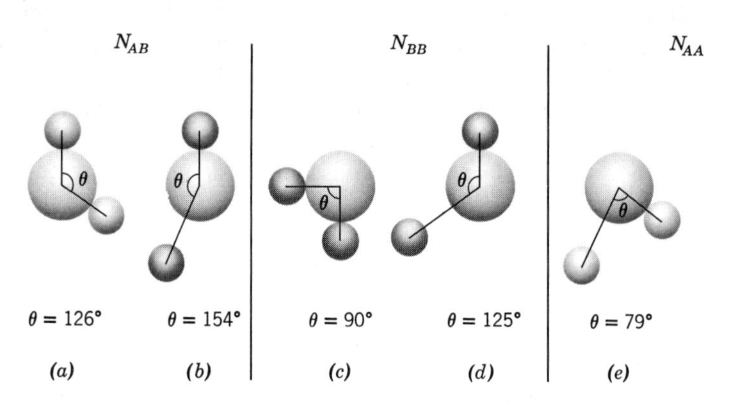

Fig. 9-3.3. Sketch of ion-pair configurations important for the superexchange interaction. Large spheres represent oxygen ions, small light ones, cations in A sites, and small dark ones, cations in B sites. (After E. W. Gorter.)

depicted in Fig. 9-3.3*a* both the angle and the distances between ions are favorable for superexchange. For all other cases, either the angle (Fig. 9-3.3*c*), one distance (Figs. 9-3.3*b*, *d*), or both (Fig. 9-3.3*e*) are unfavorable. The conclusion is that the interaction between sublattices is stronger than those within the sublattices and further that the interaction between ions in A sites is weakest of all. This result thus supports the assumption that sublattice magnetizations are antiparallel.

Rather more substantial evidence for the antiparallel spin configuration is provided by neutron diffraction experiments. In a ferrimagnet the ions in A and B sites are inequivalent crystallographically. Hence, in contrast to the usual case for antiferromagnets, no extra lines appear on cooling below T_{FN}. However, the magnetic scattering does contribute to the intensities of the lines. The magnitude of this contribution can be determined by the methods outlined in Section 8-2 and compared with that

[24] E. W. Gorter, *Philips Res. Rept.* **9,** 295 (1954).

predicted from an assumed spin configuration. In this way direct evidence for the antiparallel spin model has been obtained for Fe_3O_4[25] and $MgFe_2O_4$.[26] For some spinel solid solutions there appears[27] to be some inconsistency between the neutron diffraction and magnetic data. This may be due to a small random unobservable canting of the moments.

Neutron diffraction experiments also show that in a small number of spinels, for example $CuCr_2O_4$[28] and Mn_3O_4,[29] there is a nonantiparallel spin arrangement. The actual configuration may possibly be triangular; at this time, however, the matter is uncertain.[30] In addition, neutron diffraction studies have provided information on such things as cation site preferences,[31] precise positions of the oxygen ions,[32] and electronic ordering in magnetite at a low-temperature phase transition.[33]

We now turn to a discussion of magnetic properties in which the simple spinel compounds are considered first. The inverse of the susceptibility as a function of temperature indeed has the general shape predicted by equation 9-2.9 (see Fig. 9-3.4). The Curie constant C can be obtained from the slope of the asymptote. It is found that the experimental value is appreciably larger than that calculated, assuming only that the spin contributes to the moment.[34] This difference may be a result of a variation of the exchange energy with temperature which originates from the thermal expansion of the lattice. If it is assumed that the relationship between the molecular field constant N_{AB} and temperature is given by $N_{AB}(T) = N_{AB}(0)\,(1 + \rho T)$ and that α and β are constant, the discrepancy between the experimental and theoretical Curie constants is removed[35] if $\rho \approx 10^{-4}$ (see problem 9-6a). This value is approximately 10 times larger than the linear expansion coefficient (10^{-5}/°K) and implies that the molecular fields vary as the inverse tenth power of the cation separation.

[25] C. G. Shull, E. O. Wollan, and W. C. Koehler, *Phys. Rev.* **84,** 912 (1951).

[26] L. M. Corliss, J. M. Hastings, and F. G. Brockman, *Phys. Rev.* **90,** 1013 (1953).

[27] S. J. Pickart and R. Nathans, *Phys. Rev.* **116,** 317 (1959).

[28] E. Prince, *Acta Cryst.* **10,** 554 (1957).

[29] J. S. Kasper, *Bull. Am. Phys. Soc.* **4,** 178 (1959).

[30] I. S. Jacobs, *J. Phys. Chem. Solids* **11,** 1 (1959), **15,** 54 (1960); K. Dwight and N. Menyuk, *Phys. Rev.* **119,** 1470 (1960).

[31] G. E. Bacon and F. F. Roberts, *Acta Cryst.* **6,** 57 (1953); J. M. Hastings and L. M. Corliss, *Phys. Rev.* **104,** 328 (1956); R. Nathans, S. J. Pickart, S. E. Harrison, and C. J. Kreissman, *Proc. IEE* (*London*), *Suppl.* 5 **104-B,** 217 (1957).

[32] L. M. Corliss, J. M. Hastings, and F. G. Brockman, *Phys. Rev.* **90,** 1013 (1953); E. Prince, *Acta Cryst.* **10,** 787 (1957); S. J. Pickart and R. Nathans, *Phys. Rev.* **116,** 317 (1959).

[33] W. C. Hamilton, *Phys. Rev.* **110,** 1050 (1958).

[34] G. Foëx and A. Serres, *Compt. rend.* (*Paris*) **230,** 729 (1950); M. Fallot and P. Maroni, *J. Phys. radium* **12,** 256 (1951).

[35] L. Néel, *J. Phys. radium* **12,** 258 (1951).

The constants χ_0, σ, and θ' can be determined by fitting equation 9-2.9 to the experimental $1/\chi - T$ curve. They in turn lead to values of the molecular field constants N_{AB}, N_{AA}, and N_{BB}. It is found that these values are not sensitive to the curve fitting. Nevertheless, when the spontaneous magnetization is calculated as a function of temperature, using these values, there is rather good agreement with the experimental results. This is illustrated[36] in Fig. 9-3.4 for Fe_3O_4 and $NiFe_2O_4$. The

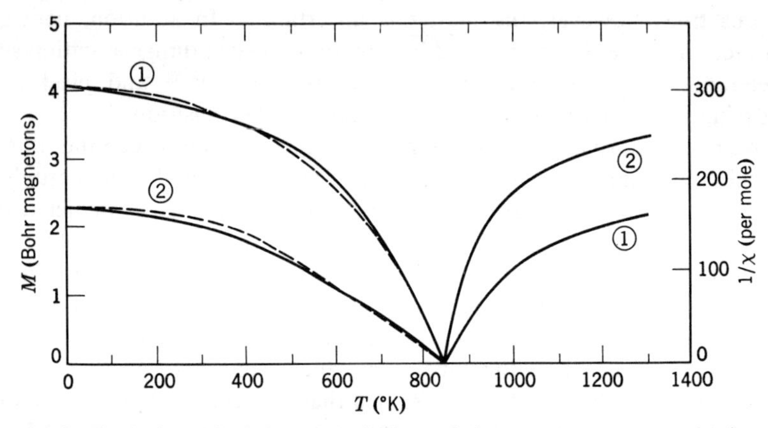

Fig. 9-3.4. The reciprocal of the susceptibility and the spontaneous magnetization as a function of temperature for (1) Fe_3O_4 and (2) $NiFe_2O_4$. The solid lines are the experimental curves; the dashed lines are theoretical curves that have been calculated with molecular field constants obtained from the paramagnetic region. [L. Néel, *Proc. Phys. Soc.* (*London*) **A-65**, 869 (1952).]

figure also serves to illustrate the typical Brillouin-type shape of the spontaneous magnetization-temperature curves which is generally observed for the simple spinels.

Data for some simple spinels are collected[37] in Table 9-3.1. The cations within the brackets are located at *B* sites, whereas those to the left of the bracket are in *A* sites. These cation distributions are made primarily on the basis of X-ray and neutron diffraction measurements. The spin-only values of the magnetic moments expected for these distributions are tabulated in Bohr magnetons for the *A* and *B* sites. The two normal spinels, $ZnFe_2O_4$ and $CdFe_2O_4$, do not exhibit ferrimagnetism. For the other spinels in the table the predicted net moment is reasonably close to

[36] L. Néel, *Proc. Phys. Soc.* (*London*) **A-65**, 869 (1952); R. Pauthenet and L. Bochirol, *J. Phys. radium* **12**, 249 (1951); G. T. Rado and V. J. Folen, *J. Appl. Phys.* **31**, 62 (1960).

[37] An extensive list of references may be found in J. Smit and H. P. J. Wijn, *Ferrites*, John Wiley and Sons, New York (1959), Ch. 8.

Table 9-3.1. Data for Some Simple Spinels

Magnetic moments are in Bohr magnetons per formula unit for 0°K; K_1 and λ_s are for $T = 293$°K

Spinel	Moment A Site	Moment B Site	Net moment Theoretical	Net moment Experimental	T_{FN} (°K)	Anisotropy Constant K_1 (ergs/cm³)	Magnetostriction Constant λ_s ($\times 10^{-6}$)
$Zn[Fe_2]O_4$	0	5 − 5	0 (anti)	para	—	—	—
$Cd[Fe_2]O_4$	0	5 − 5	0 (anti)	para	—	—	—
$Fe^{3+}[Fe^{2+}Fe^{3+}]O_4$	5	4 + 5	4	4.1	858	-110×10^3	+40
$Fe[NiFe]O_4$	5	2 + 5	2	2.3	858	-62×10^3	−26
$Fe[CuFe]O_4$	5	1 + 5	1	1.3	728	-60×10^3	−10
$Fe[CoFe]O_4$	5	3 + 5	3	3.7	793	$+1.8 \times 10^6$	−110
$Fe^{3+}[\square_{1/3}Fe_{5/3}{}^{3+}]O_4$	5	8.3	3.3	3.2	1020	-2.5×10^5	
$Fe[Li_{0.5}Fe_{1.5}]O_4$	5	0 + 7.5	2.5	2.6	943		−8
$Mn^{2+}_{0.8}Fe^{3+}_{0.2}[Mn^{2+}_{0.2}Fe^{3+}_{1.8}]O_4$	5	5 + 5	5	4.6	573	-28×10^3	−5
$Mg_{0.1}Fe_{0.9}[Mg_{0.9}Fe_{1.1}]O_4$	4.5	5.5	1.0	1.0	713	-25×10^3	−6

the observed value. The deviations may arise from any one of the causes listed in Section 9-2 (see page 498). For example, for $NiFe_2O_4$ and $CoFe_2O_4$ the orbital momentum probably is not completely quenched. For $MnFe_2O_4$ it has been suggested that Mn^{3+} and Fe^{2+} ions also occur simultaneously. For $CuFe_2O_4$ it appears that the Cu ions are not all actually in *B* sites. It has been established that $MgFe_2O_4$ has a partly inverted structure, as indicated in the table. In $Fe^{3+}[\square_{1/3}Fe^{3+}_{5/3}]O_4$, which is usually written γ-Fe_2O_3, some of the usually occupied *B* sites are vacant. In terms of Fig. 9-1.1 γ-Fe_2O_3, $MgFe_2O_4$, and $Li_{0.5}Fe_{2.5}O_4$ are examples of (*a*), whereas the other inverse spinels listed are examples of (*c*). $MnFe_2O_4$ is in a sense an example of (*c*). The easy directions of the crystalline anisotropy lie along the cube-body diagonals for all the simple spinels except $CoFe_2O_4$; then they lie along the cube edges. The linear magnetostriction is negative for all of these spinels except magnetite (Fe_3O_4).

It may be possible to obtain some desired property, not possessed by the simple spinels, with a solid solution. For example, $Co_{0.03}Ni_{0.97}Fe_2O_4$ has essentially zero crystalline anisotropy. The ferrimagnetic Néel temperature may also be varied in this way. Since solid solutions of the simple spinels may be formed in any proportions, there is a wide variety of possibilities.

Solid solutions of simple ferrimagnetic spinels with the normal spinel $ZnFe_2O_4$ are of interest. Some data for these spinels, which have the general chemical formula $Zn_{\delta}^{2+}Fe_{1-\delta}^{3+}[M_{1-\delta}^{2+}Fe_{1+\delta}^{3+}]O_4$, are plotted[38] in Fig. 9-3.5. Here M^{2+} is a metal ion such as Mn, Co, Ni, etc. Since Zn ions occupy *A* sites, the moment of the *A* sublattice decreases with increasing Zn content. As a consequence, the net moment of the spinel increases. In the limit in which $\delta = 1$ a moment of 10 Bohr magnetons per molecule might be expected. Indeed, for small δ, the moment versus δ curves have a slope, indicated by the dashed line in Fig. 9-3.5, which is consistent with

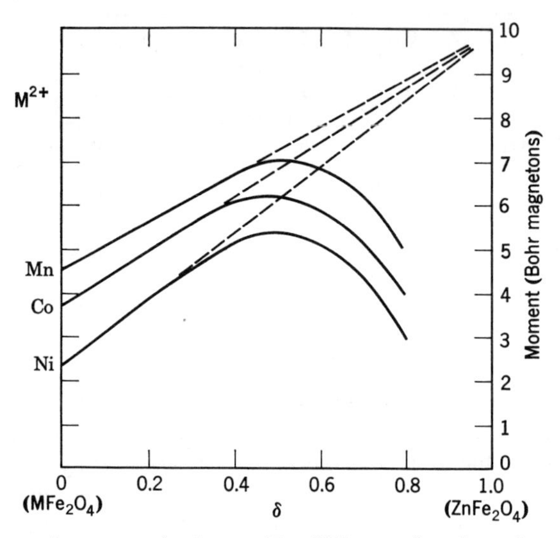

Fig. 9-3.5. Saturation magnetization at $T = 0°K$ as a function of composition for spinels with formula $M_{1-\delta}^{2}Zn_{\delta}Fe_2O_4$. The dashed lines converge toward the value of 10 Bohr magnetons. (After C. Guillaud.)

this expectation. However, for larger values of δ, the slope of the curves becomes less until finally the net moment decreases with δ. The reason is that the interaction between the sublattices decreases as the number of Fe^{3+} ions on *A* sites decreases. The molecular field within the *B* sublattice then assumes increasing importance. As a result, some sort of triangular, antiferromagnetic, or other spin arrangement occurs within the *B* sublattice and therefore the net moment decreases.

Magnetization versus temperature curves of the types shown in Fig. 9-2.2 have also been observed. A compensation temperature was first

[38] C. Guillaud, *J. Phys. radium* **12,** 239 (1951), *Revs. Mod. Phys.* **25,** 64 (1953); C. Guillaud and H. Creveaux, *Compt. rend.* (*Paris*) **230,** 1458 (1950); C. Guillaud and M. Sage, *Compt. rend.* (*Paris*) **232,** 944 (1961); E. W. Gorter, *Philips Res. Rept.* **9,** 321 (1954); J. Smit and H. P. J. Wijn, *Ferrites,* John Wiley and Sons, New York (1959).

observed[39] for $Li_{0.5}Cr_{\delta}Fe_{2.5-\delta}O_4$, where $1.0 < \delta < 1.6$. The reversal of the net magnetization was actually demonstrated by making measurements of the remanence. The variation of magnetization with temperature for various compositions of the solid solutions of $NiFe_2O_4$ and $NiFeAlO_4$ is plotted[40] in Fig. 9-3.6. The substitution of Al^{3+} for Fe^{3+}

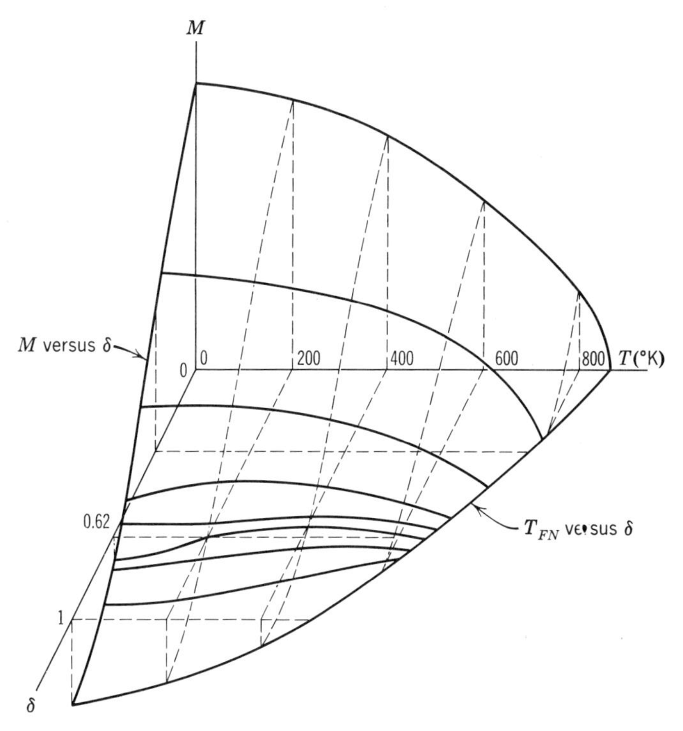

Fig. 9-3.6. Magnetization versus temperature curves for various compositions of the $NiAl_{\delta}Fe_{2-\delta}O_4$ system. [E. W. Gorter, *Proc. IRE* **43,** 1945 (1955).]

occurs primarily on *B* sites in the $NiAl_{\delta}Fe_{2-\delta}O_4$ system. Therefore, as shown by the curve in the $(T = 0)$-plane, the magnetization decreases as the percentage of Al is increased. For $\delta = 0.62$ the magnetization is zero. The Néel temperature T_{FN} also decreases with increasing δ, as indicated by the curve in the $(M = 0)$-plane. Clearly, all the magnetization-temperature curves shown in Fig. 9-2.3 are exhibited by this system. Of the simple spinels only Co_2VO_4 has so far been found to possess a compensation temperature.[41]

[39] E. W. Gorter and J. A. Schulkes, *Phys. Rev.* **90,** 487 (1953).

[40] E. W. Gorter, *Proc. IRE* **43,** 1945 (1955).

[41] N. Menyuk, K. Dwight, and D. G. Wickham, *Phys. Rev. Letters* **4,** 119 (1960).

Other nonmagnetic ions that substitute for ions on the *A* sites include Cd^{2+}, Ga^{3+}, and In^{3+}; Sc^{3+} and the combination ($Ni^{2+} + Ti^{4+}$) tend to go into *B* sites. Spinels in which S^{2-} or Se^{2-} replace the O^{2-} ions are also known.

Some spinels deviate slightly from a cubic lattice. For example, Mn_3O_4 and $CuFe_2O_4$ have a tetragonal structure, whereas Fe_3O_4 below $\approx 119°K$ is orthorhombic.

In addition to neutron diffraction experiments, evidence for non-antiparallel spin arrangements in ferrimagnets can be obtained from measurements of the magnetic moment. If the applied field is comparable to the

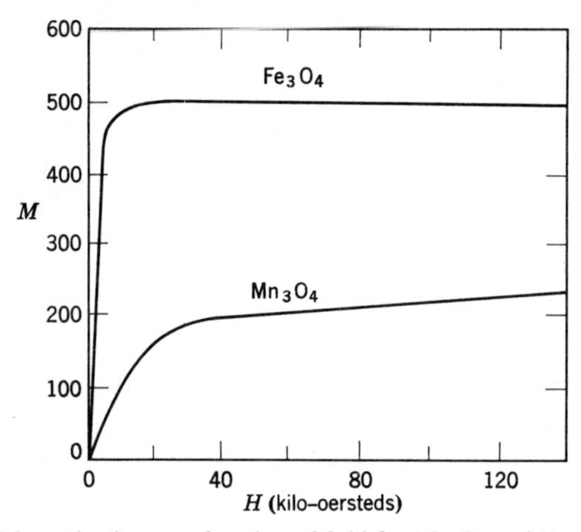

Fig. 9-3.7. Magnetization as a function of field for Mn_3O_4 and Fe_3O_4. Note the linear increase in magnetization with field for Mn_3O_4 compared to the almost constant value for Fe_3O_4 (for $H > 20$ koe.). [I. S. Jacobs, *J. Phys. Chem. Solids* **11,** 1 (1959).]

molecular field, the large torques will be sufficient to change the angles between the canted moments appreciably. As a result, once the specimen is a single domain the net magnetization will increase linearly with the applied field[42] even at $T = 0°K$. This effect has been observed for Mn_3O_4, $CoMn_2O_4$, $ZnMn_2O_4$, $MgMn_2O_4$, and their solid solutions.[42,43] In Fig. 9-3.7 the results for Mn_3O_4 are contrasted to those for Fe_3O_4 whose magnetization changes little in high fields. Although there is no doubt that the sublattice magnetizations of Mn_3O_4 are not antiparallel, the actual spin configuration is not certain at present.[44] It should also be mentioned that transitions

[42] I. S. Jacobs, *J. Phys. Chem. Solids* **11,** 1 (1959).
[43] I. S. Jacobs, *J. Phys. Chem. Solids* **15,** 54 (1960).
[44] K. Dwight and N. Menyuk, *Phys. Rev.* **119,** 1470 (1960).

between spin arrangements have been inferred from changes in the shape of magnetization versus temperature curves—an example[45] is CoV_2O_4.

4. Garnets

Garnets, which are well known as semiprecious gems when found in nature, have the general chemical formula $P_3Q_2R_3O_{12}$. The basic crystal structure is cubic with eight formula units per unit cell, that is, with 96 oxygen ions. The cations occupy interstitial sites, but in contrast to

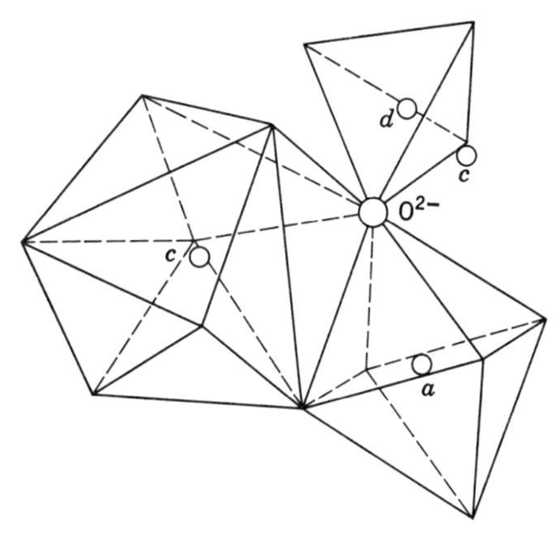

Fig. 9-4.1. The positions of the cations in *a*, *c*, and *d* sites that are nearest a given oxygen ion. Oxygen ions are also located at the vertices of the polyhedra. The dodecahedron for one of the *c*-site ions has been omitted in order to simplify the diagram. (After M. A. Gilleo and S. Geller.)

the spinels there are now three types. The cations denoted by Q and R occupy octahedral or *a* sites and tetrahedral or *d* sites, respectively. The other metal ions, P, are surrounded by eight oxygen ions located at the corners of a skewed cube, or, as it is often called (although incorrectly), a dodecahedron. These interstitials are called *c* sites. The *a*, *c*, and *d* notation for the sites is derived from crystallography. As for the spinels, when a number of different cations are present it is sometimes convenient to indicate their site location with brackets. The notation $\{P_3\}[Q_2]R_3O_{12}$ is used.

Each oxygen ion lies at a vertex that is common to four polyhedra of oxygen, one octahedron, one tetrahedron, and two dodecahedra, as indicated[46] in Fig. 9-4.1. The orientations of the polyhedra vary throughout the unit cell, although the type of symmetry for each is retained; this

[45] N. Menyuk, A. Wold, D. Rogers, and K. Dwight, *J. Appl. Phys.* **33,** 1144 (1962).
[46] M. A. Gilleo and S. Geller, *Phys. Rev.* **110,** 73 (1958).

feature is best seen with the aid of a three-dimensional model.[47] A two-dimensional view of the positions of the cations in a unit cell is given in Fig. 9-4.2. The unit cell of a typical garnet has an edge that is approximately 12.4 Å long. All the octahedral and tetrahedral sites present in the lattice are occupied by cations.

Although a number of different cations are found in the garnets that occur naturally, the R ions of $P_3Q_2R_3O_{12}$ are always tetravalent silicon, Si^{4+}. The P ions are then divalent, such as Ca^{2+}, Mn^{2+} or Fe^{2+}, and the Q

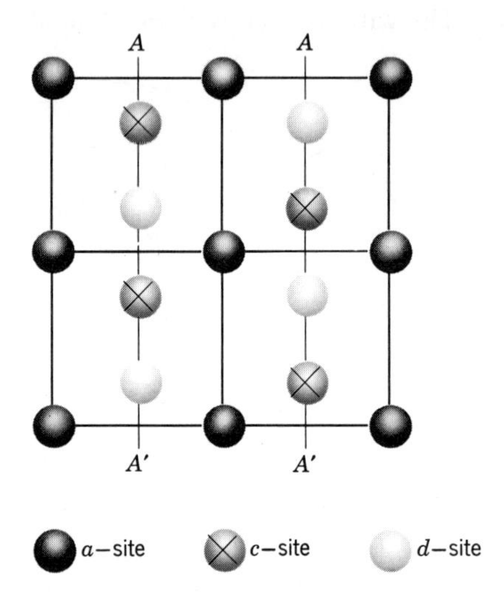

Fig. 9-4.2. The cations in one plane of the garnet lattice. The *a*-site ions form a body-centered cubic lattice. In the parallel plane that has *a*-site ions at the body-center positions the *A-A'* rows are rotated by 90°. *A-A'* rows also lie perpendicular to the plane of the paper.

ions are trivalent, such as Al^{3+}, Fe^{3+} or Cr^{3+}.[48] It has been found possible to replace Si^{4+} with Ge^{4+} in synthetic garnets.[49] However, none of the garnets with these tetravalent ions appears to be ferrimagnetic, at least above liquid helium temperatures.

The divalent ions in *c* sites and the tetravalent ions in *d* sites can also be replaced simultaneously by certain trivalent ions. The first such garnet synthesized[50] was $Y_3Al_2Al_3O_{12}$, but it was not until several years later that

[47] Obtainable for example from A. S. Lapine and Co., Chicago 29, Ill.

[48] G. Menzer, *Z. Krist.* **69,** 300 (1929).

[49] A. Tauber, E. Banks, and H. H. Kedesy, *J. Appl. Phys.* **29,** 385 (1958); S. Geller, C. E. Miller, and R. G. Treuting, *Acta Cryst.* **13,** 179 (1960); R. M. Bozorth and S. Geller, *J. Phys. Chem. Solids* **11,** 263 (1959).

[50] H. S. Yoder and M. L. Keith, *Am. Min.* **36,** 519 (1951).

the discovery of ferrimagnetic garnets was announced.[51] Perhaps the most important, and certainly the one investigated most extensively to date, has the chemical formula $\{Y_3\}[Fe_2]Fe_3O_{12}$ (also written $Y_3Fe_5O_{12}$); it is commonly known as YIG.[52] The Y^{3+} ions have a closed shell electronic configuration and, of course, are diamagnetic. In the simple or rare earth garnets the Y^{3+} ions are replaced by one of the trivalent rare earth ions, starting with Sm and running to Lu (Appendix III). The rare earth garnets are commonly called RIG's; specific members are similarly abbreviated, for example, gadolinium iron garnet is written GdIG. The radii of other rare earth ions (from La to Pm) apparently are too large to replace Y^{3+} ions; some, nevertheless, will enter the lattice when mixed with appropriate numbers of smaller ions. Examples are $\{La_{1.5}Y_{1.5}\}[Fe_2]Fe_3O_{12}$[53] and $\{Nd_{1.5}Y_{1.5}\}[Fe_2]Fe_3O_{12}$.[51] A number of other ions can be substituted for the ions of the rare earth garnets;[53] some of these compounds are discussed briefly at the end of this section.

In YIG there are magnetic ions only on the *a* and *d* sublattices. The distances from the octahedral and tetrahedral sites to the common oxygen ion are 2.00 and 1.88 Å, respectively, and the $a - O^{2-} - d$ angle is 126.6°[54] (see Fig. 9-4.1). This ionic arrangement is similar to that found for *A* and *B* sites in the spinels (see Fig. 9-3.3*a*); hence the superexchange interaction is likely to be large. On the other hand, the distances and angles for ions within either the *a* or *d* sublattices are unfavorable for appreciable superexchange. It is reasonable to expect that the magnetizations of the *a* and *d* sublattices will be antiparallel. Neutron diffraction experiments on YIG have confirmed this expectation.[55]

In the rare earth garnets the trivalent rare earth ions occupy *c* sites. Now the distances between the *c* sites and the common oxygen ion (shown in Fig. 9-4.1) are 2.43 and 2.37 Å, that is, approximately 2.4 Å, and the $c - O^{2-} - d$, $c - 0^{2-} - a$, and $c - O^{2-} - c$ angles are 122°, 100°, and 104.7°[54], respectively. On the basis of this information, it is expected that the *c* sublattice would experience the largest superexchange interaction with the *d* sublattice but that this interaction would be much smaller than that between the *a* and *d* sublattices. Therefore, although the couplings between the rare earth ions and the iron sublattices will probably be weak, it is likely that the magnetizations of the *c* and *d* sublattices will be antiparallel.

[51] F. Bertaut and F. Forrat, *Compt. rend. (Paris)* **242,** 382 (1956); S. Geller and M. A. Gilleo, *Acta Cryst.* **10,** 239 (1957).

[52] The chemical formula $3Y_2O_3{\cdot}5Fe_2O_3$ is also employed sometimes, particularly in the French literature. This formula is double the unit used here.

[53] S. Geller, *J. Appl. Phys.* **31,** 30S (1960).

[54] S. Geller and M. A. Gilleo, *J. Phys. Chem. Solids* **12,** 111 (1957).

[55] F. Bertaut, F. Forrat, A. Herpin, and P. Mériel, *Compt. rend. (Paris)* **243,** 898 (1956); E. Prince, *Acta Cryst.* **10,** 787 (1957).

As will be seen later, magnetic measurements support these conclusions, at least for most of the rare earth garnets.

Garnets are in many ways better materials for the study of ferrimagnetism than the spinels. The rare earth garnets possess only trivalent cations, hence there is no longer any uncertainty concerning the ionic states and their distribution over the lattice sites. Further, single crystals of these garnets that are virtually stoichiometric and large in volume[56] can be grown. Finally, polycrystalline garnets can easily be made with

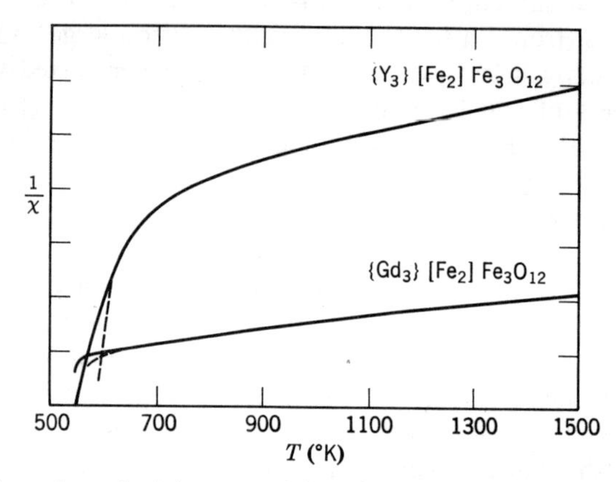

Fig. 9-4.3. The reciprocal of the susceptibility as a function of temperature above T_{FN} for YIG and GdIG. The solid lines represent the experimental data. The dashed curves show where the fitted theoretical curves deviate from the data. [R. Aléonard and J. C. Barbier, *J. Phys. radium* **20,** 378 (1959).]

about 99% of the single-crystal density; with spinels, a density above 90% is difficult to achieve.

Above the ferrimagnetic Néel temperature the inverse susceptibility versus temperature curve of YIG (Fig. 9-4.3)[57] is found to have essentially the form predicted by equation 9-2.9 for two sublattices. As for spinels, the Curie constant obtained from the slope of the asymptote is larger than the value calculated from theory. As before, this discrepancy is removed if N_{ad} is assumed to vary linearly with temperature, namely as $N_{ad}(T) = N_{ad}(0)(1 + \rho T)$. A value of $\rho \approx 10^{-4}$ is again found[58].

When there are three magnetic sublattices, it can be shown from equations 9-2.16 and 9-2.17 that the relationship between susceptibility and

[56] J. W. Nielsen and E. F. Dearborn, *J. Phys. Chem. Solids* **5,** 202 (1958); J. W. Nielsen, *J. Appl. Phys.* **29,** 390 (1958), **31,** 51S (1960).

[57] R. Aléonard and J. C. Barbier, *J. Phys. radium* **20,** 378 (1959).

[58] R. Aléonard and J. C. Barbier, *Compt. rend.* (*Paris*) **245,** 831 (1957).

temperature has the form[59] (problem 9-6*b*)

$$\frac{1}{\chi} = \frac{T}{C} - \frac{1}{\chi_0} - \frac{\sigma T + d}{T^2 - \theta' T + f}, \qquad (9\text{-}4.1)$$

where χ_0, σ, θ', d, and f are functions of the six molecular field constants and $C = C_a + C_c + C_d$. Since, for the rare earth garnets, the couplings of the rare earth sublattice are likely to be small, it is not necessary to consider all the $1/\chi - T$ curves possible for different values of the molecular field coefficients. In the limiting case, in which $N_{ac} = N_{ad} = N_{cc} = 0$, equation 9-4.1, of course, reduces to that for two sublattices (equation 9-2.9). Experimental results for GdIG are shown in Fig. 9-4.3, and, as expected, the curve is rather similar to that for YIG and other two-sublattice materials. If the coefficients N_{ad}, N_{aa}, and N_{dd}, obtained from the YIG data, are assumed to hold also for the rare earth garnets, then N_{ac}, N_{ad}, and N_{cc} can be calculated. The values of the latter coefficients taken from the data on GdIG are indeed small.[60] Similar experimental results have been obtained for terbium, dysprosium, holmium, erbium, thulium, and ytterbium garnets.[60]

The spontaneous magnetization of YIG arises solely from trivalent iron ions which have a moment of five Bohr magnetons each. Since there are three ferric ions on *d* sites for every two on *a* sites, the magnetic moment expected at $T = 0°$K is 15–10 = $5\mu_B$ per formula unit. This is in excellent agreement with the value of $4.96\mu_B$ found experimentally.[61] However, for LuIG, which should be a similar case in the sense that Lu^{3+} ions have no permanent dipole moment, the reported experimental value[62] is $4.2\mu_B$. It would appear to be worthwhile to repeat this measurement on a single crystal of high purity. For the other rare earth garnets the magnetization of the *c* sublattice has been assumed to be antiparallel to the resultant of the two iron sublattices. If both the orbital and spin angular momenta contribute to the magnetic moment according to Hund's rules, the spontaneous magnetization of a rare earth ion at $T = 0°$K would be $gJ\mu_B$ or $(L + 2S)\mu_B$ in which g is the Landé factor of equation 2-2.11. If only the spin contributes, the corresponding ionic moment would be $2S\mu_B$. Hence the net magnetization expected for these two extremes would be $|3(L + 2S) - 5|$ and $|3(2S) - 5|$ per formula unit, respectively. These values, and the experimental ones, are tabulated in Table 9-4.1. The agreement is satisfactory only for GdIG, presumably because Gd^{3+} ions have $L = 0$. For the other rare earth garnets the observed moments generally lie

[59] R. Pauthenet, *Ann. phys. (Paris)* **3,** 424 (1958).
[60] R. Aléonard and J. C. Barbier, *J. Phys. radium* **20,** 378 (1959).
[61] M. A. Gilleo and S. Geller, *Phys. Rev.* **110,** 73 (1958).
[62] Pauthenet, *loc. cit.*

Table 9-4.1. Data for Some Simple Garnets

Magnetic moments in Bohr magnetons per formula unit at 0°K

Garnet	Moment c Site $3(2S)$	Moment c Site $3(L+2S)$	Theoretical Net Moment $\lvert 3(2S)-5\rvert$	$\lvert 3(L+2S)-5\rvert$	Experimental Moment	T_{FN} (°K)
$\{Y_3\}[Fe_2]Fe_3O_{12}$	0	0	5	5	4.96	560
$\{Sm_3\}[Fe_2]Fe_3O_{12}$	15	0	10	5	4.7	578
$\{Eu_3\}[Fe_2]Fe_3O_{12}$	18	9	13	14[a]	2.6	566
$\{Gd_3\}[Fe_2]Fe_3O_{12}$	21	21	16	16	15.2	564
$\{Tb_3\}[Fe_2]Fe_3O_{12}$	18	27	13	22	15.7	568
$\{Dy_3\}[Fe_2]Fe_3O_{12}$	15	30	10	25	16.2	563
$\{Ho_3\}[Fe_2]Fe_3O_{12}$	12	30	7	25	13.7	567
$\{Er_3\}[Fe_2]Fe_3O_{12}$	9	27	4	22	11.6	556
$\{Tm_3\}[Fe_2]Fe_3O_{12}$	6	21	1	16	1.0	549
$\{Yb_3\}[Fe_2]Fe_3O_{12}$	3	12	2	7	0	548
$\{Lu_3\}[Fe_2]Fe_3O_{12}$	0	0	5	5	4.2	549

[a] Since it is the spins of the rare earth and iron ions that are antiparallel, and further since J and S have opposite signs for the first half of the rare earth series, the moments of the c and d sublattices are parallel for this case.

between the two that are calculated. The probable explanation is that the orbital momentum is partly quenched by the crystalline fields. A few calculations of this effect have been made.[63] A non-antiparallel spin arrangement, however, is another possibility.

The spontaneous magnetization of YIG is plotted as a function of temperature[64] in Fig. 9-4.4. The type of analysis employed for the two-sublattice spinels are applicable here. The magnetization-temperature curve calculated with molecular field constants obtained from the paramagnetic region (Fig. 9-4.3) agrees well with the experimental results. The values calculated for the thermal variation of the spontaneous magnetizations of the individual a and d sublattices are indicated[65] in Fig. 9-4.5. The magnetization-temperature curve of LuIG is similar in shape to that of YIG.

The ferrimagnetic Néel temperature of the rare earth garnets is close to that of YIG ($T_{FN} \approx 560$°K), as may be seen by inspection of Table 9-4.1. This adds further credence to the idea that the two iron sublattices

[63] Y. Ayant and J. Thomas, *Compt. rend. (Paris)* **248,** 1955 (1959), **250,** 2688 (1960); W. P. Wolf, *Proc. Phys. Soc. (London)* **74,** 665 (1959); R. Pauthenet, *J. Phys. radium* **20,** 388 (1959); R. L. White and J. P. Andelin, *Phys. Rev.* **115,** 1435 (1959).

[64] R. Pauthenet, *Ann. phys. (Paris)* **3,** 424 (1958); M. A. Gilleo and S. Geller, *Phys. Rev.* **110,** 73 (1958).

[65] Pauthenet, *loc cit.*

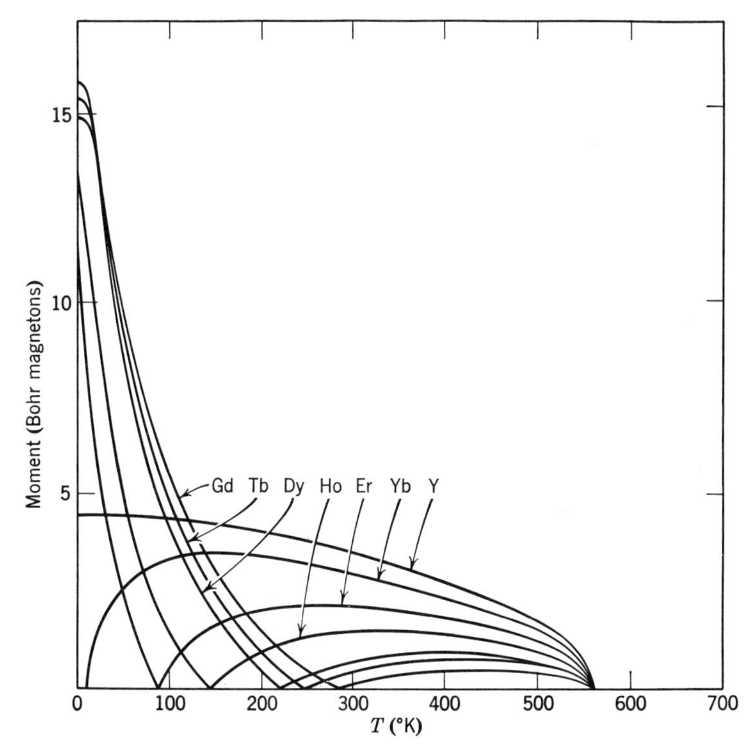

Fig. 9-4.4. Experimental values of the spontaneous magnetization of various simple garnets as a function of temperature. The formula unit is $\{P_3\}[Fe_2]Fe_3O_{12}$, where P is indicated for each curve. (After R. Pauthenet.)

are not influenced much by the magnetic rare earth sublattice; that is, the c sublattice coupling is weak. The assumption that the thermal variation of the magnetization of the iron sublattices in the RIG's is identical to that of YIG greatly simplifies the analysis.

The spontaneous magnetization versus temperature curves observed for several of the rare earth garnets are shown[65] in Fig. 9-4.4. All of the curves possess a compensation temperature. The physical reason for this is simple enough. At $T = 0°K$ the c sublattice magnetization is large and determines the direction of the net magnetization. However, at higher temperatures the magnetization of the c sublattice drops quickly because of the weak coupling. At the compensation temperature the c sublattice magnetization is equal to the resultant of the a and d sublattices. Above T_C the net magnetization is parallel to that of the d sublattice.

The thermal variation of the c sublattice's magnetization can be calculated if the molecular field constants appropriate to this sublattice are known (last of equations 9-2.17). One simplification made is the assumption that the coupling between the rare earth ions and the ensemble of iron ions can be described by one molecular field constant N_{cad}. This coefficient can then be evaluated at the compensation temperature. The magnetization of the gadolinium sublattice of GdIG thus obtained is shown as a function of temperature in Fig. 9-4.5. As expected, $M_c(T)$ decreases rapidly with temperature. Further, the sum of M_a, M_c, and M_d yields a curve that agrees well with the observed one (Fig. 9-4.4). For the

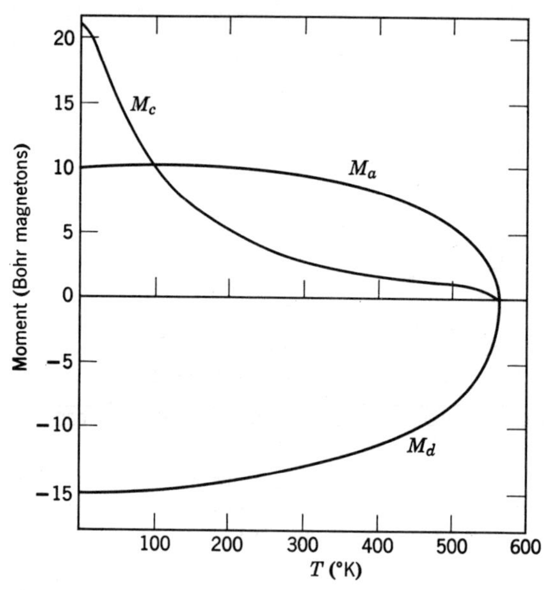

Fig. 9-4.5. The magnetization of the GdIG sublattices per formula unit as a function of temperature. (After R. Pauthenet.)

other garnets with rare earths from the second half of the series the calculations are equally successful. The data may also be analyzed in terms of the two coefficients N_{ac} and N_{dc}. The calculated values of the sublattice magnetizations of GdIG are in good accord with those found by the simpler method just described and plotted in Fig. 9-4.5.[65]

Europium and samarium iron garnets are rather different in that their magnetization versus temperature curves do not possess a compensation point. The reason is $g - 1 < 0$ for the rare earths in the first part of the series. Here g is the Landé factor of equation 2-2.11. Therefore, although the coupling between the rare earth and iron sublattices is antiferromagnetic, the rare earth moments are parallel to the resultant of the iron ions. As mentioned in Section 2-9, samarium and europium ions are unusual in that their multiplet spacing is small; EuIG and SmIG require special theoretical treatment.[66]

When magnetic fields of the order of 5000 oe or higher are applied to the rare earth garnets below T_{FN}, the magnetization M_T can be given by an expression of the form

$$M_T = M + \chi_c H, \tag{9-4.2}$$

where M is the spontaneous magnetization. The susceptibility χ_c is, of course, a paramagnetic term arising essentially from the loosely coupled

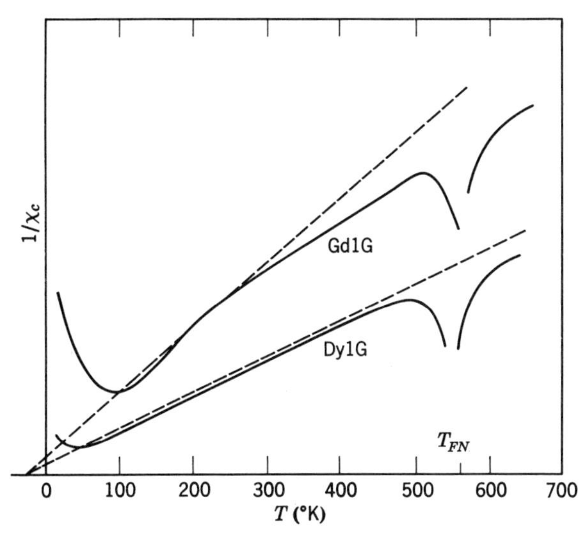

Fig. 9-4.6. Reciprocal of the susceptibility as a function of temperature for garnets of Gd and Dy below and just above the ferrimagnetic Néel temperature. The full curve represents the experimental data and the dashed lines, the Curie-Weiss law. [R. Pauthenet, *Ann. phys.* (Paris) **3,** 424 (1958).]

rare earth sublattice. The reciprocal of this susceptibility as a function of temperature is shown for GdIG and DyIG in Fig. 9-4.6;[67] the $1/\chi_c - T$ curves for the other RIG's are generally similar.[67] Now from equations 9-2.17 the magnetization of the c sublattice is given by

$$M_c = N_c g \mu_B J_c B_{J_c}(x_c). \tag{9-4.3}$$

Since the spontaneous magnetization of the c sublattice is already relatively

[66] J. H. Van Vleck, *J. Phys. Soc. Japan, Suppl. B-I* **17,** 352 (1962); W. P. Wolf and J. H. Van Vleck, *Phys. Rev.* **118,** 1492 (1960).

[67] R. Pauthenet, *Ann. phys. (Paris)* **3,** 424 (1958).

small at $T = 100°K$ (see Fig. 9-4.5), this equation can be approximated by the expression (see equation 6-2.10)

$$M_c = \frac{C_c}{T}(-N_{ca}M_a - N_{cd}M_d - N_{cc}M_c + H) \qquad (9\text{-}4.4)$$

for $T > 100°K$. Here C_c is the usual Curie constant (equation 6-2.9), defined by

$$C_c = \frac{N_c g^2 \mu_B^2 J_c(J_c + 1)}{3k}.$$

The paramagnetic susceptibility is given by $\chi_c = dM_c/dH$ and therefore from equation 9-4.4 it follows that

$$\chi_c = \frac{C_c}{T + N_{cc}C_c}, \qquad (9\text{-}4.5)$$

which is just a form of the Curie-Weiss law (equations 6-2.9 and 8-3.6). As illustrated in Fig. 9-4.5, plots of this equation calculated with $J = L + S$ do fit the data over the middle part of the 0° to T_{FN} temperature range. The deviations at low temperatures are primarily due to saturation effects. It can be shown (problem 9-7) that a curve calculated from equation 9-4.3 without making any approximations agrees well with the low temperature observations.[68] At temperatures close to T_{FN} it has been suggested that the deviations may arise from temperature-independent paramagnetism and from paramagnetism of the iron sublattices.

The crystalline anisotropy constant K_1 of YIG at room temperature is approximately equal to -6×10^3 erg/cm^3. The rare earth garnets have almost the same anisotropy constant at this temperature; in other words, the rare earth ions do not contribute much to K_1. However, this is no longer true at low temperature, and very large anisotropies have been reported. The probable reasons for this behavior will be discussed briefly in Section 9-6.

Finally, we turn to a consideration of solid state solutions of the simple garnets as well as various ionic substitutions that can be made. By making solutions of YIG with small amounts of a RIG, commonly called rare earth doping of YIG, it has been possible to study in detail how the rare earth ions influence the anisotropy and resonance line widths.[69] It turns out that even small amounts of rare earth ions cause a large increase in the anisotropy and line width at low temperatures. These effects can be interpreted by considering the energy levels of the rare earth ions. Materials

[68] W. P. Wolf, *Rept. Progr. Phys.* **24,** 212 (1961).

[69] J. F. Dillon and J. W. Nielsen, *Phys. Rev. Letters* **3,** 30 (1959), *Phys. Rev.* **120,** 105 (1960); J. F. Dillon and L. R. Walker, *Phys. Rev.* **124,** 1401 (1961).

with particular compensation temperatures can be obtained with appropriate solid state solutions. This feature is of some engineering importance, since it is desirable to operate some devices over most of the temperature range between T_C and T_{FN}, in which the spontaneous magnetization is low and almost constant.

There are more restrictions on the substitutions that can be made in garnets than in spinels. Nevertheless, a large number of substituted compounds have been made that have yielded interesting information.[70] The iron ions can be partially substituted by Sc^{3+}, In^{3+}, and Cr^{3+} and completely substituted by Al^{3+} and Ga^{3+} ions. By partial substitution with these ions the ferrimagnetic Néel temperature and the magnetic moment can be changed.[71] Also such materials provide additional knowledge of the sublattice interactions. Measurements on the completely substituted garnets, RAlG and RGaG, as well as on LuAlG and LuGaG doped with rare earth ions, give information on the crystalline fields acting on the rare earth ions. The properties of the ferrimagnetic garnets can then in turn be considered in the light of these results. It has been suggested that the rare earth ions in the RIG's are subdivided into six sublattices whose moments are appreciably canted with respect to the magnetization of the iron sublattices.[72] The tetravalent ions Ti^{4+}, Zr^{4+}, Hf^{4+}, and Sn^{4+}, as well as the two mentioned earlier, Si^{4+} and Ge^{4+}, can at least be partially substituted for iron ions. The charge balance is maintained with Ca^{2+} ions on *c* sites. For small substitutions some of these garnets have an appreciable ferrimagnetic Néel temperature.[73]

5. Other Ferrimagnetic Materials

Of the other known ferrimagnetic materials, none of which is cubic, those with the hexagonal crystal structure have received the greatest attention. The simplest hexagonal ferrimagnet has the same structure as the mineral magnetoplumbite[74] whose approximate chemical composition is $PbFe_{7.5}Mn_{3.5}Al_{0.5}Ti_{0.5}O_{19}$. Ferrimagnetic magnetoplumbite has the general chemical formula $PO{\cdot}6Fe_2O_3$, in which P may be Ba^{2+}, Sr^{2+}, or Pb^{2+}. The ferric ions can also be partly replaced by Al^{3+}, Ga^{3+}, or a combination of divalent and tetravalent ions such as Co^{2+} plus Ti^{4+}. The material with the formula $BaO{\cdot}6Fe_2O_3$, also written $BaFe_{12}O_{19}$, has the trade name

[70] For a summary see S. Geller, *J. Appl. Phys.* **31,** 30S (1960).

[71] M. A. Gilleo and S. Geller, *Phys. Rev.* **110,** 73 (1958); R. Pauthenet, *J. Appl. Phys.* **29,** 253 (1958).

[72] W. P. Wolf, M. Ball, M. J. Hutchings, M. J. M. Leask, and A. F. G. Wyatt, *Proc. Phys. Soc. Japan, Suppl. B-I* **17,** 443 (1962).

[73] Geller, *loc. cit.*

[74] V. Adelskold, *Arkiv Kemi, Min. Geol.* **12-A,** 1 (1938).

Ferroxdure;[75] it is also convenient to call this compound M for the purpose of comparison with the more complex hexagonal structures discussed later.

The atomic arrangement in a plane parallel to the magnetoplumbite c-axis is illustrated[76] in Fig. 9-5.1. The structure is built of alternate spinel (S) and hexagonal (H) blocks. The [111]-direction of the cubic S block lies along the magnetoplumbite c-axis. A barium ion replaces an oxygen ion in the middle layer of the hexagonal block. The blocks marked with an asterisk are rotated 180° about the c-axis with respect to the unmarked

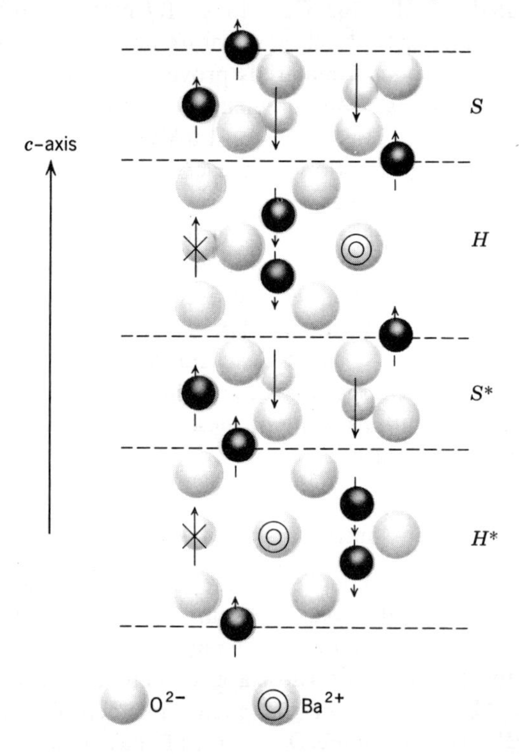

Fig. 9-5.1. A cross-sectional view of one unit cell of the magnetopumbite (M) structure. The c-axis is vertical. The structure consists of spinel (S) and hexagonal (H) blocks. The small spheres represent cations and the arrows, the direction of their magnetic moments. The small light and dark spheres indicate tetrahedral and octahedral sites, respectively. The small spheres circumscribing a cross indicate hexahedral sites, that is, sites with five oxygen nearest neighbors that form a bipyramid. The vertical rows of oxygen ions have trigonal symmetry. (After P. B. Braun.)

[75] J. J. Went, G. W. Rauthenau, E. W. Gorter, and G. W. van Oosterhout, *Philips Tech. Rev.* **13,** 194 (1951).

[76] P. B. Braun, *Philips Res. Rept.* **12,** 491 (1957).

blocks. For a unit cell, which consists of two formula units, the length of the c-axis is 23.2 Å, whereas that of the a-axis is 5.88 Å.

There are three types of cation site: tetrahedral, octahedral, and hexahedral (see Fig. 9-5.1). For the spinel block S it is reasonable to assume that the moments have the same arragement as those in the usual spinels. Thus in each S block of the unit cell the moments of the four ions in octahedral sites would be antiparallel to those of the two ions in tetrahedral sites. The moment arrangement to be expected in the hexagonal block can be deduced from a detailed consideration of the distances and angles between the cations. It is concluded[77] that the moments of the ion in the hexahedral site and three of the ions in octahedral sites would be parallel to the net moment of the spinel block, whereas the moments of the other two ions in octahedral sites would be antiparallel. Hence for $BaO{\cdot}6Fe_2O_3$, in which all the iron ions are trivalent, the predicted moment is $[(4\text{-}2) + (1 + 3 - 2)] \times 5\mu_B = 20\mu_B$ per formula unit. This is precisely the value observed experimentally[77] at 20°K. The spontaneous magnetization is plotted as a function of temperature[77] in Fig. 9-5.2. The unusual shape of this curve appears to spring from the complexity of the interactions between the different sublattices.[78]

The c-axis of $BaO{\cdot}6Fe_2O_3$ is the easy direction of the crystalline anisotropy. The anisotropy constant K_1' of equation 6-9.2 is large—about 3.3×10^6 ergs/cm^3 at room temperature. This feature makes Ferroxdure valuable as a permanent magnet material (see Table 7-4.1). The thermal variation of K_1' and the equivalent anisotropy field $H_K(=2K_1'/M)$ are also shown[79] in Fig. 9-5.2.

A large variety of other hexagonal compounds with more complex structures than magnetoplumbite have been synthesized.[80] The most interesting are called W, Y, and Z compounds. Their chemical composition and relationship to M and S compounds are indicated on the three-phase diagram of Fig. 9-5.3. There are several compounds that lie on the lines MS and MY. The crystal structure of the complex hexagonal ferrimagnets can be considered to consist of spinel (S) and hexagonal (H) blocks. The W compound for example consists of the successive blocks $SSHS^*S^*H^*$, that is, $W = M + 2S$. The hexagonal block (H_2) of the Y compound ($BaO{\cdot}QO{\cdot}6Fe_2O_3$) consists of four layers of oxygens; barium ions replace some of the oxygen ions in the two middle layers. The unit cell of the Z compound is composed of the blocks $H^*SH_2{}^*SHS^*H_2S^*$

[77] J. Smit and H. P. J. Wijn, *Ferrites*, John Wiley and Sons, New York (1959).

[78] G. Heimke, *J. Appl. Phys.* **31,** 271S (1960).

[79] Smit and Wijn, *loc. cit.*

[80] H. P. J. Wijn, *Nature* **170,** 707 (1952); G. H. Jonker, H. P. J. Wijn, and P. B. Braun, *Philips Tech. Rev.* **18,** 145 (1956); P. B. Braun, *Philips Res. Rept.* **12,** 491 (1957).

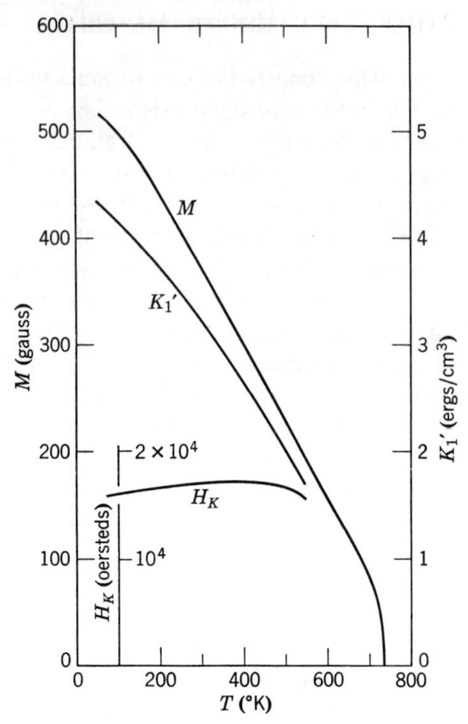

Fig. 9-5.2. Temperature dependence of the spontaneous magnetization M, the crystalline anisotropy constant K_1', and the equivalent anisotropy field $H_K(= 2K_1/M)$ of $BaO{\cdot}6Fe_2O_3$. (After J. Smit and H. P. J. Wijn.)

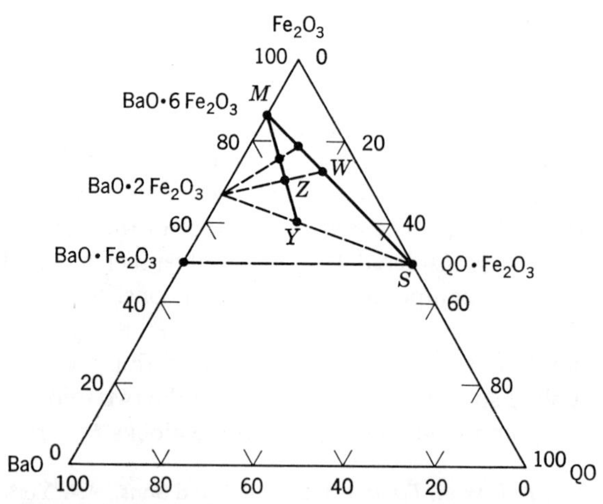

Fig. 9-5.3. The chemical composition of the hexagonal and spinel compounds indicated on a triangular coordinate system; Q represents a divalent cation. [J. S. Smit and H. P. J. Wijn, *Ferrites*, John Wiley and Sons, New York (1959), p. 177.]

in that order; that is $Z = M + Y$. This information and other crystallographic data are summarized in Table 9-5.1.

In order to indicate the type of divalent ion present, it is convenient to use the notation Q_2W, Q_2Y, and Q_2Z; specific examples are Mn_2W, Zn_2Y, and Co_2Z. It appears that most of the divalent ions occupy the same sites in the spinel blocks as they do in the ordinary spinels. At least for most of the compounds the saturation magnetization expected at

Table 9-5.1. Data for Some Hexagonal Compounds

(Q represents the divalent ions Mg^{2+}, Mn^{2+}, Fe^{2+}, Co^{2+}, Ni^{2+}, Zn^{2+}, Cu^{2+} and their combinations)

Symbol	Chemical Formula	Crystallographic Block Stacking	Chemical Relationship	Number of Formula Units per Unit Cell	Length of c-Axis of Hexagonal Unit Cell Å
S	$QO{\cdot}Fe_2O_3$	S		8	
M	$BaO{\cdot}6Fe_2O_3$	SHS^*H^*		2	23.2
W	$BaO{\cdot}2QO{\cdot}8Fe_2O_3$	$SSHS^*S^*H^*$	$M + 2S$	2	32.8
Y	$2BaO{\cdot}2QO{\cdot}6Fe_2O_3$	$SH_2SH_2SH_2$		3	53.5
Z	$3BaO{\cdot}2QO{\cdot}12Fe_2O_3$	$H^*SH_2^*SHS^*H_2S^*$	$M + Y$	2	52.3

$T = 0°K$ is in good agreement with the observed values.[81] The ferrimagnetic temperature can be varied by controlling the kind of divalent ions in the lattice.

The Y compounds possess unusual crystalline anisotropies in that their magnetization prefers to lie in the plane perpendicular to the c-axis at room temperature. The c-axis itself is the hard direction; that is, K_1' is negative. If both the anisotropy constants K_1' and K_2' must be employed, then $K_1' + 2K_2' < 0$. A typical value of $K_1' + 2K_2'$ is of the order of -10^6 ergs/cm^3; for Co_2Y, $K_1' + 2K_2' = -2.6 \times 10^6$ ergs/cm^3 at $T = 300°K$. Materials with this property are sometimes called Ferroxplana. Some of the other compounds also possess an easy plane of magnetization; for example, $Fe_{0.5}Co_{0.75}Zn_{0.75}W$, Co_2Z, and $CoZn_{0.5}Fe_{0.5}Z$. Within the basal plane there is a small crystalline anisotropy with sixfold symmetry.

It can be shown that if $0 < -K_1' < 2K_2'$ the easy direction is any straight line lying on the surface of the cone with the apex angle $\theta = \sin^{-1}(-K_1'/2K_2')^{1/2}$ (problem 9-8). Both Co_2Y and Co_2Z have an easy cone for magnetization at temperatures below about 215°K.

[81] For a summary of the details, see J. Smit and H. P. J. Wijn, *Ferrites*, John Wiley and Sons, New York (1959), Ch. 9.

Finally, it is of interest to mention that strontium-rich compounds of the system $(BaSr)_2Zn_2Fe_{12}O_{22}$, $(Zn_2 Y)$, appear to have helical spin configuration.[82]

Ferrimagnetism has been observed in other crystal structures,[83] such as those of $NiMnO_3$ and $CoMnO_3$. These compounds have rhombohedral structures identical to that of the mineral ilmenite ($FeTiO_3$) and similar to that of hematite (α-Fe_2O_3). A ferrimagnetic fluoride, $Na_5Fe_3F_{14}$, similar to the mineral chiolite ($Na_5Al_3F_{14}$), has been discovered.[84] $Na_5Fe_3F_{14}$ has a high-temperature form that is tetragonal and a low-temperature form that is monoclinic. The mineral pyrrhotite, which has the chemical formula Fe_7S_8 and a monoclinic structure, is also ferrimagnetic.[85] It seems certain that this list will grow in the future.

6. Some Quantum Mechanical Results

Hitherto ferrimagnetism has been discussed within the framework of the molecular field theory. It is obviously of interest to examine the progress made in the application of quantum mechanics to this subject. The theories that have been developed are analogous to those in the other branches of magnetism. Since almost all ferrimagnets are poor conductors, there is the advantage, already mentioned in Section 9-1, that the wave functions can be assumed with reasonable certainty to be localized. As in antiferromagnetism, the presence of more than one sublattice usually represents a formidable theoretical complication. Generally speaking, the theoretical calculations are complex and difficult. Consequently, the details will not be reproduced here. Instead, some of the assumptions underlying the theories and a brief survey of the more pertinent results will be presented.

The molecular field in ferrimagnetic insulators has its origin in an indirect exchange interaction. The general theory is the same as that already discussed in Section 9-5 for antiferromagnetism. In the application to specific ferrimagnetic materials the crystal structure must be considered. The magnitude of the superexchange interaction depends on the distance and angles between the cations; some of the details for spinels and

[82] U. Enz, *J. Appl. Phys.* **32,** 22S (1961).

[83] W. H. Cloud, *Phys. Rev.* **111,** 1046 (1958); E. F. Bertaut and F. Forrat, *J. Appl. Phys.* **29,** 247 (1958); T. J. Swoboda, J. D. Vaughan, and R. C. Toole, *J. Chem. Phys. Solids* **5,** 293 (1958).

[84] K. Knox and S. Geller, *Phys. Rev.* **110,** 771 (1958).

[85] F. Bertaut, *Compt. rend. (Paris)* **234,** 1295 (1952); L. Néel, *Revs. Mod. Phys.* **25,** 58 (1953); F. K. Lotgering, *Philips Res. Rept.* **11,** 190 (1956); E. Hirahara and M. Murakami, *J. Phys. Chem. Solids* **7,** 281 (1958); J. T. Sparks, W. Mead, A. J. Kirschbaum, and W. Marshall, *J. Appl. Phys.* **31,** 356S (1960).

garnets have already been given in Sections 9-3 and 9-4. Other indirect exchange mechanisms proposed for particular ferrites are discussed in the literature.[86]

The first application of the series expansion method to ferrimagnetism[87] has employed the approach developed by Heisenberg (see Section 6-4). This, of course, yields only the lowest order terms. The results were substantially the same as those from the molecular field theory. More recently, the susceptibility of a two-sublattice ferrimagnet has been calculated as a power series in $1/T$ by employing exact series expansion methods.[88] The coefficients have been evaluated through to a_5 for arbitrary S. In this order of approximation there are appreciable deviations from the molecular field calculations. For example, when $S_A = S_B = \frac{1}{2}$, the values predicted for J_e/kT_{FN} from the series and the molecular field theory are 0.357 and 0.235, respectively.

The Bethe-Peierls-Weiss method has been applied to ferrimagnetism[89] for the case in which all the magnetic atoms have a spin quantum number equal to $\frac{1}{2}$. The only interactions considered were between the two sublattices. The $1/\chi - T$ curves calculated from this and the molecular field theory differ appreciably for $T_{FN} < T < 4T_{FN}$; at higher temperatures, however, they approach one another. For spinels the theory predicts $J_e/kT_{FN} = 0.297$.

A number of calculations have been made, both semiclassical and quantum mechanical, concerning the spin waves in ferrimagnets. Since ferrimagnetic materials possess a spontaneous magnetization, it might be guessed that the dispersion relationship would be quadratic. The theories confirm[90] that for the lowest branch, which is the important one at low temperatures, $\omega \propto k^2$; k is the wave number. It then follows that at low temperatures the spontaneous magnetization has the same functional dependence on temperature as in the ferromagnetic case, namely,

$$M(T) = M(0)(1 - A_{\text{exp}}T^{3/2}).$$

Here A_{exp} depends on the exchange interaction between and within the sublattices and on the number of nearest neighbors. It has been found

[86] D. G. Wickham and J. B. Goodenough, *Phys. Rev.* **115,** 1156 (1959).

[87] L. Valenta, *Czech. J. Phys.* **9,** 29 (1959).

[88] P. J. Wojtowicz, *J. Appl. Phys.* **31,** 265S (1960).

[89] J. S. Smart, *Phys. Rev.* **101,** 585 (1956).

[90] H. Kaplan, *Phys. Rev.* **86,** 121 (1952); E. I. Kondorskii, A. S. Pakhomov, and T. Skiklosh, *Dokl. Akad. Nauk SSSR* **109,** 931 (1956) [trans. *Soviet Phys.-Doklady* **1,** 501 (1956)]; T. A. Kaplan, *Phys. Rev.* **109,** 782 (1958). A linear dispersion law has been found by S. V. Vonsovsky and Y. M. Seidov and reported in *Izvest. Acad. Nauk SSSR* **18,** 319 (1954); T. A. Kaplan, however, has shown that there is a mistake in their calculation.

experimentally[91] that $MnFe_2O_4$ does indeed follow a $T^{3/2}$ law up to $T/T_{FN} \approx 0.7$. For the spinels $NiFe_2O_4$, $CoFe_2O_4$, and Fe_3O_4 a better fit with a T^2 law is found.[91]

Finally, the origin of the crystalline anisotropy will be considered. Now, the usual expressions employed to describe the crystalline anisotropy energy (equations 6-9.1 and 6-9.2), refer to the direction of the resultant magnetization of the sublattices. On the macroscopic level, however, this anisotropy arises from the anisotropies of the individual sublattice magnetizations, that is, the total anisotropy energy F_K is actually the sum of the sublattice anisotropy energies. It is found that the equations for F_K provide a good description of the experimental results for many ferrimagnetic materials. The reason is that the sublattice anisotropies are small, at least compared to the exchange interactions. The free energy for each sublattice can then be expanded into a simple rapidly converging series, whose terms add to give the familiar expressions for F_K. Sometimes these expressions do not fit the data; the rare earth garnets at low temperatures are examples. In these cases it may be appropriate to include terms in F_K that have other forms. To illustrate, it has been shown on theoretical grounds that the form

$$F = \sum_{i=1}^{3} |[A^2 + (B^2 - A^2)\alpha_i^2]^{1/2}|$$

may be appropriate for some of the rare earth ions in garnets at low temperatures,[92] whereas the form

$$F_{\text{Co}} = A \sum_{i=1}^{4} \log \cosh (B \cos \theta_i)$$

where θ_i is the angle the magnetization makes with a [111]-direction may apply to the Co^{2+} ions in cobalt doped magnetite.[93] Here A and B are constants.

The thermal variation of anisotropy constants, such as K_1, may be unusual compared to the ferromagnetic case because of the manner in which each sublattice contributes. For example, K_1 for magnetite changes sign[94] at $T \approx 130°K$; just below this temperature the cube edges are easy directions, whereas above the body diagonals are the easy directions. It would be of interest to establish whether the anisotropy of a sublattice depends in a simple way on its magnetization, say via a tenth power law (see equation 6-9.11). Experiments have been conducted on spinels in

[91] R. Pauthenet, *Ann. phys.*, (*Paris*) **7,** 710 (1952).
[92] W. P. Wolf, *Proc. Phys. Soc.* (*London*) **74,** 605 (1959).
[93] J. C. Slonczewski, *Phys. Rev.* **110,** 1341 (1958).
[94] L. R. Bickford, *Phys. Rev.* **110,** 1341 (1958).

which the effect of varying the number of magnetic cations on the two sublattices has been studied. It was found[95] that at a given temperature K_1 could be written as

$$K_1 = N_A K_{1A} + N_B K_{1B},$$

where K_{1A} and K_{1B} are the anisotropy constants of the A and B sublattices, respectively, and N_A and N_B are the number of magnetic ions on the corresponding two sublattices. This result implies that in these materials at least the sublattice anisotropies are independent of one another.

On the microscopic level theories of anisotropy based on the one-ion model[93,96] mentioned in Section 6-9 have been reasonably successful. In this model the anisotropy of a single ion is calculated. This anisotropy depends on the kind of ion and the strength and symmetry of the crystalline field. The total anisotropy is then given by a suitable summation of the individual ionic anisotropies. In the analysis it is important to have available values of the anisotropy parameters in the spin-Hamiltonian. These may be obtained from electron paramagnetic resonance experiments on doped diamagnetic single crystals that are isomorphous with the ferrimagnetic compound. Particularly large anisotropies are expected when the orbital degeneracy is not completely removed by the crystalline field. This is undoubtedly the origin of the large anisotropy in spinels containing Co^{2+} ions.[97] At low temperatures the rare earth garnets exhibit what have been called giant anisotropy anomalies. It has been suggested that they occur for certain orientations of the magnetization vector because the two lowest energy levels of the rare earth ions cross over or almost cross over under the action of the combined crystalline and exchange fields.[98] In $BaO \cdot 6Fe_2O_3$ it has been shown that the classical dipole-dipole coupling makes an appreciable contribution to the anisotropy.[99]

7. Soft Ferrimagnetic Materials

The general principles that were developed for soft ferromagnetic materials in several sections of Chapter 7 apply also, of course, to

[95] V. J. Folen and G. T. Rado, *J. Appl. Phys.* **29,** 438 (1958). See also H. S. Belson and C. J. Kriessman, *J. Appl. Phys.* **30,** 170S (1959); R. F. Pearson, *J. Phys. radium* **20,** 409 (1959).

[96] G. T. Rado and V. J. Folen, *Bull. Am. Phys. Soc.* **1,** 132 (1956); K. Yosida and M. Tachiki, *Prog. Theor. Phys. (Kyoto)* **17,** 331 (1957); W. P. Wolf, *Phys. Rev.* **108,** 1152 (1957).

[97] J. C. Slonczewski, *J. Appl. Phys.* **32,** 253S (1961).

[98] C. Kittel, *Phys. Rev.* **117,** 681 (1960).

[99] J. Smit and A. J. W. Duyvesteyn, *J. Phys. radium* **70,** 368 (1959).

ferrimagnetic ones. Some properties however, are peculiar to soft ferrimagnetic materials and are worthy of specific comment.

Since polycrystalline ferrimagnetic materials are ceramics, the samples are prepared by a sintering process. The product therefore has a certain porosity. The pores may occur both in the grain boundaries and in the crystallites. Because uncompensated poles appear on the pore surfaces, there may be appreciable associated magnetostatic energy which profoundly influences the magnetic properties. For example, a domain wall in a grain may be pinned at the pores. When a magnetic field is applied, the wall motion may consist of bulging, that is, only the unpinned regions may move. In such a movement uncompensated poles would, of course, appear on the domain wall's surface, and this effect would limit the amount of bulging. Although some interesting visual studies of domain configurations have been made, many details are still uncertain.

The hysteresis loop is altered by the demagnetization fields produced at the pores. The coercive force, according to equation 7-7.10, will vary linearly with the volume of the pores. This predicted behavior has been observed in a nickel zinc spinel[100] for relative volumes of pores of less than about 40%. The remanence is, of course, reduced because of the internal fields. The initial permeability is found to increase when the porosity is reduced.[101] The initial permeability is also greater if the crystallites are larger than a certain minimum size; for manganese zinc spinels this size is of the order of 5 microns.[102] Presumably there is then a contribution from wall motion as well as from magnetization rotations.

A uniaxial anisotropy is induced in some spinels containing small amounts of Co^{2+} ions when these ferrites are cooled in the presence of a magnetic field.[103] The origin of this uniaxial anisotropy is not certain, but it seems likely either that Co^{2+} ion pairs form (see Section 7-9)[104] or that population of sites with a certain uniaxial anisotropy occurs[105] during the diffusion at elevated temperatures. Vacancies may play an important role in the process.[106] Another possible mechanism is the formation of acicular-shaped particles of a second phase.

The hysteresis loops of materials with an induced uniaxial anisotropy tend to have more of a rectangular shape than otherwise. A single preferred direction can also be introduced by the application of a tension

[100] J. Smit and H. P. J. Wijn, *Ferrites*, John Wiley and Sons, New York (1959), p. 301.

[101] F. Brown and C. L. Gavel, *Phys. Rev.* **97,** 55 (1955).

[102] C. Guillaud, *Proc. IEE* (*London*), *Suppl.* 5 **104-B,** 165 (1957).

[103] R. M. Bozorth, E. F. Tilden, and A. J. Williams, *Phys. Rev.* **99,** 1788 (1955); R. F. Penoyer and L. R. Bickford, *Phys. Rev.* **108,** 271 (1957).

[104] L. Néel, *Compt. rend.* (*Paris*) **237,** 1468 (1953).

[105] J. C. Slonczewski, *Phys. Rev.* **110,** 1341 (1958).

[106] S. Iida, *J. Appl. Phys.* **31,** 251S (1960).

or compression. This method is particularly effective if the crystalline anisotropy is small. Thus rectangular loops have been obtained by melting glass onto a nickel zinc spinel ring and cooling.[107] Although similar studies on garnets have been much less extensive to date, an induced uniaxial anisotropy has been induced in YIG and LuIG in the neighborhood of liquid nitrogen temperatures by cooling in a magnetic field.[108] It has been suggested that an electronic ordering process is involved.

Ferrimagnetic materials with rectangular loops have important applications as memory elements and as multiaperture devices for digital computers. Therefore any design principles that lead to rectangular loop materials are of special technical interest. A manganese magnesium spinel was the first such material prepared[109]; since then a number of other spinels with this property have been manufactured. It is not always clear why their hysteresis loops are rectangular. In some cases a uniaxial anisotropy produced by either crystalline fields or stresses may be the origin. In others, the field necessary to nucleate domain walls may be larger than that required for irreversible wall motion.

When spinels that contain some cobalt are cooled in the absence of a magnetic field, their minor hysteresis loops frequently have a wasp waist near the origin[110] (for an example of a wasp-waisted loop see curve (*a*) for Permivar in Fig. 7-9.3). On the other hand, their major loops have no constriction near the origin. Presumably the diffusion of the Co^{2+} ions that occurs during cooling produces an extremely stable domain structure; that is, certain regions become deep energy minima for the domain walls. It is then difficult to change the magnetization by wall motion in low fields. This stable domain structure is removed during traversal of a major loop. The whole effect, of course, is closely related to the production of rectangular loops obtained by cooling in a field.

When time-varying magnetic fields are applied to spinels, it is found that equation 7-10.5 for the total losses in ferromagnetic materials,

$$\frac{W_T}{{B_1}^2} = aB_1 + ev + c$$

[107] G. W. Rathenau and G. W. van Oosterhout, quoted in E. W. Gorter, *Proc. IRE* **43,** 1945 (1955).

[108] B. A. Calhoun, *J. Appl. Phys.* **30,** 293S (1959); D. J. Epstein, B. Frackiewicz, and R. P. Hunt, *J. Appl. Phys.* **32,** 270S (1961).

[109] E. Albers-Schoenberg, *J. Appl. Phys.* **25,** 152 (1954).

[110] E. W. Gorter and C. J. Esveldt, *Proc. IEE (London), Suppl.* 5 **104-B,** 418 (1957); O. Eckert, *Proc. IEE (London), Suppl.* 5 **104-B,** 428 (1957); N. Kosnetzki, J. Brackmann, and J. Frei, *Naturwiss.* **42,** 482 (1955); J. Smit and H. P. J. Wijn, *Ferrites*, John Wiley and Sons, New York (1959), pp. 310–317.

is no longer valid.[111] For one thing, the eddy current losses are negligible because of the large resistivity; that is, $e \approx 0$. For another, the residual losses consist of several types, such as dimensional and relaxation losses; they cannot be described by one constant. Finally, the hysteresis losses cannot generally be expressed by a simple constant except perhaps in fields of small amplitude. The magnetization curve of spinels with large initial permeabilities depends on the frequency.[112] The reversal of the magnetization of a spinel by a pulse (switching) is a subject of considerable practical importance. If the magnetization reversal occurs by means of irreversible domain-wall motions, the switching time τ of many spinels is related to the amplitude H of the pulse by the equation

$$(H - H_k)\tau = s, \tag{9-7.1}$$

where H_k and s are constants of the materials.[113]

Polycrystalline spinels have many low-frequency applications in addition to their use, already mentioned, as memory elements of computers. Components in which these materials have been employed include antenna cores, tuners, loading coils, filter inductors, wide band and other transformers, deflection yokes, and recording heads.[114]

8. Some Topics in Geophysics

Rocks may be described as agglommerates of grains of various minerals. Some of these grains may possess a spontaneous magnetization. An important constituent of rocks with this property is magnetite (Fe_3O_4), either pure or doped with cations such as Ti^{4+}, Mg^{2+}, Mn^{2+}, Al^{3+}, or V^{3+}. The magnetic oxides of greatest importance belong to the system $FeO—Fe_2O_3—TiO_2$. Some of the end products of this system besides magnetite are ulvöspinel (Fe_2TiO_4), maghemite (γ-Fe_2O_3), hematite (α-Fe_2O_3), and ilmenite ($FeTiO_3$). Ulvöspinel has the spinel structure and usually occurs as an intergrowth with magnetite; γ-Fe_2O_3 is not commonly found in rocks. Both hematite and ilmenite have the rhomohedral structure. Although α-Fe_2O_3 is antiferromagnetic, it has a small spontaneous magnetization that originates from the canting of the sublattice moments (Section 8-9). Some iron sulfide grains are also ferrimagnetic.

It is reasonable to presume that the earth's magnetic field played an important role in the production of a rock's remanent magnetization.

[111] For further details see J. Smit and H. P. J. Wijn, *Ferrites*, John Wiley and Sons, New York (1959).

[112] H. P. J. Wijn and J. J. Went, *Physica* **17,** 976 (1951).

[113] N. Menyuk and J. B. Goodenough, *J. Appl. Phys.* **26,** 8 (1955); J. K. Galt, *Bell Syst. Tech. J.* **33,** 1023 (1954).

[114] For a summary see C. D. Owens, *Proc. IRE* **44,** 1234 (1956).

Igneous rocks are produced by the fast or slow cooling of a molten mass. The final magnetization acquired by such a rock on cooling in a small magnetic field is called the thermoremanent magnetization (TRM). When a magnetic grain is cooled just below its critical (Néel or Curie) temperature, a small magnetization along the direction of the earth's magnetic field will be induced. On further cooling, the small magnetization will usually increase reversibly, although its direction will remain fixed. At normal surface temperatures ($\sim 0°C$), the magnitude of the remanent magnetization may be relatively large. Indeed, to produce a similar remanence isothermally at 0°C would require the application of a much greater field than that of the earth, namely one of the order of the coercive force. The situation is rather different for sedimentary rocks. The grains already have a remanent moment, and this moment is oriented by the geomagnetic field at the time of settling. There may, of course, be perturbing effects produced by the wind and water currents, but changes in orientation from these mechanisms appear to be small. Metamorphic rocks have been transformed by high pressures and temperatures. In these rocks the relationship of the direction of the remanence and the geomagnetic field is more obscure.

The small remanent magnetization of a rock can be measured, for example, by rotating the specimen quickly in a pick-up coil and by employing a suitable method of amplification. From these data attempts can be made to infer the direction of the earth's magnetic field at the time of the rock's formation. The history of the geomagnetic field is known from direct observations for the last 300 or 400 years. Hence for this period of time a direct test of the interpretations of rock studies can be made. For prehistoric times various other tests have been devised; however, none is conclusive. Probably the best test is simply the consistency of the results from all over the world for rocks produced at the same period of time.

The general subject is also of archaeological interest. Some bricks and baked clays of ancient origin have a thermoremanent magnetization,[115] presumably acquired when the materials were cooled in the earth's field during the last stage of the manufacturing process. A knowledge of the temporal variation of the earth's field then permits the time of manufacture to be estimated and, of course, vice versa. Complementary information can be obtained with the radioactive isotope (C^{14}) dating method.

An analysis of the TRM data on rocks must consider any changes in orientation of the original formation, for example, from uplifting or folding, as well as any physical or chemical changes in the magnetic grains themselves. The magnetization processes involved will depend on

[115] E. Thellier, *Ann. Inst. Phys. Globe, J. Phys. radium* **12,** 205 (1951).

the size of the magnetic grain. Most magnetite grains are larger than the critical single-domain size (Section 7-2). Therefore theories of small multidomain particles apply (Section 7-5). Hematite grains are usually smaller and, in terms of their ferromagnetic component, either single domains or superparamagnetic. Some single-domain grains may have a blocking temperature between 0°C and the Néel temperature; that is, on heating they become superparamagnetic (Section 7-3). Discussions of the various magnetization processes as they apply to rock magnetism may be found in the literature.[116]

Rock magnetism studies appear to lead to two main and essentially independent conclusions. First, the axis of the earth's dipole has appreciably changed its direction with respect to the continents in ancient times. Polar wandering occurs when the earth's crust or even the whole earth moves with respect to the axis of rotation. In addition, the continents may have moved in relation to one another (continental drift). Although the evidence is often fragmentary, it has been possible to suggest the direction of the axial dipole for the last 400 million years.[117] The second result is more surprising. At various intervals the direction of the earth's magnetization has reversed, with about equal time being spent in the two polarization directions. The earth's dipole has had its present (normal) direction for roughly the last million years. The intervals of time in which the dipole has assumed a given direction has varied generally from 1 to 10 million years; for the Permian the dipole was apparently reversed from the normal direction for the entire period of approximately 250 million years.[118] The time taken for reversal is relatively short, being about 10^4 years. Attempts have been made to explain these results in terms of models that consider the earth's interior; a prominent example is the dynamo theory.[119] However, the origin and properties of the earth's dipole are not yet understood.

Several mechanisms, other than the reversal of the earth's field, that could produce a reversed rock magnetization have been proposed. Because of the importance of the subject and the intrinsic interest of these processes, some of the mechanisms suggested will now be enumerated and discussed briefly.

1. If a ferrimagnetic material possesses a compensation temperature, the spontaneous magnetization will become reversed to the direction of

[116] T. Nagata, *Rock Magnetism*, Maruzen Co., Tokyo, 2nd ed. (1961); L. Néel, *Advan. Phys.* **4,** 191 (1955); F. D. Stacey, *Advan. Phys.* **12,** 45 (1963).

[117] S. K. Runcorn, *Advan. Phys.* **4,** 244 (1955).

[118] P. M. S. Blackett, *J. Phys. Soc. Japan, Suppl. B-I* **17,** 699 (1962).

[119] W. M. Elasasser, *Phys. Rev.* **72,** 821 (1947), *Revs. Mod. Phys.* **28,** 135 (1956); E. C. Bullard and H. Gellman, *Phil. Trans. Roy. Soc.* **A-247,** 213 (1954).

the earth's field on cooling (Section 9-3). Examples of such materials include the $NiAl_\delta Fe_{2-\delta}O_4$ system (Fig. 9-3.6), $Li_{0.5}Cr_\delta Fe_{2.5-\delta}O_4$, and some of the garnets (Fig. 9-4.4). Such substances have not yet been found in nature.

2. For a ferrimagnet with, say, two sublattices, diffusion of ions from one site to another may cause magnetization reversal. For example, the spontaneous magnetization of a quenched sample of the spinel $NiAl_\delta Fe_{2-\delta}O_4$ with $\delta \approx 1$ will, on annealing, change direction.[120] This effect is caused by the diffusion of Al^{3+} ions from A to B sites on heating. The diffusion could probably occur at surface temperatures in the millions of years available because of the geological time scale involved. Again, it has not been established that grains with the necessary chemical composition occur naturally.

3. Lightning and perhaps mechanical stresses may produce reversals. In lightning the local heating and associated magnetic fields may change the magnetization direction. Such effects are likely to be local in scale.

4. Reversal is also possible if the rocks have two magnetic constituents, A and B, with appreciably different ferrimagnetic Néel temperatures.[121] Suppose grains A have the higher Néel temperature, $T_{FN(A)}$. On cooling in the earth's field, the A grains, at a temperature a little below $T_{FN(A)}$, acquire a spontaneous magnetization in the field direction. The B grains, just below their Néel temperature, will experience a field that is the resultant of the earth's field and the demagnetizing field of the A grains. This resultant field may be in the opposite direction to the earth's field. If on further cooling to surface temperatures the TRM of B is larger than that of A, the rock's remanent magnetization will be reversed with respect to the direction of the earth's field. Of course, the details of the process depend on the shape and size of the grains and their spatial distribution as well as on their chemical composition.

The important interaction between the grains may be exchange rather than magnetostatic. This is most likely if the two constituents are in intimate contact, as, for example, in an intergrowth. Exchange interaction at an interface between two components has been discussed in Section 8-11.

Reversals from this cause have been observed in volcanic rocks found in Japan,[122] and there is evidence that an indirect negative exchange interaction between ordered and disordered phases of the ilmenite-hematite system[123] is operating.

[120] E. W. Gorter, *Philips Res. Rept.* **9**, 295 (1954).
[121] L. Néel, *Ann. Geophys.* **7**, 90 (1951).
[122] T. Nagata, S. Akimoto, and S. Uyeda, *J. Geomag. Geoelec.* **5**, 1681 (1953).
[123] S. Uyeda, *J. Geomag. Geoelec.* **9**, 61 (1957); W. H. Meiklejohn and R. E. Carter, *J. Appl. Phys.* **31**, 164S (1960); Y. Ishikawa and Y. Syono, *J. Phys. Soc. Japan, Suppl. B-I* **17**, 714 (1962).

5. Reversal may be the result of changes in composition, which may arise from oxidation during weathering, exsolution, and other alterations. For example, a solid solution may segregate into two phases on cooling. The exchange fields acting may then change and cause a rearrangement of the spin directions. As a consequence, the direction of the net magnetization may be reversed. It has been suggested, although not proved, that this process may occur in the ilmenite-hematite system.[124] Another example[125] is the oxidation of magnetite to γ-Fe_2O_3. The latter (*B* grains) may then become magnetized in the demagnetization field of the former (*A* grains). Ultimate demagnetization of the magnetite would lead to a reversal in the remanence. *B*-type grains could also form as the result of exsolution at surface temperatures of $\approx 0°C$. Finally, consider a partial reversal in a two-grain system, that is, in which the TRM of *A* is larger than that of *B*. Destruction of the magnetization of *A* by chemical change would produce complete reversal.

Few rocks, whether reversed in nature or not, exhibit magnetization reversal in the laboratory after heating and then cooling in a small magnetic field. A somewhat larger, although still small, fraction shows partial reversal; some of these rocks after additional heat treatment show complete reversal. Of course, this does not preclude the possibility that physical and chemical changes occurring after the rock's formation have produced the reversal. However, systematic chemical and physical differences between normal and reversed rocks have seldom been found. The current concensus favors the hypothesis that the earth's dipole has undergone many reversals in direction in the past.

Problems

9-1. (a) Show that the expression for the susceptibility of a ferrimagnet (equation 9-2.9) reduces to that for an antiferromagnet (equation 8-3.6) when the sublattices are identical, that is, if $C_A = C_B = C$ and $N_{AA} = N_{BB} = N_{ii}$. (b) Give a geometrical interpretation to the θ' that appears in equation 9-2.9. In a $1/\chi - T$ diagram plot both branches of the hyperbola that this equation represents. (c) Discuss the relative sizes and signs of the sublattice magnetizations M_A and M_B above the ferrimagnetic Néel temperature.

9-2. Show that the limiting condition for the occurrence of ferrimagnetism ($T_{FN} > 0$) is given by $\alpha\beta = 1$ where α and β are defined by equations 9-2.2.

9-3. Show that if $\alpha > M_B(0)/M_A(0)$ then $M_A < M_A(0)$ at absolute zero, whereas if $\beta > M_A(0)/M_B(0)$, $M_B < M_B(0)$ at $T = 0°K$.

[124] L. Néel, *Advan. Phys.* **4,** 191 (1955).

[125] J. W. Graham, *J. Geophys. Res.* **58,** 243 (1953).

9-4. For the two sublattice model of ferrimagnetism establish that the molecular fields H_{mA} and H_{mB} are equal in magnitude when $M_A(0)/M_B(0) = (N_{AB} + N_{AA})/(N_{AB} + N_{BB})$. Verify that $dM/dT \to 0$ as $T \to T_{FN}$ for $M_A(0)/M_B(0) = g_B(S_B + 1)/g_A(S_A + 1) \times (N_{AB} - N_{BB})/(N_{AB} - N_{AA})$.

9-5. Calculate the ferromagnetic Curie temperature T_f when there is a positive (ferromagnetic) interaction between the A and B sublattices. Find the condition that $T_f > 0$. Show that the results of problem 9-3 also hold for the present case.

9-6. (a) If $N_{AB}(T) = N_{AB}(0)(1 + \rho T)$ and α and β are constant, show that the relationship between the observed Curie constant C_e and that calculated theoretically (C) is

$$\frac{1}{C_e} = \frac{1}{C} + \frac{\rho}{\chi_0}.$$

(b) For a system with three magnetic sublattices prove that the susceptibility above T_{FN} depends on temperature as given by equation 9-4.1. For $N_{ac} = N_{ad} = N_{cc} = 0$ show that this equation reduces to that for two sublattices (equation 9-2.9).

9-7. (a) By starting with equation 9-4.3, derive an expression for the susceptibility χ_c without making any approximations. Calculate the susceptibility of the Gd sublattice in GdIG by employing the data for M_c given in Fig. 9-4.5. (b) Obtain a numerical estimate of N_{cc} from the fit of the Curie-Weiss law to the $1/\chi_c - T$ data for GdIG given in Fig. 9-4.6. (c) Prove that $\chi_c = 1/N_{cad}$, where N_{cad} is the coefficient for the molecular field produced by the ensemble of iron ions at a c site.

9-8. Assume that the crystalline anisotropy energy of material with a unique axis is given by $F_K = K_1' \sin^2 \theta + K_2' \sin^4 \theta$. Show that when $0 < -K_1' < 2K_2'$ the easy directions are any lines that make an angle $\theta = \sin^{-1}(-K_1'/2K_2')^{1/2}$ with the symmetry axis and therefore trace out a cone of revolution.

Bibliography

Fairweather, A., A. J. E. Welch, and F. F. Roberts, "Ferrites," *Rept. Progr. Phys.* **15,** 142 (1952).

"Ferrites Issue," *Proc. IRE* **44,** 1229 (1956).

Gorter, E. W., "Saturation Magnetization and Crystal Chemistry of Ferrimagnetic Oxides," *Philips Res. Rept.* **9,** 295 (1954).

"International Conference on Magnetism," *J. Phys. radium* (March 1951), (Feb.-Mar. 1959); *J. Phys. Soc. Japan, Suppl. B-I* (March 1962).

Hellwege, K. H., and A. M. Hellwege, *Magnetic Properties I,* J. Springer, Berlin (1962).

Lax, B., and K. J. Button, *Microwave Ferrites and Ferrimagnetics,* McGraw-Hill Book Co., New York (1962).

"London Conference on Ferrites," *Proc. IEE (London), Suppl.* 5 **104-B,** 127 (1957).

Nagata, T., *Rock Magnetism,* 2nd ed., Maruzen Co., Tokyo (1961).

Néel, L., "Magnetic Properties of Ferrites, Ferrimagnetism and Antiferromagnetism," *Ann. phys.* (*Paris*) **3,** 137 (1948).

"Rock Magnetism Issue," *Advan. Phys.* **4,** 113 (1955).

Smit, J., and H. P. J. Wijn, *Ferrites*, John Wiley and Sons, New York (1959).

Soohoo, R. F., *Theory and Applications of Ferrites*, Prentice-Hall, Englewood Cliffs, N.J. (1960).

Standley, J. K., *Oxide Magnetic Material*, Oxford University Press, Oxford (1962).

"Symposiums on Magnetism and Magnetic Materials," *Am. Inst. Elect. Engrs.* (Feb. 1957), *J. Appl. Phys. Suppl.* (March 1958), (April 1959), (May 1960), (March 1961), (March 1962), (March 1963), (March 1964).

Wolf, W. P., "Ferrimagnetism," *Rept. Progr. Phys.* **24,** 212 (1962).

10

Resonance in Strongly Coupled Dipole Systems

1. Introduction

Paramagnetic resonance, both electronic and nuclear, has already been discussed in Chapters 3 and 4. When there is a large exchange interaction between the magnetic atoms there exists an analogous phenomena called *ferromagnetic resonance*. When the sublattice structure is important, the phenomena is called either *ferrimagnetic* or *antiferromagnetic resonance*, depending on which is appropriate to the particular spin arrangement. A number of interesting discoveries have emerged from studies in this area, such as magnetostatic modes, high power effects, spin-wave line broadening, and direct observation of spin waves. Besides these topics, others that are discussed include magnetomechanical effects, domain-wall resonance, wave propagation in insulators, nuclear resonance in strongly coupled dipole systems, and the Mössbauer effect.

2. Magnetomechanical Effects

For a ferromagnetic material there is a relationship between the physical rotation of the specimen and a change in magnetization that can be measured with accuracy. Two main types of experiment have been devised. In one, known as the Einstein-de Haas effect, the rotation of the sample produced by a change in magnetization is determined.[1] In the

[1] A. Einstein and W. J. de Haas, *Verh. deut. phys. Ges.* **17,** 152 (1915), **18,** 173, 423 (1916); O. W. Richardson, *Phys. Rev.* **26,** 248 (1908).

other, known as the Barnett effect, the change in magnetization produced by a rotation of the sample is measured.[2] In addition, there are other possible types of experiment,[3] some of which have been attempted.[4]

The majority of experiments and all those since 1950 have employed the Einstein–de Haas method. This experimental arrangement is outlined in Fig. 10-2.1. The ferromagnetic rod or ellipsoid is suspended as a torsional pendulum in a solenoid which serves to change the magnetization. Since the angular momentum changes are small, various techniques have been developed to improve the sensitivity. The angular displacement is increased by changing the direction of the magnetization at the same frequency as the natural one of the torsional pendulum. The best results are obtained if the magnetization is changed when the pendulum crosses its zero position.[5] Extreme care must be taken to cancel out the earth's field, for otherwise the torque produced is many times larger than that to be measured.[6]

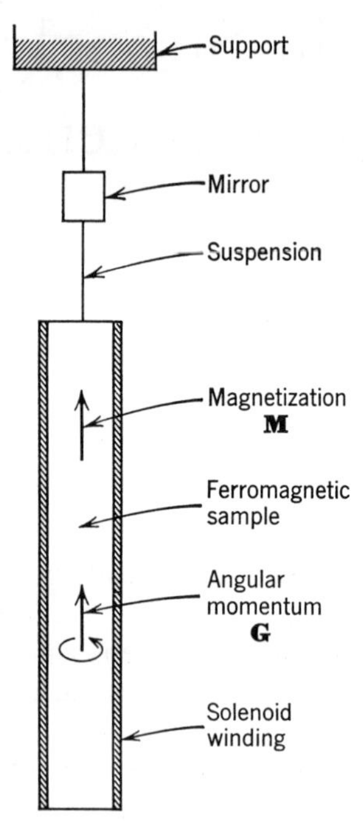

Fig. 10-2.1. Schematic of the Einstein-de Haas experiment.

The experiment measures the magneto-mechanical ratio, defined as

$$g' \frac{e}{2mc} = \frac{\text{magnetization change}}{\text{angular momentum change}},$$

where g' is known as the magnetomechanical factor or simply as the g'-factor. The purpose of the prime is to distinguish it from the analogous but different g-factor employed in ferromagnetic resonance experiments. Now the magnetization change is the result of both spin, ΔM_{spin}, and orbital, ΔM_{orbit}, contributions. On the other hand, the total angular momentum G is given by the sum

$$G = G_{\text{orbit}} + G_{\text{spin}} + G_{\text{lattice}}.$$

[2] S. J. Barnett, *Phys. Rev.* **6,** 171 (1915), **10,** 7 (1917), *Proc. Amer. Acad. Arts. Sci.* **60,** 125 (1925), **75,** 109 (1944).

[3] O. von Auwers, *Naturwiss.* **13,** 202 (1935).

[4] For a summary see L. Bates, *Modern Magnetism*, 4th ed., Cambridge University Press, Cambridge (1962).

[5] F. Coeterier and P. Sherrer, *Helv. Phys. Acta* **5,** 217 (1932).

[6] G. G. Scott, *Rev. Sci. Instr.* **28,** 270 (1957).

Since there are no external torques, the change in the total angular momentum ΔG is zero, hence

$$-\Delta G_{\text{lattice}} = \Delta G_{\text{orbit}} + \Delta G_{\text{spin}}.$$

Therefore we have

$$g' \frac{e}{2mc} = \frac{\Delta M_{\text{orbit}} + \Delta M_{\text{spin}}}{-\Delta G_{\text{lattice}}} = \frac{\Delta M_{\text{orbit}} + \Delta M_{\text{spin}}}{\Delta G_{\text{orbit}} + \Delta G_{\text{spin}}}$$

or, on omitting the Δ's,

$$g' = \frac{L + 2S}{L + S}, \tag{10-2.1}$$

where L and S refer to the expectation values of the atomic orbital and spin angular momenta, respectively.

The relationship between g and g' will now be derived. The g-factor in ferromagnetic resonance is defined as for paramagnetic resonance; that is, it is the splitting factor necessary when the total angular momentum is described only by the spin quantum number (Section 3-7). Then, since the energy per unit volume of the split levels, $gN\mu_B SH$, is also equal to MH, the magnetization may be written

$$M = gN\mu_B S = N\mu_B(L + 2S).$$

This yields

$$g = \frac{L + 2S}{S}. \tag{10-2.2}$$

From equations 10-2.1 and 10-2.2 it follows immediately that[7]

$$\frac{1}{g} + \frac{1}{g'} = 1. \tag{10-2.3}$$

If the orbital angular momentum is almost quenched, that is, if $M_{\text{orbit}} < M_{\text{spin}}$, it is easy to show (problem 10-1) that[8]

$$g - 2 = 2 - g'. \tag{10-2.4}$$

If only the spin contributes to the angular momentum, then, of course, $g' = g = 2$. If there is a small orbital contribution, the foregoing equations show that $g' < 2$ and $g > 2$.

The averages of the most recent and, presumably, most accurate determinations of g' for various materials are compiled in Table 10-2.1. The values range from 1.84 to 1.99 and thus clearly show that there is a small but significant orbital contribution. The nearness of these values to 2 probably provides the best justification for the frequently employed

[7] J. Smit and H. P. J. Wijn, *Ferrites*, John Wiley and Sons, New York (1959), p. 90.
[8] C. Kittel, *Phys. Rev.* **76**, 743 (1949); J. H. Van Vleck, *Phys. Rev.* **78**, 266 (1950).

Table 10-2.1. Data from Magnetomechanical and Ferromagnetic Resonance Experiments[a]

Material	g'	g	$\frac{g}{g-1}$
Fe	1.92	2.10	1.91
Co	1.85	2.25	1.83
Ni	1.84	2.21	1.83
50Fe50Ni	1.91	2.12	1.90
50Co50Ni	1.84	2.18	1.85
Supermalloy (16Fe79Ni5Mn)	1.91	2.09	1.91
Cu_2MnAl	1.99	2.01	1.99
MnSb	1.98	2.10	1.91
$NiFe_2O_4$	1.85	2.19	1.84

[a] G. G. Scott, *Revs. Mod. Phys.* **34,** 102 (1962). This paper contains an extensive list of references.

assumption that $g' = g = 2$. Averages of the g-factor obtained from ferromagnetic resonance experiments are also listed in Table 10-2.1. Values of g' calculated from g with the aid of equation 10-2.3 are, with the exception of MnSb, in good agreement with those measured in the magneto-mechanical experiments (Table 10-2.1).

There is no reason, in principle, why magnetomechanical experiments cannot be performed on paramagnetic materials. Because of the lack of strong coupling between the dipoles, however, there are sensitivity problems. Nevertheless, Sucksmith in the early 1930's succeeded in making such experiments work.[9] For salts with iron group ions he found generally that $g' \approx 2$. For the rare earth ions g' often deviated considerably from 2. These results are, of course, to be expected, since it is well known that quenching of the orbital angular momentum is large for the iron group ions and small for rare earth ions (Chapters 2 and 3). Hopefully, continued refinements in the experimental techniques will permit more extensive and accurate determinations of the g'-factor of paramagnets to be made in the future.

3. Ferromagnetic Resonance

The main difference between ferromagnetic and paramagnetic resonance is the presence of strong exchange forces. As a result, when a steady magnetic field is applied, the atomic moments will tend to precess

[9] W. Sucksmith, *Proc. Roy. Soc. (London)* **A-128,** 276 (1930), **A-133,** 179 (1931), **A-135,** 276 (1932).

coherently about their equilibrium orientation. It will be assumed that this steady magnetic field is large enough to remove the domain structure; unsaturated specimens are discussed later in Section 10-12. Then the process can be viewed as the precession of the total magnetic moment. Resonance will occur when the frequency of an alternating field applied transversely is equal to the precession frequency. Alternatively, the phenomena can be considered in terms of the quantization of the total spin angular momenta of the specimen. Since the smallest change in angular momentum corresponds to the reversal of a single electronic spin, the difference in energy between adjacent states is given by $g\,|\mu_B|\,H_e$. Here H_e is the effective magnetic field whose value will be calculated later and g may differ from 2 because of the spin-orbit interaction. Energy will then be absorbed at a frequency given by the equation

$$\hbar\omega = -g\mu_B H_e, \tag{10-3.1}$$

which is formally the same as that for the paramagnetic case (equation 3-7.11).

Ferromagnetic resonance may be treated microscopically with a spin-Hamiltonian that includes terms for the Zeeman, dipole-dipole, and exchange interactions, as well as one for the crystalline anisotropy energy. The equations of motion are then obtained by the usual procedure of calculating the commutator of the spin-Hamiltonian and the spin angular momentum (see Section 3-7, page 110). It is reassuring that with certain minor approximations the quantum theory leads to the usual macroscopic equations of motion[10]

$$\frac{d\mathbf{M}}{dt} = \gamma\mathbf{M} \times \mathbf{H}_T + \text{damping} \tag{10-3.2}$$

(equation 3-7.3). Since this equation is simpler to handle, common practice treats ferromagnetic resonance problems macroscopically, the approach followed here. In equation 10-3.2 the magnetic field is the resultant of the applied, demagnetizing, crystalline anisotropy, and any other fields that may be present. A phenomenological term is usually employed to describe the damping. Some of these terms are listed at the end of this section. For the sake of simplicity, the damping term will be neglected for the present.

For a specimen in the shape of an ellipsoid the resultant field may be written

$$\mathbf{H}_T = \mathbf{H} + \mathbf{H}_1(t) - \mathsf{D}\cdot\mathbf{M} + \mathbf{H}_K \tag{10-3.3}$$

where $\mathbf{H}$ represents the steady magnetic field, $\mathbf{H}_1(t)$, the alternating field,

[10] J. M. Luttinger and C. Kittel, *Helv. Phys. Acta* **21,** 480 (1948); J. M. Richardson, *Phys. Rev.* **75,** 1630 (1949); D. Polder, *Phil. Mag.* **40,** 99 (1949); J. H. Van Vleck, *Phys. Rev.* **78,** 266 (1950).

$-\mathbf{D} \cdot \mathbf{M}$, the demagnetizing field, and $\mathbf{H}_K$, the anisotropy field. Without any loss in generality, the steady field may be taken to be applied along the z-axis. Although the alternating field is usually applied perpendicular to the steady field, there are no complications to the approximations that will be employed if it is assumed that

$$\mathbf{H}_1(t) = H_{1x}(t)\mathbf{i} + H_{1y}(t)\mathbf{j} + H_{1z}(t)\mathbf{k}.$$

It will also be supposed that $|\mathbf{H}_1(t)| \ll |\mathbf{H}|$ and that the time dependence is exponential, that is, $\mathbf{H}_1(t) = \mathbf{H}_1 e^{i\omega t}$. In the expression for the demagnetizing field $\mathbf{D}$ is a tensor. For simplicity, the principal axes of the ellipsoid will be made to coincide with the reference axes. Then $\mathbf{D}$ is diagonal; otherwise it is not. This arrangement represents an important restriction, as will be seen later. The anisotropy field will depend on the crystal symmetry and the orientation of the easy and hard directions with respect to the coordinate axes. Some particular cases are specified.

The magnetization $\mathbf{M}$ may be expressed as the sum of a steady and an alternating component. The magnitude of the steady magnetization will be just the saturation magnetization M_s, which is virtually equal to the spontaneous magnetization. Provided that (*a*) the shape and crystalline anisotropies are negligible, (*b*) the easy direction of these anisotropies lie along the z-direction, or (*c*) the intermediate or hard directions lie along the z-direction and the applied field is equal or larger than the anisotropy fields, $\mathbf{M}_s$ will, to the first order of approximation, lie along the direction of the steady field $\mathbf{H}$. This follows since $d\mathbf{M}_s/dt = 0 = \gamma \mathbf{M}_s \times \mathbf{H}$ if $H_1 \ll H$. When the easy directions make some arbitrary angle with the z-axis, $\mathbf{M}_s$ is still almost parallel to $\mathbf{H}$ if H is very large compared to the equivalent anisotropy fields. Otherwise, they are not parallel, as may be clearly seen from a review of the principles for the static case outlined in Section 7-2. The problem then becomes difficult and can best be handled with the formalism to be developed in Section 10-4. The approximation that $\mathbf{M}_s$ and $\mathbf{H}$ are parallel, frequently used in ferromagnetic resonance calculations, will be assumed to apply in the present case. We may then write $\mathbf{M} = \mathbf{M}_s + \mathbf{M}_1(t) = M_{1x}(t)\mathbf{i} + M_{1y}(t)\mathbf{j} + [M_s + M_{1z}(t)]\mathbf{k}$, where $\mathbf{M}_1(t) = \mathbf{M}_1 e^{i\omega t}$ is the alternating component of the magnetization.

Substitution for $\mathbf{M}$ and $\mathbf{H}_T$ in equation 10-3.2 yields

$$\frac{d\mathbf{M}_1(t)}{dt} = \gamma[\mathbf{M}_s \times \mathbf{H}_1(t) + \mathbf{M}_1(t) \times \mathbf{H} + M_{1y}(t)(D_y - D_z)M_s\mathbf{i} - M_{1x}(t)(D_x - D_z)M_s\mathbf{j} + \mathbf{M} \times \mathbf{H}_K], \quad (10\text{-}3.4)$$

since $\mathbf{M} \times \mathbf{D} \cdot \mathbf{M} = M_s\mathbf{k} \times \mathbf{D} \cdot \mathbf{M}_1(t) - \mathbf{M}_1(t) \times \mathbf{D} \cdot \mathbf{M}_s\mathbf{k}$ when second-order quantities are neglected and $\mathbf{D} \cdot \mathbf{M}_1(t) = D_x M_{1x}(t)\mathbf{i} + D_y M_{1y}(t)\mathbf{j} + D_z M_{1z}(t)\mathbf{k}$. The anisotropy field will depend on anisotropy constants such as K_1, K_2, etc. In order to obtain a general expression now, the anisotropy field will be written in the form[11] $\mathbf{H}_K = -\mathbf{D}_K \cdot \mathbf{M}$. Again, for simplicity, the easy and hard directions will be assumed to coincide with the reference axes, hence $\mathbf{D}_K$ will be a diagonal tensor (see problem 10-2). Thus we have[12] $\mathbf{H}_K = -D_{Kx}M_{1x}(t)\mathbf{i} - D_{Ky}M_{1y}(t)\mathbf{j} - D_{Kz}M_{1z}(t)\mathbf{k}$. The components of equation 10-3.4 then are

$$\frac{i\omega M_{1x}}{\gamma} = M_{1y}[H + (D_y + D_{Ky} - D_z - D_{Kz})M_s] - M_s H_{1y},$$

$$\frac{i\omega M_{1y}}{\gamma} = -M_{1x}[H + (D_x + D_{Kx} - D_z - D_{Kz})M_s] + M_s H_{1x},$$

$$M_{1z} = 0. \tag{10-3.5}$$

The resonance frequency may be found from the condition that M_{1x} and M_{1y} have nontrivial solutions when $H_{1x} = H_{1y} = 0$. This condition is

$$\begin{vmatrix} \dfrac{i\omega_r}{\gamma} & -[H_r + (D_y + D_{Ky} - D_z - D_{Kz})M_s] \\ H_r + (D_x + D_{Kx} - D_z - D_{Kz})M_s & \dfrac{i\omega_r}{\gamma} \end{vmatrix} = 0,$$

where ω_r is the resonance frequency and H_r is the corresponding steady applied field. On solving, we find

$$\begin{aligned} \omega_r &= -\gamma\{[H_r + (D_y + D_{Ky} - D_z - D_{Kz})M_s] \\ &\qquad \times [H_r + (D_x + D_{Kx} - D_z - D_{Kz})M_s]\}^{1/2} \\ &= -\gamma H_e, \end{aligned} \tag{10-3.6}$$

where H_e is the effective field of equation 10-3.1.

Let us evaluate the resonance frequency for some particular cases. First, suppose the crystalline anisotropy is small compared to the shape anisotropy and may be neglected. Then for a disk or flat plate, with H in the plane of the specimen, $D_x = 4\pi$, $D_y = 0$, and $D_z = 0$ approximately and

$$\begin{aligned} \omega_r &= -\gamma[H_r(H_r + 4\pi M_s)]^{1/2} \\ &= -\gamma(BH_r)^{1/2}. \end{aligned} \tag{10-3.7}$$

For the steady field applied perpendicular to the plane of the sample

$$\omega_r = -\gamma(H_r - 4\pi M_s). \tag{10-3.8}$$

[11] C. Kittel, *Phys. Rev.* **73,** 155 (1948).

[12] The anisotropy field is usually chosen so that $D_{Kz} = 0$.

For a long circular cylinder with H applied along the axis of symmetry, $D_x = D_y = 2\pi$, $D_z = 0$, and

$$\omega_r = -\gamma(H_r + 2\pi M_s). \tag{10-3.9}$$

For a sphere the demagnetizing constants $D_x = D_y = D_z = 4\pi/3$ and do not appear in the equation for the resonance frequency. It is no longer permissible to neglect the crystalline anisotropy. Indeed, this is a convenient sample shape for the determination of the crystalline anisotropy constants of single crystals. For a uniaxial crystal with H applied along the symmetry axis it follows from equations 6-9.2 and 7-11.5 that, provided $K_1' \gg K_2'$, $D_{Kx} = D_{Ky} = 2K_1'/M_s^2$, $D_{Kz} = 0$, and then[13]

$$\omega_r = -\gamma\left(H_r + \frac{2K_1'}{M_s}\right). \tag{10-3.10}$$

For a cubic crystal it can be shown[13] (problem 10-3) that

$$\omega_r = -\gamma\left(H_r + \frac{2K_1}{M_s}\right) \quad \text{for } H \text{ parallel to [100]-directions,}$$

$$\omega_r = -\gamma\left(H_r - \frac{2K_1}{M_s}\right)\left(H + \frac{K_1}{M_s} + \frac{K_2}{2M_s}\right)$$
$$\text{for } H \text{ parallel to [110]-directions,}$$

$$\omega_r = -\gamma\left(H_r - \frac{4K_1}{3M_s} - \frac{4K_2}{9M_s}\right) \quad \text{for } H \text{ parallel to [111]-directions.} \tag{10-3.11}$$

If the crystalline anisotropy is small enough, M_s will remain almost parallel to H even when H is not applied along a principal direction. Then it can also be shown[14] that for H lying in a (110) plane (problem 10-3)

$$D_{Kx} = \frac{K_1}{M_s^2}(2 - \sin^2\theta - 3\sin^2 2\theta) + \frac{K_2}{2M_s^2}\sin^2\theta(6\cos^4\theta - 11\sin^2\theta\cos^2\theta + \sin^4\theta),$$

$$D_{Ky} = \frac{K_1}{M_s^2}(2 - 4\sin^2\theta - \tfrac{3}{4}\sin^2 2\theta) - \frac{K_2}{2M_s^2}\sin^2\theta\cos^2\theta(3\sin^2\theta + 2),$$

$$D_{Kz} = 0, \tag{10-3.12}$$

[13] C. Kittel, *Phys. Rev.* **73,** 155 (1948), **76,** 743 (1949).

[14] L. R. Bickford, *Phys. Rev.* **78,** 449 (1950); W. A. Yager, J. K. Galt, F. R. Merritt, and E. A. Wood, *Phys. Rev.* **80,** 744 (1950); J. F. Dillon, S. Geschwind and V. Jaccarino, *Phys. Rev.* **100,** 750 (1955).

where θ is the angle between H and the [001]-direction. Hence, when $K_1 \gg K_2$,

$$\omega_r = -\gamma\left[H_r + \frac{K_1}{M_s}(4 - 5\sin^2\theta - \tfrac{15}{4}\sin^2 2\theta)\right]. \qquad (10\text{-}3.13)$$

A typical example of the variation of H_r with θ in the (110) plane for a fixed ω_r is shown[15] in Fig. 10-3.1. The foregoing equations involving the crystalline anisotropy are, of course, valid only if the sample is a single

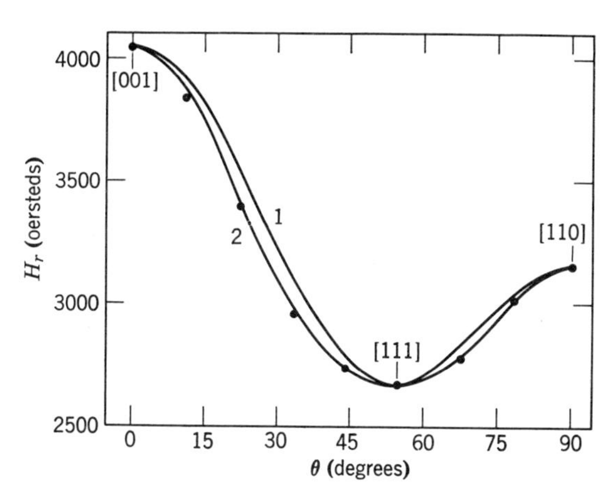

Fig. 10-3.1. Applied field required for resonance at a frequency of 9.198 Gc plotted as a function of the angle in the (110) plane of a cubic crystal with a spherical shape. All three principal directions lie in this plane; θ is the angle between H and the [001]-direction. Curve 1 is calculated by using equations 10-3.6 and 10-3.12 with $K_1/M_s = -399$ and $K_2/M_s = -121$ oe. The experimental points are for an MnZn spinel at $T = 77°$K. For curve 2 the small angle between $\mathbf{M}_s$ and $\mathbf{H}$ is taken into account. (After J. O. Artman.)

crystal. The behavior of polycrystalline samples is discussed in Section 10-6.

We now return to equations 10-3.5 and solve for M_{1x} and M_{1y}. We get

$$M_{1x} = \frac{\gamma^2 M_s[H + (D_y + D_{Ky} - D_z - D_{Kz})M_s]}{\omega_r^2 - \omega^2} H_{1x} - \frac{i\omega\gamma M_s}{\omega_r^2 - \omega^2} H_{1y},$$

$$M_{1y} = \frac{i\omega\gamma M_s}{\omega_r^2 - \omega^2} H_{1x} + \frac{\gamma^2 M_s[H + (D_x + D_{Kx} - D_z - D_{Kz})M_s]}{\omega_r^2 - \omega^2} H_{1y}. \qquad (10\text{-}3.14)$$

[15] J. O. Artman, *Proc. IRE* **44**, 1284 (1956).

This equation may be written

$$\mathbf{M}_1 = \boldsymbol{\chi}_e \cdot \mathbf{H}_1$$

where

$$\boldsymbol{\chi}_e = \begin{pmatrix} \chi_{11} & \chi_{12} & 0 \\ \chi_{21} & \chi_{22} & 0 \\ 0 & 0 & 0 \end{pmatrix}. \tag{10-3.15}$$

The elements of the tensor susceptibility are then given by

$$\chi_{11} = \frac{\gamma^2 M_s[H + (D_y + D_{Ky} - D_z - D_{Kz})M_s]}{\omega_r^2 - \omega^2},$$

$$\chi_{22} = \frac{\gamma^2 M_s[H + (D_x + D_{Kx} - D_z - D_{Kz})M_s]}{\omega_r^2 - \omega^2}, \tag{10-3.16}$$

$$\chi_{12} = -\chi_{21} = \frac{i\omega\gamma M_s}{\omega_r^2 - \omega^2}.$$

Since $\boldsymbol{\chi}_e$ relates the alternating magnetization to the externally applied alternating magnetic field, it may be called the *extrinsic* susceptibility. The high frequency susceptibility may also be expressed in terms of the alternating field inside the sample; it is then called the *intrinsic* susceptibility $\boldsymbol{\chi}_i$. It can be shown (problem 10-4) that

$$\boldsymbol{\chi}_i = \boldsymbol{\chi}_e(1 + \mathbf{D} \cdot \boldsymbol{\chi}_i). \tag{10-3.17}$$

It is convenient to employ the permeability rather than the susceptibility tensor in many calculations. The notation usually used is

$$\tilde{\boldsymbol{\mu}} = \begin{pmatrix} \mu_1 & -iK & 0 \\ iK & \mu_2 & 0 \\ 0 & 0 & 1 \end{pmatrix}, \tag{10-3.18}$$

where the subscript has been omitted, since a similar form holds for both the intrinsic and extrinsic. Since $\tilde{\boldsymbol{\mu}} = \boldsymbol{\delta}_{ij} + 4\pi\boldsymbol{\chi}$, where $\boldsymbol{\delta}_{ij}$ is the unit matrix, it follows that

$$\begin{aligned} \mu_1 &= 1 + 4\pi\chi_{11}, \\ \mu_2 &= 1 + 4\pi\chi_{22}, \\ iK &= -4\pi\chi_{12}. \end{aligned} \tag{10-3.19}$$

The off-diagonal elements that are nonzero give rise to nonreciprocal effects, which are the basis for a number of important device applications (see Section 10-11).

Finally, it is necessary to consider the question of damping. The ultimate objective is, of course, to obtain a quantitative microscopic theory of damping. Progress that has been made in this direction will be discussed in Section 10-8. For the purpose of describing the experimental results, a phenomological expression may be employed in equation 10-3.2. The prerequisite for such a term is that it contain one or more adjustable parameters and have a direction that tends to restore equilibrium. Several suitable damping terms have been proposed.

The damping terms employed by Bloch in paramagnetic resonance, namely (equations 3-4.8 and 3-6.5),

$$-\frac{M_x}{\tau_2}, -\frac{M_y}{\tau_2}, \text{ and } -\frac{M_z - M_0}{\tau_1}, \text{ with } M_0 = M_s, \tag{10-3.20}$$

are one obvious possibility; they were first applied to the ferromagnetic resonance case by Bloembergen.[16] Two parameters, τ_1 and τ_2, represent the relaxation times for the longitudinal and transverse directions, respectively. Another expression,

$$-\frac{\lambda}{M^2}\mathbf{M} \times (\mathbf{M} \times \mathbf{H}_T), \tag{10-3.21}$$

where λ is the adjustable parameter, was introduced much earlier by Landau and Lifshitz.[17] The vector $\mathbf{M} \times (\mathbf{M} \times \mathbf{H}_T)$ is perpendicular to $\mathbf{M}$ and directed toward $\mathbf{H}_T$ in the $M - H_T$-plane, that is, in the proper direction to restore equilibrium. The final form for the damping considered in this section

$$-\frac{\alpha}{M}\mathbf{M} \times \frac{d\mathbf{M}}{dt} \tag{10-3.22}$$

was proposed by Gilbert.[18] It is analogous to the Rayleigh dissipation function used almost universally in classical calculations.

It is of interest to determine the conditions under which the various damping terms are equivalent. To the approximation used in this section, $M_z - M_0 \simeq M_z - M_s \simeq 0$, and the last term of the Bloembergen-Bloch damping is zero. Then, on setting $\tau_2 = \tau$, the terms

$$-\frac{M_x}{\tau} \text{ and } -\frac{M_y}{\tau}$$

remain. If the $-(M_z - M_0)/\tau$ term is retained, it is necessary to set τ_1 equal to τ in order to establish equivalence.

[16] N. Bloembergen, *Phys. Rev.* **78,** 572 (1950).
[17] L. Landau and L. Lifshitz, *Physik Z. Sowjet.* **8,** 153 (1935).
[18] T. L. Gilbert, *Phys. Rev.* **100,** 1243 (1955).

In the Landau-Lifshitz form $\mathbf{H}_T$ can be replaced by $\mathbf{H}(t) = \mathbf{H} + \mathbf{H}_1(t)$, provided the shape and crystalline anisotropy fields are proportional to $\mathbf{M}$. Expansion of the triple cross product yields

$$-\lambda\left[\frac{\mathbf{H}(t)\cdot\mathbf{M}}{M^2}\mathbf{M} - \mathbf{H}(t)\right].$$

Since $\tau = M_0/\lambda H \simeq M_s/\lambda H$, $\mathbf{H}(t)\cdot\mathbf{M} \simeq HM_s$, and $M = M_0 \simeq M_s$ (problem 10-6), this becomes

$$-\frac{1}{\tau}\left[\mathbf{M} - \frac{M_s}{H}\mathbf{H}(t)\right].$$

Neglect of the H_{1x} and H_{1y} components of the last term leaves just the Bloembergen-Bloch components, namely,

$$-\frac{M_x}{\tau},\ -\frac{M_y}{\tau},\quad \text{and}\quad -\frac{M_z - M_s}{\tau}\left(\frac{H + H_{1z}}{H}\right) \simeq -\frac{M_z - M_s}{\tau} \simeq 0.$$

Actually, there is some merit in retaining the H_{1x} and H_{1y} components as well as $-(M_z - M_0)/\tau$, since the expression would then describe a relaxation of the magnetization toward the instantaneous value rather than the static value of the magnetic field.

The Gilbert form may be obtained by starting with the Landau-Lifshitz damping term in the equations of motion written as

$$\frac{d\mathbf{M}}{dt} = \gamma(\mathbf{M}\times\mathbf{H}_T) + \frac{\gamma\alpha}{M}[\mathbf{M}\times(\mathbf{M}\times\mathbf{H}_T)],$$

where $\alpha = \lambda/\gamma M$. The cross product of $\mathbf{M}$ with this equation yields

$$\mathbf{M}\times\frac{d\mathbf{M}}{dt} = \gamma[\mathbf{M}\times(\mathbf{M}\times\mathbf{H}_T)] + \frac{\gamma\alpha}{M}\{\mathbf{M}\times[\mathbf{M}\times(\mathbf{M}\times\mathbf{H}_T)]\}.$$

Elimination of $\mathbf{M}\times(\mathbf{M}\times\mathbf{H}_T)$ from these two equations leads to

$$\frac{d\mathbf{M}}{dt} = \gamma(\mathbf{M}\times\mathbf{H}_T) - \frac{\alpha}{M}\left(\mathbf{M}\times\frac{d\mathbf{M}}{dt}\right) + \gamma\alpha^2(\mathbf{M}\times\mathbf{H}_T),$$

since $\mathbf{M}\times[\mathbf{M}\times(\mathbf{M}\times\mathbf{H}_T)] = -M^2(\mathbf{M}\times\mathbf{H}_T)$. If α^2 is small, which is the usual case, the last term may be neglected. The desired result is then obtained.

Although one certain damping term may be the most suitable for a particular case, for the present discussion it is convenient to specify the Gilbert form.[19] The solution of the equation of motion

$$\frac{d\mathbf{M}}{dt} = \gamma(\mathbf{M}\times\mathbf{H}_T) - \frac{\alpha}{M}\left(\mathbf{M}\times\frac{d\mathbf{M}}{dt}\right) \tag{10-3.23}$$

[19] S. Iida, *J. Phys. Chem. Solids* **24,** 625 (1963).

may be obtained with the general method already presented in this section. The elements of the susceptibility tensor will now, of course, be complex and give rise to a power loss. For a spherical sample with negligible crystalline anisotropy, it is easy to show (problem 10-7) that

$$\chi_{11} = \chi_{22} = \frac{\gamma M_s[(\gamma H + i\omega\alpha) + (4\pi/3)\gamma M_s]}{(\omega_r + i\omega\alpha)^2 - \omega^2}$$

and

$$\chi_{21} = -\chi_{12} = \frac{i\omega\gamma M_s}{(\omega_r + i\omega\alpha)^2 - \omega^2}. \tag{10-3.24}$$

It is worth emphasizing that the analysis in this section has assumed that the alternating field was uniform over the volume of the sample. Effects that occur when this is not true are discussed in Sections 10-5 and 10-7.

4. Energy Formulation of the Equations of Motion

The equations of motion may be expressed in terms of the total free energy F_T (equation 7-1.2) instead of effective fields. One way of recasting the equations is to write the torques as derivatives of the free energy and to equate them to the time change of the angular momenta.[20] Another way is to form a Lagrangian and then use either Hamilton's principle or Lagrange's equations of motion.[21] When carried out correctly, all approaches, of course, yield identical results. There are, however, several advantages to the energy formulation of the equations of motion. First, they are then rather more general in form. Second, when the free energy terms are complicated functions of the angular coordinates, they can be introduced routinely into the equations. By contrast, considerable physical intuition may be required to obtain the equivalent magnetic fields. Third, when the easy directions of the various anisotropies have different orientations, the computations are more straightforward with the energy equations.

Suppose the magnetization $\mathbf{M}_s$ of a ferromagnetic specimen has the static equilibrium direction described by the azimuth ϕ_0 and the polar angle θ_0, as in Fig. 10-4.1. A Lagrangian function, $\mathscr{L}$, consistent with the accepted equations of motion (equation 10-3.2) can be obtained by considering the magnetic system as a classical top with principal moments of inertia (0, 0, C) and Euler angles ϕ, θ, and χ. Here χ is the angle that describes a rotation of the top about its symmetry axis. The angular

[20] J. Smit and H. G. Beljers, *Philips Res. Rept.* **10,** 113 (1955).

[21] T. L. Gilbert, *Phys. Rev.* **100,** 1243 (1955), *Supp. Report, Project No. A059*, Armour Research Foundation, May 1, 1956; W. F. Brown, Jr., *Micromagnetics*, Interscience Publishers, New York (1963).

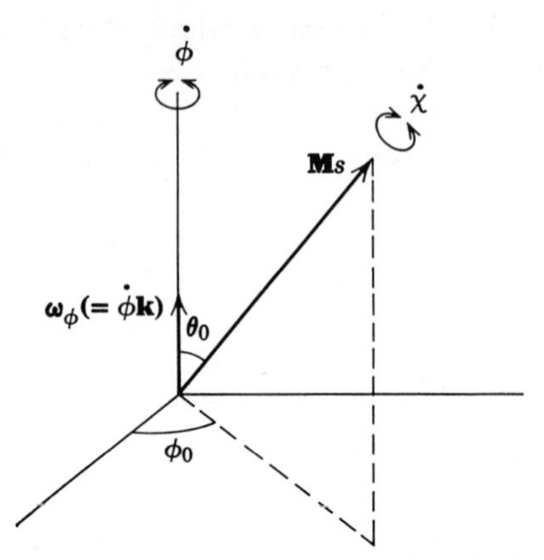

Fig. 10-4.1. To illustrate pertinent angular velocities of classical top. Static equilibrium direction of $\mathbf{M}_s$ is also shown.

velocity of the top about this symmetry axis is then $\dot{\chi}$ plus the component of $\boldsymbol{\omega}_\phi$, namely $\phi \cos \theta$ (Fig. 10-4.1). The kinetic energy T of the top is therefore given by[22]

$$T = \tfrac{1}{2}C(\dot{\chi} + \phi \cos \theta)^2$$

and the Lagrangian function by $\mathscr{L} = T - V(\phi, \theta)$, where V is the potential energy. Now Lagrange's equations of motion have the general form[22]

$$\frac{d}{dt}\frac{\partial \mathscr{L}}{\partial q_\alpha} - \frac{\partial \mathscr{L}}{\partial q_\alpha} = 0,$$

where q_α is a coordinate. Application of this equation to the present problem yields

$$\frac{d}{dt}[C(\dot{\chi} + \phi \cos \theta) \cos \theta] + \frac{\partial V}{\partial \phi} = 0,$$

$$\frac{d}{dt}(0) + C(\dot{\chi} + \phi \cos \theta)\dot{\phi} \sin \theta + \frac{\partial V}{\partial \theta} = 0,$$

and

$$\frac{d}{dt}[C(\dot{\chi} + \dot{\phi} \cos \theta)] = 0.$$

From the last of these equations it follows that

$$C(\dot{\chi} + \dot{\phi} \cos \theta) = \text{constant};$$

in other words the angular momentum conjugate to χ is conserved. In order to return to the magnetic system, it is necessary only to recognize

[22] H. Goldstein, *Classical Mechanics*, Addison-Wesley Publishing Co., Reading, Mass. (1959).

that this angular momentum should be set equal to M_s/γ, a quantity that is also a constant of the motion in this approximation. The equations then become

$$\frac{M_s}{\gamma}\dot{\theta}\sin\theta = \frac{\partial F_T}{\partial\phi}$$

and (10-4.1)

$$-\frac{M_s}{\gamma}\dot{\phi}\sin\theta = \frac{\partial F_T}{\partial\theta},$$

since $V = F_T$.

The precessional motion of $\mathbf{M}_s$ will be assumed to be confined to small deviations from the equilibrium orientation (ϕ_0, θ_0) and described by the small angles $\Delta\phi$ and $\Delta\theta$. Thus we may write

$$\phi = \phi_0 + \Delta\phi$$
$$\theta = \theta_0 + \Delta\theta,$$

hence

$$\sin(\theta_0 + \Delta\theta) = \sin\theta_0\cos\Delta\theta + \cos\theta_0\sin\Delta\theta$$
$$\simeq \sin\theta_0 + \cos\theta_0\,\Delta\theta.$$

For exponential time variation of $\Delta\phi$ and $\Delta\theta$ it follows that

$$\dot{\phi} = i\omega\,\Delta\phi$$

and

$$\dot{\theta} = i\omega\,\Delta\theta.$$

The total free energy may be expanded about the equilibrium position, for which $\partial F_T/\partial\phi = \partial F_T/\partial\theta = 0$:

$$F_T = F_{T0} + \tfrac{1}{2}[F_{T\phi\phi}(\Delta\phi)^2 + 2F_{T\phi\theta}\,\Delta\phi\,\Delta\theta + F_{T\theta\theta}(\Delta\theta)^2 + \cdots,$$

where $F_{T\phi\phi} = \partial^2F_T/\partial\phi^2$, $F_{T\phi\theta} = \partial^2F_T/\partial\phi\,\partial\theta$, and $F_{T\theta\theta} = \partial^2F_T/\partial\theta^2$ are derivatives evaluated at the static equilibrium position. Substitution of these expressions into equations 10-4.1 then yields to first order

$$\frac{i\omega M_s}{\gamma}\sin\theta_0\,\Delta\theta = F_{T\phi\phi}\,\Delta\phi + F_{T\phi\theta}\,\Delta\theta$$

and (10-4.2)

$$-\frac{i\omega M_s}{\gamma}\sin\theta_0\,\Delta\phi = F_{T\phi\theta}\,\Delta\phi + F_{T\theta\theta}\,\Delta\theta.$$

This set of homogeneous equations for $\Delta\phi$ and $\Delta\theta$ has a nontrivial solution only when the determinant of the coefficients equals zero. Application of this condition then yields

$$\omega_r = \frac{-\gamma}{M_s\sin\theta_0}(F_{T\phi\phi}F_{T\theta\theta} - F^2_{T\phi\theta})^{1/2}, \quad (10\text{-}4.3)$$

where ω_r is the resonance frequency.

The introduction of a Lagrangian to obtain this result may seem rather artificial. There is, however, the merit that any general theorems concerning Lagrangians also apply to this one. For example, when T is a quadratic function of the generalized velocities, the quantity $T + V$ is conserved. Now, since $T = (1/2C)(M_s/\gamma)^2$, both the kinetic and potential energies are constants of the motion. The precessional motion of the magnetization $\mathbf{M}_s$ occurs therefore along a path of constant F_T.

The foregoing analysis assumes, of course, that there is no dissipation of energy. A damping term may be added directly to the Lagrangian or to the equations of motion[23] (equations 10-4.1). The magnetization vector then tends to return to the static equilibrium position; that is, F_T decreases. The high-frequency susceptibility can be calculated, provided the torques produced by the applied alternating magnetic fields are added to the equations of motion[23] (equations 10-4.1).

The resonance condition given by equation 10-4.3 of course reduces to that obtained in the preceding section (equation 10-3.6) when the appropriate expression for F_T is inserted (problem 10-8). Cases in which the magnetization $\mathbf{M}_s$ is not parallel to the steady applied field can be treated readily with equation 10-4.3. A simple yet important example is the following. Suppose that a spherical crystal with uniaxial anisotropy has its easy direction along the z-axis and that a steady field is applied along the x-axis. The total free energy is then

$$F_T = K_1' \sin^2 \theta - HM_s \sin \theta \cos \phi,$$

where ϕ and θ are the usual polar coordinates. At static equilibrium $\partial F_T/\partial \phi = \partial F_T/\partial \theta = 0$ and this yields $\phi = 0$ and

$$\sin \theta = \frac{HM_s}{2K_1'} \quad \text{for } H < \frac{2K_1'}{M_s}$$

or

$$\theta = \frac{\pi}{2} \quad \text{for } H \geqslant \frac{2K_1'}{M_s},$$

results already obtained in Section 7-2 (see equation 7-2.10). The second derivatives, calculated at the equilibrium position, are

$$\frac{\partial^2 F_T}{\partial \phi^2} = \frac{H^2 M_s^2}{2K_1'}, \qquad \frac{\partial^2 F_T}{\partial \theta^2} = 2K_1' - \frac{H^2 M_s^2}{2K_1'}, \qquad \frac{\partial^2 F_T}{\partial \phi\, \partial \theta} = 0,$$

when $\phi = 0$, $\theta = \sin^{-1} HM_s/2K_1'$, and

$$\frac{\partial^2 F_T}{\partial \phi^2} = HM_s, \qquad \frac{\partial^2 F_T}{\partial \theta^2} = HM_s - 2K_1', \qquad \frac{\partial^2 F_T}{\partial \phi\, \partial \theta} = 0,$$

[23] E. P. Valstyn, J. P. Hanton, and A. H. Morrish, *Phys. Rev.* **128**, 2078 (1962).

when $\phi = 0$, $\theta = \pi/2$. From equation 10-4.3 the resonance frequency is therefore

$$\omega_r^2 = \gamma^2\left(\frac{4K_1'^2}{M_s^2} - H_r^2\right) \quad \text{for } H < \frac{2K_1'}{M_s}$$

and

$$\omega_r^2 = \gamma^2 H_r\left(H_r - \frac{2K_1'}{M_s}\right) \quad \text{for } H \geqslant \frac{2K_1'}{M_s}. \tag{10-4.4}$$

These equations are plotted in Fig. 10-4.2. For a fixed frequency ω that is less than $2K_1'/M_s$, indicated for example, by the dashed horizontal line

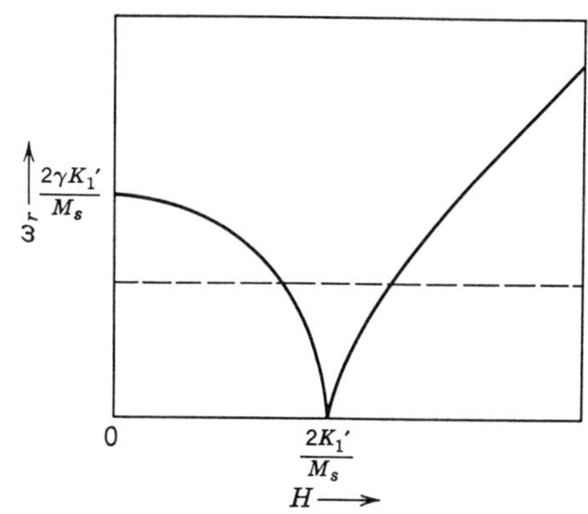

Fig. 10-4.2. Resonance frequency plotted as a function of steady magnetic field applied perpendicular to easy direction of a uniaxial crystal. Intersection of the horizontal line with curve indicates fields at which two peaks occur in the absorption spectrum for a given frequency. [J. Smit and H. G. Beljers, *Philips Res. Rept.* **10,** 113 (1955).]

in the figure, there are two steady fields at which resonance occurs. Two peaks in the absorption spectrum have been observed in Ferroxdure[24] ($BaFe_{12}O_{19}$), a hexagonal crystal with a large uniaxial crystalline anisotropy (Section 9-5).

When the symmetry is cubic, the calculations are more complex and it is usually best to use a computer. The theory shows that cubic materials, especially those with large crystalline anisotropies, may exhibit multiple resonance peaks.[25] This behavior has been observed in cobalt spinels.[26]

In Section 10-3 the field required for resonance in the (110) plane of a particular cubic crystal was calculated by assuming that $\mathbf{M}_s$ and $\mathbf{H}$ were parallel. When the angle between $\mathbf{M}_s$ and $\mathbf{H}$ is taken into account by

[24] J. Smit and H. G. Beljers, *Philips Res. Rept.* **10,** 113 (1955); M. T. Weiss, *IRE Natl. Conv. Record* **3,** Part 8, 95 (1955).

[25] H. Suhl, *Phys. Rev.* **97,** 555 (1955); J. O. Artman, *Proc. IRE* **44,** 1284 (1956).

[26] P. E. Tannenwald, *Phys. Rev.* **99,** 463 (1955).

using equation 10-4.3, curve 2 of Fig. 10-3.1 is obtained.[27] It is seen that the error introduced by the assumption, although small, is significant.

5. Resonance in Ferromagnetic Metals and Alloys

The skin depth in a good electrical conductor is only 10^{-4} or 10^{-5} cm at microwave frequencies. Consequently, the alternating magnetic field will be uniform over the volume of a ferromagnetic metal only if the sample consists of very small particles. Although colloidal suspensions of ferromagnetic particles have been studied,[28] most of the experiments have been concerned with plane or disk samples which often form one wall of the resonant cavity. As will be seen later, the shape of the absorption curve is then modified.

The energy absorbed by the plane sample will be proportional to the real part of the input impedance. Suppose the z-axis is along the normal to the sample's surface. The impedance Z is defined[29] for a plane wave as $Z = -E_y/H_x$. It can be calculated easily from Maxwell's equations (equations 1-8.1 to 1-8.4). In equation 1-8.4 $\mathbf{j} = \sigma\mathbf{E}$ by Ohm's law; σ is the electrical conductivity. For a good conductor this term is much larger than $\epsilon \dfrac{\partial \mathbf{E}}{\partial t}$; hence the latter term may be neglected. When operating with the curl on both sides of this equation, employing equation 1-8.3, and introducing exponential time dependence, it follows that

$$\frac{\partial^2 H_x}{\partial z^2} = \frac{4\pi\sigma\mu i\omega}{c^2} H_x$$

$$= k^2 H_x.$$

The solution shows that when the wave propagates into the metal the magnetic field component is attenuated as e^{-kz}. From the y-component of equation 1-8.4 we also have

$$\frac{\partial H_x}{\partial z} = \frac{4\pi\sigma}{c} E_y$$

or

$$-kH_x = \frac{4\pi\sigma}{c} E_y.$$

[27] J. O. Artman, *Proc. IRE* **44,** 1284 (1956).

[28] D. M. S. Bagguley, *Proc. Phys. Soc.* (*London*) **A-66,** 765 (1953), *Proc. Roy. Soc.* (*London*) **A-228,** 549 (1955).

[29] See, for example, S. Ramo and J. R. Whinnery, *Fields and Waves in Modern Radio*, John Wiley and Sons, New York (1953), p. 279.

On substituting for k, we find

$$Z = \left(\frac{i\omega\mu}{4\pi\sigma}\right)^{1/2}.$$

Now, since the permeability is complex, $\mu = \mu' - i\mu''$, the surface impedance may be written

$$Z = \left(\frac{\omega}{8\pi\sigma}\right)^{1/2}[(|\mu| + \mu'')^{1/2} + i(|\mu| - \mu'')^{1/2}], \qquad (10\text{-}5.1)$$

where $|\mu| = (\mu'^2 + \mu''^2)^{1/2}$.

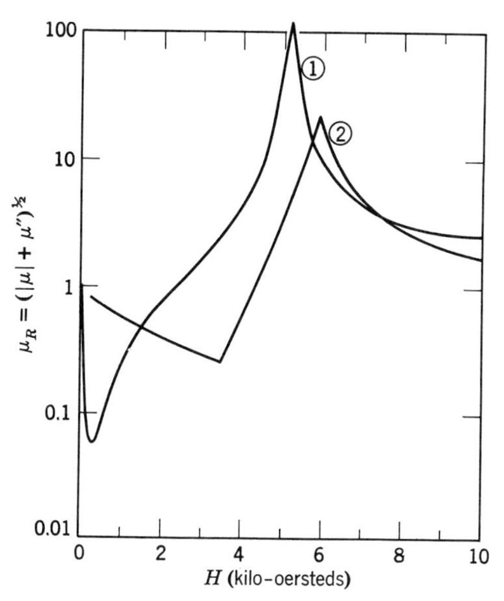

Fig. 10-5.1. Absorption spectra of (1) Supermalloy and (2) a 9.75Cu90.25Ni alloy. The ferromagnetic resonance data was obtained at room temperature with K-band frequencies ($\nu \approx$ 24 Gc) and the steady magnetic field applied parallel to the plane of the sample. [(1): N. Bloembergen, *Phys. Rev.* **78,** 572 (1950); (2): K. J. Standley and K. H. Reich, *Proc. Phys. Soc.* (*London*) **B-68,** 713 (1955).]

Therefore, the power absorption is proportional to $\mu_R = (|\mu| + \mu'')^{1/2}$; that is, it depends on both μ' and μ''. As a result, the absorption spectrum will be a combination of two curves which have the general forms shown in Figs. 3-5.1 or 3-8.10. Actual absorption lines observed for two nickel-rich alloys are shown in Fig. 10-5.1.[30] These curves are typical in that they possess a minimum as well as a maximum. When the power absorption is a minimum, the sample is sometimes said to be at antiresonance.

[30] N. Bloembergen, *Phys. Rev.* **78,** 572 (1950); K. J. Standley and K. H. Reich, *Proc. Phys. Soc.* (*London*) **B-68,** 713 (1955).

In the ideal case antiresonance occurs when the high-frequency flux B_x and B_z are zero; in other words the components of the permeability in the x- and y-plane equal zero.

The antiresonance condition can be derived simply in the following way. Suppose the alternating field is applied along the x-axis. Then from equation 10-3.14 we have

$$M_{1x} = \frac{\gamma^2 M_s[H + (D_y - D_z)M_s]}{\omega_r^2 - \omega^2} H_{1x}, \tag{10-5.2}$$

provided that the crystalline anisotropy is negligible. Also, at antiresonance $\mu_1 = 1 + 4\pi\chi_{11} = 0$ in the notation of equations 10-3.18 and 10-3.15, hence $\chi_{11} = M_{1x}/H_{1x} = -1/4\pi$. Substitution of this value into equation 10-5.2 yields

$$\omega_a = \omega_r\left\{1 + \frac{4\pi\gamma^2 M_s}{\omega_r^2}[H + (D_y - D_z)M_s]\right\}^{1/2}$$

for the antiresonance frequency ω_a. When the value of ω_r (equation 10-3.6) is inserted, this becomes

$$\omega_a = -\gamma\{[H_a + (D_y - D_z)M_s][H_a + (D_x - D_z + 4\pi)M_s]\}^{1/2}. \tag{10-5.3}$$

This relationship has been derived by assuming no damping. It is essentially correct, provided the eddy current damping is small, the sample is magnetized to the saturation value M_s, and μ'' is essentially constant in the vicinity of the minimum.

Particular cases follow immediately. For a sphere, $D_x = D_y = D_z = 4\pi/3$ and

$$\omega_a = -\gamma(B_a H_a)^{1/2}. \tag{10-5.4}$$

For a flat plate with H parallel to the plane surface, $D_y = 4\pi$, $D_x = D_z = 0$, and

$$\omega_a = -\gamma B_a. \tag{10-5.5}$$

For the steady field applied perpendicular to the flat surface, $D_z = 4\pi$, $D_x = D_y = 0$, and

$$\omega_a = -\gamma[H_a(H_a - 4\pi M_s)]^{1/2}. \tag{10-5.6}$$

An expression for antiresonance is especially useful if M_s as well as g of the material is unknown. This expression and the one for the resonance provide two equations that serve to determine M_s and g.

Average values of g for some representative ferromagnetic metals and alloys have been tabulated in Table 10-2.1. For alloys the observed g-values are approximately a mean of the g-values of the constituent metals. For example, the Ni-Fe alloys have an effective g lying between the values found for pure iron and nickel.[31,32] It has been observed that the addition of some diamagnetic metals to a ferromagnetic metal leaves the g-value unchanged; examples are Ni-Cu, Ni-Al, and Ni-Sb alloys, which have the same g-value as pure nickel.[31] The g-value is apparently independent of temperature below the Curie point.[33] It appears that the g-value may be a function of the frequency of the microwave field.[34] It has been suggested that this may result from the Zener-Vonsovsky type of coupling between the $3d$ and $4s$ electrons (Section 6-7).[35]

The crystalline anisotropy constants of a small number of single crystals have been determined with the aid of equations developed in Section 10-3. The value of $K_1 = -4.6 \times 10^4$ ergs/cm^3 found for nickel[36] is in good agreement with static measurements (Table 6-9.1). For single crystals of FeSi alloys the agreement between the resonance[37] and static results are similarly satisfactory.

Analysis of the data yields the imaginary part of the complex permeability, μ''. These rather broad lines generally have half-widths of several hundreds of oersteds, which correspond to relaxation times τ of the order of 10^{-9} sec (equation 10-3.20). Much narrower line widths have been observed for iron whisker crystals.[38] Since whiskers are the nearest things to perfect crystals, this result suggests that imperfections play a prominent role in line broadening. Relaxation processes are discussed in Section 10-8.

6. Ferromagnetic Resonance of Poor Conductors

The presence of eddy currents complicates the study of ferromagnetic resonance. Obviously, these currents are virtually eliminated if samples with small electrical conductivity are investigated. Unfortunately, few ferromagnetic insulators are known. One of the rare examples is $CrBr_3$; ferromagnetic resonance has been observed in this ionic compound.[39]

The great majority of ferrimagnetic materials are poor conductors. Moreover, many properties, such as hysteresis curves, can be analyzed in

[31] K. J. Standley and K. H. Reich, *Proc. Phys. Soc.* (*London*) **B-68,** 713 (1955).

[32] G. Asch, *Compt. rend.* (*Paris*) **249,** 1483 (1959).

[33] N. Bloembergen, *Phys. Rev.* **78,** 572 (1950); D. M. S. Bagguley and N. J. Harrick, *Proc. Phys. Soc.* (*London*) **A-67,** 1052 (1954).

[34] R. Hoskins and G. Wiener, *Phys. Rev.* **96,** 1153 (1954); G. S. Barlow and K. J. Standley, *Proc. Phys. Soc.* (*London*) **B-69,** 1052 (1956).

[35] C. Kittel and A. H. Mitchell, *Phys. Rev.* **101,** 1611 (1956).

[36] K. H. Reich, *Phys. Rev.* **101,** 1647 (1956).

[37] A. F. Kip and R. D. Arnold, *Phys. Rev.* **75,** 1556 (1949); Barlow and Standley, *loc. cit.*

[38] D. S. Rodbell, *J. Appl. Phys.* **30,** 187S (1959).

[39] J. F. Dillon, *J. Appl. Phys.* **33,** 1191 (1962).

terms of their resultant magnetization. The same is true for most of their resonance properties. However, some resonance phenomena that originate from the sublattice structure do exist; it is then appropriate to talk about ferrimagnetic resonance. Otherwise, it is essentially the ferromagnetic resonance of ferrimagnetic materials that is being studied. The use of this nomenclature, although reasonable, is not universal.

Spherical samples are often used in the experiments on ferrimagnetic materials. The resonance conditions are then simplified, for the shape

Table 10-6.1. Observed Effective *g*-Values of Some Ferrimagnets ($T = 300°K$)

Compound	g	Reference
$MnFe_2O_4$	2.004	(a)
Fe_3O_4	2.12	(b)
$NiFe_2O_4$	2.19	(c)
YIG	2.004	(d)

[a] J. F. Dillon, S. Geschwind, and V. Jaccarino, *Phys. Rev.* **100,** 750 (1955).
[b] L. R. Bickford, *Phys. Rev.* **76,** 137 (1949).
[c] W. A. Yager, J. K. Galt, F. R. Merritt, and E. A. Wood, *Phys. Rev.* **80,** 744 (1950).
[d] G. P. Rodrigue, H. Meyer, and R. V. Jones, *J. Appl. Phys.* **31,** 376S (1960).

anisotropy no longer enters explicitly. It is important that the samples be truly spherical if the shape anisotropy is to be neglected. Even small deviations from sphericity have an appreciable effect on the resonance frequency, as may be readily confirmed by an inspection of the demagnetization factors listed in Appendix II.

The observed g-value will be an average of the g-values of the individual ions, provided there is a parallel-antiparallel moment arrangement (for two sublattices see Figs. 9-1.1*a* to *e*). The effective g-value is given by

$$g = \frac{2mc}{e}\frac{\sum_i \mathbf{M}_i}{\sum \mathbf{S}_i} = \frac{\sum_i g_i \mathbf{S}_i}{\sum_i \mathbf{S}_i}, \tag{10-5.1}$$

where the summation is over all ions of each sublattice per formula unit. Some g-values are listed in Table 10-5.1. For $MnFe_2O_4$ and YIG all the cations are in S states and therefore $g \approx 2$ is expected and found. For Fe_3O_4 and $NiFe_2O_4$, which have the inverse spinel structure, the contributions from the Fe^{3+} ions cancel. An effective g-value that is characteristic of the Fe^{2+} and Ni^{2+} ions, respectively, hence is anticipated. The measured values are indeed close to those observed for these ions in paramagnetic salts (Table 3-10.2). The reverse procedure can be followed; that is, when reliable values of g_i are known, the ionic distribution over the different lattice sites can be inferred from the effective g-value.[40] For many spinels the g-value is independent of temperature. A change occurs if the average spin angular momenta of the sublattices have different temperature variations; examples are $CuFe_2O_4$ below 300°K[41] and $Ni_{0.75}Fe_{2.25}O_4$.[42] The g-values appear to be independent of frequency for single crystals. Sublattice effects are important near compensation temperatures or if there is strong damping within one sublattice; these topics are discussed in Section 10-13. Further, equation 10-5.1 is not valid for triangular[43] or spiral arrangements.

Many of the anisotropy constants quoted for ferrimagnetic materials in Chapter 9 were obtained from resonance experiments. When there are both ferrous and ferric ions present in spinels, it is found that the static and microwave measurements of K_1 differ.[44] It is believed that this effect is related to electron transfers from ferrous to ferric ions. A different value of the anisotropy would occur when the period of a microwave becomes less than the electronic transition time. There is good agreement between static and resonance data for YIG and some of the rare earth garnets.[45] For other rare earth garnets, such as YbIG and ErIG, there are differences, particularily at lower temperatures. It seems that these results may be due to fast relaxation in the rare earth sublattice.

The line widths of ferrimagnetic specimens are generally large, (of the order of 100 oersteds) and depend on temperature and crystal orientation. YIG is an exceptional case; a line width of less than 1 oersted has been observed.[46] Various relaxation mechanisms which have been proposed are discussed in Section 10-8.

The foregoing remarks have referred to single crystals. Actually, many studies have been made on polycrystalline samples, partly because of the difficulty in growing good single crystals and partly because such materials are employed in devices. In spite of the increased complexity, it is possible with proper analysis to extract data concerning fundamental quantities.

[40] E. W. Gorter, *Nature (London)* **173,** 123 (1954); J. S. Smart, *Phys. Rev.* **94,** 847 (1954).

[41] T. Okamura and Y. Kojima, *Phys. Rev.* **86,** 1040 (1952).

[42] W. A. Yager, J. K. Galt, and F. R. Merritt, *Phys. Rev.* **99,** 1203 (1955).

[43] A. Eskowitz and R. K. Wangness, *Phys. Rev.* **107,** 379 (1957); R. K. Wangness, *Phys. Rev.* **119,** 1496 (1960).

[44] R. M. Bozorth, B. B. Atlin, J. K. Galt, F. R. Merritt, and W. A. Yager, *Phys. Rev.* **99,** 1898 (1955).

[45] R. F. Pearson, *J. Appl. Phys.* **33,** 1236 (1962).

[46] R. C. LeCraw, E. G. Spencer, and C. S. Porter, *Phys. Rev.* **110,** 1311 (1958).

It is found experimentally that resonance data on polycrystalline samples can often be fitted to the equation

$$\omega_r = -\gamma(H + H_i) \tag{10-6.2}$$

in which H_i is an effective internal field.[47] This equation is valid either when the microwave frequency applied to a given sample is varied or when samples with different porosities are measured at a given frequency.[48] An equation with this form can be derived theoretically, provided several simplifing assumptions are made. First, suppose the applied steady field is strong enough to align the magnetizations of each grain along its direction. Then, for a system of randomly oriented noninteracting spherical crystallites with cubic anisotropy, a first-order approximation yields the result[49]

$$\omega_r = -\gamma\left(H - \frac{K_1}{2M_s}\right); \tag{10-6.3}$$

that is, $H_i = -K_1/2M_s$. For a spherical sample with a spherical hole and no crystalline anisotropy the internal field is $(4\pi/3)M_s(v/V)$, in which v and V are the volumes of the cavity and sample, respectively.[49]

The line width will also be increased as a result of the internal fields adding or subtracting to the effect of the applied steady field. As an example, for independent crystallites with cubic anisotropy the increase in line width ΔH will be approximately equal[49] to $2H_i$; that is, $\Delta H \approx |K_1|/M_s$. For nonmagnetic inclusions such as pores the lines are broadened[50] by an amount proportional to v/V. When the saturation magnetization is comparable to the internal fields, the dipole interaction tends to couple the crystallites. As a result, the line width is narrower than it is when the crystallites are independent.[51]

When the crystalline anisotropy is large, $\mathbf{M}_s$ and $\mathbf{H}$ may no longer be parallel. Then, in analogy to the case illustrated by Fig. 10-4.2, the absorption curve may have more than one peak. It has been predicted that cubic spinels with $K_1 < 0$ will have two maxima[52] and that hexagonal compounds with threefold symmetry in the basal plane will have three.[53] Both effects have been observed.[52,53]

[47] T. Okamura, Y. Torizuka, and Y. Kojima, *Phys. Rev.* **88,** 1425 (1952).

[48] Y. Kojima, *Sci. Rept. Tôhoku Univ.* **A-6,** 614 (1954); T. R. McGuire, *Proc. Conf. Mag. and Mag. Materials, Pittsburgh* **T-78,** 43 (1955).

[49] E. Schlömann, *Proc. Conf. Mag. and Mag. Materials, Boston* **T-91,** 600 (1956).

[50] P. E. Seiden and J. G. Grunberg, *J. Appl. Phys.* **34,** 1696 (1963).

[51] S. Geshwind and A. M. Clogston, *Phys. Rev.* **108,** 49 (1957); E. Schlömann, *J. Phys. Chem. Solids* **6,** 242 (1958).

[52] E. Schlömann and J. R. Zeender, *J. Appl. Phys.* **29,** 341 (1958).

[53] E. Schlömann and R. V. Jones, *J. Appl. Phys.* **30,** 177S (1959).

7. Magnetostatic Modes

In the discussion on ferromagnetic resonance up to this point it has been assumed that the atomic dipoles precess in phase. Such a precession is called the uniform, or Kittel, mode. It is possible, however, to excite other precessional modes. The type of mode that is considered in this section can be activated by applying an inhomogeneous alternating magnetic field.

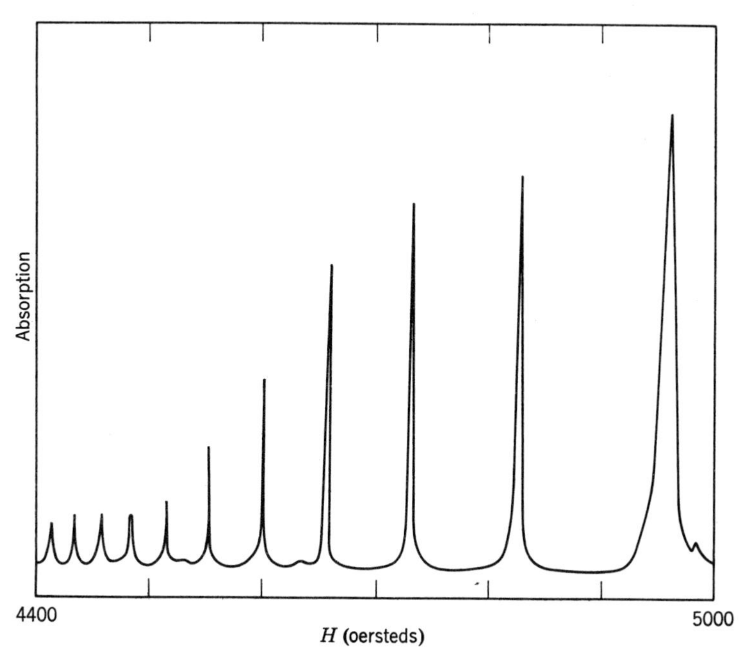

Fig. 10-7.1. Spectrum of a disk of YIG in a uniform rf field and a steady field applied perpendicular to the plane of the disk. The magnetostatic modes appear to the left of the uniform mode. (After J. F. Dillon.)

This can be accomplished either by placing an ellipsoidal sample in a location in the microwave cavity, where the alternating field is nonuniform, or by using a sample with a nonellipsoidal shape. Small samples are employed in order to ensure that propagation effects are small or negligible. The observation of these nonuniform modes was first reported in 1956;[54] the spectrum observed for a disk of YIG placed in a uniform microwave field is shown[55] in Fig. 10-7.1.

[54] R. L. White, I. H. Solt, and J. E. Mercereau, *Bull. Am. Phys. Soc.* **1,** 12 (1956); J. F. Dillon, *Bull. Am. Phys. Soc.* **1,** 125 (1956); R. L. White and I. H. Solt, *Phys. Rev.* **104,** 56 (1956).

[55] J. F. Dillon, *J. Appl. Phys.* **31,** 1605 (1960).

The fact that an inhomogeneous alternating field is required means that the atomic dipoles must be precessing out of phase in different parts of the sample. Further, the peaks of the spectrum are almost independent of the sample size, provided the size is not too large. They do, however, depend strongly on the sample shape and the saturation magnetization. These experimental data are consistent with the idea that these modes have a long wavelength that is comparable to the dimensions of the sample. Under these circumstances the increase in exchange energy will be small and essentially negligible compared to the classical dipole-dipole interaction energy. Since there is little or no propagation of the waves, the problem is essentially a magnetostatic one. This has led to the name magnetostatic modes, although they are also sometimes called Walker modes, since a pertinent theory was first developed by Walker.[56]

In order to simplify the problem, consider an ellipsoid of revolution with steady field applied along the polar axis and negligible crystalline anisotropy. It follows from the equations of motion that the alternating magnetization is given by (equations 10-3.4)

$$M_{1x} = \chi_{11} H_{1x} + \chi_{12} H_{1y}$$

and (10-7.1)

$$M_{1y} = \chi_{21} H_{1x} + \chi_{22} H_{1y},$$

where $\chi_{12} = -\chi_{21}$ and $\chi_{11} = \chi_{22}$, since for this case $D_x = D_y$. If there is no propagation, Maxwell's equations (Section 1-8) reduce to the magnetostatic ones, namely,

$$\nabla \times \mathbf{H}_1 = 0$$

and (10-7.2)

$$\nabla \cdot \mathbf{B}_1 = \nabla \cdot (\mathbf{H}_1 + 4\pi \mathbf{M}_1) = 0.$$

It is convenient to introduce the scalar potential φ, defined as $\mathbf{H}_1 = -\nabla\varphi$. Equations 10-7.1 become

$$-M_{1x} = \chi_{11} \frac{\partial \varphi}{\partial x} + \chi_{12} \frac{\partial \varphi}{\partial y}$$

and (10-7.3)

$$-M_{1y} = \chi_{21} \frac{\partial \varphi}{\partial x} + \chi_{22} \frac{\partial \varphi}{\partial y},$$

whereas from equations 10-7.2 we find

$$\nabla^2 \varphi - 4\pi \nabla \cdot \mathbf{M}_1 = 0. \tag{10-7.4}$$

[56] L. R. Walker, *Phys. Rev.* **105**, 390 (1957).

Elimination of $\mathbf{M}_1$ between equations 10-7.3 and 10-7.4 shows that φ satisfies the differential equation

$$(1 + 4\pi\chi_{11})\left(\frac{\partial^2\varphi}{\partial x^2} + \frac{\partial^2\varphi}{\partial y^2}\right) + \frac{\partial^2\varphi}{\partial z^2} = 0 \qquad (10\text{-}7.5)$$

inside the ellipsoid and, of course, Laplace's equation

$$\nabla^2\varphi = 0 \qquad (10\text{-}7.6)$$

inside the ellipsoid.

The usual boundary conditions at the surface of the ellipsoid, the continuity of φ and the normal component of $\mathbf{B}_1$, must be applied; as a consequence of the latter condition the susceptibility χ_{12} re-enters the equations. Further, if the ellipsoid is situated far from the cavity walls, it can be assumed that φ goes to zero at infinity. The result is an eigenvalue problem with solutions for φ that depend on associated Legendre polynomials $P_n^{\ m}$ and $e^{im\phi}$, in which ϕ is the usual azimuthal coordinate.[57] It follows from the $e^{im\phi}$ factor that the field configurations are circularly polarized, hence rotate together. It is convenient to label the modes with the integral indices (*nmr*). Here *n* and *m* indicate the periodicity in the alternating magnetization with respect to the z- and ϕ-coordinates, respectively, whereas *r* specifies the root of a characteristic equation that yields the resonance frequency. In this notation the uniform precessional mode has the index (110). The alternating magnetization for the (430)-mode in a sphere is indicated[58] in Fig. 10-7.2. The magnetic fields calculated[58] for the resonance of some modes in a sphere at a given microwave frequency are plotted in Fig. 10-7.3. Some peaks expected in the spectrum when $\omega/4\pi M_s\gamma = 0.80$ are given by the intersection of the curves with the nearly vertical line. This value of $\omega/4\pi M_s\gamma$ corresponds to YIG at room temperature and an X-band frequency. The nonuniform modes may occur on either the low or high field side of the uniform precession line. However, for an ellipsoid and a fixed microwave frequency the theory predicts that the modes always occur within a range of $2\pi M_s$ in the steady magnetic field. The theory is in good accord with experiments.[59] By passing light through a $CrBr_3$ sample undergoing resonance it has been found possible to observe some magnetostatic modes visually.[60] The regions with a large amplitude of precession appear as bright spots. It is of interest to mention that magnetostatic modes have also been observed in

[57] L. R. Walker, *Phys. Rev.* **105,** 390 (1957).
[58] L. R. Walker, *J. Appl. Phys.* **29,** 318 (1958).
[59] P. C. Fletcher, I. H. Solt, and R. O. Bell, *Phys. Rev.* **114,** 739 (1959).
[60] J. F. Dillon, H. Kamimura, and J. R. Remeika, *J. Appl. Phys.* **34,** 1240 (1963).

paramagnetic materials; examples are $CuSO_4 \cdot 5H_2O$,[61] $CuK_2(SO_4)_2 \cdot 6H_2O$,[61] and $CuK_2Cl_2 \cdot 2H_2O$.[62]

The magnetostatic modes yield valuable information[63] concerning the various magnetic parameters such as the crystalline anisotropy, the g-value, the saturation magnetization, and the relaxation times. For

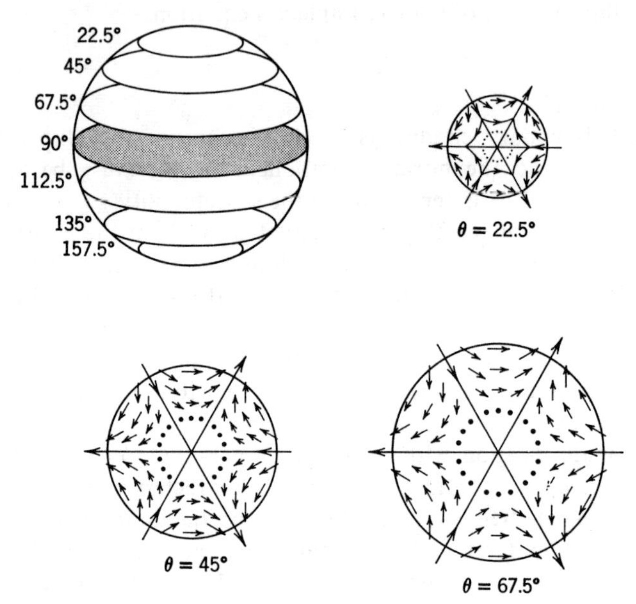

Fig. 10-7.2. To illustrate the (430) magnetostatic mode of a sphere, the arrows indicate the direction of the alternating magnetization in planes perpendicular to the applied steady field for a suitably chosen rotating system of coordinates. The transverse magnetization is 180° out of phase in the corresponding planes in the lower hemisphere. (After L. R. Walker.)

example, the mode spacing for a spherical sample depends on the magnetization, as indicated in Fig. 10-7.4. The magnetization can, of course, be determined from measurements on the uniform mode alone, but then it is necessary to employ two different microwave frequencies. Further, if there are propagation effects, they may go unrecognized.

Electromagnetic propagation, present in samples that are relatively large, produces two effects. First, the fields required for resonance are shifted

[61] G. Seidel and I. Svare, *Paramagnetic Resonance*, Vol. II., W. Low, Editor, Academic Press, New York (1963), p. 468.

[62] H. Abe, H. Morigaki, and K. Koga, *Paramagnetic Resonance*, Vol. II, Academic Press, New York (1963), p. 557.

[63] R. L. White, *J. Appl. Phys.* **31,** 86S (1960).

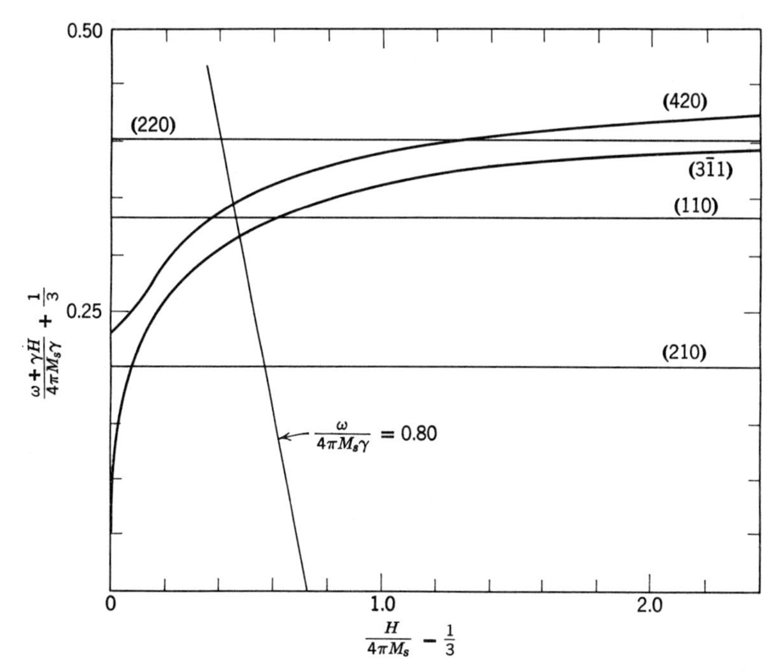

Fig. 10-7.3. The resonance condition for some magnetostatic modes in a sphere as a function of applied steady magnetic field. (After L. R. Walker.)

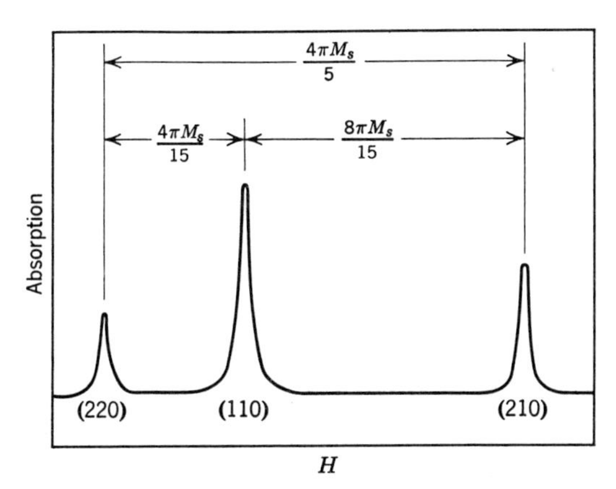

Fig. 10-7.4. Spectrum showing some of the important magnetostatic modes in a sphere and their spacing. [R. L. White, *J. Appl. Phys.* **31,** 86S (1960).]

upward for all modes. The amount of the shift has been calculated[64] for a given radius of sample; hence the data can be corrected. Second, the modes are no longer independent of each other; that is, there is a coupling present between certain modes. For modes that have almost the same energy (Fig. 10-7.3), the second effect may result in an anomalous intensity and a large shift in position of the line.[65] Empirical selection rules have been established from the data.[65]

8. Relaxation Processes

The shape of a ferromagnetic absorption line is, as in the paramagnetic case (equations 3-4.6 and 3-8.10), proportional to χ'', the imaginary part of the complex susceptibility. Expressions for the tensor susceptibility are given by equations 10-3.24 and in problem 10-7*c*. The susceptibility may also be derived for an elliptically or circularly polarized field that is rotating in the same direction as the precession; the analogous calculation for paramagnetic materials is given in Section 3-8. Then χ'' can be obtained from the appropriate expression for $\boldsymbol{\chi}$. As usual, the half-width of the line, ΔH, is defined as the difference between the applied fields corresponding to absorptions equal to half the peak value when the microwave frequency is kept fixed. It is easy to show (problem 10-9) that the damping factor α (equation 10-3.23) is related to ΔH by

$$\frac{1}{\alpha} = \frac{2\omega}{\gamma\,\Delta H}. \tag{10-8.1}$$

Alternatively, the Bloembergen-Bloch relaxation time τ (equation 10-3.20) is given in terms of ΔH by

$$\tau = \frac{2\omega}{\gamma^2 H\,\Delta H}, \tag{10-8.2}$$

since $\tau = 1/\alpha\gamma H$. For a typical ferromagnetic absorption line $\Delta H \approx 100$ oe, and this leads to $\alpha \approx 0.07$ and $\tau \approx 10^{-9}$ sec. A few ferrimagnetic materials have exceptionally narrow line widths, the most notable being YIG for which $\Delta H \approx 0.2$ oe.[66] Other examples are $Ni_{0.96}Fe_{2.04}O_4$ with $\Delta H \approx 15$ oe[67] and $Ba_2Zn_2Fe_{12}O_{22}$ (Zn *Y*) with $\Delta H \approx 8$ oe.[68] However, these results were obtained only by preparing the samples with special

[64] R. A. Hurd, *Can. J. Phys.* **36,** 1072 (1958); J. E. Mercereau, *J. Appl. Phys.* **30,** 184S (1959).

[65] P. C. Fletcher and I. H. Solt, *J. Appl. Phys.* **30,** 181S (1959).

[66] R. C. LeCraw and E. G. Spencer, *Phys. Rev. Letters* **4,** 130 (1960).

[67] H. Sekizawa and K. Sekizawa, *J. Phys. Soc. Japan* **17,** 359 (1962).

[68] A. Tauber, S. Dixon, and R. O. Savage, *J. Appl. Phys.* **35,** 1008 (1964).

techniques. In general, the line widths observed in ferromagnetic resonance are large compared to those in paramagnetic resonance.

It is clear that some mechanism or mechanisms are permitting the dipoles to relax extremely well. Of course, for good electrical conductors there are always the eddy currents. However, the main difference between ferro- or ferrimagnetic and paramagnetic materials is the presence of strong exchange forces in the first two. It has therefore long been recognized that spin waves must play a role in the relaxation; the details of the mechanisms operating emerged more slowly. Actually, a large number of relaxation processes have been proposed, including that produced by the strong coupling of paramagnetic ions to the lattice and by changes of ionic valencies. Since a wide variety of materials has strongly coupled dipoles, it is likely that different mechanisms apply for particular compounds. Thus, even though the line widths can usually be described adequately by one damping constant, such as α, λ, or τ, the microscopic origin of this parameter may be complex and depend on the chemical composition, the crystallographic structure, and the imperfections. Various relaxation mechanisms will now be examined.

Relaxation via spin waves in insulators. The excitation of spin waves by thermal agitation was discussed in Section 6-6. The dispersion relationship (equation 6-6.17)

$$\omega_k = \frac{J_e a^2}{\hbar} k^2$$
$$= -\gamma H_m a^2 k^2$$

was derived. Here H_m is the exchange field given by $H_m = J_e/\gamma\hbar$. In the presence of a static field H this expression becomes

$$\omega_k = -\gamma(H + H_m a^2 k^2). \qquad (10\text{-}8.3)$$

The uniform mode, for which $k = 0$, is the ground state. Hence the transfer of energy from the uniform mode to a spin-wave mode requires a source of additional energy. Such transitions would have a low probability of occurrence. However, the point is that equation 10-8.3 was derived for an infinite medium. In ferromagnetic experiments small samples are employed. Demagnetizing fields then enter the problem. For simplicity consider an ellipsoid of revolution with demagnetization factors D_a and D_b and the polar axis coincident with the direction of the static field. For a spin wave propagating along the polar, or z-, axis the dispersion relationship is given by

$$\omega_{kz} = -\gamma(H - D_a M_s + H_m a^2 k^2). \qquad (10\text{-}8.4)$$

The propagation of a spin wave along any other than the z-direction, say at an angle θ with respect to the polar axis, requires more energy. This extra energy, however, does not originate in the transverse demagnetizing fields. This follows since the uncompensated positive and negative poles on the surface alternate with a spacing of approximately $\lambda(=2\pi/k)$. The fields arising from these poles propagate a distance only of the order of λ, and hence, provided k is large enough, will be restricted to a thin surface layer.

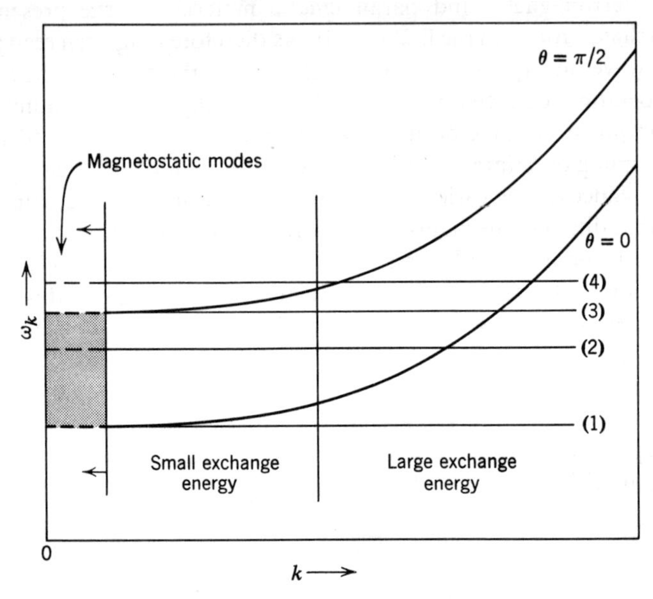

Fig. 10-8.1. Schematic diagram of the dispersion relation for an ellipsoid of revolution (equation 10-8.5). The spin-wave spectrum lies between the $\theta = 0$ and $\theta = \pi/2$ limits. The horizontal lines (1) to (4) indicate the spin waves that are degenerate in energy ($\hbar\omega_k$) with the uniform modes of various frequencies. The ω_k and k scales are different for each of the cases (1) to (4).

The additional energy is necessary because the shape and static field make the z-axis the easy direction. The movement of a spin wave at some angle θ is therefore more difficult. It is reasonable to expect that the extra energy would be proportional to the difference in the principal demagnetizing fields of a disk, namely $4\pi M_s$, and would also depend on θ. Calculation[69] shows that the actual equation for the angular frequency ω_k is

$$\omega_k^2 = \gamma^2(H - D_a M_s + H_m a^2 k^2)(H - D_a M_s + H_m a^2 k^2 + 4\pi M_s \sin^2 \theta). \tag{10-8.5}$$

Plots of this equation for $\theta = 0$ and $\pi/2$ are sketched schematically in Fig. 10-8.1. Curves for all other values of θ lie between these two limits,

[69] A. M. Clogston, H. Suhl, L. R. Walker, and P. W. Anderson, *J. Phys. Chem. Solids* **1**, 129 (1956); H. Suhl, *Proc. IRE* **44**, 1270 (1956), *J. Phys. Chem. Solids* **1**, 209 (1957).

and the region between may appropriately be called the spin-wave manifold. For large k the k^2 term dominates and ω_k increases rapidly for a constant θ. For intermediate values of k the curves are almost horizontal. For k less than about 2000 cm^{-1} the exchange term is small, if not negligible. The uncompensated poles on the surfaces now are important, however, and the spin waves are no longer the correct modes. The magnetostatic modes, discussed in Section 10-7, are, in fact, the proper ones. Hence the dispersion relation is not valid in this region. For higher k the spin waves differ from the true modes only because of the small effect produced by the densely spaced surface poles. The influence of these poles is limited to a thin outer shell of the material; hence the spin waves at the interior are almost unaffected.

Since crystalline anisotropy is being neglected here, the resonance frequency of the uniform mode is given by (equation 10-3.6)

$$\omega_r = -\gamma[H + (D_b - D_a)M_s]. \qquad (10\text{-}8.6)$$

The resonance frequency ω_r can be varied in relation to the spin-wave manifold by changing either the sample's shape,[70] hence D_b and D_a, or the applied field H.[71] In general, the uniform mode will have the same energy as a large number of spin waves (see the horizontal lines (2), (3), and (4) in Fig. 10-8.1). The flat disk is an exception (line 1 in Fig. 10-8.1); this is the lowest value of ω_r possible. Comparison of equation 10-8.6 with equation 10-8.4 or 10-8.5 shows that the equivalence of energy arises from the term for the transverse demagnetizing field when $k = 0$, compensating for the appearance of the exchange terms when $k \neq 0$.

Energy will be transferred from the uniform mode to spin waves with the same frequency, provided some type of coupling exists. The spin waves, in turn, transfer this energy to the lattice, with which they presumably interact strongly. Hence the spin-lattice relaxation proceeds indirectly via the agency of the spin-wave manifold.

If all other things are equal, the greatest line width is expected when the resonance frequency of the uniform mode is degenerate with the largest number of spin waves. This situation corresponds to case 3 in Fig. 10-8.1; here $\omega_r = \gamma(8\pi/3)M_s$ if the sample is a sphere. For YIG at room temperature this equation yields the frequency $\omega_r/2\pi \approx 3.4$ Gc. In experiments on polycrystalline YIG a maximum in the line width was actually found at a frequency of 3.6 Gc. (Fig. 10-8.2);[72] this represents satisfactory agreement. As expected, thin disks (case 1 of Fig. 10-8.1) have relatively narrow lines.[73]

[70] A. S. Risley and H. E. Bussey, *J. Appl. Phys.* **35,** 896 (1964).

[71] C. R. Buffler, *J. Appl. Phys.* **30,** 172S (1959).

[72] *Ibid.*

[73] J. F. Dillon, *J. Appl. Phys.* **31,** 1605 (1960).

The mechanisms proposed for the coupling between the uniform and spin-wave modes all have their origin in some kind of imperfection or disorder. Thus for inverse spinels the disorder of the *B*-site ions leads to local fluctuations in the spin-orbit interaction[74] and in the anisotropy.[75] Vacancies and grain boundaries may produce significant variations in the crystalline and exchange fields. It has been demonstrated experimentally that surface irregularities on a macroscopic scale profoundly affect the

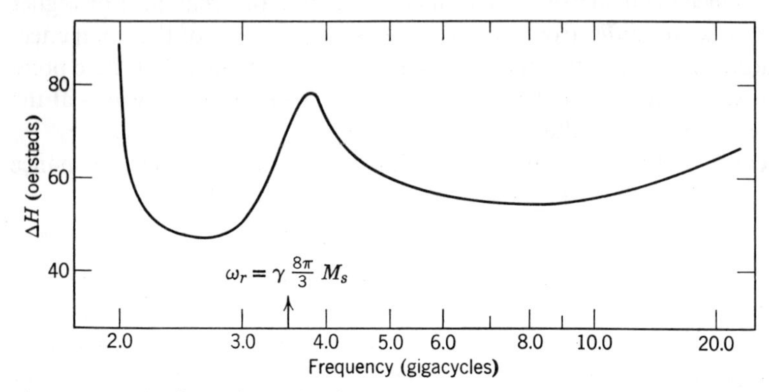

Fig. 10-8.2. The line width of polycrystalline YIG as a function of frequency for a spherical sample at $T = 300°\text{K}$. In the neighborhood of 2 Gc the sample is not saturated, and the line width is due to the domain structure. (After C. R. Buffler.)

line width of YIG single crystals[76] (Fig. 10-8.3). Presumably the fluctuations in the demagnetizing field near the surface give rise to the coupling. Single crystals of YIG are particularly suitable for such studies because of the high degree of perfection within their volume. Regardless of origin, the coupling is expected in general to be a function of the wavelength of the spin wave.

The spin waves excited from the uniform mode interact with the thermal spin waves to establish an equilibrium spin temperature. Various intrinsic processes that involve several magnons as well as phonons have been considered in the literature.[77]

[74] H. B. Callen and E. Pittelli, *Phys. Rev.* **119,** 1523 (1960).

[75] A. M. Clogston, H. Suhl, L. R. Walker, and P. W. Anderson, *J. Phys. Chem. Solids* **1,** 129 (1956).

[76] R. C. LeCraw, E. G. Spencer, and C. S. Porter, *Phys. Rev.* **110,** 1311 (1958).

[77] A. Akheiser, *J. Phys. (USSR)* **10,** 217 (1946); A. I. Akheiser, V. G. Bar'yakhtar, and S. V. Peletminskii, *J. Exptl. Theoret. Phys. (USSR)* **35,** 288 (1958) [trans. *Soviet Phys.-JEPT* **8,** 157 (1959)]; V. G. Bar'yakhtar and G. I. Urushadze, *J. Exptl. Theoret. Phys., (USSR)* **39,** 355 (1960) [trans. *Soviet Phys.-JEPT* **12,** 251 (1961)]; M. I. Kagonov and V. M. Tsukernik, *J. Exptl. Theoret. Phys. (USSR)* **36,** 224 (1959) [trans. *Soviet Phys.-JEPT* **9,** 151 (1959)]; J. Morkowski, *Acta Phys. Polon.* **19,** 3 (1960); T. Kasuya and R. C. LeCraw, *Phys. Rev. Letters* **6,** 223 (1961); P. Pincus, M. Sparks, and R. C. LeCraw, *Phys. Rev.* **124,** 1015 (1961); E. Schlömann, *Phys. Rev.* **121,** 1312 (1961).

There is also the possibility of the direct interaction between the uniform mode and the lattice. When ions with degenerate orbital ground states are present in the lattice, this direct relaxation is likely to be important (see page 574). Otherwise, relaxation of this type proceeds via some intrinsic process, such as the dipole-dipole interaction,[78] and is relatively slow. However, for highly polished and very pure YIG, which has a line

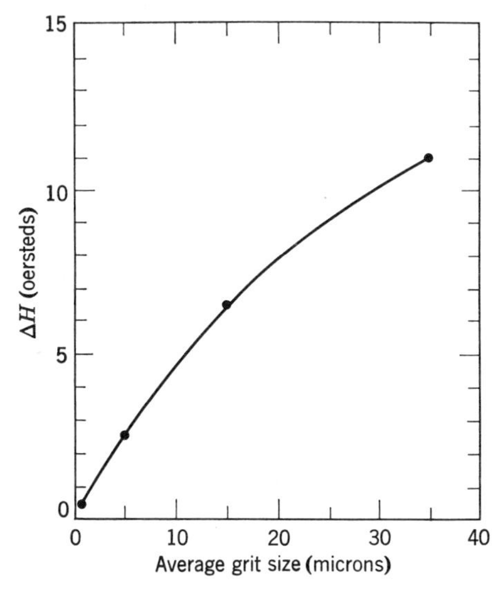

Fig. 10-8.3. Line width of YIG as a function of surface polish at 9.3 Gc. The abscissa represents the mean grit size of the polishing paper. [R. C. LeCraw, E. G. Spencer, and C. S. Porter, *Phys. Rev.* **110,** 1311 (1958).]

width of less than 0.1 oe, it seems probable that this is the relaxation mechanism.

Parameters whose purpose is to distinguish between the different microscopic relaxation processes have been introduced into the equations of motion.[79,80] For example, separate times for the uniform mode-lattice, uniform mode-spin waves, and the spin-wave mode-lattice relaxations have been defined.[81] Attempts have been made to measure these individual quantities by varying the properties of the sample and employing frequency modulation[81] and a parallel pumping technique[82] (see Section 10-9).

[78] M. Sparks, R. Loudon, and C. Kittel, *Phys. Rev.* **122,** 791 (1961).
[79] H. B. Callen, *J. Phys. Chem. Solids* **4,** 256 (1958).
[80] R. C. Fletcher, R. C. LeCraw, and E. G. Spencer, *Phys. Rev.* **117,** 955 (1960).
[81] *Ibid.*
[82] R. C. LeCraw and E. G. Spencer, *J. Phys. Soc. Japan, Suppl. B-I* **17,** 401 (1962).

Finally, it should be mentioned that experiments in which high microwave powers are applied to the samples have yielded much important information on relaxation processes and the role of the spin-wave manifold. These matters are discussed in Section 10-9.

Relaxation via spin waves in conductors. In metals or alloys with dimensions large compared to the skin depth the amplitude of the microwave field decreases with penetration into the sample. As a result, the magnetization in the skin-depth region is nonuniform and the exchange energy may be appreciable. Eddy currents, present because of the conductivity and the exchange, then give rise to energy dissipation. Line broadening by the eddy current damping of spin waves will be most important when the skin depth and internal field are small. The line widths of a Permalloy with 66% Ni[83] and iron whiskers[84] are believed to develop primarily from this mechanism.

The effects of exchange on line widths have been calculated with the aid of certain simplifying assumptions.[83,85] In the course of the analysis the general boundary condition

$$\left(\frac{2A}{M_s^2}\right)\mathbf{M} \times \left(\frac{\partial \mathbf{M}}{\partial n}\right) + \mathbf{L}_s = 0 \qquad (10\text{-}8.7)$$

was derived.[86] Here A is an exchange constant (see Section 10-9) and $\mathbf{L}_s$ is a torque due to a surface anisotropy. If $\mathbf{L}_s = 0$, this dynamical equation becomes identical in form to that for the static case (equation 7-1.14).

Fast relaxation via paramagnetic ions. Paramagnetic ions with degenerate orbital ground states are expected to be strongly coupled to the lattice, hence to have short relaxation times. This idea finds confirmation in the large line widths observed for these ions in dilute paramagnetic salts. Examples are all the rare earth ions except Gd^{3+}, the iron group ions Fe^{2+}, Co^{2+} and Mn^{2+} in octahedral sites, and Fe^{2+}, Ni^{2+} and Mn^{3+} in tetrahedral sites.

In ferrimagnetic materials these ions are strongly coupled by the exchange interaction to the ions in S states. Thus the ions with unquenched orbital angular momentum play an intermediate role in the transfer of energy from the otherwise relatively undamped spin systems to the lattice. In more picturesque language they provide a thermal path of high conductivity.

[83] G. T. Rado and J. R. Weertman, *J. Chem. Phys. Solids* **11,** 315 (1959).

[84] D. S. Rodbell, *J. Appl. Phys.* **30,** 187S (1959).

[85] J. R. MacDonald, *Phys. Rev.* **103,** 280 (1956); P. Pincus, *Phys. Rev.* **118,** 658 (1960).

[86] G. T. Rado and J. R. Weertman, *loc. cit.*

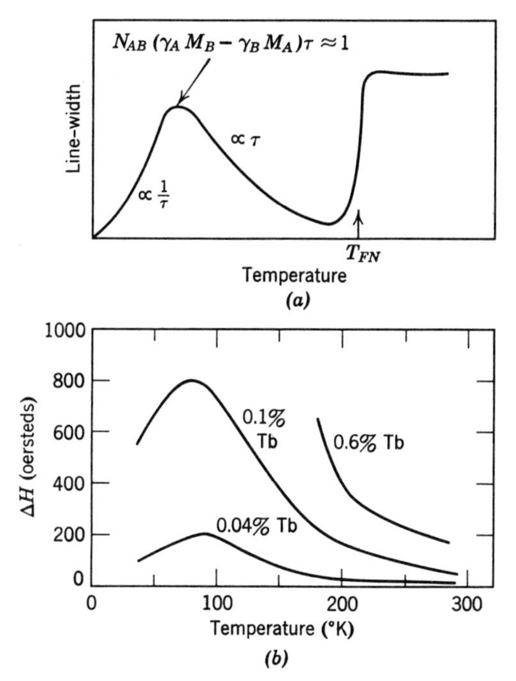

Fig. 10-8.4. Variation of line width with temperature (*a*) predicted by the fast relaxation model and (*b*) observed for YIG doped with various amounts of terbium. [(*a*): C. Kittel, *J. Appl. Phys.* **31,** 115 (1960); (*b*): J. F. Dillon and J. W. Nielsen, *Phys. Rev. Letters* **3,** 30 (1959).]

This relaxation mechanism is supported by various experimental evidence. When YIG is doped with 0.6 atomic per cent Tb, the line width at room temperature increases from a fraction of an oersted to about 200 oe[87] (Fig. 10-8.4).[88,89] The substitution of Sm in YIG leads to a linear increase in ΔH of about 22 oe/atomic per cent of Sm.[90] The rare earth garnets have line widths of hundreds or even thousands of oersteds.[91] Spinels with non-S state ions, notably Fe^{2+}, have large line widths.[92]

Additional evidence that favors this model has been obtained from studies of the line widths as a function of temperature. According to

[87] J. F. Dillon and J. W. Nielsen, *Phys. Rev. Letters* **3,** 30 (1959).

[88] C. Kittel, *J. Appl. Phys.* **31,** 11S (1960).

[89] Dillon and Nielsen, *loc. cit.*

[90] M. H. Sirvetz and J. E. Zniemer, *J. Appl. Phys.* **29,** 431 (1958).

[91] G. P. Rodrigue, J. E. Pippin, W. P. Wolf, and C. L. Hogan, *Trans. IRE Microwave Theory Tech.* **6,** 83 (1958).

[92] L. R. Bickford, *Phys. Rev.* **78,** 499 (1950); J. K. Galt, W. A. Yager, and F. R. Merritt, *Phys. Rev.* **93,** 1119 (1954); P. E. Tannenwald, *Phys. Rev.* **100,** 1713 (1955); A. D. Schnitzler, V. J. Folen, and G. T. Rado, *J. Appl. Phys.* **31,** 348S (1960).

macroscopic and microscopic theories,[93] the line width below the ferrimagnetic Néel temperature will vary directly as τ until $\omega_{AB}\tau \approx 1$, in which ω_{AB} is an exchange frequency given by

$$\omega_{AB} = N_{AB}(\gamma_A M_B - \gamma_B M_A).$$

Since τ increases rapidly at low temperatures for most paramagnetic ions, ΔH will also increase. ΔH has a maximum at $\omega_{AB}\tau \approx 1$, since for lower temperatures $\Delta H \approx 1/\tau$.[94] Above T_{FN} the line width depends on a fluctuation process. The variation of ΔH with temperature predicted by this theory is indicated schematically in Fig. 10-8.4*a*. Experimental data for YIG doped with various amounts of terbium indeed exhibit[95] the general behavior expected below T_{FN} (Fig. 10-8.4*b*). A maximum in the line width at low temperatures has also been observed for impure YIG[96] and for MnFe[97] and NiFe[98] spinels that contain significant amounts of ferrous ions.

The presence of fast relaxing ions can also have a profound influence on the *g*-value.[99] This effect is illustrated with a simple two-sublattice model.[100] Suppose that the *A* sublattice consists of ions with negligible damping and the *B* sublattice of ions with essentially infinite damping ($\tau \to 0$). The equation of motion for the magnetization $\mathbf{M}_A$ of the *A* sublattice is

$$\frac{d\mathbf{M}_A}{dt} = \gamma_A \mathbf{M}_A \times (\mathbf{H} - N_{AB}\mathbf{M}_B), \tag{10-8.8}$$

where $(\mathbf{H} - N_{AB}\mathbf{M}_B)$ is the effective field acting on the *A* ions, since the anisotropy, shape, and intrasublattice fields are assumed to be very small. The magnetization $\mathbf{M}_B$ of the *B* sublattice will instantaneously assume the direction of the effective field $\mathbf{H} - N_{AB}\mathbf{M}_A$ that acts on the *B* ions, hence

$$\mathbf{M}_B = \frac{\mathbf{H} - N_{AB}\mathbf{M}_A}{|\mathbf{H} - N_{AB}\mathbf{M}_A|} M_B$$

$$\approx (\mathbf{H} - N_{AB}\mathbf{M}_A)\frac{M_B}{N_{AB}M_A}. \tag{10-8.9}$$

[93] P.-G. de Gennes, C. Kittel, and A. M. Portis, *Phys. Rev.* **116,** 323 (1959).
[94] *Ibid.*
[95] J. F. Dillon and J. W. Nielsen, *Phys. Rev. Letters* **3,** 30 (1959).
[96] E. G. Spencer, R. C. LeCraw, and A. M. Clogston, *Phys. Rev. Letters* **3,** 32 (1959).
[97] P. E. Tannenwald, *Phys. Rev.* **100,** 1713 (1955).
[98] W. A. Yager, J. K. Galt, and F. R. Merritt, *Phys. Rev.* **99,** 1203 (1955).
[99] R. K. Wangness, *Phys. Rev.* **111,** 813 (1958); C. Kittel, *Phys. Rev.* **115,** 1587 (1959).
[100] C. Kittel, *J. Appl. Phys.* **31,** 11S (1960).

Substitution of equation 10-8.7 into (10-8.6) yields

$$\frac{d\mathbf{M}_A}{dt} = \gamma_A\left(1 - \frac{M_B}{M_A}\right)\mathbf{M}_A \times \mathbf{H}. \tag{10-8.10}$$

It follows immediately that the effective g-value is given by

$$g_{\text{eff}} = g_A\left(\frac{M_A - M_B}{M_A}\right). \tag{10-8.11}$$

If $g_A \approx 2$, then g_{eff} may be considerably less than 2. The rare earth garnets appear to be good examples. Here the ensemble of ferric ions would constitute the A sublattice and the rare earth ions, the B sublattice. Some of the effective g-values observed for the RIGs at $T = 500°K$ are:[101] 1.58 for ErIG, 1.36 for HoIG, and 1.9 for YbIG and SmIG.

Slow relaxation via electron redistribution. Quite a different model which forms an alternative to that of fast relaxation has been suggested. It was first proposed for spinels with both ferric and ferrous ions on the B sites.[102] On thermodynamic grounds the arrangement of these ions at equilibrium is expected to depend on the direction of the magnetization. Changes in the valency of the Fe^{2+} and Fe^{3+} ions can take place simply by the transfer of electrons. The approach to equilibrium is described by a relaxation time τ, which is assumed to be comparable to the reciprocal of the microwave frequency. Even though YIG doped with rare earth ions or the rare earth ions themselves have only trivalent cations, a variation of this model can still be applied.[103] Magnetization changes in the ferric sublattices will lead to time-varying exchange fields acting on the rare earth ions. The splitting of the energy levels of the rare earth ions will then oscillate, provided there is a component of the alternating exchange field parallel to the static splitting field (for the analogous paramagnetic case see Section 3-5). The populations of the energy levels will then tend toward equilibrium with a relaxation time τ. For either the spinels or the garnets the finite time required to produce equilibrium gives rise to a phase lag in the magnetization and hence to energy losses and line broadening. This mechanism is referred to as slow relaxation.

For slow relaxation the peak in the line width versus temperature curve is expected to occur at a different temperature when the microwave frequency

[101] G. P. Rodrigue, H. Meyer, and R. V. Jones, *J. Appl. Phys.* **31,** 376S (1960).

[102] J. K. Galt, W. A. Yager, and F. R. Merritt, *Phys. Rev.* **93,** 1119 (1954); J. K. Galt, *Bell System Tech. J.* **33,** 1023 (1954); H. P. J. Wijn and H. van der Heide, *Revs. Mod. Phys.* **80,** 744 (1953); A. M. Clogston, *Bell System Tech. J.* **34,** 739 (1955).

[103] R. W. Teale and K. Tweedale, *Phys. Letters* **1,** 298 (1962); J. H. Van Vleck and R. Orbach, *Phys. Rev. Letters*, **11,** 65 (1963).

is changed; the magnitude of the peak, however, will be independent of frequency. Just the reverse is predicted by the fast relaxation theory. Experimentally it is found that for the spinel $Ni_{0.75}Fe_{2.25}O_4$ the peak in the $\Delta H - T$ curve is a function of frequency.[104] Similar support for the slow relaxation theory has been obtained from observations on YIG doped with a number of the rare earths.[105]

9. Nonlinear Effects

A study of the response of ferrimagnetic materials to high microwave powers has yielded significant information concerning relaxation processes in general and spin waves in particular. In addition, these investigations have led to the development of a number of important high-frequency devices. Two main experimental arrangements are used for fundamental studies. In one the microwave field is applied perpendicular to the static magnetic field, whereas in the other the microwave field is applied parallel to the static field. Both are discussed in turn.

In the first experiments two effects were observed. The peak of the absorption line decreased for sufficiently large microwave fields and a subsidiary absorption peak appeared on the low field side. These are illustrated in Fig. 10-9.1.[106,107] Of course, a decrease in the susceptibility at the main peak is expected because of saturation. For saturation the usual restriction of ferromagnetic resonance that M_x and M_y (or M_{1x} and M_{1y}) be small and $M_z \approx M$ is removed. The analogous paramagnetic case has been considered in detail in Section 3-8. From equations 3-8.4 and 3-8.6 it follows that at resonance

$$\frac{M_z(H_1)}{M_0} = \frac{\chi''(H_1)}{\chi''(0)} = \frac{1}{1 + \gamma^2 H_1^{\,2} \tau_1 \tau_2}, \qquad (10\text{-}9.1)$$

where $\chi''(0)$ is the susceptibility when $H_1 \approx 0$. Then saturation will set in when $\gamma^2 H_1^{\,2} \tau_1 \tau_2 \approx 1$. If $\tau_1 \approx \tau_2 \approx 1/\gamma\, \Delta H$, where ΔH is the line width, the threshold for the onset of saturation is $H_1 \approx \Delta H$. It is easy to show that a similar result is obtained for the corresponding ferromagnetic case. The surprising thing in the high power experiments is that χ'' decreased when H_1 was very much smaller than ΔH (Fig. 10-9.1*a*). Further, in contradiction to the predictions of equation 10-9.1, M_z changed but little

[104] J. K. Galt and E. G. Spencer, *Phys. Rev.* **127,** 1572 (1962).
[105] J. F. Dillon, *Phys. Rev.* **127,** 1495 (1962); B. H. Clarke, R. F. Pearson, and K. Tweedale, *J. Appl. Phys.* **34,** 1269 (1963).
[106] N. Bloembergen and S. Wang, *Phys. Rev.* **93,** 72 (1954).
[107] R. W. Damon, *Revs. Mod. Phys.* **25,** 239 (1953).

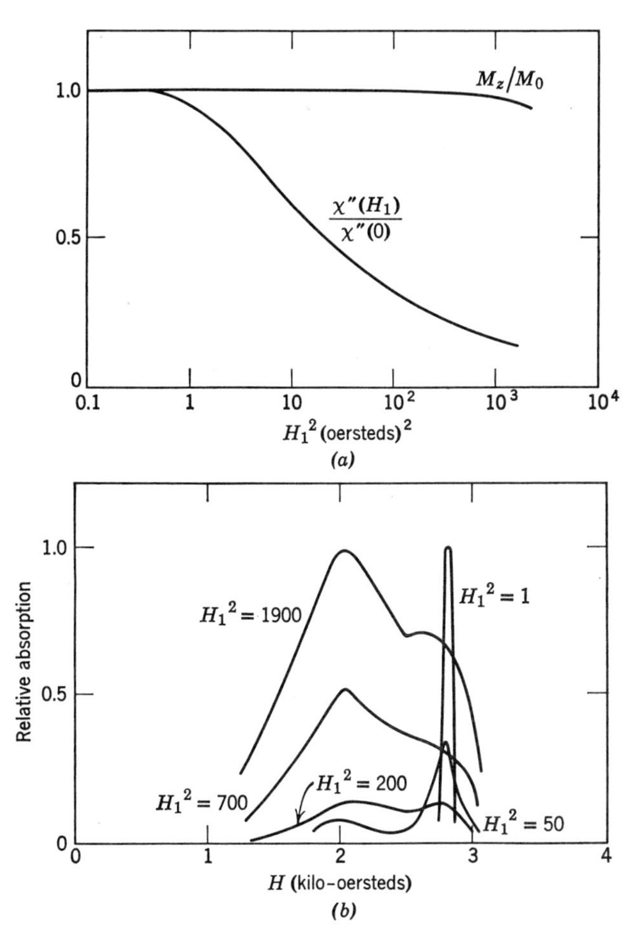

Fig. 10-9.1. The response of a single crystal of $NiFe_2O_4$ to a microwave signal of large amplitude in a resonance experiment ($T = 300°K$). In (*a*) the normalized values of χ'' and M_z at the peak of the main resonance are plotted as a function of H_1^2. In (*b*) the absorption spectra is shown for various microwave power levels. The ordinates of the curves for $H_1^2 = 1$ and 50 oe differ from those for the other microwave fields. Note the increase of the low field peak at the expense of the principal peak. [N. Bloembergen and S. Wang, *Phys. Rev.* **93,** 72 (1954).]

at the highest signal levels employed. In other words, the precession angle does not increase very much with the microwave power.

The answer to this problem is found by retaining higher order terms involving products of the spin-wave modes (including $k = 0$). These terms are present even if the crystal specimen is chemically and crystallographically perfect. They are obtained from the equations of motion when the sum of the applied field, the demagnetizing fields, and the exchange fields is substituted for H_T.[108] It turns out that only two of the nonlinear terms are important, one that couples the uniform mode and a spin-wave mode of wave number k and the other that involves the square of the uniform mode and a spin-wave mode k. The terms containing the uniform mode are the important ones because this mode has a large amplitude induced by the microwave field. These two terms lead to effects that are generally referred to as nonlinear first- and second-order processes, respectively. Although the theoretical analysis is essentially straightforward, a considerable amount of mathematical detail is required. Because of this the discussion here is essentially qualitative.

It is the first-order process that gives rise to the subsidiary absorption peak of Fig. 10-9.1a. Because of the conservation of momentum the spin-waves excited by the coupling to the uniform precession have the wave numbers k and $-k$. From the conservation of energy it follows that

$$\omega_{\mathbf{k}} + \omega_{-\mathbf{k}} = \omega$$

or, since $\omega_{\mathbf{k}}$ is an even function of $\mathbf{k}$,

$$\omega_{\mathbf{k}} = \frac{\omega}{2}. \qquad (10\text{-}9.2)$$

Here ω is the frequency of the uniform precession and also of the signal.

It is convenient to describe the damping or relaxation of the spin wave with wave number k by the line width ΔH_k. When the amplitude of the uniform mode is small, only a little energy that is consistent with the dissipation ΔH_k will be transferred to the spin waves. However, when the amplitude of the uniform mode becomes large enough to overcome this dissipation, the critical spin waves will grow exponentially. Hence at higher signal powers the energy above this threshold will tend to go into the spin waves. Further, the amplitude of the uniform mode will thereafter increase only slightly. The situation is similar to that of the circuit oscillator. There, when sufficient power (or, in circuit language, sufficient negative resistance) is added to overcome the resistance in the tank circuit, instability results and the circuit breaks into oscillation.

The value of the threshold field required for the first order instability, H_{1t}, is a function of the applied field and is given approximately by[109]

$$H_{1t} \simeq + \frac{\Delta H_k}{4\pi\gamma M_s} g(\theta)[(\omega - \omega_r)^2 + \gamma^2(\Delta H)^2]^{1/2}; \qquad (10\text{-}9.3)$$

[108] H. Suhl, *Proc. IRE* **44,** 1270 (1956); P. W. Anderson and H. Suhl, *Phys. Rev.* **100,** 1788 (1955); H. Suhl, *J. Phys. Chem. Solids* **1,** 209 (1957).

[109] Suhl, *Proc. IRE, loc. cit.*

provided

$$H < +\frac{\omega}{2\gamma} + D_a M_s. \qquad (10\text{-}9.4)$$

Here θ is the angle between $\mathbf{k}$ and the z-axis, $g(\theta)$ is a function of θ equal to $+\infty$ for $\theta = 0$ and $\theta = \pi/2$ and possesses a minimum between these limits, and ω_r is the resonance frequency of the uniform mode for an ellipsoid of revolution with negligible crystalline anisotropy; that is (equation 10-3.6),

$$\omega_r = -\gamma[H + (D_b - D_a)M_s]. \qquad (10\text{-}9.5)$$

When the static field reaches the value

$$H = \frac{\omega}{2\gamma} + D_a M_s \qquad (10\text{-}9.6)$$

the threshold field becomes infinite. Equation 10-9.6 is obtained from the dispersion relationship (equation 10-8.3) with $\omega_{\mathbf{k}} = \omega/2$, $k^2 = 0$, and $\theta = 0$. For static fields larger than $(\omega/2\gamma) + D_a M_s$ the first order instability is no longer possible.

The subsidiary absorption first appears at the static field for which the threshold is a minimum; this occurs when H is somewhat less than the limiting value of equation 10-9.6. In common experimental situations the subsidiary absorption is located a few hundred oersteds below the main resonance. As an example, the predicted ratio of the static field at the subsidiary resonance to that at the main resonance for a sphere of $NiFe_2O_4$ ($4\pi M_s = 3{,}320$ oe) at an X-band frequency is 0.83. Experimentally (Fig. 10-9.1*b*) the ratio 0.73 is found. In general, the agreement between theory and experiment for this ratio and for the value of the threshold field, H_{1t}, is fairly good. The unstable spin waves generated at the subsidiary absorption usually have $|\mathbf{k}| = 0$ and $0 < \theta < \pi/3$. Thus the analysis should actually employ the magnetostatic modes. It appears, however, that the error involved is not serious.[110]

For low signal frequencies and suitable sample shapes and saturation magnetizations it is possible to have the subsidiary and main resonances coincide. From equations 10-9.6 and 10-9.5 it is easy to see that the static field required for an infinite threshold will lie above the resonance field of the uniform mode if

$$\frac{\omega}{2} < \gamma D_b M_s. \qquad (10\text{-}9.7)$$

[110] H. Suhl, *J. Appl. Phys.* **29**, 416 (1958).

When this condition applies, coincidence can be obtained. From equation 10-9.3 the threshold field is then given by

$$H_{1t} \approx \frac{\Delta H_k \, \Delta H}{4\pi M_s}, \tag{10-9.8}$$

since $g(\theta) \approx 1$ at its minimum. This value of H_{1t} is exceptionally small. At resonance no additional absorption appears above threshold. Instead,

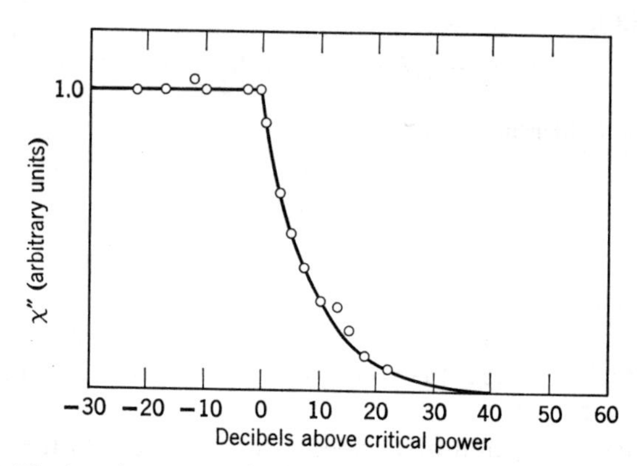

Fig. 10-9.2. The imaginary part of the susceptibility as a function of H_1^2 (signal power) when the subsidiary and main resonances coincide. The open circles represent the experimental data obtained from a disk of $MnFe_2O_4$ with the static field in the plane of the disk and a frequency of 4040 Mc/sec. The agreement with theory (solid line) is good. Above the threshold χ'' declines as (signal power)$^{-1/2}$, that is, as H_1^{-1}. [H. Suhl, *Proc. IRE* **44,** 1270 (1956).]

because of the additional losses suffered by the uniform mode, the susceptibility decreases. There is good agreement between the theory and the experimental data for an $MnFe_2O_4$ disk (Fig. 10-9.2)[111]; for nonstoichiometric spinels, however, the agreement is poor.[112,113] For YIG, which has an extremely narrow line width, instability is produced by a signal field of only a few millioersteds;[114] this is in good accord with the predictions of equation 10-9.8.

When the inequality (10-9.7) is reversed, that is, if

$$\frac{\omega}{2} > -\gamma D_b M_s,$$

[111] H. Suhl, *Proc. IRE* **44,** 1270 (1956).

[112] *Ibid.*

[113] E. Schlömann, J. H. Saunders, and M. H. Sirvetz, *Trans. IRE Microwave Theory Tech.* **8,** 96 (1960).

[114] R. C. LeCraw, E. G. Spencer, and C. S. Porter, *J. Appl. Phys.* **29,** 326 (1958).

it is no longer possible for the subsidiary and main resonance to coincide. The decline in the susceptibility of the main peak, such as that in Fig. 10-9.1, can then be caused by the second-order process. Since the square of the uniform mode, hence $e^{i2\omega t}$, is involved, the conservation of energy now requires

$$\omega_{\mathbf{k}} + \omega_{-\mathbf{k}} = 2\omega_{\mathbf{k}} = 2\omega$$

or

$$\omega_{\mathbf{k}} = \omega. \tag{10-9.9}$$

The threshold for the second-order process is[115]

$$H_{1t} = \Delta H\left(\frac{2\Delta H_k}{4\pi M_s}\right)^{1/2}, \tag{10-9.10}$$

which is appreciably larger than that for the first-order process (equation 10-9.8) but still much smaller than ΔH, the value predicted by the Bloch equations. The unstable spin waves excited in this second-order process propagate along the z-axis ($\theta = 0$) and have a wavelength of the order of 50 lattice spacings.[115]

Imperfections, which are always present in any specimen, also couple the uniform mode to spin waves with the same frequency in the manner discussed in Section 10-8. Coupling by imperfections is not important when the nonlinear first-order process applies, for then $\omega_{\mathbf{k}} = \omega/2$ (equation 10-9.2). As a result of the additional coupling the threshold for the second-order process is reduced. Further, on the basis of a microscopic theory it has been shown that[116] the susceptibility initially varies linearly with the square of the microwave level; that is

$$\frac{\chi''}{\chi''(0)} \approx 1 - C\left(\frac{H_1}{\Delta H}\right)^2. \tag{10-9.11}$$

This predicted behavior has been observed for several spinels and garnets;[117] in particular the data for polycrystalline YIG plotted in Fig. 10-9.3 supports the theory.[118] The constant C depends on composition, grain size, and the location of the uniform mode re the spin-wave manifold. Although C is usually positive, it can be negative for part of the range of H_1. A maximum in the χ'' versus $H_1{}^2$ curve was actually observed for $Gd_{1.5}Y_{1.5}Fe_2Fe_3O_{12}$.[119] At powers somewhat above those corresponding to the threshold of equation 10-9.10 a decline in χ'' as $1/H_1$ is generally

[115] H. Suhl, *Proc. IRE* **44,** 1270 (1956).
[116] E. Schlömann, *Phys. Rev.* **116,** 828 (1959); H. Suhl, *J. Appl. Phys.* **30,** 1961 (1959).
[117] J. J. Green and E. Schlömann, *Trans. IRE Microwave Theory Tech.* **8,** 100 (1960).
[118] P. E. Seiden and H. J. Shaw, *J. Appl. Phys.* **31,** 224S (1960).
[119] E. Schlömann, J. J. Green, and U. Milano, *J. Appl. Phys.* **31,** 386S (1960).

found (Fig. 10-9.3). This decline is the same as that in the nonlinear first-order process and follows because $\chi'' = |M_+/H_1| = \theta M_s/H_1$, where θ is now the precession angle of the uniform mode. Above the threshold there is little increase in the precessional angle, since the additional signal power goes into the unstable spin waves, hence $\chi'' H_1 \approx$ constant.

High resolution experiments have revealed[120] a fine structure in the χ''-H_1 curves. This feature apparently is present because the spin-wave

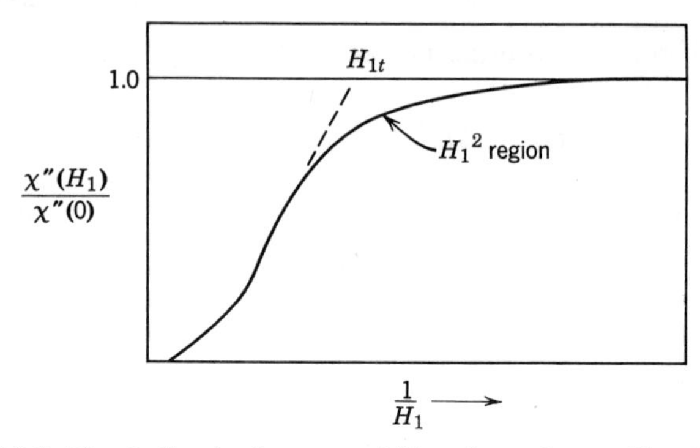

Fig. 10-9.3. The decline in the susceptibility of a polycrystalline YIG sample at the main resonance peak as a function of the reciprocal of the alternating field amplitude. Extrapolation of the linear part of the decline yields the threshold for the second-order instability. [P. E. Seiden and H. J. Shaw, *J. Appl. Phys.* **31,** 225S (1960).]

spectrum is actually discrete and not continuous as usually assumed.[121] Some Ni-Zn spinels have the interesting property that the limiting value of $\gamma D_b M_s$ (equation 10-9.7) lies close to X-band frequencies. In this case the first-order process goes over to the second order as the power is increased[119] because the decrease in M_z produces an increase in $\omega_{\mathbf{k}}$.

Other effects that have been observed at high power levels when $\mathbf{H}_1$ is perpendicular to $\mathbf{H}$ include the dependence of the threshold on the crystallographic orientation[122] and various phenomena involving low-frequency oscillations.[123] Of particular interest are the magneto-acoustic resonances[124] found in polished spheres of YIG and other garnets. Acoustic vibrations in the range of 0–30 Mc can be induced in YIG samples that

[120] E. Schlömann and J. J. Green, *Phys. Rev. Letters* **3,** 129 (1959).

[121] E. Schlömann, *Phys. Rev.* **116,** 828 (1959).

[122] J. I. Masters, *J. Appl. Phys.* **31,** 41S (1960); P. Gottleib, *J. Appl. Phys.* **31,** 2059 (1960).

[123] M. T. Weiss, *Phys. Rev. Letters* **1,** 239 (1958), *J. Appl. Phys.* **30,** 146S (1959); T. S. Hartwick, E. R. Peressini, and M. T. Weiss, *J. Appl. Phys.* **32,** 223S (1961).

[124] E. G. Spencer and R. C. LeCraw, *Phys. Rev. Letters* **1,** 241 (1958), *J. Appl. Phys.* **30,** 149S (1959); R. C. LeCraw and T. Kasuya, *Phys. Rev.* **130,** 50 (1963).

are close to ferromagnetic resonance at microwave frequencies. The coupling between the spin waves and the acoustic modes is provided by the magnetostriction. It turns out that YIG possesses a high acoustic Q of the order of 10^5.

In order to test the validity of the threshold formulas (equations 10-9.3, 10-9.8, and 10-9.10), it is necessary to employ a value for ΔH_k. Usually it is assumed either that $\Delta H_k \approx \Delta H$ or that ΔH_k can be calculated from a measurement of, say, the spin-lattice relaxation time. Conversely, the experimental determination of H_{1t}, together with the appropriate threshold equation, yields a value for ΔH_k. Of course, there are complications to be met; for example, extrapolation may have to be made to obtain H_{1t} (Fig. 10-9.3). It is found that ΔH_k is independent of a single crystal's surface conditions[125] and of the porosity.[126] Further, for high power levels ΔH_k is observed to be directly proportional to the amount of rare earth doping in YIG.[127] This is reasonable, since the rare earth ions are expected to lead to faster relaxation. Similar results are found for increasing amounts of cobalt in a Ni-Co spinel.[128] All in all, there is considerable merit in the study of ΔH_k; certainly it is a better measure of intrinsic loss processes than ΔH.

As first pointed out by Schlömann et al.[128] and Morgenthaler,[129] spin waves and their line widths can be investigated directly by applying the microwave field parallel to the static field. This experimental arrangement is the same as that employed for the study of spin-lattice relaxation in paramagnets (Section 3-5). In the ideal ferromagnetic case there is no microwave power absorbed until a certain threshold required for the growth of some spin waves is reached. The process depends on the tip of the precessing magnetization vector following an elliptical rather than a circular path; that is, the anisotropy along the x- and y-axis must be different for some of the spin waves. A simple example is the uniform magnetization of a disk when the two fields are applied in the plane of the sample (problem 10-10). The general technique is often called the *parallel pump* method.

It has been shown[128] that the threshold field for spin-wave instability when the steady and alternating fields are parallel to the z-axis is given by

$$H_{1t} = \frac{\omega}{4\pi M_s \gamma} \min \frac{\Delta H_k}{\sin^2 \theta}, \qquad (10\text{-}9.12)$$

[125] R. C. LeCraw and E. G. Spencer, *J. Appl. Phys.* **30,** 185S (1959).
[126] J. J. Green and E. Schlömann, *Trans. IRE Microwave Theory Tech.* **8,** 100 (1960).
[127] E. Schlömann, J. J. Green, and U. Milano, *J. Appl. Phys.* **31,** 386S (1960).
[128] *Ibid.*
[129] F. R. Morgenthaler, *J. Appl. Phys.* **31,** 95S (1960).

where the minimum is computed under the condition

$$\omega_{\mathbf{k}} = \frac{\omega}{2}. \qquad (10\text{-}9.13)$$

Here θ is the angle between the z-axis and the propagation direction ($\mathbf{k}$) of the spin wave. Below a certain limiting field, H_{lim}, the unstable spin waves have $k \neq 0$ and $\theta = \pi/2$; that is, they are directed perpendicular to the magnetic fields. The limiting field occurs when $k = 0$ and $\theta = \pi/2$; it follows from the dispersion relationship that[130]

$$H_{\text{lim}} = -2\pi M_s + \left[(2\pi M_s)^2 + \left(\frac{\omega}{2\gamma}\right)^2\right]^{1/2} + D_z M_s. \qquad (10\text{-}9.14)$$

For $H > H_{\text{lim}}$, $k = 0$ and $\theta < \pi/2$ on the basis of spin-wave theory Actually, magnetostatic modes should be employed in this range; then, however, the modifications required in the results appear to be minor.[131]

The threshold as a function of the applied static field observed in experiments on a YIG sphere[132,133] is shown in Fig. 10-9.4; the minimum threshold field occurs for $H = H_{\text{lim}}$. The data provide information concerning the dependence of ΔH_k on k. Further, particular spin waves that become unstable can be chosen by the proper selection of ω and H.

It is to be noted in Fig. 10-9.4 that increased thresholds occur with peaks at the static fields labeled H_a and H_b. These localized maxima are believed to be caused by a magnetoelastic coupling between the spin waves and phonons with the same frequency ($\omega/2$) and the same k. Measurements of the magnitude, width, and location of the interaction line lead to values of the acoustic Q, one of the magnetoelastic constants (b_2), and the exchange energy constant.[134] Thus magnetoelastic interactions can also be studied directly by the parallel pump method.

The interesting nonlinear effects have a variety of important practical applications. Components of the high-frequency magnetization that are integral multiples of the signal frequency permit frequency multiplication

[130] E. Schlömann, J. J. Green, and U. Milano, *J. Appl. Phys.* **31,** 386S (1960); I. Brady and E. Schlömann, *J. Appl. Phys.* **33,** 1377 (1962).

[131] E. Schlömann and R. I. Joseph, *J. Appl. Phys.* **32,** 165S (1961).

[132] E. H. Turner, *Phys. Rev. Letters* **5,** 100 (1960).

[133] See also W. E. Courtney and P. J. B. Clarricoats, *J. Appl. Phys.* **34,** 1279 (1963); J. J. Green and B. J. Healy, *J. Appl. Phys.* **34,** 1285 (1963).

[134] R. C. LeCraw and L. R. Walker, *J. Appl. Phys.* **32,** 167S (1961); F. R. Morgenthaler, *J. Appl. Phys.* **34,** 1289 (1963); F. A. Olson, *J. Appl. Phys.* **34,** 1281 (1963).

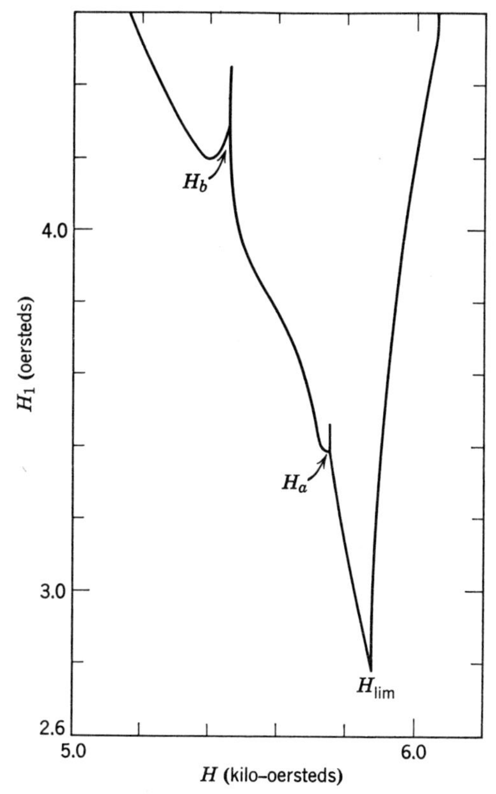

Fig. 10-9.4. The threshold field as a function of the static field for a YIG sphere in a parallel pump experiment. The microwave frequency is 34.627 Gc and the two magnetic fields are applied along the [111]-direction. [E. H. Turner, *Phys. Rev. Letters* **5**, 100 (1960).]

and mixing to be achieved.[135] The ferromagnetic amplifier is a form of the parametric amplifier and depends on the introduction of energy from the uniform mode to small k spin-wave or magnetostatic modes and to microwave cavity modes.[136] With sufficient pumping, the amplifier will in principle become an oscillator. The threshold properties have made

[135] W. P. Ayres, P. H. Vartanian, and J. L. Melchor, *J. Appl. Phys*, **27,** 188 (1956); J. E. Pippin, *Proc. IRE* **44,** 1054 (1956); E. N. Skomal and M. A. Medina, *J. Appl. Phys.* **29,** 423 (1958), **30,** 161S (1959); A. S. Risley and I. Kaufman, *J. Appl. Phys.* **33,** 1296 (1962).

[136] H. Suhl, *Phys. Rev.* **106,** 384 (1957), *J. Appl. Phys.* **28,** 1225 (1957), **29,** 416 (1958); M. T. Weiss, *Phys. Rev.* **107,** 317 (1957), *J. Appl. Phys.* **29,** 421 (1958); W. L. Whirry and F. B. Wang, *J. Appl. Phys.* **30,** 150S (1959); R. T. Denton, *Proc. IRE* **48,** 937 (1960), *J. Appl. Phys.* **32,** 300S (1961); B. A. Auld and C. H. Holmes, *J. Appl. Phys.* **34,** 1277 (1963).

power-limiting devices possible.[137] Thresholds arising from both the first- and second-order nonlinear processes have been utilized. Ferromagnetic limiters have a number of advantages, including immediate recovery, long lifetime, and a large range of power limits.

10. Spin-Wave Spectra of Thin Films

Kittel[138] first suggested that a uniform microwave field should be able to excite standing spin waves in thin films. In order that microwave energy

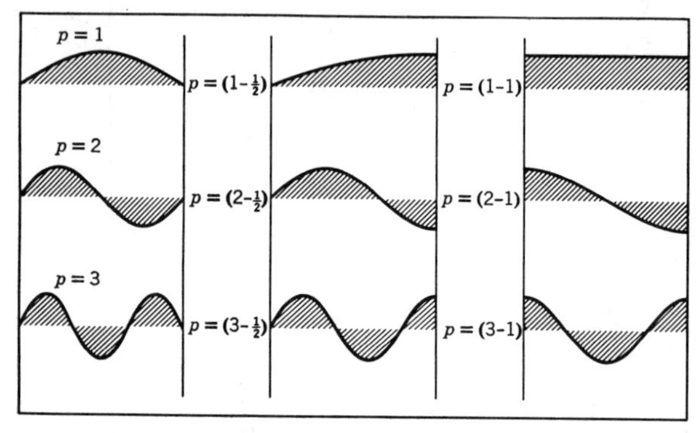

Fig. 10-10.1. The possible standing spin-wave modes in a thin film for different boundary conditions. For (*a*) there are nodes at both boundaries, for (*b*) there is a node at one boundary and an antinode at the other, and for (*c*) there are antinodes at both boundaries.

may be absorbed, it is necessary that the spins be pinned to some degree at one boundary of the film at least. The reason for these conditions can be appreciated by considering Fig. 10-10.1.[139] The three cases shown represent (*a*) complete pinning at both surfaces, (*b*) complete pinning at one surface and none at the other, and (*c*) no pinning at either surface of the film. Only the lowest order modes are shown for each case. Let $f(z)$

[137] G. S. Uebele, *Trans. IRE Microwave Theory Tech.* **7,** 18 (1959); F. R. Arams, M. Grace, and S. Okwit, *Proc. IRE* **49,** 1308 (1961); R. W. DeGrasse, *J. Appl. Phys.* **30,** 155S (1959); F. C. Rossol, *Proc. IRE* **49,** 1574 (1961); W. F. Krupke, T. S. Hartwick, and M. T. Weiss, *Trans. IRE Microwave Theory Tech.* **9,** 472 (1961); F. J. Sansalone and E. G. Spencer, *Trans. IRE Microwave Theory Tech.* **9,** 272 (1961); J. Brown and G. R. Harrison, *Proc. IRE* **49,** 1575 (1961); J. Clark and J. Brown, *J. Appl. Phys.* **33,** 1270 (1962); E. Stern, *J. Appl. Phys.* **32,** 315S (1961); F. S. Hickernell, B. H. Auten, and N. G. Sakiotis, *J. Appl. Phys.* **33,** 1272 (1962); R. L. Comstock and W. A. Dean, *J. Appl. Phys.* **34,** 1275 (1963).

[138] C. Kittel, *Phys. Rev.* **110,** 1295 (1958).

[139] C. W. Searle, A. H. Morrish, and R. J. Prosen, *Physica* **29,** 1219 (1963).

represent the amplitude of a particular spin-wave mode as a function of z, the coordinate normal to the film's surface. The energy coupled from the microwave field will be proportional to the integral

$$\int_0^L f(z)\,dz, \tag{10-10.1}$$

where L is the film's thickness. For (a) there will be a net absorption of energy only for the modes with an odd number of half wavelengths, that is,

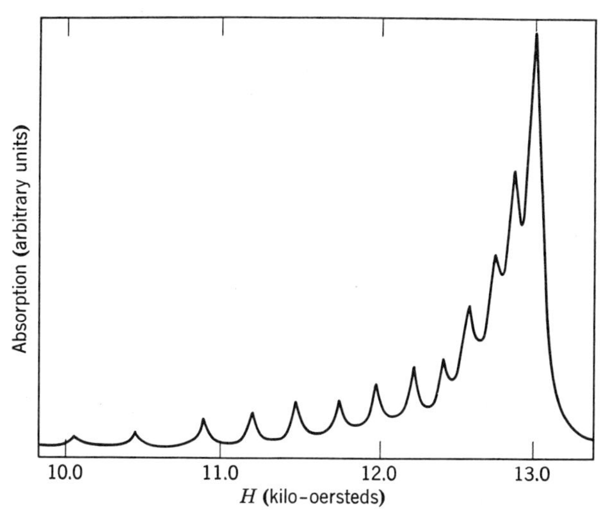

Fig. 10-10.2. Spin-wave spectrum of a Permalloy film. The static field is applied perpendicular to the plane of the film.

for $p = 1$, 3, etc. For (b) all the modes will absorb energy, and for (c) only the uniform mode ($\lambda = \infty$, $k = 0$) couples with the microwaves. Hence it is clear why some pinning is required. In order that sufficient energy may be absorbed to be observable, one dimension of the film should be comparable to the wavelength of a standing spin wave. Thin films with thicknesses in the range of 1000 to 10,000 Å satisfy this requirement.

Standing spin waves in thin films were first detected by Seavey and Tannenwald.[140] A well-defined spectrum observed[141] for a Permalloy film is shown in Fig. 10-10.2. Most experimental investigations have been made on Permalloy films with about 80Ni20Fe, primarily because this material

[140] M. H. Seavey and P. E. Tannenwald, *Phys. Rev. Letters* **1,** 168 (1958).
[141] At the University of Minnesota.

has negligible magnetostriction. Other metallic films such as Ni and Co have also been examined. The metallic films studied have thicknesses considerably less than the skin depth; hence to a good degree of approximation the microwave field is uniform throughout the film. It is difficult to manufacture films of ferrimagnetic materials; there have been no reports of the observation of standing spin waves in nonmetallic films.

The resonance condition for standing spin waves may be found very simply by starting with the equations of motion (equation 10-3.2) and introducing an equivalent exchange field. A convenient expression for the exchange free energy per unit volume (equation 7-1.12) is

$$F_e = A[(\nabla\alpha_1)^2 + (\nabla\alpha_2)^2 + (\nabla\alpha_3)^2]$$
$$= \frac{A}{M^2}[(\nabla M_x)^2 + (\nabla M_y)^2 + (\nabla M_z^2)]$$

where $A = 2J_eS^2/a$. By use of the vector identity derived from $\nabla^2(\mathbf{u}\cdot\mathbf{u}) = 0$ (see just before equation 6-3.32), the exchange free energy becomes

$$F_e = -\frac{A}{M^2}\mathbf{M}\cdot\nabla^2\mathbf{M}. \tag{10-10.2}$$

This free energy may also be written $F_e = -\frac{1}{2}\mathbf{M}\cdot\mathbf{H}_m$ (compare with equation 1-6.6), in which $\mathbf{H}_m$ is the equivalent exchange field and the factor of $\frac{1}{2}$ enters because F_e represents a self-energy. Hence we have

$$\mathbf{H}_m = \frac{2A}{M^2}\nabla^2\mathbf{M}. \tag{10-10.3}$$

When the static field is applied perpendicular to the film's plane ($\mathbf{H}_\perp$), that is, along the z-direction, the expression for the total field is

$$\mathbf{H}_T = \mathbf{H}_\perp - 4\pi\mathbf{M} + \frac{2A}{M^2}\nabla^2\mathbf{M},$$

since $D_x = D_y = 0$, $D_z = 4\pi$, and it is unnecessary to include the microwave field when only the resonance frequency is desired. If losses are neglected, the equations of motion become (see equation 10-3.4)

$$\frac{1}{\gamma}\frac{d\mathbf{M}}{dt} = \mathbf{M}\times\left(\mathbf{H}_\perp - 4\pi\mathbf{M} + \frac{2A}{M^2}\nabla^2\mathbf{M}\right).$$

It is reasonable to assume that $\mathbf{M} = \mathbf{M}_s + \mathbf{M}_1(t, k) = M_s\mathbf{k} + \mathbf{M}_1e^{i(\omega t - kz)}$, where $\mathbf{M}_1$ lies in the x-y plane. The more general form $e^{i(\omega t - \mathbf{k}\cdot\mathbf{r})}$ for plane waves is not employed, since for thin films only waves traveling in the direction perpendicular to the plane of the film are of interest. On substitution for $\mathbf{M}$ in the equations of motion, we find to first order

$$\frac{i\omega}{\gamma}\mathbf{M}_1(t, k) = \mathbf{M}_1(t, k) \times \mathbf{H}_\perp - 4\pi\mathbf{M}_1(t, k) \times \mathbf{M}_s - \frac{2Ak^2}{M_s^{\,2}}[\mathbf{M}_s \times \mathbf{M}_1(t, k)]$$

or

$$\frac{i\omega}{\gamma}\mathbf{M}_1(t, k) = \mathbf{M}_1(t, k) \times \mathbf{H}_\perp - 4\pi\mathbf{M}_1(t, k) \times \mathbf{M}_s + \frac{2Ak^2}{M_s^{\,2}}[\mathbf{M}_1(t, k) \times \mathbf{M}_s].$$

For clarity, this may be rewritten

$$\frac{i\omega}{\gamma}\mathbf{M}_1(t, k) = \mathbf{M}_1(t, k) \times \left(H_\perp - 4\pi M_s + \frac{2Ak^2}{M_s}\right)\mathbf{k}. \quad (10\text{-}10.4)$$

Now the vectors denoted by the left- and right-hand sides of this equation are parallel, since i serves as an operator that produces a rotation of 90° in the x-y plane and the right-hand side represents a vector perpendicular to $\mathbf{k}$ (or $\mathbf{M}_s$) and $\mathbf{M}_1(t, k)$. Hence the magnitudes of the two vectors of equation 10-10.4 must be equal, and we get

$$\omega_r = -\gamma\left(H_\perp - 4\pi M_s + \frac{2Ak^2}{M_s}\right), \quad (10\text{-}10.5)$$

which is the resonance condition. If the static field is applied parallel to the plane of the film ($H_\parallel$) it is easy to show (problem 10-11) that the natural modes of oscillation are

$$\omega_r = -\gamma\left[\left(H_\parallel + \frac{2Ak^2}{M_s}\right)\left(H_\parallel + 4\pi M_s + \frac{2Ak^2}{M_s}\right)\right]^{1/2}. \quad (10\text{-}10.6)$$

The boundary conditions require the establishment of standing spin waves with a sinusoidal spatial variation; this restricts the values that can be assumed by k. If p represents the number of half wavelengths in L, the film's thickness, it is necessary that

$$k = \frac{p\pi}{L}. \quad (10\text{-}10.7)$$

In general p is nonintegral. Integral numbers can be assigned to all modes, however, by letting

$$p = n - \delta, \quad (10\text{-}10.8)$$

where n is an integer and δ is bounded by

$$0 \leqslant \delta \leqslant 1. \quad (10\text{-}10.9)$$

The exact value of δ depends on the pinning conditions at the boundaries. For example, in Fig. 10-10.1 δ equals 0, $\frac{1}{2}$, and 1 for (*a*), (*b*), and (*c*), respectively.

Spin-wave spectra of thin films have special significance for a number of reasons. First, such spectra probably represent the most direct observation of spin waves. Second, the spin-wave resonance data permit the deduction of what are probably the best values of the exchange constants (A, J_e, etc.). It is found, for example, that $A = 1.22 \times 10^{-6}$ erg/cm for Permalloy (80Ni20Fe) at room temperature.[142] Third, the variation of the exchange constant, and therefore also the magnetization, as a function of temperature can be determined with considerable accuracy. The Bloch $T^{3/2}$ law for the magnetization is found[143] to be essentially valid below but not above 80°K. Fourth, the interaction between spin waves and phonons can be studied by depositing the film on a quartz rod. Phonons can be excited in quartz by the piezoelectric effect. Interesting information on how the interaction depends on the wavelengths of the magnons and phonons has been obtained.[144]

The mechanism responsible for the pinning of the spin waves remains to be discussed. Calculations show that the discontinuity in the lattice at the surface produces too small a pinning to be of consequence.[145] The following two possibilities have been proposed.

1. Films may have a thin antiferromagnetic surface layer, for example FeO or NiO; the ferromagnetic-antiferromagnetic exchange interaction at the interface (Section 8-11) may then cause pinning.[146]
2. The magnetization of a film's surface layer may differ from the value in the body of the film, perhaps because of differences in chemical composition. The spins in the surface layers will resonate at a slightly different frequency compared to the spins in the bulk of the film. These "off-resonance" surface spins may pin the resonant spins by exchange coupling.[147]

Support for both mechanisms has been found, and indeed it is conceivable that the first mechanism applies for some films and the second for others. It is even possible that both mechanisms act in some films. However, the most convincing evidence has been obtained for the second or dynamic pinning model; this evidence will now be described briefly.

Portis[148] has assumed that the saturation magnetization M_s varies parabolically, with the smallest values occurring at the film's surfaces.

[142] C. W. Searle and A. H. Morrish, *Phys. Letters* **7**, 229 (1963).

[143] R. Weber and P. E. Tannenwald, *J. Phys. Chem. Solids* **24**, 1357 (1963).

[144] H. E. Bömmel and K. Dransfield, *Phys. Rev. Letters* **3**, 83 (1959); M. Pomerantz, *Phys. Rev. Letters* **7**, 312 (1961); M. F. Lewis, T. G. Philips, and H. M. Rosenberg, *Phys. Letters* **1**, 198 (1962); C. F. Kooi, *Phys. Rev.* **131**, 1070 (1963).

[145] R. F. Soohoo, *J. Appl. Phys.* **32**, 148S (1961).

[146] C. Kittel, *J. Phys. radium* **20**, 81 (1959).

[147] P. E. Wigen, C. F. Kooi, M. R. Shanabarger, and T. D. Rossing, *Phys. Rev. Letters* **5**, 206 (1962).

[148] A. M. Portis, *Appl. Phys. Letters* **2**, 69 (1963).

When the static field is applied perpendicular to the plane of the film, the alternating component of the magnetization, $\mathbf{M}_1$, satisfies a harmonic oscillator type of equation. It then follows that the mode positions will depend linearly on p, instead of quadratically, as in equation 10-10.5. Some films exhibit almost a linear spectrum.[149] When the static field is applied parallel to the plane of the film, a harmonic oscillator equation is again obtained for $\mathbf{M}_1$, except that now the sign of the potential is reversed.[150] Hence there will be no pinning in this orientation. This conclusion is confirmed by experiment; for $H_{\parallel}$ the spectrum is observed to consist only of the uniform mode, or possibly two uniform modes associated with two regions of different M_s.[151] Further, the spin-wave spectrum is expected to collapse when the static field makes some critical angle lying between the extreme $H_{\perp}$ and $H_{\parallel}$ orientations; this effect is also observed.[152] However, when the magnetization M_s increases near the film's surfaces, for example, when Fe is deposited on the boundary surfaces of a Permalloy film, the theory predicts spin-wave modes in the parallel orientation and no distinct modes in the perpendicular orientation; that is, just the reverse of the situation just described. This behavior has been observed.[153] When the variation in M_s is slight, only small deviations from the square law of equation 10-10.6 are expected and indeed this is what is found for most films.[153]

11. Electromagnetic Wave Propagation in Gyromagnetic Media

When the material containing strongly coupled dipoles is large compared to the wavelength of the microwaves inside the sample, propagation effects become important. Of course, appreciable propagation will occur only in samples with a low electrical conductivity; most ferrimagnetic materials, however, satisfy this criterion. The analysis of propagation in gyromagnetic media will involve both the equations of motion (equation 10-3.2) and Maxwell's equations (equations 1-8.1 to 1-8.4). Two cases are of main interest, one in which the direction of propagation is parallel to the applied static field (hence to the static magnetization) and the other in which the direction of propagation is perpendicular to $\mathbf{H}$.

First, consider an infinite medium consisting of some appropriate ferrimagnetic material and a plane wave propagating along the z-direction,

[149] M. Nesinoff and R. W. Terhune, *J. Appl. Phys.* **35,** 806 (1964).

[150] C. W. Searle and A. H. Morrish, *Phys. Letters* **7,** 229 (1963).

[151] D. Chen and A. H. Morrish, *J. Appl. Phys.* **33,** 1146 (1962).

[152] P. E. Wigen, C. F. Kooi, M. R. Shanabarger, and T. D. Rossing, *Phys. Rev. Letters* **5,** 206 (1962).

[153] C. W. Searle, A. H. Morrish, and R. J. Prosen, *Physica* **29,** 1219 (1963).

which is also the direction of **H**. It can be shown (problem 10-12) that the solution consists of two plane circularly polarized waves. According to the result of problem 10-5, the waves rotating counterclockwise and clockwise will experience the permeabilities $\mu_+ = \mu_{11} + K$ and $\mu_- = \mu_{11} - K$, respectively. From equations 10-3.16 and 10-3.19 it follows that

$$\mu_+ = 1 + \frac{4\pi\gamma M_s}{\gamma H + \omega}$$

and

$$\mu_- = 1 + \frac{4\pi\gamma M_s}{\gamma H - \omega}, \tag{10-11.1}$$

since the demagnetizing factors are all zero.

A linearly polarized plane wave may be considered as two circularly polarized waves rotating clockwise and counterclockwise. Because of the difference in the permeabilities μ_- and μ_+, the velocities v_- and v_+ of the two circularly polarized waves will differ, for $v = c/\sqrt{\epsilon\mu}$ in general for an electromagnetic wave. In a distance l the two circular components will rotate through the angles θ_- and θ_+ in relation to some fixed axis. This behavior will produce a net rotation θ in the linearly polarized wave, given by

$$\begin{aligned}\theta &= \frac{\theta_- - \theta_+}{2} \\ &= \frac{l}{2}(k_- - k_+) \\ &= \frac{l\omega}{2}\left(\frac{1}{v_-} - \frac{1}{v_+}\right),\end{aligned} \tag{10-11.2}$$

where k_- and k_+ are the wave numbers or phases and no or equal attenuation of the two circular components is assumed. This effect is known as the Faraday rotation. The substitution for $\mu_\pm$ from equation 10-11.1 then yields

$$\theta = \frac{l\omega\sqrt{\epsilon}}{2c}\left[\left(1 + \frac{4\pi\gamma M_s}{\gamma H - \omega}\right)^{1/2} - \left(1 + \frac{4\pi\gamma M_s}{\gamma H + \omega}\right)^{1/2}\right]. \tag{10-11.3}$$

If $\omega \gg \gamma H$ and $\omega \gg 4\pi\gamma M_s$, equation 10-11.3 becomes

$$\theta = \frac{2\pi l\gamma M_s\sqrt{\epsilon}}{c}. \tag{10-11.4}$$

Thus in this approximation the rotation is independent of the frequency and proportional to the saturation magnetization.

When a spinel or garnet sample with finite dimensions is inserted into a waveguide, the boundary conditions modify the propagation behavior considerably. In general, the mathematical analysis is complicated.[154] Some solutions have been obtained with the aid of computers or with perturbation methods.[155] A formal treatment of the problem can be made with scattering matrix analysis;[156] this method does not consider the details of the propagation in the ferrimagnetic material itself but instead

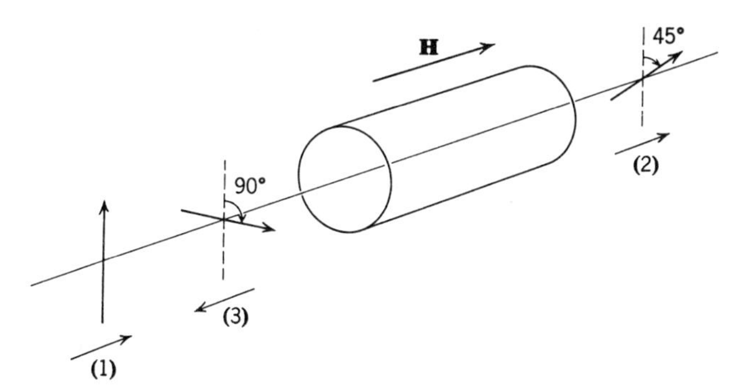

Fig. 10-11.1. The Faraday rotation isolator.

describes only the final amplitudes of the waves reflected and transmitted by the ferrimagnet.

The propagation properties of ferrimagnetic materials has led to a number of important microwave devices. A full analysis of these devices requires the consideration of a finite specimen and its location in the waveguide. The operating principles, however, can be understood qualitatively by referring to the action of the infinite medium.

Suppose a ferrimagnetic material inside a circular waveguide is of such length that the plane of polarization of a linearly polarized wave is rotated by 45°, as illustated by (1) and (2) in Fig. 10-11.1. If the microwave energy is fed into and out of the circular guide via rectangular guides rotated at 45° with respect to one another, there is little attenuation. A reflected wave, moving from right to left, has its polarization plane rotation by 45° in the same direction, as indicated by (3) in Fig. 10-11.1. It is to be noted that actually the rotation of the polarization with respect to the direction of propagation is reversed; that is, the rotation is counterclockwise instead

[154] H. Suhl and L. R. Walker, *Phys. Rev.* **86,** 122 (1952), *Bell System Tech. J.* **33,** 579 (1954); H. Gamo, *J. Phys. Soc. Japan* **8,** 176 (1952); A. A. van Trier, *Appl. Sci. Res.* **B-3,** 305 (1953); R. A. Waldron, *J. Brit. IRE* **18,** 677 (1958).

[155] H. Suhl and L. R. Walker, *Bell System Tech. J.* **33,** 1133 (1954), B. Lax, *Proc. IRE* **44,** 1368 (1956).

[156] G. S. Heller, *Proc. IRE* **44,** 1386 (1956).

of clockwise. Since the reflected wave is rotated by 90°, it will not be accepted by the rectangular guide at the input. In order to absorb the reflected energy, resistance vanes with appropriate orientations are inserted at the entrance and exit of the circular guide. Hence the device is nonreciprocal, for energy is transmitted in one direction and not the other. The device serves to isolate the energy source (klystron) from changes in the load and indeed is called a Faraday rotation isolator.[157]

Suppose the rectangular guide at the input to the circular guide has a 90° twist and the ferrimagnetic material now produces a 90° Faraday rotation. A wave traveling from right to left will be rotated 180°, whereas one traveling from left to right will be rotated 0° (90–90°). This device is called a *gyrator*.[157] A variation of a gyrator or isolator leads to a *circulator*,[157] a device with three or four ports that permits microwave energy to be transmitted only between the ports in the sequence 1 to 2 to 3 to 4 to 1, etc.

In Faraday rotation devices the static field applied is relatively small and the ferrimagnetic sample is usually below saturation. Energy absorption from ferromagnetic resonance is therefore slight. If the applied field **H** is produced by a coil (solenoid), this field, hence the magnetization of the ferrimagnet, can be changed by varying the current in the coil. The device can then be used to produce amplitude or frequency modulation and variable phase shifts.[158] If the magnetization direction is reversed, the device acts as a switch.[158] Various design considerations for Faraday rotation devices may be found in the literature.[159]

In the other experimental arrangement of importance the steady field is applied perpendicular to the direction of propagation of the plane wave. For the infinite ferrimagnetic medium the solution consists of two linearly polarized waves. The time-varying magnetic field for one of these waves is parallel to the static applied field (the z-direction); that is, $H_{1x} = H_{1y} = 0$. There is then no interaction with the magnetic dipoles of the ferrimagnet, hence the permeability has the free space value. The time-varying magnetic field for the other wave, however, lies in the plane perpendicular to the static applied field. There is now an interaction with the spin system and the permeability is given by (problem 10-12)

$$\mu_{\perp} = \frac{\mu_{11}^2 - K^2}{\mu_{11}} = \frac{\gamma^2(H + 4\pi M_s)^2 - \omega^2}{\gamma^2 H(H + 4\pi M_s) - \omega^2}. \qquad (10\text{-}11.5)$$

[157] C. L. Hogan, *Bell System Tech. J.* **31,** 1 (1952), *Revs. Mod. Phys.* **25,** 253 (1953).

[158] R. C. LeCraw, *J. Appl. Phys.* **25,** 678 (1954); A. C. Brown, R. S. Cole, and W. N. Honeyman, *Proc. IRE* **46,** 722 (1958); E. H. Turner, *Trans. IRE Microwave Theory Tech.* **6,** 300 (1958); C. Bowness, *Microwave J.* **1,** 13 (1958).

[159] J. H. Rowen, *Bell System Tech. J.* **32,** 1333 (1953); A. G. Fox, S. E. Miller, and M. T. Weiss, *Bell System Tech. J.* **34,** 5 (1955); B. Lax, *J. Appl. Phys.* **26,** 919 (1955); H. N. Chait and N. G. Sakiotis, *Trans. IRE Microwave Theory Tech.* **6,** 38 (1959).

Media with this property are said to exhibit *birefringence*,[160] and the phenomena is sometimes known as the Cotton-Mouton effect. Because of the different phase retardation for the two polarizations, an incident plane polarized wave will become elliptically polarized. It is to be noted that $\mu_\perp$ does not depend on the direction of propagation. A ferrimagnetic rod placed at the center of a rectangular guide will then act as a reciprocal phase shifter[161] and can also serve as a modulator.[162] It should, however, be mentioned that the most useful reciprocal ferrimagnetic phase shifter

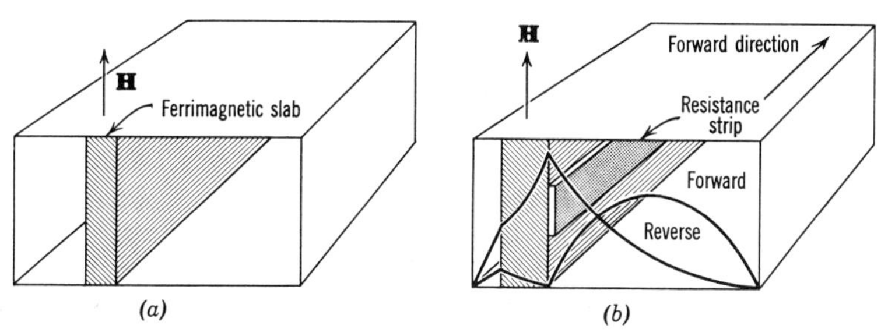

Fig. 10-11.2. (*a*) Construction of resonance isolator or nonreciprocal phase shifter. The ferrimagnetic slab which is transversely magnetized is placed at a position in the rectangular waveguide in which the microwave magnetic field is approximately circularly polarized. (*b*) To illustrate the field-displacement isolator, the magnitude of the microwave electric field for the forward and reverse directions of propagation are indicated.

employs a longitudinal static field, that is, one that lies along the direction of propagation.[162] The operation of this device depends[163] on concentration of the field by the dielectric constant and the high-frequency permeability that can be varied by changing the longitudinal field **H**.

A number of nonreciprocal devices in which the field **H** is applied transversely to the direction of propagation have been developed and are generally in wider use than those with longitudinal fields. The physical principles underlying some of the important transverse devices will now be discussed briefly.

Consider a rectangular waveguide with a ferrimagnetic rod or slab located at a region in which the microwave magnetic field is approximately circularly polarized (Fig. 10-11.2*a*). For a wave propagating from

[160] D. Polder, *Phil. Mag.* **40,** 99 (1949); F. K. du Pré, *Philips Res. Rept.* **10,** 1 (1955).

[161] M. T. Weiss and A. G. Fox, *Phys. Rev.* **88,** 146 (1952); J. Cacheris, *Proc. IRE* **42,** 1242 (1954).

[162] D. Bush, *Proc. IEE* (*London*), *Suppl.* 5 **140-B,** 368 (1957); F. Reggia and E. G. Spencer, *Proc. IRE* **45,** 1510 (1957).

[163] K. J. Button and B. Lax, *Proc. IEE* (*London*), *Suppl.* 21 **109-B,** 56 (1962).

left to right the rotation of the alternating magnetic field $\mathbf{H}_1$ will be counterclockwise, say, whereas for propagation from right to left the rotation will be clockwise. Suppose the transverse field $\mathbf{H}$ is adjusted to satisfy the ferromagnetic resonance condition. Then the microwave field will rotate in the opposite sense to the precessing dipoles for propagation in one direction, for example, from right to left, hence little absorption of energy will result. For the reverse direction of propagation the microwave field and dipoles will rotate in the same sense, and appreciable ferromagnetic resonance absorption will occur. The device, called a *resonance isolator*,[164] thus permits one-way transmission. Its directional properties may be improved by attaching a dielectric material to the ferrimagnetic slab;[164] this combination concentrates more microwave energy in the ferrimagnet. The *nonreciprocal phase shifter* also has the construction shown in Fig. 10-11.2*a*. Its action, however, now depends on the dispersive rather than the absorptive part of the complex susceptibility of the ferrimagnet. The phase shift differs for the two directions of propagation.[165] If the phase difference is made 180°, the phase shifter can be used to build a circulator.[166] Another form of circulator has a transversely magnetized ferrimagnetic post placed at the junction of three rectangular waveguides.[167] The operation of this device depends on the nonreciprocal scattering of the microwaves by the ferrimagnetic post.

By adjusting the thickness and position of the ferrimagnetic slab indicated in Fig. 10-11.2*a* and adding a suitable value of $\mathbf{H}$ it is possible to achieve the microwave electric field distribution shown in Fig. 10-11.2*b*. For the forward direction of propagation, say from left to right, the electric field at one face of the ferrimagnet is almost zero, whereas for the reverse direction of propagation the electric field at the same face is appreciable. If a resistance strip is placed at the interface, there will be considerable attenuation in the reverse direction but little in the forward direction. This

[164] B. Lax, *Proc. IRE* **44,** 1370 (1956), *Trans. IRE Microwave Theory Tech.* **6,** 5 (1958); M. T. Weiss, *Trans. IRE Microwave Theory Tech.* **4,** 240 (1956); W. W. Anderson and M. E. Hines, *Trans. IRE Microwave Theory Tech.* **9,** 63 (1961); G. S. Heller and G. W. Catuna, *Trans. IRE Microwave Theory Tech.* **6,** 97 (1958); E. Stern and R. E. Mangiaracina, *Trans. IRE Microwave Theory Tech.* **7,** 11 (1959); J. L. Melchor and P. H. Vartanian, *Trans. IRE Microwave Theory Tech.* **7,** 15 (1959).

[165] B. Lax, K. J. Button, and L. M. Roth, *J. Appl. Phys.* **25,** 1413 (1954); M. L. Kales, H. N. Chait, and N. G. Sakiotis, *J. Appl. Phys.* **24,** 816 (1953); J. Roberts, C. M. Srivastava, *Proc. IEE* (*London*), *Suppl.* 5 **104-B,** 340 (1957); S. Weisbaum and H. Boyet, *J. Appl. Phys.* **27,** 519 (1956).

[166] A. G. Fox, *Proc. IEE* (*London*), *Suppl.* 5 **104-B,** 371 (1957).

[167] B. A. Auld, *Trans. IRE Microwave Theory Tech.* **7,** 238 (1959); H. N. Chait and T. R. Curry, *J. Appl. Phys.* **30,** 152 (1959); M. Grace and F. R. Arams, *Proc. IRE* **48,** 1497 (1960); H. Bosma, *Proc. IEE* (*London*), *Suppl.* 21 **109-B,** 137 (1962).

device is called a nonreciprocal *field-displacement isolator*.[168] It has the advantage of requiring much smaller static fields than the resonance isolator. It is also possible to construct such a device by using a circular waveguide and a longitudinal field.[169]

12. Resonance in Unsaturated Samples

If the magnetic field applied is too small to produce saturation or if, in the extreme case, no field is applied, samples possessing a spontaneous magnetization will usually have some sort of domain structure. An exception would be an assembly of single-domain particles. The presence of domains and domain walls leads to new resonance phenomena, some of which will be discussed briefly.

Resonances in addition to any expected for single-domain samples have been observed for static fields applied along the hard directions of anisotropic single crystals. These extra resonances occur at low fields in which a domain structure is likely. In general, one resonance peak is expected for each type of domain. Since only one, or sometimes two, additional resonances are usually found, it is reasonable to assume a rather simple configuration consisting of two types of domain. For example, the domains in uniaxial $BaFe_{12}O_{19}$ are considered to be thin slabs or laminas with magnetizations lying alternately along the positive and negative easy directions. In the case of cubic crystals with **H** along the [110]-direction the domain walls are pictured as running perpendicular to the applied field and the magnetizations of the lamellar domains as lying in sequence along one of the two easy directions. Any component of the static magnetization perpendicular to the domain walls is continuous. However, the time-varying magnetizations perpendicular to the wall may be out of phase and thus lead to uncompensated poles at the walls. These poles produce demagnetizing fields with demagnetizing factors 4π which couple the two magnetization vectors together. The calculations[170] show that inclusion of this demagnetizing field in the equations of motion leads to two resonance conditions, one for the microwave field perpendicular to **H**

[168] A. G. Fox, S. E. Miller, and M. T. Weiss, *Bell System Tech. J.* **34,** 5 (1955); S. Weisbaum and H. Seidel, *Bell System Tech. J.* **35,** 877 (1956); S. Weisbaum and H. Boyet, *Proc. IRE* **44,** 554 (1956), **5,** 194 (1957); K. J. Button, *Trans. IRE Microwave Theory Tech.* **6,** 303 (1958).

[169] J. L. Melchor, W. P. Ayres, and P. H. Vartanian, *J. Appl. Phys.* **27,** 72 (1956); P. H. Vartanian, J. L. Melchor, and W. P. Ayres, *Trans. IRE Microwave Theory Tech.* **4,** 8 (1956).

[170] T. Nagamiya, *Prog. Theor. Phys. (Kyoto)* **10,** 72 (1953); J. Smit and H. G. Beljers, *Philips Res. Rept.* **10,** 113 (1955); J. O. Artman, *Proc. IRE* **44,** 1284 (1956), *Phys. Rev.* **102,** 1008 (1957).

and the other for the microwave field parallel to **H**. Of course, at higher applied fields the sample will be a single domain, and the usual resonance condition will then hold; it may or may not be possible to satisfy this condition, depending on the frequency being used. Resonances believed to arise from the domain structure have been observed in $BaFe_{12}O_{19}$,[171] in Co and Mn spinels,[172] and in rare earth garnets.[173]

A somewhat different although analogous effect to that produced by domain structure occurs in the low temperature form of magnetite. Above about 119°K the crystal structure of magnetite is cubic, whereas below it is orthorhombic. In this orthorhombic phase the *c*- and *a*-axes are the easy and hard directions, respectively, of the crystalline anisotropy, and the *b*-axis is intermediate. When a steady magnetic field is applied along, say, the [001]-direction of a single crystal of magnetite as it is cooled through the transition temperature, this particular cube edge is selected as the *c*-axis of the orthorhombic structure.[174] The orthorhombic *a*- and *b*-axes lie along the two cube face diagonals perpendicular to the cubic [001]-direction. Even though the *c*-axis is uniquely determined, the (*a*, *b*)-axes system has two possible orientations (1 and 2 in Fig. 10-12.1*a*), and the crystal will be "twinned." Suppose that the two orientations of the (*a*, *b*)-axes are arranged as a series of laminas parallel to the *c*-axis (Fig. 10-12.1*b*). When a field **H** is applied along the *c*-axis, the crystal is, in a static sense, a single domain. However, the amplitude of the time-varying magnetization in any direction in the basal plane will be different in the two regions, and a high-frequency pole density will appear on the interface. As a result, the resonance spectrum is expected to consist of two lines; this is indeed what is observed[175] (Fig. 10-11.1*c*). The twinning of the crystal may be removed by applying a static field along one of the cubic face diagonals in the temperature range $80° < T < 90°K$; this direction becomes the *b*-axis. When this is done, the two resonance lines merge into a single one at an intermediate value of **H** (Fig. 10-12.1*c*).

When no static field is applied, resonance phenomena can still be observed by varying the frequency of the alternating field. The complex initial permeability, $\mu = \mu' - i\mu''$ is actually measured as a function of frequency. Although a number of measurements have been made on ferromagnetic metals and alloys, studies in recent years have focused on polycrystalline ferrimagnetic materials; then at least the eddy current and

[171] J. Smit and H. G. Beljers, *Philips Res. Rept.* **10,** 113 (1955).

[172] P. E. Tannenwald, *Phys. Rev.* **99,** 463 (1955), **100,** 1713 (1955).

[173] R. V. Jones, G. P. Rodrigue, and W. P. Wolf, *J. Appl. Phys.* **29,** 434 (1958).

[174] C. H. Li, *Phys. Rev.* **40,** 1002 (1932); L. R. Bickford, *Phys. Rev.* **78,** 449 (1950); H. J. Williams, R. M. Bozorth, and M. Goertz, *Phys. Rev.* **91,** 1107 (1953); B. A. Calhoun, *Phys. Rev.* **94,** 1577 (1954).

[175] D. B. Bonstrom, A. H. Morrish, and L. A. K. Watt, *J. Appl. Phys.* **3,** 272S (1961).

skin depth complications are eliminated. Data[176] for a commercial ferrite, consisting primarily of Mg spinel, are shown in Fig. 10-12.2. It is to be noted that there are two resonances, that is, two peaks in the curve

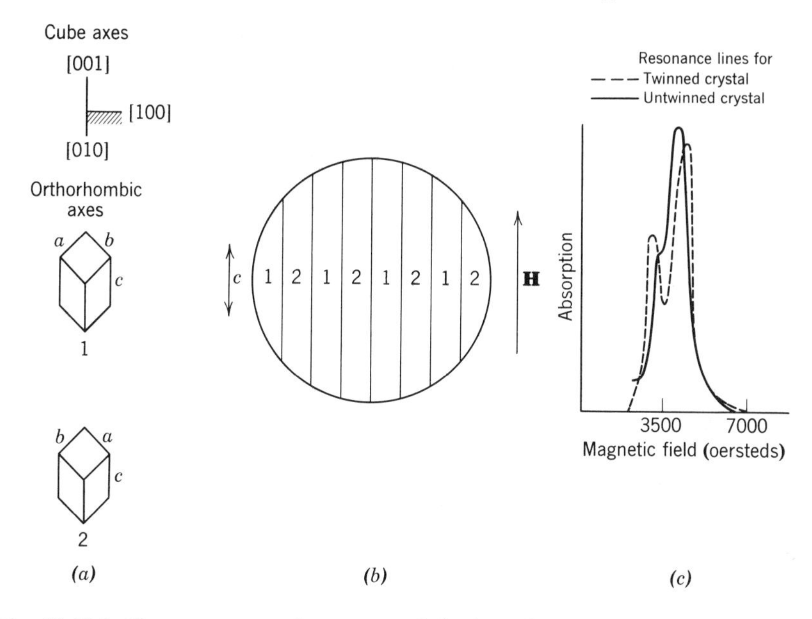

Fig. 10-12.1. The structure and resonance behavior of the orthorhombic form of magnetite (Fe_3O_4), (*a*) the possible (*a*, *b*)-axes twins when the *c*-axis is unique; (*b*) postulated lamillar structure in a twinned crystal; (*c*) resonance lines observed in a twinned and untwinned magnetite crystal with **H** parallel to the *c*-axis and at a frequency of 24 Gc.

for the imaginary part of the permeability. For some other spinels only one resonance has been observed; examples are $NiFe_2O_4$,[177] various NiZn and MnZn spinels,[178,179] Li and Cu spinels,[179] and a number of hexagonal oxides with the basal plane as the preferred plane of magnetization.[179] Two peaks in μ'' have also been observed in hexagonal oxides with the

[176] G. T. Rado, R. W. Wright, and W. H. Emerson, *Phys. Rev.* **80,** 273 (1950); see also, G. T. Rado and A. Terris, *Phys. Rev.* **83,** 177 (1951); G. T. Rado, R. W. Wright, W. H. Emerson, and A. Terris, *Phys. Rev.* **88,** 909 (1952); G. T. Rado, *Revs. Mod. Phys.* **25,** 81 (1953); G. T. Rado, V. J. Folen, and W. H. Emerson, *Proc. IEE* (*London*), *Suppl.* 5 **104-B,** 198 (1957); A. J. E. Welch, P. F. Nicks, A. Fairweather, and F. F. Roberts, *Phys. Rev.* **77,** 403 (1950).

[177] J. L. Snoek, *Philips Tech. Rev.* **8,** 353 (1946), *Nature* (*London*) **160,** 90 (1947); *Physica* **14,** 207 (1948); H. G. Beljers and J. L. Snoek, *Philips Tech. Rev.* **11,** 313 (1949).

[178] F. Brown and C. L. Gravel, *Phys. Rev.* **97,** 55 (1955).

[179] J. Smit and H. P. J. Wijn, *Ferrites,* John Wiley and Sons, New York (1959).

c-axis as the preferred direction[179] and in yttrium and other iron garnets;[180] three peaks have been found in some Mn spinels.[181] Other parameters that have been varied are the sample density and the temperature. Curves of μ'' versus frequency for polycrystalline YIG samples of various densities are plotted[182] in Fig. 10-12.3. These samples were prepared by grinding

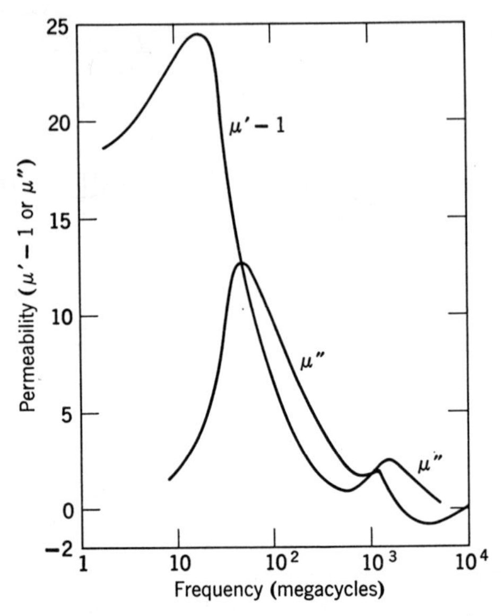

Fig. 10-12.2. Initial permeability spectrum of a sintered ferrimagnetic material consisting mainly of Mg spinel. [G. T. Rado, R. W. Wright, W. H. Emerson, and A. Terris, *Phys. Rev.* **88**, 909 (1952).]

pure single crystals into a powder and then pressing and sintering. The densest sample has two well-defined absorption peaks. The separation between these two peaks decreases as the sample density decreases until for 72% density only one peak is observed.

The origin of the peaks is obviously an important question requiring an answer. When no steady field is applied, ferromagnetic resonance is still possible if some strong orienting force is present. This force may be considered as an effective field H_e which takes the place of the externally applied field. The resonance frequency is then given by $\omega_r = \gamma H_e$. In

[180] R. D. Harrington and A. L. Rasmussen, *Proc. IRE* **47,** 98 (1959); D. J. Epstein and B. Frackiewicz, *J. Appl. Phys.* **30,** 295S (1959); E. E. Anderson, J. R. Cunningham, and G. E. McDuffie, *Phys. Rev.* **116,** 624 (1959); G. E. McDuffie, J. R. Cunningham, and E. E. Anderson, *J. Appl. Phys.* **31,** 47S (1960).

[181] S. E. Harrison, C. J. Kreissman, and S. R. Pollack, *Phys. Rev.* **110,** 844 (1958).

[182] J. D. Holm and A. H. Morrish, *J. Appl. Phys.* **35,** 894 (1964).

sintered materials the crystalline anisotropy is likely to be the primary source of the effective field ($H_e = H_K$). However, uncompensated poles on the surfaces of pores resulting from voids and inclusions and on the boundaries of domains will modify this effective field. The extreme values

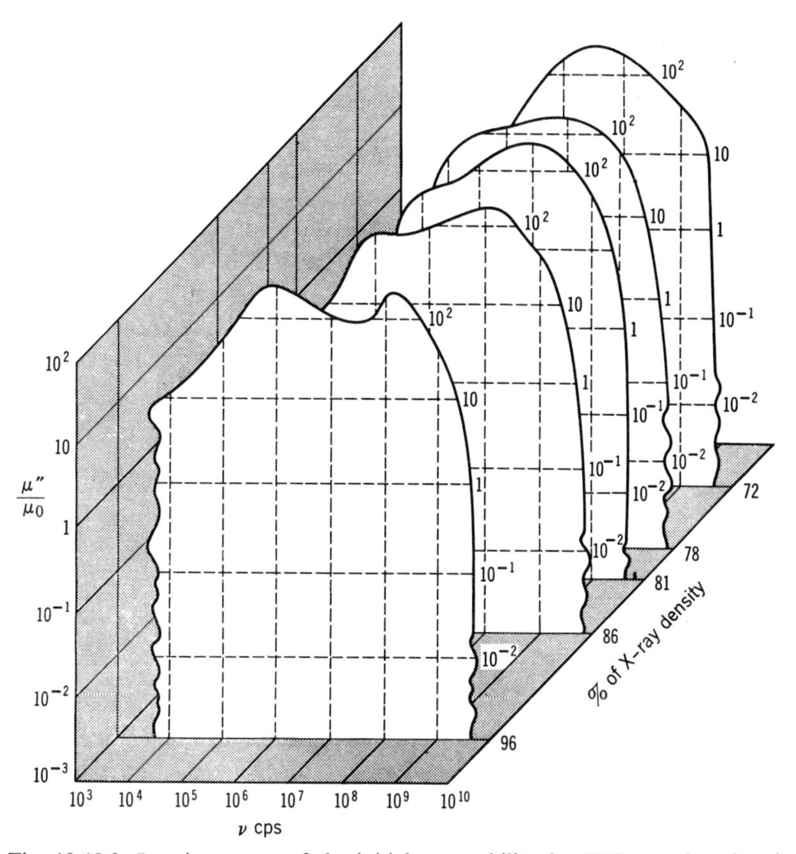

Fig. 10-12.3. Imaginary part of the initial permeability for YIG samples of various densities.

to be expected for the resonance frequency are $\omega_r = \gamma(H_K + 4\pi M_s)$[183] for a disk-shaped grain with a lamellar domain structure and $\omega_r = \gamma(H_K - 4\pi M_s)$ for a disk-shaped grain that is a single domain. Different parts of the sample then presumably resonate at different frequencies. Thus broad absorption lines are expected and indeed are usually observed. In samples exhibiting two absorption peaks, the high-frequency peak sometimes corresponds approximately to the effective field H_K expected on the basis

[183] D. Polder and J. Smit, *Revs. Mod. Phys.* **25,** 89 (1953).

of the known values of the anisotropy constants (K_1 or K_1'). When the high-frequency peak is displaced appreciably from the frequency γH_K, the interpretation is that demagnetizing fields are also playing an important role. For spectra with one peak ferromagnetic resonance is considered to be at least partly and perhaps entirely the mechanism responsible. The investigations have not been sufficiently detailed to ascertain whether the two resonance peaks observed in some samples now merely overlap to produce one peak (see Fig. 10-12.3) or whether the mechanism causing the low-frequency peak is truly absent.

The low-frequency line of a spectrum with two peaks is generally attributed to a domain-wall resonance; the idea is the following. As will be shown later, a domain wall moves as though it had an effective mass m_w. If the wall's motion is confined to the vicinity of a local minimum in the energy F_{wt} (Fig. 7-7.1), to a good degree of approximation the restoring force acting on the wall is proportional to the displacement x from the equilibrium position. Further, the relaxation processes will damp the wall motion. Therefore the equation of motion for a 180° wall is the same as that for a damped harmonic oscillator, namely

$$m_w\ddot{x} + \beta\dot{x} + \alpha_w x = 2M_s H_1(t). \tag{10-12.1}$$

Here β is a damping constant, α_w is the restoring force constant, and $2M_s H_1(t)$ is the driving force that results from the application of the oscillatory field $H_1(t)$. It follows from this equation that a resonance occurs at the frequency

$$\omega = \sqrt{\alpha_w/m_w}, \tag{10-12.2}$$

provided $\beta^2 \ll \alpha_w m_w$.

We will show that an effective mass can be attributed to a moving domain wall.[184] Consider a 180° wall and let ϕ be the angle between an atomic moment in the wall and an easy axis, as in Fig. 7-5.2. When the wall moves with a velocity $v = \dot{x}$, the atomic moments will rotate with the angular velocity

$$\omega = \frac{d\phi}{dt} = -v\frac{d\phi}{dx}.$$

This rotation may be considered to be produced by an effective field H_e directed perpendicular to the wall's surface and defined by

$$\begin{aligned} H_e &= \frac{-\omega}{\gamma} \\ &= +\frac{v}{\gamma}\frac{d\phi}{dx}. \end{aligned} \tag{10-12.3}$$

[184] W. Döring, *Z. Naturforsch.* **3a**, 374 (1948); C. Kittel, *Phys. Rev.* **80**, 918 (1950); R. Becker, *J. Phys. radium* **12**, 332 (1951); G. T. Rado, *Phys. Rev.* **83**, 821 (1951).

The additional energy F_{wv} arising from the wall's motion is then (equation 1-6.11)

$$F_{wv} = \frac{1}{8\pi}\int_{-\infty}^{+\infty} H_e^{\,2}\, dx$$
$$= \frac{v^2}{8\pi\gamma^2}\int_{-\infty}^{+\infty}\left(\frac{d\phi}{dx}\right)^2 dx \tag{10-12.4}$$

per unit area of wall. Since this energy is proportional to v^2, it may be identified as kinetic energy and set equal to $\frac{1}{2}m_w v^2$. From equations 7-5.5, 7-5.7, and 7-5.12, the integral is equal to $(K_1/A)^{1/2}$, where A is an exchange constant (equation 7-1.12). Hence the effective mass is

$$m_w = \frac{1}{4\pi\gamma^2}\left(\frac{K_1}{A}\right)^{1/2} \tag{10-12.5}$$

or, since the width δ of a wall is approximately equal to $\pi(A/K_1)^{1/2}$ (equation 7-5.13),

$$m_w = \frac{1}{4\gamma^2\delta}\,.$$

Typical values of m_w are 10^{-10} to 10^{-11} gm/cm^2.

If the relaxation processes associated with small amplitude motions of domain walls are the same as those occurring in ferromagnetic resonance, then it is meaningful to derive expressions relating the viscous damping constant β to one of the phenomenological damping constants employed in the precessional equations of motion (equations 10-3.20, 10-3.21, and 10-3.22). In particular, in terms of the Landau-Lifschitz parameter λ (equation 10-3.21) β is given by (problem 10-13)

$$\beta = \frac{\lambda}{\gamma^2}\left(\frac{K_1}{A}\right)^{1/2}$$
$$= 4\pi\lambda m_w. \tag{10-12.6}$$

However, since the domain-wall resonance frequency is usually much lower than that for ferromagnetic resonance, it is conceivable that different relaxation mechanisms are important. For example, slow relaxation, described in Section 10-8, may make a significant contribution at one frequency and not at the other.

From equation 7-7.1 it follows that the restoring force constant is given by

$$\alpha_w = \frac{d^2F_{wt}}{dx^2}, \tag{10-12.7}$$

where the derivative is evaluated at the wall's equilibrium position. Hence α_w has its origin in the same imperfections that determine F_{wt} (see Section 7-7). A typical value of α_w is about 10^5.

If $\beta \gg m_w$, as it is for large λ, and if the wall acceleration is not too large, equation 10-12.1 reduces to

$$\beta\dot{x} + \alpha_w x = 2M_s H_1(t). \tag{10-12.8}$$

Although the imaginary part of the initial permeability still has a maximum, the real part now has no peak. This kind of permeability versus frequency curve is sometimes called the relaxation type; by contrast, when equation 10-12.1 applies, the corresponding curve is said to be the resonance type. This nomenclature leads to possible confusion and is therefore unfortunate. The initial permeability data obtained from single crystals of Fe_3O_4[185] and $NiFe_2O_4$[186] cut in the form of picture frames (see Fig. 7-7.5) are described very well by equation 10-12.8.

In fields larger than those necessary to move a domain wall over the maxima in the dF_{wt}/dx versus x curve (Fig. 7-7.1*c*), the wall motion is irreversible and it is no longer appropriate to retain the restoring force term $\alpha_w x$ in the equation of motion. Further, if the wall velocity $\dot{x}$ is essentially constant, a result confirmed by experiment provided the applied field is not too large,[185,187] equation 10-12.1 reduces to

$$\beta\dot{x} = 2M_s(H - H_0). \tag{10-12.9}$$

Here H_0 is a field that is necessary to produce large amplitude displacements and is a measure of the influence of the imperfections; H_0 has a magnitude of the order of the coercive force. Experiments of this nature have been conducted on picture-frame specimens of single crystals of silicon iron,[188] some spinels,[189,190] and YIG.[191] These experiments should yield good values of β; for Fe_3O_4 and $Ni_{0.75}Fe_{2.25}O_4$ β was found to be equal to 0.44 and 0.023, respectively, at room temperature.[188] These values agree well with those expected on the basis of the ferromagnetic resonance line width.[190] However, for YIG which has a very narrow line width, there is an appreciable discrepancy.[191]

Domain structure is eliminated if the sample consists of an assembly of single-domain particles. The initial permeability of several powders has

[185] J. K. Galt, *Phys. Rev.* **85,** 664 (1952).

[186] J. K. Galt, B. T. Matthias, and J. P. Remeika, *Phys. Rev.* **79,** 391 (1950).

[187] J. K. Galt, *Bell System Tech. J.* **33,** 1023 (1954).

[188] H. J. Williams, W. Shockley, and C. Kittel, *Phys. Rev.* **80,** 1090 (1950); K. H. Stewart, *J. Phys. radium* **12,** 325 (1951).

[189] Galt, *loc. cit.*

[190] J. F. Dillon and H. E. Earl, *J. Appl. Phys.* **30,** 202 (1959).

[191] F. B. Hagedorn and E. M. Gyorgy, *J. Appl. Phys.* **32,** 282S (1961).

been measured; usually only one resonance peak is observed and this occurs at microwave frequencies.[192,193] The resonance frequency of single-domain particles is expected to depend on the shape as well as on the crystalline anisotropy. This is confirmed by the spectra observed for some γ-Fe_2O_3 powders with various shape distributions[193] (Fig. 10-12.4). Theoretical permeability versus frequency curves, computed by taking into account

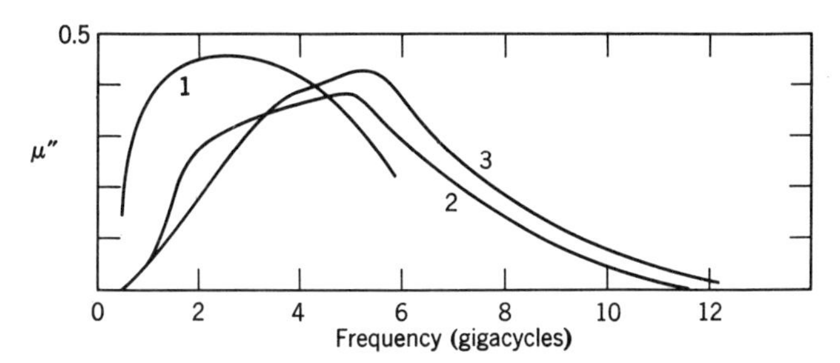

Fig. 10-12.4. Imaginary part of the permeability spectrum for an assembly of single-domain particles. For curve (1) the particles are roughly spherical; for curves (2) and (3) the particles are acicular with mean axial ratios of about 3:1 and 6:1, respectively.

the particle-shape distributions obtained with an electron microscope, the crystalline anisotropy, and the particle interactions are in respectable agreement with the experimental ones.[193]

13. Ferrimagnetic Resonance

Hitherto, resonance in ferrimagnetic materials has been treated as though it were identical to the ferromagnetic case, except, of course, with g or γ replaced by effective values (equation 10-5.1). This premise will now be examined more carefully. In addition, resonance conditions that depend explicitly on the sublattice structure will be derived.

The analysis is restricted to the simplest situation, namely that of two sublattices, A and B, with antiparallel saturation magnetizations, $\mathbf{M}_{As}$ and $\mathbf{M}_{Bs}$. Suppose the single crystal specimen is uniaxial, with equivalent anisotropy fields $\mathbf{H}_{KA}$ and $\mathbf{H}_{KB}$ (equations 8-7.2) acting along the easy z-direction; in general, $H_{KA} \neq H_{KB}$. The results will also apply to cubic lattices, provided that the magnetizations lie in the vicinity of one of the

[192] J. B. Birks, *Nature* (*London*) **158,** 671 (1946), *Proc. Phys. Soc.* (*London*) **60,** 282 (1948), *Proc. Phys. Soc.* (*London*) **B-63,** 65 (1950); G. T. Rado, R. W. Wright, and W. H. Emerson, *Phys. Rev.* **80,** 273 (1950); J. Bluet, I. Epelboin, and D. Quivy, *Compt. rend.* (*Paris*) **246,** 245 (1962).

[193] W. F. Brown, J. P. Hanton, and A. H. Morrish, *J. Appl. Phys.* **31,** 214 (1960); E. P. Valstyn, J. P. Hanton, and A. H. Morrish, *Phys. Rev.* **128,** 2078 (1962).

easy directions. In order to neglect demagnetizing fields, it will be assumed that the sample is spherical. Then, if the exchange interaction is represented by the molecular fields $\mathbf{H}_{mA}$ and $\mathbf{H}_{mB}$ (equations 9-2.1), the equations of motion are given by

$$\frac{d\mathbf{M}_A}{dt} = \gamma_A[\mathbf{M}_A \times (\mathbf{H} - N_{AB}\mathbf{M}_B + \mathbf{H}_{KA})]$$

and

$$\frac{d\mathbf{M}_B}{dt} = \gamma_B[\mathbf{M}_B \times (\mathbf{H} - N_{AB}\mathbf{M}_A + \mathbf{H}_{KB})], \tag{10-13.1}$$

in which the damping terms have been omitted, again for simplicity. The alternating magnetic field is not included on the right-hand side of these equations, because the calculation is confined to the derivation of the resonance conditions (however see problem 10-14). Suppose that the static field $\mathbf{H}$ is applied along the easy direction parallel to $\mathbf{H}_{mA}$ and antiparallel to $\mathbf{H}_{mB}$ (see equations 8-3.9 and 8-3.10); this, of course, implies that $|\mathbf{M}_A| > |\mathbf{M}_B|$. Also, let the sublattice magnetizations be written

$$\begin{aligned}\mathbf{M}_A &= \mathbf{M}_{As} + \mathbf{M}_{A1}(t)\\ &= M_{A1x}(t)\mathbf{i} + M_{A1y}(t)\mathbf{j} + [M_{As} + M_{A1z}(t)]\mathbf{k}\end{aligned}$$

and

$$\begin{aligned}\mathbf{M}_B &= \mathbf{M}_{Bs} + \mathbf{M}_{B1}(t)\\ &= M_{B1x}(t)\mathbf{i} + M_{B1y}(t)\mathbf{j} + [M_{Bs} + M_{B1z}(t)]\mathbf{k}\end{aligned} \tag{10-13.2}$$

where $\mathbf{M}_{A1}(t) = \mathbf{M}_{A1}e^{i\omega t}$ and $\mathbf{M}_{B1}(t) = \mathbf{M}_{B1}e^{i\omega t}$. The components of equations 10-13.1 are then

$$\frac{i\omega M_{A1x}}{\gamma_A} = (H + N_{AB}M_{Bs} + H_{KA})M_{A1y} + N_{AB}M_{As}M_{B1y} \tag{10-13.3}$$

$$-\frac{i\omega M_{A1y}}{\gamma_A} = (H + N_{AB}M_{Bs} + H_{KA})M_{A1x} + N_{AB}M_{As}M_{B1x} \tag{10-13.4}$$

$$\frac{i\omega M_{B1x}}{\gamma_B} = (H - N_{AB}M_{As} - H_{KB})M_{B1y} - N_{AB}M_{Bs}M_{A1y} \tag{10-13.5}$$

$$-\frac{i\omega M_{B1y}}{\gamma_B} = (H - N_{AB}M_{As} - H_{KB})M_{B1x} - N_{AB}M_{Bs}M_{A1x} \tag{10-13.6}$$

together with $M_{A1z} \approx 0$ and $M_{B1z} \approx 0$. Although a solution can readily be obtained from the foregoing component equations, it is rather more convenient to rewrite them in terms of the magnetization amplitudes in the x-y plane, namely, $M_{A1\pm} = M_{A1x} \pm iM_{A1y}$ and $M_{B1\pm} = M_{B1x} \pm iM_{B1y}$ (compare with Section 3-8). This is so because, as a result of the choice of axial symmetry, the precessional motion will be circular. Thus equations 10-13.3 to 10-13.6 become

$$\left[\frac{\omega}{\gamma_A} \pm (H + N_{AB}M_{Bs} + H_{KA})\right]M_{A1\pm} \pm N_{AB}M_{As}M_{B1\pm} = 0$$

and

$$\mp N_{AB}M_{Bs}M_{A1\pm} + \left[\frac{\omega}{\gamma_B} \pm (H - N_{AB}M_{As} - H_{KB})\right]M_{B1\pm} = 0. \tag{10-13.7}$$

The resonance conditions, found by equating the determinant of the coefficients in these equations to zero, are

$$\left[\frac{\omega_{r\pm}}{\gamma_A} \pm (H + N_{AB}M_{Bs} + H_{KA})\right]\left[\frac{\omega_{r\pm}}{\gamma_B} \pm (H - N_{AB}M_{As} - H_{KB})\right] + N_{AB}{}^2M_{As}M_{Bs} = 0, \tag{10-13.8}$$

where ω_{r+} and ω_{r-} are the resonance frequencies of (M_{A1+}, M_{B1+}) and (M_{A1-}, M_{B1-}), respectively. Equation 10-13.8 is a hyperbola with two roots for each of ω_{r+} and ω_{r-}. From a consideration of the general solution of a quadratic, it is obvious that the two roots of ω_{r+} have opposite signs and are equal in magnitude but opposite in sign to the roots of ω_{r-}.

It is anticipated that one solution will correspond to the ferromagnetic resonance frequency; the fields $\omega_{r\pm}/\gamma_A$ and $\omega_{r\pm}/\gamma_B$ will then be small compared to the molecular field. For this mode, therefore, it is reasonable to retain only those products that involve the molecular field. In this approximation equation 10-13.8 becomes

$$\mp \frac{\omega_{r\pm}}{\gamma_A} N_{AB}M_{As} \pm \frac{\omega_{r\pm}}{\gamma_B} N_{AB}M_{Bs} \mp HN_{AB}(M_{As} - M_{Bs}) \mp N_{AB}(H_{KA}M_{As} + H_{KB}M_{Bs}) = 0;$$

that is, N_{AB} disappears from the equation. It follows that

$$\omega_{r\pm} = \mp\gamma_{\text{eff}}(H + H_{K\text{eff}}), \tag{10-13.9}$$

where

$$\gamma_{\text{eff}} = \frac{M_{As} - M_{Bs}}{M_{As}/\gamma_A - M_{Bs}/\gamma_B} \tag{10-13.10}$$

and

$$H_{K\text{eff}} = \frac{H_{KA}M_{As} + H_{KB}M_{Bs}}{M_{As} - M_{Bs}}. \tag{10-13.11}$$

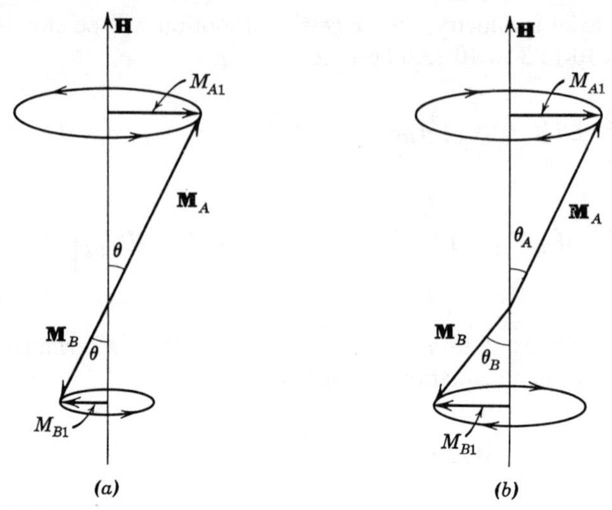

Fig. 10-13.1. Two ferrimagnetic resonance modes. In (*a*), the ferromagnetic resonance mode, the sublattice magnetizations are collinear and precess counterclockwise viewed from above. In (*b*), the exchange resonance mode, the precessional angles of the two sublattice magnetizations are unequal and the precession is clockwise viewed from above. If $|\gamma_A| > |\gamma_B|$, then $m_{A1} > m_{B1}$; otherwise if $|\gamma_A| < |\gamma_B|$, then $m_{A1} < M_{B1}$. Usually, $\theta_A < \theta_B$, but it is conceivable that for widely different values of g_A and g_B that $\theta_B < \theta_A$ (see problem 10-14).

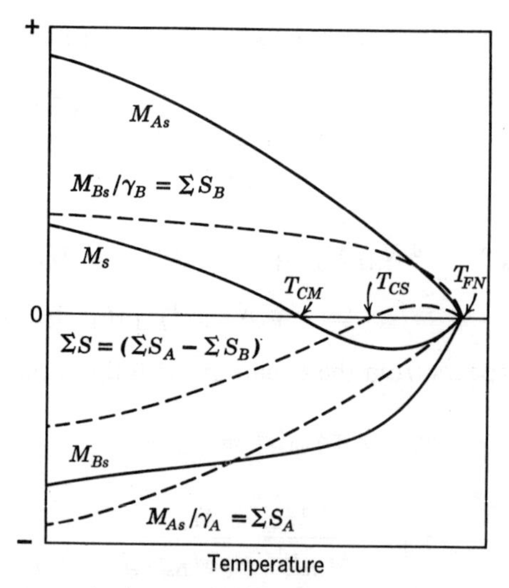

Fig. 10-13.2. The magnetizations and angular momenta of a hypothetical ferrimagnet with two sublattices plotted as a function of temperature. Compensation points in the resultant magnetization and angular momentum occur at the temperatures T_{CM} and T_{CS}. [R. K. Wangness, *Am. J. Phys.* **24,** 60 (1956).]

Since $M_{As} > M_{Bs}$, $H_{K\text{eff}} > 0$. Equation 10-13.10 may also be rewritten in terms of an effective g-value, namely (compare with equation 10-5.1),

$$g_{\text{eff}} = \frac{M_{As} - M_{Bs}}{M_{As}/g_A - M_{Bs}/g_B}. \tag{10-13.12}$$

Equation 10-13.9 indeed has the form of the usual ferromagnetic resonance relation but with a generalized gyromagnetic ratio and crystalline anisotropy field. As indicated in Fig. 10-13.1a, the sublattice magnetizations $\mathbf{M}_A$ and $\mathbf{M}_B$ remain collinear as they precess about the static field (problem 10-14). The frequency ω_{r+} represents a counterclockwise precession of the net magnetization $\mathbf{M}$ $(= \mathbf{M}_A - \mathbf{M}_B)$ about $\mathbf{H}$; the other frequency ω_{r-} may be ruled out on physical grounds.

Generally the properties of this ferrimagnetic resonance mode are similar to those found in ferromagnetic resonance, as may be seen from the discussion in Section 10-6. However, unusual behavior does occur near the temperature of the compensation point. At T_{CM}, $M_{As} = M_{Bs}$, and from equations 10-13.10 and 10-13.12 both γ_{eff} and g_{eff} equal zero. Further, since g_A is rarely equal to g_B, the angular momentum of the two sublattices will become equal at some temperature close to but not equal to T_{CM}; then the denominators of equations 10-13.10 and 10-13.12 will equal zero and γ_{eff} and g_{eff} will become infinite. Actually, approximations have been employed to obtain the equations for γ_{eff} and g_{eff}; a more careful calculation shows that although γ_{eff} and g_{eff} become very large at the angular momentum compensation point they do not become infinite. The two compensation points at T_{CM} and T_{CS} are shown in Fig. 10-13.2. It will be recalled that for any sublattice the angular momentum vector is always antiparallel to the magnetization vector. It is therefore interesting to note that for temperatures between the magnetization and angular momentum compensation points the net angular momentum is parallel to the net magnetization, that is, in this range $\gamma_{\text{eff}} > 0$.

It is reasonable to suspect that the other solution of the resonance condition (equation 10-13.8) corresponds to a mode in which the precession angles θ_A and θ_B of the sublattice magnetizations are appreciably different. In this event the exchange-field torque will be much larger than those of the applied and anisotropy fields. Therefore, as a first approximation, it is legitimate to neglect H, H_{KA}, and H_{KB}, compared to $N_{AB}M_{As}$ and $N_{AB}M_{Bs}$; equation 10-13.8 then becomes

$$\left[\frac{\omega_{r\pm}}{\gamma_A} \pm N_{AB}M_{Bs}\right]\left[\frac{\omega_{r\pm}}{\gamma_B} \mp N_{AB}M_{As}\right] + N_{AB}^{\,2}M_{As}M_{Bs} = 0.$$

It follows that

$$\omega_{r\pm} = \pm N_{AB}(\gamma_B M_{As} - \gamma_A M_{Bs})$$

$$= \pm \gamma_A \gamma_B N_{AB}\left(\frac{M_{As}}{\gamma_A} - \frac{M_{Bs}}{\gamma_B}\right). \qquad (10\text{-}13.13)$$

Since this solution represents a precession about the exchange field, the mode is often referred to as *exchange resonance.*[194] Since $\omega_{r+} < 0$, the precession is clockwise, that is, opposite to the corresponding ferromagnetic mode (Fig. 10-13.1*b*). Again the ω_{r-} frequency is not a physical possibility. For ferrimagnetic materials with $T_{FN} > 100°K$, ω_{r+} is a very high frequency of the order of 10^{11} to 10^{12} c/sec; such radiation has a wavelength in the millimeter or submillimeter range, that is, in the far infrared part of the spectrum. The one exception to this statement occurs at or near the angular momentum compensation point, for here $M_{As}/\gamma_A \approx M_{Bs}/\gamma_B$ and, according to equation 10-13.13, $\omega_{r+} \approx 0$. Actually, the approximations employed to derive equation 10-13.13 are no longer valid; that is $N_{AB}(\gamma_B M_{As} - \gamma_A M_{Bs})$ is no longer very large compared to H, H_{KA}, and H_{KB}. A more detailed calculation shows that, provided that the exchange and anisotropy fields are not too large, both the exchange and ferromagnetic modes can be observed near the compensation point with microwaves in the centimeter wavelength range[195] (problem 10-15).

The calculations may be extended in a straightforward way to include crystalline anisotropies of other symmetries, shape anisotropy, and damping; however, the details then become much more complex. In the case of damping it is interesting to note[196] that the Landau-Lifshitz damping term (equation 10-3.21) provides a much better description of the experimental results than the Bloembergen-Bloch term (equation 10-3.20). The line width is expected to be greatest near the compensation point.

A short survey of the experimental results will now be given. Anomalous values of g_{eff} (or γ_{eff}) for the ferromagnetic mode have been observed at temperatures near the compensation points in LiCr spinel,[197] NiAl spinel,[198] and GdIG.[199] The results for the LiCr spinel are shown in Fig. 10-13.3. For the spinels it is interesting to note that the values obtained

[194] J. Kaplan and C. Kittel, *J. Chem. Phys.* **21,** 760 (1953); R. K. Wangness, *Phys. Rev.* **91,** 1085 (1953), **93,** 68 (1954), **95,** 339 (1955), **98,** 1200 (1955); B. Dreyfus, *Compt. rend.* (*Paris*) **241,** 552 (1956), **241,** 1270 (1956); J. Paulevé and B. Dreyfus, *Compt. rend.* (*Paris*) **242,** 127 (1956).

[195] S. Geschwind and L. R. Walker, *J. Appl. Phys.* **30,** 163S (1959).

[196] R. K. Wangness, *Phys. Rev.* **111,** 813 (1958), *Bull. Am. Phys. Soc.* **3,** 43 (1958).

[197] J. S. Van Wieringen, *Phys. Rev.* **90,** 488 (1953).

[198] T. R. McGuire, *Phys. Rev.* **93,** 682 (1954).

[199] B. A. Calhoun, J. Overmeyer, and W. V. Smith, *Phys. Rev.* **107,** 993 (1957), *J. Appl. Phys.* **29,** 427 (1958).

lie in the range $1 < g_{eff} < 7$. No minimum, only a maximum in g_{eff} was found for GdIG. The line width of these materials is a maximum at the compensation temperature, in agreement with the theoretical predictions.

The appearance of two peaks in the absorption spectra near a compensation point was first observed in a polycrystalline sample of $Li_{0.5}Fe_{1.25}Cr_{1.25}O_4$.[200] Since the low-field peak disappeared at higher temperatures, it presumably was the exchange-resonance absorption.

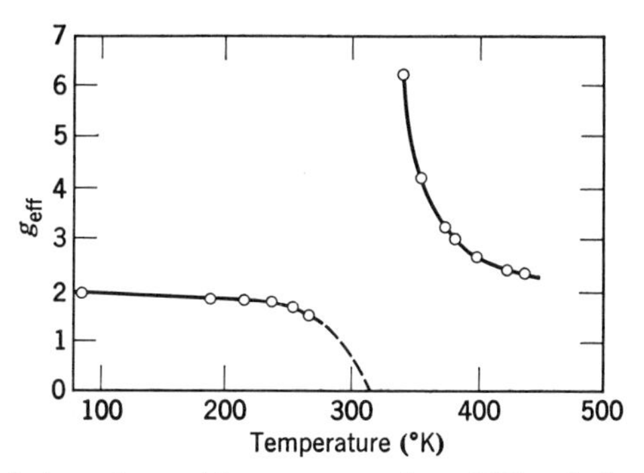

Fig. 10-13.3. Variation of g_{eff} with temperature for a LiCr spinel at the vicinity of the compensation points. [J. S. Van Wieringen, *Phys. Rev.* **90,** 488 (1953).]

Unfortunately, with polycrystalline samples, additional peaks can occur because of the manner in which the randomly oriented crystalline anisotropy contributes to the line width (see end of Section 10-6).[201] It is therefore advisable to use single crystals for these studies. Extensive resonance investigations of GdIG single crystals near T_{CM} have been reported[202] (Fig. 10-13.4). The sense of precession of the two modes was identified unambiguously by employing circularly polarized microwaves. A theoretical calculation of the resonance field for each mode as a function of temperature was in excellent agreement with the data. A new complication was discovered at temperatures within one degree of T_{CM}. Apparently, it is energetically favorable for domains to be oriented with their magnetizations along the various [111]-directions. Very close to T_{CM} the magnetization is small; indeed the domains are almost antiferromagnetic. Hence the magnetostatic energy is essentially zero when there is a domain structure. Further, when the applied field makes a large angle to the sublattice magnetizations of a domain, a larger net magnetization along the direction of the field may result than when the applied field

[200] T. R. McGuire, *Phys. Rev.* **97,** 831 (1955).
[201] E. Schlömann and J. R. Zeender, *J. Appl. Phys.* **29,** 341 (1958).
[202] S. Geschwind and L. R. Walker, *J. Appl. Phys.* **30,** 163S (1959).

is parallel to the sublattice magnetizations. The domain structure manifests itself in the appearance of additional peaks in the absorption spectrum. As many as five peaks have been observed; a maximum of six is expected.[202]

Radiation of very high frequency is required to observe the exchange resonance mode at temperatures that are not close to the compensation point. To be specific, wavelengths approximately in the range of 0.1 to 1 mm (100 to 1000 microns), corresponding to frequencies of 100 to

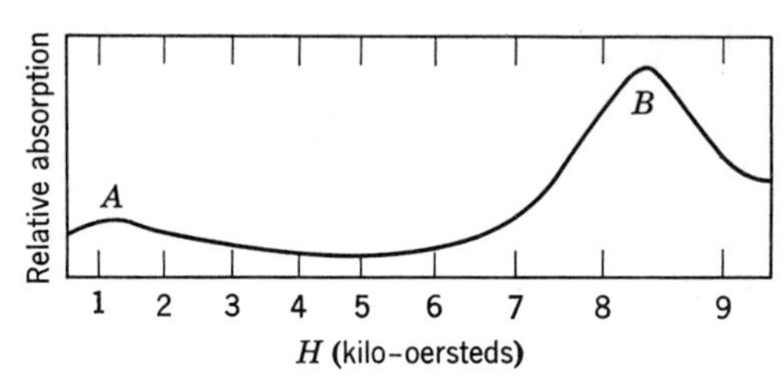

Fig. 10-13.4. Two ferrimagnetic resonance modes in a single crystal of GdIG at $T = 6.81°C$; H is along the [111]-direction. The compensation temperature of GdIG is $T = 13°C$. The normal or ferromagnetic mode (B) has a much greater absorption than the exchange mode (A). The modes were observed with circularly polarized radiation with opposite senses of rotation. [S. Geschwind and L. R. Walker, *J. Appl. Phys.* **30,** 163S (1959).]

10 cm^{-1}, are needed. This region of the spectrum is referred to as the far infrared. Although such sources present a number of problems, it nevertheless has been found possible to construct a suitable far infrared monochromator[203] and to detect the exchange resonance in a number of rare earth garnets. The absorption spectrum of YbIG is shown[204] in Fig. 10-13.5. The exchange absorption line can be distinguished from the other lines because its frequency varies with temperature. This follows since the magnetization of the loosely coupled rare earth sublattice, M_A, is a sensitive function of temperature (equation 10-13.13). The frequencies of the exchange resonance of several rare earth garnets are listed[205] in Table 10-13.1. None is observed for GdIG; since the line intensity depends on $(g_A - g_B)^2$, the reason presumably is that $g_A \approx g_B \approx 2.00$ because both Gd^{3+} and Fe^{3+} ions have S ground states.

Other lines with low infrared frequencies have been identified as transitions in single rare earth ions. The two lines for YbIG at 23.4 and 26.4 cm^{-1} are examples (Fig. 10-13.5). The transition is between the

[203] R. C. Ohlmann and M. Tinkham, *Phys. Rev.* **123,** 425 (1961).
[204] A. J. Sievers and M. Tinkham, *Phys. Rev.* **124,** 321 (1961).
[205] A. J. Sievers and M. Tinkham, *Phys. Rev.* **129,** 1995 (1963).

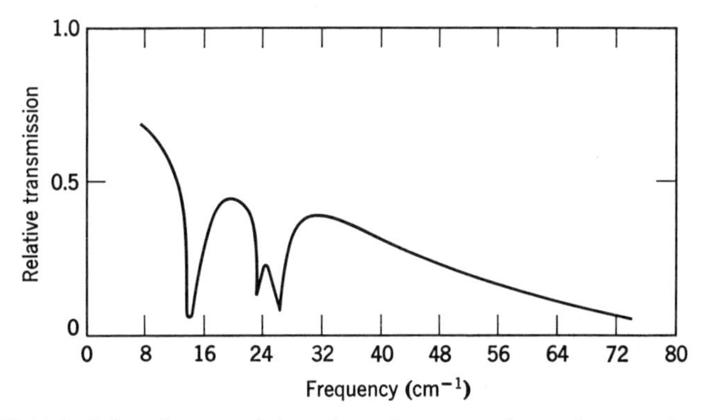

Fig. 10-13.5. Infrared transmission through a pressed powder sample of YbIG at $T = 2°K$. The coherent exchange resonance absorption occurs at 14.0 cm^{-1} (716 microns). [A. J. Sievers and M. Tinkham, *Phys. Rev.* **124,** 321 (1961).]

two levels of the lowest Kramers doublet, split by the crystalline electric field and the exchange field of the iron sublattice.[206] These transitions, which are normally forbidden by the selection rules, are allowed because of the anisotropy of the transverse g-values. The frequency of the single-ion transitions is given by

$$\omega_{si} = \gamma_A N_{AB} M_B; \tag{10-13.14}$$

hence it is almost temperature independent. Here M_B is the magnetization of the iron sublattice and both γ_A and N_{AB} are anisotropic. Two lines are observed for GdIG because there are two magnetically inequivalent sites when the spontaneous magnetization is along the [111]-easy direction. Several such lines observed for HoIG and ErIG develop from transitions

Table 10-13.1. Exchange Resonance Data for Rare Earth Iron Garnets[a]
$(T = 2°K)$

	Exchange Resonance Mode	
Garnet	Frequency (cm^{-1})	Wavelength (microns)
SmIG	33.5	298
GdIG	None observed	
HoIG	38.5	260
ErIG	10.0	1000
YbIG	14.0	716

[a]A. J. Sievers and M. Tinkham, *Phys. Rev.* **129,** 1995 (1963).

[206] M. Tinkham, *Phys. Rev.* **124,** 311 (1961), *J. Appl. Phys.* **33,** 1248 (1962).

to higher lying energy levels. No single-ion lines are found for GdIG; this is to be expected, for the anisotropy for Gd^{3+} ions is probably negligible. When static magnetic fields are applied, the frequency of single-ion lines shifts as a result of Zeeman splitting. Studies of this nature have provided valuable information concerning the fields acting on the rare earth ions in garnets.

14. Antiferromagnetic Resonance

A theoretical study of resonance in a two-sublattice antiferromagnet was initiated[207] before the development of the ferrimagnetic case. For our purposes it is more convenient to start with the results already obtained for ferrimagnetic resonance and to apply the particular conditions that are peculiar to antiferromagnetism. Specifically, for two identical sublattices, $\gamma_A = \gamma_B = \gamma$ and $H_{KA} = H_{KB} = H_K$. Further, at absolute zero, $M_{As} = M_{Bs} = M_s$, where M_s is the spontaneous magnetization of *one* sublattice. Equation 10-13.8, which, it is to be recalled, holds only when the magnetic field **H** is applied along the easy direction, then becomes

$$\frac{\omega_{r\pm}^2}{\gamma^2} \pm \frac{2\omega_{r\pm}}{\gamma} H + H^2 - 2H_K N_{AB} M_s - H_K^2 = 0. \qquad (10\text{-}14.1)$$

The roots of this quadratic equation are

$$\omega_{r\pm} = \mp\gamma H \pm \gamma[H_K(H_K + 2N_{AB}M_s)]^{1/2}. \qquad (10\text{-}14.2)$$

The two sets of roots, ω_{r+} and ω_{r-}, correspond to the field **H** applied parallel and antiparallel to the A sublattice magnetization, respectively. Since the sublattices are indistinguishable physically, only one set, say $\omega_{r+} = \omega_r$ need be considered.

For $H = 0$ it follows that

$$\omega_r = \pm\gamma[H_K(H_K + 2N_{AB}M_s)]^{1/2}; \qquad (10\text{-}14.3)$$

that is, the modes are degenerate in frequency but precess in opposite senses. This is illustrated in Fig. 10-14.1 (see also problem 10-16*a*). When $H \neq 0$, the modes are no longer degenerate but continue to precess in the opposite sense until $\omega_r = 0$ for one of them; that is

$$\begin{aligned} H &= [H_K(H_K + 2N_{AB}M_s)]^{1/2} \\ &\approx (2H_K N_{AB} M_s)^{1/2} \\ &\approx (2H_K H_m)^{1/2}, \end{aligned} \qquad (10\text{-}14.4)$$

[207] C. Kittel, *Phys. Rev.* **82,** 565 (1951); T. Nagamiya, *Prog. Theor. Phys.* (*Kyoto*) **6,** 342 (1951).

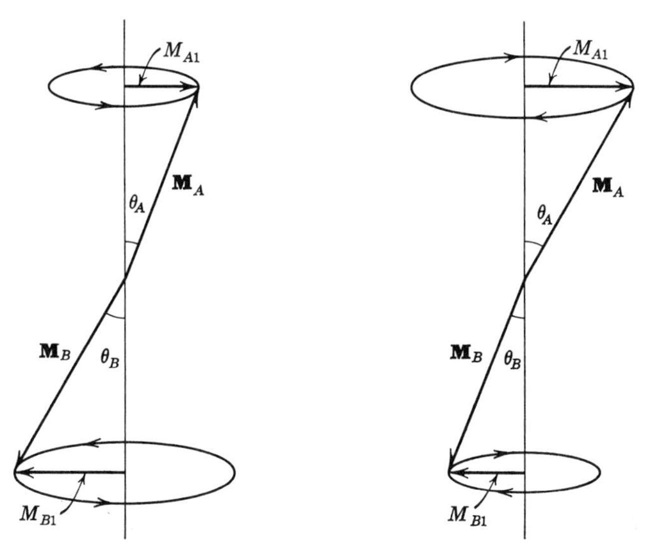

Fig. 10-14.1. Normal modes for a two-sublattice antiferromagnet when $H = 0$. The difference between M_{A1} and M_{B1} has been exaggerated.

where $H_m = N_{AB}M_s$ is the intersublattice molecular field. Equation 10-14.4 gives precisely the value of H required to produce spin flopping, that is, a reorientation of the sublattice magnetizations to a direction perpendicular to the easy axis (compare equation 8-7.12 and problem 8-8*c*). The resonance condition (equation 10-14.1) is, of course, no longer valid. When equation 10-14.4 is exactly satisfied, ($H = H_f$), resonance absorption can occur; this phenomena is sometimes called orientation resonance.[208] For $H > H_f$ it can be shown that (problem 10-16*b*)

$$\omega_r = \pm\gamma(H^2 - 2H_K H_m)^{1/2}. \tag{10-14.5}$$

At temperatures above absolute zero $\mathbf{M}_{As}$ is slightly larger than $\mathbf{M}_{Bs}$ when $\mathbf{H}$ is applied parallel to $\mathbf{M}_{As}$. Equation 10-13.8 then becomes

$$\left[\frac{\omega_r}{\gamma} + (H + N_{AB}M_{Bs} + H_K)\right]\left[\frac{\omega_r}{\gamma} + (H - N_{AB}M_{As} - H_K)\right] + N_{AB}{}^2 M_{As}M_{Bs} = 0.$$

The roots of this quadratic are

$$\omega_r = -\gamma\left[H + \frac{N_{AB}(M_{Bs} - M_{As})}{2}\right] \pm \gamma\left[-H_K{}^2 + H_K N_{AB}(M_{As} + M_{Bs}) + \frac{N_{AB}{}^2}{4}(M_{Bs} - M_{As})^2\right]^{1/2} \tag{10-14.6}$$

[208] T. Nagamiya, *Prog. Theor. Phys. (Kyoto)* **11,** 309 (1954).

If H_K is small compared to the molecular field, H_K^2 may be neglected in the second term on the right-hand side. Further, to a good degree of approximation $N_{AB}(M_{As} + M_{Bs}) = 2H_m$. Finally, the difference between M_{As} and M_{Bs} may be written in terms of α, defined as $\alpha = \chi_{\parallel}/\chi_{\perp}$, where $\chi_{\parallel} = (M_{As} - M_{Bs})/H$ and $\chi_{\perp} = 1/N_{AB}$ (Chapter 8). Equation 10-14.6 reduces to

$$\omega_r = -\gamma H\left(1 - \frac{\alpha}{2}\right) \pm \gamma\left[2H_K H_m + \left(\frac{\alpha}{2}\right)^2 H^2\right]^{1/2}; \qquad (10\text{-}14.7)$$

this result is valid, of course, only if $H < H_f$.

The case in which the field $\mathbf{H}$ is applied perpendicular to the easy axis is also of interest. The sublattice magnetizations then precess about the static equilibrium positions given in Fig. 8-3.3. Since the sublattice magnetizations are tilted equally toward $\mathbf{H}$, they are equal in magnitude at any given temperature; hence one complication present in the earlier calculation is absent.

In order to derive the resonance conditions, we begin with equations 10-13.1 and set $\gamma_A = \gamma_B = \gamma$ and $\mathbf{H}_{KA} = \mathbf{H}_{KB} = \mathbf{H}_K$. As usual, it is the alternating components of the magnetizations $\mathbf{M}_A$ and $\mathbf{M}_B$ in the planes perpendicular to the static equilibrium axes that are important. Now, however, these planes make an angle 2ϕ with each other. It is therefore appropriate to introduce the axes x' and x'', as indicated in Fig. 10-14.2, and to consider the magnetization components $M_{A1x'}$, M_{A1y}, $M_{B1x''}$, and M_{B1y}. The components of the equations of motion are

$$\frac{i\omega M_{A1x'}}{\gamma} = (N_{AB}M_{Bs}\cos 2\phi + H_K\cos\phi)M_{A1y} + N_{AB}M_{As}\cos 2\phi M_{B1y}, \qquad (10\text{-}14.8)$$

$$-\frac{i\omega M_{A1y}}{\gamma} = (H\sin\phi + N_{AB}M_{Bs}\cos 2\phi + H_K\cos\phi)M_{A1x'} + N_{AB}M_{As}\cos 2\phi M_{B1x''}, \qquad (10\text{-}14.9)$$

$$\frac{i\omega M_{B1x''}}{\gamma} = -N_{AB}M_{Bs}\cos 2\phi M_{A1y} - (+N_{AB}M_{As}\cos 2\phi + H_K\cos\phi)M_{B1y}, \qquad (10\text{-}14.10)$$

$$-\frac{i\omega M_{B1y}}{\gamma} = -N_{AB}M_{Bs}\cos 2\phi M_{A1x'} - (H\sin\phi + N_{AB}M_{As}\cos 2\phi + H_K\cos\phi)M_{B1x''}, \qquad (10\text{-}14.11)$$

together with $M_{A1z'} \approx 0$ and $M_{B1z''} \approx 0$. There is a lack of symmetry in these equations, and it is no longer possible simply to combine them and

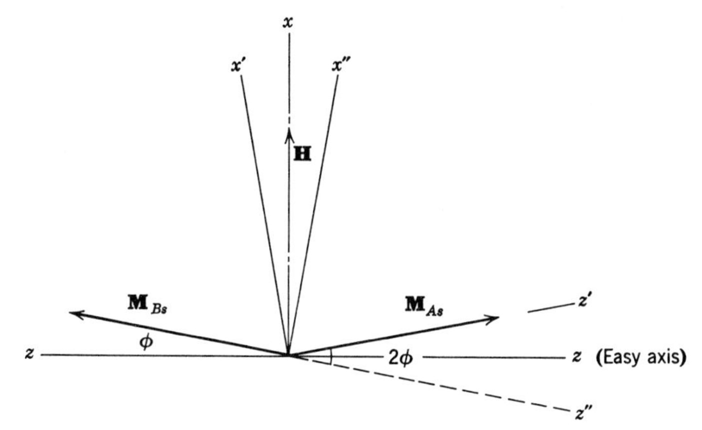

Fig. 10-14.2. The calculation of the resonance conditions when H is applied perpendicular to the easy axis.

to introduce M_{A1+} and M_{B1+}. Instead, two different resonance equations will be obtained, one for each pair of components. First, it will be recalled that $iM_{A1x'} = -M_{A1y}$, $iM_{A1y} = M_{A1x'}$, $iM_{B1x''} = -M_{B1y}$, $iM_{B1y} = M_{B1x''}$, and $M_{As} = M_{Bs} = M_s$. Then, from equations 10-14.9 and 10-14.11, the resonance condition, obtained by setting the determinant of the coefficients equal to zero, is

$$\begin{vmatrix} \dfrac{\omega_r}{\gamma} + (H \sin\phi + N_{AB}M_s \cos 2\phi + H_K \cos\phi) & N_{AB}M_s \cos 2\phi \\ -N_{AB}M_s \cos 2\phi & \dfrac{\omega_r}{\gamma} - (H \sin\phi + N_{AB}M_s \cos 2\phi + H_K \cos\phi) \end{vmatrix} = 0.$$

If it is assumed that $H < N_{AB}M_s$, $H_K < N_{AB}M_s$, and ϕ is small, the determinant yields

$$\omega_r = \pm\gamma(2HN_{AB}M_s\phi + 2H_KN_{AB}M_s)^{1/2}.$$

It has already been shown that at static equilibrium $\phi = H/2N_{AB}M_s$ (equation 8-3.15); therefore

$$\begin{aligned} \omega_r &= \pm\gamma(H^2 + 2H_KN_{AB}M_s)^{1/2} \\ &= \pm\gamma(H^2 + 2H_KH_m)^{1/2}, \end{aligned} \qquad (10\text{-}14.12)$$

where $H_m = N_{AB}M_s$. This mode will be excited when a microwave field with the proper frequency is applied along the y-direction, that is, perpendicular to **H**.

The other components (equations 10-14.8 and 10-14.10) yield the resonance condition

$$\begin{vmatrix} \dfrac{\omega_r}{\gamma} + (N_{AB}M_s \cos 2\phi + H_K \cos\phi) & N_{AB}M_s \cos 2\phi \\ -N_{AB}M_s \cos 2\phi & \dfrac{\omega_r}{\gamma} - (N_{AB}M_s \cos 2\phi + H_K \cos\phi) \end{vmatrix} = 0.$$

Under the same approximations as before, this leads to

$$\begin{aligned} \omega_r &= \pm\gamma(2H_K N_{AB} M_s)^{1/2} \\ &= \pm\gamma(2H_K H_m)^{1/2}. \end{aligned} \tag{10-14.13}$$

This mode is unusual in that it is excited only when the microwave field is applied along the x-axis, that is, when $\mathbf{H}_1$ is *parallel* to $\mathbf{H}$.

The general case in which $\mathbf{H}$ is applied at an angle θ to the easy direction has been analyzed in the literature. It is found[209] that

$$\begin{aligned} \left(\frac{\omega_r}{\gamma}\right)^2 &= 2H_K H_m + \tfrac{1}{2}H^2[1 + (1+\alpha)^2\cos^2\theta] \\ &\quad \pm \tfrac{1}{2}H\{8H_K H_m(2-\alpha)^2\cos^2\theta \\ &\quad + H^2[\sin^4\theta + \cos^2\theta(2-\alpha)^2(2\sin^2\theta + \alpha^2\cos^2\theta)]\}^{1/2}. \end{aligned} \tag{10-14.14}$$

When shape anisotropy is included in the calculation with $\mathbf{H}$ parallel to the easy axis, the results for the general ellipsoid with $D_x \neq D_y$ predict two sets of resonance frequencies. Even for a sphere, the demagnetizing factor $D(= 4\pi/3)$ enters formally into the resonance condition; for example, at $T = 0°\text{K}$ it has been shown that[210]

$$\omega_r = -\gamma[H \pm H_K(2H_m + H_K + 2DM_s)]^{1/2}.$$

Crystalline anisotropies with other than uniaxial symmetry are of interest; the case of the orthorhombic lattice has been considered in detail.[211] Damping has been taken into account by introducing either the Bloembergen-Bloch[212] (equations 10-3.20) or the Landau-Lifshitz[213] (equations 10-3.21) terms into the equations of motion. As for the ferrimagnetic exchange mode, the Landau-Lifshitz term provides the best fit with the experimental data.

Since the product $(H_K H_m)^{1/2}$ appears in the expressions developed for ω_r, it is anticipated that a very high frequency will usually be required for resonance. An exception occurs for antiferromagnets with low Néel temperatures. The outstanding example is orthorhombic $CuCl_2 \cdot 2H_2O$, which has $T_N = 4.3°\text{K}$; then the absorption spectra has been observed

[209] T. Nagamiya, *Prog. Theor. Phys. (Kyoto)* **6,** 342 (1951); F. Keffer and C. Kittel, *Phys. Rev.* **85,** 329 (1952).

[210] R. K. Wangness, *Phys. Rev.* **86,** 146 (1952).

[211] T. Nagamiya, K. Yosida, and R. Kubo, *Advan. Phys.* **4,** 1 (1955).

[212] H. J. Gerritsen and M. Garber, *Physica* **22,** 481 (1956); Yu. M. Seiidov, *Fiz. Metal i Metalloved* **7,** 443 (1959).

[213] E. S. Dayhoff, *J. Appl. Phys.* **29,** 344 (1958); A. M. Portis and D. Teaney, *Phys. Rev.* **116,** 838 (1959); G. S. Heller, J. J. Stickler, and J. B. Thaxter, *J. Appl. Phys.* **32,** 307S (1961); R. C. Ohlmann and M. Tinkham, *Phys. Rev.* **123,** 425 (1961).

at frequencies of 4, 9.4,[214] and 35[215] Gc. Both the modes of equation 10-14.2 were identified. Antiferromagnetic resonance has also been detected in $CuBr_2{\cdot}2H_2O$($T_N = 6°K$) and $CuCl_2{\cdot}2D_2O$ at X-band frequencies.[216]

The development of microwave spectrometers in the millimeter range has permitted the investigation of antiferromagnets with moderate Néel temperatures; examples include Cr_2O_3($T_N = 307.5°K$), MnF_2($T_N = 67.7°K$),

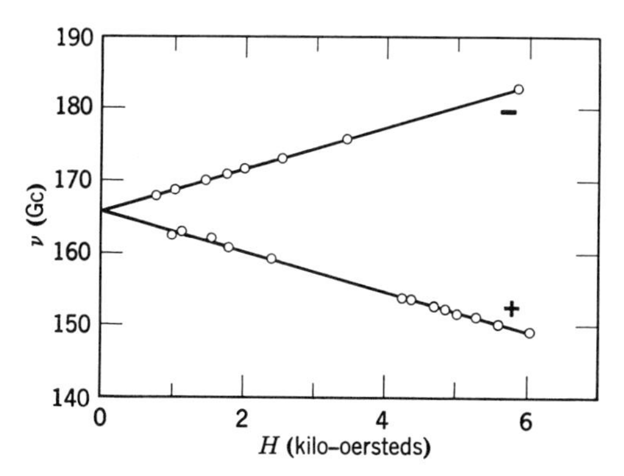

Fig. 10-14.3. The resonance frequency ν_r ($= \omega_r/2\pi$) as a function of the magnetic field **H** applied along the easy (c) axis of a single crystal of Cr_2O_3. $T = 77°K$. [G. S. Heller, J. J. Stickler, and J. B. Thaxter, *J. Appl. Phys.* **32,** 307S (1961).]

and $MnTiO_3$ ($T_N \approx 60°K$). The zero-field resonance of Cr_2O_3 at 77°K occurs at about 166 Gc. By varying both ω and H it has been found possible to observe both the modes predicted by equation 10-14.2. This data is shown in Fig. 10-14.3,[217] clearly the resonance frequencies for the modes of opposite polarization diverge linearly with H. The use of large fields will greatly reduce the frequency required for resonance of the mode with positive circular polarization. Thus an absorption line has been observed in Cr_2O_3 at 35 and 70 Gc by applying a pulsed field.[218] The microwave frequency necessary for resonance can also be reduced by

[214] N. J. Poulis, J. van den Handel, J. Ubbink, J. A. Poulis, and C. J. Gorter, *Phys. Rev.* **82,** 552 (1951); J. Ubbink, J. A. Poulis, H. J. Gerritsen, and C. J. Gorter, *Physica* **18,** 361 (1952).

[215] H. J. Gerritsen, R. F. Okkes, B. Bölger, and C. J. Gorter, *Physica* **21,** 629 (1955); H. J. Gerritsen, *Physica* **21,** 639 (1955).

[216] M. Date, *Phys. Rev.* **104,** 623 (1956), *J. Phys. Soc. Japan* **12,** 1168 (1957).

[217] G. S. Heller, J. J. Stickler, and J. B. Thaxter, *J. Appl. Phys.* **32,** 307S (1961).

[218] S. Foner, *J. Phys. radium* **20,** 336 (1959).

working near the Néel temperature, since then $H_m(= N_{AB}M_s)$ is comparatively small. This idea has led to successful experiments[219] on Cr_2O_3 at frequencies as low as 24 Gc.

MnF_2 has been investigated between 100 and 200 Gc in zero applied field[220] and at 35 and 70 Gc with pulsed fields.[221] It was found that the field $(2H_KH_m)^{1/2}$ varied with temperature as a Brillouin function with $S = \frac{5}{2}$, a result expected if H_K, as well as H_m, is proportional to M_s. Actually a Brillouinlike temperature dependence for $(2H_KH_m)^{1/2}$ has been observed for most of the antiferromagnetic materials studied to date; an exception, however, is Cr_2O_3. One of the interesting conclusions that have emerged from studies of $MnTiO_3$ is that pertinent data can be obtained from powdered samples.[222]

The line widths in antiferromagnetic resonance are generally large and usually range from a few hundred to several thousand oersteds. The line widths are observed to increase rapidly in the vicinity of the Néel temperature.[223] It is to be hoped that the production of purer and higher quality crystals will lead to a marked reduction in the line widths and permit more meaningful studies of relaxation processes in antiferromagnets.

The use of far infrared radiation has led to the detection of antiferromagnetic resonance in a number of other compounds; these include FeF_2,[224] CoF_2,[225,226] NiF_2,[225,226] $KNiF_3$,[225] MnO,[227] NiO,[228] and $MnCO_3$.[226] In addition, the results obtained for MnF_2 with microwaves have been confirmed with optical techniques.[225] The large values of

$$H_f\{=[H_K(H_K + 2H_m)]^{1/2}\}$$

and ΔH, the line width, make it difficult to resolve the two modes by the application of ordinary magnetic fields. However, it has been found possible to observe the individual modes in some of the compounds by employing either very low temperatures or superconducting solenoids or

[219] T. R. McGuire, E. J. Scott, and F. H. Grannis, *Phys. Rev.* **102**, 1000 (1956); E. S. Dayhoff, *Phys. Rev.* **107**, 84 (1957).

[220] F. M. Johnson and A. H. Nethercot, *Phys. Rev.* **104**, 847 (1956), **114**, 705 (1959).

[221] S. Foner, *Phys. Rev.* **107**, 683 (1957).

[222] J. J. Stickler and G. S. Heller, *J. Appl. Phys.* **33**, 1302 (1962).

[223] F. M. Johnson and A. H. Nethercot, *Phys. Rev.* **114**, *loc. cit.*

[224] R. C. Ohlmann and M. Tinkham, *Phys. Rev.* **123**, 425 (1961).

[225] P. L. Richards, *J. Appl. Phys.* **34**, 1237 (1963).

[226] P. L. Richards, *J. Appl. Phys.* **35**, 850 (1964).

[227] F. Keffer, A. J. Sievers, and M. Tinkham, *J. Appl. Phys.* **32**, 65S (1961); M. Tinkham, *Phys. Rev.* **124**, 311 (1961); A. J. Sievers and M. Tinkham, *Phys. Rev.* **124**, 321 (1961).

[228] H. Kondoh, *J. Phys. Soc. Japan* **15**, 1970 (1960); M. Tinkham, *J. Appl. Phys.* **33**, 1248 (1962); A. J. Sievers and M. Tinkham, *Phys. Rev.* **129**, 1566 (1963).

both.[224,226] The theoretical analysis required for MnO and NiO is somewhat different, inasmuch as their moments lie in the (111) plane; that is, the [111]-axis is the hard direction. FeF_2 has an exceptionally large anisotropy field, with $H_K \approx 10^5$ oe; as a result, this material has the highest resonance frequency measured so far. CoF_2 is piezomagnetic; the application of a stress decreases the resonance frequency.[229] Just as

Table 10-14.1. Resonance Data on Antiferromagnets at $T \approx 0°K$

Compound	Resonance with $H = O$ Frequency	Wavelength	$H_f = \frac{\omega_r}{\gamma}$ (oersteds)	H_K (oersteds)	H_m (oersteds)	T_N (°K)
$CuCl_2 \cdot 2H_2O$	17.5 Gc	1.7 cm	7×10^3			4.3
$RbMnF_2$[a]				5	7.1×10^5	77
Cr_2O_3	158 Gc	1.9 mm	6×10^4	9×10^2	2×10^6	307.5
$MnTiO_3$	150 Gc	2 mm	5×10^4	10^3	10^6	65
MnF_2	8.7 cm^{-1}	1.150 mm	9×10^4	9×10^3	5×10^5	67.7
FeF_2	52.8 cm^{-1}	0.190 mm	5×10^5	2×10^5	5×10^5	78.4
CoF_2	28.5 and 36 cm^{-1}	0.351 and 0.278 mm	3×10^5 3.5×10^5			37.7
NiF_2	31.1 cm^{-1}	0.322 mm	3×10^5			73.2
$KNiF_3$	48.7 cm^{-1}	0.205 mm	4.6×10^5			
MnO	27.6 cm^{-1}	0.362 mm	3×10^5	3×10^4	10^6	120
NiO	36.5 cm^{-1}	0.274 mm	4×10^5	4×10^4	4×10^6	523
$MnCO_3$	4.1 cm^{-1}	2.440 mm	4×10^4			31.6

[a] D. J. Teaney, M. J. Freiser and R. W. H. Stevenson, *Phys. Rev. Letters* **9**, 212 (1962).

in microwaves, the line widths of the infrared transmission spectra are described relatively well by the Landau-Lifshitz damping parameter. A summary of various data obtained from antiferromagnetic resonance experiments is given in Table 10-14.1.

The subject of antiferromagnetic resonance in materials with canted or spiral spin arrangements is also of considerable interest. Canting can be produced either by different crystalline fields acting on magnetic ions in two nonequivalent sites or by an antisymmetric superexchange interaction (Section 8-9). The first mechanism is believed to act, for example, in NiF_2. Calculations for this case show[230] that two resonance frequencies that are nondegenerate even in zero field are expected. The lowest frequency does not depend on the exchange interaction, hence may be considered as a ferromagnetic mode. Two absorptions which have indeed

[229] P. L. Richards, *J. Appl. Phys.* **35**, 850 (1964).
[230] T. Moriya, *Phys. Rev.* **117**, 635 (1960).

been observed in NiF_2 at 31.1 cm^{-1} and 3.33 cm^{-1} have been identified as the antiferromagnetic and ferromagnetic modes, respectively.[231]

The other mechanism gives rise to the Dzialoshinski-Moriya Hamiltonian, $\mathcal{H}_{\mathfrak{D}} = \mathfrak{D} \cdot (\mathbf{S}_i \times \mathbf{S}_j)$ (equation 8-9.1). The resonance conditions may be calculated by inserting the effective Dzialoshinski-Moriya field, $\mathbf{H}_{\mathfrak{D}A} = \mathfrak{D} \times \mathbf{M}_B$ or $\mathbf{H}_{\mathfrak{D}B} = \mathfrak{D} \times \mathbf{M}_A$ into the equations of motion. For crystals with a rhombohedral structure it can be shown (problem 10-17) that the resonance frequencies are given by

$$\omega_r = -\gamma[H(H + H_{\mathfrak{D}}) + 2H_m H_K']^{1/2} \qquad (10\text{-}14.15)$$

and

$$\omega_r = -\gamma[H_{\mathfrak{D}}(H + H_{\mathfrak{D}}) + 2H_m H_K]^{1/2}, \qquad (10\text{-}14.16)$$

where H_K and H_K' are the anisotropy fields along the [111]-direction and within the (111) plane, respectively.[232] Since H_K' is usually small, equation 10-14.15 represents a much lower frequency than equation 10-14.16. The low-frequency mode has been observed in α-Fe_2O_3[233] and $MnCO_3$,[234] whereas to date the high frequency mode has been detected only in $MnCO_3$[235,236] (see Table 10-14.1).

The resonance condition in materials with a spiral or helical spin configuration will depend on the details of the spin arrangement, the anisotropies, and the orientation of the applied fields.[237,238] Generally very high frequencies are required for resonance[238]; however, it is possible that in certain situations for some materials with small anisotropies resonance will occur at ordinary microwave frequencies. Apparently resonance has not yet been observed in materials with helical spin configurations.

15. Nuclear Magnetic Resonance in Ordered Magnetic Materials

Nuclear magnetic resonance has been observed in ferromagnetic, antiferromagnetic, and ferrimagnetic materials. Signals have been detected

[231] P. L. Richards, *J. Appl. Phys.* **35,** 850 (1964).

[232] P. Pincus, *Phys. Rev. Letters* **5,** 13 (1960); A. S. Borovik-Romanov, *J. Exptl. Theoret. Phys. (USSR)* **36,** 766 (1959) [trans. *Soviet Phys.-JEPT* **9,** 539 (1959)]; E. A. Turov and N. G. Gusseinov, *J. Exptl. Theoret. Phys. (USSR)* **38,** 1326 (1960) [trans. *Soviet Phys.-JEPT* **11,** 955 (1960)].

[233] P. W. Anderson, F. R. Merritt, J. R. Reimika, and W. A. Yager, *Phys. Rev.* **93,** 717 (1954); H. Kumagai, H. Abe, K. Ono, I. Hayashi, J. Shimada, and K. Iwanaga, *Phys. Rev.* **99,** 1116 (1955); A. Tasaki and S. Iida, *J. Phys. Soc. Japan* **18,** 1148 (1963).

[234] M. Date, *J. Phys. Soc. Japan* **15,** 2251 (1960).

[235] Richards, *loc. cit.*

[236] H. J. Fink and D. Shaltiel, *Phys. Rev.* **130,** 627 (1963).

[237] A. Yoshimori, *J. Phys. Soc. Japan* **14,** 807 (1959).

[238] B. R. Cooper, R. J. Elliott, S. J. Nettel, and H. Suhl, *Phys. Rev.* **127,** 57 (1962); B. R. Cooper and R. J. Elliott, *Phys. Rev.* **131,** 1043 (1963).

from the nuclei of both diamagnetic and transition group ions. The subject began with the discovery of proton resonance in the antiferromagnetic compound $CuCl_2 \cdot 2H_2O$.[239] Resonance of the nuclei in paramagnetic ions was first observed for Co^{59} in face-centered cubic cobalt metal.[240] Since then the list of metallic and dielectric materials in which nuclear resonance has been found has increased rapidly.

The hyperfine structure (hfs) produced by the interaction between the nuclear and electronic magnetic moments has already been discussed in Section 3-9. This interaction produces an internal field H_n at the nucleus. The nuclear resonance data provides a direct numerical estimate of the magnitude of H_n. Although the sign of H_n can also be determined under certain circumstances from resonance data, this information is usually obtained more readily from the Mössbauer effect to be discussed in the next section. It is worth mentioning in passing that experimental values for the internal field H_n can also be obtained from specific heat measurements at low temperature and from observations of the anisotropy in the γ-ray emission from radioactive nuclei.[241]

The internal field H_n is, of course, not to be confused with the exchange field which causes the ordering of the electronic moments. Nevertheless, the exchange field does play an important role in nuclear resonance phenomena, as will become clearer later. The internal field may be expressed in terms of the hyperfine interaction ($\mathcal{A}\mathbf{I} \cdot \mathbf{S}$) of the spin-Hamiltonian, namely,

$$H_n = \frac{\mathcal{A}\bar{S}}{g_n \mu_{B_n}} \tag{10-15.1}$$

where $\bar{S}$ is the expectation value of the electron spin. The internal field has several sources. For a $3d$ transition ion there will be the usual dipole-dipole interaction between the $3d$ electronic moment and the nuclear moment. The inner $2s$ and $3s$ electrons, which have nonzero wave functions at the nucleus, contribute a field as a result of the contact interaction (equation 3-11.2). This term arises because the $3d$ electrons couple differently to $2s$ and $3s$ electrons in the two possible spin states. The different densities of the s electrons of opposite spin at the nucleus lead to a large *negative* internal field. For ferromagnetic metals the $4s$ conduction electrons contribute to H_n partly because they are polarized and partly from the admixture with $3d$ electrons. There is also a further contribution from Lorentz, demagnetizing, and applied fields. However,

[239] N. J. Poulis and G. E. G. Hardeman, *Physica* **19,** 391 (1953).
[240] A. C. Gossard and A. M. Portis, *Phys. Rev. Letters* **3,** 164 (1959).
[241] N. Kurti, *J. Appl. Phys.* **30,** 215S (1959).

in many experiments the resonance is observed in zero applied static field; the demagnetizing field is then often also zero. Numerical estimates of the various contributions to H_n have been made.[242] Their sum in the case of cobalt, for example, does indeed agree in magnitude with the observed resonance value of 217 kilo-oe at $T \approx 0°K$. However, the theory predicts that H_n is positive, whereas Mössbauer and other experiments show that it is negative. This means that H_n has the opposite direction to the magnetization of the material. This result appears to imply that the $2s$-$3s$ electron contact term, which is apparently the only one that is negative, is much larger than expected. A re-examination of the theory of the internal field has been made[243] which confirms that the inner contact term is indeed very large.

In ferromagnetic alloys there are additional possible sources of the internal field at the nucleus of a solute atom. The $4s$ electrons may be polarized by the $3d$ electrons of both the solute and solvent atoms. A transfer of a $3d$ electron between the solvent and solute atoms may also occur which would affect the $2s$ and $3s$ as well as the $4s$ electron polarizations. Resonance measurements of H_n for both diamagnetic and paramagnetic solute atoms indicate the importance of the $4s$ polarization.[244]

The nuclear magnetic resonance signal is found to be remarkably intense. Measurements of the saturation of the resonance line for face-centered cubic cobalt show that the rf field seen by the nuclei is about 1000 times larger than the applied rf field.[245] This large enhancement can occur at the nuclei situated within a domain wall as may be seen by considering the following simple model. Suppose a Co particle with the shape of an ellipsoid of revolution has one domain wall lying parallel to the polar axis, assumed to be an easy direction of crystalline anisotropy, as shown in Fig. 7-2.3b. When an external rf field with the magnitude H_1 is applied parallel to the wall's surface, the wall will be displaced along a direction x lying perpendicular to the wall (see Fig. 7-7.1a). If the effect of internal imperfections is negligible, the maximum wall displacement x_m will be limited by the increase in demagnetizing energy. Hence, if the equatorial radius of the particle is b, then

$$\frac{x_m}{b} \approx \frac{H_1}{\frac{1}{2}D_a M_s},$$

[242] W. Marshall, *Phys. Rev.* **110,** 1280 (1958).

[243] R. E. Watson and A. J. Freeman, *J. Appl. Phys.* **32,** 118S (1961), *Phys. Rev.* **123,** 2027 (1961), *J. Appl. Phys.* **33,** 1086 (1962).

[244] T. Kushida, A. H. Silver, Y. Koi, and A. Tsujimura, *J. Appl. Phys.* **33,** 1079 (1962); V. Jaccarino, *J. Appl. Phys.* **32,** 1023 (1961); L. H. Bennett and R. L. Streever, *J. Appl. Phys.* **33,** 1093 (1962).

[245] A. M. Portis and A. C. Gossard, *J. Appl. Phys.* **31,** 205S (1960).

where D_a is the longitudinal demagnetizing factor. In the displacement x_m each moment is rotated approximately through the angle

$$\theta = \frac{\pi x_m}{\delta},$$

where δ is the width of the 180° wall. The magnitude of the rf field at the nucleus H_{1n} will be

$$H_{1n} = \theta H_n \approx \frac{2\pi}{\delta} \frac{H_1 b}{D_a M_s} H_n.$$

The enhancement factor η is therefore

$$\eta = \frac{H_{1n}}{H_1} = \frac{2\pi}{\delta} \frac{b}{D_a M_s} H_n. \tag{10-15.2}$$

For $\delta \approx 10^{-5}$ cm, $b \approx 10^{-4}$ cm, $H_n = 2.2 \times 10^5$ oe, $M_s = 1400$ gauss, and $D_a \approx 2\pi$, equation 10-15.2 gives $\eta \approx 1600$, in reasonable accord with the experimental findings for Co.

Additional confirmation of this idea[246] is obtained by applying a static magnetic field. When this static field is large enough to remove the domain structure partly, the signal intensity drops rapidly. Close to saturation of the electronic magnetization, the signal is only about 1% of its original value.

An enhancement is also produced when the magnetization within a domain rotates coherently under the action of the rf field. Consider a single-domain particle with negligible crystalline anisotropy and suppose the rf field is applied perpendicular to the magnetization M_s. At equilibrium the torques produced by the rf and demagnetizing fields will be equal; that is

$$H_1 M_s = D_a M_s^2 \theta,$$

where θ is the maximum angle M_s makes with the easy direction during the rotation and it is assumed that $\sin\theta \approx \theta$. Hence

$$H_{1n} = \theta H_n = \frac{H_1}{D_a M_s} H_n$$

[246] A. C. Gossard and A. M. Portis, *Phys. Rev. Letters* **3,** 164 (1959); J. I. Budnick, L. J. Bruner, R. J. Blume, and E. L. Boyd, *J. Appl. Phys.* **32,** 120S (1961); R. G. Shulman, *J. Appl. Phys.* **32,** 126S (1961); R. L. Streever, L. H. Bennett, R. C. LaForce, and G. F. Day, *J. Appl. Phys.* **34,** 1050 (1963).

and the enhancement η is

$$\eta = \frac{H_n}{D_a M_s}. \tag{10-15.3}$$

For $H_n \approx 2 \times 10^5$ oe and $D_a M_s \approx 5 \times 10^3$, $\eta \approx 40$. Therefore this type of enhancement is appreciably less than when the nuclei are situated within domain walls. Nevertheless, resonances from nuclei located within domains have been observed in the ferromagnetic metals Fe, Co, and Ni,[247] in ferrimagnets such as magnetite (Fe_3O_4),[248] and in the canted antiferromagnet NiF_2.[249] In antiferromagnetic materials the nuclear resonance is enhanced because it is driven by an antiferromagnetic resonance mode.[250]

Additional nuclear resonance lines are observed in some materials. Some extra lines found in the spectra of cobalt powders are believed to develop from nuclei located at various stacking faults in the cubic phase.[251] The resonance of nuclei in the hexagonal phase of cobalt has also been detected.[251] In alloys the additional lines observed appear to originate from nuclei that surround a solute atom.[252] In some instances the change in the internal field is too small to produce resolvable lines and instead only a line broadening occurs.

In several experiments the temperature dependence of the nuclear resonance frequency, and hence also of H_n, has been determined. According to equation 10-15.1, H_n is proportional to $\bar{S}$, the average value of the spin per atom. Thus, provided $\mathcal{A}$ is independent of temperature, the measurements yield the temperature variation of the magnetization associated with particular ions. Data of this sort have been obtained for metallic iron,[253] some alloys,[254] the Mn^{2+} ions on the A sublattice of the spinel $MnFe_2O_4$,[255] the Fe^{3+} ions on both the a and d sublattices of YIG,[256]

[247] M. Weger, E. L. Hahn, and A. M. Portis, *J. Appl. Phys.* **32,** 124S (1961).

[248] E. L. Boyd and J. C. Slonczewski, *J. Appl. Phys.* **33,** 1077 (1962).

[249] R. G. Shulman, *Phys. Rev.* **121,** 125 (1961).

[250] A. Narath, *Phys. Rev.* **131,** 1929 (1963); A. Nakamura, V. Minkiewicz, and A. M. Portis, *J. Appl. Phys.* **35,** 842 (1964); A. J. Heeger and D. T. Teaney, *J. Appl. Phys.* **35,** 846 (1964).

[251] W. A. Hardy, *J. Appl. Phys.* **32,** 122S (1961); R. Street, D. S. Rodbell, and W. L. Roth, *Phys. Rev.* **121,** 84 (1961).

[252] R. C. LaForce, S. R. Ravitz, and G. F. Day, *Phys. Rev. Letters* **6,** 226 (1961); Y. Koi, A. Tsujimura, T. Hihara, and T. Kushida, *J. Phys. Soc. Japan* **16,** 574 (1961).

[253] J. I. Budnick, L. J. Bruner, R. J. Blume, and E. L. Boyd, *J. Appl. Phys.* **32,** 120S (1961).

[254] T. Kusida, A. H. Silver, Y. Koi, and A. Tsujimura, *J. Appl. Phys.* **33,** 1079 (1962); L. H. Bennett and R. L. Streever, *J. Appl. Phys.* **33,** 1093 (1962).

[255] A. J. Heeger and T. Houston, *J. Appl. Phys.* **35,** 836 (1964).

[256] C. Robert, *Compt. rend.* (*Paris*) **251,** 2685 (1960).

and others. However, experiments on iron in which the pressure was varied show that $\mathcal{A}$ does depend on the second power of the temperature.[257] This effect may be caused by the excitation of electrons higher into the $3d$ band.

Both spin-spin and spin-lattice relaxation times have been measured in a number of ordered materials by a variety of standard techniques. For the ferromagnetic metals τ_1 is of the order of 10^{-3} sec.[258] The results indicate that a spin diffusion analogous to that observed for liquids (Section 4-4) occurs.[259] It is likely that the spins ultimately relax to the lattice by coupling to the conduction electrons.[258] It has been proposed that an important contribution to the line width comes from a long-range coupling between the nuclear dipoles via the electronic spin-waves.[260] In this process the nuclear spins, respectively, emit and absorb electronic spin-waves through the hyperfine interaction. Since the initial and final electronic states are identical, the spin waves are called virtual. This mechanism, for example accounts for the line width observed in the canted antiferromagnet $KMnF_3$.[261] The virtual spin waves also produce a shift in the resonance frequency.[262] This frequency pulling may be the origin of the nuclear spin-lattice relaxation in the spinel $MnFe_2O_4$.[263]

The double resonance method, discussed in Section 4-10, has also been applied to materials with strongly coupled dipoles. Substances in which the nuclear resonance has been detected via the electron resonance include the ferrimagnetic spinel $MnFe_2O_4$[264] and the antiferromagnetic compounds $KMnF_3$[265] and $MnCO_3$.[266]

16. The Mössbauer Effect

The magnetic field at the nucleus H_n may also be determined by the Mössbauer effect. This method consists of the emission and subsequent

[257] G. B. Benedek and J. Armstrong, *J. Appl. Phys.* **32,** 106S (1961).

[258] A. M. Portis and A. C. Gossard, *J. Appl. Phys.* **31,** 205S (1960); M. Weger, E. L. Hahn, and A. M. Portis, *J. Appl. Phys.* **32,** 124S (1961).

[259] A. M. Portis, *J. Phys. Soc. Japan, Suppl. B-I* **17,** 81 (1962).

[260] H. Suhl, *Phys. Rev.* **109,** 606 (1958); T. Nakamura, *Prog. Theor. Phys. (Kyoto)* **20,** 542 (1958).

[261] A. Nakamura, V. Minkiewcz, and A. M. Portis, *J. Appl. Phys.* **35,** 842 (1964).

[262] P.-G. DeGennes, P. A. Pincus, F. Hartmann-Bouton, and J. M. Winter, *Phys. Rev.* **129,** 1105 (1963); A. M. Portis, G. Witt, and A. J. Heeger, *J. Appl. Phys.* **34,** 1052 (1963); A. J. Heeger and D. Teaney, *J. Appl. Phys.* **35,** 846 (1964).

[263] A. J. Heeger, T. G. Blocker, and S. K. Ghosh, *J. Appl. Phys.* **35,** 840 (1964).

[264] A. J. Heeger, S. K. Ghosh, and T. G. Blocker, *J. Appl. Phys.* **34,** 1034 (1963).

[265] Portis, Witt, and Heeger, *loc. cit.*; A. J. Heeger, A. M. Portis, D. T. Teaney, and G. L. Witt, *Phys. Rev. Letters* **9,** 212 (1962); K. Lee, A. M. Portis, and G. L. Witt, *Phys. Rev.* **132,** 144 (1963).

[266] D. Shaltiel and H. J. Fink, *J. Appl. Phys.* **35,** 848 (1964).

absorption of a gamma ray with extremely narrow line width. Resonance fluorescence has been known for a long time for photons with optical frequencies. When an atom decays from an excited state (2) to the ground state (1), a photon with frequency ω_{21} is emitted. This photon, on passing through a gas of the same type of atom, may be absorbed, hence may excite one of these atoms into the state (2).

For the corresponding process with gamma rays a problem exists. When a gamma ray is emitted from a free atom at rest, the atom recoils with an energy $R = p^2/2M = E_{12}{}^2/2Mc^2$, where $E_{12} = E_1 - E_2$. As a result, the energy E of the emitted γ-ray is $E = E_{12} - R$, whereas for absorption the energy $E_{12} + R$ is required since the target atom also recoils. In spite of the appreciable line broadening by the Doppler effect arising from the thermal motion, there is little overlap of the emission and absorption spectra; hence the resonance fluorescence is usually negligible.

The situation can be quite different when the source atoms are bound in a solid. If the γ ray is very energetic, there is the possibility that the recoiling atom will be dislodged from its lattice site. Usually, however, the recoil energy will appear as lattice vibrations or phonons. Suppose the γ-ray energy is smaller than the Debye energy, $\hbar\omega_D$ or $k\theta_D$, where ω_D is the maximum frequency allowed by the Debye theory and θ_D is the corresponding Debye temperature. Then there is an appreciable probability that no phonon will be emitted in a particular γ-ray emission. Momentum, of course, will still be conserved by the recoil of the solid as a whole. Because of the solid's large mass, this recoil energy is extremely small. The conclusion is that the γ ray has a good chance of being emitted with the full energy E_{21}. The same situation holds for the absorption process. Quantum mechanical calculations confirm these qualitative statements.[267]

In recoilless emission the line width of the γ ray in the ideal case is determined by the lifetime τ of the excited state. If $\tau \approx 10^{-7}$ sec, it follows from the uncertainty principle that $\Delta E \approx h/\tau \approx 10^{-8}$ eV. The ultimate resolving power for a 10 keV γ ray is then $\Delta\nu/\nu = \Delta E/E \approx 10^{-12}$. Although other factors may enter to increase the line width, the resolution attainable is far greater than that ever before achieved in spectroscopy. As a consequence, recoilless nuclear resonance absorption has found many applications. These include the measurement of the gravitational red shift, investigations of biophysical interest, the analysis of lattice dynamics, and the study of hyperfine interactions. Only the last topic is discussed here.

[267] W. E. Lamb, *Phys. Rev.* **55,** 190 (1939); R. L. Mössbauer, *Naturwiss.* **45,** 538 (1958); W. M. Visscher, *Ann. Phys. (New York)* **9,** 194 (1960); H. J. Lipkin, *Ann. Phys. (New York)* **9,** 332 (1960); K. S. Singwi and A. Sjolander, *Phys. Rev.* **120,** 1093 (1960); C. Tzara, *J. Phys. radium* **22,** 303 (1961).

Recoilless nuclear resonance absorption was first discovered by Mössbauer[268] for the 129 keV γ ray from Ir^{191}. It is curious that the Mössbauer effect was not discovered sooner, since X-ray diffraction by crystals is an analogous phenomena. In addition, a similar effect occurs in the absorption of neutrons by a solid, as was shown implicitly in an early theoretical paper.[269] Finally, a theoretical paper by Dicke[270] showed

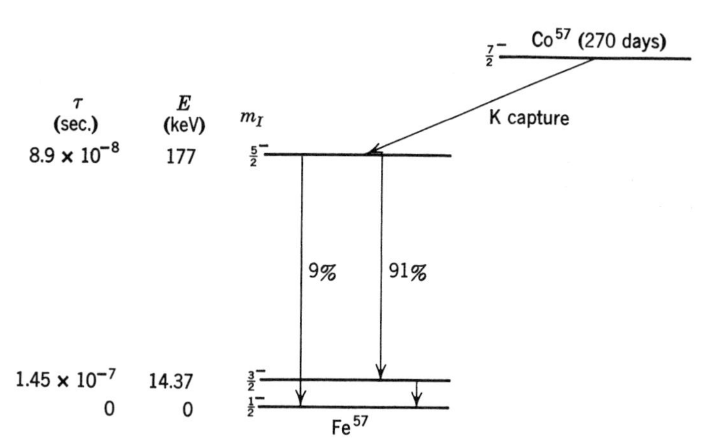

Fig. 10-16.1. Decay scheme of Fe^{57}. Pertinent nuclear data are indicated.

that when a gas is confined in a box an appreciable fraction of the emitted radiation is not broadened by the Doppler effect.

The majority of Mössbauer experiments performed up to 1964 have employed the 14.4 keV transition in Fe^{57}; its decay scheme is indicated in Fig. 10-16.1. Reasons that make this isotope particularly suitable for Mössbauer experiments include the low energy of the γ ray, the relatively long lifetime of the transition's excited state, and the long lifetime of the parent isotope (Co^{57}). For magnetic studies another advantage is that iron is a common constituent of magnetic materials. The particular isotope Fe^{57} has a natural abundance of 2.14%. Sometimes materials are manufactured with iron enriched in this isotope.

The nuclear energy levels will be split by a magnetic field. The fields normally available in the laboratory produce a splitting that is too small to yield well-resolved spectra. Use is therefore made of the large magnetic field H_n at the nucleus that exists in ordered materials. The splitting of

[268] R. L. Mössbauer, *Z. Physik* **151,** 124 (1958), *Naturwiss.* **45,** 538 (1958), *Z. Naturforsch.* **14a,** 211 (1959).

[269] W. E. Lamb, *Phys. Rev.* **55,** 190 (1939).

[270] R. H. Dicke, *Phys. Rev.* **89,** 472 (1953).

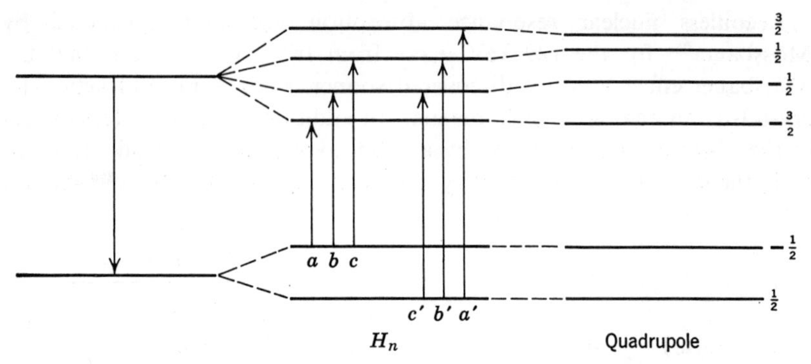

Fig. 10-16.2. Ground and first excited states of Fe^{57}. The splitting of the energy levels by the internal field H_n and by the quadrupole interaction are shown. The allowed transitions, $\Delta m_I = 0, \pm 1$, are also indicated.

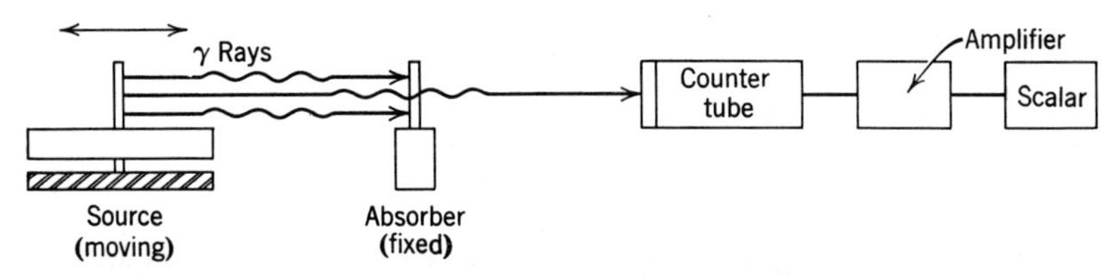

Fig. 10-16.3. Experimental arrangement for observation of the Mössbauer effect. Cryogenic apparatus is added when other than room temperature is desired for the source or absorber.

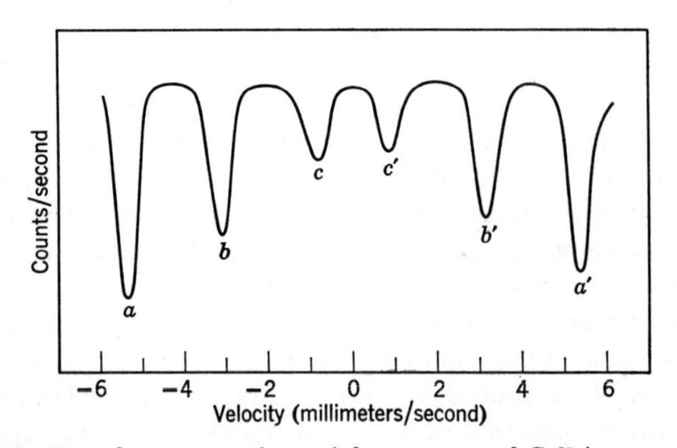

Fig. 10-16.4. Hyperfine spectra observed for a source of Co^{57} in unmagnetized iron and an absorber of stainless steel. The absorption line arising from a particular transition shown in Fig. 10-16.2 is labeled with corresponding letters. There is no quadrupole interaction because the lattice is cubic. [C. E. Johnson, quoted in A. J. F. Boyle and H. E. Hall, *Rept. Prog. Phys.* **25,** 441 (1962).]

the lowest nuclear energy levels of Fe^{57} by H_n is shown diagrammatically in Fig. 10-16.2.

It is not necessary that either the source or the absorber have a monochromatic line,[271] but it is a convenience. The ferro-cyanides, which are diamagnetic (Table 2-12.1), and hence have no field at the iron ions,[272] have found use as the absorber. Monochromatic sources have been made by diffusing Co^{57} into some stainless steels at elevated temperatures.[273] It follows that these steels must not be ordered.

If a monochromatic source is then employed, the hyperfine spectra of, say, iron can be observed by giving the source a velocity. The energy of the emitted γ ray will be shifted because of the Doppler effect. When the energy of the γ ray coincides with one of the possible transitions in the iron, there will be resonance absorption that can be readily detected with a proportional or scintillation counter. The velocities required are not great, being only of the order of 1 cm/sec. The experimental arrangement just described is sketched in Fig. 10-16.3. Alternatively, the source may be fixed and the absorber given a velocity. In another arrangement the source may consist of Co^{57} in iron and the absorber of stainless steel. The absorption spectrum observed[274] for this arrangement is shown in Fig. 10-16.4. The positive and negative velocities refer to the source moving toward and away from the absorber, respectively. According to theory, the line intensities will be proportional to the Clebsch-Gordan coefficients.[275] A calculation for iron shows that the lines a, b, and c should be in the ratio 3:2:1; the same is true for the lines a', b', and c'. This is indeed the case for the results shown in Fig. 10-16.4; however, frequently the predicted ratio is not observed because the absorber is too thick.

The shift of the energy levels is

$$\Delta E_1 = -m_{I1} g_{n1} H_n$$

for the excited state and

$$\Delta E_2 = -m_{I2} g_{n2} H_n$$

for the ground state. Knowledge of the nuclear moment of the ground state, which can be obtained, for example, from a beam experiment,

[271] S. S. Hanna, J. Herberle, C. Littlejohn, G. J. Perlow, R. S. Preston, and D. H. Vincent, *Phys. Rev. Letters* **4,** 177 (1960); S. S. Hanna, J. Heberle, G. J. Perlow, R. S. Preston, and D. H. Vincent, *Phys. Rev. Letters* **4,** 513 (1960).

[272] S. L. Ruby, L. M. Epstein, and K. H. Sun, *Rev. Sci. Instr.* **31,** 580 (1960).

[273] G. K. Wertheim, *J. Appl. Phys.* **32,** 110S (1961).

[274] C. E. Johnson, quoted in A. J. F. Boyle and H. E. Hall, *Rept. Prog. Phys.* **25,** 441 (1962).

[275] M. E. Rose, *Elementary Theory of Angular Momentum*, John Wiley and Sons, New York (1957).

(see Section 4-11), then yields both the magnetic field at the nucleus and the magnetic moment of the excited state. Values of the internal field determined experimentally are listed in Table 10-16.1; they are of the order of 10^5 oe. The sign of H_n can be determined by applying a magnetic field and noting whether the hyperfine splitting decreases or increases. The sign of H_n for cases in which it has been determined is specifically

Table 10-16.1. The Magnetic Field at a Nucleus as Determined by the Mössbauer Effect

Nucleus	Host Lattice	$H_n(\times 10^5$ oe)	T(°K)
Fe^{57}	Fe	−3.30	room
Fe^{57}	Co	3.12	room
Fe^{57}	Ni	2.80	room
Co^{59}	Fe	3.00	4.2
Ni^{61}	Ni	−1.70	room
$Fe^{57(3+)}$	YIG (tetrahedral site)	3.90	room
$Fe^{57(3+)}$	YIG (octahedral site)	4.80	room
$Fe^{57(3+)}$	Fe_3O_4 (tetrahedral site)	5.00	room
$Fe^{57(2+)}$	Fe_3O_4 (octahedral site)	4.50	room
$Dy^{161(3+)}$	DyIG (tetrahedral site)	3.90	room
$Dy^{161(3+)}$	DyIG (octahedral site)	4.80	room
Fe^{57}	FeF_2	3.40	room
Sn^{119}	Fe	−0.88	100
Sn^{119}	Co	−0.21	100
Sn^{119}	Ni	+0.185	100
Au^{197}	Fe	2.80	room

indicated by a + or − symbol in Table 10-16.1. For most materials H_n is negative; the significance of this result has already been discussed in Section 10-15.

The field at the Fe^{57} nucleus does not depend greatly on whether the host is iron or any of the other ferromagnetic metals or alloys. It appears then that H_n is produced primarily by the electrons of the iron nucleus in question and is not influenced greatly by the crystalline fields and magnetization of the host lattice. The spectra of compounds is more complicated when there are nonequivalent lattice sites. For example, when the magnetization of YIG is along the [111]-direction, there is one pattern from the ions in tetrahedral sites and two patterns from the ions in the octahedral sites.[276] For Fe_3O_4 three superposed patterns from Fe^{3+} ions in A sites, Fe^{3+} ions in B sites, and Fe^{2+} ions in B sites appear.[276] The

[276] G. K. Wertheim, *J. Appl. Phys.* **32,** 110S (1961).

variation of H_n with temperature has also been determined in some cases. For Fe^{57} in iron H_n has the same temperature dependence as the saturation magnetization.[277]

Of course, since the early days of the subject, many other nuclei have been found to be suitable for Mössbauer studies. These include Ni^{61}, Sn^{119}, Dy^{161}, Tm^{169}, and Au^{197}; the internal fields at some of these nuclei are listed in Table 10-16.1. Actually, most rare earth elements possess nuclei with first excited states of low energy and with associated γ rays with relatively narrow line widths. It is therefore likely that much work in the future will concentrate on rare earth metals and compounds.

It is interesting to consider the possibility of studying the field H_n in paramagnetic salts with the Mössbauer effect. When the relaxation time for transitions between the electronic levels is short, the field at the nucleus has a time average of zero[278]; hence the magnetic hyperfine spectrum will not be observable. At low temperatures the relaxation times will be longer and, further, only the lowest energy levels will be populated. A nonzero H_n may then result; however, no observation of the magnetic hyperfine spectra in paramagnetic salts yet appears to have been reported.

When the nucleus has an electric quadrupole moment, the interaction with the electric field gradient at the nucleus will lead to a shift of the nuclear energy levels, as discussed in Sections 3-9 and 4-8. For the case in which the crystal has axial symmetry the energy shift according to equation 3-9.10 will be

$$\Delta E = \mathscr{P}[m_I^2 - \tfrac{1}{3}I(I + 1)]. \tag{10-16.1}$$

For the first excited state of Fe^{57}, $I = \frac{3}{2}$, and it follows immediately from this equation that $\Delta E_{3/2} = \Delta E_{-3/2} = -\Delta E_{1/2} = -\Delta E_{-1/2}$. This shifting of the energy levels is indicated in Fig. 10-16.2. The ground state levels are not changed, for the quadrupole moment is zero when $I \leqslant \frac{1}{2}$. The result is that the resonance absorption spectrum will now be shifted; as an example, the resonance spectrum observed for α-Fe_2O_3 is shown[279] in Fig. 10-16.5. Measurements of such spectra yield values for the quadrupole interaction, hence lead to information concerning the crystalline field gradient at the nucleus.

There is still another hyperfine interaction of importance. The volume occupied by the nucleus is different for the ground and excited states. The interaction with the s-state electrons is a function of the overlap of the nuclear and electronic wave functions and consequently the nuclear

[277] D. E. Nagle, H. Frauenfelder, R. D. Taylor, D. R. F. Cochran, and B. T. Matthias, *Phys. Rev. Letters* **5**, 364 (1960).

[278] R. Orbach, *Proc. Phys. Soc. (London)* **77**, 821 (1961).

[279] O. C. Kistner and A. W. Sunyar, *Phys. Rev. Letters* **4**, 412 (1960).

energy levels are shifted by different amounts. In addition, the electronic charge density at the nucleus depends on the chemical nature of the material. For example, metallic iron has 4*s* conduction electrons that are not present in ionic compounds. Further, Fe^{2+} and Fe^{3+} ions differ in the number of 3*d* electrons and therefore the shielding of the inner *s* electrons is altered. As a result, the γ-ray energy differs from compound to compound. In a Mössbauer experiment, the position of the absorption line

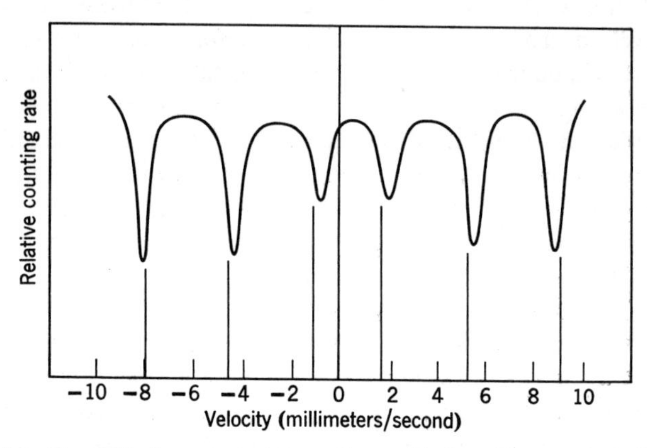

Fig. 10-16.5. The Mössbauer spectrum of $\alpha - Fe_2O_3$. The vertical lines indicate where the greatest absorption would have occurred if the quadrupole interaction had been zero. The source is Co^{57} bound in stainless steel. [O. C. Kistner and A. W. Sunyar, *Phys. Rev. Letters* **4,** 412 (1960).]

for a given source will depend on the chemical form of the absorber. As an illustration, the lines of α-Fe_2O_3 in Fig. 10-16.5 are all shifted upward by 0.5 mm/sec.[280] This effect is known as the *isomer* shift, since nuclei with the same atomic weight and number but in different states are known as isomers. Measurement of the isomer shift permits changes in the *s* electron wave functions at the nucleus to be studied.

Problems

10-1. Show that if the orbital angular momentum is almost quenched $g - 2 = 2 - g'$. Under what condition is equation 10-2.3 identical to equation 10-2.4?

10-2. Suppose that for the anisotropy field $\mathbf{H}_K = -\mathbf{D}_K \cdot \mathbf{M}$ all the off-diagonal elements of $\mathbf{D}_K$ are zero except D_{Kxy} and D_{Kyx}. Calculate the resonance frequency corresponding to equation 10-3.6.

10-3. Derive equations 10-3.11 and 10-3.12.

[280] O. C. Kistner and A. W. Sunyar, *Phys. Rev. Letters* **4,** 412 (1960).

10-4. Show that the high-frequency extrinsic and intrinsic susceptibility are related by $\boldsymbol{\chi}_i = \boldsymbol{\chi}_e (1 + \mathbf{D} \cdot \boldsymbol{\chi}_i)$ (equation 10-3.16). Find the similar relationship that holds for the permeability.

10-5. Consider the alternating magnetic field as composed of two circularly polarized fields, $H_{1+} = H_1 e^{i\omega t}$ and $H_{1-} = H_1 e^{-i\omega t}$ rotating counterclockwise and clockwise, respectively, about H_{1z} (see Section 3-7). Show that the permeability tensor is then given by

$$\tilde{\boldsymbol{\mu}} = \begin{pmatrix} \mu_{11} + K & 0 & 0 \\ 0 & \mu_{11} - K & 0 \\ 0 & 0 & 1 \end{pmatrix}$$

under conditions where $\mu_{11} = \mu_{22}$.

10-6. Show that M may be set equal to M_0 in the Landau-Lifshitz damping term by proving that M is conserved when this damping expression is used in the equations of motion. Also show that M is not conserved with Bloembergen-Bloch damping.

10-7. (a) Start with equation 10-3.23 and find the high-frequency susceptibility elements for a spherical sample with zero crystalline anisotropy (equation 10-3.24). (b) If the tensor elements are written $\chi_{11} = \chi_{11}' - i\chi_{11}''$ and $\chi_{12} = -\chi_{12}'' - i\chi_{12}'$, compute the real and imaginary parts. (c) Repeat the calculation for a sample with ellipsoidal shape; hence show that

$$\chi_{11} = \frac{\gamma M_s[(\gamma H + i\omega\alpha) + D_y \gamma M_s]}{\omega_r^2 - \omega^2 + 2\omega\alpha\gamma i\{H + [(D_x - D_y)/2]M_s\}},$$

$$\chi_{22} = \frac{\gamma M_s[(\gamma H + i\omega\alpha) + D_x \gamma M_s]}{\omega_r^2 - \omega^2 + 2\omega\alpha\gamma i\{H + [(D_x - D_y)/2]M_s\}},$$

and

$$\chi_{21} = -\chi_{12} = \frac{i\gamma\omega M_s}{\omega_r^2 - \omega^2 + 2\omega\alpha\gamma i\{H + [(D_x - D_y)/2]M_s\}}.$$

10-8. Consider an ellipsoidal specimen situated in a steady magnetic field applied along the z-direction. Suppose this field is large enough to align the magnetization along the z-direction also. Show that the resonance frequency as given by equation 10-4.3 reduces to the Kittel formula (equation 10-3.6).

10-9. Show that the Bloembergen-Bloch relaxation time τ is related to the half-width of the absorption line by

$$\tau = \frac{2\omega}{\gamma H \, \Delta H}.$$

10-10. Set up the equations of motion for the case in which the static and microwave fields are parallel and lie in the plane of a disk. Show that the threshold field for the instability of spin waves with wave number $\mathbf{k}$ is given by

$$H_{1t} = \frac{\omega \, \Delta H_k}{4\pi M_s \gamma},$$

provided that $\omega = 2\omega_{\mathbf{k}}$.

10-11. Consider a thin film with the static field applied parallel to the film's plane. Show that the resonance frequency is given by equation 10-10.6.

10-12. Consider a plane wave with field vectors varying as $e^{i\omega t - \Gamma(\mathbf{n}\cdot\mathbf{r})}$, where Γ is the complex propagation constant and $\mathbf{n}$ is a unit vector normal to the wavefront. If ϕ is the angle between the direction of propagation and the steady magnetic field $\mathbf{H}$, show that

$$\Gamma_{\pm}^2 = -\omega^2\epsilon\left\{\frac{(\mu_{11}^2 - \mu_{11} - K^2)\sin^2\phi + 2\mu_{11} \pm [(\mu_{11}^2 - \mu_{11} - K^2)^2\sin^4\phi + 4K^2\cos^2\phi]^{1/2}}{2(\cos^2\phi + \mu_{11}\sin^2\phi)}\right\}.$$

Find the propagation constant for $\phi = 0$ and $\pi/2$ and compare with the results quoted in Section 10-11.

10-13. Find the relationship between the viscous damping constant β employed in the theory of domain-wall resonance and the Landau-Lifshitz damping parameter λ (equation 10-12.6).

10-14. For a two-sublattice system undergoing ferromagnetic resonance obtain expressions for the components of the alternating magnetizations $M_{A1\pm}$ and $M_{B1\pm}$. Consider the ratio of $M_{A1\pm}/M_{B1\pm}$ for both the ferromagnetic and exchange modes. Make reasonable approximations and justify the diagrams of Fig. 10-13.1.

10-15. (a) Derive a better approximation than that of equation 10-13.13 for the high-frequency mode of ferrimagnetic resonance. (b) Derive a resonance equation for a two-sublattice ferrimagnet that is valid in the vicinity of the compensation temperature.

10-16. (a) Find the relationship between the alternating components of the sublattice magnetizations in antiferromagnetic resonance for the case in which $T = 0°K$ and $\mathbf{H}$ is parallel to the easy direction. From this establish the mode configurations in Fig. 10-14.1. (b) Derive the resonance conditions given by equations 10-14.5. (c) Show that equation 10-14.14 for $\theta = 0$ and 90° reduces to the resonance conditions derived in the text.

10-17. Show that the resonance frequencies of the two modes for α-Fe_2O_3 are given by equations 10-14.15 and 10-14.16.

10-18. (a) Calculate the magnetic field at the Fe^{57} nucleus in metallic iron from the data plotted in Fig. 10-16.4. (b) Deduce from Fig. 10-16.5 the shift in the nuclear-energy levels produced by the quadrupole interaction and the isomer shift in α-Fe_2O_3 when the source is Co^{57} in stainless steel.

Bibliography

Artman, J. O., "Microwave Resonance Relations in Anisotropic Single Crystal Ferrites," *Proc. IRE* **44,** 1284 (1956).

Bagguley, D. M. S., and J. Owen, "Microwave Properties of Solids," *Rept. Prog. Phys.* **20,** 304 (1957).

Bloembergen, N., "Magnetic Resonance in Ferrites," *Proc. IRE* **44,** 1259 (1956).

Boyle, A. J. F., and H. E. Hall, "The Mössbauer Effect," *Rept. Prog. Phys.* **25,** 441 (1962).

Clarricoats, J. B., *Microwave Ferrites*, Chapman and Hall, London (1961).

Conferences on the Mössbauer Effect, John Wiley and Sons, New York (1962); *Revs. Mod. Phys.* (January, part 2, 1964).

Epstein, P. S., "Theory of Wave Propagation in a Gyromagnetic Medium," *Revs. Mod. Phys.* **28,** 3 (1956).

Frauenfelder, H., editor, *The Mössbauer Effect*, W. A. Benjamin, New York (1962).

"International Conferences on Magnetism," *J. Phys. radium* (March 1951) (February-March 1959), *J. Phys. Soc. Japan, Suppl. B-I* (March 1962).

Katz, H. W., *Solid State Magnetic and Dielectric Devices*, John Wiley and Sons, New York (1959).

Lax, B., and K. J. Button, *Microwave Ferrites and Ferrimagnetics*, McGraw-Hill Book Co., New York (1962).

Rado, G. T., "Ferromagnetic Resonance," *J. Appl. Phys.* **32,** 129S (1961).

Rado, G. T., and H. Suhl, editors, Magnetism, Vol. 1, *Magnetic Ions in Insulators, their Interactions, Resonances and Optical Properties*, Academic Press, New York (1963).

Smit, J., and H. P. J. Wijn, *Ferrites*, John Wiley and Sons, New York (1959).

Soohoo, R. H., *Theory and Applications of Ferrites*, Prentice-Hall, Englewood Cliffs, N.J. (1960).

"Symposiums on Magnetism and Magnetic Materials," *J. Appl. Phys. Suppl.* (March 1958), (April 1959), (May 1960), (March 1961), (March 1962), (April 1963), (March 1964).

Wolf, W. P., "Ferrimagnetism," *Rept. Prog. Phys.* **24,** 212 (1961).

APPENDIX I

Systems of Units

In terms of the dimensions length, mass, and time (LmT), two systems of units have become standard; one is the centimeter, gram, and second or cgs system and the other is the meter, kilogram, and second or mks system. It should be emphasized, however, that the choice of units and dimensions is completely arbitrary. To illustrate, in certain theoretical calculations the quantities c and h are set equal to unity and chosen to be dimensionless; all the physical quantities can then be expressed in terms of one dimension, usually taken to be the length L ($L = T = m^{-1}$).

There are several cgs systems of units, but the one in common use today is the gaussian. It, in turn, is based on two earlier systems, electrostatic units (esu) and electromagnetic units (emu). Electrostatic units were defined from the expression describing the force between two electrical point charges, which is analogous to equation 1-2.1 for magnetic charges. The constant of proportionality in the equation was set equal to unity. Electromagnetic units may be defined from the force equation for magnetic charges (equation 1-2.1) or, as is more usual, from the expression either for the force between two current-carrying wires or for the field produced by a current (equation 1-7.1). The ratio between the quantity of charge, or, equivalently, the current density, in the esu and emu systems turns out to be equal to c, the velocity of light in vacua; that is,

$$\frac{q_{\mathrm{esu}}}{q_{\mathrm{emu}}} = \frac{j_{\mathrm{esu}}}{j_{\mathrm{emu}}} = c.$$

In the gaussian system it is arranged that **E** and **B** have the same dimensions; this is the origin of the $1/c$ factor on the right-hand side of equation 1-8.3.

The phrase rationalization of units refers to the positioning of the factor 4π. By proper definition of the proportionality constants, the 4π that appears in some equations can be eliminated. It should be clearly recognized that a factor 4π will then appear in other equations. Hence all that is really achieved is the choice of the equations in which the 4π is located. In the gaussian system rationalization is accomplished by putting the proportionality constants in the electric and magnetic force equations (Coulomb's law) equal to $1/4\pi$. Then the relationship between **B**, **H**, and **M** becomes $\mathbf{B} = \mathbf{H} + \mathbf{M}$ and there is no 4π in any of Maxwell's equations.

This is known as the Heaviside-Lorentz system and is but little used nowadays.

In the mks system rationalization of the units is almost universal. Further, the electric charge q is introduced as an additional dimension, with the coulomb being defined as the charge transported by a steady current of one ampere in the time of one second. Since the proportionality constant in Coulomb's electrostatic force equation is written $1/4\pi\epsilon_0$ so that

$$\mathbf{F} = \frac{1}{4\pi\epsilon_0}\frac{q_1q_2}{r^2}\mathbf{r}_0,$$

it follows that ϵ_0, which is called the permittivity of free space, has the dimensions $m^{-1}L^{-3}T^2q^{+2}$. If we assume $c = 3.00 \times 10^8$ m/sec, it is found that $\epsilon_0 = 1/36\pi \times 10^{-9}$, or 8.854×10^{-12} farad/meter. For the magnetic case the proportionality constant in a force equation is numerically defined. Thus, in equation 1-2.1 $k = 1/4\pi\mu_0$, where μ_0 is called the permeability of free space and is equal to $4\pi \times 10^{-7}$ henry/meter. In the equation relating **H**, **B**, and **M** an additional choice is possible. If we write $\mathbf{B} = \mu_0\mathbf{H} + \mathbf{M}$, then **M** and **B** have the same units; this is the logical point of view if the uncompensated poles are considered to be the source of the magnetic induction **B**. Alternatively, we may write $\mathbf{B} = \mu_0(\mathbf{H} + \mathbf{M})$ if we wish **H** and **M** to have the same units, as is appropriate if amperian currents are considered as the sources of **B**.

One of the main advantages of the mks rationalized system is the inclusion of practical units such as volts, amperes, and ohms. Another advantage is that some frequently used equations do not contain factors of 4π, c, ϵ_0, or μ_0. In particular, Maxwell's equations become

$$\nabla \cdot \mathbf{D} = 0, \qquad \nabla \times \mathbf{E} = -\frac{\partial \mathbf{B}}{\partial t},$$

$$\nabla \cdot \mathbf{B} = 0, \qquad \nabla \times \mathbf{H} = \mathbf{j} + \frac{\partial \mathbf{D}}{\partial t}.$$

The overwhelming majority of papers quoted in this book employ gaussian or some form of cgs units. Also, in a text devoted to magnetism there are drawbacks, from a pedagogical viewpoint, to the assignment of a permeability to free space. These are the principal reasons why the gaussian system has been adopted. Nevertheless, on certain occasions, for example, when reference is made to a current in a solenoid, practical units are used. The relationship between gaussian and rationalized mks units is spelled out in the following conversion table.

Relationship between Gaussian and Rationalized MKS Units

To convert gaussian into mks rationalized ones, multiply the number of gaussian units by the conversion factor given in the table. The inverse conversion, of course, is accomplished by employing the reciprocal of the tabulated conversion factor.

Quantity	Symbol	Gaussian Unit	Conversion Factor	MKS Rationalized Unit
Length	l, L	centimeter (cm)	10^{-2}	meter (m)
Mass	m, M	gram (gm)	10^{-3}	kilogram (kg)
Time	t, T	second (sec)	1	second (sec)
Force	$\mathbf{F}$	dyne	10^{-5}	newton
Energy	E, U, W, F	erg	10^{-7}	joule
Energy density	F/V, etc.	erg/cm³	10^{-1}	joule/m³
Power	dW/dt	erg/sec	10^{-7}	watt
Magnetic pole	m	cgs pole strength	$4\pi \times 10^{-8}$	weber
Magnetic field	$\mathbf{H}$	oersted (oe)	$\frac{1}{4\pi} \times 10^{3}$	ampere-turns/meter
Magnetic dipole moment	μ	cgs magnetic dipole moment	$4\pi \times 10^{-10}$	weber-meter
Magnetization	$\mathbf{M}$	gauss	$4\pi \times 10^{-4}$	weber/m²
Magnetic induction	$\mathbf{B}$	gauss	10^{-4}	weber/m²
Magnetic flux	Φ	maxwell	10^{-8}	weber
Magnetic scalar potential: magnetomotive force	φ	gilbert	$\frac{1}{4\pi} \times 10^{-3}$	ampere-turn
Demagnetization factor	D		$\frac{1}{4\pi}$	
Electric charge	q	statcoulomb	$\frac{1}{3} \times 10^{-9}$	coulomb (coul)
Electric charge density	ρ	statcoulomb/cm³	$\frac{1}{3} \times 10^{-3}$	coulomb/m³
Current	i	statampere	$\frac{1}{3} \times 10^{-9}$	ampere (amp)
Current density	$\mathbf{j}$	statamp/cm²	$\frac{1}{3} \times 10^{-5}$	amp/m²
Electric field	$\mathbf{E}$	statvolt/cm	3×10^{4}	volt/m
Displacement vector	$\mathbf{D}$	statvolt/cm	$\frac{1}{12\pi} \times 10^{-5}$	coul/m²
Polarization	$\mathbf{P}$	statcoul/cm²	$\frac{1}{3} \times 10^{-5}$	coul/m²
Electrical potential or emf	$\varphi_E, \mathscr{E}$	statvolt	3×10^{2}	volt
Conductivity	σ	sec⁻¹	$\frac{1}{9} \times 10^{-9}$	mho/m
Resistance	R	sec/cm	9×10^{11}	ohm
Capacity		cm	$\frac{1}{9} \times 10^{-11}$	farad
Inductance		cgs (es) units	9×10^{11}	henry

Relationship between Gaussian and Rationalized MKS Units

APPENDIX II

Demagnetization Factors for Ellipsoids of Revolution

$$D_b = \tfrac{1}{2}(4\pi - D_a)$$

$$m = \frac{b}{a}$$

m	D_a	m	D_a
0.00	12.57	1.40	3.13
0.01	12.38	1.50	2.92
0.02	12.17	1.60	2.75
0.03	11.98	1.80	2.44
0.04	11.81	2.00	2.18
0.05	11.62	2.50	1.70
0.06	11.46	3.0	1.37
0.07	11.30	3.5	1.13
0.08	11.13	4.0	0.95
0.09	10.97	4.5	0.81
0.10	10.82	5.0	0.70
0.125	10.42	6.0	0.54
0.167	9.84	7.0	0.44
0.200	9.41	8.0	0.36
0.25	8.84	9.0	0.30
0.30	8.30	10.0	0.25
0.40	7.39	15.0	0.13
0.50	6.61	20.0	0.085
0.60	5.98	30.0	0.043
0.70	5.42	40.0	0.026
0.80	4.95	50.0	0.018
0.90	4.54	70	0.0098
1.00	4.18	100	0.0053
1.10	3.96	200	0.0015
1.20	3.59	500	0.00029
1.30	3.34		

APPENDIX III

Periodic Table of the Elements[a]

(showing outer electronic configuration and ground state term; the principal transition elements are also indicated.)

1 H $1s$	$^2S_{1/2}$	3 Li $2s$	$^2S_{1/2}$	11 Na $3s$	$^2S_{1/2}$	19 K $3p^6 4s$	$^2S_{1/2}$	37 Rb $4p^6 5s$	$^2S_{1/2}$	55 Cs $5p^6 6s$	$^2S_{1/2}$	87 Fr $6p^6 7s$	—
		4 Be $2s^2$	$^1S_0$	12 Mg $3s^2$	$^1S_0$	20 Ca $3p^6 4s^2$	$^1S_0$	38 Sr $4p^6 5s^2$	$^1S_0$	56 Ba $5p^6 6s^2$	$^1S_0$	88 Ra $6p^6 7s^2$	$^1S_0$
						21 Sc $3d4s$	$^2D_{3/2}$	39 Y $4d5s^2$	$^2D_{3/2}$	57 La $5p^6 5d6s^2$	$^2D_{3/2}$	89 Ac $6d7s^2$	—
										58 Ce $4f5d6s^2$	$^3H_4$	90 Th $6d^2 7s^2$	—
										59 Pr $4f^3 6s^2$	—	91 Pa $5f^2 6d7s^2$	—
										60 Nd $4f^4 6s^2$	—	92 U $5f^3 6d7s^2$	—
										61 Pm $4f^5 6s^2$	—	93 Np $5f^4 6d7s^2$	—
										62 Sm $4f^6 6s^2$	$^7F_0$	94 Pu $5f^6 7s^2$	—
										63 Eu $4f^7 6s^2$	$^8S_{7/2}$	95 Am $5f^7 7s^2$	—
										64 Gd $4f^7 5d6s^2$	$^9D_2$	96 Cm $5f^7 6d7s^2$	—
										65 Tb $4f^9 6s^2$	—	97 Bk $5f^8 6d7s^2$	—
										66 Dy $4f^{10} 6s^2$	—	98 Cf $5f^{10} 7s^2$	—
										67 Ho $4f^{11} 6s^2$	—	99 Es $5f^{11} 7s^2$	—
										68 Er $4f^{12} 6s^2$	—	100 Fm $5f^{12} 7s^2$	—
										69 Tm $4f^{13} 6s^2$	—	101 Md $5f^{13} 7s^2$	—
										70 Yb $4f^{14} 6s^2$	—	102 No $5f^{14} 7s^2$	—
										71 Lu $4f^{14} 5d6s^2$	$^2D_{3/2}$	103 Lw $5f^{14} 6d7s^2$	—

Elements 58–71: $4f$, rare earth or lanthanum transition group

Elements 90–103: $5f$–$6d$ or actinium transition group

			3d or iron transition group	4d or paladium transition group	5d or platinum transition group
			22 Ti $3d^24s^2$ $^3F_2$	40 Zr $4d^25s^2$ $^3F_2$	72 Hf $5d^26s^2$ $^3F_2$
			23 V $3d^34s^2$ $^4F_{3/2}$	41 Nb $4d^45s$ $^6D_{1/2}$	73 Ta $5d^36s^2$ —
			24 Cr $3d^54s^1$ $^7S_3$	42 Mo $4d^55s$ $^7S_3$	74 W $5d^46s^2$ $^5D_0$
			25 Mn $3d^54s^2$ $^6S_{5/2}$	43 Tc $4d^65s$ —	75 Re $5d^56s^2$ $^6S_{5/2}$
			26 Fe $3d^64s^2$ $^5D_4$	44 Ru $4d^75s$ $^5F_5$	76 Os $5d^66s^2$ —
			27 Co $3d^74s^2$ $^4F_{9/2}$	45 Rh $4d^85s$ $^4F_{9/2}$	77 Ir $5d^9$ —
			28 Ni $3d^84s^2$ $^3F_4$	46 Pd $4d^{10}$ $^1S_0$	78 Pt $5d^96s$ $^3D_3$
			29 Cu $3d^{10}4s$ $^2S_{1/2}$	47 Ag $4d^{10}5s$ $^2S_{1/2}$	79 Au $5d^{10}6s$ $^2S_{1/2}$
			30 Zn $3d^{10}4s^2$ $^1S_0$	48 Cd $4d^{10}5s^2$ $^1S_0$	80 Hg $5d^{10}6s^2$ $^1S_0$
	5 B $2s^22p$ $^2P_{1/2}$	13 Al $3s^23p$ $^2P_{1/2}$	31 Ga $4s^24p$ $^2P_{1/2}$	49 In $5s^25p$ $^2P_{1/2}$	81 Tl $6s^26p$ $^2P_{1/2}$
	6 C $2s^22p^2$ $^3P_0$	14 Si $3s^23p^2$ $^3P_0$	32 Ge $4s^24p^2$ $^3P_0$	50 Sn $5s^25p^2$ $^3P_0$	82 Pb $6s^26p^2$ $^3P_0$
	7 N $2s^22p^3$ $^4S_{3/2}$	15 P $3s^23p^3$ $^4S_{3/2}$	33 As $4s^24p^3$ $^4S_{3/2}$	51 Sb $5s^25p^3$ $^4S_{3/2}$	83 Br $6s^26p^3$ $^4S_{3/2}$
	8 O $2s^22p^4$ $^3P_2$	16 S $3s^23p^4$ $^3P_2$	34 Se $4s^24p^4$ $^3P_2$	52 Te $5s^25p^4$ $^3P_2$	84 Po $6s^26p^4$ —
	9 F $2s^22p^5$ $^2P_{3/2}$	17 Cl $3s^23p^5$ $^2P_{3/2}$	35 Br $4s^24p^5$ $^2P_{3/2}$	53 I $5s^25p^5$ $^2P_{3/2}$	85 At $6s^26p^5$ —
2 He $1s^2$ $^1S_0$	10 Ne $2s^22p^6$ $^1S_0$	18 Ar $3s^23p^6$ $^1S_0$	36 Kr $4s^24p^6$ $^1S_0$	54 Xe $5s^25p^6$ $^1S_0$	86 Rn $6s^26p^6$ $^1S_0$

[a] J. A. Cotton and G. Wilkinson, *Advanced Inorganic Chemistry*, Interscience Publishers, New York (1962); G. T. Seaborg, *Man Made Transuranic Elements*, Prentice-Hall, Englewood Cliffs, N.J. (1963).

APPENDIX IV

Numerical Values for Some Important Physical Constants[a]

Electronic charge e	-4.80298×10^{-10} cm$^{3/2}$ gm$^{1/2}$ sec^{-1}
Electron rest mass m	9.1091×10^{-28} gm
Proton rest mass M_P	1.67252×10^{-24} gm
Velocity of light c	2.997925×10^{10} cm/sec
Planck's constant h	6.6256×10^{-27} erg-sec
$\hbar$	1.05450×10^{-27} erg-sec
Charge to mass ratio for electron e/m	5.27274×10^{17} cm$^{3/2}$ gm$^{-1/2}$ sec^{-1} (emu); 1.758796×10^{7} cm$^{1/2}$ gm$^{-1/2}$ (esu)
Bohr magneton μ_B	-9.2732×10^{-21} erg/oe
Nuclear Bohr magneton μ_{B_n}	5.0505×10^{-24} erg/oe
Electron volt eV	1.60210×10^{-12} erg; 3.8291×10^{-20} calorie
Associated "temperature"	1.16049×10^{4} °K
Associated wave number	8065.73 cm^{-1}
Boltzmann's constant k	1.38054×10^{-16} erg/°K
Gas constant R	8.3143×10^{7} erg/°K-mole
Avogadro's number N (number atoms/mole)	6.02252×10^{23}/mole
Loschmidt's number N (number atoms/cm^3)	2.687444×10^{19}/cm^3
Bohr radius a_0	0.529167×10^{-8} cm

[a] *Physics Today* **17**, 48 (1964).

APPENDIX IV

Numerical Values for Some Important Physical Constants[a]

Author Index

Subject Index